Strategic Management of Technology and Innovation

The Irwin Series in Management and The Behavioral Sciences

Strategic Management of Technology and Innovation

Robert A. Burgelman
Stanford University

Modesto A. Maidique
Florida International University

1988

Homewood, Illinois 60430

Case material of the Harvard Graduate School of Business Administration is made possible by the cooperation of business firms and other organizations which may wish to remain anonymous by having names, quantities, and other identifying details disguised while maintaining basic relationships. Cases are prepared as the basis for class discussion rather than to illustrate either effective or ineffective handling of administrative situations.

This book was set in 10 point Garamond Light Roman by Bi-Comp, Inc.
The editors were William Bayer, Doris Hill, Paula M. Buschman, Joan A. Hopkins.
The production manager was Carma W. Fazio.
The drawings were done by Tom Mallon.
Arcata Graphics/Kingsport was the printer and binder.

ISBN 0-256-03481-8

Library of Congress Catalog Card No. 87–80949

Printed in the United States of America

11 K 5 4 3

In memory of our parents,
Robert, Louisa,
Hilda, and Modesto

Preface

This book is a major by-product of our 25 years of collective experience as researchers, teachers, consultants, and practitioners in the management of technology field. The main objective of the book, and the associated teaching manual for instructors, is to articulate a course concept which we jointly developed as part of our teaching and course development responsibilities at Harvard University, Stanford University, and New York University.

The book brings together those cases and readings that in our case method work with over 1,000 M.B.A.s, engineers, and executives have proven to be most effective in communicating the basic principles of our field. Most of the cases and many of the articles we have included were developed as part of the research and teaching work of the authors at Stanford University and the Harvard Business School. All of the cases and readings we have included have been tested by the authors in class. On the other hand, the book includes several previously unpublished cases, articles, and notes and an introductory section that provides a lead for each section of the book.

The structure of the book has been designed to stimulate the reader to perform multiple levels of analysis. The cases address both content and process issues (the what and the how) within progressively more complex units of analysis (the project, the business unit, the corporate context, and the overall industry environment). The cases also address longitudinal changes, interactions within product families, and the organizational evolution that occurs as technology twists and turns. In summary, our objective is to provide a rich exposure to a range of organizational, contextual, and longitudinal problems in which technology plays an important role.

We believe the subject of the book, the management of technological innovation, is a timely one for students of American management. The United States is no longer the dominant technological innovator in the world. It now faces challenges to its technological prowess from both the West and the East. For this reason, business scholars have increasingly turned their attention, during the past decade, to America's competitive position in a dynamically changing world environment. Most of those that have critically analyzed the current situation conclude that there is considerable room for improvement in current U.S. management practice, and that in particular, our ability to manage technology will be an increasingly important competitive dimension during the remainder of this century.

Yet, despite the numerous books that assess our technological position and diagnose the causes of America's shift in competitive position, and a few excellent books of readings on technological innovation, only one casebook has been developed to teach students and practitioners how to better deal with basic issues of R&D management and how to integrate technology into corporate strategy. But, *Research, Development, and Technological Innovation* (Richard D. Irwin, 1967), Jim Bright's pioneering book, is now 20 years old, and though it is replete with ageless insights, it does not reflect those truths through the mirror of current reality.

It is from Jim Bright's work that the inspiration came to prepare a book of cases and readings on the management of technology two decades after his initial effort. Much, however, has changed since Jim Bright selected the materials for his book of cases and readings. Several new industries have been born: computer-integrated manufacturing, genetic engineering, minicomputers, personal computers, solar heating, and integrated circuits. There are cases and readings in the book that deal with each of these important developments. Secondly, there have been numerous academic contributions to the field that have helped us understand how to better analyze technological developments and how to manage technology competitively: Porter's concept of industry and competitor analysis; von Hippel's findings regarding the role of the user in the innovative process; Allen's development of the gatekeeper concept; Abernathy and Utterback's model of business unit evolution patterns; Maidique and Zirger's model of the new product development process as a learning cycle; and Burgelman's view of the strategic process as a dyad consisting of induced and autonomous components. These, as well as other important new contributions to the field, are reflected in our choices of cases and clarified in the accompanying readings.

The book is structured into four parts which embody a *course concept* that has evolved from the authors' close collaboration over the past five years. It is providing a course concept which we hope will be the major contribution of this book.

The book begins with a case that illustrates the issues faced by a manager who has to make general management decisions in a context that is shaped by technological considerations. This case serves as an introduction to the topics that will be discussed in the remainder of the text.

The second part of the book provides a framework that brings together those decisions that constitute the firm's *technological strategy*. This part includes one or more cases on each of the major classes of strategic technological decisions faced by the top managements of technology-intensive firms, such as technological choices, design trade-offs, licensing, purchase of technology, timing and entry into the marketplace, and the organization and management of research and development activities.

The third part focuses on a central objective of businesses that compete in industries that are periodically destabilized by new waves of technology: the development of new products and new product families. This section starts with cases that focus on the link between R&D management and product development and progress from the development of new products to groups of related products, that is, to entirely new businesses based on new market or technological developments.

The fourth and last part of the book deals with the problems of technological entrepreneurship in the large diversified corporation. The book closes with cases and readings that illustrate the most recent attempts to develop entrepreneurial systems within large complex organizations.

In preparing this book, we have benefited

greatly from the work of many of our colleagues, as is evident by the breadth and variety of the materials we have included. On the other hand, a few of our colleagues—in addition to Jim Bright—have specially influenced what we teach and how we teach it. Foremost amongst these are the late Bill Abernathy of the Harvard Business School; Ed Roberts, Jim Utterback, and Eric von Hippel at the Massachusetts Institute of Technology; Richard Rosenbloom and Robert Hayes at Harvard University; Steve Wheelwright, Nate Rosenberg, and Hank Riggs at Stanford University; Leonard Sayles and Mike Tushman at Columbia University; and James Brian Quinn at Dartmouth University.

While, of course, any inadequacies, omissions, and errors in our book are our own, whatever value it does have has been enhanced significantly by our study of the work of these scholars and our personal interaction with them over the past 10 years. To all of them and to numerous other colleagues from whom we have learned much during our careers, our profound thanks.

We also want to acknowledge the generous support for our course development efforts received from Stanford University since 1981. For Burgelman, this support has come from the Strategic Management Program of the Graduate School of Business; for Maidique, from the Department of Industrial Engineering and Engineering Management.

As anyone who has attempted (and completed) a book-length manuscript knows, the final product is a team effort. Margaret Talt, Andrea Reisman, and Jeanie McGuire, our secretaries at different stages in the development of the manuscript, typed most of the early drafts of the manuscript. This book would never have been completed without the collaboration of B. J. Zirger, a Stanford M.B.A. and now a doctoral candidate at Stanford University, who, with our assistance, was largely responsible for preparing the teaching guide that significantly enhances the value of the book as a useful teaching tool. Furthermore, B. J.'s tireless enthusiasm and her willingness to help out in a variety of ways at crucial points during the book's preparation was an essential ingredient in achieving our deadlines. Tom Kosnik, now a faculty member at Harvard Business School, coauthored several of the cases, notes, and teaching notes during his tenure as a Stanford Business School doctoral candidate, under the supervision of one of the authors. Martine van den Poel, research associate at the Stanford Business School during 1984–85 and now a research program director at Insead, and John Ince, during his tenure as a research assistant at the Harvard Business School, also contributed to several of the materials included in this book. In addition to Tom, B. J., Martine, and John, we are grateful to the hundreds of Stanford and Harvard M.B.A. and engineering students who, through their class participation and written case analyses, helped us to test and improve the materials in the course.

Finally, a word of thanks to Rita and Anna for their unwavering support throughout this venture.

Robert A. Burgelman
Modesto A. Maidique

Acknowledgments for Case Materials

W. J. Abernathy	"Patterns of Industrial Innovation"; "Silicon Valley Specialists, Inc. (A)"
R. M. Atherton	"Hewlett-Packard: A 1975–1978 Review"
J. L. Bower	"PC&D, Inc."
I. C. Bupp	"Texas Instruments' 'Speak and Spell' Product"
E. T. Christiansen	"PC&D, Inc."
D. M. Crites	"Hewlett-Packard: A 1975–1978 Review"
J. M. Crowe	"Biodel, Inc. (A)"
J. S. Gable	"Apple Computer (A)"
G. Greenberg	"Hewlett-Packard: A 1975–1978 Review"
A. L. Frevola	"Data Net"; "Technological Strategy"
R. G. Hamermesh	"PC&D, Inc."
C. Huntington	"Claire McCloud"
J. F. Ince	"The Grumman Corporation"; "Grumman Energy Systems"
A. L. Jakimo	"Texas Instruments' 'Speak and Spell' Product"
S. Koreisha	"Golden Gate Semiconductor"
T. J. Kosnik	"Toward an Innovative Capabilities Audit Framework"; "Software Architects (A)"; "Medical Equipment (A)"
P. Lawrence	"Aerospace Systems (D)"
R. S. Rosenbloom	"Advent Corp. (C)"
A. Ruedi	"Aerospace Systems (D)"
C. C. Swanger	"Apple Computer: The First Ten Years"
P. H. Thurston	"Silicon Valley Specialists (A)"
S. Tylka	"Apple Computer (A)"
M. van den Poel	"Toward an Innovative Capabilities Audit Framework"; "Medical Equipment (C)"; "Control Data Corporation"

The authors would also like to acknowledge B. J. Zirger, Julio Gonzalez, and Al Frevola for all their work in preparing the text for publishing.

Contents

I Integrating Technology and Strategy

Technology and innovation must be managed. That much is generally agreed upon by thoughtful management scholars. But can technology management be taught, and if so, how? What concepts, techniques, facts, and management processes result in successful technological innovations? The answers to these questions and to several other related ones are an important part of the agenda of those academics and practitioners who concern themselves with organizations in which technology and technological innovation play an important role.

These concerns are, however, of relatively recent vintage. For decades after World War II it appeared that America would reign indefinitely as the world's technological superpower. This fantasy has been shattered by the industrial recovery of Europe and the rise of the Japanese as an economic and technological superpower.

Both Japan and Europe have made major inroads into industries that were once considered unassailable U.S. strongholds. At first, it seemed that the challenge was mainly in the traditional capital-intensive, heavy manufacturing industries such as steel and automobiles and in consumer electronics. In the 1980s, the challenge broadened to include all aspects of electronics, including semiconductors, as well as aerospace products and telecommunications. These developments have made effective technology management a high-priority issue for U.S. business.

Notwithstanding the importance of technology as a determinant of domestic and international competitive advantage, very few courses on the subject exist in U.S. business schools. While a great deal of emphasis is given to training students in finance, accounting, marketing, and strategy, few courses focus on technology. Yet, ironically, it is often technology that is the driving variable behind major industrial change.

Even if the importance of the role of technology is taken as a given, there are widely differing views as to what needs to be known to manage a business unit for which technology is a key strategic variable. According to one school of thought, it is enough to understand the parameters that are transformed by the technological "black box," e.g., the computer or the instrument in question. In effect—so goes this mode of thinking—it is enough to know *what* the technological device or system does, not *how* it does it. An alternative view is to argue that unless one understands the functioning of a technological device and the laws that delineate its limitations, one cannot make effective judgments regarding the shaping of the relevant technologies into successful products.

Part I of the book focuses on this fundamental issue. The first case situation finds Claire McCloud, a recently minted M.B.A. who, as a newly appointed general manager of a high-technology business with little technological education or experience, has to determine what she needs to know to function effectively in her new job. Coming to terms with this issue leads Claire to define the need for a framework that defines that set of technology-related strategic decisions with which she, as a general manager, will have to deal. This line of inquiry leads naturally to Part II of the book which is devoted to the development of the concept of technological strategy.

Technology and the Manager

Case I–1
Claire McCloud

C. Huntington and M. A. Maidique

"Think it over, Claire. You've worked closely with me for two years, and I think you're now ready to take on a division of your own. You're smart, determined, and you've got good business sense.

"We need an executive at the fiber optics division who can solve the present operating problems and at the same time get it established firmly in the fiber optics market. Give it serious thought, Claire, and I'll come by in the morning to hear your decision." With that, Mr. W. H. Walton, "the Colonel," chairman and founder of Walton International, closed the door behind him as he left Claire McCloud's office. As usual, Mr. Walton, a physicist who was unusually proud of his stint as a communications officer in Korea, had made his case forcefully and explicitly.

McCloud sat back to consider Walton's offer. Walton International had had great hopes for the fiber optics business when Optical Wavelength Specialists had been acquired three years before in 1976, but the division had not performed as well as had been expected. The fiber optics industry, however, was expected to grow rapidly as it expanded its participation in the telecommunication equipment market and thus it presented a major opportunity for the firm. McCloud also knew that Walton was personally committed to establishing the company's position in the fiber optics industry. If she accepted the job of managing the division, it could be a unique and exciting opportunity. But she wondered if her experience and abilities were suited for the job.

Prior to joining Walton International, McCloud had received a B.A. in economics magna cum laude from Stanford, worked four years as an economist for a large private investment management firm, and spent two years at a well-known eastern business school where she received her M.B.A. degree. During the past two years as "the Colonel's" assistant, she had gained experience in working with people and in dealing with company problems from the top. She had worked closely with Walton in the step-by-step reorganization of the components division and the instruments division, and had studied and initiated changes in the pension administration policies of the company.

Harvard Business School case 9-680-030

Through this involvement, she had become intimately familiar with the aggressive, highly technical top management group of Walton International.

But perhaps what was of most relevance to her current decision was her direct involvement with the fiber optics division. Claire had evaluated several project proposals from the fiber optics division. These proposals had ranged from requests for new capital equipment to a formal proposal for the acquisition of a wire and cable company. While her principal contributions in these project evaluations had been largely in the areas of finance and marketing, she felt, thanks to her persistent inquisitiveness, that she had also gained a fairly good understanding of the technical aspects of the fiber optics business from frequent conversations with the fiber optics division technical staff. But she wondered if this level of technical understanding was sufficient for the position Walton had proposed. She also wondered what particular tools she would need to be an effective manager of a high-technology business such as fiber optics. She had the rest of the day to reach a decision.

FIBER OPTICS[1]

The principle of sending signals by light was first demonstrated in 1880 by Alexander Graham Bell. By 1978, the fiber optics industry was still embryonic but was expected to evolve into one of the most exciting and fruitful growth industries of the 1980s. A conservative estimate by an industry analysis firm placed the market at over $500 million by 1987. A breakdown of one of these projections is given in Exhibit 1.

[1] This section is based on three main sources: (1) G. R. Elion and H. A. Elion, *Fiber Optics in Communication Systems* (New York: Marcel Dekker, Inc., 1978); (2) "The Fiber Optics Industry in 1978: Competition," Case Clearing House #1-379-139; and (3) "The Fiber Optics Industry in 1978: Products, Technology, and Markets," Case Clearing House #1-379-136.

EXHIBIT 1 Market Segmentation Projection for Fiber Optic Communication Systems, 1987

Estimated Expenditures on Fiber Optics (dollars in millions)

	1987
Commercial telecommunications	$350.0
CCTV and broadcast TV	3.0
Data communications	2.0
Computer applications	75.0
Other (industrial, office equipment, instrumentation)	15.0
Military and aerospace	100.0
Total	$545.0

SOURCE: Frost and Sullivan Report #415.

The invention of the laser by Bell Laboratories in 1960 and other recent developments in electronics had paved the way for the application of optical fiber technology to communication equipment. According to many observers, fiber optics had the potential to revolutionize the communications industry. Claire explained the elements of fiber optics technology:

> In principle, the technology offers many advantages over conventional copper cable for voice, data, and video communication. Because of the unique aspects of optical physics, a glass or plastic fiber can transmit light signals much faster and can carry much more information than copper cable. A fiber optic cable is also much smaller, lighter, and more flexible, thus simplifying and reducing the cost of shipping and handling. A further significant but not yet completely realized advantage is that of manufacturing cost; with volume production, fiber optic cable is expected to cost at least 1/10 that of premium coaxial cable and, in addition, requires fewer peripheral structures such as repeaters and grounding equipment that would otherwise add to cost.[2]

[2] Repeaters are used to boost the signals on long-length cable runs.

EXHIBIT 2 Block Diagram of a Fiber Optic System

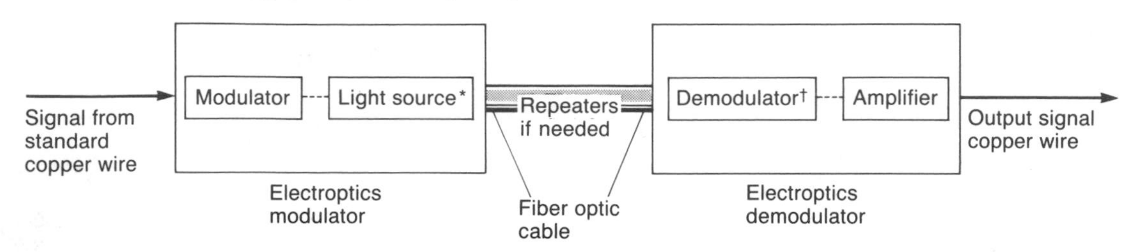

* Typically an LED (light-emitting diode) or laser. At present, LEDs usually employed for lasers are less durable and more expensive. It was expected, however, that lasers would ultimately dominate the fiber optics market due to their greater bandwidth potential.

† Typically accomplished via a standard semiconductor device known as a PIN diode.

SOURCE: Casewriter's drawing.

Optical systems also offer many advantages that are not attainable in metal or wire cables because of the physical properties of the fiber. For example, they are immune from electrical or electromagnetic interference. They can be made to have total electrical isolation, and are unaffected by extreme weather or temperature conditions. Another advantage is that the system can be upgraded easily, so technological improvements can translate immediately into better performance.

The components of a fiber optic system are described in Exhibit 2, and consist of a power source, a modulator, a light source such as a laser or light-emitting diode (LED), the optical fiber encased in a cable, and a demodulator and amplifier for receiving the signal.[3] The modulator and demodulator are used to modulate the signals transmitted between the fiber optics cable and the copper cable.

By 1978, fiber optic technology had undergone radical changes. Significant improvement had taken place in the four major aspects of system design: 1) information-carrying capacity, 2) the distance over which signals could be carried, 3) reliability, and 4) cost. Information-carrying capacity, measured in bandwidth, had improved significantly and had become extremely competitive with other communications methods by 1978, as shown in Exhibits 3–A and 3–B. While the distance over which fibers could commercially operate was still fairly short—very few systems had been tested at over five kilometers—there has been rapid improvement. Technological developments had focused on reducing attenuation losses (weakening in the light signal as it travels through the fiber). In eight years, from 1970 to 1978, attenuation losses had improved from 20 decibels per kilometer (db/km) in Corning Glass Works' initial low-loss fiber to a fiber with a loss of only 1 db/km. Finally, costs were declining, and by 1978, optical fiber had become cost competitive with conventional cable. Prices of fibers had fallen from $25 per meter for fiber with an attenuation loss of 30 db/km to $1 per meter for fibers with losses of 6 to 10 db/km. Corning projected that fiber prices would eventually reach 15 cents per meter.

By 1978, there were several techniques for fabricating optical fibers. The major processes were double crucible, stratified melt, chemical vapor deposition, rod-in-tube, and direct fiber drawing. All required ultrapurified raw materials. The main differences among the various

[3] A light-emitting diode (LED) is a semiconductor device that emits light when it is powered by an electric voltage.

EXHIBIT 3–A Information-Carrying Capacity of Selected Communications Systems

	Twisted Pair Wire	Coaxial Cable	Radio	TV	Microwave	Fiber Optics
Usable bandwidth (Hz)*	5×10^3 5×10^3	5×10^3 2×10^8	4.2×10^3	4.2×10^6	4.2×10^7	4.2×10^{11}
Number of usable voice channels†	0–1	0–10^4	1.0	10^3	10^4	10^8
Number of usable video channels‡	0	0–10	0	1	10	10^5

* The frequency of the signal source determined the nominal frequency. Inefficiencies within the system resulted in a usable bandwidth that was somewhat less than normal frequency.
† Voice channels required 4.2×10^3 Hz per channel.
‡ Video channels required 4.2×10^6 Hz per channel.
SOURCE: Industry analyst's estimates. Reprinted with permission from page 15, ICCH #379-136.

EXHIBIT 3–B Coaxial versus Optical Underwater Cable Systems

System Parameter	Coaxial Cable	Optical Cable	Ratio: Coaxial to Optical
Cable diameter (cm)	3.2	1.6	2.0
Cable weight (kg/km)	1,250	625	2.0
Total cable weight (kg)	6×10^6	3×10^6	2.0
Total cable volume (m^3)	4,021	1,005	4.0
Cost of cable ($M)	35	25	1.4
Number of repeaters	150	500	0.3
Cost of each repeater ($K)	60	10	6.0
Cost of all repeaters ($M)	9	5	1.8
Total cable cost ($M)	44	30	1.5
Number of channels	300	600	0.5
Cost per channel ($K)	147	50	2.9

SOURCE: G. R. Elion and H. A. Elion, *Fiber Optics in Communications Systems* (New York: Marcel Dekker, Inc., 1978).

processes involved the extent to which the process was batch or continuous and the way in which the fiber was drawn. Walton International used the double crucible method, in which the "core" and "cladding" materials are drawn together from separate crucibles in a continuous process. See Exhibit 4 for a simplified diagram of the process. The advantages of this method are that it allows continuous production and provides a superior core/cladding interface. However, it can have the disadvantage of minor variations in the fiber's diameter which can limit its potential for low attenuation. Manufacturing problems were also sometimes experienced in the melt of the core and cladding materials. By 1978, fibers produced by this method had achieved attenuation losses of about 5 db/km.

The fiber fabrication process could be contrasted with the electronics manufacturing process used by OWS for demodulator and modulator instruments. Very few of the components

EXHIBIT 4 Double Crucible Fiber Drawing Process

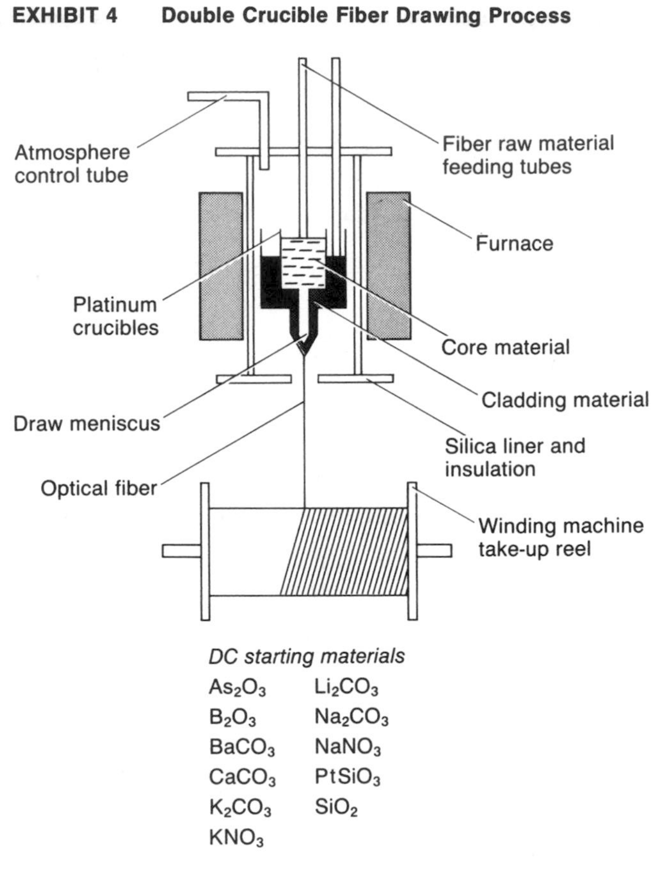

SOURCE: G. R. Elion and H. A. Elion, *Fiber Optics in Communications Systems* (New York: Marcel Dekker, Inc., 1978), p. 23, by courtesy of Marcel Dekker, Inc.

used in assembly of fiber optic instruments were made by Walton International. Lasers, diodes, transformers, and other components were purchased from other manufacturers. Instruments were assembled in batches of five and carefully tuned and inspected by an individual technician in the last stage of the process.

As fiber optic technology becomes increasingly sophisticated and as production costs decline, the market for optical fiber is expected to expand significantly. Current applications include: telephone loops, trunks, terminals and exchanges; internal and external computer links; cable TV; space vehicles; military and commercial aircraft; ships; submarine cable; security and alarm systems; electronic instrumentation systems; medical systems; satellite ground stations; and industrial automation and process control. Of all these, telecommunications and computer applications are expected to account for as much as three fourths of the market in the early 1980s.

By 1978, a diverse multitude of players had become involved in producing one or more components, and an even greater number of

EXHIBIT 5 Selected Manufacturers of Fiber Optic Products

Optical Fibers and Cables (*includes international*)

American Optical
Corning Glass
Fiber Optic Cable
General Cable
ITT
Pilkington Brothers
Quartz Products
Siemens
Valtec

Fiber Optic Connectors or Splices

Bell-Northern
Elecro-Fiberoptics
Fujitsu
ITT
Thomson & CSF

LED or Laser Light Sources

AEG-Telefunken
Hewlett-Packard
Hitachi
ITT
Laser Diode Laboratories
National Semiconductor
Tektronics
Texas Instruments
Thomson & CSF

PIN or APD Photodetectors

Bell-Northern
Bell & Howell
EG&G
Ferranti
Fujitsu
General Electric
General Instrument
Hewlett-Packard
Motorola Semiconductor
Plessey
Raytheon
RCA

SOURCE: G. R. Elion and H. A. Elion, *Fiber Optics in Communications Systems.*

potential participants were rumored to be waiting in the wings. Corning Glass Works and Bell Laboratories had established themselves as the technological leaders in manufacturing optical fiber, but many smaller companies produced components or complete systems. The participants came from many other industries: electronics, glass, wire and cable, chemicals, and others with peripheral connections to fiber optic technology. The industry, although embryonic, showed great potential. Exhibit 5 provides a partial list of competitors in the industry in 1978.

THE COMPANY

Walton International was established in 1958 by W. H. Walton, shortly after he was honorably discharged from the U.S. Air Force, to supply specialized electronic devices to the U.S. military. By 1978, the company had grown to $300 million in annual revenues and had diversified into supplying a variety of specialized electronic instruments and components to both industrial and military markets. Walton was known to be a technological leader in many of its markets and this leadership had become a source of pride to "the Colonel." Several years ago, he had established a corporate R&D lab to supplement R&D programs conducted in the divisions; he claimed that his focus on R&D was one of the reasons behind the company's success. A summary organization chart and financial history of the company are presented in Exhibits 6 and 7.

In 1975, Walton International acquired Optical Wavelength Specialists (OWS) in the belief that it could combine its electronic capabilities with experience in fiber optics technology to gain a foothold in what was thought to be a rapidly growing and potentially lucrative business. At the time, OWS had annual revenues of close to $30 million. OWS's main products were a broad line of fiber optic modulators and demodulators using both LEDs and lasers as

EXHIBIT 6 Organization Chart

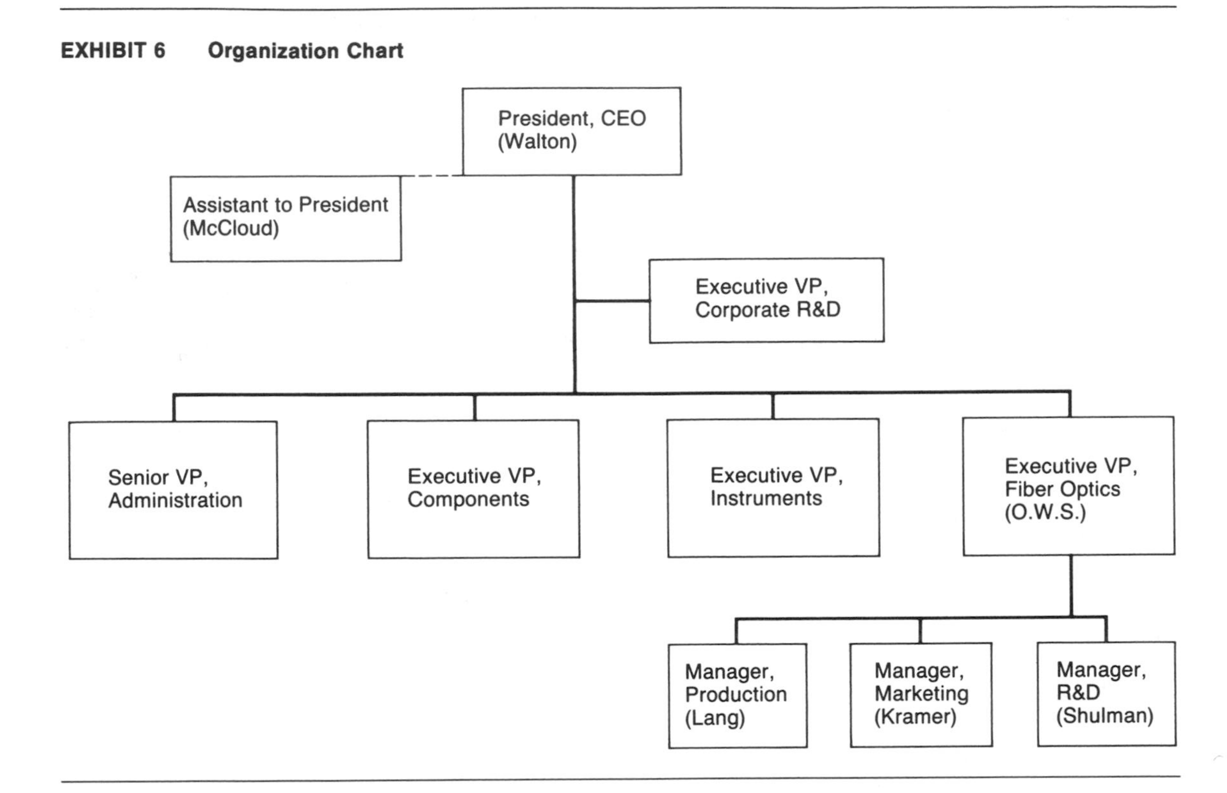

optical sources. The Walton International corporate R&D labs had also produced pilot quantities of soda-boro-silicate fiber optic cable.

Since its acquisition, however, the division had not lived up to expectations. Sales had remained relatively flat; over the last three years, its leadership position in low-loss short-length fiber manufacture and design had deteriorated. Sources and detectors now accounted for about four fifths of sales.[4] An additional $3 million came from cable sales. The division had just been awarded a $3 million contract from the U.S. government. This was the first major government contract for advanced sources and detectors that the division had won.

EXHIBIT 7 Statement of Income (millions of dollars)

	1976	1977	1978
Revenues	$235	$270	$300
Cost of goods	165	190	210
Gross margin	70	80	90
SG&A	31	35	41
Research and development	15	17	19
Profit before tax	24	28	30
Income taxes	12	14	15
Profit after tax	$ 12	$ 14	$ 15

OWS executives were asked to describe the division. All had been with OWS before the acquisition and had remained with the division. OWS's founder, Allan Gray, was one of the few principals in OWS who did not remain with Walton International after the acquisition.

John Lang, production manager for the division and 56 years old, had been with Corning

[4] A *source* was another term for *modulator*.

Glass Works before he joined OWS in 1972. He described the situation as follows:

> When Walton acquired us in 1975, most of us figured we would get the best of all worlds—we would get the financing we needed because Walton seemed so committed to the business, and we would get independence in running our own affairs. Well, it didn't work out. We had all the backing we could have wanted from Walton, but he didn't know how to run this kind of business and neither did the guy he sent in as division manager. We wanted to get into higher volume production because we feel we have the know-how and the market for our products, but too many people want their fingers in the pie.
>
> R&D, however, keeps changing the design specs on some of our fiber runs and coming up with so many new ideas—most of which don't work—that I can never get my shop under control. When we were just a small business and there were only a few of us around, everything worked fine.
>
> Now we've got these new sources and detectors which appear to be pretty good products, but combined with the rest of the disorganization around here they just complicate my job.

John Kramer, marketing manager and 32 years old, saw the situation in a different light:

> The business is evolving so rapidly now—you blink and someone's made a lower loss fiber or more efficient laser source. . . . The competition is fierce and believe me, it's technology that counts. At present, if you have a couple of first-rate components, you can make money. But five years from now, it may be that you need to have whole systems to sell in order to be considered a viable competitor. Anyway, aside from the long-term situation, what concerns me here is how we seem to be missing the boat on some of this technology.
>
> When Allan [Gray] was here, he knew exactly who was doing what and where R&D should be putting its best efforts. He *should* have known—he was a materials science engineer with a Ph.D. in physics from MIT, a real R&D man. But since 1975, even though we have some good new lines, we seem to lack strong direction. This is unfortunate because the industry is really taking off now.

Dr. Larry Shulman, 33 years old and the manager of R&D, had been acting division manager for three months following his predecessor's unexpected resignation. Dr. Shulman was considered one of the nation's foremost authorities on fiber optic electronic systems. In addition to a Ph.D. in physics from Princeton, he holds an M.S. in electrical engineering from MIT. He also holds 20 patents, about half of them on fiber optic systems. He had joined Walton International from Bell Labs the same year that Claire McCloud finished her M.B.A. At Bell, he had done research on fiber optic materials and systems. He stated:

> This business depends on its research talent. Without new ideas, a fiber optic company wouldn't last long because of the speed of technological change. Of course, there is also the problem of getting too many ideas, and that must be managed as well. No clear strategy was outlined and no direction was given to the division for R&D. We don't lack bright and talented people here; that's not the problem. We have one of the strongest technical teams in the industry. But R&D must be a first priority for this division, and we have got to organize around it.

THE DECISION

McCloud knew she had a great deal to consider that afternoon before she could make a decision on whether or not to take the division manager's job. She decided her first step should be to prepare a list of the important issues she felt she had to address.

Her first question concerned the technology itself. What did she have to know about fiber optics to be an effective manager? How much would she have to know to keep current of the advances taking place? She pulled from her files a recent list of papers on the industry (see Exhibit 8). Would the journals listed there keep her adequately informed? Even though she had had little prior training in engineering, physics, or electronics, she had picked up quite

EXHIBIT 8 Walton International: Recent Papers on Fiber Optics

Bodem, F., et al. "Investigation of Various Transmission Properties and Launching Techniques of Plastic Optical Fibers Suitable for Transmission of High Optical Power." *Optical and Quantum Electronics,* vol. 7, no. 5, 1975.

Brandt, G. B. "Two Wavelength Measurements of Optical Waveguide Parameters." *Applied Optics,* vol. 14, no. 4, 1975.

Eickhoff, W., et al. "Measuring Method for the Refractive Index Profile of Optic Glass Fibers." *Optical and Quantum Electronics,* vol. 7, no. 2, 1975.

Ford, S. G., et al. "Principles of Fiber Optical Cable Design." *Proceedings of IEE,* vol. 123, 1976.

Inada, K., et al. "Experimental Consideration of the Pulse Spreading of Multimode Optical Fibers." *Electronics Communication Engineers of Japan,* vol. 75, 1976.

Justice, B. "Strength Considerations of Optical Wavelength Fibers." *Fiber and Integrated Optics,* vol. 1, 1977.

Kapron, F. P. "Maximum Information Capacity of Fiber Optic Waveguides." *Electronic Letters,* vol. 13, no. 3, 1977.

Kawana, A., et al. "Fabrication of Low-Loss Single-Mode Fibers." *Electronic Letters,* vol. 13, no. 7, 1977.

Kuroha, T., et al. "Optical Fiber Drawing and Its Influence on Fiber Loss." International Conference on Integrated Optics and Optical Fiber Communication, July 1977, Tokyo, Japan.

Lussier, F. M. "Widening Choices in Fiber Optics." *Laser Focus,* June 1977.

Marcatili, E. A. J. "Factors Affecting Practical Attenuation and Dispersion Measurements." Topical Meeting on Optical Fiber Transmission, February 1977, Williamsburg, Virginia.

Marcuse, D. "Scattering and Absorption Losses of Multimode Optical Fiber Lasers." *Bell System Technical Journal,* vol. 55, no. 10, 1976.

Maurer, R. C. "Strength of Fiber Optical Waveguides." *Applied Physics Letters,* vol. 27, 1975.

Mims, F. M. "Measuring LED Power Distribution." *Electro-Optical Systems Design,* June 1976.

Titchmarsh, J. G. "Fiber Geometry Control with the Double Crucible Technique." Second European Conference on Optical Fiber Communication, September 1976, Paris, France.

a bit of general knowledge on the subject during her two years as Walton's assistant. She also felt familiar with the basic principles of fiber optics. Would this be enough?

What framework would she use to manage the division? What were the unique aspects of a fiber optics business that required a special management framework? In business school, she had developed an overall familiarity with technological innovation and some of the strategies for managing the innovation process; she thought these would give her a start. For example, one decision she felt she would have to make is where the division should position itself technologically in the fiber optics market. Did it have the resources and talent to become a technological leader, and if so, what kind of products should it emphasize—modulators, demodulators, or the basic fibers themselves? What would the organizational and recruitment implications be for being a technological leader? Where could she find the additional technical know-how it required?

This brought her to another question. How should new products be evaluated? R&D clearly had problems right now in its ability to manage new product ideas and evaluate those ideas. What system should she use to choose between proposals with apparently equal financial potential?

A further question was how the division should be organized. Should R&D really be the focal point for this kind of a business as Shulman recommended? Once she had developed a "technological strategy," she thought the organization problem would fall into place.

Nevertheless, she knew she would still have to choose among a variety of organizational structures. Should she organize functionally in a matrix structure or form new venture groups? How closely should OWS's R&D group work with the corporate R&D office? Her choice should, of course, also be very sensitive to the people currently in the division.

She felt that the people management issue was critical to her framework. How would these individuals be most productive and most satisfied? She knew that it would take delicate leadership to get the division working again as a "team." Not the least of her concerns was how a group of technically trained people would accept an economist—and a young one at that—as their boss.[5] How closely would she need to work with the vice presidents of R&D, marketing, and production? To what extent should she rely on their judgments? To what extent should she seek the counsel of other technical experts in the organization?

Finally, she would have to plan for her first priorities and would have to anticipate what her first problems would be. She had much more thinking to do before she could give Walton an answer in the morning.

[5] Claire McCloud was 30 at the time of the case.

II Designing a Technological Strategy

Technological strategy is an important but often ignored link in the strategic formulation chain. This is not to say that technology is altogether absent from the strategic plans of technology-intensive companies, but it generally appears in a fragmented, piecemeal fashion as part of other functional strategies such as marketing. In part this is to be expected. The concept of technological strategy is relatively new and is not yet part of the standard tool kit of the M.B.A. or the practicing manager. The objective of this part of the book is to address this gap by proposing and illustrating a framework that defines a business unit's technological strategy.

Functional strategies are the building blocks of business strategy. In turn, functional strategies are themselves defined by a set of interrelated decisions that define the business unit's posture toward financial, manufacturing, technological, and marketing related issues. Marketing strategy, for instance, deals with four principal variables: pricing, distribution, product planning, and promotion and advertising. This set of variables is sometimes referred to as the marketing mix or the "four *p*'s of marketing."[1]

[1] In order to make the four dimensions alliterate, purveyance is sometimes (and somewhat awkwardly) substituted for distribution.

For many firms, and especially for those competing in industries characterized by rapid technological change, developing a sharply defined and self-consistent set of policies to deal with technology is of paramount importance. To develop an effective technological strategy, several major issues should be addressed:

- Deciding on generic strategies for different technology-based businesses in the corporate portfolio.
- Choosing product-market combinations in the light of their evolving technological requirements.
- Understanding sources of technologically based synergies and technological leverage.

The technological strategy that evolves from these considerations will generally serve as the basis for several fundamental decisions:

1. Which technologies should be the basis of our business?
2. In which technologies should we become especially proficient? What distinctive technological competencies does our business require?
3. How should those technologies be embodied into products? (What criteria or guidelines should we use to design our products?)

4. Where should we obtain the requisite technology?
5. How much should we invest in technology development or purchases?
6. How should we organize and manage technology and innovation?
7. When should the technology be introduced to the market?

Just as marketing strategy deals with choices in *distribution, pricing, product planning,* and *promotion,* technological strategy deals with choices in *technology, product design and development, sources of technology,* and *R&D management and funding.*

There are eight cases in Part II of the book. Each case addresses one or more of these interrelated issues. The two closing cases in this part give the student the opportunity to integrate the factors discussed above into a coherent technological strategy which can then be tested for consistency with business strategy and for its competitive effectiveness.

Distinctive Technological Competences

Case II–1
Advent Corporation (C)

R. S. Rosenbloom

Early in November 1970, Henry Kloss was reviewing the progress Advent Corporation had made in the preceding months. The September profit and loss statement had registered a net profit of almost $30,000, against a cumulative loss of nearly $165,000 in the preceding 10 months. The new Advent cassette recorder, Model M200, had just completed its third month on the market. The M200 recorder, with its sophisticated circuitry, was felt to represent real potential as a replacement for the phonograph as the central element in any home entertainment system. With the financial turnaround, Mr. Kloss felt confident that a sales level of $40 to $50 million was achievable by Advent within five years. His problem was how to organize for continuing innovation.

Harvard Business School case 9-674-027

INTRODUCTION

Mr. Kloss was a well-known figure in consumer electronic product design and manufacturing. Prior to Advent, he had participated in the founding and operation of Acoustic Research, Inc. (AR), and later, KLH Corporation. He had been the mind behind the products at KLH, an organization which was renowned for its very high-quality, slightly odd-ball electronic products. He left KLH in 1967 after 10 years as president.

The formation of AR had originated during the Korean crisis. While stationed in New Jersey, Mr. Kloss was able to attend the City College of New York, where he was a student of Edgar Vilchur. He and Vilchur had mutual interests in an acoustic suspension speaker because of its immense reproductive advantages over conventional mechanical speaker systems and its small size. With Mr. Kloss providing some capital and a "garage," Acoustic Research, Inc., was formed. Financial guidance of the business was provided by Anton (Tony) Hofmann, who was later to become a principal of KLH, and then treasurer of Advent.

Mr. Kloss and other active management sold their share of AR, Inc., after irreparable dis-

EXHIBIT 1

Advent Corporation (C)
Balance Sheet
As of September 26, 1970

Assets		
Current assets:		
Cash		$ 64,488.34
Accounts receivable		650,226.68
Less: Reserve for bad debts		(10,000.00)
Advance to employees		(650.00)
Inventory:		
Material	$ 375,486.13	
Labor	37,076.97	
Manufacturing overhead	38,189.91	450,753.01
Prepaid insurance and other assets		10,958.21
Total current assets		1,165,770.24
Property, plant, and equipment	221,030.07	
Less: Accumulated depreciation	(57,524.71)	163,505.36
Deferred financing expense		5,450.00
Advent television system		205,085.92
Total assets		$1,539,811.52
Liabilities		
Current liabilities:		
Accounts payable		$ 347,449.36
Notes payable, bank		666,714.00
Due officers		0.00
Loans, other		50,000.00
Accrued debenture interest		4,002.65
Accrued payroll		17,441.39
Royalties payable		(2,000.00)
Accrued royalty expense		7,584.80
Accrued audit and legal fees		20,628.38
Accrued taxes and fringe benefits		37,083.67
Accrued promotion and discount allowances		65,312.20
Miscellaneous accounts		17,471.29
Total current liabilities		1,231,687.74
Long-term debt:		
8% convertible debentures		200,000.00
Stockholders' equity:		
Common stock (10¢ par value)	$ 45,925.10	
Additional paid-in capital	821,866.29	
Retained earnings deficit to 10/31/69	(595,130.66)	
Deficit 11/1/69 to date	(164,536.95)	
Total stockholders' equity		108,123.78
Total liabilities and stockholders' equity		$1,539,811.52

agreements with Vilchur over company policies. KLH was initiated shortly thereafter with $60,000 in capital and Mr. Kloss as president, Malcolm Low as manager of sales, and Mr. Hofmann as financial manager. After seven years and a series of innovative audio products that were producing a $4 million level of sales, KLH was sold because of sheer tiredness of the managers and uncertainties associated with KLH's growing size. With the sale, Mr. Kloss agreed to remain as president for three years, and left in 1967.

EXHIBIT 2

Advent Corporation (C)
Statement of Profit and Loss
As of September 26, 1970

	Current Month			November 1, 1969, to Date
Gross sales:	Units	Amount		Amount
Regular speakers	1,561	$115,222.44		$ 685,003.10
Utility speakers	278	17,838.58		46,653.19
F.B.C.	161	23,148.34		182,995.73
M 100	303	50,481.83		260,995.63
M 101	295	24,139.50		68,485.25
M 101 Advocate	146	11,826.00		13,284.00
M 200	988	170,718.00		245,960.30
CC-1	6	100.02		363.40
WC-1	138	1,603.12		3,371.03
Parts	—	605.00		2,757.16
Crolyn tape	1,824	3,997.44		11,108.20
Total		419,680.27		1,520,976.99
Less: Provision for promotional and quantity discounts		21,489.78		86,385.59
Net sales		398,190.49	(100%)	1,434,591.40
Cost of sales:				
Material		196,431.82		663,770.07
Labor		41,366.25		199,930.01
Manufacturing overhead		45,908.85		232,392.33
Royalties		3,182.88		13,222.26
Total cost of sales		286,889.80	(72%)	1,109,314.67
Gross profit		111,300.69	(28%)	325,276.73
Operating expenses:				
Sales		47,517.13		242,799.78
General and administrative		13,753.68		91,570.51
Research and development		14,371.13		195,877.20
Total operating expenses		75,641.94	(19.0%)	530,247.49
Operating profit (loss)		35,658.75	(9.0%)	(204,970.76)
Other income (expense)		(6,428.79)	(−1.6%)	(34,652.11)
Capitalization of Advent TV system, (included in R&D above)				75,085.92
Net profit		$ 29,229.96		$ (164,536.95)

Advent Corporation was incorporated by Mr. Kloss in May 1967 for the purpose of manufacturing specialized electronic products for home entertainment use. The actual justification for forming the company was to do work in television, especially to create an organization which would support the R&D and marketing of a large screen (4′ × 6′) color television system. Formal development work on the television system had been suspended in 1970.

With the formation of Advent Corporation, Mr. Kloss embarked on a plan to see what a big company could do. He felt that growth was always a primary goal, always desirable, but that one had to think in terms of what was realizable without beating his head against the wall. Mr. Kloss sought to retain strong financial control of the company, having sold his share of Acoustic Research, Inc., under duress and his share of KLH Corporation with mixed feelings. He had this to say to the case researchers about financial policies:

> The size one desires is really only limited by the dollars available for working capital. There's a firm intention to reach the middle tens of millions of dollars certainly in less than five years; one anticipates a faster accumulation of staff, faster than the 30 percent one might be able to do from profits, so the question becomes how fast does one dribble out equity if you're not staff limited?

Mr. Kloss continued,

> Eighteen months ago, there was a small private offering of 12 percent of the company in which we offered 20 units consisting of $10,000 in 8 percent convertible debentures, and 300 shares of equity common at $7.50 per share, 10 cents par value. I retained 75 percent control; company directors and others have 13 percent. It was simply that circumstances warranted our doing that. In addition, we have a $1.15 million line of credit, of which $600,000 is revolving and $550,000 open, secured by the directors and pegged to 80 percent of the accounts receivable. I will not offer any further equity until a really big push (for which the sales are guaranteed) requires it, and when a price several times the $7.50 price per share is attainable. Beyond that, we are working hard to slash overhead and to build profits.

Financial data regarding the operations of Advent Corporation are given in Exhibits 1 and 2.

CURRENT OPERATIONS

In the fall of 1970, Advent Corporation manufactured and sold five products for home entertainment use: the Advent loudspeaker; the Advent Frequency Balance Control, which allowed the listener to alter the relative musical balance in any audible octave; two models of the Advent Noise Reduction Unit, which allowed virtually hiss-free tape recording and playback; and the new Advent Tape Deck, which also featured noise-free recording and playback. These products, as well as a special recording tape that Advent sold under license from Du Pont, are described in detail in Exhibit 3, in a piece of Advent promotional literature.

Several specific policies of Advent Corporation served to interlock the company with the consumer electronics market. Most important, perhaps, was product policy. Mr. Kloss felt that there were several repugnant aspects to direct competition with the industry giants such as Zenith, Magnavox, and Motorola. Advent sought to turn to specialized areas of the audio market, the 5 percent or so where no competition existed, where whole new classes of products might be developed. Quality was an important Advent byword; to make the most efficient piece of equipment at the lowest possible price to the consumer was the primary objective. Such product sanctity was not protected by patent but rather by the product itself, which had a real name, which gathered equity as it was seen and became known, and which hopefully represented the perfect low-price product. Even though the entry fee was low, Advent anticipated specializing upon a base product already determined by the major suppliers (e.g., tape

EXHIBIT 3 A Progress Report from Advent on Loudspeakers, Cassette Recorders, a New Kind of Tape, and Other Matters

After more than a year in business, we (Advent Corporation) think it's time for an accounting of where we are and why.

We began, you may remember, with the intention of making products that would differ significantly from other people's—products that would fill special needs others weren't filling, explore genuinely new ways of doing things, and keep testing accepted limits of performance and value.*

One of the products we had in mind was a new kind of color television set, a high-performance system with a screen size several times the present limit for home use. We are happy to report that it's coming along nicely (and slowly, as such things do), and that the present prospects for prerecorded video material make it look more appealing than ever.

Audio, however, was where we could do the most the quickest, and our first product was:

The Advent Loudspeaker

Anybody who knew us might have predicted that we would make a loudspeaker system pretty early in the game, but few would have predicted that we'd make just *one,* call it simply The Advent Loudspeaker, and say flatly that it was the best we could offer for a long way into the future.

The reason for that was, and is, that it had become possible to design a speaker system as good as anyone would ever need for home listening—one as good in every measurable and audibly useful way as any speaker system of any size or price—at a cost slightly below what most people consider the "medium price" category. Our prior experience in design and manufacturing techniques convinced us that this could be done, and we did it.

We will be happy to send you full particulars on The Advent Loudspeaker, including its reviews. But we believe its sound will tell you quickly enough why it has become, in its first year, one of our industry's all-time best sellers.

(To avoid surprises in a showroom, we should note that our one speaker system comes in two styles of cabinet: the original walnut model, priced at $125†, and a "utility" version that is actually in a rather handsome vinyl finish that looks like walnut, priced at $105.† Both sound the same.)

All of the first year's reviews of The Advent Loudspeaker finished by saying that it was an auspicious beginning for a company. But it represented only one of our immediate directions. The next was:

The Advent Frequency Balance Control

One of the things to be learned in the design of speaker systems is that "flat" frequency response is in the ear of the beholder and virtually nowhere else. True, there are amplifiers and tuners with straight-line frequency response, but practically everything else—recordings, listening rooms, cartridges, loudspeakers—is anything but flat. Different things sound different, not because of basic differences in quality or performance in many instances, but because a recording engineer, or speaker designer, or room plasterer had a slightly special view of the world.

There is nothing wrong with those differences, in our view. And one of the challenges for a speaker designer is to accept and cope with them by designing for an octave-to-octave musical balance that sounds "right" with the widest variety of present recording techniques. But there is no single perfect balance, and that lack is a source of discomfort to a number of critical listeners. It causes many listeners with really superb (and really expensive) sound equipment to keep trading for new and more expensive equipment in the hope that it will sound "perfect" for everything from Deutsche Grammophon's conception of the Berlin Philharmonic's sound to Columbia's notions about Blood, Sweat, and Tears.

Anyone who keeps pursuing that ideal, and many who don't, would be well advised to investigate our Frequency Balance Control, a unique device that enables listeners to alter the relative musical balance of any octave in the audible fre-

EXHIBIT 3 (*continued*)

quency spectrum. It is uniquely flexible and uniquely effective in dealing with sonic differences between recordings, equipment, and even the placement of speakers in a room—and in making things sound subjectively "right" more consistently than could be accomplished any other way.

The FBC, designed around our own experience with subjective judging of sound quality, is worth investigation by anyone who can't just sit back and listen, accepting the bad with the wonderful. At $225‡, it is a far better, more pertinent investment than most changes of components.

One of the special abilities of the FBC is the reclaiming of many recordings from an unlistenable state. The need for another kind of recording reclamation led to another kind of product:

The Advent Noise Reduction Units (Models 100 and 101)

Background noise in tape recording—specifically, tape hiss—is a far bigger enemy of sound quality than most listeners realize. One reason it isn't properly identified (and vilified) is that few people have heard tape recordings without it. Lacking the standard of blessed silence is something like never having seen a television picture without "snow." If you don't know it isn't supposed to be there, you just look or listen past it and accept it as part of the medium. But once you see—or hear—things free of interference, life is different.

Getting rid of tape noise is a prime function of the now-famous Dolby® System of noise reduction, which in its professional version is in use in virtually every major recording studio in the world.§ We became interested in the Dolby System not only because it helps rid even the best conventional tape recordings of background noise, but because it had even greater possibilities when applied to low-speed home tape recording. Home recording at 3¾ and 1⅞ ips has been plagued by the problem of really excessive tape hiss—which manufacturers have chosen either to tolerate or to "reduce" by giving up frequency and dynamic range in recording at those speeds. The Dolby System makes it possible to remove that problem and get first-class performance at the low speeds best suited, from the standpoint both of economy and convenience, to home recording.

So we designed a product that would make the Dolby System available—in a version designed by Dolby Laboratories exclusively for home recording and prerecorded tapes—for use with any good tape recorder. The product was our Model 100 Noise Reduction Unit, a flexible and effective piece of equipment that can make any recorder sound better and can do wonders in opening up the world of low-speed recording to the home user.

The Model 100 combines the Dolby System with a recording control system that supersedes a recorder's own and provides a recording accuracy and simplicity seldom seen in home tape equipment. One crucial advantage of that control system, which provides separate input level controls (with input-mixing) *and* a master record-level control, is that it gets stereo recording balance right and does so easily. Improper balance, almost guaranteed with many tape recorders, is the chief reason for recordings (on even the best recorders) that don't sound like the original. It is, in other words, the chief reason for many people's dissatisfaction with their recorders.

The Model 100, at $250, is a required investment for anyone who takes recording very seriously and measures the results critically. But since some people won't need its tremendous flexibility, we also decided to offer the Model 101—which, at $125, provides identical performance at half the price. To make that possible, we omitted the input-mixing provided with the Model 100, supplied slightly less flexible recording controls (it takes a bit longer to get stereo balance just right), and provided one Dolby circuit per channel instead of two. (As in the professional studio Dolby System, you switch the Model 101's two circuits to function first for stereo recording and then for stereo playback, but not for both at the same time.) The result, again, was performance identical to the more elaborate unit, at a price that makes sense for serious recordists on tight budgets.

EXHIBIT 3 *(continued)*

While designing the Noise Reduction Units, we became interested in what the Dolby System and other factors might do for a kind of tape recording that no one was taking seriously enough. The result was:

The Advent Tape Deck (Model 200)

We had known before, and confirmed in our work on the Model 100, that tape hiss was the underlying reason for the compromised, Am-radio kind of sound quality that people had come to associate with cassette recording. Because the hiss was present in a quantity that made wide-range recording unpleasant to listen to on cassettes, it had effectively set an upper limit on quality—giving manufacturers little incentive to optimize *any* aspect of cassette recording, including mechanical performance.

We realized that once you used the Dolby System to get rid of the noise, you would then have reason to go on to improve all the performance areas that nobody was really attending to. So, to show just how good cassette recording could be, we optimized everything we could around a good cassette transport, added our Noise Reduction Unit, and held a demonstration for the press. The reaction, even though we couldn't demonstrate everything we wanted to in a rigged-together unit, was that we had proved that cassette performance could be as good as, and in some ways better than, the standard for records.

In the meantime, we worked on our own cassette recorder—which was to include not only the Dolby System and the necessary improvements in all areas of performance, but also the means, not given to our knowledge with any previous cassette recorder, to make really superb recordings. That meant effective and precise controls for setting balance and recording levels, including a VU meter that read both stereo channels simultaneously and indicated the louder of the two at a given moment.

We felt that calling the resulting tape machine a cassette recorder wouldn't fully indicate our conviction that it was probably the single best choice among *all* kinds of recorders for most serious listeners who want to tape records and broadcasts. So we called it The Advent Tape Deck (Model 200) and let its being a cassette machine speak for itself. At $260, it is a new kind of tape machine that we hope will prove the key, given "Dolbyized" commercial cassette releases, to making cassettes the medium most serious listeners prefer for most listening.

About midway in our development of The Advent Tape Deck, we became convinced that the Dolby System's contribution to performance would become even greater if it were combined with the use of DuPont's chromium-dioxide tape in cassettes. Lots of people had been talking about DuPont's "Crolyn," but nobody had hard facts on what it could do in cassette recordings.§ So we got samples, experimented with its characteristics, and were convinced that we had to supply a means to use it on our recorder. That meant a special switch on The Advent Tape Deck to provide the right recording and playback characteristics (a good bit different from those of other tape formulations) for its use. It also meant another product:

Advocate Crolyn Tape

Although DuPont's Crolyn tape was being used extensively in critical video recording applications, and justifying its advance press notices, no one had made the leap to marketing it for audio purposes for home use. We decided to do so because we felt that Crolyn was necessary for the very best in potential cassette performance.

We are, then, marketing Crolyn tape under the "Advocate" brand in cassettes. One of our hopes in doing so is to get others to market chromium-dioxide tape as well.

There is no doubt in our mind that it's worth the trouble. Chromium dioxide has the ability to put greater high-frequency energy on tape than other oxide formulations, and is also increasingly sensitive as frequency goes up. Those are ideal characteristics for cassette recording, making possible a still greater signal-to-noise ratio in conjunction with the Dolby System and better overall high-frequency performance than any other tape we know of.

EXHIBIT 3 *(concluded)*

The Advent Packet

At this writing, we can't predict exactly what product is going to follow Advocate Crolyn tape. As you probably have noted by now, we develop products in what might be thought of as organic style, letting each product stand on its own. We don't sit down and decide to manufacture a "line" of speakers or amplifiers or tape recorders.

We are into other things at this point, and hope that they will be firm enough to talk about soon. In the meantime, we invite you to write us at the address below for any information you would like, including a list of Advent dealers.‖

If you like, ask for "The Advent Packet." That will bring you everything we have on all of our products, and will also—unless you specify otherwise—put you in jeopardy of getting future informational mailings from us.

So much for the first year.

---

* Having helped found two successful companies previously, and having prior credit for some of audio's most significant products (including something like half of the loudspeakers in use in music systems and serious radios and phonographs in this country), our president, Henry Kloss, had some pretty firm notions about what he wanted to do now.

† Slightly lower in some parts of the country.

‡ Slightly higher in some parts of the country.

§ "Dolby" is a trademark of Dolby Laboratories. "Crolyn" is a trademark of Du Pont.

‖ Advent Corporation, 377 Putnam Ave., Cambridge, Mass. 02139.

decks), which had an appeal to a broad spectrum of the market.

Production operations of Advent Corporation were closely supervised by Mr. Kloss, although there was a production manager for all but the M200 line. Speakers were manufactured in a separate 12,000-square-foot plant in Cambridge, Massachusetts. Major operations of the company took place in a 20,000-square-foot, three-story building also in Cambridge, which Mr. Kloss leased upon forming the company. A move was being planned to consolidate the operations of the company in the spring of 1971 into a 64,000-square-foot building also in Cambridge, which had already been leased.

Production itself was typical of the small manufacturers in the industry. Approximately 130 production workers formed the products in a specified sequence of assembly steps that was usually determined by Mr. Kloss. He also carried out "time and motion" studies to determine an appropriate production rate. No significant economies of scale existed in the industry. Mr. Kloss felt very strongly that higher overhead would destroy any advantages to be gained by mechanization. In addition, it seemed that after a quantity order of 100 per week or more, no important savings could be gained from higher quantity parts orders. It had been found that direct labor ran about one half of material cost over a wide range of products. With manufacturing overhead being determined as a percent of direct labor, cost of sales could easily be forecast for any given product, and a price determined on the basis of a typical margin percentage. Pricing policy, therefore, was also dependent upon the emphasis on making an excellent low-cost product, and not on selling products at a what-the-market-may-bear level.

Marketing management at Advent was a relatively autonomous activity. Vice president of sales was Mr. Stan Pressman, who had performed similar duties at KLH before coming to Advent. Nationwide distribution was maintained through 150 dealers across the country, who were carefully selected on their ability to sell and service Advent products intensively. Shelf space was originally attained by contacting each dealer personally and promising him a succession of useful and high-quality products, with which it would be valuable for him to be associated. The reputation of Mr. Kloss was also emphasized. Finally, exposure to the trade press and to the public had been attained through press conferences designed to place the Advent audio products in sink-or-swim competition with similar offerings then on the market. Response had been overwhelmingly favorable.

Under pressure to reach the marketplace with successful products and to improve profitability, Advent had expanded on a day-to-day functional basis. Emphasis on "continually optimizing its position" rather than responding to a long-range plan had placed substantial importance upon production efficiency and rapid response to daily marketing problems. As a result, current operating managers were expected to monitor the functions of their departments in fine detail.

INNOVATION AT ADVENT

Both Acoustic Research and KLH had demonstrated the ability to recognize changing product and consumer trends and to respond quickly in a dynamic marketplace. Henry Kloss had been able to achieve similar success during the initial life of Advent. Mr. Kloss was unable to explain why Advent had succeeded in accomplishing responses to market needs in advance of other companies in the industry. He discussed this phenomenon at some length during conversations with the case researchers in his president's office, a room that was bewilderingly cluttered with all sorts of electronic gear. His desk was laden with trade journals and other papers reporting the current developments in home electronics. Only a few feet from his desk was the door that led to the R&D section, which was never seen closed. Mr. Kloss said,

> Perhaps a recent example will highlight what I mean. Du Pont Company, which is really not concerned with products at all, I mean, their basic formulations are raw materials or processes to make raw materials, recently developed a way of making a material which is simply a process kind of thing. That was chromium dioxide, which can be used in the manufacture of magnetic tape and which results in a really quite important product. But Du Pont stopped short very early in the process. They'll sell you all the chromium dioxide you want. But their involvement with the resulting product (Crolyn tape) was absolutely nil. A lot of time was lost until Advent recognized the product and did something about it. They (Du Pont) had no market for it at all. And they are extremely grateful to us for it now. I really didn't think a big company could be so pleasant to work with.

Casewriter

Are you suggesting that product innovation is primarily characterized by observation that a need or a market exists for that product and then going after it, after that specific product?

Mr. Kloss

Yes. And from the process innovation, which is a new way of making something, or some new combination of things. Often a new process could have a connection with a new product, but it doesn't tell.

All of *our* working has been backwards from the person. Others work hard to find a physical phenomenon, or to develop a new bearing, and then work hard to find a market. This is to work completely in isolation, with no connection to the product at all.

Nobody asked at Du Pont, "In what way can this new process make a higher quality result?" At the same time, we were asking, "In what way can this be used irrespective of presently established systems of using tape—what are the limitations inherent within this tape on its ability to produce music for the listener at home?" And we found that it had a distinct and strong advantage, and this had not even been done by the Du Pont people. You know, it's really hard to believe! I'm not trying to boost Advent, or knock Du Pont, but their detachment from this thing in terms of people was absolutely complete!

Exemplifying the kind of reasoning that went on at Advent prior to a product decision, Mr. Kloss mentioned the following incident:

> Somebody came around the other day with a way of making a very high-powered amplifier that requires only a very small size and bulk. Any normal amplifier wastes up to half its power at any one moment in heat loss. There is a way of making an amplifier, which we've known for some time, that is 97 percent efficient—you waste almost no power in the amplifier itself. Now since the size of the amplifier is largely determined by the need to dissipate power, clearly the size here could be reduced. One has known it can be done; it's called Class D circuitry. This size might make possible a whole new class of things; whether we do this in a year or so is uncertain. But it's a possible kind of thing, which we didn't go to invent, and which has been around for years and years and years, but which might become practical to do, if you do the rest of the things to get all the merit out of it, such as creating a small power supply and all, which calls for minor invention on our part. We've had a feeling that exceedingly small kinds of things were worthwhile; when something like this comes up, you notice it more sharply than somebody else, who looks at it only as just a cheaper way of getting a high-power amplifier.

Formal market research at Advent Corporation was never mentioned. Mr. Kloss had the following remarks to offer when the case researchers asked him about it:

Mr. Kloss

Oh! One never does market research! The only test of the market that there will ever be is to fully commit to a product itself; one is never going to make any test marketing or any asking of anything. And it will be done whenever it's the product that will most certainly, most quickly, give a certain amount of money here. It's just a matter of priority of products; one could, within a couple of months' time, make a noise reduction unit and turn it into a product and sell it. That had to be done first.

Casewriter

But with all due respect, you must feel that it will go, that when people see it, they're going to buy it?

Mr. Kloss

Well, yes. But there's no way of proving this before you spend the money to produce it, that I know of.

Casewriter

Experience and intuition tell you that it will go?

Mr. Kloss

This is about, yes, all that one has. Experience that my intuition has been right gives me a little more confidence, maybe.

THE DOLBYIZED CASSETTE RECORDER

Critical to Advent's recovery from unprofitable development operations was the successful manufacture and marketing of the Dolbyized cassette recorder (Advent Tape Deck, M200), described briefly in Exhibit 3. The way the idea of noise reduction recording became a product for Advent, and at just the right time, is indicative of the whole Advent innovation process.

Mr. Kloss had noted very early in the company's history that it was possible, in theory, to do something like noise reduction. That is, he noted that at any moment in the recording process, the normal recording methods from basic information theory resulted in great waste. He noted that there ought to be some way of continuously optimizing the recording technique. However, his investigation stopped there. He knew it was possible, but he did not embark then and there upon a process of invention. Instead, Mr. Kloss became sensitized to noticing if somebody else had really done it. All of the Advent products began in familiar fashion. Mr. Kloss commented,

> The things that I have done have never started from noticing something was important and then working backwards to the fundamental way to do it. You know, "Gee, it would really be desirable to have instant photographs," and then work hard to do it. I don't know if that's what Land did or not. But that has never been our particular way of doing things. All of the work has been to think about things that would be desirable to do, and then be continuously looking around to see what things are possible to do, perhaps with minor invention on our part, which would satisfy a perceived need in the market and begin to define a product. Only when the need in the marketplace simultaneously matches the knowledge of the technology does one spend more than a few minutes thinking about it.
>
> . . . So any product I think of for longer than a few minutes is already one that I know can be made. . . . You want to constantly have in mind, stored with very short access time, the different technologies. You sort of somehow keep aware of what kind of things can be done. When several of these come together to form a product, that can result in your deciding to make that product. You have to have, at any moment, a moderate-sized number of floating possibilities of things that you can do.
>
> But there's a cost to this floating process of having all these pieces of information available which makes it very hard to expand to a large group of people.

In 1967, Mr. Kloss heard about Ray Dolby, a man who had been making professional noise reduction systems in England and was just starting to sell them in the United States. That was just at the conclusion of Mr. Kloss's presidency at KLH. He negotiated an agreement between Dolby and KLH for KLH to have the rights to incorporate that system in a tape deck. Mr. Kloss agreed to manufacture that tape deck for KLH, to help KLH introduce the Dolby system to the world.

For many reasons, the product, which was envisioned as a $600 reel-to-reel machine with Dolby circuitry, never got made. By May 1969, Mr. Kloss personally had suffered a loss of $265,000, largely through design and production problems. At that time, Advent began manufacturing the Advent loudspeaker to support further development work on the large-screen television. Simultaneously, KLH had renegotiated a manufacturing contract with a Japanese firm, Nakamichi, to build a $250 reel-to-reel machine with Dolby circuitry. Such a product was on-line by the fall of 1969, when Nakamichi offered KLH a similar deal on a cassette recorder with Dolby circuitry. Mr. Kloss described the events that followed:

> Even though KLH had a selling reel-to-reel machine with Dolby, they decided not to make the cassette machine. There were many reasons for this: they were having trouble with the Nakamichi machine they had, they had had gross trouble with my deck, and they had just gotten a new president who was against expansionist moves. So they just backed off the whole thing, just when the right product was there. Advent's contribution to the process was really a floating knowledge of the benefits of chromium dioxide tape, the Dolby circuit, and a manufacturer of heads who knew about Dolby. It was gathering these things together into a product and bringing them to people's attention that Advent accomplished.

Within hours, before the Nakamichi representatives had returned to Japan, Mr. Kloss had negotiated an agreement granting Advent the productive capacity to employ Nakamichi heads in an Advent deck. The Dolby system it uses is described in Exhibit 4. While the new product received numerous adulations in the press, by October 1970 Mr. Kloss felt that the primary shift toward central cassette recording that he had expected with the marketing of the Advent Tape Deck was not occurring as fast as he had hoped. He felt that the primary reason for this deficiency was the inherent difficulty of depending on a dealer organization to push Advent products that incorporated sophisti-

EXHIBIT 4 How the Dolby System Works

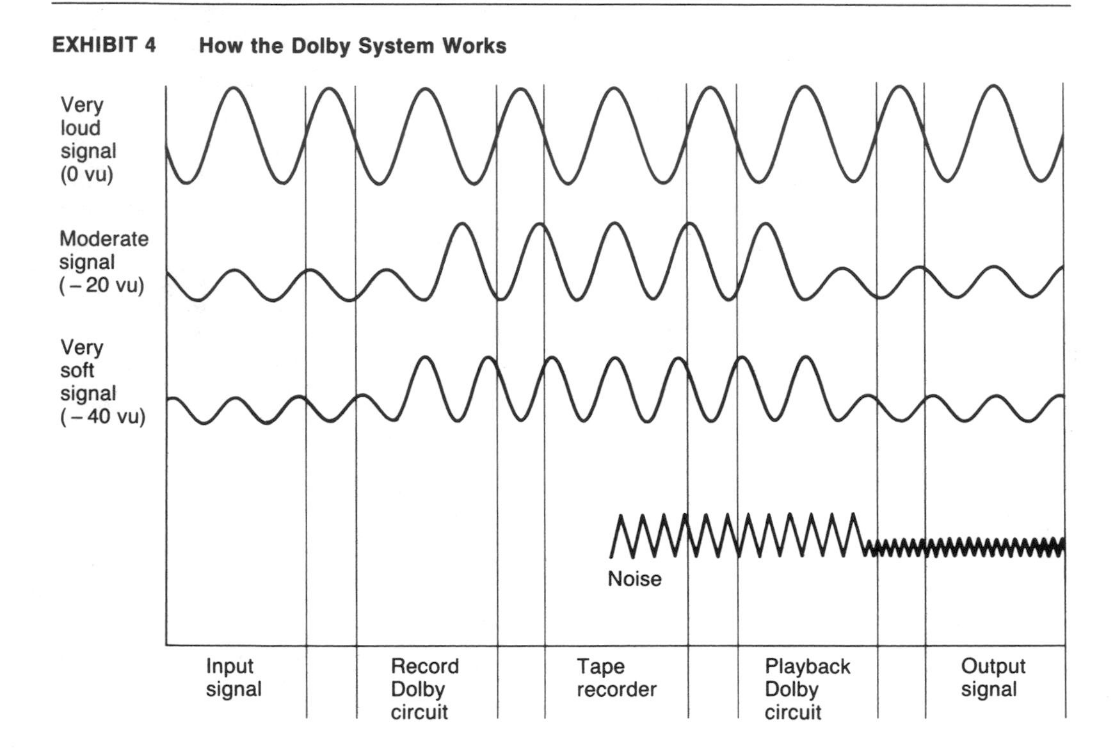

1. The signal being recorded passes through the record Dolby circuit *first.* The Dolby circuit operates on the higher ("hiss") frequencies in a predetermined manner, depending on their loudness level. The loudest signals (0vu) pass unaffected through the circuit. Signals of moderate intensity (−20vu) are boosted moderately, while the very soft signals (−40vu) receive maximum boost.

2. After being thus "Dolbyized," the signal is recorded onto the tape. It is at this point that tape hiss makes its appearance. You can see on the diagram how the record Dolby circuit's action has made the low-level signal louder than usual, relative to the tape hiss.

3. On playback, the signal from the tape is passed through the playback Dolby circuit, which is an exact "mirror-image" of the record Dolby circuit. The playback Dolby *lowers* the previously boosted parts of the signal, by precisely the same amount they had been boosted. The tape hiss—which made its appearance between the record and playback halves of the Dolby System—is automatically lowered at the same time by a very substantial amount, effectively 10 db or 90 percent. At the same time, because of the precise "mirror-image" playback action, the Dolby System causes no other change in the signal relative to the original source that was recorded.

cated innovations, features that had to be understood by the consumer before he made the logical choice of an Advent product. Consequently, he and Mr. Pressman were spending considerable time in attempting to find a solution to this problem, the final step in completing the innovation of the Dolbyized recorder.

TELEVISION

Although Mr. Kloss had suspended formal development work on Advent's large-screen color television set, he continued to make minor modifications to it when time was available. Several experimental sets functioned without major problems in the Advent plant and homes of employees, but decisions remained as to the exact design the set would have and the marketing approach to be used. Mr. Kloss estimated that the first production models would be available for sale six to nine months after the "go" decision was made, and that the decision would be made "whenever it's the product that most certainly, most quickly, can give the right amount of money here."

Describing the product's origins, Kloss said:

> I was vaguely interested in TV as an important medium. One reads a magazine article that points out a way to make projection television. All you had to do was read that article and see that it could apply to a screen this size [four and one-half feet by six feet]. And then you quickly ask the question "Is this worthwhile?"; you make a guess that it might be worthwhile at the right price.

In 1964, Kodak announced the development of a screen which could effectively increase the amount of perceived reflected light by a factor of five over ordinary mat screens. This development suggested that it might be possible to diffuse light from a projection tube over a larger screen of this type and still retain satisfactory brightness. However, Mr. Kloss said that he would have built a high-quality television set even if the large screen had not been possible.

> If there never was a big screen, we'd be in television anyway because you can do a high-quality small set. So our interest in TV is not restricted to the big screen, though it's much more fun because there's no comparison available.

Mr. Kloss believed he could discern in color TV the typical product life cycle of consumer electronics products working to the advantage of new producers with sufficient marketing skill. During the late 1950s and early 1960s, color TV quality improved as bugs were worked out of it, but by 1966 short-cut production methods were reducing overall quality. Mr. Kloss observed that

> NTSC [National Television Standards Committee] standards permit a very high quality to be broadcast which is usually badly degraded by a set at home. What you see on a regular picture is not what you would see on a really high-quality set.

Evaluation of demand, though not verbalized, suggested to Mr. Kloss that larger screen TV was an inherently desirable thing. The evaluation was not, he said, an extrapolation of the popularity of larger screens in ordinary television sets. Nor could he isolate any one other factor which dominated his evaluation except that it was the kind of thing he would like to have in his home. There was no way to extrapolate from sales of expensive large-screen sets. He said:

> There is absolutely no experience on large-screen television for consumer use. . . . Yes, I feel that large-screen TV will be popular but there's no way to prove this until you spend the money to produce it, that I know of.
>
> A lot of people go to the flicks. The whole business is to bring things up close, large and important. . . . This is doing that and there's that kind of rightness about it. That's about the only defense one has. It just doesn't have any connection with television as one thinks about it. Once

> you say television, somebody brings to mind almost repugnant kinds of images. They don't do it for books though. You talk about books and they think about great books and the University of Chicago, and this kind of thing. They don't think about the kinds of things they sell down on Washington Street.
>
> . . . And for big screen, there's no expressed desire for anybody to want a big screen. . . .
>
> Exactly what's happening out in the store, where people are expressing what they want, sure I get some information on that from somebody else. But this sort of shapes the end features of products. People are not out there expressing a new kind of thing that they would like to have—a compacter for kitchen garbage; I've never heard anybody say that they ever wanted something like this. I think maybe some people do; we'll see. . . . The kind of products that people might want are not limited to what people have said they want or what people, when you knock of their door, say that they will want. In the first case, it's too late if people express the desire for what they want. In the second case, the answers are invalid when you ask about it.

Development

Shortly after organizing Advent Corporation in May 1967, Mr. Kloss began working on the television set. Though he was confident that the idea was technically feasible, there were many questions yet to be answered regarding design. For example:

> The way of finishing mirrors at a very low cost—it's been used in the eyeglass industry; it's not used in making lenses; it's not used in telescopic work. But the technology to make very low-cost kinds of mirrors exists in the trade. And we sort of know that technology is there and go and use it. If it had required our finding a very low-cost way of making a lens which hadn't been developed yet, I would have cut out from any of our consideration the making of a low-cost projection television. . . . It maybe would have been a very fruitful investigation, but it would have been the kind of thing for which you couldn't be absolutely certain of finding an answer. We've always avoided the kind of investigation where the answer had some reasonable chance of being negative.

The major cost in operating the Advent large-screen TV was expected to be cathode-ray tube replacement. Phosphor life (and therefore tube life) was expected to lie between 700 to 2,500 playing hours. The projection tube had been used in some of the earliest television sets, but the large screen desired would put new demands on it for maximizing total light output. Thus, an RCA commercial projection system with the mirror and corrector lens outside the tube was rejected as too inefficient and troublesome.

Rights to produce the Kodak screen had been given to Advent with no guarantee of the practicality of doing so in a large size. It was concave toward the audience and leaned forward slightly. These two factors required that the screen extend about a foot out from a normal wall. Brightness fell off rapidly as the viewer moved about 70° off an axis perpendicular to the screen. While satisfactory viewing required the room to be no brighter than would be required to read a newspaper with strain, a bright light could be situated to the side of the screen without seriously degrading the image. Mr. Kloss believed the Kodak screen was the best presently available, but hoped to develop a proprietary flat screen which could be patented. It would be composed of many elements which would each direct light in the optical direction.

In a conventional color receiver, the electronics assembly feeds information to a single picture tube which contains three electron guns. The Advent system was based on similar electronic circuitry, but the video image was projected on the screen by three separate cathode-ray tubes, one each for the red, green, and blue color constituents. The Advent tube is diagramed in Exhibit 5. Within each tube, a stream of electrons of varying intensity was beamed toward the positively charged internal anode,

EXHIBIT 5 Projection Television Tube

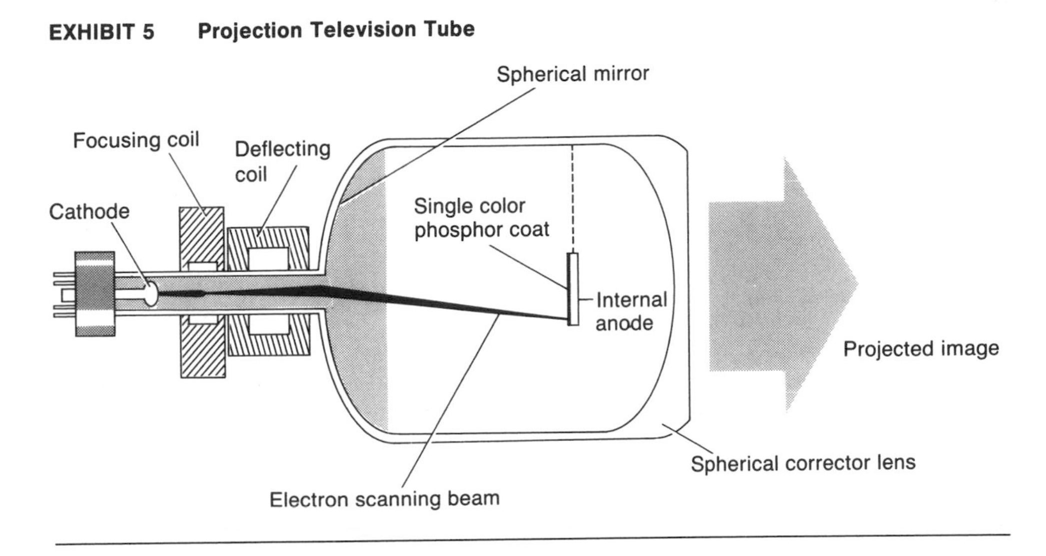

coated with a phosphor that generated one of the three colors to be projected. This beam was accelerated, focused, and deflected in a rapid horizontal scan of 15°, with the U.S. standard 525 sweeps for each vertical transit. This stream of electrons hitting the anode recreated the transmitted picture for that color. The internal spherical mirror reflected this image and focused it through the corrector lens on the external screen where the three colors were superimposed.

Tests of experimental models had shown that this system, based on three projection tubes with internal optics, could produce large images of amazingly high quality. Internal optics (mirror and corrector lens within the tube) were superior to external optics which required exact positioning of mirrors, greater light wastage, and attendant problems in keeping the optics clean. Internal optics had been used by the U.S. Navy many years previously and were not patentable. Mr. Kloss commented, "We may very well have been in error in the past in not getting some nominal patents to make it easier to sit down with somebody and sell some of this technology." He did not feel that protection was the primary value in patents "because the reluctance of manufacturers to want to get into any new field is really quite surprising. It's unfamiliar, sort of strange; they would like to buy it. We wouldn't mind, but it's always a mess to do manufacturing for someone else."

Competition

There were no large-screen TV sets on the market which would compete directly with the Advent set in the home market. The Eidophore system, developed in Switzerland in the early 1950s, used an electron beam physically to change the surface of an oil film. A light projected through this film and onto a screen provided a much brighter and larger picture. The Eidophore set, however, required an operating technician and cost about $40,000 for a monochrome version and over $100,000 for color. General Electric had produced a modified version which was more easily operated and cost about $29,000 in monochrome and $38,000 in color. People who had seen this set reported to Mr. Kloss that its brightness and resolution were inferior to Advent's though he felt that improvement might be obtained by using a higher gain screen.

Very little of the current discussion in trade journals about future trends in television centered on large screens. More was aimed at miniaturized TV or at development of a flat screen which could be installed in a wall. Mr. Kloss commented that "there might be a message there. You talk about desirable things of the future; nobody talks about larger screen television at all." During 1968, Sony Corporation demonstrated an experimental set which was flat and large (eight feet diagonally). This set was essentially a board of 26,000 elements of one red, one blue, and one green light each. This compared with 350,000 elements of phosphor in the typical shadow mask tube and resulted in a picture of noticeably poor resolution. The problems to be solved were to decrease the size of individual lights and obtain more rapid switching of the lights. While many years of development work remained before this set could be competitive, Mr. Kloss felt that the ultimate and best TV of the future would in some way generate light on the screen itself.

Although projection television was well recognized in the industry as a means of obtaining both a large and flat screen, the immediate objection was that it could not be made bright enough for viewing in a lighted room. Several large companies had experimented with it nonetheless. Mr. Kloss described one such effort in explaining why the large manufacturers were unlikely to provide competition soon after he began production. A large military and consumer electronics firm had shown him a cathode-ray tube under development for projection use. A configuration of optics very similar to Advent's had been adopted, but a maximum falloff in brightness at the edges of the screen of 10 percent to 20 percent had been specified by someone. Conventional knowledge of human eye sensitivity would suggest that humans would be insensitive to falloffs of 200 percent to 300 percent. However, this error caused engineers to design an aspherical corrector lens which would disperse light nearly perfectly on the screen and cost many thousands of dollars. Similar mistakes eventually caused the system to reach a height of nearly six feet and to require a sealed refrigeration unit for cooling. RCA had estimated about a year earlier that it could develop a large-screen TV within several years for from $5 to $50 million (the exact figure was not recalled). Mr. Kloss believed that a radically improved system could not be designed and built within 5 years and more likely 10 years. Competitive projection systems would probably require two years minimum after the Advent set was introduced.

Production and Costs

Production was expected to be carried out in Advent's new 64,000-square-foot plant. Receiver units were the same as those required in ordinary sets except that less deflection power was needed due to the decreased deflection angle. No decision had yet been made whether to make or buy the receiver units. Projection tubes were expected to cost about $50 each produced on a small scale and involved no unusual technology. Some equipment had already been purchased to produce test models. The cost could probably be reduced to $100 for a set of three tubes on an automated line. Production of the screen involved handwork to stretch and mold a thin aluminum foil, apply backing, and construct a frame and stand.

Mr. Kloss did not envision a highly automated line. He believed from experience at KLH and Acoustics Research that cost penalties would be only 10 percent to 20 percent if as few as 100 sets a week were produced. Electronic components were priced the same to all buyers if ordered in quantities over approximately 1,000, and the inflexibility and high fixed costs of an automated line would prevent great economies of scale. Tooling costs for a line adequate for 100 sets per week would be "many tens of thousands of dollars." Production costs were expected to be similar to those for other Advent products, with direct labor costing about one half as much as materials. Mr. Kloss believed that production costs of shadow mask tubes were about $100 and far outweighed cabinet and receiving equipment costs.

Concessions were made to simplify design, and production of the Advent set included replacement of electrostatic focusing with the less common and more expensive static magnetic focusing. This decision would result in a selling price about $200 higher than it would otherwise be. Similar concessions were expected in screen design.

Distribution and Marketing

Mr. Kloss believed that most of the expected technical problems which could become customer complaints had been effectively ironed out of the TV design. He felt that sales personnel could cause complaints, however, by creating or allowing unreasonable expectations. He said that "the expected kind of troubles are that we just haven't anticipated somebody's attitude toward this or his expectations. This comes from rather recent learning in noise reduction systems where you have a difficult time explaining to someone."

Although the set would eventually be designed for installation by the customer, the first installations would require a technician:

> It will be exactly equivalent to what the early color sets had with technicians running around . . . . The whole thing was mechanically fragile and fussy. . . . The beginning of any new kind of thing is troublesome. You can't even tell how you finally want to make it until you go through this manufacturing process.

Retail price had originally been estimated at $1,500 to $2,000, but had been revised upward to $2,000 to $2,500 based upon estimated costs and normal margins.

Mr. Kloss expected to engage in enough advertising to identify the product in consumer minds as reasonably priced and to lock in a portion of the market. The only scheme he had which might help lock in the consumer was to give the product a simple name and then not change it. He would "never engage in what might be interpreted as an annual model change." He felt that this strategy had given KLH an advantage over other companies which introduced new products and consequently destroyed their equity in the name.

Mr. Pressman, marketing vice president of Advent, said he had purposely avoided having his attention diverted by the TV which was still at least several months from introduction. He did feel that video products were exciting and had a greater long-range potential for Advent than audio did. He thought, as well, that 5,000 unit sales a year sounded possible. Though attempts by other video product manufacturers to distribute through audio dealers had proven unsuccessful, he had not eliminated the possibility of trying it again. No opinions regarding advertising and promotional strategies had been formed.

When the Advent large-screen television was conceived, Mr. Kloss envisioned its use in the home as a high-quality display medium and believed that the increasing quantity of broadcast materials would lead to proportionally increased quality programs for which a large screen would be preferred. Video tape recording units being brought out by several manufacturers would permit quality programs to be recorded or purchased, which might encourage more intensive viewing of programs at convenient times on a unit like the Advent TV. However, recorders would have to have sufficient capability to reproduce most of the information content of the signal broadcast so that playback on a large screen would be of adequate quality. Mr. Kloss had no immediate plans to produce complementary products except the improved screen, though he did wish to broaden Advent's product line over the long term.

The possibility of selling the television set to another company was not considered because:

> If a product that I developed and sold to somebody else did not succeed, I would be free to blame somebody else. And that's an unsatisfactory position. I have to have the complete responsibility. I really, honestly wouldn't know whether it was their fault or mine; so I have no way of knowing whether I've done anything worthwhile or not if I don't have complete knowledge of the total process. So to me it would be very unsatisfactory to invent things and sell them to somebody. If they continually and regularly were successful, I'd, after a period of time, be satisfied . . . with my contribution. This probably wouldn't happen.

THE FUTURE

Despite several problems with Advent that were apparent in late 1970, the company's future promised to be an exciting one. Mr. Kloss especially looked forward toward making the decisions necessary to reach his stated sales goal, a level of sales which he felt confident of reaching. Specifically, Mr. Kloss felt that a $50 million sales level could be reached within the $2.25 billion audio equipment market and the $2.5 billion television market, without sacrificing Advent's policy of operating within a specialized and protected market niche. Beyond that point, however, it was uncertain whether such a position could be maintained. Mr. Kloss commented,

> If one grows in an established market area, then there can be a succession of products that are based on a careful and sensitive reading of what people in the marketplace express that they want, and what competent engineering can produce, and this may well be an important part of Advent's future. . . . I've no objection to growing in the regular kind of way, and that's the kind of thing that can be happily delegated to somebody else. In fact, to delegate enough of that to make a strong, growing company, and yet continue in the company, would be highly desirable. How strong you have to be before you can have the luxury of doing "me too" kinds of products, though, I don't know.
>
> I think a perfectly honorable way is to continue to grow making products which, on the strength of the market position, are salable. . . . Up until now one has restricted their attention to things which are fundamentally better and different than anything else. But there is nothing wrong with growing doing ordinary kinds of products . . . . The idea of making products which continually add to the volume of Advent may well be completely done by someone else. I'd be happy to have that done. That would leave me increasingly free to think longer about things which were different in kind, new kinds of products.
>
> There's an ideological inclination to want to make broad spectrum kinds of products. The interest is to get back to where one was at KLH. The cassette recorder with Dolby, I envision that as not nearly as broad spectrum at the present time as it was planned to be. All the products that would grow out of the fact of the cassette being the primary music listening medium for a lot of people in the home, this isn't happening so fast.

One issue of great concern to Mr. Kloss was the institutionalization of the Advent innovation process. On the one hand, Mr. Kloss felt it would be possible to find a full-time administrator who could work closely with him in handling the company's growing management responsibilities, while he could continue to devote his major efforts to the very enjoyable work of conceiving new products and staying abreast of consumer electronics technology. On the other hand, Mr. Kloss felt that it was possible to institutionalize the product conception function, but he was unsure how best to proceed. In the current situation, he personally perceived market needs, was able to match those needs with the technological state of the art, and was further capable of completing the product conception that fulfilled the market-technology match. As the company grew, Mr. Kloss recognized that some division of these functions would have to take place. Should he separate the more routine R&D functions from the esoteric, or should he attempt to pool the efforts of a large number of people in order to arrive at an effective product conception function? In late 1970, Mr. Kloss could not see how the latter plan might work.

Concerning his role as Advent grew, Mr. Kloss mentioned his admiration for the situation Edwin Land was reputed to have at Polaroid, namely, the situation of ready access to any level of R&D. Mr. Kloss commented, "To contribute to it or direct it without interfering with its normal process. That to me is a really very desirable kind of thing. And it can't frequently be achieved." Mr. Kloss felt that he might be on the way toward such a situation already, toward an Advent that could carry on, increasing a bit in his absence, but to which he could contribute substantially.

Reading II–1
Toward an Innovative Capabilities Audit Framework

R. A. Burgelman, T. J. Kosnik, and M. van den Poel

INTRODUCTION

Over the years, firms have become accustomed to the concept of an audit and have used it extensively for accounting, marketing, and strategy-formulation purposes. Innovation audits, on the other hand, are rarely conducted. In this paper, a framework for auditing the *innovative capabilities* of a firm is provided. This is an area that must be addressed by both small and large firms in order to manage and improve the existing potential for innovation in their businesses, as well as to plan for future strategic developments.

At least three key questions must be addressed in the context of an innovative capabilities audit:

1. How has the firm been innovative in the areas of product/service offerings and/or production/delivery systems?
2. How good is the fit between the firm's current business and corporate strategies and its current innovative capabilities?
3. What are the firm's needs in terms of innovative capabilities to support its long-term business and corporate competitive strategies?

The first section of this note defines key concepts used in the innovation literature for our specific purposes. The second section discusses technology as one major component in an innovative capabilities audit. The third section presents comprehensive innovative capabilities audit frameworks for the business unit and corporate levels of analysis. The fourth section briefly discusses the use of these frameworks.

I. DEFINING SOME KEY CONCEPTS

Turning to the innovative capabilities audit, it is useful to consider some related key concepts which when used elsewhere may have a variety of meanings, and to define them for purposes of this paper. These concepts are innovation, entrepreneurship, invention/discovery, technology, and scientific research.

Innovation

By *innovation,* we usually mean particular new things like large-sized tennis rackets (product), dry cleaning (service), assembly line (production system), or computer stores (delivery system). Such innovations are the outcome of the *innovation process:* The combined activities leading to a new marketable product/service and/or new production/delivery system. The strategic management of innovation requires an understanding of the innovation process. *Incremental* innovations (from the perspective of the producing firm) involve the adaptation, refinement, and/or enhancement of existing product/service categories and/or of production/delivery systems. *Radical* innovations, on the other hand, involve entirely new product/service categories and/or production/delivery systems. The criteria for success in innovation are commercial (economical) rather than technical. A successful innovation is one that returns the original investment plus some additional returns to the developing organizations. This implies that there be a market for the product/service that justifies the cost of the original development.

Entrepreneurship

Entrepreneurship refers to activities involved in creating new resource combinations that did not previously exist. Entrepreneurship is a fundamental element of the innovation process. It can rest with one individual (individual entrepreneurship) or can involve the combined activities of multiple participants in an organization (corporate entrepreneurship). Successful entrepreneurship—and hence successful innovation—involves both the capacity to *identify* opportunities for combining resources in a novel and commercially viable way, and the capacity to *exploit* such opportunities by deploying resources to capitalize on the opportunities. Note that entrepreneurship as defined here is more than arbitrage, and does not necessarily imply personal financial risk taking.

Invention/Discovery

At the origin of the innovation process are often *inventions* and/or *discoveries*. As Webster points out, "We discover what before existed, though to us unknown; we invent what did not before exist." Both invention and discovery are the results of creative processes which are often serendipitous in nature and thus very difficult to organize and predict. Researchers in universities and industrial and government laboratories following the canons of modern science, as well as idiosyncratic tinkerers in a garage, play a role in these processes. Through *patents* or *trade secrets,* invention and discovery sometimes allow their originators to establish a potential for economic rents. However, the criteria for success regarding inventions/discoveries are technical rather than commercial.

Technological Development

It is useful to distinguish invention/discovery from *technological development*. Technological development refers to the activities involved in putting inventions/discoveries to practical use. For example, three main inventions in the semiconductor industry—transistor (1947), integrated circuit (1959), and microprocessor (1971)—gave rise to a myriad of successive new generations of technology in areas such as data processing, memory devices, etc. Not all inventions/discoveries can be put to practical use and become technologies. Also, there may be a substantial delay before an invention/discovery can be turned into a technology (genetic engineering and lasers are good examples). Generally speaking, then, *technology* refers to the practical knowledge, know-how, skills, and artifacts that can be used to develop a new product/service and/or new production/delivery system. Technology can be embodied in people, materials, cognitive and physical processes, plant, equipment, and tools. The criteria for success regarding technology are also technical (can it do the job?) rather than commercial (can it do the job profitably?). In Section II of this paper, we shall elaborate on the role of technology in the innovation process.

Scientific Research

It is also useful to distinguish invention/discovery and technology from scientific research. *Basic* scientific research refers to the activities involved in generating new knowledge ("truth") about physical, biological, and social nature. The cumulative body of knowledge resulting from research forms the substratum for many inventions, but not for all (e.g., the wheel was not the result of scientific research). There is normally a significant time lag (often 10 years or more) between doing basic scientific research and the possibility of using research results for creating innovations. *Applied* scientific research is geared toward solving particular technical problems and results often in new technology. *Development* work serves to embody technology into new product/services.

Activities and Results

From our discussion in this section, it is clear that one can distinguish *activities* and *result* aspects for the various key concepts. Figure 1 represents a simplified picture of the connections between the various key concepts that are used to describe the innovation process.

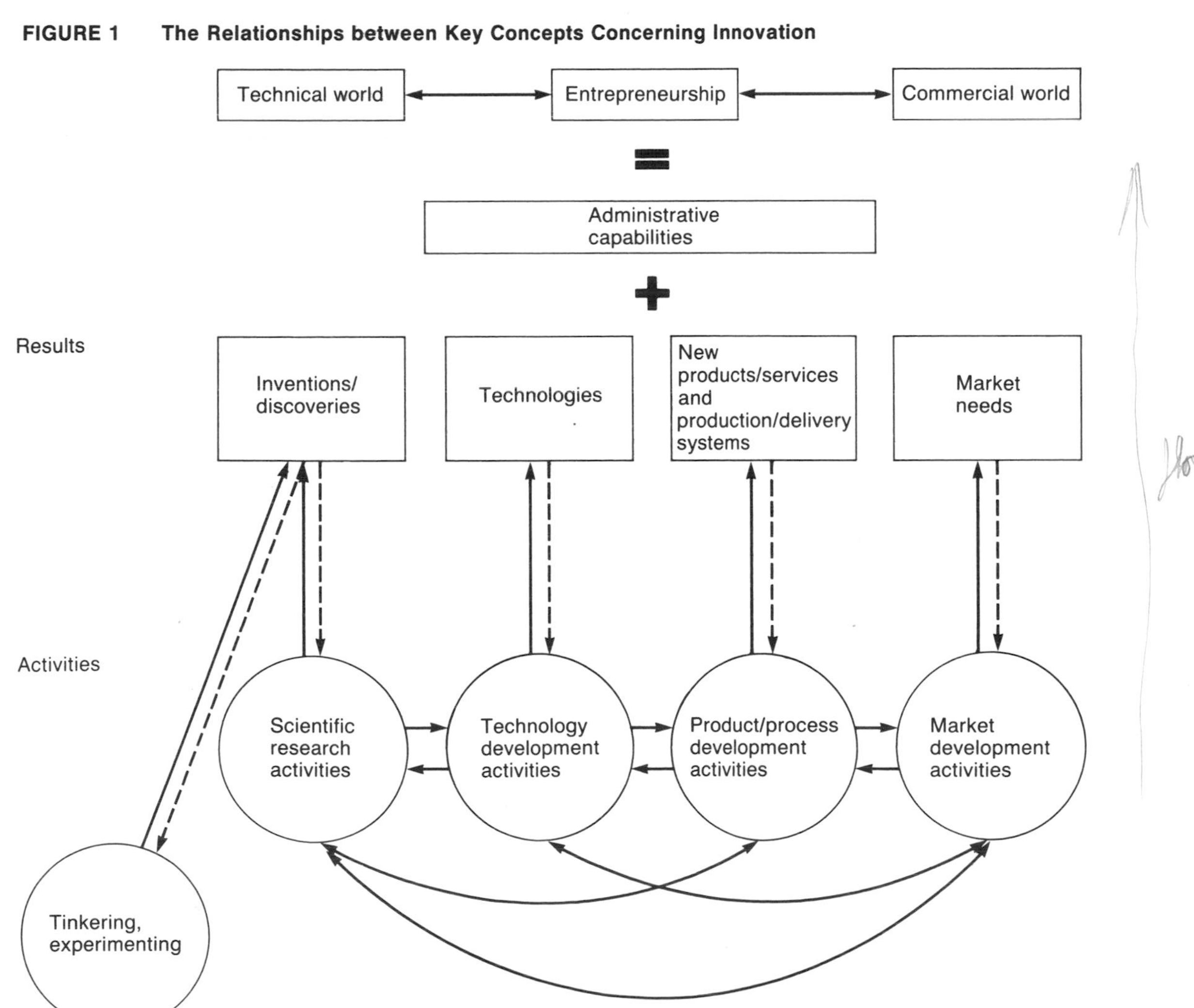

FIGURE 1 The Relationships between Key Concepts Concerning Innovation

It is important to realize that the direction of causality in Figure 1 can go from market needs to scientific research as well as from scientific research to market needs. The *innovation process* in real-life situations will almost always be iterative and simultaneous rather than unidirectional and sequential.

II. TECHNOLOGY: ONE ELEMENT OF INNOVATIVE CAPABILITIES

In the broadest sense, a firm's technology encompasses the *set* of technologies that are used in the different aspects of its activities. A firm's technology has thus often been decomposed into its constituting technologies. One useful framework to do so examines the various technologies used in the different parts of a firm's value chain. Figure 2 shows a simplified version of this framework.

It is important to realize that key elements of a firm's technologies often exist in *embedded* form (e.g., trade secrets based on know-how). Key parts of the technologies may exist only in *tacit* form rather than be expressed or codified in manuals, routines and procedures, recipes, rules of thumb, or other explicit articulations. "Craftsmanship" and "experience" usually have a large tacit component.

At any given time, a firm (or other productive unit) has a stock of technologies which are

FIGURE 2 Representative Technologies in a Firm's Value Chain

Inbound logistics	Operations	Outbound logistics	Marketing sales	Service
Transportation technology	Basic product technology	Transportation technology	Media technology	Diagnostic and testing technology
Material handling technology	Materials technology	Material handling technology	Audio and video recording technology	Communication system technology
Storage and preservation technology	Machine tool technology	Packaging technology	Communication system technology	Information system technology
Communication system technology	Material handling technology	Communication system technology	Information system technology	
Testing technology	Packaging technology	Information system technology		
Information system technology	Maintenance methods			
	Testing technology			
	Building design operation technology			
	Information system technology			

SOURCE: Adapted with permission of the Free Press, a division of Macmillan, Inc. from *Competitive Advantage: Creating and Sustaining Superior Performance* by Michael E. Porter. Copyright © 1985 by Michael E. Porter (New York: Free Press, 1985).

to a greater or lesser extent embodied in its products/services and production/delivery systems. There is also a rate of change in this stock of technologies which is driven by internal and external technological development efforts.

In the so-called "high-tech" firms, technology is the major driving force of the innovation process. Such firms spend a significant portion of their resources (usually well above 3 percent of sales) on technology-related activities (including R&D). Both high-tech and other firms have, in recent years, become more aware of the critical role of technology in strategic decisions, and of integrating them into their strategic management process.

At a more micro level, it may be useful to focus on the technologies embodied in a product (or product group) and/or service (or service group). Each product or service can be decomposed in its constituting technologies, and an assessment can be made of the relative strength (degree of distinctive competence) that the firm has with respect to that technology. Figure 3 shows the outline for a product/technology matrix.

Maidique and Patch (see references) have proposed that a *technology strategy* encompasses a number of dimensions: technology choice, level of technical competence, sourcing to technology, level of R&D funding, competitive timing of technology introduction, organization of the technology function. In recent years, various consulting firms have developed an approach to technology strategy development. SRI International, for example, has its Innovation Search Technique seminars involving a company management team (including CEO/COO), in which SRI attempts to relate the current position of the company to the specific technologies, its competitors, and its overall strategy, followed by market studies and strategy development. Arthur D. Little tries to identify the critical technologies, analyzes the industry maturity vis-à-vis the technologies, and the subsequent trade-offs and risks from both business unit and corporate level perspectives. Booz-Allen investigates the match (or mismatch) of the business portfolio and the technology portfolio of a firm and the resulting technology investment priorities.

FIGURE 3 The Product/Technology Matrix

	Product A	Product B	•••	Product N
Technology 1	(*)			
Technology 2				
•				
•				
•				
Technology K				

Note: Each entry (*) should establish the firm's relative strength vis-à-vis the state of the art.

SOURCE: Adapted from A. Fusfeld, "How to Put Technology into Corporate Planning," *Technology Review*, May 1978. Reprinted with permission from *Technology Review*, MIT Alumni Association, © 1978.

An audit focusing on the technology component will examine *what* the characteristics of the firm's technology strategy are. It will also look at *how* technology strategy is determined.

Technological Evolution and Forecasting

Technological change is one of the most important forces affecting a firm's competitive position, and research suggests that firms find it difficult to respond to such changes. An audit of the technological capabilities should, therefore, address the issue of how well the firm understands the dynamics of the life cycle of the various technologies that it employs. Figure 4 shows ways in which the relative importance of the various technologies of the firm and their evolution can be assessed.

An important element in developing a technology strategy is the capacity to perform systematic technological forecasting that goes beyond simple expert opinion. At this point, this is still more art than science. Twiss (see references) discusses useful techniques such as technological progress functions, trend extrapolation, the Delphi method, and scenarios. A proper audit should assess how well top management has integrated forecasting efforts into its strategic planning process.

Underlying the capacity to forecast and, perhaps more importantly, to see relationships between technologically significant events is the effort to gather data systematically and continuously. Maintaining a log book for just this purpose is often an effective way of doing such data collection. An audit of the technology forecasting capacity should examine the extent to which individuals and groups collect and interpret data on technically relevant events.

III. INNOVATIVE CAPABILITIES AUDIT FRAMEWORK (ICAF)

Bringing a new product or service to the market (i.e., innovation) depends not only on the technological capabilities but on other critical functional capabilities of the organization (marketing and distribution, manufacturing, human resources management, financing). For instance, a technological strategy designed to achieve product performance needs to be enhanced with a technically trained sales force which can educate the customer regarding performance advantages, and with a manufacturing system which has built-in adequate quality controls.

Innovative Capabilities

Innovative capabilities can simply be defined as the *comprehensive set of characteristics of an organization that facilitate and support its innovation strategies.*

It is useful to differentiate between the innovative capabilities at a *business unit* level and at

FIGURE 4 Technology Life Cycle and Competition Advantages

Stages in Technology Life Cycle	Importance of Technologies for Competitive Advantage
I. Emerging technologies	Have not yet demonstrated potential for changing the basis of competition.
II. Pacing technologies	Have demonstrated their potential for changing the basis of competition.
III. Key technologies	Are embedded in and enable product/process. Have major impact on value-added stream (cost, performance, quality). Allow proprietary/patented positions.
IV. Base technologies	Minor impact on value-added stream; common to all competitors; commodity.

a *corporate* level. A business unit is a unit for which a particular strategy and resource commitment posture can be defined because it has a distinct set of product markets, competitors, and resources. An innovative capabilities audit tries to identify the critical variables that influence the innovation strategies at the business unit level.

The corporate level deals with a number of business units (i.e., number of product/markets, processes, technologies) and an innovative capabilities audit tries to identify the critical variables that influence both the relationships *between* corporate level and business unit level in terms of innovative capabilities and the formulation of *overall* corporate innovative strategy.

An innovative capabilities audit should not only address the *formulation* of business and corporate level innovation strategies; it should also address the firm's capacity to *implement* such strategies.

Business Unit Level Audit

In general, the innovative strategies at the business unit level can be characterized in terms of:

- Timing of market entry with new products/services and/or production/delivery systems.
- Technological leadership/followership in new products/services and/or production/delivery systems.
- Scope of innovativeness in product/market portfolio.
- Rate of innovativeness in particular categories of products/services and/or production/delivery systems.

Five important categories of *variables* influence the innovation strategies of a business unit:

1. Resources available for innovative activities.
2. Capacity to understand competitor strategies and industry evolution with respect to innovation.
3. Capacity to understand technological developments relevant to the business unit.
4. Structural and cultural context of the business unit affecting entrepreneurial behavior.
5. Strategic management capacity to deal with entrepreneurial initiatives.

Figure 5 shows the five categories of variables influencing the business unit level innovation strategies.

Resources, capacity to understand competitor strategies and industry evolution, and capacity to understand relevant technological developments are important inputs for the *formulation* of business unit innovation strategies.

Structural and cultural context, and strategic management capacity, are important system characteristics which determine the *implementation* of business unit innovation strategies.

Exhibit 1 lists some critical issues for auditing each of the five categories. Clearly, these lists are not exhaustive; additional items may be added to reflect the particulars of different situations.

The combination of the five categories determines the relative strength of the business unit for formulating and implementing innovation strategies. So, the audit should address this as well. A business unit may, for example, have ample resources for new product development but lack the strategic management capacity to channel those resources both within the unit and in relation to competitors' moves. Conversely, the necessary resources and strategic management capacity might bring about a new product whose technological advantage is already obsolete.

Corporate Level Audit

In today's economy, many large, established firms comprise a number of different but par-

FIGURE 5 **Innovative Capabilities Audit Framework—Business Unit Level**

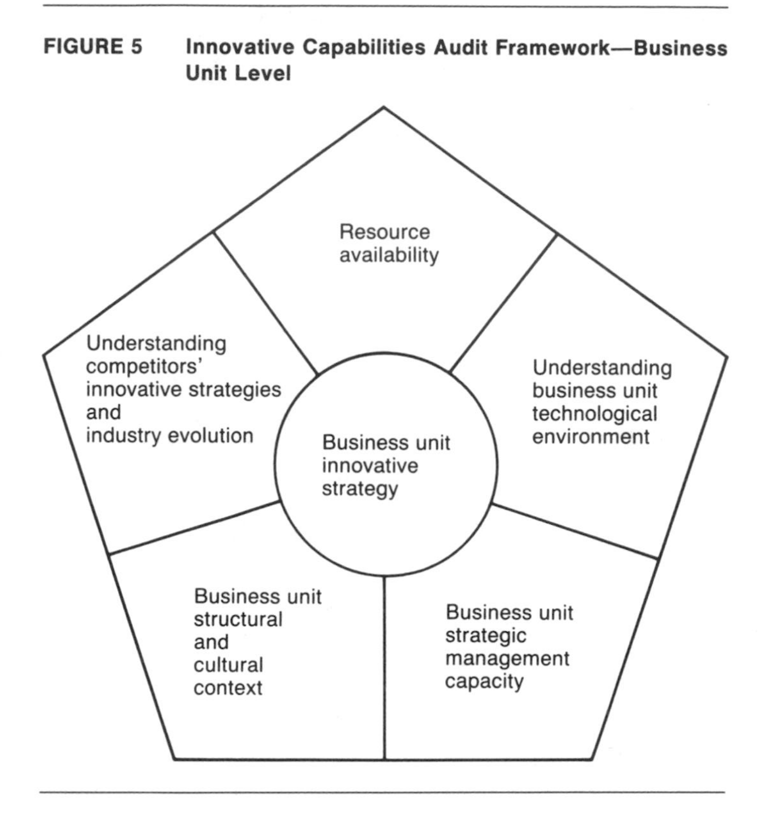

tially related businesses. The *raison d'être* of such complex corporations is based on the capacity of corporate strategic management to identify and exploit "synergies"—operational and/or financial—which would not be, or less completely, realized if the various businesses were to operate independently of each other.

An audit of the innovative capabilities at the corporate level thus brings in an additional dimension. Here, it is necessary to examine *how* the corporate innovative capabilities enhance the innovative capabilities at the business level. In other words, it is necessary to investigate whether and how the total corporate innovative capabilities are larger than the sum of the innovative capabilities of the business units.

In general, the innovative strategies at the corporate level can then be characterized in terms of:

- Scope and rate of development of *new* product/services and/or production/delivery systems that are derived from combining innovative capabilities across existing business units.
- Scope and rate of new business development based on *corporate* technology development efforts.
- Timing of entry with respect to the above.

Five categories of *variables* are again considered. These are represented in Figure 6.

Each of the five categories at the corporate level corresponds to an equivalent category at the business unit level, but has now a somewhat different emphasis: The capacity to do

EXHIBIT 1 Innovative Capabilities Audit Framework—Business Level

1. *Resource Availability and Allocation*
 Level of R&D funding and evolution:
  In absolute terms.
  As percentage of sales.
  As percentage of total firm R&D funding.
  As compared to main competitors.
  As compared to leading competitor.
 Breadth and depth of skills at business unit level in R&D, engineering, market research.
 Distinctive competences in areas of technology relevant to business unit.
 Allocation of R&D to:
  Existing product/market combinations.
  New product development for existing product categories.
  Development of new product categories.
2. *Understanding Competitors' Innovative Strategies and Industry Evolution*
 Intelligence systems and data available.
 Capacity to identify, analyze, and predict competitors' innovative strategies.
 Capacity to identify, analyze, and predict industry evolution.
 Capacity to anticipate facilitating/impeding external forces relevant to business unit's innovative strategies.
3. *Understanding the Business Unit's Technological Environment*
 Capacity for technological forecasting relevant to business unit's technologies.
 Capacity to assess technologies relevant to business unit.
 Capacity to identify technological opportunities for business unit.
4. *Business Unit Structural and Cultural Context*
 Mechanisms for managing R&D efforts.
 Mechanisms for transferring technology from research to development.
 Mechanisms for integrating different functional groups (R&D, engineering, marketing, manufacturing) in the new product development process.
 Mechanisms for funding unplanned new product initiatives.
 Mechanisms for eliciting new ideas from employees.
 Evaluation and reward systems for entrepreneurial behavior.
 Dominant values, business unit mythology, definition of "success."
5. *Strategic Management Capacity to Deal with Entrepreneurial Behavior*
 Business unit level management capacity to define a substantive development strategy.
 Business unit level management capacity to assess strategic importance of entrepeneurial initiatives.
 Business unit level management capacity to assess relatedness of entrepreneurial initiatives to unit's core capabilities.
 Capacity of business unit level management to coach product champions.
 Quality and availability of product champions in the business unit.

more in each area than what the business unit would be able to do on its own.

1. Resource availability and allocation (e.g., corporate R&D, cash availability for risky projects).
2. Capacity to understand multi-industry competitive strategies and evolution (e.g., corporate strategic planning for innovation).
3. Capacity to understand technological developments (e.g., multi-industry scanning and technological forecasting).
4. Structural and cultural context (see separate section below).
5. Strategic management capacity (e.g., ex-

ploitation of synergies in innovation through "horizontal strategies"; internal corporate venturing and acquisition strategies).

The five categories, again, can be classified as inputs (#1, 2, 3) or system characteristics (#4, 5).

Exhibit 2 lists critical issues to be addressed in each category to carry out the corporate level audit.

Like for the business unit level audit, the combination of the five categories should be considered in the corporate audit.

Organizational Designs for Innovation

A subaudit at the corporate level relates to the use of various organization designs to facilitate corporate entrepreneurship and innovation. This can be done on an internal basis or external basis. On an *internal* basis, a number of alternative organizational designs can be envisioned. These can range from mechanisms for directly integrating innovative projects in the mainstream operations to independent business units. (Exhibit 3 lists a number of internal organization designs.) On an *external* basis, arrangements can range from acquisitions, joint ventures, to venture capital investments, R&D agreements, university contracts. (Exhibit 4 lists a number of external arrangements.)

These internal and external designs have varying implications for control and management and bring different value to the firm. A joint venture arrangement, for example, to develop and market a new product may indeed

FIGURE 6 Innovative Capabilities Audit Framework—Corporate Level

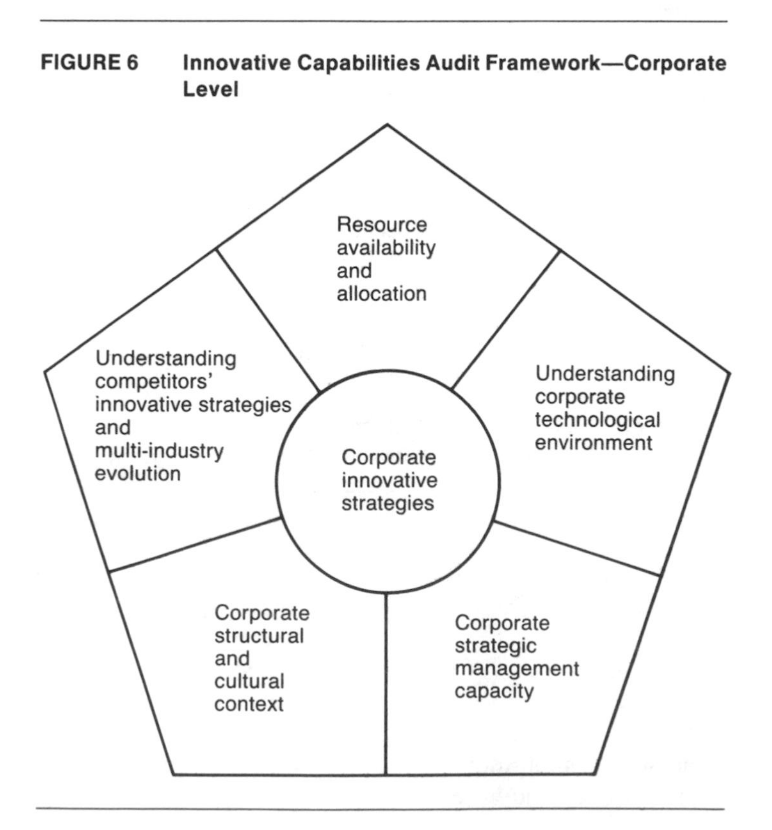

EXHIBIT 2 Innovative Capabilities Audit Framework—Corporate Level

1. *Resource Availability and Allocation*
   Corporate R&D funding level and evolution:
     In absolute terms.
     As percentage of sales.
     As compared to average of main competitors.
     As compared to leading competitor.
   Breadth and depth of skills of corporate level personnel in R&D, engineering, market research.
   Distinctive competences in areas of technology relevant to multiple business units.
   Corporate R&D allocation to:
     Exploratory research.
     R&D in support of mainstream business.
     R&D in support of new business definition.
     R&D in support of new business development.
2. *Understanding Competitors' Innovative Strategies and Multi-Industry Evolution*
   Intelligence systems and data available.
   Capacity to identify, analyze, and predict competitors' innovative strategies spanning multiple industries.
   Capacity to develop scenarios concerning evolution of interdependencies between multiple industries.
   Capacity to anticipate facilitating/impeding external forces relevant to firm's innovative strategies.
3. *Understanding the Corporate Technological Environment*
   Capacity for technological forecasting in multiple areas.
   Capacity to forecast cross-impacts between areas of technology.
   Capacity to assess technologies in multiple areas.
   Capacity to identify technological opportunities spanning multiple areas.
4. *Corporate Context (Structural and Cultural)*
   Mechanisms to share technologies across business unit boundaries.
   Mechanisms to define new business opportunities across business unit boundaries.
   Internal and external organization designs for managing new ventures (see Exhibits 3 and 4).
   Mechanisms to fund unplanned initiatives.
   Evaluation and reward systems for entrepreneurial behavior.
   Movement of personnel between mainstream activities and new ventures.
   Dominant values, corporate mythology, definition of "success."
5. *Strategic Management Capacity to Deal with Entrepreneurial Behavior*
   Top management capacity to define a substantive long-term corporate development strategy.
   Top management capacity to assess strategic importance of entrepeneurial initiatives.
   Top management capacity to assess relatedness of entrepreneurial initiatives to the firm's core capabilities.
   Middle-level management capacity to work with top management to obtain/maintain support for new initiatives (organizational championing).
   Middle-level management capacity to define corporate strategic framework for new initiatives.
   Middle-level management capacity to coach new venture managers.
   New venture managers' capacity to build new organizational capabilities.
   New venture managers' capacity to develop a business strategy for new initiatives.
   Availability of product champions to identify and define new business opportunities outside of mainstream activities.

EXHIBIT 3 Internal Arrangements for Innovation

	Higher	Mechanisms for direct integration
Degree of control		New product/business departments in existing divisions
		Special business units/new divisions
		Micro new venture department
		New venture division (corporate level)
	Lower	Independent business units

EXHIBIT 4 External Arrangements for Innovation

	Higher	Acquisition of other companies
		Joint ventures between companies
Degree of control		Licensing agreements
		R&D agreements between companies
		Venture capital investments
		Limited R&D partnership
		Nurturing and contracting with spin-offs
	Lower	Research contracts with universities

diminish the firm's needs for R&D and bring economies of scale and/or enhanced distribution networks, but may result in more limited value appropriation to the firm than an internally generated venture.

IV. USE OF THE INNOVATIVE CAPABILITIES AUDIT FRAMEWORK (ICAF)

What Frame of Reference Should Be Used?

The audit should provide to the user a profile of how the firm stands in terms of innovative capabilities. To make the assessment meaningful, the firm's current situation must be put in a specific frame of reference. One frame of reference concerns the historical *time* dimension through analysis of past, current, and projected strategic position of the firm; another one concerns the firm's position relative to current competitors, customers, and suppliers at a particular moment in time. Both allow to identify major variances with respect to desired positions.

Who Should Conduct the ICAF?

The first issue with respect to who should conduct the ICAF is whether to use only insiders or to enlist the aid of outsiders such as consultants. Actors from *inside* the organization will likely have an advantage in understanding the firm's structural and cultural context, the resource availabilities and constraints, and the product-market portfolio currently served by the various business units. On the other hand, actors from the *outside* may be able to more objectively assess the firm's strategic management capacity to deal with new ventures, or its capacity to understand, develop, and acquire new technologies. Depending upon their prior experience with or access to the firm's customers, competitors, and suppliers, outsiders may be able to conduct a more complete analysis of the industry and competitive environment.

The major disadvantage of using insiders is the possibility for a narrow or biased perspective. The major disadvantage in using outsiders is the possibility of misunderstanding internal realities and delivering impractical recommendations. The choice of staffing the audit must weigh the advantages and disadvantages of each approach.

While the audit could be undertaken by the corporate strategic planning department, the audit would provide more valuable insights if an ad hoc audit team would be set up with representatives from corporate strategic planning, corporate R&D, new business development, and business unit new product managers. The timing of the audit should precede the cycles of strategic planning processes and budget reviews to which it should form a major input.

V. CONCLUSION: WHY SHOULD AN ICAF BE UNDERTAKEN?

The Innovative Capabilities Audit Framework which has been described above requires a substantial investment of the time and attention of key actors in the organization. What are the benefits of an ICAF which might justify the effort involved?

First, the ICAF can provide management at the business unit and corporate unit levels with an opportunity to discover the firm's most promising opportunities for innovation and renewal. Management of any organization must make difficult decisions about which opportunities will receive its attention and resources, and which will not be pursued. Insights about each business unit's potential for strategic renewal in response to its environment are crucial to making effective strategic choices in allocating scarce resources.

Second, the ICAF can explicitly identify the most critical barriers to the implementation of innovation. By focusing management attention on the key roadblocks, the ICAF makes it possible to take effective proactive steps to remove or circumvent these barriers to success.

Finally, the ICAF provides a vehicle for communication between management at the business unit and corporate level, and between technologists and business professionals throughout the firm. By pinpointing the five categories of variables which affect innovative strategies, ICAF offers a frame of reference that can facilitate the discussions about how to make the firm "more innovative." This shared perspective may result in more timely consensus about what concrete steps to take during the evolutionary process of strategic renewal.

SELECTED REFERENCES

Abell, D. *Defining the Business*. Englewood Cliffs, N.J.: Prentice-Hall, 1980.

Abernathy, W. *The Productivity Dilemma*. Baltimore: The Johns Hopkins Press, 1978.

Abernathy, W.; K. Clark; and A. Kantrow. *Industrial Renaissance*. New York: Basic Books, 1983.

Booz-Allen and Hamilton. "The Strategic Management of Technology." *Outlook*, Fall–Winter 1981.

Burgelman, R. "Corporate Entrepreneurship and Strategic Management: Insights from a Process Study." *Management Science*, December 1983.

Burgelman, R., and L. Sayles. *Inside Corporate Innovation*. New York: Free Press, 1986.

Cooper, A., and D. Schendel. "Strategic Responses to Technological Threats." *Business Horizons*, February 1976.

Drucker, P. *Innovation and Entrepreneurship*. New York: Harper and Row, 1985.

Hayes, R., and S. Wheelwright. *Regaining Our Competitive Edge*. New York: John Wiley & Sons, 1984.

Horwitch, M., ed. *Technology in the Modern Corporation*. New York: Pergamon Press, 1986.

"Innov-aha-tion: Or How to Make New Things Happen." Prepared by the editors of *Technology Review* with the assistance of Myron J. Exelbert and the MIT Alumni Center of New York, 1976.

Kanter, R. *The Change Masters*. New York: Basic Books, 1983.

Lawrence, P., and D. Dyer. *Renewing American Industry*. New York: Free Press, 1983.

Little, A. D. "The Strategic Management of Technology." European Management Forum, 1981.

Maidique, M., and P. Patch. "Corporate Strategy and Technological Policy." Boston: Harvard Business School Clearing House #9-679-033, 1980.

Miller, W. *Technology*. Menlo Park, Calif: Stanford Research Institute, October 1983.

Porter, M. *Competitive Advantage*. New York: Free Press, 1985.

Reich, R. *The New American Frontier*. New York: Times Books, 1983.

Rosenberg, N. *Inside the Black Box*. Cambridge: Cambridge University Press, 1982.

Rosenbloom R. S., ed. *Research on Technological Innovation, Management, and Policy,* Volumes 1–3. Greenwich, Conn.: Jai Press, 1981–1986.

Rothberg, R., ed. *Corporate Strategy and Product Innovation*. New York: Free Press, 1981.

Tushman, M., and W. Moore, eds. *Readings in the Management of Innovation*. Boston: Pitman, 1982.

Twiss, B. *Managing Technological Innovation*. London: Longman, 1982.

Williamson, O. *The Economic Institutions of Capitalism*. New York: Free Press, 1985.

Wilson, J. *The New Venturers*. Reading, Mass.: Addison-Wesley Publishing, 1985.

Technology Choice

Case II–2
The Grumman Corporation

J. F. Ince and M. A. Maidique

INTRODUCTION

"Solar power is on the threshold of a boom that may well transform it into a key energy source by the turn of the century." So wrote *Business Week* shortly after Ron Peterson of the Grumman Corporation returned from a year of study at MIT on a Sloan Fellowship. In 1971, prior to Peterson's departure, Joseph Gavin, president of Grumman Aerospace, Grumman's largest subsidiary, had put together a team of four engineers from space programs to decide how Grumman should redirect its efforts and use space technology in the future. Upon Peterson's return, Gavin asked him to review the work of this team and make recommendations for coordination of the various energy efforts going on within Aerospace. These efforts included engineering services in energy conservation, development and study of solar thermal heating systems, wind power, photovoltaics, ocean thermal, solar satellite power stations, and some work in nuclear.

Later in 1973, Gavin formed the "Energy Program Department" within Grumman Aerospace, which he asked Peterson to head up. In this capacity, Peterson began an in-depth look at the solar market potential. During his research, he saw growing support in Congress and state legislatures for solar subsidies and came to expect billions of dollars to be pumped into the industry.

Yet at the same time, Peterson saw a fragmented solar industry with technology in a state of flux. He saw questionable user economics and numerous plumbing, installation, and construction problems in solar heating. He saw climatic limitations in many areas and institutional barriers in the form of antiquated building codes. Peterson knew of numerous corporate giants (Exxon, Shell, Westinghouse, GM, PPG, GE, IBM) entering various aspects of the field who perhaps were better suited for solar. And he was aware of powerful sectors of the government and country who were extremely skeptical about solar's potential. A 1973 task force of the National Petroleum Council

Harvard Business School case 9-680-032

summarized these views in one such statement:

> Because it is so diffuse and intermittent when it reaches the earth, solar energy can be put to no foreseeable large-scale use over the next 15 years even with appreciable improvements in technology.

In early 1975, Peterson felt solar possibilities had received adequate study and now was faced with the decision on whether or not to recommend a larger commitment on the part of Grumman into solar, and if so which of the several promising areas to commit to.

THE GRUMMAN CORPORATION

In the early 1930s, Leroy Grumman, an aeronautical engineer, launched the Grumman Aircraft Engineering Corporation with no product, no plant, no customers, and 21 employees. That was shortly after the stock market crash, but despite all the obstacles, they built the FF-1, then the fastest fighter plane in the world. During World War II, their Wildcat and Hellcat carrier-based fighters helped defeat such Axis planes as the renowned Japanese Zero and helped the Allied powers control the skies. In the 1960s and 70s Grumman participated heavily in the design and production of lunar modules and in production of the space shuttle wings. So from modest beginnings Grumman has grown into one of the foremost aeronautical and related product manufacturing companies in the world.

In the early 1970s, Grumman changed its name to the present one and reorganized, creating five wholly owned subsidiaries: Aerospace, Data Systems, Allied Industries, Ecosystems, and International. In the early 1970s, Grumman's principal product was the advanced F-14 fighter, and in 1972 when they ran into contract problems it created a severe cash drain. As a consequence, board chairman Towle, in 1973, wrote in his annual letter to stockholders, "These 12 months were the most critical in Grumman's history." Also at the time the first phase of the space program was drawing to a close and the rate of growth of the aerospace industry was declining. (See Exhibits 1, 2, and 3 for financial history.) When, in 1973, Grumman lost its bid to serve as prime contractor in the space shuttle, the arguments for diversification became compelling.

Grumman's expertise in aerospace manufacturing involved many areas that were fertile grounds for expansion: metallurgy, laser

EXHIBIT 1

THE GRUMMAN CORPORATION
Consolidated Statement of Income*
(in millions)

	Year Ended December 31	
	1974	1973
Sales	$1,112	$1,082
Other income	16	5
	1,129	1,087
Costs and expenses:		
Wages, materials, and other costs	1,080	1,045
Interest	11	7
Minority interests	—	1
Provision for federal income taxes	16	16
	1,109	1,071
Income before gain on debenture exchange and extraordinary item	20	16
Gain on debenture exchange	9	—
Income before extraordinary item	29	16
Extraordinary item—federal income tax benefit of operating loss carryover	3	11
Net income	$ 32	$ 28

* Totals may not add due to rounding.
SOURCE: Grumman annual reports 1973 and 1974.

EXHIBIT 2

THE GRUMMAN CORPORATION
Consolidated Balance Sheet*
(in millions)

Assets	December 31 1974	1973
Current assets:		
Cash	$ 13	$ 12
Certificates of deposit and U.S. government securities (at cost)	6	31
Accounts receivable	124	88
Inventories, less progress payments	182	103
Prepaid expenses	2	2
Total current assets	329	237
Long-term lease contract receivable	5	4
Property, plant, and equipment (at cost):		
Buildings	85	80
Machinery and equipment	170	164
Leasehold improvements	15	15
	271	261
Less: Accumulated depreciation and amortization	174	162
	97	98
Construction in progress	2	1
Land	7	6
Net property, plant, and equipment	107	107
Other assets and deferred charges	2	3
Total assets	$445	$353

Liabilities and Shareholders' Equity

	1974	1973
Current liabilities:		
Notes payable to banks and others, including current installments on long-term debt	$ 11	$ 11
Accounts payable	75	70
Federal income and other taxes	16	—
Accrued wages and employee benefits	21	18
Progress payments and customer deposits	13	24
Other liabilities	16	12
Total current liabilities	$156	$138

EXHIBIT 2 *(concluded)*

	1974	1973
Long-term debt:		
Subordinated loan—Bank Melli Iran	$ 75	$ —
Bank loans—revolving credit agreement	17	—
Advances	—	24
4¼% convertible subordinated debentures due September 1, 1992	24	49
8% convertible subordinated debentures due September 1, 1999	15	—
Other debt	24	39
	156	112
Deferred investment tax credit	2	2
Minority interests in subsidiaries	4	5
Shareholders' equity:		
Preferred stock—authorized 10 million shares of $1 par value; issued 170,651 shares of 80 cents convertible preferred	2	—
Common stock—authorized 20 million shares of $1 par value; issued 1974, 7,277,793 shares; 1973, 7,267,956 shares	34	34
Retained earnings	99	69
	135	103
Less: Cost of common stock in treasury; 477,028 shares	9	9
Total shareholders' equity	126	94
Total liabilities and shareholders' equity	$445	$353

* Totals may not add due to rounding.
SOURCE: Grumman annual reports 1973 and 1974.

equipment, work in new aircraft materials such as composites, flight stabilizing equipment, electronic devices, thrust reversers. With these and other areas as bases, Grumman gradually sought to establish a wider product base to cushion the cycles of aerospace. A newly created Grumman Data Systems Corporation pur-

EXHIBIT 3 The Grumman Corporation: Financial Review

Sales and Other Income (in millions)

	1974	1973	1972	1971	1970
Military aircraft and space systems	$ 963	$ 947	$586	$729	$920
General aviation	82	86	67	44	55
Commercial products (nonaerospace)	54	44	34	27	21
Data processing systems and services	50	37	35	31	29
Other operations	38	17	9	8	7
	1,189	1,133	733	842	1,034
Intergroup eliminations	59	45	46	41	38
Consolidated totals	$1,129	$1,087	$686	$800	$995

Income (Loss) before Federal Taxes, Gain on Debenture Exchange, and Extraordinary Item (in millions)

	1974	1973	1972	1971	1970
Military aircraft and space systems	$ 30	$ 23	$(114)	$(36)	$34
General aviation	8	12	2	(2)	—
Commercial products (nonaerospace)	2	2	2	1	—
Data processing systems and services	1	1	—	—	2
Other operations and unallocated corporate expenses	(4)	(5)	(3)	(3)	(2)
	37	34	(111)	(40)	35
Intergroup eliminations and minority interests	1	1	—	(4)	(2)
Consolidated totals	$ 36	$ 32	$(111)	$(36)	$37

SOURCE: Grumman annual reports 1970–74.

chased design and manufacturer's right for a printer-controller which provided non-IBM computer users with a way to use IBM high-speed printers. The new Grumman Ecosystem Corporation began several large municipal construction projects including a solid waste treatment facility in West Palm Beach and five packaged sewage treatment plants on Long Island. The Grumman Houston Corporation, formed in 1973, obtained contracts for oil exploration control units, electrical cables, test and calibration equipment, and electromechanical products. Grumman Allied Industries, Inc., entered the recreational vehicle, ground transportation, and health services businesses with such products as Pearson yachts, mini-buses, ambulances, and special-purpose vehicles. Allied also acquired rights and patents for storage and transportation systems of fresh foods, produce, and perishables using a unique water-saturated air flow technique.

Grumman also had interests in energy fields: an environmental data acquisition contract for offshore siting of nuclear power facilities and an extensive energy conservation program on its own facilities. That conservation program carried out in 1973 was especially enlightening to Grumman management. By using high-technology analyses to uncover excessive energy use, and by modifying heating, ventilating, and air conditioning systems, Grumman saved nearly a million gallons of fuel per year, enough to heat 1,200 single-family homes. Based on that success, Grumman, in 1974, created the Energy Programs Department within Grumman Aerospace to provide similar services to outside clients, including state governments and schools. They also made preliminary studies of power conversion by wind generators and solar collectors.

Ron Peterson

Ron Peterson received his B.S. in aeronautical engineering from St. Louis University in 1959 and joined Grumman Aerospace in August 1959 as an aerodynamic performance analyst. In 1964, he was selected to participate in Grumman Engineering's professional development program during which time he held assignments in operations analysis, business development, and program management. In 1966, he was made an assistant to the vice president for aircraft programs, and in 1968 promoted to program manager for the KA-6D, an aerial version of the A6A intruder aircraft. Then, in 1973 he was selected by Grumman to attend MIT as a Sloan Fellow.

Peterson is reluctant to categorize himself as a managerial type. "I'm basically a retread engineer, and tend to be informal in my managerial style. I haven't written a memo in eight years." He emphasized that in his experience a strong technical background was not the key to success. "Technical decisions are relatively easy," he explained. "You're dealing with basic laws of science. It's management and personnel decisions that are really tough."

He characterizes the management of innovation as an especially challenging role. He explained, "I've just spent a year at MIT studying management and I wrote my thesis on Internal Entrepreneurship. It's all been very influential on my thinking about managing innovation." Peterson has had extensive management responsibility in the product design, manufacturing, and to a lesser extent, marketing interface. The solar group would be his first opportunity to combine all these areas under his span of authority.

Joseph Gavin

Joseph Gavin received B.S. and M.S. degrees from MIT. After serving in the navy from 1942 to 1946, he joined Grumman as an aeronautical engineer. In 1957, he was appointed chief of the missile and space program and made a major contribution to the development of the lunar module for NASA's Apollo Space Programs in 1968. He was promoted to senior vice president of Grumman Aerospace in 1970, president in 1972, and chairman of the board in 1973. He is presently president of Grumman Corporation. He is also a director of the Grumman Corporation and various other organizations including the North East Solar Energy Center.

SOLAR ENERGY

Solar energy is a general term that incorporates many different technologies, some high, some low, some of immediate potential, some long term, all united by two common elements: their source is the sun, and they are renewable. Solar energy incorporates ideas ranging from wood-burning stoves to science fiction-type space satellites, from hydroelectric generating facilities to windmills; these are all methods of harnessing solar energy.

The Department of Energy (DOE) has categorized solar energy technologies under eight possibilities.

- Solar heating (and cooling).
- Wind (as from windmills).

- Photovoltaics (solar cells).
- Solar space satellites.
- Ocean thermal electric.
- Biomass (plant matter and animal waste).
- Agricultural and industrial process heating.
- Hydropower (hydroelectric dams).

Ron Peterson selected the first four of DOE's categories for Grumman's careful consideration.

Solar Heating

A solar heating system is a device that captures energy from the sun, converts it to heat, stores it, and transfers it to a point of usage. Solar heating systems conjure up images of banks of collectors, vast storage tanks, and other paraphernalia, but many effective solar systems rely simply on architectural design and have no moving parts. These are called passive solar and include such elementary ideas as south-facing windows, overhangs, thermal walls, double-glazed windows, reflectors, and U-shaped design. Natural principles of convection, conduction, thermal air movement, and optics can be employed to significantly reduce a home's heating load.

Peterson, although interested in passive possibilities, saw more immediate potential in those systems involving moving parts, or active systems. Two main avenues of opportunity presented themselves to Peterson here: space heating and domestic hot water. The principles of design of each were similar, but scale and costs were quite different. Typically, space heating systems range in cost from $5,000 to $10,000 while domestic hot water systems range from $1,000 to $2,000. Both must be supplemented by conventional heat.

Active solar heating systems consist of a collector, a heat transmission fluid (gas or liquid), a storage device, a pump or air circulator. (See Exhibit 4.) The collector (see Exhibit 5 for collector diagram) works on the principle of the "greenhouse effect." Greenhouses are actually huge solar collectors. Light waves, which have a high electromagnetic frequency, can easily penetrate the glass walls, but once inside, the light strikes plants (in greenhouses) or copper absorber plates (solar collector) and is converted to heat. Heat waves of lower frequency cannot penetrate the glass so easily and are trapped. Once trapped, water, a circulating fluid, antifreeze, or air transfers the heat to a water storage tank, a bed of rocks, gravel, or some other storage device, where it remains until that night or whenever it is needed.

Perhaps Peterson's greatest misgiving about the near-term potential of solar heat concerned user economics. Although it's very difficult to find figures on which experts will agree, most estimates of the total energy bill for a typical American family fall in the range of $1,000 to $1,500. Of that amount, 50 percent to 60 percent is attributable to space heating and from 20 percent to 25 percent to hot water needs.

Both supply of solar heat and demand fluctuate widely over the course of a day and over the span of a year. The largest fluctuation in demand (see Exhibit 6) is attributable to space heating demands. In general, the higher the percentage of a total heating load a solar system is designed to meet, the greater the unused capacity and the less economical is the system. Estimates of the percentage of home heating needs that can be provided by solar heat vary widely depending on climatic conditions, and claims of collector efficiency (defined as the percentage of entering sun's energy that is captured and transferred by the collector), heat loss in transmission, and heat loss in storage. Calculations are further complicated by the fact that collector efficiency is critically dependent on two other factors: the gradient between ambient temperature and the temperature inside the collector and the absolute temperature inside the collector. That is, a given collector can be highly efficient in summer and involve great heat loss in winter.

Another very significant factor in performance of a solar system is that of ground water

EXHIBIT 4 The Grumman Corporation

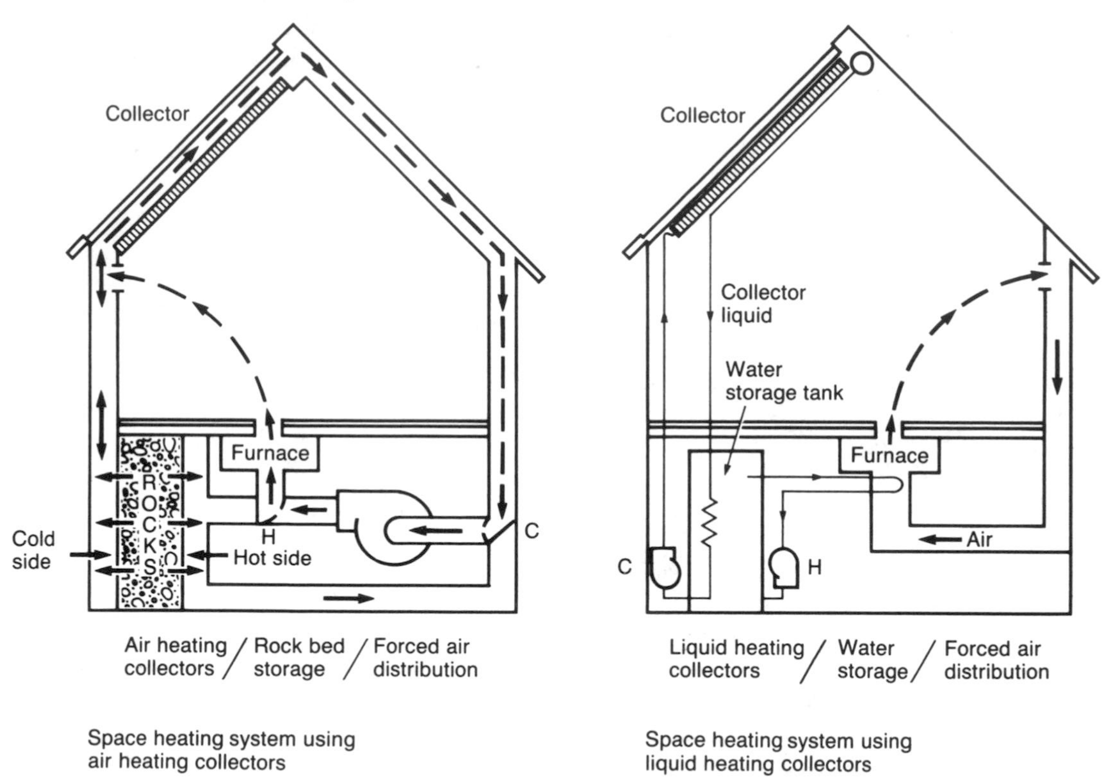

SOURCE: *Survey of the Emerging Solar Industry*, Solar Energy Information Services, 1977, p. 51.

temperature. If one operates on colder northern ground water, one is using more energy to achieve a given temperature than if one is in the south with warmer ground water.

Inexpensive low-technology collectors ($20 per square foot) typically average 40 percent efficiency while more expensive—$35 per square foot (evacuated tube)—collectors can increase that to 60 percent. A typical solar system can provide 1 million BTUs of energy per year for every $200 of collector. Every gallon of oil has an energy content of 140,000 BTUs; a typical oil heating system will operate at 65 percent efficiency.[1]

[1] In 1978, the price of home heating oil was about 58 cents a gallon.

The potential market for active heating systems was enormous. Unlike passive solar, it was not constrained by the pace of new construction. By "retrofitting," old homes can be converted to solar. Estimates of the percentage of the 55 million American homes suitable for retrofitting vary from 25 percent to 50 percent depending on judgments about the extent of exposure to the sun required to make solar pay. To be sure, shadings of trees and other buildings render much retrofitting uneconomical.

Potential solar markets include single-family and multifamily dwelling units, office buildings, industrial situations, schools, hospitals, garages, warehouses, and agricultural use for crop drying. Solar systems are also possible for

EXHIBIT 5 The Grumman Corporation

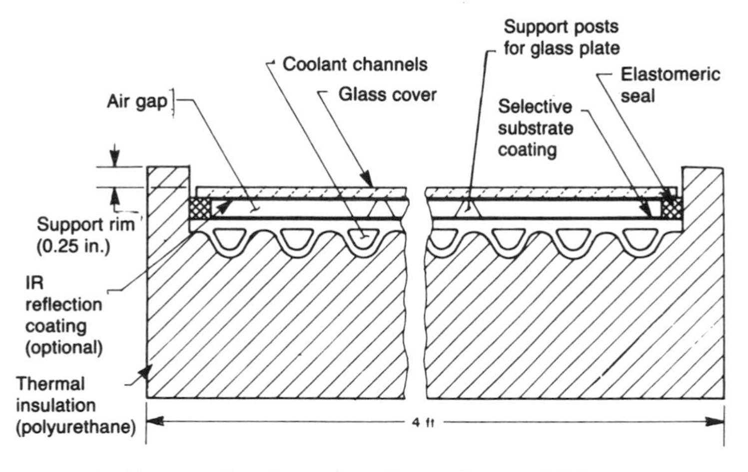

Prototypical low-cost flat plate solar collector. Source: (NSF)

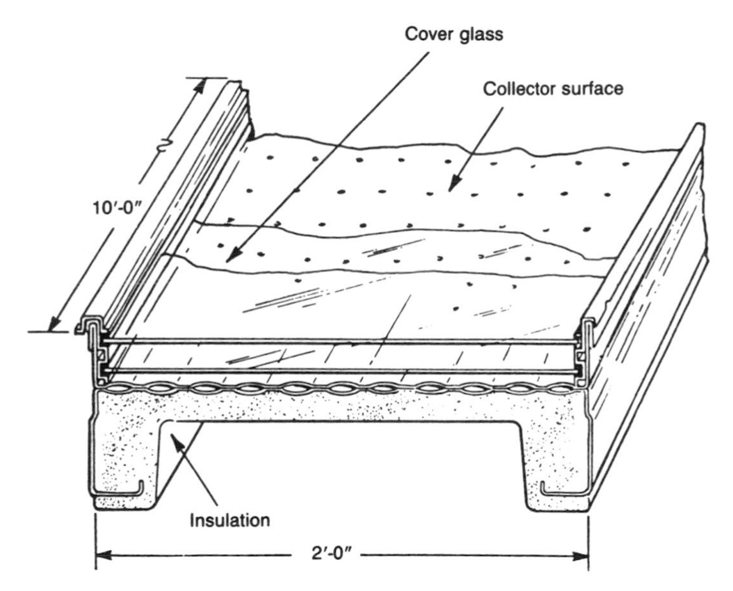

Structurally integrated solar collector unit. Source: (MOO)

SOURCE: Bruce Anderson, *Solar Energy, Fundamentals in Building Design* (New York: McGraw-Hill Book Company, 1977), p. 167.

a wide range of commercial structures, especially places like shopping centers, banks, and restaurants, where the publicity value of having a solar system might make a marginal economic situation look attractive.

Of these markets, the new single-family detached home is the most basic market. Hot water, space, and swimming pools all have great potential. Photovoltaics may also play a large role here. The greatest drawback to single-family market development is the fact that each installation is highly site specific. Because of

EXHIBIT 6 The Grumman Corporation

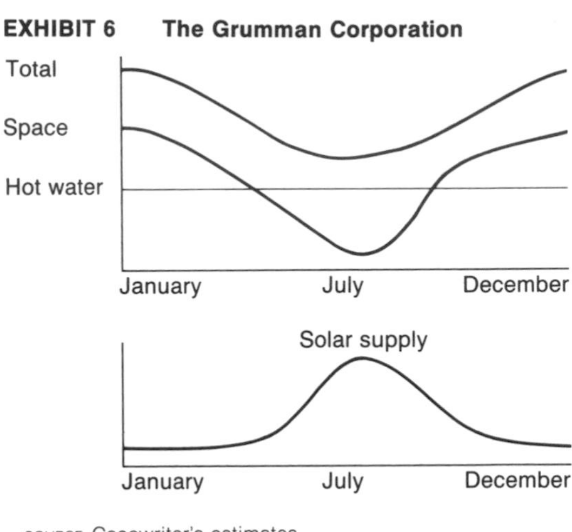

SOURCE: Casewriter's estimates.

this, little standardization is possible and payback periods remain long. To date, passive systems are more prevalent than active in new construction.

Today one tends to encounter resistance among developers of new townhouses to solar because plans are often tight and there is less room for ducts and solar storage. In situations where there are high roofs, costs of installation are increased and unless a townhouse has individual metering, there is little personal incentive for solar. In urban residential and office environments, the small roof area means solar can only make a minimal contribution. Hospitals and schools offer attractive markets in that federal financing is often available. Peterson was aware that architectural construction and financing problems peculiar to different markets often were critical to the purchase decision.

Market growth was largely dependent on oil and gas prices, governmental incentives, and advertising and marketing outlays.

The pace of diffusion of solar technology is in many ways incalculable because of the trendish nature of the American consumer. Many people were, in Peterson's words, "curious," or "intrigued by the romance of solar," but were waiting for a major technological breakthrough. Peterson wondered whether it would be possible to dispel some of the mythology concerning solar with a concerted consumer campaign.

Because solar power is largely nonpolluting, ecologically sound, without waste disposal problems, by and large safe and plentiful, it had captured the public imagination. But Peterson does not necessarily see this as a plus. "There's a real danger that poor initial performance or exaggerated expectations could dispel the high measure of public enthusiasm—cause a severe setback for several years." To Peterson, then, timing of entry is a key consideration.

Another issue of paramount concern to Peterson and the solar heating industry is the role of utilities. During the 1940s in Florida, solar energy was a booming business, but the advent of cheap gas and low utility energy rates effectively killed it. Peterson saw a similar threat today. Utilities, although not unaware of the benefits of solar, view the new industry as a threat. Their reasoning is based on simple economics. Utilities have tremendous capital investments in plant and it is to their advantage to operate facilities at capacity or near capacity as much of the time as they can. If energy demand varies widely on a daily or weekly or seasonal basis, utilities are left with overtaxed facilities at some times and underutilized facilities at others.

Solar energy is in most abundant supply during the day when demand is lowest on utilities—and is often unavailable in the evening and in winter when peak utility demands occur. Thus, simple solar energy devices can have the net effect of aggravating the utilities' peaking problem and reducing the base demand. If utilities offer off-peak rates to entice users to use energy during low-capacity periods (evenings, for instance), the already tenuous user economics of solar could become prohibitively low.

Another serious concern to Peterson is the unpredictability of the technology in solar devices. Dr. William Shurcliffe, a Harvard physicist, articulates Peterson's concern in an article in the *Bulletin of Atomic Scientists*:

> Is there one most promising avenue in solar design? Are there two or three clearly superior avenues? Unfortunately not. Perhaps there will be in five years. But today, to the dismay of newcomers to the field, there are literally hundreds of approaches and the race is neck and neck. No one knows which of the many approaches will gain a decisive lead.

The solar heating industry, like most industries in the early stages of growth, was fragmented. No one manufacturer had more than 3 percent of the market. Dr. Shurcliffe offers several categorizations of the solar heating industry:

> Investigations are divided on the issue of collector efficiency. They have divided into high-technology types and low-technology types: they have divided them into those that receive huge government grants and those that do not, those that produced glamorous and highly uneconomic systems and those who have produced drab but economic systems. The two groups are scarcely on speaking terms.

Total sales for solar heating, including installation, in year ending 1974 were $25 million. Peterson expected growth rates in the vicinity of 200 percent for the next few years. His projections of growth, however, were clouded by the uncertainty surrounding governmental incentives. "The government," he declared, "is actually hurting the industry by simply discussing subsidies. People are holding off purchases, waiting for the government action."

Although fragmented, many large and well-financed companies were moving cautiously ahead in the solar field. Mobil Oil Company had formed a joint venture, Mobil-Tyco, to investigate opportunities in the photovoltaics field. The Exxon subsidiary Daystar was involved in solar collector manufacture. Heating firms like Honeywell and Lenox were also entering the field. In addition, hundreds of small firms were rushing into the field.

Solar heating faces numerous obstacles beyond the strict economics. There remain unresolved questions as to legal status of rights to the sun. Can they be deeded? Also, rigid building codes in many areas have acted as barriers to solar installations.

Wind

People have used wind to move ships and power windmills since the beginning of recorded history. By 1910, wind conversion systems generated usable amounts of electricity in Denmark. Recently, a wind system in Ohio, sponsored by ERDA and built by the National Aeronautic and Space Administration, operating in a wind velocity of 14 miles per hour could generate enough electricity for 30 average-size homes. This has prompted several utilities in windy coastal regions, the Midwest, and Hawaii to look closely at wind generators in the 100 to 1,000-kilowatt range.

A wind energy system consists of three smaller systems: an aerodynamic, a mechanical, and an electrical subsystem. Each interacts to affect the other's performance. U.S. technology is highly developed in all three areas, but the most serious challenge is in the aerodynamic-structural interface.

Wind generators consist of a rotating electrical generator[2] turned by a propeller which, in turn, is pushed by the force of the wind. The amount of electricity generated is dependent on four things: the amount of wind, the diameter of the propeller, the size of the generator, and the efficiency of the system. With an eight-

[2] An electrical generator is a device which transforms rotary mechanical energy into electrical energy. (293 kwhr = 1 million BTUs = approximately the energy derived from 11 gallons of oil.) Oil costs approximately 58 cents per gallon and residential oil furnaces operate at an efficiency of about 65 percent.

foot propeller, blades with an aerodynamic efficiency of 70 percent, connected to a generator with an efficiency of 70 percent capable of delivering 1,000 watts, a 5 miles per hour breeze will produce 10 watts; a 15 miles per hour breeze will produce 250 watts; and 20 miles per hour will produce 600 watts. Presently, electricity in the northeast costs approximately 5 cents per kilowatt-hour.[3] Government agencies are also experimenting with nonpropeller systems including a vertical axis turbine system which looks like the lower part of an egg-beater.

Although atmospheric scientists agree the raw power available in the wind is vast, three obstacles stand in the way of widespread use of wind energy: relative costs, intermittance of wind, and storage systems. Although government research is bringing the cost down, today wind systems are only economically feasible in remote areas. Showcase windmills flourish, however. Congressman Henry Reuss of Wisconsin installed a windmill in his home at Chenague at a cost of $5,000. He gets about 75 percent of his power needs from the windmill.

In a recent study, the Department of the Interior stated, "The energy shortage seen imminent compels the recognition of any possible opportunities to develop major amounts of electricity." Their report suggests that so-called "wind farms may soon be such an opportunity." Exhibit 7 shows an ERDA map with potential farm sites.

EXHIBIT 7 The Grumman Corporation: Sites for Wind Farms

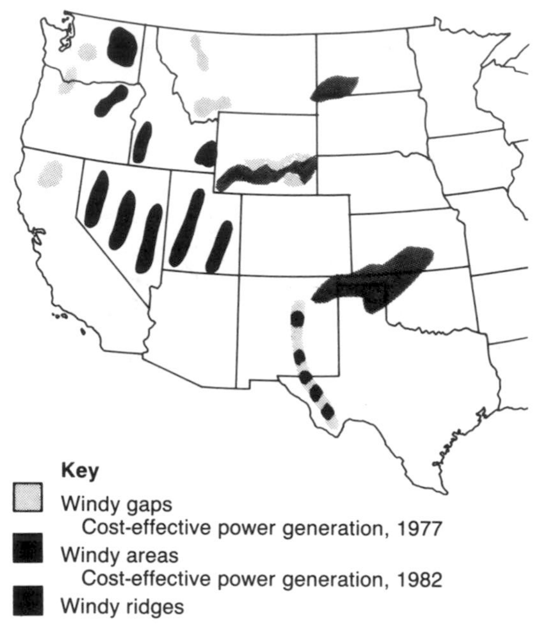

SOURCE: U.S. Department of the Interior, August 1977.

The intermittances of wind is a serious but not insoluble problem. Since no lighting systems and few appliances can accommodate themselves to a power source where the output is constantly fluctuating, virtually every system then must include some means of stabilizing the electrical current. A synchronous inverter is a device that can perform this function; however, it requires commercial power. At remote sites, the most practical solution is to run power into a battery—which will also store the energy.

Because of these obstacles, wind, as one executive puts it, is a "fledgling industry." Wide commercialization is believed to be five to seven years away. But despite this, there were 22 manufacturers of wind machines on the market in 1975. Capacity of these machines ranged from 50 watts to 15 kilowatts. Installed costs ranged from $3,000 to $15,000 per kilowatt. ERDA's goal is to bring the installed cost per kilowatt down to less than $1,000. The development emphasis according to one executive should be on small machines (under 100 kilowatts). Within this category he sees three markets:

- 1 kilowatt high reliability machines for use in remote areas (e.g., navigation buoys).

[3] On a national basis, it ranges from 3 cents to 6 cents per kilowatt-hour.

- 8 kilowatt machines designed for residential use.
- 30 kilowatt machines suitable for use in agricultural or industrial processes or for small groups of homes.

Grumman Aerospace had made a preliminary investment in wind energy. This work resulted in the invention of the Tornado Wind Energy System. According to one industry source, it is the only revolutionary idea in windmills in 20 years.

Dr. Yen, the inventor, discusses his concept and why it is revolutionary:

> Firstly, we can reach much higher wind densities. We propose a stationary surface to collect wind—nothing moving. Then we add concentration action—like vortex of a tornado, which we harnessed in a small turbine generator. We propose a stationary tower that would be much larger than ordinary wind systems—300 feet tip to tip. We also propose a system 1,000 to 2,000 feet in height—that could generate 100 megawatts.
>
> Our system can also be powered by heat. If heat is supplied at the bottom of the vortex, it creates an updraft. So if there is some wind we reinforce the vortex and if there is no wind we create it. We can utilize low-grade air like exhaust fumes from the power plants.
>
> This design would enable us to capture part of the immense amounts of heat wasted by utilities. It has real commercial significance.

Peterson, although enthusiastic about the Yen mill, would expect a phased move into wind power. He explains:

> Grumman would probably want to enter the wind field initially in the small wind generator area as it has the most near-term potential. Also, I'd expect that there would be significant government contracts in this area.

Photovoltaics

Silicon, the principal raw material used in the manufacture of photovoltaic cells, is the most abundant solid element on earth. And photovoltaic cells, alone, convert sunlight into one of the most highly prized energy forms—electricity.

Photovoltaic cells are made with the same advanced manufacturing processes used by the semiconductor industry to manufacture transistors and integrated circuits. Windows are chemically etched on a protective oxide[4] layer covering a wafer[5] of silicon. Through these windows, the pure silicon is selectively "contaminated" with different gaseous impurities in high-temperature, closely controlled furnaces. These impurities define electrical boundaries between the elements of cells that are connected by a vacuum evaporating a microscopically thin layer of aluminum on the circuit. The resulting cells will be photovoltaic. The term *photovoltaic* literally means voltage created by photons or light waves from the sun. The sun excites the silicon producing charged carriers that travel around the circuit and generate voltage. The theory underlying this phenomenon was first developed by Albert Einstein in a 1905 Ph.D. paper.

The early photovoltaic cells were used to provide electricity for orbiting space satellites. Clusters of the cells covered the butterflylike wings that opened when the satellite was in space. Cost was not a problem. Reliability, technical performance, and sophistication were the key considerations.

These early cells were manufactured using perfect silicon crystals and the same tightly controlled and costly processes used for making transistors for aircraft equipment. They were, indeed, very expensive. Early systems cost as much as $1 million per peak kilowatt—1,000 to 2,000 times more than the cost of producing electricity by conventional means.

But what happened to the transistor industry is now happening to photovoltaics. The space

[4] S_iO_2 (silicon dioxide), an insulator.

[5] Silicon crystals are produced in three-inch circular "wafers."

program's need for lightweight, reliable electronic amplifiers created a constant, strong stream of federal purchases for what were then often $200 transistors. The rapid fall of semiconductor prices is now legend. A commercial version of the same transistor can today be purchased for 20 cents.

Prices for photovoltaic cells have also been dropping dramatically. The average commercial price for photovoltaic systems in 1975 was slightly over $15,000 per peak kilowatt, but some of the latest quotes for new installations are as low as $3,000 per peak kilowatt.

Even though present price levels reflect a decrease in cost to less than a hundredth of the original satellite cells, they restrict the current market for photovoltaics to special applications, such as unmanned railroad crossings, remote microwave repeaters, marine battery recharging, and pipeline leak warning systems. The market for these photovoltaic applications grew by 50 percent in 1975 while shipments were about $10 million.

The continued reduction of costs could open up a vast market of conventional electric power generation via photovoltaics. At the heart of the cost reduction problem is the high cost of single crystal silicon. As one semiconductor executive explained, "Eighty percent of the cost of a photovoltaic system is in the materials and 80 percent of the material cost is in the silicon." However, it was expected that there would continue to be substantial reductions in photovoltaic costs.

There is a wide consensus that in order to penetrate the broader market and obtain longer term learning curve projections, the industry must move away from single crystal silicon. A reduction in cost to 1/10 of present levels (to 30 to 50 cents per peak watt) is necessary to make photovoltaics competitive with conventional fuels. Reaching these cost levels will probably require a radical move to very thin cells (up to 100 times thinner than the thickness of today's cells) or to amorphous (uncrystallized) semiconductors. Researchers at RCA and at Energy Conversion Devices have already demonstrated the feasibility of thin film, amorphous cells with moderate efficiencies. Stamford Ovshinsky, the inventor of the amorphous semiconductor, estimates that electricity generated by amorphous cells could be cheaper than conventionally generated electricity in a few years.

Assuming that the cost barrier is crossed, then the potential for photovoltaics is very large. But in what form? Photovoltaic cells could be used in two distinct ways, with quite different effects on the American energy supply system. The cells could be used on-site like solar heating. But photovoltaics, as a generator of electricity, would have greater effect than solar heating on the role of utilities, for if on-site power generation by photovoltaics became widespread, then the utility industry would have to redefine its business, emphasizing power distribution instead of power generation. Utilities would purchase power from on-site installations, mark it up, and distribute it to where it is needed. Unless low-cost means of electrical power storage were developed, the utilities would still supply backup power. Such structural changes would require imaginative solutions on the part of regulatory commissions. Moreover, they would require the support of the utilities themselves.

Recently, a Harvard physicist concluded in a report that photovoltaic cells could be expected to produce 1 percent of the nation's electricity by the year 2000 and ultimately could be as significant as coal or nuclear power in generating electricity. An ERDA official has estimated that photovoltaics could contribute up to 3 percent of the nation's energy by the turn of the century. Other projections are considerably more optimistic.

Grumman had recently been approached by a reputable physicist to form a joint research effort. That physicist, who had been involved in semiconductor manufacture, believes he has a cell with a unique thin film device that not only converts solar energy into electricity but serves as a storage device as well. Peterson, however, is viewing this idea cautiously and investigating the possibilities of outside funding for the idea.

Solar Satellites

Solar satellites are giant satellites which would capture the sun's energy with photovoltaic cells and beam it back to the earth using microwaves. The power from solar satellites would be sold to utilities and the high-technology firms of Boeing, Martin-Marietta, and Lockheed. Construction would proceed along the lines of aerospace mission programs. Indeed, the solar satellites fit well with the technical and managerial practices and expertise of the nation's research establishment. Many of those directing government energy R&D programs—coming out of the Atomic Energy Commission, the National Aeronautics and Space Administration (NASA), and the Department of Defense—are accustomed to dealing with major high-technology companies in this way. The microwave solar power satellite concept, for instance, would require orbiting space platforms with literally square miles of photovoltaic collectors and could cost upwards of $70 billion per system.

Even though the concept is only at the feasibility stage, critics have been quick to point out the military risks of obtaining a sizable amount of our power from orbiting satellites. Ecologists are concerned about the impact of bombarding patches of several square miles of the earth's surface with microwaves, and the accompanying risks to airline travel. Unfriendly powers can be expected to react negatively to the potential military value of hundreds of controllable microwave patches. Against these negatives, supporters point out that by rotating in phase with the sun the satellites could supply round-the-clock power and would convert the sun's rays before they lose energy passing through the atmosphere.

Grumman could become involved in solar satellites in two different ways. Grumman could bid individually on some of the components of the satellite program such as structural sections or control systems or it could attempt to become the overall contract manager for the program. In either case, the bids would be made on a cost-plus basis. Depending on the magnitude and complexity of the items Grumman chose to bid on, the investment could range from one million dollars to tens of millions of dollars. Returns would be expected to be compatible with returns from Grumman's other aerospace operations.

CONCLUSION

Before Peterson submitted his proposal to Gavin, he had second thoughts about rejecting other solar technologies without any serious consideration. Some experts saw a potential for burning wood and other plant matter of up to the equivalent of 3 million barrels of oil equivalent per day, over three times the present contribution of nuclear.[6] Large-scale use of wood, the principal biomass fuel, would require construction of chipping or pelletizing plants for easier transportation. This offered a possible opportunity, as did the sale of wood stoves and the construction of municipal solid and liquid waste recycling plants. Small hydropower systems and ocean-thermal plants for electricity generation both appeared to have significant potential and Peterson was wondering if he should take the time to investigate them further. Beyond these technologies, a myriad of lesser known solar technologies had been proposed by a variety of scientists and inventors.

Peterson and Gavin both saw tremendous potential for solar. Increasingly, an awareness of the serious energy constraints facing the country and the world were filtering down to the everyday worker. People were coming to

---

[6] Total U.S. energy consumption was 37 million barrels of oil equivalent per day. In 1978, world consumption was about 100 million barrels of oil equivalent per day.

see the nation's energy crisis in personal terms. The implication of a sound national energy policy on the balance of payments problem on the environment, the economy, and the international situation were more and more becoming matters of national awareness. The extent to which this awareness could be translated into sound business opportunities was the question facing Peterson and Gavin in mid-1975 as they pondered whether or not to make a major commitment in solar and if so, to which area(s) to commit their resources.

Reading II–2
Technological Forecasting for Decision Making

B. C. Twiss

Thus in policy research we are not only concerned with anticipating future events and attempting to make the desirable ones more likely and the undesirable less likely. We are also trying to put policymakers in a position to deal with whatever future actually arises, to be able to alleviate the bad and exploit the good.

Herman Kahn

THE NEED TO FORECAST

All we know for certain about the future is that it will be different from the present. Thus the products, organizations, skills, and attitudes, which today serve a business well, may have little relevance to the conditions of tomorrow. If a business is to survive it must change. And the changes must be timely and appropriate to meet the needs of the future. The difficulty arises in prognosticating what these needs will be. Nevertheless, an attempt must be made to do so and, however imperfect forecasts may be, managers cannot afford to take decisions affecting the future of their organization without examining any clues they can find to the best of their ability.

Forecasts are important inputs to the process of strategy formulation and planning. They have been used to gain a better understanding of the threats and opportunities likely to be faced by established products and markets and, consequently, of the nature and magnitude of the changes needed. Thus, anticipation enables the business to be steered into the future in a purposeful fashion, in contrast to belated reaction to critical events. Nowadays, technical lead times are often so long that a market can be lost before a proper response is made. While this approach may be satisfactory conceptually, it is of little value unless one can make sufficiently accurate forecasts to aid the practical manager in his decision making. During recent years, numerous techniques for technological forecasting have been developed with the object of enabling the manager to obtain the maximum use from the information available to him.

Since technology is responsible for many of the most important changes in our society, forecasting future advances in technology and their impact can be as vital for top management in the formulation of corporate strategy as it is for the technologist reviewing his R&D program. For technological changes may sometimes result in a major redefinition of an industry or market. We have seen, for example, that the producer of metal cans may find his major threats coming not from direct competitors in can manufacture, but from other packaging technologies such as glass, plastics, or paper, or from different forms of preservation such as freeze drying. However, the area of consideration might need to be widened even further to take in factors such as: the size of the future market for convenience goods upon which the relative economic merits of alternative packag-

---

Managing Technological Innovation, 2d ed. (New York: Longman, 1980), pp. 206–234. Reprinted with permission.

ing technologies may depend; the availability and future costs of packaging materials; customer tastes in convenience products; health and hygiene standards; and a multitude of similar items. Similarly, the manufacturer of razor blades, whilst aware of the direct alternative of the electric razor, may eventually suffer more dangerous competition from the development of effective depilatory creams. But in order to assess this threat, it is necessary not only to forecast the technological possibility of a satisfactory product being developed but also to predict whether it would be socially acceptable to the market.

It follows that the field of investigation has to be very extensive if some important indication is not to be missed and it must include a wide range of social, economic, and political as well as technological factors. While sometimes identification of the threat or opportunity may be sufficient, more frequently the real implication can only be properly evaluated by a detailed study. When it is realized that each factor is associated with a high degree of uncertainty, the task of the forecaster might seem well nigh impossible. Yet, however daunting the difficulties may appear, the manager cannot avoid making decisions which will be proved good or bad by future events. He must take a view of the future. If forecasting techniques can enable the manager to obtain a more accurate picture of the future and in consequence improve his decision making, the effort devoted to it is justified. This can be the only real justification for forecasting.

In any consideration of forecasting, it is important not to lose sight of the high degree of uncertainty in the outcome. The future will never be predictable. Forecasting can assist in the formation of managerial judgment but will never replace it. Many of the critics of forecasting condemn it for its failures, of which there are abundant examples. Perhaps the forecasters themselves are largely to blame for encouraging others to place more reliance on their forecasts than they merit. We must learn to accept that many forecasts will remain poor, but appreciate that if conducted conscientiously, they will lead to fewer errors and the avoidance of some of the most costly mistakes.

With this perspective in mind, it can be seen that there are limitations which must set some bounds to the resources devoted to forecasting. For like any other managerial activity, the investment of time and resources can only be justified in terms of the benefit expected to derive from it. This criterion must be used to prevent excessive expenditure on forecasts which do not aid decision making, however interesting they may be in themselves. As Drucker comments:

> Decisions exist only in the present. The question that faces the long-range planner is not what we should do tomorrow, it is: What do we have to do today to be ready for an uncertain tomorrow? The question is not what will happen in the future. It is what futuristics do we have to factor into our present thinking and doing; what time spans do we have to consider, and how do we converge them to a simultaneous decision in the present?[1]

In some industries, today's decisions cover a long time scale. This is true of fuel, power generation, and communications where the effects of what is decided now will still be felt several decades hence. In the case of a power station, for example in the choice of the reactor system for a nuclear plant, the design and construction stages may take 5 to 6 years to be followed by an operating life of 20 years or more. Although the importance of the more distant dates loses significance because of financial discounting, it may still be necessary to make forecasts for a period of 20 or more years.

By contrast, the planning horizons for most companies are much shorter—of the order of 5 to 10 years. It is still important to forecast, for many significant changes can occur in a dec-

---

[1] P. F. Drucker, *Technology, Management, and Society* (London: Heinemann, 1970).

ade. Thus the only meaningful determinant of the time period for which forecasts are necessary would appear to be the planning horizon of the company. This is a function of the rate at which the company's activities can be made to respond to changes rather than the rate at which the environment itself is changing.

Hall of International Computers, Ltd., writes:

> In the field of product development, however, close control is necessary to ensure commercial viability, control of costs, and particularly of time scales. In this area, we can in the computer field respond to outside stimuli within about five years for any normal project and seven years for the very largest. . . . The practical range of technological forecasting required, therefore, and on which whole attention needs to be focused, is about five to seven years, this corresponding both to our response time and to the generation period of computer development.[2]

Summarizing, we can see that forecasting can assist business decision making in the following ways:

1. Wide-ranging surveillance of the total environment to identify developments, both within and outside the business's normal sphere of activity, which could influence the industry's future and, in particular, the company's own products and markets.
2. Estimating the time scale for important events in relation to the company's decision making and planning horizons. This gives an indication of the urgency for action.
3. The provision of more refined information following a detailed forecast in cases where an initial analysis finds evidence of the possibility of a major threat or opportunity in the near future, but where this evidence is insufficient to justify action,

   OR

   continued monitoring of trends which, while not expected to lead to the necessity for immediate action, are, nevertheless, likely to become important at some time in the future and must consequently be kept under review.
4. Major reorientation of company policy to avoid situations which appear to pose a threat or to seek new opportunities by:

   a. Redefinition of the industry or the company's business objectives in the light of new technological competition.
   b. Modification of the corporate strategy.
   c. Modification of the R&D strategy.

5. Improving operational decision making, particularly in relation to:

   a. The R&D portfolio.
   b. R&D project selection.
   c. Resource allocation between technologies.
   d. Investment in plant and equipment, including laboratory equipment.
   e. Recruitment policy.

LEVEL OF INVESTMENT IN TECHNOLOGICAL FORECASTING

There is a common assumption that technological forecasting (TF) is for the larger companies, since it is only they who can afford to undertake the exhaustive process of data collection and analysis required by the more sophisticated techniques. Technological forecasters often point to the dangers of forecasts based on inadequate analysis and argue that useful results can only be expected when information is meticulously gathered and carefully analyzed. This view is supported by Nicholson, who states:

> The useful analysis of present information to predict the future is likely to be more than a

[2] P. D. Hall, "Technological Forecasting for Computer Systems," *Proceedings of National Conference on Technological Forecasting*, University of Bradford, 1968.

spare-time occupation for research or marketing staff. There needs to be a nucleus of planning staff who are familiar with the subject and who can act as a focus. Specialist staff can then be brought in temporarily to form teams for particular studies as required. In current practice, these teams are attached either to the headquarters of a company or to the central R&D department. Scientists and engineers have to date predominated in such activities, but economists and hopefully, sociologists, should be able to play a considerable part. In forecasting, one needs a wide range of thought processes brought to bear as well as a variety of methods.[3]

TF, then, appears to be an expensive activity employing a multidisciplinary team in detailed analysis. Without doubt, effort on this scale is normally essential if the most accurate forecasts are to be prepared. But this is a counsel of perfection. Small companies equally face an uncertain future and need to build some form of forecasting into their decision making even if it is based only on the judgment of the chief executive. If his judgment can be assisted by simple forecasting techniques, then they should be used, even though it is realized that more accurate forecasts might have resulted from the investment of greater effort in sophisticated techniques. The question should not be whether or not to forecast, but:

- To what extent is it necessary to forecast?
- Which are the most appropriate techniques to use, given the limitations of what can be afforded?

When attention is focused on need rather than the ability to pay, it can be seen that the importance of forecasting is much more a matter of the business environment than the size of the company. The large firm in a mature industry is unlikely to be overtaken suddenly by a technological development which will have a catastrophic effect. For such a company, TF can be confined to monitoring the environment to give early warning of the occasional advance; only then does it need more detailed forecasts. It was not, for example, St. Gobain's ignorance of the float glass process which led to its late investment in the new technology, but its lack of adequate response to the threat. TF might have helped the company's understanding but its effect upon the decision-making process is likely to have been negligible.

By contrast, one can see how TF could have been applied both by IBM and its competitors when the IBM 360 series was being evaluated. Computer technology was advancing rapidly and forecasts showed a very close correlation between a computer's performance and the date of its introduction.[4] The rate of technological change in an industry should, therefore, be a much better criterion than size of company as a determinant of the need for TF, although as we have seen earlier, the speed at which a company can react to change is also important.

The obscurity of the precise nature of a threat also has an important bearing on the degree of detail required in a TF. Sometimes the threat is obvious, as in the case of the manufacturer of lead additives for petrol—the only area of doubt is in the time scale of antipollution legislation. On the other hand, the maker of lead acid batteries has a much more complex problem to resolve. Firstly, he has to consider whether the conventional reciprocating petrol engine will be totally banned, or whether the petroleum or motor industry will find an acceptable solution to exhaust pollution. If banned, it might be replaced by the

[3] R. L. R. Nicholson, *Technological Forecasting as a Management Technique* (London: HMSO, 1968).

[4] R. U. Ayres, "Envelope Curve Forecasting," in *Technological Forecasting for Industry and Government,* ed. James R. Bright (Englewood Cliffs, N.J.: Prentice-Hall, 1968). Ayres shows a remarkably high degree of correlation between computer performance measured as $\frac{\text{capacity (bits)}}{\text{add time (secs)}}$ and date for the period 1945–70. Hall has applied the Delphi technique to forecasting computer applications for up to 40 years from 1968. (Hall, "Technological Forecasting for Computer Systems.")

Wankel engine, gas turbine, steam, or electric propulsion. Even if his analysis shows a high probability of electric traction being widely adopted, he will be aware that major motor manufacturers are engaged in research into new forms of electric power storage. It is, therefore, not immediately obvious whether he is facing a threat or an opportunity. In this situation, detailed technological forecasting could be of invaluable assistance.

An offensive R&D strategy exposes a company to much greater risks than a defensive strategy, particularly in relation to the state of the technological art and timing. Greater steps into the unknown are involved. In this situation, opportunities are being sought, rather than the avoidance of threats. Thus, a company adopting an offensive strategy should devote correspondingly more resources to TF.

In summary, we can conclude that:

1. All companies should undertake some form of technological forecasting.
2. The amount of effort devoted to TF should take into account:

    a. The rate of change in the environment.
    b. The planning horizon determined by the technological and marketing lead times for new products or processes.
    c. The complexity of the underlying problems.
    d. The R&D strategy.
    e. The size of the company only in so far as the availability of resources limits the choice of techniques which can be afforded.

THE DEFINITION OF TECHNOLOGICAL FORECASTING

Although there is nothing new in attempting to forecast the future trends of technology, it is only during the last decade or so that the range of techniques known collectively as technological forecasting has been developed. They differ from informal methods in their systematic analysis of data within a formalized structure. The main features of what we know by TF can be seen by examining several definitions.

Prehoda defines TF as:

> The description of prediction of a foreseeable technological innovation, specific scientific refinement, or likely scientific discovery, that promises to serve some useful function, with some indication of the most probable time of occurrence.[5]

or Bright:

> Forecasting means systems of logical analysis that lead to common quantitative conclusions (or a limited range of possibilities) about technological attributes and parameters, as well as technical economic attributes. Such forecasts differ from opinion in that they rest upon an explicit set of quantitative relationships and stated assumptions, and they are produced by a logic that yields relatively consistent results.[6]

or Cetron who adds:

> A prediction, with a level of confidence, of a technological achievement in a given time frame with a specified level of support.[7]

Let us now look at some of the phrases used in these definitions to obtain a clearer picture of the underlying basis for the techniques of TF.

"Foreseeable"

This reservation is not so restrictive as might be thought, for although it is true that forecast-

---

[5] R. W. Prehoda, *Designing the Future—The Role of Technological Forecasting* (London: Chilton Books, 1967).

[6] J. R. Bright, ed., *Technological Forecasting for Industry and Government: Methods and Applications* (Englewood Cliffs, N.J.: Prentice-Hall, 1968.)

[7] M. J. Cetron, *Technological Forecasting—A Practical Approach* (London: Gordon and Breach, 1969).

ing is not an occult science, the fact emerges from the accumulated body of forecasting knowledge that unexpected breakthroughs are much rarer than commonly supposed. The much quoted example of penicillin is the exception rather than the rule. Most of the innovations of the next 20 years will be based upon scientific and technological knowledge existing now. The difficulty lies in identifying what is of real significance. With hindsight, what today appears obscure will tomorrow seem remarkably clear. The role of TF is to evaluate today's knowledge systematically, thereby identifying what is achievable and more particularly, how one technological advance, perhaps in conjunction with another, could fulfil a human need.

"Specific Scientific Refinement"

Most writers lay heavy stress upon the need to focus attention upon a specific development or discovery. The complex interplay of environmental factors escalates a detailed forecast rapidly even when the subject of the exercise is defined in specific terms. Without this restriction, the study would quickly get out of hand. Admittedly the scenarios for the year 2000 of the "think tank" investigators such as Kahn explore every aspect of human activity, but these are far removed from the practical consideration of industrial forecasters, who are concerned with decision making in respect of specific threats, opportunities, or innovations.[8]

Nevertheless, we have seen already that the initial screening of the environment should be wide ranging, otherwise there is a danger of becoming distracted by detailed forecasting of the wrong phenomenon. Thus, while accepting that detailed forecasts must be focused upon a specific development, they must be preceded by a much broader investigation to ensure that the effort is being allocated to the right task.

". . . that promises to serve some useful function"

The importance of a market orientation in successful technological innovation has already been stressed. As a corollary, one might assume that technological progress will respond to market needs. But is this so? It is certainly not true of science where progress stems from the desire of scientists to push back the frontiers of knowledge. Technology, however, is different. It advances through industrial investment motivated by the desire to reap specific benefits, financial in the case of business, though they may be social or political where government expenditure is involved. Thus, it is the market which ultimately determines the pace of progress.

Corroborative evidence comes from an extensive study of the growth and application of technology carried out in respect of defense projects in the United States under the title of "Project Hindsight." Isenson, reporting on this project, writes, "The first conclusion of the study is that real needs result in accelerated technological growth. Conversely, in the absence of real needs, technological growth is inhibited."[9]

The importance of practical objectives for technological advance is self-evident. Nevertheless, it is easy for technological forecasters, wrapped up in the elegance of their techniques, to lose sight of this simple truth. There is a clear distinction to be drawn between *what could happen*, because of man's capabilities, and *what is likely to occur*, because of what he wants.

---

[8] H. Kahn and A. J. Weiner, *The Year 2000: A Framework for Speculation on the Next Thirty-Three Years* (New York: Macmillan, 1967).

[9] C. W. Sherwin and R. S. Isenson, *First Interim Report on Project Hindsight,* Office of the Director of Defense Research and Engineering Clearing House for Scientific and Technological Information, no. AD. 642-400, 1966. See also Bright, *Technological Forecasting for Industry and Government.*

"Quantitative Conclusions": "Levels of Confidence"; "A Given Time Frame"

Since the purpose of a forecast is to provide a useful basis for decision making, it must be expressed in terms the manager can apply. Vague qualitative statements are of little use to him. He requires specific quantitative information with an indication of its reliability. The uncertainties inherent in any forecast must also be reflected in its presentation.

Whereas the statement: "The computer will eventually eliminate the need for cheques in financial transactions" is unhelpful since it is unquantified, the alternative: "By 1990, 80 percent of the financial transactions currently carried out by cheque will be replaced by the computer," is misleading, because although quantified, it gives no indication of the uncertainties involved. This second statement needs to be associated with a probability distribution, albeit subjective, relating the time of occurrence of the event to the probability of it happening by that date.

"Specific Level of Support"

We have already noted that technology advances in response to a market need. The rate of progress, however, depends upon the scale of effort devoted to it, in itself a function of the importance attached to the need. In many cases, the need will be stimulated by events largely unconnected with technology. One example is provided by the programme to land a man on the moon's surface. The prime objective was national prestige rather than the furtherance of science. For the origins one must look to the "space race" between the United States and Russia. Thus international technopolitical competition created conditions enabling the U.S. government to provide NASA with funds on a massive scale. Although technological feasibility was a precondition for success, the date when the mission could be accomplished was closely geared to the level of financial support.

Similarly, the development of motor car antipollution devices is not the result of direct pressure from individual owners unwilling to poison their fellow citizens; it is the outcome of social pressure which subsequently manifests itself in political action resulting in legislation. This determines the scale of the motor manufacturer's investment in the development of antipollution equipment. Another example is found in the development of, and investment in, automation equipment where the supply and cost of labour provides the incentive rather than the availability of suitable technologies.

Figure 1 shows this interlinking chain of events. The timing of a specific innovation depends upon the investment available for the development and the technological complexity. But the level of financial support or investment is in turn determined by the importance of the need as perceived by those people responsible for making available the resources; usually this will be corporate management, though it may also be the government or another sponsoring organization. However, it is a convergence of environmental pressures which gives rise to the need and their strengths which determines its importance. Thus, it follows that technological forecasting must look behind the need at the whole of the environment if any useful attempt is to be made to gauge the likely level of support and consequently the timing of an innovation. Furthermore, technology cannot be looked at in isolation, since it is necessary to forecast the interaction of these economic, political, social, industrial, and technological forces.

". . . produced by a logic that yields relatively consistent results"

This phrase goes to the heart of the matter. For TF is a logical, systematic examination of data which substantially eliminates human judgment from the processing of the informa-

FIGURE 1 Technological Forecasting and the Innovation Chain

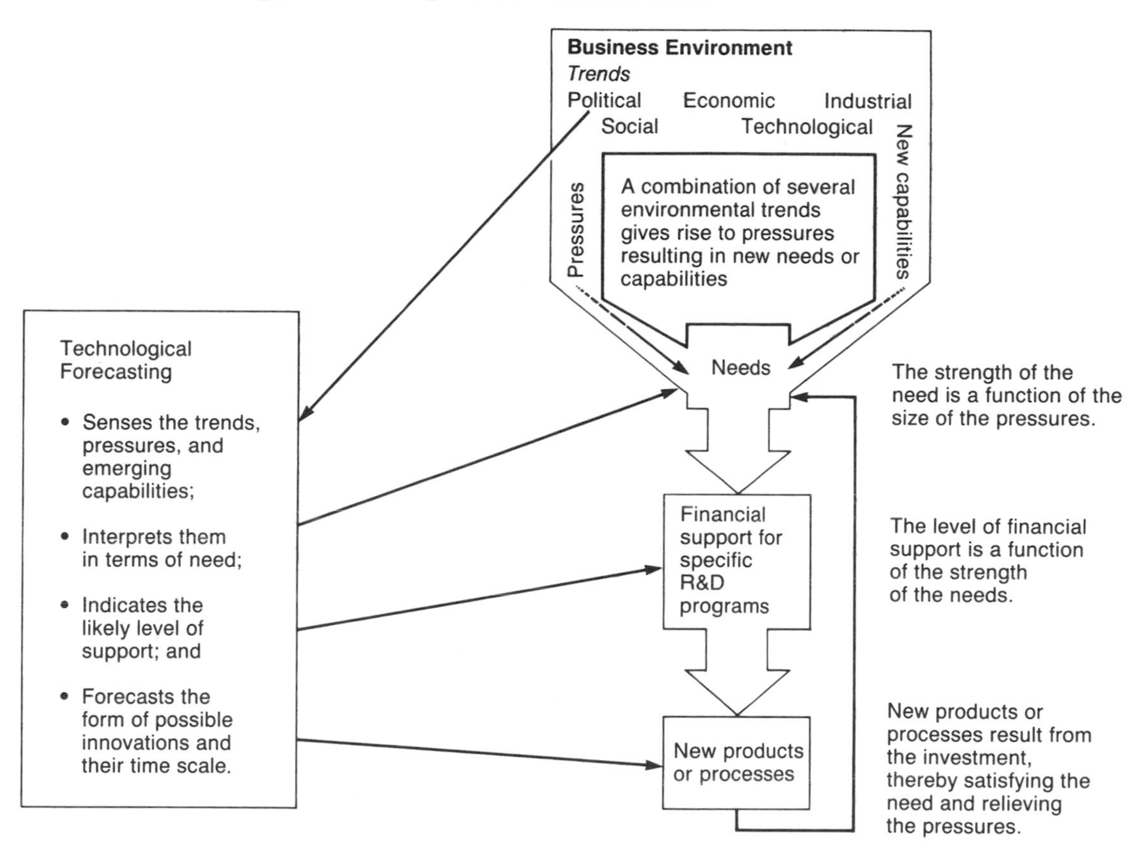

tion. Judgment enters in the selection of the area to be examined (e.g., technological parameter), the collation of data, and the interpretation of the results. The stage in the middle, the application of the technique, should, however, be little affected by opinion—two people using the same technique on identical data should produce forecasts different only in detail.

INPUTS TO THE FORECASTING SYSTEM

Like any other procedure for systematic analysis, forecasting can only be as accurate as the information fed into it. The inadequacy of the input data cannot be compensated, though it might easily be hidden, by highly sophisticated quantitative analysis. And what we have at our disposal is limited to:

1. Information from the past.
2. Knowledge of the present.
3. The ability of the human intellect—logical thought process, insight, and judgment.

These are the only resources to be marshaled and interpreted within a forecasting system. Discussion of TF inevitably centers on the detail of the techniques themselves, but this must not be allowed to distract attention from the two factors that determine the usefulness of the results—the quality of the input data and the caliber of the minds applied to the task.

CLASSIFICATION OF FORECASTING TECHNIQUES

The techniques of TF are commonly classified under the headings of "exploratory" or "normative." The term *exploratory* covers those techniques based upon an extension of the past through the present and into the future. They look forward from today taking into account the dynamic progression which brought us to today's position.

A normative approach starts from the future. The mind is projected forward by postulating a desired or possible state of events represented by the accomplishment of a particular mission, the satisfaction of a need, or state of technological development. It is then necessary to trace backwards to determine the steps necessary to reach the end point and assess the probability of their achievement. In the elaborate scenarios developed by the research of "think tanks" such as the Hudson Institute, alternative futures for the whole of mankind are developed. These workers claim that the scenarios present a choice from which mankind can select the most desirable of the alternative futures enabling the formulation of policies and the allocation of resources to travel the path to reach the desired end. However, these prognostications are beyond the concern of the industrial manager whose objectives are specific and limited. Nevertheless, they do underline that there is always a choice to be made.

In a practical forecasting exercise, it is usually necessary to employ a combination of techniques, some of which may be exploratory and some normative. For in dealing with great uncertainty, it is essential to examine every clue from as many angles as possible.

Too much stress must not be placed on "making the future happen." It is rare that one company or series of decisions has a profound influence on the future of technology, although the consequences of a decision may be of great significance for the company concerned. If one firm does not proceed with a potential innovation, there is a high probability that another will do so within a short space of time. The history of technology contains many examples where similar innovations occurred almost simultaneously in several places. This is not chance, but the result of a combination of the advances in several technologies necessary for the achievement of an innovation. It was not, for example, the discovery of the principle of the gas turbine, which had been known for many years, that made possible its practical realization in the late 1930s, but the availability of high-temperature materials. Project Hindsight provides further evidence and concludes, "Engineering design of improved military weapon systems consists primarily of skillfully selecting and integrating a large number of innovations from diverse technological areas so as to produce systematically the high performance achieved." This statement is supported by evidence from several examples investigated during the research program:

> As an example, transistor technology is credited with being responsible for size reduction in electronic equipment; however, without the development of such ancillary technologies as tantalytic capacitors, high-core permeability chokes and transformers, printed circuits, dip soldering, nickel cadmium batteries, and silicon cell power supplies, the electronic equipment chassis would be only marginally (perhaps 10 percent) smaller than a vacuum tube version.[10]

Thus, if a forecasting exercise is to give useful guidance to decision makers regarding a specific innovation, it must take into account not only a wide range of technological advances but also their mutual interactions. Fur-

---

[10] R. S. Isenson, "Technological Forecasting Lessons from Project Hindsight," in *Technological Forecasting for Industry and Government: Methods and Applications*, ed. J. R. Bright (Englewood Cliffs, N.J.: Prentice-Hall, 1968).

thermore, timing is of the essence. Before a certain date, the convergence of technological capabilities is insufficient to support the innovation. Once that stage has been reached, competitive forces are likely to ensure a limited time advantage to the company that seizes the initiative. TF can assist in deciding when to start. But the competitive edge so gained will be quickly eroded by an R&D program poorly conceived or conducted without vigor.

Most businesses are living in the type of environment we have just examined, where opportunities will emerge from marginal improvements in a number of advancing and converging technologies. In this situation, exploratory techniques of TF are likely to be the most helpful.

Mission-oriented programs are rarer. Frequently, they result from socially or politically motivated needs and are likely to involve major investments. The end point of the project will be postulated—landing a man on the moon, low toxicity motor car exhausts, a supersonic airliner, or an antiballistic missile system. Normative techniques, usually in conjunction with exploratory methods, can be of assistance here in forecasting the likely time of achieving the desired outcome with different levels of support, evaluating the alternative paths and selecting the best, and estimating the probabilities of each component of the overall program being satisfied within a given time scale.

FIGURE 2 History of a Developing Technology Where Rate of Advance Does Not Follow a Regular Path

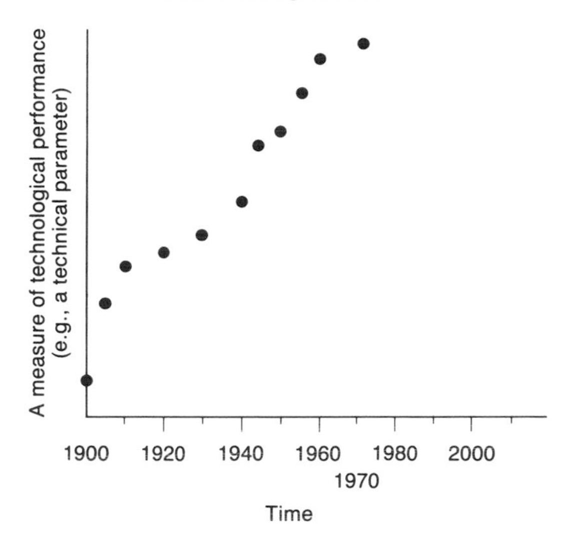

Note: 1. Although there has been an advance of performance with time, it has not followed a regular pattern. With this information it is not possible to fit a curve to give a meaningful forecast for the future.

2. If technological parameters followed such an erratic path, forecasting would not be possible. Fortunately this is rarely the case.

THE PATHS OF TECHNOLOGICAL PROGRESS

If technological progress consisted of a succession of random events (Figure 2) where it was not possible to establish any relationship between the rate of technological advance and time, any attempt to forecast would be impossible. Fortunately, however, analysis of historical data from a considerable number of phenomena shows that progress is not random and discontinuous, but follows a regular pattern when a selected attribute, such as functional performance (e.g., aircraft speed), a technical parameter (e.g., the tensile strength to density ratio for a material), or economic performance (e.g., cost per kwhr for electrical generation) is plotted against time. Characteristically, one finds an S-curve pattern (Figure 3).

The S-curve is similar to a product life cycle in that we observe a slow initial growth, followed by a rapid rise of approximately exponential growth, which slows down as it approaches asymptotically an upper limit normally set by some physical property.

Two important guidelines for R&D management can be inferred from observation of the

FIGURE 3 The S-Curve

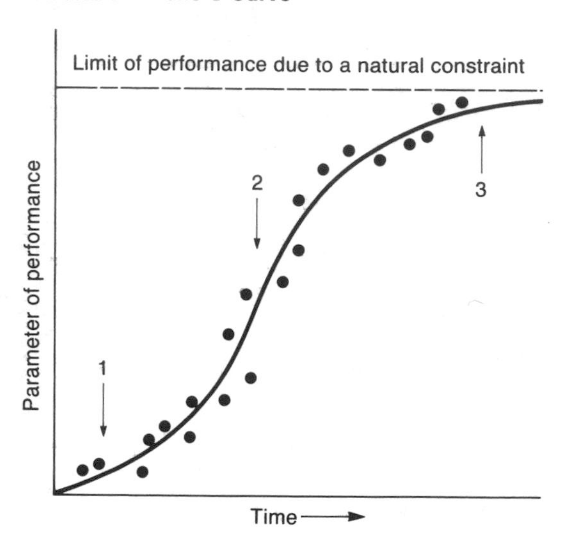

1 Period of slow initial growth.
2 Rapid exponential growth.
3 Growth slows as performance approaches a natural physical limit asymptotically.

shape of the S-curve:

1. The human intellect, with its linear thought patterns, may seriously underestimate the rapidity of the potential progress when establishing design specifications during the exponential midphase.
2. The decreasing managerial returns which may be expected from investment in a technology as its physical limit is approached.

It must not be assumed, however, that there is a predetermined path which progress inevitably follows. No progress occurs without human investment decisions. Thus, although a technology follows an S-curve, the actual path it takes will be one of a family of such curves. If the stimulus to invest in reaching a higher performance is low, the curve described is likely to be of the form OB (Figure 4). By contrast, OC_2 exhibits the much greater rate of progress to be found when market needs have led to a high expenditure in R&D. It is also possible to modify the shape of the curve; for example, a technology having reached the point A_1 on curve OA, at time t_1 may receive a stimulus from a needs-oriented mission which causes it to follow the path $A_1C_1C_2$; there is, of course, a limit to the maximum slope of the curve set by the size of the resources available or the rate at which they can be usefully employed. From the viewpoint of a forecaster at time t_1 attempting to predict the future with the knowledge that progress to date has followed the path OA_2, the indications are that the curve A_1A_2 has a high probability of describing the immediate future,

FIGURE 4 Possible Paths of Progress

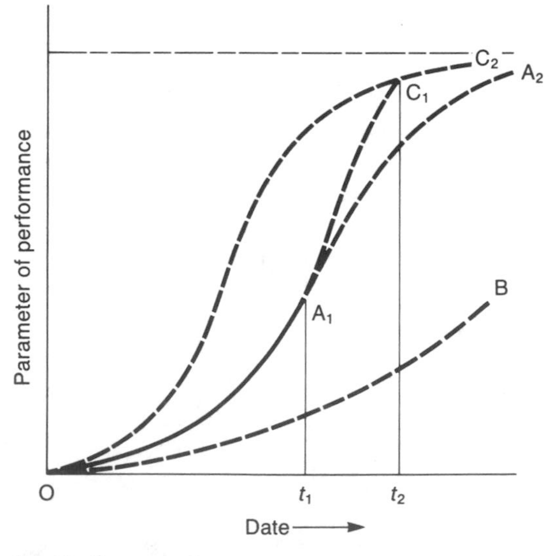

OA Medium growth rate
OB Low growth rate
OC Fast growth rate

The rate of growth is largely determined by the technological effort devoted to it. A technology which has advanced along OA_1 to time t_1 may be expected to progress along A_1A *unless* an identifiable outside factor causes an acceleration along A_1C_1.

unless he can positively identify some factor likely to cause a discontinuity producing a departure along a different path. In practice, such occurrences are infrequent.

An approaching physical limit does not remove the need to forecast. For it is at such a time that a new technology may emerge. This new technology will have a different physical limit and a potential for further progress in performance. Eventually it is likely to supplant the existing technology in a wide range of applications, particularly when they call for high performance. This situation frequently gives rise to a succession of S-curves (Figure 5) which can be contained within an "envelope" curve, also of the familiar S-shape. Ayres has plotted the efficiency of the external combustion engine from the time of Savery (1698) and Newcommen (1712) to the high pressure turbines of the present day, spanning seven different technological approaches.[11] In every case, he observes the same pattern. For example, between 1820 and 1850, the efficiency of the Cornish type engine rose rapidly from less than 5 percent to over 15 percent. Between 1850 and 1880, progress was slow until the triple expansion engine appeared on the scene, raising the efficiency to 22 percent by 1910, at which date the Parsons Turbine brought about another rapid rise. In the total period of 270 years, efficiencies have risen from virtually zero to over 50 percent and the "envelope" curve describing them has closely followed the S-shape. On the much shorter time scale of 30 years, Ayres has plotted computer performance on semilog paper, showing a remarkably close straight line relationship between computer performance and time spanning four technologies—valves, transistors, hybrids, and integrated circuits.

FIGURE 5 Effect of an Emergent New Technology on an Established Technology

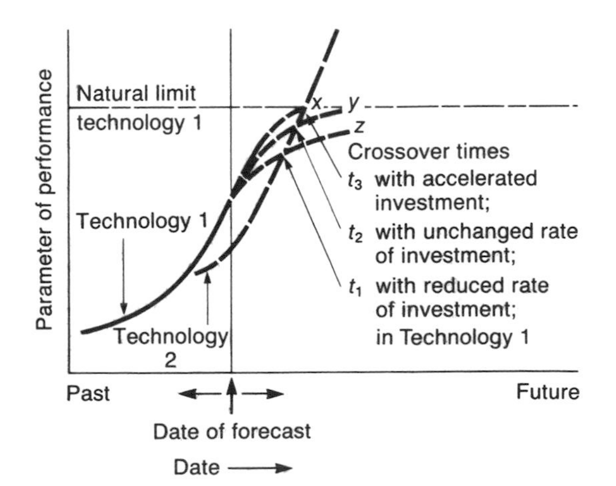

An emerging technology 2 is likely to replace the existing technology 1. The question is, when? What effect will the threat from 2 have on investment in 1? And how will this influence the date of substitution of 2 for 1?

As would be expected from the S-curve, an emergent new technology at first grows slowly. This is because its early performance is likely to be inferior to the highly developed existing technology. But when the performance of the two technologies approaches the same level, the greater potential of the newcomer attracts increasing investment, particularly once it has taken the lead when it begins to grow exponentially.

No natural law governs the emergence of a new technology. There is a physical limit even to the discovery of new technologies as evidenced by the flattening of the "envelope" curve. This phenomenon does, however, present the R&D manager with some particularly tricky judgments. The experience of history suggests that a slowing down or the approach of a natural limit should alert him to the possible appearance of something new. But his correct reaction to the identification of a substitute technology is not straightforward.

The initial expectation is that R&D effort will be transferred and that attempts to approach closer to the upper performance limit of the

[11] Ayres, "Envelope Curve Forecasting."

established technology will be reduced. However, in practice, the reaction may be the reverse. A great deal of capital is invested in the existing system. The threat may call for a defensive response and an increased rather than reduced investment when some development potential remains. This can be seen in Figure 5 where an established technology (1) is threatened by progress in technology (2). Extrapolating the past trends suggests a crossover point at time t_2 where T represents the present. But in spite of the approaching natural limit for (1) it is still possible to invest more heavily in it to follow the path TX rather than TY thereby delaying the crossover point to time t_3. In the perspective of history, the delay represented by t_2–t_3 is insignificant, but it may be of great importance to the short-term strategy of a business heavily invested in (1). This effect is often observed in practice. A good example is provided by the substitution of nuclear for conventional fuels in electric power generation. The forecasts of the economic crossover point for these two systems made 20 years ago have been shown to be grossly optimistic. One important reason for this has been an unexpected improvement in the performance of conventional power stations stimulated by the threat of nuclear power.

We can see that study of the past progress of technology confirms the existence of regular patterns, and provides a framework within which forecasting can be undertaken. It also indicates the types of information that the R&D manager would like to extract from forecasts, as:

1. The performance levels likely to be reached within his decision-making horizons in the absence of any new stimuli.
2. The identification of stimuli which may change the level of technological support and the effect this would have on the rate of progress.
3. The approach of a natural or physical limit for a technological parameter and its impact on the rate of progress.
4. The emergence of new technologies and the identification of crossover points:
   a. without additional investment in the established technology.
   b. with changed levels of investment.
5. The determination of performance milestones in several technologies the achievement of which would together make feasible a specific innovation.

THE TECHNIQUES OF TECHNOLOGICAL FORECASTING

It is beyond the scope of this book to explain in detail the many techniques for technological forecasting now available and fully described in the literature.[12] The following brief descriptions of a few of the most widely used techniques will not attempt to provide the reader with a working basis for their practical application; they are intended to show how the principles described previously can be applied and some of the major problems likely to be encountered. But it is worth reiterating that the techniques are not an end in themselves, and their successful application must rest heavily upon the technological experience and insights of the managers using them.

Trend Extrapolation

The extrapolation of past trends into the future is a technique economic forecasters have used for many years. At first sight, this seems to present the easy exercise of applying a mathematical curve-fitting technique to past data and

---

[12] E. Jantsch, *Technological Forecasting in Perspective,* Organization for Economic Cooperation and Development (OECD), 1967. Also, H. Jones and B. C. Twiss, *Forecasting Technology for Planning Decisions* (New York: Macmillan, 1978).

extrapolating into the future. However, the technological forecaster will encounter many practical difficulties and traps into which he may fall. Considerable judgment is required both in the choice and use of data.

Selection of the appropriate parameters or attributes to plot is critically important. The wrong choice will lead to the wrong conclusion. Bright, for example, suggests that one reason for the late entry of U.S. aeroengine manufacturers into gas turbines for civil air transport was their preoccupation with specific fuel consumption as the criterion of aeroengine performance. Consideration of the performance of the aircraft system in terms of passenger miles per unit of cost would have shown the potential of the gas-turbine powered aircraft in spite of the high specific fuel consumption of the power units. With the benefit of hindsight, it is easy to see how this error could occur. For several decades, the development of propeller-driven aircraft powered by piston engines had been progressing along one S-curve. Marginal improvements in engine fuel consumption which did not significantly modify the system were the main concern. Thus, the thinking of the aeroengine manufacturers was focused on their own technology. But the emergence of the gas turbine called for a reexamination of the engine in relation to the total system of which it formed only a part. The technological orientation of the traditional aeroengine manufacturers caused them to lose sight of the needs of the market for the system in which their product was embodied.

Although it is desirable for the parameter being plotted to be an independent variable, this is not always possible. Lenz quotes the example of aircraft passenger capacity, passenger miles, load factor, and total aircraft miles flown.[13] These are clearly interdependent. Whereas the graphs for passenger miles and seating capacity follow a rising pattern, that for aircraft miles flown begins to fall after about 1965 because of the impact of increased aircraft size. This fall is unlikely to be predicted from consideration of this factor in isolation and the simple fitting of a curve to past data. It is incorrect, therefore, to select aircraft miles flown as a parameter for trend extrapolation; it should be derived from extrapolation of the other factors (i.e.,

$$\text{Aircraft miles flown} = \frac{\text{Passenger miles}}{\text{Aircraft seating capacity} \times \text{Load factor}})$$

This again underlines the dangers of attempting to forecast mechanically without a proper understanding of the underlying technologies and their interrelationships.

One of the most frequent sources of difficulty arises from the absence or inadequacy of data upon which to base the forecast. Whereas the economic forecaster has at his disposal a wealth of relatively reliable statistics, systematically gathered, this is rarely the case with historical technological data. Furthermore, the data available to the economist—population and demographic information, GNP figures, income distribution, etc.—is unambiguous and requires the minimum of judgment in selection. The technological forecaster attempting to establish the trend for, say, aeroengine specific fuel consumption faces a much more difficult problem. Where does he find the data from the past? What do we mean by the term—is it the maximum achieved on the test bed? In the latest military application? Or in civil use? Unless such questions are resolved satisfactorily, the data plotted will not be strictly comparable. In any case, he will be lucky if the data he eventually unearths does not present him with a difficult curve-fitting problem. Thus, he would be wise to make several projections, perhaps using different curve-fitting techniques, although he should avoid the temptation of offering so

[13] R. C. Lenz, "Forecasts of Exploding Technologies by Trend Extrapolation," in Bright, *Technological Forecasting for Industry and Government.*

wide a range of possibilities that, while providing himself an alibi for the future, he gives the decision maker little useful guidance.

The inadequacy of data about the past available today indicates a first priority when initiating TF within a company. A system for collecting, interpreting and recording contemporary data should be established to provide future forecasters with an improved data base. In this way, forecasting accuracy can be steadily improved.

In spite of the difficulties of accurate trend extrapolation, it provides one of the most useful and widely used techniques. However, confidence will diminish with the time period of the forecast and this technique is consequently of greatest value in the short term, say 5 to 15 years. In this period, the accuracy of the forecast is likely to fall within limits imposed by the quality of the input data, unless some distorting factor, a new technology, or a mission-oriented program emerges.

FIGURE 6 Speed Trends of Combat versus Transport Aircraft, Showing Lead Trend Effect

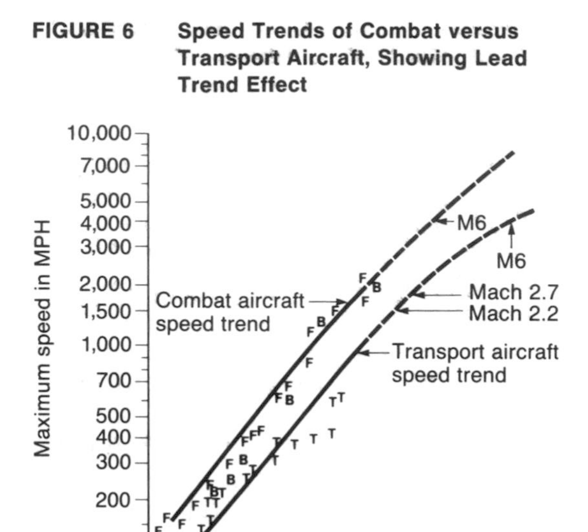

SOURCE: Ralph C. Lenz, "Forecasts of Exploding Technologies by Trend Extrapolation."

Precursor Trends—Curve Matching

Most industrial R&D is conducted in areas where the commercial application of a new technology or innovation has been preceded by its use where special attributes have outweighed the possible shortcomings. Thus, in scientific research or military applications, high cost or doubtful reliability may be less important than performance. Aerospace and electronics are two industries where this is particularly true.

The trends for both the first application and the commercial adoption are likely to follow each other, but with a regular lag (Figure 6). Lenz has shown that the trend line for the speed of transport aircraft has lagged behind that for combat aircraft by a period which has slowly extended from 9 years in 1930 to 11 years by 1970. The value of this knowledge to the commercial aerospace forecaster is obvious. If he can identify a precursor application which follows a regular trend behind which his own application follows, then he has a good indication of when today's advanced technologies are likely to appear in commercial products. The example of aircraft speed also shows how careful one has to be in attempting to use past data to forecast the future. Whereas the correlation was good up to the mid-1960s, we may query whether it will still hold for the future. This arises from two factors. Firstly, the importance to passengers of reduced air travel times as a proportion of actual travel time is decreasing. There is, of course, a natural limit of zero travel time. But more importantly, the size of the investment required to develop supersonic airliners is so large that it has been removed from the realm of industrial commercial decision making to that of politics and governments. This change of the industrial environment may well cause a departure from the trend line produced from past data.

In the frequency of radio transmissions, three stages of development can be traced—research, amateur and military, and commercial applications. In all three, a good correlation has been observed from the time of Marconi's first propagation experiment to the present day, the time lag between research and commercial use being about 30 years, with the military/amateur application appearing about halfway through this period.

This discovery of a precursor relationship gives the forecaster additional confidence for he now has two "fixes" for a point in the future, that obtained from extrapolating the curve for his own application, and a cross-reference from the precursor. It may also provide him with a useful clue to the date when a new substituting technology (e.g., the laser), at present in use in the research laboratory, is likely to achieve commercial exploitation.

Technological Substitution

Frequently a new technology substitutes for an existing technology. Study of a large number of historical examples indicates that the growth pattern follows the S-shaped curve of Figure 3. Fisher and Pry have shown that when the proportion of the new product has reached about 5 percent, the dynamics of the substitution process is likely to be well established. Thus, the remainder of the curve may be predicted with reasonable confidence in the absence of any discontinuities which may modify the rate of substitution.

Delphi

The opinion of experts can give important insights into the future, particularly in the identification of potential innovations likely to disturb the path of progress away from the extrapolated trend. Traditionally, expert opinion has been brought to bear through the medium of committee meetings. The Delphi technique was developed by Helmer at the Rand Corporation to overcome the weaknesses of the committee by using the individual judgments of a panel of experts working systematically and in combination, divorced from the distortions introduced by their personalities.

The committee method suffers from a variety of shortcomings. In the first place, geographical dispersion and the full diaries of prominent experts severely limit the membership and opportunities to bring them together. Once assembled, the committee process may not lead to a conclusion representing the unbiased views of all its members. Some people who are persuasive or articulate in discussion have a greater influence because of these characteristics rather than the strength of their case.

Other members may obtain a better hearing because of a position of authority or a high scientific reputation. Furthermore, there is a natural reluctance to change publicly a view which has previously been expressed strongly.

Another major weakness of the committee is the "bandwagon" effect produced by the disinclination of an individual to disagree with a majority view, in spite of his own judgment. This phenomenon is illustrated with great force in the work of the psychologist Asch.[14] In a series of experiments, all but one of a group of students were briefed to support an erroneous view unbeknown to the remaining member. He described one of his experiments in which the subjects were shown two cards, one bearing a standard line, and the other three lines of which only one was the same length as the standard. When asked to select which of these three was the same length as the standard, Asch found that, "whereas in ordinary circumstances, individuals matching the lines will make mistakes less than 1 percent of the time, under group pressure, the minority subjects swing to acceptance of the misleading majori-

[14] S. E. Asch, "Opinions and Social Pressures," *Scientific American,* November 1955. Reprinted in *Readings in Managerial Psychology,* eds. H. J. Leavitt and L. R. Pondy (Chicago: University of Chicago Press, 1964).

ty's wrong judgment in 36.8 percent of the selections." When such distortions can be created in a straightforward situation, one can appreciate the possible influence of this effect where there is genuine uncertainty as is inevitable when considering the future. Nevertheless, the minority view might well be valid and its expression should not be suppressed.

Delphi attempts to eliminate these problems by using a questionnaire technique circulated to a panel of experts who are not aware of the identity of their fellow members. The procedure is as follows:

1. *Round 1* Circulate the questionnaire by post to the panel.
2. *Round 2* After analysis of the Round 1 replies, recirculate, stating the median and interquartile range of the replies. Respondents are asked to reconsider their answers and those whose replies fall outside the interquartile range are invited to state their reasons—these may result from lack of knowledge, or more importantly, from some specialist information unknown to the other members.
3. *Round 3* Analysis of replies from Round 2 are recirculated, together with the reasons proferred in support of the extreme positions, in the light of which the panel members are asked once more to reconsider their replies.
4. Further rounds for additional clarification may be employed if thought necessary.

Selection of the panel members is a task of the utmost importance, for the value of the forecast is a function of the caliber and expertise of the individual contributors as well as the appropriateness and comprehensiveness of the areas of knowledge they represent. Questionnaire formulation also requires considerable skill to ensure the right questions are asked and that they are framed in specific, quantified, and unambiguous terms.

Delphi is most widely used for longer range forecasts. Figure 7 is part of a typical Delphi presentation for the computer industry. Another interesting published study has been conducted by Smith, Kline, and French Laboratories, into future developments in medicine.[15]

The evidence supports the contention that Delphi studies result in a gain of consensus either by the rejection of extreme positions or by shifting the median as a consequence of specialist knowledge introduced by one or more members of the panel. But how accurate are the resulting forecasts? The technique is of too recent origin to provide a great deal of validation. There are some indications, however, that there may be a tendency to err in an optimistic direction in the short term due to an underestimation of development times. By contrast, long-term forecasts may well be pessimistic because of the mind's inability to appreciate fully the effects of exponential growth.

Rand Corporation has, however, conducted experiments to validate the methodology. These show clearly that when experts have been asked about current phenomena where it is possible to establish a factual answer, consensus grows as the study progresses, and moves towards the correct answer. This suggests possible applications outside TF such as in cost estimation and R&D project selection.

Scenarios

Scenario writing has become most widely known through the work of American "think tanks" such as the Hudson Institute. The scenarios describe a possible future situation based upon a wide-ranging environmental

[15] A. D. Bender, et al. "Delphic Study Examines Developments in Medicine," *Futures,* June 1969.

FIGURE 7 Part of Delphi Forecast for Computer Systems—International Computer, Ltd.

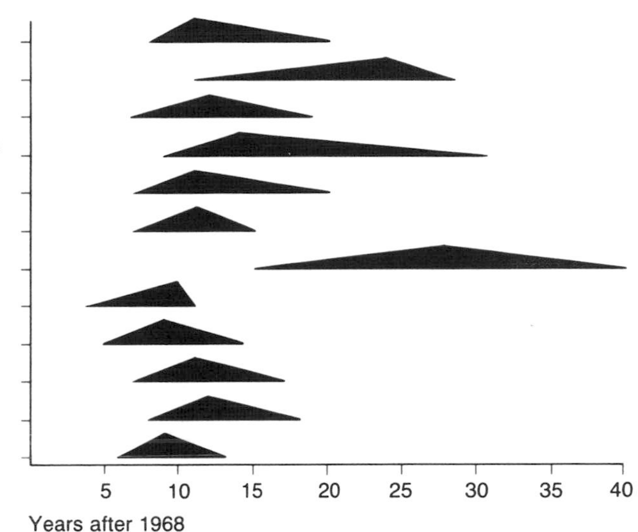

SOURCE: P. D. Hall, "Technological Forecasting for Computer Systems," Proc. National Conference on Technological Forecasting, University of Bradford, 1968.

analysis. Frequently several scenarios or alternative futures are prepared supported by detailed research using a wide variety of TF techniques. Scenario writing is based upon the recognition that it is not always possible to choose between alternative sets of assumptions.

In recent years, interest in scenario writing has increased at both the national and industrial level. Energy forecasting is a field where this approach has been used extensively.[16] A number of practical techniques for industrial scenario writing have been developed to enable the consideration of the mutual interactions of a wide range of environmental factors both upon themselves and upon an organization's strategic objectives.[17] Thus, this is an approach which extends beyond technological planning and enables top management to review their strategic assumptions and the consequences flowing from them.

The use of scenarios in decision making is illustrated diagrammatically in Figure 8. Three alternative futures have been identified: A, B, and C. Although it is possible to attach probabilities to the likelihood of occurrence of each of the alternatives, they are not likely to be very reliable. It may, of course, emerge that all the evidence points in one direction with a high probability of the future being similar to that described by Scenario A. In this case, the

[16] *Energy Research and Development in the UK*, Energy Paper No. 11, HMSO, 1976. Also, P. Chapman, *Fuel's Paradise: Energy Options for Britain* (New York: Penguin Books, 1975).

[17] Jones and Twiss, *Forecasting Technology*. Also, C. A. R. MacNulty, "Scenario Development for Corporate Planning," *Futures*, April 1977.

FIGURE 8 A Minimum Risk Policy for Possible Alternative Futures

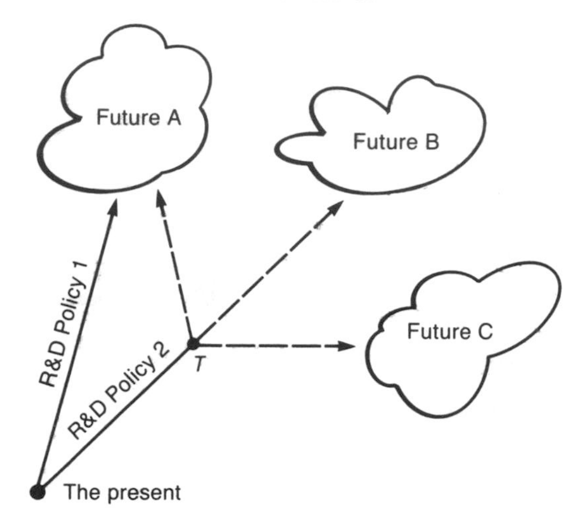

Futures A, B and C are three scenarios of alternative futures. The forecast may result in R&D action as follows:

Policy 1 Future A is considered so probable that decisions are made assuming the forecast is correct.

Policy 2 A minimum risk policy permitting an advance to *T*, without precluding any of the scenarios. At time *T* a decision cannot be deferred any longer. In this way the decision has been delayed to the latest possible date.

long-term R&D programs (Policy 1) could be based on this assumption. Policy 1 might, however, be such that it has no relevance to Futures B or C. In the more likely case when it is not possible to choose clearly between A, B, or C, a program (Policy 2) which keeps the options open is likely to be the best compromise. At a later date, the program can be reoriented in the light of future events. Thus, a minimum risk strategy can be adopted as a conscious act of policy.

Most writers stress the interaction between the decisions resulting from forecasts and the determination of the future. If Scenario A is thought to be likely, then following Policy 1 as a result of the forecast could make it a self-fulfilling prophecy. Such freedom to shape the future is open to few companies. But one can see how scenarios could be used to shape government policy. If, for example, Scenario C described a highly undesirable future for mankind, then, given the correct political conditions, government policies could ensure that the path leading to it is sealed off.

Relevance Trees

The purpose of a relevance tree is to determine and evaluate systematically the alternative paths by which a normative objective or mission could be achieved. How this is done can best be seen by examining the abbreviated version illustrated in Figure 9.

The objective (O) is the starting point—let us say a pollution-free road transport system. At the next level, the problem can be broken down into alternative solution concepts A, B, C, etc. (e.g., electric car, Wankel Engine, "clean" petrol piston engine, etc.), or by functions to be performed (e.g., for a moon landing there would be launch, midcourse flight and control, lunar landing, lunar take-off, earth reentry). For a particular solution, a variety of systems will be necessary (e.g., fuel systems, combustion, and exhaust), each of which in turn may be satisfied in a variety of ways involving alternative subsystems (e.g., mechanical, electrical, or catalytic exhaust cleaners).

Thus, starting from the desired end result, it is possible to examine exhaustively all the alternative paths by which it can be reached, working backwards through a hierarchy ending with detailed and limited R&D project objectives. The next stage is to investigate each step in greater depth including feasibility, resources required, probability of success, and time scale. This involves the employment of other TF techniques such as Delphi and trend extrapolation.

FIGURE 9 Relevance Tree

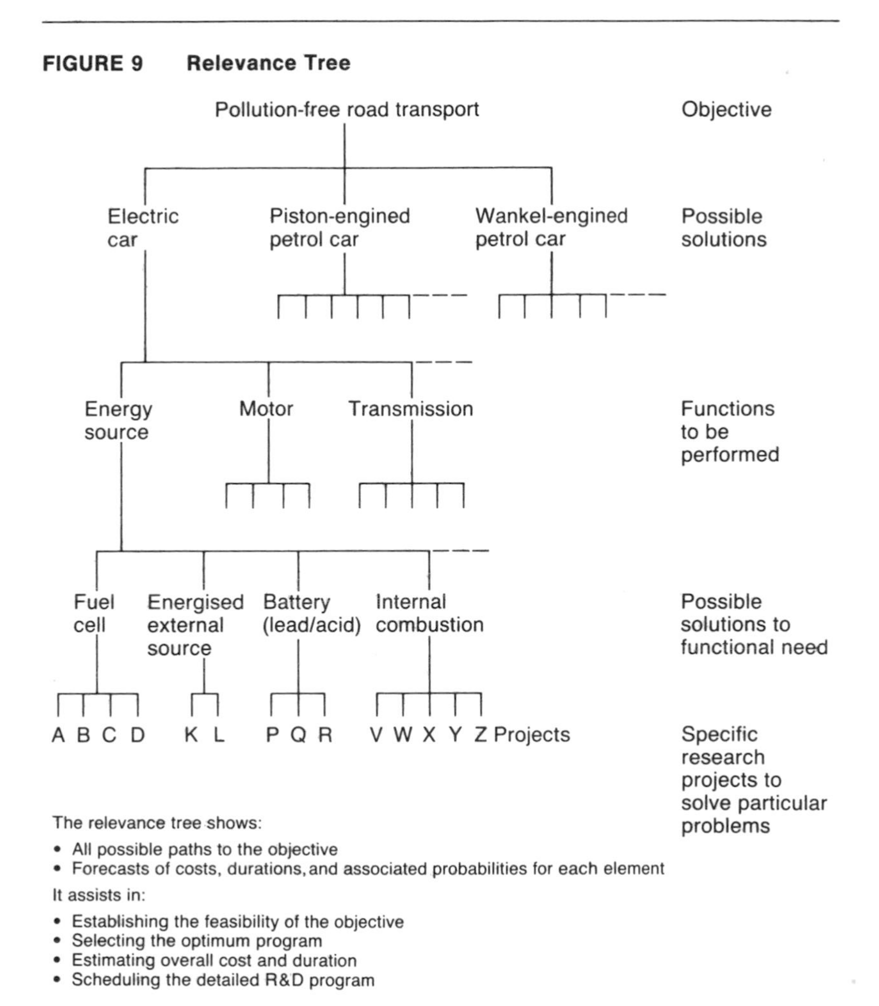

Relevance trees can be useful to the R&D manager in:

1. Establishing the feasibility of a technological mission. If no feasible path can be found, the mission cannot be achieved with the technological capability of the present or the foreseeable future.
2. Determining the optimum R&D program, i.e., the paths to be followed up the hierarchy. This could vary with the program objective. Early achievement of the objective may call for a different program (i.e., path) from that for minimum cost.
3. Selecting and planning the initiation of detailed research projects.

This technique can lead to highly complex and sophisticated computer-based approaches. In the Honeywell Pattern program, all military and aerospace activities in which the company was or could be involved were covered. The cost of setting up this program has been estimated at $250,000 to $300,000 with a further annual running cost of $50,000. Relevance trees have also been used by the British Post Office R&D department for relating the choice of projects to the organization's objectives.[18] The scale of effort involved in Pattern should not however deter the industrial R&D manager with limited resources at his disposal. The basic relevance tree methodology in a much simplified form can still be a useful planning tool.

Technology Monitoring

We have noted previously that innovation frequently depends upon the convergence of advances in several technologies and, furthermore, that the period between the emergence of a technological advance or a new technology and its practical application may span a number of years. Well-informed judgment and insight are also important. Most managers receive their information inputs haphazardly, from reading, discussions, conferences, etc. If judgment is to be based upon good and comprehensive information, the gathering of these inputs should be organized, so far as possible. Those of potential future value should be stored to enable retrieval without relying upon the vagaries of human memory, and their significance noted.

Bright has proposed monitoring the environment on a systematic basis. He writes:

> Note that monitoring includes much more than simply "scanning." It includes search, consideration of alternative possibilities and their effects, and a conclusion based on evaluation of progress and its implications. The feasibility of monitoring rests on the fact that it takes a long time for a technology to emerge from the minds of men into economic reality, with its resulting social impacts. There are always some identifiable points, events, relationships, and other types of "signals" along the way that can be used in an analytical framework. If a manager can detect these signals, he should be able to follow the progress of the innovation relative to time, cost, performance, obstacles, possible impacts, and other considerations. Then he will have two more important inputs to his decisions:
>
> (*a*) Awareness of new technology and its progress;
>
> and
>
> (*b*) Some thoughtful speculation about its possible impact.[19]

The monitoring process is based upon a journal in which significant events are recorded. Bright suggests four column headings: Date; Event and Technical Economic Data; Possible Significance; and Things to Consider. The events may be likened to the pieces of a jigsaw. Over a period of time, the number of pieces we have increases. Not all may be useful, but by careful selection and assembly, a picture of one part of the future may slowly emerge. Sometimes a vital piece may be missing; this could lead to the deduction—"if advance X were made, then innovation Y would be possible," with the corollary "we must monitor the environment carefully for signs of X being achieved since we shall then know that Y is feasible." The reporting of X might appear first in an obscure scientific paper in any part of the world.

The attraction of monitoring is that it can be performed by any individual manager for his own information. It is surprising how much one man can glean from systematically processing the information received daily. The richness of the information and the deductions are obviously enhanced if organized on a departmental or interdepartmental basis.

---

[18] H. Beastall, "The Relevance Tree in Post Office R&D," *R&D Management* 1, no. 2, 1971.

[19] J. R. Bright, "Evaluating Signals of Technological Change," *Harvard Business Review,* January/February 1970.

Cross-Impact Analysis

The techniques we have discussed so far have centered on forecasting the future behavior of specific attributes or likely events in isolation, although we have seen that in fact many of them are interdependent not only within the technological sphere but also between technology and other environmental factors. Gordon of the Institute of the Future classifies these interconnective modes as *enhancing* (enabling or provoking) or *inhibiting* (denigrating or antagonistic). Thus, our forecasts for automotive antipollution legislation enhance the probability of the development of catalytic exhaust cleaners, but the development of noncatalytic exhaust cleaners would inhibit the probability of developments of new processes for purifying platinum, the future demand for which is closely related to the potential demand for catalysts.

The cross-impact analysis technique has been devised to permit these interrelationships to be recognized and reflected in the forecasts. The procedure adopted by the Institute of the Future is briefly as follows:

a. Establish the principal possible events associated with the problem under consideration.

b. Assign to each of these events individual probabilities, e.g., by using Delphi or trend extrapolation.

c. Determine the interactions between events in terms of their interconnective mode and strength, identifying in some cases predecessor–successor relationships.

d. After programming into a computer, select an event from the predecessor group. Use random numbers to determine whether an event occurred, in which case adjust the other items accordingly. Repeat this process for all items.

e. Repeat stage (*d*) up to 1,000 times.

ORGANIZATION FOR TECHNOLOGICAL FORECASTING

Projects such as PATTERN presuppose a heavy investment of organizational and financial resources in TF. For the majority of companies, however, the employment of more than two or three full-time technological forecasters is the exception. In most of the companies where the techniques have been studied, the involvement has been largely confined to a part-time activity of a few interested staff in the R&D department.

The best person to make a forecast is the manager who has to make a decision which is future related. Usually, however, he will need assistance in two areas:

1. The provision of data.
2. The application of TF techniques.

Much of the data required for TF is of the type which is needed in several areas of the company (e.g., economic and statistical data, market research) and is most conveniently collected and stored centrally within the organization, perhaps in the economics or corporate planning section. Other data which is of a specialist nature, of interest only to the departments preparing the particular forecast, needs to be collected at departmental level. The value of such data storage grows with time since early efforts at TF are frequently frustrated by the absence of information. Thus, the first and critical step in adopting TF is to lay the foundation for future forecasting by organizing the collection and recording of data.

Few managers have been trained in TF and it is impracticable to provide the training for all those who might be called upon to make forecasts. There is thus need for consultancy advice from a small staff who are well versed in the techniques and can advise the manager in the preparation of his forecasts and warn him of the many pitfalls. Nevertheless, it is better that the individual manager makes his own forecast, with assistance, rather than the removal of forecasting to a central service department remote from the problem area.

CURRENT STATUS OF TECHNOLOGICAL FORECASTING

Most of the initiatives in TF originally came from the United States, where several major companies invested heavily in it. There is some doubt whether the results justified the substantial investments.

In Europe, the rate of adoption has been much slower. The fuel crises of recent years have given a stimulus to industrial concern with the future and a growth in forecasting activity. It is possible to identify two levels of support; the first is concerned primarily with R&D planning decisions whereas the second focuses on corporate strategic issues. The latter is often referred to as "futures studies" rather than as forecasting. There is also a recognition that the future is more uncertain now than it was in the recent past and, furthermore, it is more likely to experience discontinuities. Thus, we are witnessing a growth in the use of scenarios and such techniques as trend-impact analysis.[20]

A major cause for the low rate of acceptance is likely to be reluctance to invest the large resources, which the literature suggests are necessary, in an untried technique. This has not encouraged more modest attempts. Nevertheless, as the pace of technological progress continues to increase, so will the need to forecast. Thus, we may expect to see a growth in the use of TF. But this growth is likely to be hindered by exaggerated claims for what it can achieve.

[20] J. Fowles, ed. *Handbook of Futures Research* (Westport, Conn.: Greenwood Press, 1978). Also, H. A. Linstone and W. H. Simmonds, *Futures Research: New Directions* (Reading, Mass.: Addison-Wesley Publishing, 1977).

SUMMARY

Since the benefits from R&D decisions are gained in the future, it is incumbent upon the R&D manager to satisfy himself, so far as is possible, that the results of his investments are relevant to the market needs and the competitive technologies at the time they reach fruition. Thus, all R&D decision makers must take a conscious view of the future. Forecasts are needed which take full account of the information available and the techniques of TF. It was seen, however, that the effort devoted to TF should be related to the characteristics of the industry, the company, and the decisions to be made, rather than to the size of the company.

Technological forecasting cannot enable the decision maker to predict the future with certainty. But it can assist him in refining his judgments. The value of the forecasts was seen to be highly dependent upon the quality of the informational inputs to the forecasting process and the caliber of the minds applied to it. Sophisticated forecasting techniques can only be aids to this process, and care should be taken to guard against TF absorbing greater resources than can be justified in economic terms.

The principles underlying technological forecasting were discussed and brief descriptions given of how they are applied in some of the most commonly used techniques—trend extrapolation, precursor trend curve matching, Delphi, scenarios, relevance trees, technology monitoring, and cross-impact analysis.

Whilst the practical application of TF is still limited, it was concluded that its use is likely to become more widespread once a better understanding is gained of what it can contribute to R&D decision making and, perhaps more important, what it cannot be expected to do. No R&D manager can afford to ignore TF, but enthusiasm for the techniques must be tempered by the realization that they alone will never remove completely the uncertainties inseparable from any consideration of the future.

ADDITIONAL REFERENCES

Arnfield, R. V., ed. *Technological Forecasting.* Edinburgh: Edinburgh University Press, 1969.

Currill, D. L. "Technological Forecasting in Six Major UK Companies." *Long-Range Planning,* March 1972.

Jantsch, E. *Technological Planning and Social Futures.* Cassell/Associated Business Programmes, 1972.

Linstone, H. A., and M. Turoff. *The Delphi Method: Techniques and Applications.* Reading, Mass.: Addison-Wesley Publishing, 1975.

Martino, J. P. *Technological Forecasting For Decision Making.* New York: Elsevier-North Holland Publishing, 1972.

Morrell, J. *Management Decisions and the Role of Forecasting.* London: Pelican Books, 1972.

Twiss, B. C. "Economic Perspectives of Technological Progress: New Dimensions for Forecasting Technology." *Futures,* February 1976.

Twiss, B. C. "The Production Manager's Need for Technological Forecasting." *Production Engineer,* August 1972.

Design Choice

Case II–3
Grumman Energy Systems

J. F. Ince and M. A. Maidique

INTRODUCTION

"We've got two years to make it or it's all over," said Ron Peterson, president of Grumman Energy System, Inc., in June 1978. GES had been formed on Grumman President Joseph Gavin, Jr.'s, and Peterson's enthusiastic initiative two and a half years earlier, on the premise that active solar energy systems and specifically solar hot water offered a lucrative business opportunity. But Mr. Gavin had set a five-year time frame for GES to become a profitable, self-sufficient operation. The first two years had been more than exciting, but the financial state of the industry was best summed up by an article in *The New York Times:*

> Although 94 percent of the persons polled by Louis Harris in June said harnessing solar energy might ease national energy problems, very few homeowners committed any cash to this belief. Sales figures are sketchy, but solar energy firms who managed to stay in business report that 1978 is a disaster.

Peterson was balancing the disappointment of the last years' performance against the long-term potential as evidenced by a survey published in *Housing Magazine.* That survey showed that the percentage of those who wanted solar hot water in various regions was as follows: 34 percent in Washington, D.C., 58 percent in Miami, 25 percent in Chicago, 77 percent in Phoenix, and 40 percent in San Diego.

Grumman's sales had been doubling every year, and the company now had the largest market share in the domestic hot water industry, but Peterson was concerned about the future of the solar markets and he was reevaluating GES's product design, production, organizational, sales, and marketing policies.

Harvard Business School case 1-680-033

THE GRUMMAN CORPORATION

In the early 1930s, Leroy Grumman, an aeronautical engineer, launched the Grumman Aircraft Engineering Corporation with no product, no plant, no customers, and 21 employees. Despite all the obstacles, they built the FF-1, then the fastest fighter plane in the world. During World War II, their Wildcat and Hellcat carrier-based fighters helped defeat such Axis planes as the renowned Japanese Zero and helped the Allied powers control the skies. Grumman in 1969 was awarded one of the most important Department of Defense contracts of the last decade, the Navy F-14 air superiority fighter. In the 1960s and 70s, Grumman participated in the design and production of the wings of the lunar modules, and they designed and supported the space shuttle missions. So from modest beginnings, Grumman has grown into one of the foremost aeronautical development and manufacturing companies in the world.

In the early 1970s, Grumman changed its name to the Grumman Corporation and reorganized, creating five major wholly owned subsidiaries: Aerospace, Data Systems, Allied Industries, Ecosystems, and International Division. At that time, one of Grumman's principal products was the advanced F-14 fighter, and in 1972 when they ran into contract problems, it created a severe cash drain. As a consequence, Board Chairman Towl, in 1973, wrote in his annual letter to stockholders, "These 12 months were the most critical in Grumman's history." Also at the time, the first phase of the space program was drawing to a close and the rate of growth of the aerospace industry was declining. (See Exhibits 1, 1*a*, and 2 for financial history.) When, in 1973, Grumman lost its bid to serve as prime contractor in the space shuttle, the arguments for diversification became compelling.

Grumman's expertise in aerospace manufacturing involved many areas that were considered fertile grounds for expansion: metallurgy, laser-equipment manufacturing of test equipment, work in new aircraft materials such as composites, flight stabilizing equipment, electronic systems, thrust reversers. With these and other areas as bases, Grumman gradually sought to establish a wider product base to cushion the cycles of aerospace.

Energy-related products were seen as one such alternative. Consequently, Grumman began to develop interests in energy fields, such as an environmental data acquisition contract for offshore siting of nuclear power facilities and an extensive energy conservation program on its own facilities. The conservation program, carried out in 1973, was especially enlightening to Grumman management. By using high-technology analyses to uncover excessive energy

EXHIBIT 1 Grumman Energy Systems

THE GRUMMAN CORPORATION
Consolidated Statement of Income*
(dollars in millions)

	Year Ended December 31	
	1977	1976
Sales	$1,552	$1,502
Other income	11	21
	1,564	1,523
Costs and expenses:		
Wages, materials, and other costs	1,492	1,463
Interest	14	20
Minority interests	—	—
Provision for federal income taxes	24	15
	1,531	1,499
Net income	$ 32	$ 23
Earnings per share:		
Primary	$4.04	$3.04
Fully diluted	3.53	2.66

**Note*: Figures may not add because of rounding off.
SOURCE: Grumman annual reports, 1976 and 1977.

EXHIBIT 1*a* Grumman Energy Systems

THE GRUMMAN CORPORATION
Consolidated Balance Sheet*
(dollars in millions)

	December 31	
	1977	1976
Assets		
Current assets:		
Cash	$ 18	$ 19
Marketable securities (at cost, approximating market)	14	11
Accounts receivable	140	122
Inventories, less progress payments	239	256
Prepaid expenses	8	5
Total current assets	421	416
Long-term financing contracts	12	9
Property, plant, and equipment (at cost)	321	303
Less: Accumulated depreciation and amortization	211	197
	110	105
Construction in progress	2	4
Land	9	9
Net property, plant, and equipment	122	119
Other assets and deferred charges	12	11
Total assets	$569	$556
Liabilities and Shareholders' Equity		
Total current liabilities	$235	$219
Long-term debt	123	157
Other:		
Deferred investment tax credit	4	3
Minority interests in subsidiaries	6	5
Other liabilities	4	3
	14	12
Shareholders' equity:		
Preferred stock	3	1
Common stock	51	47
Retained earnings	152	127
	203	176
Less: Cost of common stock in treasury	7	9
Total shareholders' equity	196	167
Total liabilities and shareholders' equity	$569	$556

* Figures may not add due to rounding.
SOURCE: Grumman annual reports, 1976 and 1977.

EXHIBIT 2 **Grumman Energy Systems**

THE GRUMMAN CORPORATION
Ten-Year Summary
(dollars are stated in thousands except per share amounts)

	1977	1976	1975	1974	1973	1972	1971	1970	1969	1968
Operations:										
Sales	$1,552,695	$1,502,058	$1,328,622	$1,112,855	$1,082,570	$ 683,457	$ 799,021	$ 993,261	$1,180,328	$1,152,571
Other income	11,522	21,064	21,932	16,940	5,378	2,823	1,718	2,149	2,295	1,466
Costs and expenses:										
Wages, materials, and other costs	1,492,915	1,463,522	1,288,725	1,080,896	1,045,684	791,464	830,019	949,363	1,132,554	1,110,473
Interest	14,084	20,086	19,455	11,092	7,955	5,756	6,805	8,277	5,313	4,645
Minority interest	821	160	1,027	878	1,366	211	105	(2)	168	182
Provision (credit) for federal income taxes	24,000	15,800	17,800	16,900	16,000	(41,125)	(18,200)	17,500	22,500	19,700
Income (loss) before extraordinary items	32,397	23,554	23,547	20,029	16,943	(70,026)	(17,990)	20,272	22,088	19,037
Extraordinary items				12,900	11,300					
Net income (loss)	32,397	23,554	23,547	32,929	28,243	(70,026)	(17,990)	20,272	22,088	19,037
Percent of sales	2.09%	1.57%	1.77%	2.96%	2.61%	(10.25)%	(2.25)%	2.04%	1.87%	1.65%
Fully diluted earnings per share:										
Income (loss) before extraordinary items	$3.53	$2.66	$2.68	$2.36	$2.07	$(9.32)	$(2.35)	$2.39	$2.51	$2.21
Cash dividends paid per common share	.95	.70	.60	.45	.15	.25	1.00	1.00	1.00	1.00
Financial position at December 31:										
Working capital	$ 185,916	$ 196,641	$ 242,860	$ 175,979	$ 101,581	$ 96,144	$ 105,192	$ 124,619	$ 104,160	$ 108,675
Net property, plant, and equipment	122,920	119,700	115,390	107,337	107,020	108,053	120,531	132,345	133,909	114,563
Long-term debt	123,304	157,389	210,490	156,260	112,780	140,432	88,464	85,442	75,589	78,277
Shareholders' equity per common share	24.18	21.29	19.09	16.60	12.63	8.98	18.53	21.88	20.12	18.27
Other data:										
Expenditures for facilities (net)	$ 22,191	$ 20,367	$ 24,162	$ 16,625	$ 8,363	$ 7,162	$ 10,056	$ 22,991	$ 36,012	$ 39,518
Number of employees	27,000	28,000	28,000	30,000	27,000	25,000	25,000	28,000	34,000	36,000
Common shares outstanding	8,095,201	7,805,246	7,561,216	7,480,841	7,470,021	7,490,261	7,511,711	7,673,294	7,847,242	7,829,430

SOURCE: Grumman annual report, 1977.

use, and by modifying heating, ventilating, and air-conditioning systems, Grumman saved nearly a million gallons of fuel per year, enough to heat 1,200 single-family homes.

SOLAR DIVERSIFICATION

One of the energy technologies that Grumman decided to explore as a possible investment opportunity was solar energy. To spearhead this task, Grumman President Gavin chose Ronald Peterson, an experienced Grumman hand who had joined Grumman 14 years earlier as a performance analyst. Peterson, an aeronautical engineer, had held various program planning and management positions at Grumman, including overall program management for a version of the A6A intruder aircraft.

In 1973, Peterson was selected by Grumman to attend MIT as a Sloan Fellow. Shortly after his return, he was tapped by Mr. Gavin to head an effort to review and coordinate the various energy efforts within the corporation. The efforts included energy conservation, development work on solar thermal heating systems, wind power, photovoltaics, ocean thermal, solar satellite power stations, and some work in nuclear. As a result of Peterson's study and recommendations, Mr. Gavin asked him to head up the new "Energy Program Department" within Aerospace. In this capacity, Peterson began an in-depth look at solar market potential and preliminary development work on a solar domestic hot water system.

By 1975, this program had grown into a viable product concept and Gavin created the "Energy System Division." Peterson today heads Grumman Energy Systems, a division of 150 employees and several million dollars in sales in Ronkonkoma, New York.

Peterson was enthusiastic about his new responsibilities. He characterized bringing to market a new innovation as an especially challenging role. He explained, "I've just spent a year at MIT studying management and I wrote my Master's thesis on Internal Entrepreneurship. It's all been very influential on my thinking about managing innovation." Peterson has had extensive management responsibility in the product design, manufacturing, and to a lesser extent, marketing interface. The solar group would be his first opportunity to combine all these areas under his span of authority.

Another factor that added to Peterson's motivation was Mr. Gavin's support. Mr. Gavin, an MIT trained aeronautical engineer, was enthusiastic about Grumman's prospects in the burgeoning field of solar energy and about solar energy in general. For these reasons, Mr. Gavin had agreed to become a director of the Northeast Solar Energy Center, an organization whose charter was to stimulate solar industry development in the Northeast.

THE SOLAR INDUSTRY

Like most emerging industries, the solar heating industry consisted of a multitude of competitors, no clear industry leader (unless it is Grumman), no dominant product design, and relatively high manufacturing costs. According to the Solar Energy Industry Association, there were some 525 manufacturers of solar equipment as of late 1977. These firms can be categorized into three major groups: a building trades group, a corporate diversification group, and a solar company group.

The *building trades group* is composed of companies which are already active in marketing a product to the residential and/or commercial construction market. Examples of companies in this group are General Electric, PPG, Revere Copper and Brass, LOF, and the Lennox/Honeywell team. These companies generally draw upon existing marketing strengths within the construction market and use present distribution channels. Installation and service can often be carried out by the present staff with minimal additional training.

Solar manufacturing operations of this group often complement existing manufacturing operations. In the case of the glass companies such as LOF and PPG, for instance, the manufacturing represents a forward integration move into systems assembly.

These companies are not showing a profit on their solar operations at this time. Public statements, annual reports, and industry contacts indicate that the best reported profit is only for very "brief periods of profitability." Most companies are reporting losses, some of which are very substantial. PPG has recently announced that they are dropping out of the solar industry, citing poor return on investment as the reason.

General Electric, one of the firms in this group, is now concentrating on large-scale commercial projects, apparently with heavy government support.

The *corporate diversification* group is composed of companies whose principal marketing and manufacturing operations lie in fields which are largely unrelated to solar energy. Representatives of this group include Grumman, Owens-Illinois, Daystar (Exxon Enterprises), Hexcell, and Northrup. These companies are building their solar organizations mostly from the ground up and their marketing focuses vary widely.

Their manufacturing is generally unsophisticated. They have entered the solar energy field with little or no volume assembly manufacturing capability. Fabrication consists of a metal bending shop and assembly area.

These companies, however, have the ability to sustain long periods of losses in hopes of being one of the survivors as the industry shakes out. Public and private sources indicate that they are also not making any profit in solar.

The third, and most numerous, group of competitors are the *solar companies.*[1] Due to the number of competitors in this segment, there is a wide diversity of market strategies. This entrepreneurial group of companies can be described as scrambling for a distribution network. The companies which are attempting to get national distribution are generally faring poorly since they are at a great disadvantage in comparison with the building trades group. Others in the solar group have chosen to concentrate geographically and are performing better. Many solar companies are offering very good warranties on their products which they may not be able to honor in the event of a large number of claims.

These companies, often charging low prices, sacrifice profitability in order to meet competition and achieve market penetration. For example: Solaron, a leader in air-based systems, had profits of $34,295 before tax on sales of $2,666,414 in 1977. American Heliothermal had losses of $441,000 on solar sales of $760,000 in 1978 after three years of operation. Inter-Technology Solar Corporation had a loss of $800,000 on sales, primarily to the government, of $1 million. In spite of this, new companies spring up to fill the void each time a company succumbs to the financial pressures of the industry.

This proliferation of small competitors is partially a result of the type of managers who start companies in the solar industry. Many see solar energy as an important social goal. Their personal goals are not simply rational profit maximization.

Grumman was most concerned with the very large companies, including General Electric, General Motors, and others. One firm that caused Peterson some concern was Daystar, a wholly owned subsidiary of Exxon Enterprises. However, in spite of the massive resources available to back up their venture, it did not appear that Daystar was taking nearly as ambitious an approach as Grumman.

THE SUNSTREAM FLAT PLATE PRODUCT LINE

Grumman Energy Systems entry strategy to the solar industry was based on a moderately

[1] Companies which derive more than 50 percent of their revenues from solar products.

priced, novel collector design backed up with massive advertising. Through this strategy, Grumman hoped to establish an early toehold in the residential solar water heating market.

Peterson was operating on the assumption that, as the industry matured, barriers to entry would grow, protecting the position of the leader. "The expenditure of a million dollars in advertising will buy us much more today than it will five years from now," said Peterson.

GES's proprietary collector design was called the "Sunstream." (See Exhibit 3 for collector diagram.) The standard Sunstream collector was a box-shaped unit measuring about 9′ × 3′. The device consisted of an aluminum housing containing a flat metallic collector plate through which liquid was circulated. The liquid was heated by the sun. The plate was insulated on its backside, and the whole assembly was topped by an arched acrylic dome. In-

EXHIBIT 3 Grumman Energy Systems

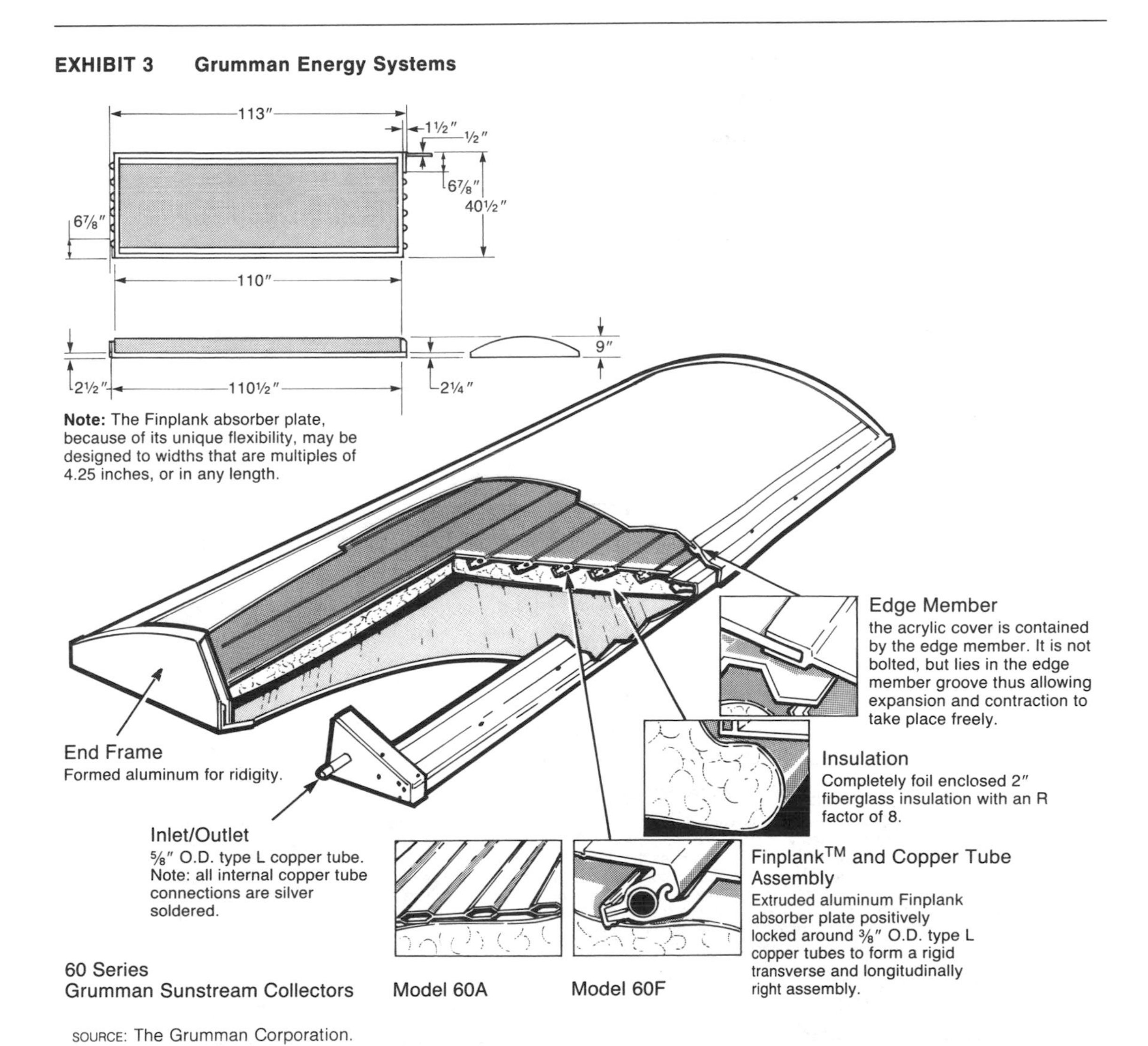

SOURCE: The Grumman Corporation.

stalled on a roof, the collector resembled a skylight.

A typical residential solar collector hot water heating system consisted of two flat-plate collectors, totaling some 50 square feet of active collector area. (See Exhibit 4 for system diagram.) These supplied solar-heated fluid to a storage tank and heat exchanger located adjacent to the home's hot water heater. The solar-heated fluid transferred its thermal energy to the domestic water supply via the heat exchanger. The existing hot water heater remained available to supplement the solar-heated water.[2] The same operating principles applied to space heating, although space heating systems were more elaborate.

[2] Such systems supplied about 60 percent of the energy necessary to heat hot water for the typical house. For the typical oil-heated home, this meant savings of $10/month in oil or about 16 gallons of oil.

EXHIBIT 4 Grumman Energy Systems

SUNSTREAM® Solar Domestic Hot Water System

Contains:
- Two model 60A collectors
- Storage tank with heat exchanger
- Circulating pump
- Thermostatic control (two sensors; one control box)
- Expansion tank
- Check valve
- Sunstream anti-freeze

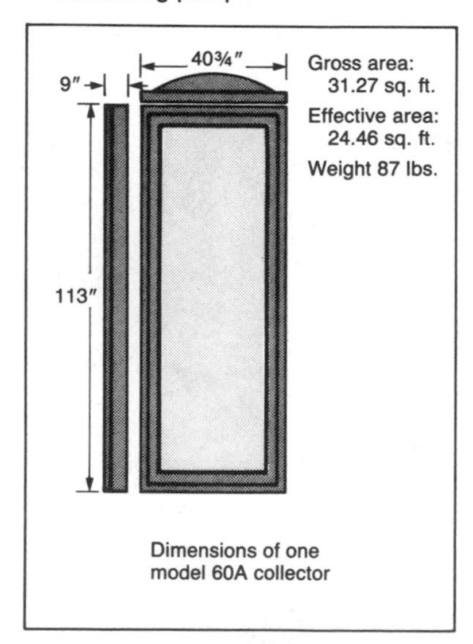

Dimensions of one model 60A collector

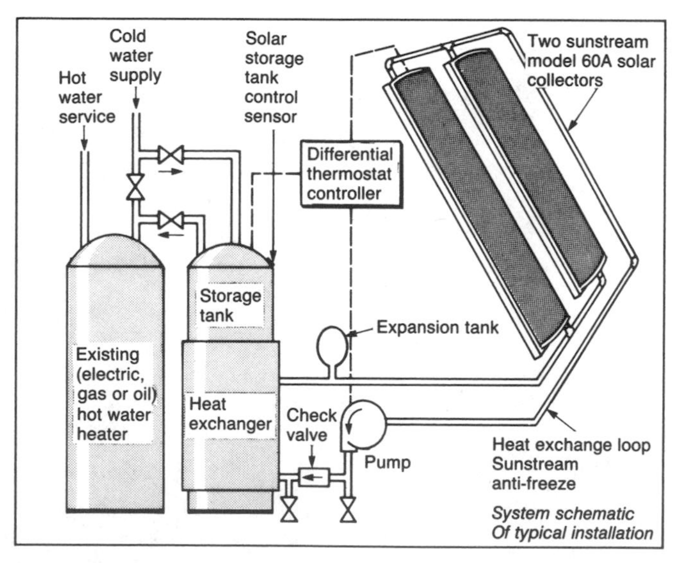

System schematic Of typical installation

How Does The Sunstream Domestic Hot Water System Work?

The heart of the Sunstream Solar Domestic Hot Water System is the collector which absorbs the sun's energy thereby heating the water. The Sunstream system is a freeze-proof year-round system. The Sunstream anti-freeze is circulated in a closed loop by the pump; goes through the collector, absorbs heat and gives it up as it goes through the heat exchanger. This process allows the water inside the storage tank to be heated. The cold water supply from the main is connected to this tank and the delivery pipe from it is connected to the existing hot water heating system. Thus, the solar system functions as a preheater to the existing system. In sunny weather, the water in the tank is hot enough that the existing heater has no need to come on. But it will be available to "top up" the hot water temperature in times of poor sun conditions or extraordinarily heavy use of hot water by the family.

GRUMMAN *Sunstream®*
4175 Veterans Memorial Highway
Ronkonkoma, L.I., New York 11779

GES marketed two basic residential collector models, the "60AS" and the "60FS." The former employed an aluminum absorber plate, and carried a suggested retail price of $1,260 as a complete system. The 60FS was mechanically the same but its absorber plate was copper, and it carried a suggested retail price of $1,460. These prices were at the upper end of the market. Variants on these two models were also available such as systems for heating swimming pools, or systems for use in nonfreezing climates. The collectors were also available by themselves (i.e., without the ancillary equipment, tank, or controls) for about 30 to 35 percent of the cost of the complete system. Installation costs would typically run $400 to $600 for a domestic hot water system. Space heating systems were several times more expensive, since they required more collectors, larger storage and heat exchanger capabilities, more elaborate ancillary equipment, and much higher installation costs. (See Exhibit 5 for a list of Grumman solar domestic hot water products and prices.)

Taking note of today's solar economics, Peterson stated, "One thing is perfectly clear. Our real competition is not the solar manufacturers, but the utilities and the Middle East." When replacing oil heat, solar hot water heating systems had paybacks ranging from 8 to 16 years, assuming present energy costs. A comparison of delivered energy costs for the various fuel sources in late 1978 is given in Exhibit 6.

Design Considerations

Although the solar industry was still in its infancy, architects, builders, solar installation specialists, and engineers were becoming increasingly sophisticated in their appraisal of collectors and their cost effectiveness. Dr. William Shurcliff, a Harvard physicist and solar expert, enumerated what he considered to be the four main considerations in solar collector and system design:

- Achievement of high-efficiency solar radiation collection.
- High durability of equipment.
- Low product cost.
- Equipment that can be successfully and economically retrofitted.

Efficiency is defined as that percentage of the sun's energy entering a collector that ultimately finds its way to the point of usage. It is most easily conceptualized if one thinks of a collector as a small building. It has heat losses just like a house. Much heat loss in a building is through windows, and, so too, in collectors—the selection of a glazing and cover plate is critical. Does one use glass or plastic? If glass, should it be tempered?

If one uses plastic—what kind? Acrylic (plexiglas, acrylite) is expensive and moves a lot, but is durable and very transparent. Polycarbonate (Lexan) is very durable and transparent, but very expensive. Tedlar film is tough but degrades due to heat; it also expands and contracts with temperature changes. Teflon film is heat resistant but not available in thickness required for outside layer.

Fiberglass-reinforced polyester resin is cheap, remaining strong and stiff under moderate temperatures, but it is only moderately transparent.

If one uses plastic, the sealing is critical. When one uses film, the problem is one of securing the material. One must concern oneself with venting to prevent condensation. And there's also an option of providing double glazing with a selective surface. But double glazing reduces the transmission of heat to the collector. Once the energy is inside the collector, one must concern oneself with the efficiency of the heat transfer fluid and transmission devices.

The choice of an absorber coating is also important. If one uses paint to absorb the heat, is it durable? One must be sure the paint is professionally applied and properly primed.

The durability objective is perhaps the most

EXHIBIT 5 Grumman Energy Systems: Product Summary

Model	Description	Dealer Price	Suggested Retail Price
Sunstream Solar Domestic Hot Water Systems			
60AS	An aluminum indirect (closed loop) system, for use in all climates, including: Two model 60A solar collectors. One 82-gallon glass-lined steel tank with external aluminum heat exchanger, uninsulated. Circulating pump. Expansion tank. Thermostatic control with sensors. Special system installation components. Four gallons Sunstream antifreeze.	$ 720.00	$ 995.00
60AST	Same as the 60AS system above with insulated, enamel jacketed tank and components factory mounted, wired, and piped.	912.00	1,260.00
60FS	A copper, indirect (closed loop) system, for use in all climates, including: Two model 60F solar collectors. One 82-gallon glass-lined steel tank with external copper heat exchanger, uninsulated. Circulating pump. Expansion tank. Thermostatic control with sensors. Miscellaneous system installation components.	845.00	1,200.00
60FST	Same as the 60FS system above with insulated, enamel jacketed tank and components factory mounted, wired and piped.	1,035.00	1,460.00
60FD	A direct (open loop) system, for nonfreezing climates, including: Two model 60F solar collectors. One 66-gallon glass-lined steel tank, insulated, enamel jacketed. Thermostatic control with sensors. Circulating pump. Miscellaneous system installation components.	785.00	1,060.00
Sunstream Solar Collectors			
60A	Aluminum Rollbond® collector plate encased in aluminum chassis with arched acrylic cover, fully assembled for use in closed loop systems.	227.00	275.00
60F	Finplank™ collector plate, featuring copper fluid passages, encased in aluminum chassis with arched acrylic cover, fully assembled for use in open or closed loop systems.	248.00	299.00
FP	Finplank™ collector plate for use in swimming pool heating.	140.00	169.00
100/200	A line of industrial quality collectors, all incorporating the Finplank™ collector plate and tempered glass; available with single or double glazing and insulation options.	Special order pricing	

Purchase of six domestic hot water systems required for dealership.
(Product specifications and prices subject to change without notice.)

SOURCE: The Grumman Corporation.

EXHIBIT 6 Grumman Energy Systems: Model Costs of Energy per Million Btu for Space Heating Systems in Individual Residences

Fuel	Capital, Operation, and Maintenance Costs per Energy Unit Supplied (dollars per million Btu)	Total Energy Cost per Million Btu Delivered (dollars per million Btu)
Solar* (low and high estimates)	12–22	12–22
Gas	1.5	4.9
Synthetic gas	1.5	9.2
Oil	2.4	7.4
Synthetic oil	2.4	11.7
Electricity:		
Resistance heating	1.2	13.2
Heat pumps	3.2	10.9

* Solar costs for hot water heating only are about one-third lower.
SOURCE: S. Schurr et al., *Energy in America's Future* (Resources for the Future, 1979), p. 338.

elusive in that it requires a very specific knowledge of the actual conditions under which the collector functions, which can vary widely in different times of the year and parts of the world. Dr. Shurcliff discusses some of the problems:

> The history of the development of solar heating is a history replete with tragedies: glass cracking, edge seals failing, tubes breaking loose from the metal sheet they had been soldered to, pipes becoming clogged, serpentine tubes failing to drain properly and subsequently freezing and bursting, joints leaking, insulation installed badly and allowing excessive heat loss, underground insulation becoming water-logged, rapid corrosion. Parts that hold up well in winter may deteriorate fast under the hot summer sun (and under glass!). In a four-day tour of 12 solar houses, the writer found that 8 of them had broken glass and leaks of water or moisture. Mistakes are seldom publicized, however.

And many other durability considerations come into play. What frame material does one use? Aluminum runs into problems in attaching it to steel. Steel may rust if not galvanized. Wood is inexpensive but must be flashed and pressure treated.

While Peterson was in agreement with Dr. Shurcliff on the need for lower product costs, he was not optimistic about cost reductions, for as he explained: "Eighty percent of the system's cost is in the materials." Similarly, he expected retrofitting costs to escalate with plumbers' and carpenters' fees.

Dean Cassell, director of manufacturing of GES, emphasized a different obstacle to cost reduction: the background of the Grumman employees and their realm of experience, which was largely in aerospace.

> You have to remember that Grumman guys are mill-spec-ed[3] to death. I've had to emphasize, drill into them, that this isn't a gold-plated airplane we are building.

Dr. Shurcliff perhaps best summarized the tension that has characterized Grumman's and other solar manufacturing thinking about design and cost consideration.

> Many newcomers [to the solar heating market] are still preoccupied with technical elegance,

[3] Military specifications—which for aerospace equipment can be very tight.

blindly trusting that, somehow, cost problems will vanish when mass production is begun.

Is this trust well founded? Will mass production drastically reduce cost? Costs can certainly be reduced below today's levels—assembly costs, especially. But whether the reduction will be substantial or only moderate is uncertain. Why? Because (*a*) components such as glass, sheet metal, tubes, insulation, and valves are already being mass produced; and (*b*) the equipment is bulky and fragile—shipping it to a site will always be expensive, and installation work will often be "special" and may remain expensive. Electric clocks and transistor radios are truly cheap; but houses and solar heating systems are a different kettle of fish. Consider the cost of modernizing a bathroom: the new equipment may cost only $300, but the total bill may reach $3,000.

The Second-Generation Collector System

Peterson and his staff were exploring two divergent ways in which to redesign GES's present collector system. One alternative was

EXHIBIT 7 Grumman Energy Systems

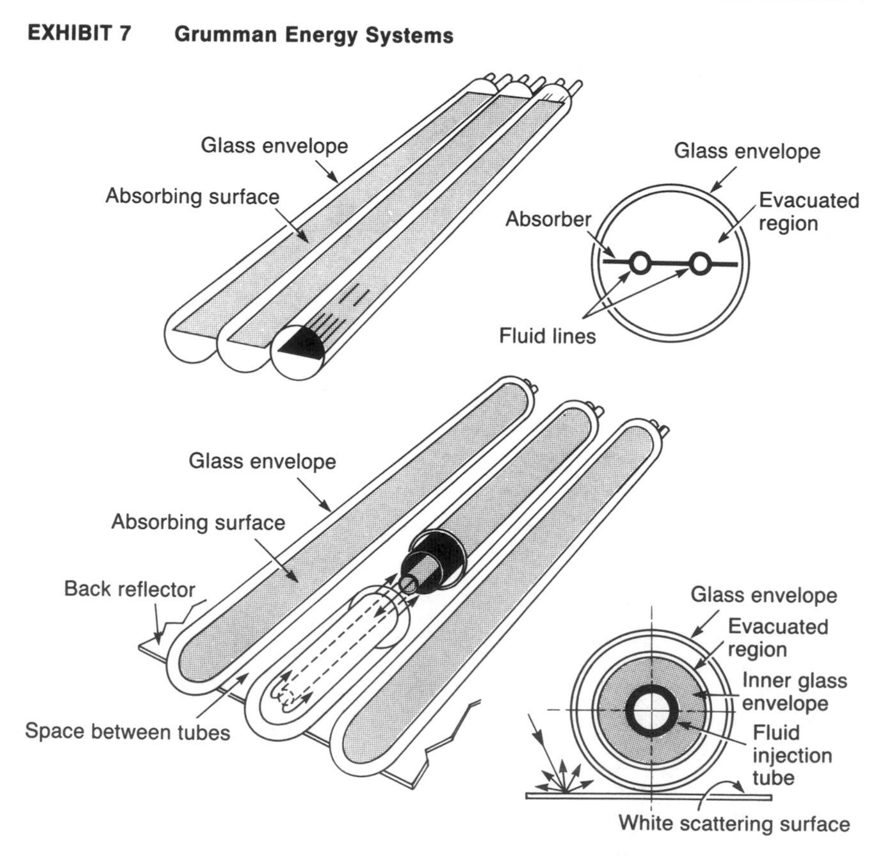

Two designs for nonconcentrating evacuated solar collectors. The top configuration is used by Corning and the bottom by Owens-Illinois.

SOURCE: Meinel and Meinel, *Applied Solar Energy—An Introduction* (Reading, Mass.: Addison-Wesley Publishing, 1977), p. 451.

to bet on a new technology (Evacuated Collectors) that promised improved collector efficiency; the other was to concentrate on refining and standardizing the GES flat plate collector system. A key issue was one of timing, for Peterson felt that for solar systems to realize their full potential they had to function and appear like appliances. In his words,

> We need to put sizzle on the stick. We need to ultimately make the solar system look like an appliance. That's the basic idea. When people hear about a solar system, they think of something that can be installed as simply as a refrigerator—just plug it in. We need to reach towards a standardized design that can incorporate much of the individualized plumbing, piping, and fittings. That's where our competitive advantage will be.
>
> We can now design a system where we are not wedded to the existing hot water tank and where we can use it horizontally. It can be a one-tank system, with internal electric elements for backup.[4] It would look like an enameled bathtub with only two input and two output connections. It will be basically foolproof.

Peterson was considering several additional product design changes. The first would eliminate the arched dome. By eliminating the arch, the company could ship 60 percent more collectors in the same space—at roughly the same cost. However, this change would eliminate the distinctiveness of the Grumman design, which Peterson felt was important to sales. The second change concerned the introduction of a new 4′ × 8′ panel for the small family. "This one panel would replace the two 3′ × 7′s," Peterson explained. "This would be another way to cut costs. Also, we would offer variable sizes so that architects could simply select the most suitable size out of our catalog."

GES was also exploring evacuated collectors, a wholly different way to design collector systems. Evacuated tubes were in an embryonic stage now, but if proven viable they could pose a major challenge to flat plate collector systems. They also could place new demands on GES's manufacturing and engineering departments.

Evacuated tube collectors had been recently characterized by several firms in the solar heating industry as a "technological breakthrough" in collector design that could potentially eclipse the Grumman Sunstream System. This design, the so-called evacuated tube design, was now in initial production at General Electric and at Owens-Illinois (see evacuated tube design in Exhibit 7). According to an article in *Science Magazine,* "evacuated tubes can potentially give twice the performance at half the cost of flat devices." See Exhibit 8 for a simplified comparison of flat plate collector and ETC efficiencies. As the president of one evacuated collector supplier saw it: "Flat plates compared to [evacuated] tubes are like a horse and buggy compared to the internal combustion engine."

Evacuated collectors consist of an inner glass cylinder, blackened to absorb sunlight, enclosed within an outer protective cylinder, with the space between the two cylinders evacuated. The heat is transferred to a fluid, either air or

EXHIBIT 8 Grumman Energy Systems: Comparison of Flat Plate and Evacuated Tube Collector Efficiencies (Simplified)

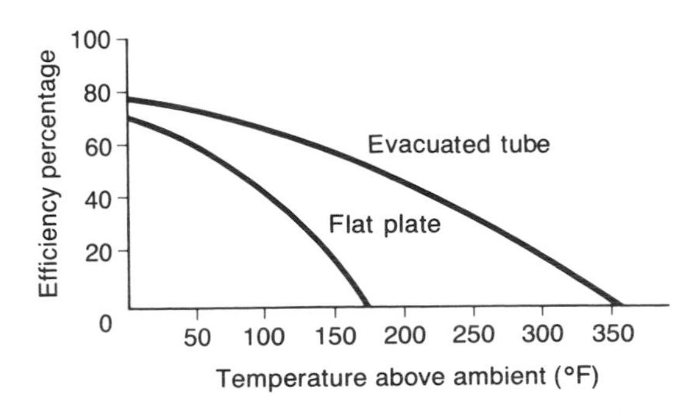

SOURCE: Casewriter's estimates.

[4] If an oil or gas water heating system is used, this normally would require another tank.

a liquid, that flows through the inner cylinder. Evacuated tubes improve performance over flat plate collectors because when the air is removed from the space between the absorber and windows, the convective loss[5] is suppressed. The evacuated tubes can yield temperatures well above 212°F at 50 to 60 percent efficiency. Residential hot water heaters operate at temperatures of about 120°F.

Cylindrical evacuated tube collectors can also absorb light coming from any direction, and they work somewhat better than flat plate collectors early and late in the day. They are essentially unaffected by either high winds or cold weather, both of which degrade flat plate performance; in fact, the output of evacuated tubes is essentially independent of the ambient temperature. Their high efficiency translates into higher temperature heat. Evacuated tube collectors usually operate at 180°F or above for space heating and providing industrial process heat [as opposed to 110°F average for flat plate collectors], and with a reflective material behind them, they can operate at 240°F, high enough to drive absorption air conditioners.[6]

The glass collectors appear to have promise for reduced costs. Today evacuated tubes generally cost between $25 and $35 per square foot, installed, compared to $20 to $25 for standard flat plate collectors. The manufacturing process for evacuated tubes is closely related to that used for producing fluorescent lights and lends itself to automated mass production. In some designs, the individual tubes are removable from the arrays, and can thus be replaced like light bulbs. They are lighter than flat plate devices, use both less glass and less metal (materials costs are reported to run about $5 per square foot), and are more resistant to corrosion. The most serious problems involved in evacuating a collector are how to seal the system for a reliable lifetime of operation, how to alleviate the mechanical pressure resulting from the surrounding atmosphere, and how to insure against breakage.

SOLAR MARKET SEGMENTATION

The residential sector dominates the solar thermal market, accounting for approximately 80 percent of manufacturer's sales of $150 million in 1977. This dominance is expected to last at least through the 1980s. Residential systems are primarily hot water systems using flat plate medium temperature collectors and low temperature pool heating systems. Exhibit 9 shows the residential sector in terms of numbers and types of systems installed during 1977 and estimated average costs per system.

[5] Loss due to heat carried by air current in the tube.

[6] This discussion of evacuated collector characteristics was adapted from Meinel and Meinel, *Applied Solar Energy* (Reading, Mass.: Addison-Wesley Publishing, 1976).

EXHIBIT 9 Grumman Energy Systems: Residential Sector Analysis 1977

System Type	Average Cost	Number Installed	Total Spent
Hot water	$ 2,000	63,000	$126 million
Heat and hot water	10,000	3,000	33 million
Pool heaters	1,100	35,000 est.	38.5 million
			$197.5 million

Note: Assumes 1977 installed system cost per square foot of collector $40 on average exclusive of pool heaters.
SOURCE: Solar Energy Industry Association.

EXHIBIT 10 Grumman Energy Systems: Sector Segmentation of Solar Thermal Market

Sector	Percentage Market Dollars	Applications
Residential	75% to 85%	Hot water, space heating, cooling, pool heating.
Commercial/institutional	15% to 20%	Hot water, heating, cooling.
Industrial	5%	Hot water, process heat, heating, cooling, crop drying.

SOURCE: Casewriter's estimates.

The nonresidential sector accounts for less than 20 percent of the solar thermal market. (See Exhibit 10.) The vast majority of the installations in this segment are solar hot water systems for institutions (hospitals, schools, nursing homes, apartment complexes, etc.). The nonresidential market is still in an embryonic state. Most projects are funded by the Department of Energy for demonstration of feasibility and to promote commercialization. The potential, however, for nonresidential applications is vast. Approximately 18 percent of the national energy consumption is nonresidential hot water, space heating, and process heat applications within the temperature range of solar thermal power (less than 350°F), and its rate of increase is projected at 5 percent per year.

THE GRUMMAN SALES STRATEGY

To date, Grumman's sales efforts had been aimed primarily at the residential, single-family domestic hot water market (and some space heating). Most of their sales efforts had been concentrated in the major metropolitan areas of the East Coast, California, and the southwestern states.

GES marketing strategy rested on two foundations. First, an intensive national advertising campaign aimed at this preselected target market. Second, the rapid development of a high-quality coast-to-coast dealer network to provide sales, installation, and service.

1. Advertising

The 1977 Sunstream ad campaign consisted of 73 separate advertisements, 61 in full color, that were inserted in 15 publications to yield over 364 million "advertising impressions." The media use included "lifestyle" magazines like *Sunset,* professional-oriented publications like *The Wall Street Journal,* and several major East Coast newspapers. Although the ad campaign was national in scope, its greatest emphasis was on cities in the Northeast, both because of that area's high energy costs and proximity to the production facility on Long Island.

The themes of the ads tended to be pragmatic, and they often showed pictures of actual installations in typical suburban residential settings. According to Peterson, "We are emphasizing that solar energy is right here *now*, and we have a positive tone in our ads." For example, one headline ran "Solar Energy: Practical, Clean, and Now." He contrasted this to the advertising of competitive energy source companies like Exxon and Mobil who are saying that solar energy will become practical at some time in the future, thereby giving the consumer the impression that they should not rush into a purchase of solar equipment. He adds, however, "When the market is ready, companies such as Exxon and Mobil, who are energy companies, will be there with a product." (See Exhibit 11 for sample ad.)

The cost of the ad campaign was about $1 million in 1977, accounting for about half of all of GES's R&D, S, and G&A expenditures. Their advertising campaign had produced about 80,000 inquiries from the public. The ads invited the reader to write to "Grumman Sunstream" headquarters in Ronkonkoma, Long Island. Inquiries received at headquarters were processed in a "Communications Center" established specifically for GES marketing. In this closely guarded operation, inquiries were recorded, and a standard brochure was mailed to the individual. The brochure was a glossy, full-color foldout piece entitled, "The Story about Using the Sun's Energy Today." It was prepared along a question and answer format, discussing the most common and pragmatic questions about solar energy, including its economics. A postpaid business reply card was included as part of this brochure. Addressed again to Grumman Sunstream, it permitted the homeowner to request that he be contacted by an authorized Grumman Sunstream dealer. (See Exhibit 12 for the brochure.) If the homeowner returned this card, the Communications Center channeled his request to the appropriate Sunstream dealer.

But Peterson had some doubts about the follow-through operations of this advertising campaign. "One ad campaign we had produced 15,000 inquiries. Ninety percent of those people never heard back from us or our dealers. Whereas, in another situation we had 6,000 inquiries and 1,200 ended up buying—and that 20 percent conversion factor included selling efforts of only 15 percent of the dealers. One dealer alone sold almost 100 systems." So Peterson was wondering how to improve the system. One possibility was to establish a toll-free "800" phone number to further encourage inquiries.

EXHIBIT 11 Grumman Energy Systems

Start collecting today.

Question: **WHY BUY NOW?**

Answer: THE COST OF CONVENTIONAL ENERGY IS ON THE RISE. It will dramatically increase over the next five to ten years. Hence, it makes sense to cut your costs today, and for the future, by using our greatest known energy resource. . . the sun. The system is an investment that may add significantly to the value of your home. It's also an investment whose use of free solar energy pays you immediate dividends that continue to increase as all other energy costs escalate.

ELECTRICITY IN DOLLARS PER MILLION BTU'S

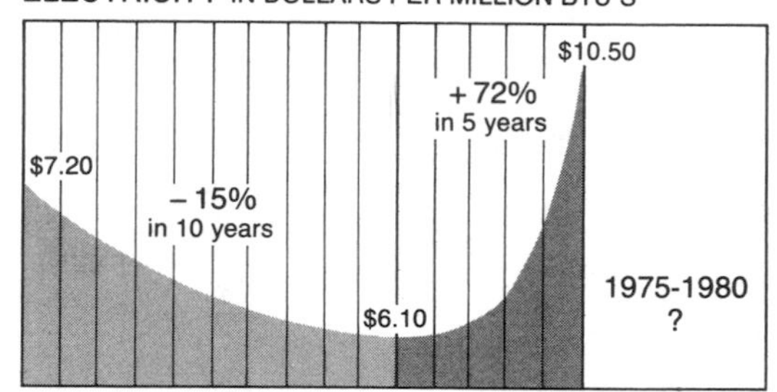

#2 FUEL OIL IN DOLLARS PER MILLION BTU'S

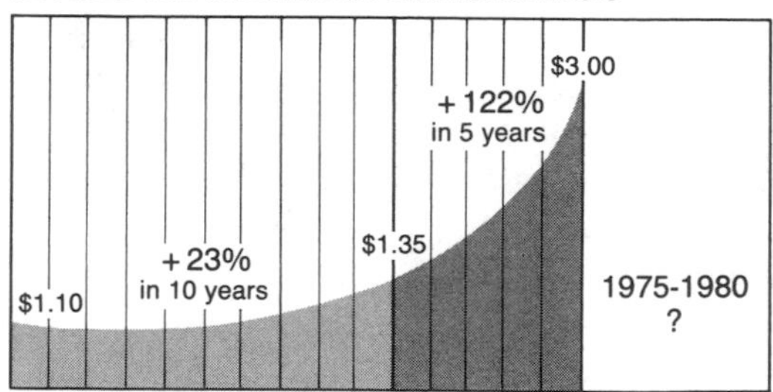

NATURAL GAS IN DOLLARS PER MILLION BTU'S

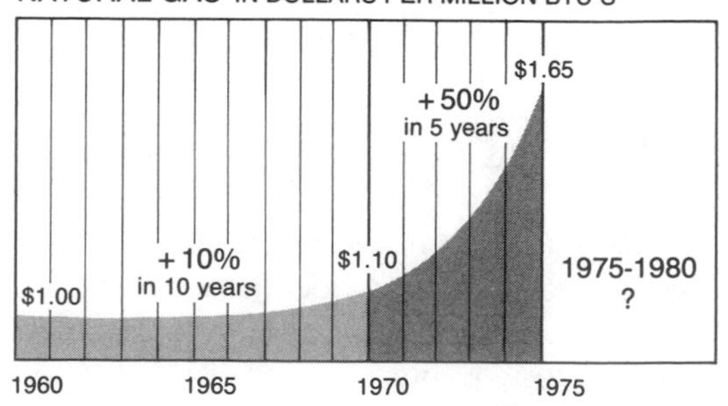

Put the sun to work for you today. A trained and authorized Grumman Sunstream™ dealer would be most happy to tell you more about how you can save energy for your home or business. Simply fill out and detach the prepaid postcard.

SOURCE: The Grumman Corporation.

EXHIBIT 12 An Information Brochure for the Potential Sunstream Dealer

The company that put the man on the moon . . .

The technological and financial resources of the Grumman Corporation are vast. The corporation whose aerospace technology produced NASA's lunar module (which landed men on the moon with a 100 percent safety record) has also engineered and constructed products that range from jet aircraft, yachts, and transportation vehicles to municipal and industrial pollution control systems. The name Grumman is, and has always been, synonymous with quality and engineering excellence. And because Grumman is the manufacturer, you can be sure that every effort has been expended to make Sunstream an outstanding line of solar products.

. . . is bringing the sun down to earth.

Not long ago fossil fuels were inexpensive and plentiful. Today, they are neither. And as fossil fuels have become ever more expensive, solar energy has come into its own as an alternate source of energy. That is why the Sunstream solar domestic hot water system has been designed and built to use the sun's energy to heat water for household needs. The system can be installed in almost any house now relying totally on electricity, oil, or gas to do the same job. Given today's cost of energy (and predictions of future increases), the system's potential value becomes apparent.

The Sunstream Product Line Addresses All the Solar Applications that Make Sense Today . . .

The solar domestic hot water system, our primary product and the one upon which dealerships are founded, is available in all aluminum as well as copper fluid passage designs; with open and closed loop (with heat exchanger) options for nonfreezing and freezing climates. In addition, our line of Sunstream solar collectors is available for sale through dealers for application in residential and commercial space heating, large domestic hot water heating systems, and swimming pool heating.

The Benefits of the Domestic Hot Water System

- *Installation versatility.* Sunstream collectors can be mounted anywhere the sun shines—on the roof of your home, garage, or on separate support structures, either vertically or horizontally.
- *Aesthetically unique.* The soft arch of the Sunstream collectors, together with the straight line of water passages in the absorber plate, provide a clean, distinctive appearance. Further, because of this shape, a "self-cleansing action" is promoted and glare is greatly reduced, making the Sunstream collector a more friendly neighbor.
- *Easily maintained.* The only moving mechanical part in the system is the circulating pump.
- *Freezeproof.* The closed loop system works year round when used with Sunstream antifreeze. The open loop system is designed for use in nonfreezing climates only.
- *Low shutdown temperature.* The collector is designed to minimize temperature extremes, allowing the use of low-cost materials. Cost and performance have been kept in balance.
- *Automatic.* The system is thermostatically controlled to run by itself.
- *Lightweight.* Reduces or eliminates the need for roof reinforcement; easily handled for installation.
- *Safety.* The "double wall" effect of the wrap-around heat exchanger provides an extra degree of protection against contamination of potable water.

EXHIBIT 12 *(continued)*

Sunstream Dealer Support Materials

In order to help you generate sales, we can provide you with the following materials and services:

1. *Training at Sunstream*
   We will train you in the installation and service of the system. These training sessions will take place at our facilities on Long Island.
2. *Consumer Literature*
   We will supply you with a variety of promotional literature to use in your selling efforts.
3. *Flip Chart*
   This sales aid is designed to help you sell on a face-to-face basis to your prospect.
4. *Door/Window Identification*
   Decals to identify you as a Sunstream dealer.
5. *Counter Displays*
   Eye-catching Sunstream counter display cards are provided to you, showing the product warranty and examples of applications.
6. *Advertising Materials*
   The following materials are available for use by you in the advertising media in your sales area: *Newspaper mats* (just add your name and address to the prepared ad); *Radio scripts* (in increments from 10 to 60 seconds).
7. *Cooperative Advertising Program*
   For every system you buy, we will credit you with $10 for your use towards any Sunstream advertising.
8. *Trade/Consumer Show Support*
   Upon formal request from you, we may be able to supply booth display material for your shows. Local inquiries resulting from Sunstream shows will be forwarded to you.
9. *Publicity*
   This is a free service for you provided by Sunstream to help you build sales in your local area. These materials include: *News releases* from Sunstream about you; *prepared news releases* for you to send to the local press; and a *newsletter* filled with a flow of ideas on how to use Sunstream to increase profits.
10. *Engineering Services*
    A free service providing sizing estimates for large domestic hot water systems, space heating, and swimming pool systems. Convenient job description forms are provided for you to submit to initiate this service.

In summary, when you become a Sunstream dealer, you can be provided with all of the materials needed to expand the profitability of your business. These materials include:

- Training
- Literature
- Flip chart
- Door/window identification
- Counter cards and ceiling display
- Advertising materials
- Cooperative advertising
- Show support
- Publicity
- Engineering services

Limited Warranty

Model 60AS/60AST/60FS/60FST/60FD systems
Model 60A/60F/FP collectors

1. What is covered and for how long:
   Grumman Energy Systems warrants to the original consumer for a period of five years from date of delivery with respect to solar collectors and one year from date of delivery with respect to other components that the components of its solar domestic hot water systems are free from defects in materials and workmanship under normal use and circumstances, when installed and used in accordance with the "Sunstream Installation and Maintenance Instructions."

EXHIBIT 12 *(concluded)*

2. What is not covered:
   This warranty does not apply to installation services nor to any components, materials, or services not supplied by Grumman Energy Systems and does not include any warranty or guaranty made by any Dealer. The warranty does not apply to components of Grumman Energy Systems' solar domestic hot water system used outside the continental United States.
3. Where and how warranty claims are made:
   All warranty claims are to be made through an authorized Grumman Energy Systems Dealer directly to Grumman Energy Systems, within warranty periods outlined in paragraph 1, above, by the original consumer. When such claim is valid, the defective component shall be repaired or, at the option of Grumman Energy Systems, replaced through the services of an authorized dealer or serviceman who will remove the defective component and repair or replace and reinstall the same free of charge, provided, however, that charges for labor will be made for repair or replacement of solar collectors on which warranty claims are received after one year from delivery.
4. Limitation on the length of implied warranties: The implied warranties of merchantability and/or fitness for a particular purpose are limited in duration to the warranty periods outlined in paragraph 1, above, from the date of delivery to the original consumer.
5. Other important information
   a. GRUMMAN ENERGY SYSTEMS DOES NOT UNDER ANY CIRCUMSTANCES, ASSUME RESPONSIBILITY FOR INCIDENTAL OR CONSEQUENTIAL DAMAGES, INCLUDING BUT NOT LIMITED TO DAMAGE TO PERSON OR PROPERTY OR LOSS OF REVENUE.
   b. The entire obligation of Grumman Energy Systems regarding the sale of its products is stated within this written warranty. Grumman Energy Systems does not authorize its dealers or any other person to make any other warranties or assume for it any other liabilities in connection with the sale of its products.

2. Distribution

In creating a dealer network, Grumman was moving just as aggressively. The GES started 1977 with about 40 Sunstream dealers, now they had some 140, and were hoping to finish the year with around 500. Also, there was no ceiling on the maximum number of dealers Sunstream wanted.

The types of enterprises GES was seeking as dealers varied from fuel oil suppliers (who probably comprised the bulk of present dealers in the Northeast) to building and HVAC (heating, ventilating, and air-conditioning systems) contractors. For example, of the Sunstream dealers on Long Island, five were fuel oil dealers, one was a plumbing contractor, and one an HVAC installer. As part of its strategy, Sunstream was encouraging fuel oil dealers to change their self-conception to that of a supplier of all forms of energy, not just fossil fuels. A separate advertising campaign was being run to solicit dealers. This campaign had brought in some 5,000 inquiries in a 60-day period.

In spite of its rapid buildup of a dealer network, GES was paying careful attention to the quality of dealer it accepted. Thus, the dealer application process was fairly involved, and only experienced and well-capitalized dealers were accepted.

The incentives for becoming a Sunstream dealer were substantial. First, the margins were very attractive. While Sunstream competitors typically allowed the dealer a margin of some

10 to 15 percent, margins on Sunstream products could range as high as 38 percent. The magnitude of profit involved was great. For fuel oil dealers in particular, the profit from the sale of just one Sunstream system could equal the profit made on a very large quantity of fuel oil. Second was the value of being associated with an organization of the stature and resources of Grumman. Grumman was by all appearances "playing to win" and could be counted on for a quality product and generous backup.

Once GES had accepted a dealer, it maintained tight control over him. All advertising had to be done in conjunction with GES plans, and all dealer ads were subject to prior approval. If a dealer made a single "bad" installation, he was required to send personnel back to GES for retraining at his own expense. If a dealer was so unfortunate to have made a second "bad" installation, he automatically lost his dealer status.

Peterson, however, was anxious to improve aspects of the distribution program. With sales offices in Chicago, Houston, Los Angeles, New York, and Atlanta, he was considering expansion, perhaps overseas. Peterson explained, "We are now doing international licensing for solar products in several countries, particularly the Netherlands. We already have the largest market share in solar domestic hot water in the United States."

Peterson had other distribution ideas. "Another major new avenue not yet pursued is opening solar stores. Places in California may very soon have local ordinances requiring solar. In such a situation, our traditional dealership arrangement might not be adequate. We would also consider selling through wholesalers."

Peterson was also considering moves to strengthen their position vis-à-vis competition in service. "If we could offer 24-hour delivery service to builders, we'd have a real advantage over the small manufacturers. That would require a purchase of at least three tractor trailers at a cost of $100,000 and the setup of a 24-hour notice system."

ORGANIZATION AND OPERATIONS

Peterson had full control over GES staff, finances, manufacturing, and marketing. (See Exhibit 13 for organization chart of GES.) One important influence figuring in the operation of GES, however, was not captured in the organization chart. That influence was simply the personal relationship between Ron Peterson and Joe Gavin, Grumman president. Gavin had been behind energy from the beginning. An example of Gavin's role involves a problem GES faced in designing one of their collector plates. Peterson related: "To give you an idea of how close Joe Gavin is to this thing, one day, when confronted with a problem we had in collector design, he developed a new concept for an adjustable length collector system made primarily of aluminum but with small copper pipes, on which they have a patent pending." Speaking about his autonomy, he said, "Every year we make an operating plan and I can approve anything that is within the plan. The plan is very detailed for one year, and is laid out in rough terms for five years. The things I can't do are to change the total capital equipment outlay or to depart from my bottom-line plan without authorization."

Peterson, however, was considering a general restructuring of the GES organization. He had several concerns about the way things were functioning now. One tension Peterson felt was in the pressure to hire from within Grumman. He elaborated: "The way it works is that all jobs at Grumman are advertised internally. This is a procedure that started some time ago. But our needs vary here at GES. The marketing vice president Rod Noonan came from Xerox and our salesmen's backgrounds vary from selling washing machines to automobiles. For instance, to do the new system design, we'll need additional mechanical engi-

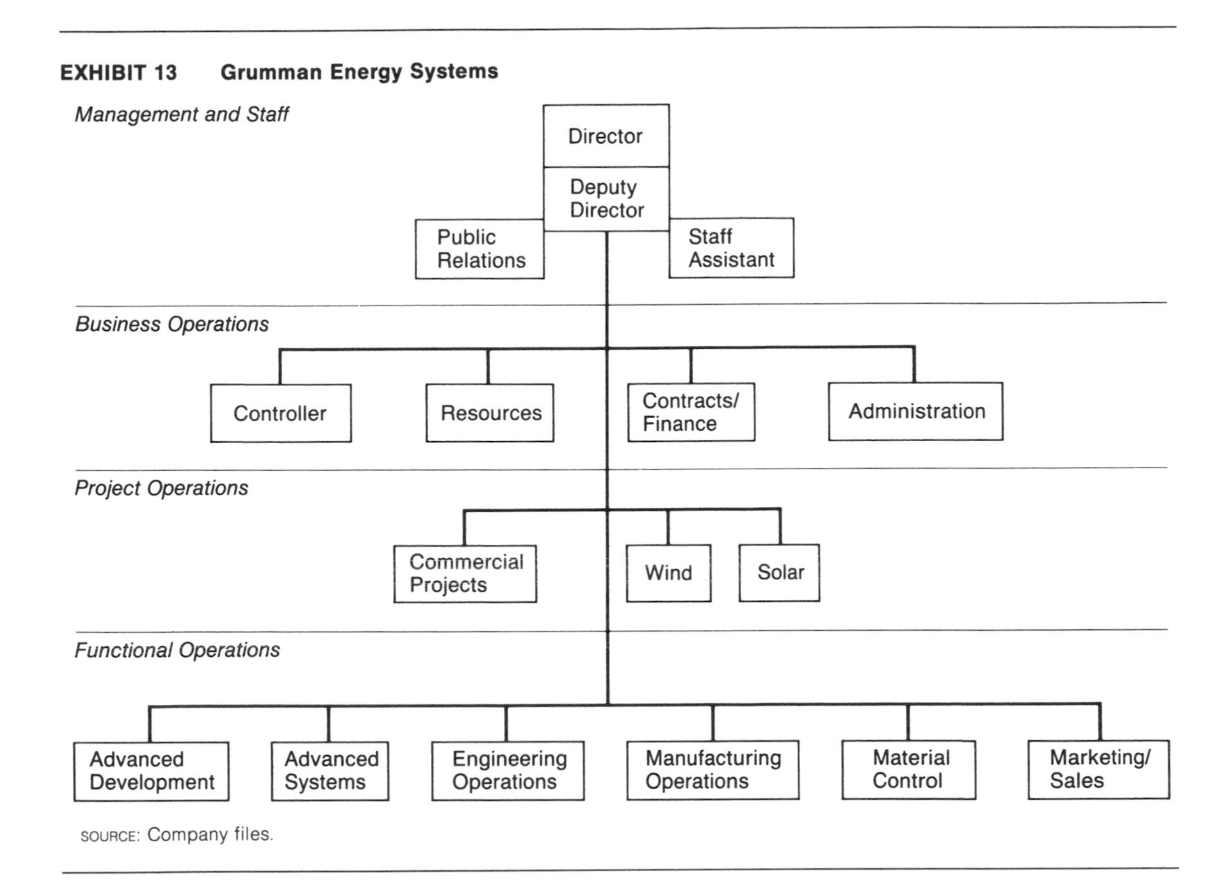

neers. We've done a lot of subcontracting from Grumman Aerospace."

PRODUCTION

Production operations at GES were centralized under Dean Cassell, a former navy aviator with an undergraduate degree in business administration. Cassell had recently graduated from the Program for Senior Executives at MIT, and had spent most of his career at Grumman. Peterson said that Cassell "probably knows Grumman better than anybody else over here, which is very important, for about 90 percent of my job is internal operations."

Grumman's manufacturing process of solar collectors is labor intensive. They use mostly low-skill labor. Dean Cassell described the fabrication process: "Basically, we have an assembly line operation. We use a roller conveyor to move tanks around."

The present system has evolved gradually from several diverse influences. Cassell speaks of the evolutionary process: "When we started, we looked at our (Grumman's) canoe manufacturing operations, figuring the processes were very similar." Then he speaks of the influence of his and Grumman's experience in airplane manufacturing on the production philosophy:

> Initially we were a packager, we bought just about everything and put it together in a system. This was very similar to making airplanes, where typically you buy at least 50 percent of the parts; here we were buying even more. Now we are

gradually moving from buy to make. For instance, right now we buy the water tank liner but we build the rest of the tank from then on.

Peterson would like to see GES integrate even farther backwards. He spoke of the future emphasis in production. "Eventually we want to get the balance changed where we are making more than we are buying. It's a matter of making commitments in capital equipment such as tooling." Peterson is hoping this shift will improve profitability. He elaborated, "Gross margins were running about 20 percent, but with improved, more automated production techniques, we expect our gross margin to go up to 40 percent." Nonetheless, Peterson acknowledged that such an increase would be difficult "because as it stands our direct production cost is primarily materials."

Another major issue Peterson faced is the location of a production facility. Presently, the primary facilities are on Long Island, and shipping costs to the western markets will cut heavily into profit. Cross-country shipping would add 10 percent to collector system cost. The fastest growing markets were in Arizona and California, and he wondered whether this trend justified a new facility. He spoke of some of the considerations involved in a move. "If we located in a plant out west we would have to develop duplicate tooling on Long Island and ship it out there. Also, we would have to find a plant manager who would be able to fit well into the California community and still know his way around Grumman." But, if the plant was located in one of the more rural areas of California, Peterson speculated: "We would realize labor savings of about one third from New York rates."[7]

The possible move to evacuated tubes was a further complication. Manufacturing expertise for the glass tubes was centered in the Northeast. Furthermore, Peterson doubted the wisdom of developing a new technology in a distant plant. An alternative was to keep the R&D in Long Island and limit the new plant to manufacturing only. But this required that R&D personnel be separated from the manufacturing engineering people at a time when there was likely to be intensive design and process development.

Dean Cassell, vice president of manufacturing for GES, on the other hand was wondering whether a new plant might lend itself to higher automation. He speculated on the possibility: "If we were to build a semiautomated plant, something like a bottling plant, we probably would have to spend $3 million over and above the million for basic plant equipment, but with such a plant we could make 12,000 systems a year on a two-shift basis." But how would these plans be affected if GES moved to EVTs?

CONCLUSION

Peterson had grown up in the aerospace industry. When he started college, propeller aircraft dominated the skies. In his early years at Grumman, he witnessed a full shift to jet power for both commercial and military aircraft. He then saw supersonic aircraft technology make major inroads in fighter aircraft design. And all the time aircraft speed continued to rise (see Exhibit 14). He wondered if this experience could serve as a model for what was occurring in the solar heating industry. Would evacuated tube collectors replace flat collectors, and then be themselves replaced by some new esoteric technology such as photovoltaics (solar cells)? If this was a reasonable forecast of the future, what were the implications for the manufacturing, marketing, and organizational decisions that he faced?

Would the commercial solar heating market technology lead the residential market as military aircraft technology has lead commercial aircraft technology? Should he hedge by maintaining twin development programs? If so, how should he split his R&D budget? These and

---

[7] Typical labor rates for hourly workers in the Long Island area were $4.50.

EXHIBIT 14 Grumman Energy Systems: Aircraft Speed Record Set under International Competition Rules

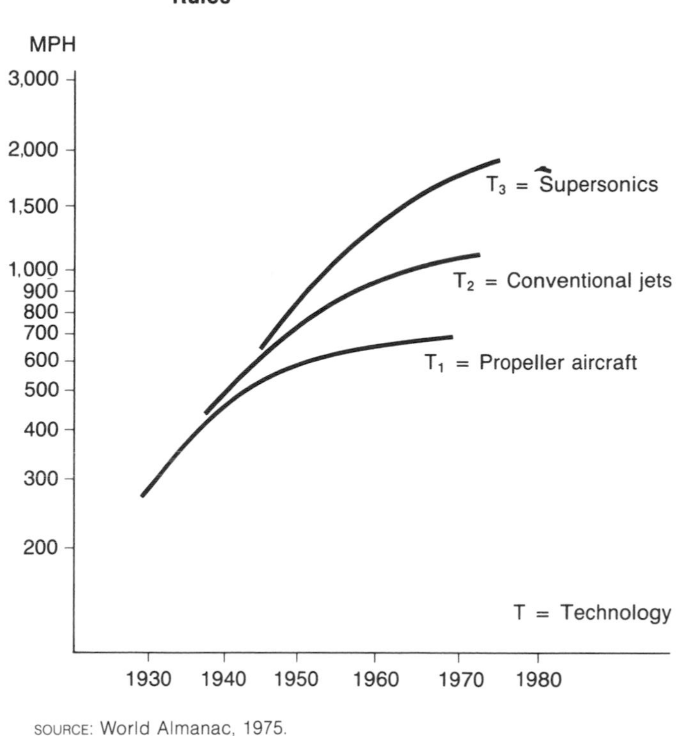

SOURCE: World Almanac, 1975.

other related questions were concerning Ron Peterson as he approached the critical next two years.

Reading II–3 Management Criteria for Effective Innovation

George R. White

Early in the corporate era, it may have been easy to believe that technological stability was the normal condition and technological innovation the occasional, fortuitous balm for problems to which we otherwise could have adapted. But since World War II, innovation has been the norm; technology-based innovations, coming in rapid sequence, have been seen as the crucial source of prosperity, the panacea for all business problems.

Now we know that this panacea is not necessarily benign. The U.S. electronics industry was in far better condition in the early 1950s, before the emergence of the transistor, than it is now with the consumer business largely penetrated by Japan. On the other hand, it is also true that a previous modest participant in electronics, the Motorola Corporation, used the transistor to expand its position in consumer

Technology Review. February 1978, pp. 21–28. Reprinted with permission from *Technology Review,* MIT Alumni Association, copyright © 1978.

electronics and then, by integration backwards, gained a significant new role as a component supplier.

When compared to those available in finance, marketing, and production, the tools available to management for assessing and directing technological innovation are rudimentary. Intrigued by this fact, Margaret B. W. Graham and I determined to study in detail two fully completed, well-documented innovations—the transistor in consumer electronics and the jet engine in subsonic jet transports. We postulated that the criteria for success in these cases could be applied to predicting the future outcomes in two immature innovations—the supersonic transport and computerized automobiles.

We conclude that we can in fact identify management criteria which effectively discriminate between profitable and unprofitable new technologies, and that these criteria have utility in appraising technological innovation in a wide variety of cases.

I. THE DETERMINANTS OF SUCCESS

We began by predicting that determinants of success could be found in both technology and business contexts. In the realm of technology, the determinants would surely depend on some appraisal of the quality and significance of the innovative concept itself: it must be new, and it must also be good. But such an innovative concept alone does not assure technical potency; there must be an embodiment for the new device, a product or system which is waiting for it. The embodiment surrounding an inventive concept has a major effect on how profitable the new technology proves to be.

Even with technical potency, a high rate of adoption and great profitability are not assured. The operational consequences of the new technology on manufacturing, marketing, and distribution must be considered.

Finally, market dynamics are extremely important—and often complex. Many industries have several dependent stages of intermediate demand; for example, transistor manufacturers sell to radio manufacturers, and radio manufacturers in turn sell to consumers. It is not enough to study the transistor market; analysis of final consumer demand is essential to understanding the outcome of transistor technology.

Balancing Old and New Constraints

Three questions turn out to be crucial in determining the technical potential of any inventive concept:

What fundamental technical constraints limiting the prior art are lifted? This is the key technical challenge: identify the core physical constraints underlying the previous technologies that have been lifted by the new invention, and assess the significance of lifting those constraints. Consider an example from the field of aircraft engines. In the piston engine, the upper limit in compression ratio is set by the detonation of the fuel charge in the cylinder. A turbine engine has no such limit; it is possible to have a higher compression ratio in a turbine engine than in a piston engine, and today's successful turbine engines do have those higher compression ratios.

What new technical constraints are inherent in the new art? The first question had to do with the credit side, and this question determines the debit side—that is, what fundamental constraints limit the effectiveness of the innovation. In jet engines, for example, the wake efficiency or the Froude efficiency of an aircraft propulsion system depends on the ratio of the velocity of the rearward stream of air to the forward velocity of the aircraft; the lower that ratio, the more efficient the propulsion. A propeller, moving a large amount of cold air slowly, has a higher wake efficiency than a jet engine moving a small amount of hot air rapidly, so a new constraint exists.

How favorable is relief of the former weighed against the stringencies of the latter? The net of the first two questions with respect to any inventive concept is a qualitative technical balance. The comparisons cannot be quantitative because they are not necessarily of similar characteristics; so this is highly judgmental balance, but it can be technically quite meaningful.

Putting Innovations in Context

The second stage of applying these management criteria is to analyze the embodiment in which the new technology will go to market. Here again the analysis takes the form of answering three questions:

Is the end product enhanced by additional technology and components required to make use of the innovation? This question calls for an analysis of the changes which must be made to a product if the innovation is to be used in it. A good example occurs in radios. Every radio must have a power supply, an R-F section, an I-F section, and an audio section. The transistor penetrated the automobile radio as a replacement for the output power tube, but the R-F and I-F sections of the radio were unaffected by the presence of a germanium power transistor instead of an output power pentode; however, because the output stage no longer required 300-volt B^+ plate potential, it was possible to eliminate the unreliable vibrator power supplies as soon as R-F and I-F tubes requiring only 12-volt B^+ potential were developed. These hybrid radios were much more reliable, but they could not have succeeded without the 12-volt tube development.

Is the inventive concept itself diluted or enhanced by the embodiment required? Now analyze the effect on the innovation itself of the changes required for its use in the product. There are favorable cases where the additional art surrounding the new invention enhances its value; that is a very happy situation. But there are many cases where the embodiment surrounding the new art dilutes it.

Does the additional embodiment offer opportunity for further inventive enhancement? Once more a balance is needed, this time between the value added to a product and that subtracted from it by the requirements of the innovation. Does it add more punch, or does the new innovation on balance decrease the acceptability of the product into which it is incorporated?

A Balance Sheet on Financing Business Operations

The answers to these two sets of questions establish the technical potency of a new innovation, but they offer no criteria in the business context from which to judge such things as profit potential. Three questions are also involved here:

What previously emplaced business operations are displaced or weakened by the new innovation? Assess the potential changes in existing business that will be brought about by the innovation. In the case of the entry of the transistor into electronics, the impact on the dealer service network is a perfect example. All tube sets were guaranteed to have tube failures and to require tube replacement. That phenomenon simply does not exist in transistor devices, and the dealer service network predictably declined in importance as transistors were introduced.

What new business operations are needed or wisely provided to support the new innovation? Assess the nature and cost of new business operations required by adoption of the new technology. The Japanese penetration of the U.S. consumer electronics market provides an example; it was simultaneously an innovation in distribution and retail marketing. Trading companies were now the distributors, and retail sales by stores, department stores, and discount houses replaced the previous pattern of selling which focused on the brand-franchise dealer.

How favorable is cessation of the former practices weighed against provision of the latter? Draw the trial balance again; we claim that this analysis can yield a qualitative business balance stemming from the new invention.

What Will Sell and What Won't

Finally, we determine a set of criteria having to do with market dynamics, on the basis of three questions:

Does the product incorporating the new technology provide enhanced effectiveness in the marketplace serving the final user? The Pilkington Float Glass Process represents a substantially more effective way of making plate glass by casting the glass against molten tin instead of grinding the surfaces; there are dramatic cost savings. But smooth glass is smooth glass, and there is no increased effectiveness resulting from its use in windows. The process is a perfect example of an innovation which made no change at all in the marketplace; it does not, unlike many others, yield economic payback because of market expansion due to enhanced effectiveness.

Does the operation reduce the cost of delivering the product or service? Taken together, this question and the one above are really the scissors of supply and demand, the first dealing with demand, the second with supply. If the answer to both of these questions is "no," we can forget the whole thing; and if it is "yes," then there need be little market uncertainty. The challenging case is where one of these is positive and the second is not.

Does latent demand expansion or price elasticity expansion determine the characteristics of the new markets? When the factor driving a market area is lower price per unit, market expansion by hundreds of percent is hard to obtain; major expansion in revenue is much more likely when the change in the market is driven by a dramatic change in product effectiveness. This final question, of course, determines the quantitative business balance.

In these criteria, we have avoided terms such as "return on investment" and "return on assets managed." Our view is that these issues are overwhelmed whenever a new inventive concept can be placed in a beneficial embodiment which will enhance its value in a major latent market with lowered operational costs. If evaluations of an innovation must be based on assumptions of narrow differences in return on investment, they are quite possibly based on fallacy. What we propose is a logic structure to identify a small class of innovations of great promise whose success will transcend the cash value of any normal investment.

II. PUTTING THE DETERMINANTS TO THE TEST OF HISTORY

Some examples from our work show how this logic structure might have served as a meaningful discriminant between successful and unsuccessful innovations in the past.

The Japanese Portable Radio Game

Consider first three aspects of the changes in the consumer electronics market wrought by the transistor. Transistor radios made their first strong showing in the U.S. portable radio market in 1956; prior to that time all portables had been tube sets. In 1956, the center of gravity of the U.S. market (according to listings in *Consumer Reports*) was in personal-size transistorized portable radios weighing an average of 20 ounces and costing an average of $57; eight models were available. This was a substantial innovation; the dominant radio in the previous year's market had been a tube-based portable weighing about six pounds. No Japanese sets were in the U.S. market in 1956.

By 1959, the U.S. industry had responded fully to the transistor as it was then applied; there were 25 portable models in the 20-ounce size, and prices were down slightly.

But in 1959, there was also a new market never populated before, filled by 11 different Japanese miniature portable sets weighing about 10 ounces (less than half as much as the smallest U.S. radios then available) and costing 10 percent less than the 20-ounce U.S. radios. The Japanese had made a dramatic innovation in the size and weight of personal portable radios and thus had opened up a new market not serviced by the U.S. industry.

The Japanese did not do this by innovating in transistors; they did it by other innovations through which they reduced the sizes of tuning capacitors, loudspeakers, battery supplies, and antennas. It was the transistor innovation supported by these additional innovations that allowed the Japanese to open up this exclusive new pocket-radio market and begin their successful penetration into U.S. consumer electronics. In fact, the U.S. industry's innovation based on the transistor was an incomplete innovation; it had not taken advantage of the embodiment surrounding the transistor as the Japanese had done.

By 1962, the market outcomes were clear. In 1955, the U.S. market for portables was 2 million sets, all tubes, and all made in the United States. By 1962, the Japanese had captured 58 percent of the market, and they had in fact captured 68 percent of the market growth made possible by the transistor.

Auto Radios and TV: Who Needs Transistors?

Compare this history with that of the second transistor innovation, which was in auto radios.

In 1955, the only auto radios were tube sets; they used high-voltage R-F, I-F, and output tubes, and of course they required vibrator power supplies.

In 1956 came a new type of auto radio with no vibrator power supply, no step-up transformer, and no high B+ potential. Germanium output transistors allowed low power drain; they were driven directly by the negative-ground 12-volt power supply of the automobile battery. In addition, an embodiment innovation provided 12-volt B+ tubes to handle R-F and I-F.

Only one year later, the first all-transistor auto radio, completely transistorized in R-F and I-F and with low-drain output transistors, all running directly on the battery supply, became available.

On average the tube sets of 1955 cost $45, the hybrid sets cost $8 more (a modest step-up), and the completely transistorized sets of 1957 cost $125, a luxury prestige item in the Cadillac Eldorado but otherwise priced out of the market.

The production of auto radios follows very closely the production of automobiles. In the mid-1950s, 67 percent of new cars went to market with radios. Since then this figure has increased steadily—no sudden changes in auto radio use have been associated with the new technologies—until now it is almost 100 percent. There were 6.86 million auto radios made in 1955 and only 6.43 million in 1960; these figures reflect almost exactly the volume of new-car demand. In this period, the transition from tube sets to the tube-plus-output-transistor hybrid set was essentially completed (fully transistorized radios were still not sold in any meaningful quantity). Yet there was no expansion at all in this market.

Turn now to color television consoles, the last of three transistor substitution innovations we have studied. All consoles marketed from 1955 through 1967 used tubes. The first transistorized console was available in 1968, but only in 1974 (almost two decades after the first transistor portable was sold) did transistorized color sets become the industry standard. Since portability and maintenance cost are dominated by the vacuum cathode-ray tube, transistor penetration was very slow.

Summing a Transistorized Balance

Portable radios, automobile radios, and color television consoles represent three different transistor innovations which were technically very similar in circuit design and in cost. Yet the business outcomes were dramatically different. Our criteria applied to this field reveal the differences in technology and its business context that led to these strikingly different outcomes.

Recall that our criteria reflected first a balancing of constraints removed against new constraints added by an innovation. In the case of portable radios, transistors meant that weight, size, and frequency of repair were all improved dramatically. For auto radios, the vibrator failure mode (which provided 60 to 80 percent of the maintenance engagement) was eliminated and with it the problem of battery drain when the radio was used without the engine running. Frequency of repair was generally reduced. In television, transistors served only to reduce the frequency of repair. As we progress from auto radios to television, the value of constraints lifted by the transistor declines.

New technology always has new problems. The audio fidelity of the little portable radios was poor, and that is a fundamental constraint. Similarly, the capture value of the tiny ferrite antennas was not as good as that of the bigger antennas. There were no particular penalties in auto radios or in television.

Drawing a technical balance on the value of these inventions, we conclude that the portable radios presented a vast increase in portability—from six to eight pounds down to shirt-pocket size. The auto radios offered a major increase in reliability. Transistorized television sets only had a slight increase in reliability going for them. We conclude that the portable radio had dramatic new value, the auto radio had substantial new value in its hybrid mode, and the television set had very little new value.

Next, what additional technology was required in the embodiments in which transistors were placed to realize their full potential? Small tuning capacitors, loudspeakers, antennas, and batteries were crucial to the reduced size of portable radios. If these had not been included, the value of the transistor innovation would have been diluted, since in fact tubes were not the fundamental limit on the size of portable radios. As the Japanese demonstrated—to the detriment of their American competitors—only if you spent extra money to miniaturize all the other components could you fully capitalize on the transistor invention itself.

On auto radios, the additional innovation required for auto radios were the 12-volt tubes for R-F and I-F sections, and these were rapidly achieved. This allowed elimination of the vibrator; without that, the transistor innovation would have been diluted and transistorized auto radios might not have been successful.

An interesting situation prevails in television. As far as we can imagine now, a television set requires a cathode-type picture tube. Even after 20 years of looking, we find the prospects dim for a cheap, all-solid-state display. Therefore, we are nailed to the low reliability, high repair cost, and large size of present picture tubes. There is essentially no opportunity for incremental enhancement in the embodiment surrounding the transistors.

As we compare these different applications of the transistor on the basis of these criteria statements, we can understand why the portable radio transition was complete and rapid, why the auto radio transition was quite rapid for one portion (hybrid) and quite slow for the other portion (fully transistorized), and why the color television transition was quite slow.

On the operational side, the transistor led to some business innovations. In Japan, a wholly new concept of low-cost mass marketing followed the design of nine-ounce radios.

The fundamental change in a key component of auto radios—the output tube became a germanium output transistor—was adopted throughout the industry and led to a new business opportunity. The Motorola Corporation, a radio manufacturer which never made tubes in its life, was effectively able to integrate backwards and started making germanium power output transistors—first for use in its own production of auto radios and also in its military equipment and later as components for sale to others. Thus this change in technology allowed Motorola to expand its role and penetration in the industry. The transistorized television set, having only slight advantages, offered no new business opportunities; essentially it is a null case.

The application of transistors to the portable radio definitely increased the effectiveness of the product to the final user; it now became a

	Effects of the Advent of the Transistor on:		
	Portable Radios	Auto Radios	Television
Inventive concept:			
Constraints lifted	Weight, size, and frequency of repair.	Vibrator failure, battery drain, and frequency of repair.	Frequency of repair.
New constraints	Low sensitivity, low fidelity.	None.	None.
Advantage	Vast increase in portability.	Major increase in reliability.	Slight increase in reliability.
Embodiment merit:			
Additional components	Condenser, speaker, antenna, and battery.	12-volt R-F/I-F tubes.	Cathode-ray tube.
Dilution or enhancement	Dilution of size and weight gains.	Dilution if vibrator required.	Dilution, since no rewards of small size or low weight.
Additional opportunity	Enhancement if above are miniaturized.	Enhancement if 12-volt tubes eliminate vibrator.	No enhancement; elimination of CRT tube seems impossible.
Operational practice:			
Displaced business operations	Dealer service no longer very important.	Service less important.	Service slightly easier.
New business operations	Low transport and inventory cost encourage wide distribution network.	New field of transistor and electronic manufacture opened.	None.
Advantage	Low-cost mass marketing opens new market to imports.	Radio makers integrate backward to transistors, auto makers to radios.	Slight if any.
Market dynamics:			
Enhanced effectiveness to final user	Great increase only in portability.	Slight.	None.
Reduced cost	Higher cost in early years.	Only slightly higher cost, due to vibrator savings.	Much higher cost in early years.
Expansion or substitution market	Expansion in miniature size only.	Substitution.	Substitution.

Assessing the impact of the transistor on consumer electronics. The author proposes that analysis of these 12 aspects of the changes wrought by the transistor demonstrate why its penetration was so instant and its revolution so complete in the portable radio industry, its penetration equally complete but its revolution insignificant in auto radios, and its penetration small and slow in the television industry.

go-with-you-anywhere radio rather than a carry-it-and-set-it-down radio. The combination of the transistor and the further innovation surrounding it led to a new product that claimed a new latent market never populated before, rather than a substitution market. In contrast, the enhanced effectiveness of the transistorized auto radio to the final user was quite slight—a little bit of reliability, a little bit of battery drain. It was a substitution market; in fact, the total demand did not increase at all. There simply was one radio (approximately) for every new car sold. The television case is our null case; nothing happened.

The Battle of the Turbines

Here is a brief review of how our criteria illuminate the different outcomes in the case of turbine aircraft. Here we are dealing with three fundamentally different types of aircraft—the wide-bodied jets (707 and DC-8), the Lockheed Electra, and the Boeing 727. The first jets, as well as the first Electras, entered service in 1958. By 1961, the Big Five U.S. airlines (these firms provided 75 percent of U.S. passenger miles in the late 1950s, and they were historically the airlines which first bought new equipment) had 177 big jets; by 1969 they were using 500 such aircraft. The Lockheed Electra went from an initial 1961 fleet of 72 down to only 28 aircraft in 1969. The 727, which was not even in the first round of purchases, turned out to be the single most effective jet aircraft; by 1969, 400 of them were in service in the United States.

The inventions on which these aircraft were based were largely similar—the substitution of rotating compressors and high-temperature gas turbines for reciprocating pistons. But these aircraft are very different in their ensemble of other elements beyond turbine engines—embodiment merit, according to the terminology of this article. To achieve the full potential for higher speed inherent in the turbine, the 707 and DC-8 used swept wings, at 35° and 30°, respectively. This was possible because Boeing and Douglas engineers solved (each in slightly different fashion) the problem of controlling a phenomenon called Dutch roll, which affected stability. The outcome was that the 707 and DC-8 were fundamentally faster than the British Comet, which was designed with a 20° wing sweep to avoid the stability problem. Thus the opportunity came not in jet engines but in solving aerodynamic problems brought into relevance by jets.

Lockheed engineers were trapped with a dilemma in designing the Electra. They concluded that they were better off moving a large mass of cold air slowly than a small mass of hot air rapidly, because of the runway length requirements of the latter. So they stuck with propellers. But propellers in fact have an ultimate speed limit, since their tips cannot easily go faster than the speed of sound, and this in turn constrains the aircraft to a top speed nearly 200 miles slower than that of a turbojet such as the 707. The Electra designers chose the old, familiar art of propellers, where no risk was entailed but no new enhancement was possible.

The 727, the third member of our jet set, has been successful because of two additional embodiment innovations—fanjet engines and high-wing-lift devices. In order to use the short runways of intermediate-range airports, the 727 had to obtain much more take-off thrust and much more lift from the wing in landing than had been possible before; yet it also required a small wing for cruising at high speed. This was achieved by using fanjet engines (with cold air flow around the hot turbine exhaust) in the rear of the aircraft so that the wing was clean, and by using triple-slotted flaps with leading edge slats which provided the equivalent of a variable-configuration wing. The innovation was effective; the 727 ended up with short-field capability that matches that of propeller aircraft, yet it cruises at speeds typical of all jets.

Market dynamics are the next criteria of importance. The 707 and DC-8 presented a great

	707/DC-8	Electra	727
Embodiment merit:			
Additional components	Swept wings required.	Propellers required.	High-lift wing devices needed.
Dilution or enhancement	Slight roll control problem.	Speed and maintenance constraint.	Clean wing with triple-slotted flaps from rear engine.
Additional opportunity	Speed advance over Comet due to high sweep angle.	None.	Short-field capability matches that of propeller aircraft.
Market dynamics:			
Final user effectiveness	Great advance in speed and comfort.	Only modest gain over piston engine in speed and vibration.	Great advance in speed and comfort.
Cost reduction	Net cost less than long-range piston.	Cost much less than piston planes.	Costs roughly equal to Electra.
Expansion or substitution market	Strong expansion.	Substitution for piston craft.	Substitution for Electra and piston craft.

The battle of the turbine-powered transports. Only 6 of the author's 12 criteria are needed to demonstrate the superiority of pure jets in the marketplace—why the 707, DC-8, and 727-type aircraft became the standard for U.S. air travel in the 1970s. He believes the same kind of analysis can be used to show why innovations leading to a supersonic transport aircraft have much less potential in the U.S. market.

advantage in comfort and speed; they flew higher and faster than any aircraft before, and it was simply more pleasant to travel. In addition, because of their high speed and capacity, they cost less per seat mile to operate than long-range piston planes. So the best of all possible worlds obtained: demand was higher and cost was lower. These aircraft led to a strong market expansion for air travel.

The Electra offered only modest improvements. Some of the vibration coming from piston engines had disappeared, but Electras could not fly as high or as fast as the jets. Operating cost per seat mile excluding depreciation was less than that of medium-range piston aircraft, but this meant a market based on substitution (slowly penetrating by doing the same function against depreciated equipment), not one based on expansion (rapidly penetrating on the basis of payback from new customers).

Finally, the 727 had the speed and comfort of the big jets and costs roughly equal to the Electra, and it could fly in and out of the smaller airports. As soon as the 727 was available, all of the intermediate-range traffic, piston or Electra, went to it.

III. FORECASTING FUTURE INNOVATION

Now that we have seen how the innovation criteria apply to the transistor case and to the case of jet aircraft, it is appropriate to make some general statements about their application to two prospective innovations, microprocessors for automobiles and supersonic transport aircraft.

Automotive Microprocessors: Everything Up

Microprocessors promise flexibility and precision of control and operation of automotive engines that are simply not available in mechanical control systems, and this is the heart of the technical advantage.

At the level of embodiment, we find that the

sensors and actuators, not the computer chips, are the most crucial components requiring further development. An automobile is an analog mechanical environment; a microprocessor is a digital electronic environment. We need either sensors, actuators, and/or analog-to-digital converters, and these are the key embodiment elements. We also know that they represent the dominant portions of system cost and are the dominant determinants of system performance. If these can be made right, there will be regulatory benefits, better driveability (which to the auto industry means desire to buy cars), and long-life, stable performance; automobiles will stay in tune and their control and performance functions will not deteriorate over the life of the car.

These will be major new design, manufacturing, and marketing opportunities, and our operational practice criteria are useful in seeing how to make a business operation out of these possibilities. Absolutely, one would expect that firms such as Bendix or TRW (with aerospace and electronic skills and a large presence in the auto industry) could take advantage of this opportunity to expand backwards (as Motorola did with auto radios) into special electronic precision sensors and actuators, seizing a key part of this technical ensemble for a long-term, stable market. We also have the possibility of car manufacturers expanding their domain of technical activities.

The market that is implied by these prospects for computerized automobiles is absolutely unique. It has been decreed by the U.S. government. Microprocessors and related control systems do not have to be evaluated against the cost of today's mechanical alternatives; they offer the most promising way we can yet envision to meet emission and economy standards mandated for motor vehicles in the 1980s. There is a billion-dollar value to manufacturers in avoiding the fines for high fuel consumption or the preemption of marketability if new automobiles fail to meet pollution standards set by federal law. The value to the auto industry hinges not on new revenue gain—because all the industry can hope to do is continue to sell high-value cars—but rather on the avoidance of loss.

The Fundamental Problem of the S.S.T.

To evaluate prospects for a supersonic transport, one can go through the same four-point check sequence on criteria. The key inventive concept, the thing that is fundamentally new on supersonic transports, is supersonic aerodynamics, the increase of aerodynamic drag at supersonic speed. Two different aerodynamic structures to deal with this problem have been examined in the United States. One (which is now a failed concept for transports) is the swing wing; the other, which survived until the entire transport project was shelved, is a wing swept back at an angle sharper than the Mach cone so the wing is subsonic while the airplane is supersonic.

This is a good concept. But it cannot deal with the fact that aircraft flying faster than the speed of sound always leave a sonic boom below, and the energy required to overcome that lost in the sonic boom results in high fuel consumption. So there is a good concept, but it has some debits.

Now we go to the embodiment criteria. The key regulatory decision was that sonic booms would not be allowed over the United States. So American supersonic aircraft must be efficient at subsonic cruise over the United States as well as at supersonic cruise over areas where boom is permitted. This requires what are called variable configuration engines, operating in bypass mode below the speed of sound and as straight jets above—a corollary innovation. The sharply swept wings present some unique control and stability problems; they lack the inherent stability of conventional wings, behaving much like classroom paper airplanes. There must be active controls, called "fly-by-wire." This is not hard; but creating reliable "fly-by-wire" equipment that will last for the 20-year life of an airplane presents a significant challenge.

A supersonic aircraft requires structures that

go beyond those we have had before, because supersonic flight causes thermal as well as aerodynamic loads. The required composite materials represent a new art which now must be mastered.

Finally, we have a question about pollution: we absolutely know that oxides of nitrogen behave differently in the meteorological system at 65,000 feet than at 25,000 feet. The problem is that we do not know how they behave. If these oxides lead to depletion of the ozone layer, we will have somehow to change engine combustion.

If we understand all these embodiments surrounding the supersonic wing innovations, we can proceed to the issues of operational practice. The key problem here is that the U.S. domestic market has underwritten the basic costs of all major air transport innovations since the DC-3. It will not do so for the supersonic transport; long-range aircraft earn value only in international travel. Our airlines and our manufacturers need to understand what it means to be primarily international; pooling of traffic and manufacturing consortia are probable.

The market outcome is not clear. Is it an expansion market or a substitution market? We already have very effective long-range aircraft; if the only problem were to fly 4,000 to 5,000 miles, supersonic travel would be a substitution market, and the economics would not be optimistic. On the other hand, if the value of time saved is substantial, it is conceivable that supersonic travel could result in market expansion.

Having followed this procedure of drawing orderly balances in the areas of inventive concepts, embodiment merit, operational practices, and market outcomes, I have concluded that, though no single constraint prohibits supersonic transports from being commercially successful, the broad array of concerns says that the mere passage of time will not assure an S.S.T. There must be some urgent national mission to override some of the constraints to their emergence.

We believe our procedure for evaluating the viability and likely outcomes of an innovation can largely account for the differentiated outcomes in high-technology businesses—businesses that are as far removed from each other as transistor radios and jet transports. When these criteria are applied to important potential future innovations, they indicate plausibility for a computerized car, given a reasonable regulatory atmosphere, and they indicate implausibility for many years for a U.S. supersonic transport. We are convinced that a similar analysis can be useful in indicating the likely future course of other projected innovations.

Product Design

Case II–4
Texas Instruments' "Speak and Spell" Product

A. L. Jakimo and I. C. Bupp

In June 1978, the semiannual Consumer Electronics Show was held in Chicago. A few days before it opened, Gene A. Frantz, program manager for speech products at Texas Instruments' Consumer Electronics Group, was putting the finishing touches on a device that closely resembled a colorful, hand-held toy radio or typewriter. It weighed about one pound and measured about 6 × 10 inches. Forty keys were arranged in 4 rows of 10 with command keys across the top and the characters arranged in alphabetical order. In spite of its toylike appearance, the product represented a major breakthrough in microelectronic technology.

Harvard Business School case 9-679-089

It was a hand-held learning aid that could talk.

Its semiconductor memory was programmed with more than 200 words considered by noted educators to be among the most frequently misspelled from beginning spelling through adulthood. Words are selected at random from one of four lists of about 50 each, graduating in degree of difficulty. The selected word is "spoken" electronically—but with human inflection and fidelity—through a small speaker at the top of the case. A child using the keyboard then attempts to spell the word. As the child presses the keys, the machine speaks each letter and displays it on the LED screen. If the child fails after two tries, "Speak and Spell" says: "That is incorrect," and goes on to spell it, pronouncing each letter and the entire word. In addition, several keys are available for word games to further stimulate and encourage learning.

The "roll-out" of Speak and Spell at the June 1978 Consumer Electronics Show appeared to herald a new era of electronically synthesized speech products.

TEXAS INSTRUMENTS INCORPORATED

Speak and Spell was developed by Texas Instruments Incorporated. The company was founded in 1930 as a geophysical exploration company. In the early 1950s, the company launched a determined effort to acquire proprietary expertise in the new field of semiconductor technology. This campaign met with early technical success: in 1956, TI was one of the first companies to produce an all-transistor radio. Unfortunately, the radio was unable to meet with similar success in the marketplace. The company, nonetheless, continued its thrust into the age of semiconductors. By the early 1960s, it had established itself as a leader in the manufacture of integrated circuits. From the mid-1960s through the late 1970s, the company's investments in semiconductor technology bore substantial fruit: in 1964, sales revenues stood at approximately $400 million; by 1978, this figure had grown to $2.5 billion. Recent financial statements are contained in Exhibits 1 and 2.

A major factor underlying TI's sustained growth through the 1960s and 1970s was overall expansion in the worldwide electronics market (see Exhibit 3). This expansion was very much fueled by continuing development of low-cost semiconductor products. But, more important, TI's growth was marked by its capture of dominant market shares through exploitation of learning curve economies and forward integration into end-user products. While virtually all successful semiconductor manufacturers were able, if not forced, to exploit the learning curve's declining manufacturing cost effect, only a handful were able to succeed at forward integration. At TI, forward integration was best exemplified by the company's move into hand-held calculators during the early 1970s; by the late 1970s, TI was the leading manufacturer of such devices. (See Appendix I for more detail on Texas Instruments integration strategy.)

THE LEARNING AIDS PRODUCT LINE

In 1975, Ron Ritchie, later to become a vice president in the Consumer Electronics Group, was instructed by a group of influential retired TI officers to explore the use of calculators in classroom environments. In response to this directive, Ritchie and his subordinates set out to develop a program of packaged activities for demonstrating the calculator as a useful educational device. Working with universities and several schools, Ritchie's group designed a package that consisted of a TI calculator and instructional materials. Like the Scientific Research Association's "SRA Reading Packages," introduced during the 1960s, the TI packages were designed to aid in classroom learning.

EXHIBIT 1 **Consolidated Financial Statements** (in thousands of dollars, except per share amounts)

	For the Year Ended December 31	
	1977	1976
Income and Retained Earnings		
Net sales billed	$2,046,456	$1,658,607
Operating costs and expenses:		
Cost of goods and services sold	1,459,490	1,185,426
General, administrative, and marketing	325,978	265,889
Employees' retirement and profit sharing plans	50,151	44,666
Total	1,835,619	1,495,981
Profit from operations	210,837	162,626
Other income (net)	9,261	23,782
Interest on loans	(9,179)	(8,310)
Income before provision for income taxes	210,919	178,098
Provision for income taxes	94,281	80,678
Net income	116,638	97,420
Retained earnings at beginning of year	530,822	458,153
Cash dividends declared on common stock ($1.41 per share in 1977; $1.08 in 1976)	(32,196)	(24,751)
Retained earnings at end of year	$ 615,264	$ 530,822
Earned per common share (average outstanding during year)	$ 5.11	$ 4.25
Changes in Financial Position		
Sources of working capital:		
Net income	$ 116,638	$ 97,420
Depreciation	$ 108,063	87,290
Provided from operations	224,701	184,710
Proceeds—common stock under options	110	1,263
Other	7,233	3,400
	232,044	189,373
Uses of working capital:		
Additions (net) to property, plant, and equipment	199,283	136,454
Dividends on common stock	32,196	24,751
Decrease in long-term debt	12,193	10,735
Purchase of common stock of the company for employee benefit plans	4,799	13,401
	248,471	185,341
Increase (decrease) in working capital	$ (16,427)	$ 4,032

EXHIBIT 1 *(concluded)*

	December 31, 1977	December 31, 1976
Balance Sheet		
Assets		
Current assets:		
Cash and short-term investments	$ 257,131	$ 293,755
Accounts receivable	334,152	282,251
Inventories (net of progress billings)	214,278	197,647
Prepaid expenses	9,337	9,520
Total current assets	814,898	783,173
Property, plant, and equipment at cost	713,787	606,380
Less: Accumulated depreciation	319,694	303,507
Property, plant, and equipment (net)	394,093	302,873
Other assets and deferred charges	46,053	41,657
Total assets	$1,255,044	$1,127,703
Liabilities and Stockholders' Equity		
Current liabilities:		
Loans payable (international subsidiaries)	$ 38,759	$ 46,103
Accounts payable and accrued expenses	287,909	211,674
Income taxes	85,792	113,421
Accrued retirement and profit sharing contributions	34,113	30,473
Dividents payable	9,582	7,540
Current portion long-term debt	10,416	9,208
Total current liabilities	466,571	418,419
Deferred liabilities:		
Long-term debt	29,671	38,169
Incentive compensation payable in future years	14,184	10,836
Total deferred liabilities	43,855	49,005
Stockholders' equity (common shares outstanding at year-end: 1977—22,814,689; 1976—22,851,443)	744,618	660,279
Total liabilities and stockholders' equity	$1,255,044	$1,127,703

See accompanying notes.

Whereas the SRA packages focused on reading, the TI packages covered mathematics.

One of the initial problems encountered by Ritchie's group was a fear on the part of teachers that calculators would make their students lazy thinkers. The teachers thought students would come to believe that all that needed to be known to add two numbers together was the sequence of buttons to push on a calculator keyboard. TI claimed that its testing experiences proved these fears ill-conceived: Where children used calculators in the absence of any packaged program, their test performance was statistically equal to that of children who had no experience with calculators. But where children used calculators as part of an organized

EXHIBIT 2 Financial Information

	Years Ended December 31									
	1977	1976	1975	1974	1973	1972	1971	1970	1969	1968
	Summary of Operations (thousands of dollars)									
Net sales billed	$2,046,456	$1,658,607	$1,367,621	$1,572,487	$1,287,276	$943,694	$764,258	$827,641	$831,822	$671,230
Operating costs and expenses	1,835,619	1,495,981	1,252,833	1,403,105	1,141,824	860,625	705,094	773,113	769,983	621,917
Profit from operations	210,837	162,626	114,788	169,382	145,452	83,069	59,164	54,528	61,839	49,313
Other income (net)	9,261	23,782	11,971	4,159	6,746	7,178	6,840	4,529	3,936	4,258
Interest on loans	(9,179)	(8,310)	(10,822)	(10,741)	(6,654)	(5,676)	(6,526)	(7,014)	(5,474)	(3,209)
Income before provision for income taxes	210,919	178,098	115,937	162,800	145,544	84,571	59,478	52,043	60,301	50,362
Provision for income taxes	94,281	80,678	53,795	73,179	62,309	36,541	25,755	22,182	26,790	24,038
Net income	$ 116,638	$ 97,420	$ 62,142	$ 89,621	$ 83,235	$ 48,030	$ 33,723	$ 29,861	$ 33,511	$ 26,324
Earned per common share (average outstanding during year)*	$5.11	$4.25	$2.71	$3.92	$3.67	$2.17	$1.53	$1.35	$1.53	$1.21
Cash dividends paid per common share*	1.320	1.000	1.000	.920	.555	.415	.400	.400	.400	400
Common shares (average shares outstanding during year, in thousands)*	22,842	22,933	22,920	22,854	22,691	22,139	22,085	22,072	21,919	21,819
	Financial Condition (thousands of dollars)									
Working capital	$348,327	$364,754	$360,722	$314,302	$306,968	$282,049	$261,398	$210,957	$189,271	$157,158
Property, plant, and equipment (net)	394,093	302,873	253,709	280,449	219,941	154,992	154,954	171,436	182,377	145,835
Long-term debt, less current portion	29,671	38,169	47,530	72,755	67,690	71,373	94,778	86,801	94,595	52,927
Stockholders' equity	744,618	660,279	585,288	541,372	469,337	369,627	328,702	303,236	281,548	253,462
Employees at year-end	68,521	66,162	56,682	65,524	74,422	55,934	47,259	44,752	58,974	46,747
Stockholders of record at year-end	24,438	22,425	21,359	18,977	16,135	15,177	16,210	17,738	17,808	18,649

* Adjusted for stock split in 1973.

EXHIBIT 3 Estimates of Total Value (at the factory level) of Goods Shipped by U.S.-Based Manufacturers (in millions of dollars)

Year	Consumer Electronics*	Industrial and Commercial	Federal Electronics	Total
1973	7,014.2	19,553.0	11,929	38,496.2
1974	6,768.5	22,227.7	12,497	41,493.2
1975	7,186.8	23,603.1	12,910	43,699.9
1976	9,425.2	26,136.0	15,659	51,220.2
1977	12,135.0	30,084.7	16,638	58,857.7
1978	14,030.0	35,048.0	18,210	67,288.0
1979	15,393.6	41,037.6	19,920	76,351.2
1981	15,767.0	50,651.0	20,754	87,172.0
1982	21,402.2	59,908.2	24,460	105,770.4

* Includes domestic-made equipment, domestic-label imports, and foreign-label imports.
SOURCE: *Electronics,* January 1975; January 1978; January 4, 1979.

program, their test performance was statistically better than that of children who had no experience with calculators.

Notwithstanding the apparent benefits of the TI packages, they could not be successfully marketed. Among the reasons cited for this failure was the existence of hundreds of highly autonomous and greatly politicized school districts. Setting up a sales network to market the packages to these districts was beyond TI's expertise. Moreover, the product life cycle of a typical product marketed to school districts was about three to four years; in contrast, the product life cycle for a typical TI product was about 18 to 24 months. Ritchie's attention thus turned from classroom packages containing TI calculators to products that could be marketed through the retail outlets with which TI was accustomed to doing business.

The first such product to which Ritchie turned his attention was a hand-held device that could help children learn the principles of arithmetic. Unlike a calculator which is passive in that it makes no inquiries to the user, the learning device was to be capable of carrying on a "dialog" with a small child. In the words of one person involved with the project, the device was to function like an "electronic flashcard machine." A product of this type would require a light-emitting diode (LED) display capable of presenting "dialog" characters, such as question marks, plus signs, and equal signs, as well as the 13 to 14 "passive" characters built into conventional calculator displays. Satisfying this requirement with 1975 LED-display technology, however, would be cost prohibitive. Not until TI achieved a breakthrough in this technology during 1976 did "dialog" characters become economically feasible. Indeed, this breakthrough allowed Ritchie to develop the "Little Professor"—a hand-held device packaged in a bright yellow plastic housing and slightly larger than a conventional calculator.

While the TI mathematics packages developed a year earlier were retail priced at $100 apiece (including the calculator and the instructional materials), Little Professor was retail priced at less than $20 apiece. More significantly, whereas the packages were designed to be marketed to school districts, Little Professor was to be marketed to consumers.

Soon after commercial production of Little Professor began in August of 1976, TI quickly discovered, much to its pleasant chagrin, that it could not build the product fast enough. Production was increased by a factor of four over a demand estimate made in June of 1976.

Within a year after Little Professor was introduced, further advances in battery, integrated circuit, and display technology allowed the design of a second mathematics learning aid: "Dataman." This product incorporated all the features of Little Professor, as well as seven additional learning activities. By Christmas of 1977, TI appeared to be on the verge of a stunning success with its two learning aids.

Near the time that shipments of Little Professor began, in the fall of 1976, Paul Breedlove, a product manager working under Ron Ritchie, conceived of an idea to build a lost-cost speech synthesizer on an integrated circuit that could be incorporated in consumer products.

Synthetic speech had been around for some time. It is produced by circuits that are essentially an electronic model of the human vocal tract. Just as human speech is created by air impulses passing through the vocal cords and the vocal tract, synthetic speech is generated by processing electronic impulses through a rapidly changing electronic filter. The result is synthetically produced speech that sounds like that heard over the telephone.

Human speech generates two basic types of sounds. The vowel sounds, referred to as voiced sounds, are produced by an air impulse from the lungs passing through the open vocal cords, causing them to vibrate. As the vibrating air impulse moves up through the vocal tract, the shape of the throat, the nasal passages, the jaw, and the lips determine what the pitch of that vowel sound is to be—high or low—and thus which vowel is sounded.

Consonants are represented by unvoiced sounds that are produced by an air impulse being forced through constricted openings along the vocal tract. This generates an air turbulence resulting in such sounds as S, SH, TH, F, etc. Where the sound waves produced by voiced vowel sounds are regularly spaced low-frequency sound waves, those produced by unvoiced sounds are erratic, randomly spaced high-frequency waves.

Speech is first converted into frames of 12 digital codes each at the rate of 40 frames per second. As each frame of a motion picture film stops the action for that instant of time, each frame of speech "stops the action" of the vocal tract for that 1⁄40 of a second or milliseconds. This "action" is then converted into 12 codes that represent the pitch for that instant of sound, the loudness, and 10 characteristics reflecting the shape of the vocal tract at the instant.

To artificially produce speech, the microcomputer within the learning aid generates 8,000 electronic signals per second that act like the air impulses generated by the lungs. At the same time, the microcomputer recalls a speech frame from the memory. The pitch characteristic selects either a periodic pitch signal to represent a voiced sound or a random noise signal for an unvoiced sound.

This signal is combined with a loudness characteristic and then processed through a unique 10-stage lattice filter where it is combined with the 10 vocal tract characteristics. In essence, the filter acts on the electronic signals to accomplish what the vocal tract does to the air signals from the lungs. The result is passed through a digital-to-analog converter and out through a small speaker.

Breedlove had in mind a major breakthrough in electronic speech synthesis. He wanted to design a single LSI chip that, in addition to the logic circuitry needed to accomplish speech synthesis, would also include memory on the order of 256K bits and circuitry from innovative new modes of user interaction. He applied for a $25,000 "IDEA grant."

"OBJECTIVES, STRATEGIES, AND TACTICS" (OST) AT TEXAS INSTRUMENTS

The organizational structure at Texas Instruments combined a traditional product-line hierarchy with an array of management systems known as "Objectives, Strategies, and Tactics" or, simply, "OST."

TI's 32 operating divisions (ranging from $50 million to $150 million each) were divided among four major groups: Digital Systems, Consumer Electronics, Equipment, and Semiconductors.

Digital Systems encompassed three major product areas: Computer Systems, Terminals and Peripherals, and Printers.

The Consumer Electronics group covered four major product areas: Specialty Products, such as hand-held learning devices; Consumer Calculators; Professional Calculators; and Time Products, such as digital watches.

The Equipment group was primarily responsible for radar and other electronic hardware for military applications.

Finally, the Semiconductor group served as TI's source of integrated circuits. This group manufactured and sold their wares to TI's other groups, as well as to outside original equipment manufacturers.

Each operating division was divided into Product Customer Centers ($10 million to $100 million each). As of 1978, there were 80 such units. Roughly equivalent to the product departments of conventional product-line structured firms, many PCCs had their own engineering, manufacturing, and marketing units; although assistance was also available from corporate-level staffs. The latter included a high-powered Corporate Research Laboratory.

Overlaid across the operating hierarchy was OST. In the early 1960s, Patrick E. Haggerty, then president of TI, diagnosed the company's susceptibility to a classic business problem: in the absence of proper incentives, operating managers will concentrate on the profitability of existing products at the expense of developing products for future growth. Haggerty conceived of OST, a management system that would hopefully immunize TI to this potential malady. In the late 1960s, his successor, Mark Shephard, Jr., turned the concept into a working reality by creating a dual reporting structure. In addition to maintaining the conventional product-line operating hierarchy, a separate OST hierarchy was established.

At the bottom of the OST pyramid were specific R&D projects, called Tactical Action Programs (TAPs). Each TAP was headed by a TAP manager. Related TAPs were grouped into Strategies, which were managed by Strategy managers. In turn, groups of related Strategies formed Objectives. Each Objective was headed by an Objective manager. In 1978, OST consisted of 250 TAPs, 60 Strategies, and 12 Objectives.

In addition to creating a dual reporting structure, Shephard split TI's annual spending budget into two separate funds: an OST fund and an operating fund. While the operating fund was managed by the conventional operating hierarchy, the OST fund was placed under the control of a corporate committee charged with the task of setting R&D spending guidelines through the use of a zero-based budgeting system. On an annual basis, each Strategy manager was required to submit a prioritized list of funding proposals for TAPs to be carried out during the following year. In March of each year, a strategic planning conference was held in Dallas. Attending this meeting were TI's 500 managers as well as the board of directors. Together, the managers and the board decided on the corporate plan and the allocation of OST funds. In March of 1979, over 400 TAPs proposals were expected. A table showing the growth of TI's R&D budget is contained in Exhibit 4.

OST was connected to the operating hierarchy in two ways. First, most TAP managers were simultaneously PCC managers. Similarly, most Strategy managers were simultaneously division

EXHIBIT 4 Research and Development and Capital Expenditures at TI (in millions of dollars)

Year	R&D	Capital
1975	50.0	67.4
1976	71.7	137.0
1977	95.7	200.0
1978	113.0	300.0

SOURCE: *Business Week*, September 18, 1978, p. 66.

heads. This link, the company contended, afforded an opportunity to avoid the classic "hands-off" problem of moving products from R&D into commercial production. Second, OST allowed a TAP to be pulled together from whatever PCCs were required; thereby affording TI a substantial amount of agility and speed for entering new businesses.

The purpose of the dual OST/operating structure was to ensure that managers would not underplay or postpone long-range strategic programs in lieu of short-range profits. Nevertheless, the OST bureaucracy had a tendency to miss ideas proposed by people deep within the corporation. In 1973, TI attempted to overcome this problem by creating IDEA—a mini-R&D program designed to avoid the massive presentation and documentation procedures required in OST. In 1978, approximately $1 million was allocated to IDEA. These funds were administered by 40 senior technical staffers; each had authority to grant up to $25,000 to employees with notions for product or process development.

DEVELOPMENT OF SPEAK AND SPELL

During the fall of 1976, Breedlove received his $25,000 IDEA grant and a design team was assigned to the program: Gene Frantz as program manager and Larry Brantingham as MOS–LSI chip designer. One of their first hurdles would be gaining the support of TI's Corporate Research Laboratory, for therein worked Richard Wiggins, one of TI's experts at designing large-scale integrated (LSI) circuits for consumer products. They could not proceed with their speech synthesis concept without Wiggins' assistance.

The design team which now included Wiggins spent December of 1976 feeling each other out. By the end of the month, they had reached an agreement: Working together they would attempt to develop Speak and Spell, a talking device designed to help children learn how to spell. Demonstrating the technical feasibility of such a product to upper management would be their first joint task.

To establish the technical feasibility of Speak and Spell, two steps were necessary. First, the sequence of steps needed to be performed by the product's circuitry would have to be well defined. For lack of a better term, we will call such a sequence of steps a "product logic." After being well-defined, the product logic would have to be proven effective. To demonstrate the effectiveness of a product logic, the latter would have to be first incorporated in a computer program for simulation by a general-purpose computer. An important objective during the computer simulation phase of the project was designing a product logic that would produce a voice capable of being understood. As Frantz noted: "Usually, a single word pronounced out of context is hard to understand when uttered by a human being, let alone an electronic device. And this is especially true with single syllable words, which become their own context."

The second objective to be achieved during the simulation phase was setting forth a convincing case to upper management that the logic used for the computer simulation could be put into a chip set consisting of only two or three chips.

At the end of March 1977, a computer simulation of Speak and Spell was presented to upper management. The first word uttered by the TI 980 computer used for the presentation was "ovoviviparous." While the simulation's voice was sufficiently clear, the probability of being able to incorporate the logic into a small, relatively inexpensive set of integrated circuits was not. The logic employed by the Brantingham-Wiggins team required 200,000 additions and 200,000 multiplications be performed in one second. Mostly on faith, management gave Frantz the go-ahead to take the project one step further: the development of a "breadboard" system. During this next stage, the Speak and Spell product logic would be converted into a circuit logic.

Once the design of a circuit logic is completed, it must be shown to work. One way of accomplishing this would be to manufacture a system of integrated circuits which would incorporate the precise circuit logic to be proven. The fabrication of such a specific set of ICs, although the ultimate goal, would be quite expensive and time consuming. Moreover, there would be no guarantee that the initially proposed circuit logic would perform according to expectation. In the result of a need for circuit modification, a completely new set of ICs would have to be fabricated, thereby costing more in precious R&D dollars and irrecoverable engineering time. An alternative to this method of "fabricate and test" is the "breadboard" system. The development of the latter calls for putting together a group of existing ICs which contain all the necessary gates but which need to be wired to each other to emulate the circuit logic. The device on which the ICs are linked to each other consists of an array of sockets which facilitate assembly and disassembly of various circuits from the group of existing ICs. This device is known by electronics engineers as a "breadboard." As characterized by Brantingham, "the gate-by-gate breadboard circuit offers the first chance to estimate the size and price of the barn."

DESIGN TO COST

In June of 1977, the Speak and Spell project was formally transferred from the IDEA program to the OST system. At that point, Frantz was at work trying to answer two interrelated questions: First, what price could TI hope to draw from consumers for the Speak and Spell? This information had a direct impact on the design of the circuit logic, since different designs would entail varying manufacturing costs. Second, would the Speak and Spell chip system consist of two chips, or three chips?

> The design-to-cost concept must be made to permeate Texas Instruments. Its application must become second nature not only to those who design, but to those who manufacture and market. It must be applied not only to high-volume products, but—in adapted form—to limited-volume products. And it must be applied not only to the products and services we sell, but to our internal functions as well. Design-to-cost principles, creatively adapted, will make a major contribution to the achievement of TI's objectives.[1]

Design-to-cost is an engineering discipline that establishes, as a primary design parameter, a timetable of unit production costs derived from a timetable of unit prices at which the product can win and retain the desired market share.

Frantz knew that three features of this statement were of special importance: First, unit cost is a primary design parameter, equal in importance to performance. Second, in the case of high-volume markets, this cost parameter takes the form of a timetable of steadily decreasing costs spanning the entire lifetime of the product. This cost timetable is established long before the product goes into production. But since experience curves are volume dependent rather than time dependent, these cost goals are subjected to continual revision as the

[1] J. Fred Bucy, president, Texas Instruments. (See Appendix II for further remarks by Mr. Bucy.)

future unfolds and as volume requirements become more predictable.

The third feature is the heart of the concept. The cost timetable is derived from a price elasticity curve. Preparation of this curve begins with an exhaustive analysis of the number of units the market will be willing to buy at specific prices during the life of the product.

Frantz also knew that while the design-to-cost concept was simple common sense, its application was uncommon. Too often in high-technology areas, design practice is merely to advance performance capabilities, total up the costs, mark these up by the normal margin, and then hand the product over to marketing with instructions to "sell it" (Appendix I). But customers are not interested in technology for technology's sake. Customers want quality, reliability, and performance at the lowest feasible price.

Normally the first step of the design-to-cost approach would be to determine basic performance requirements: What does the potential customer want the product to do? Then, when dealing with markets where the volume is elastic, it is necessary to try to determine price elasticity: How will the market expand as the price of the product is lowered—that is, how much are how many customers willing to pay for this product?

After this careful market study has established prices and volumes, profit margins necessary to finance growth and reward stockholders can be applied. In this way, working backwards from price, unit production cost objectives are established. The price elasticity curve also serves as a basis for determining the production volume and price at which a significant share of market can be obtained. Taken together, these estimates of performance, price, cost, and volume give a definite target for product design.

The next step is to apply experience-curve theory to determine future manufacturing costs. Working with prices and volumes from the price elasticity curve, required costs versus cumulative production are plotted. These points will rarely plot as a straight line, so the experience curve must be drawn through two important checkpoints: The cost of the first significant manufacturing run, and the required cost of the product at maturity.

Finally, the experience curve is translated into target costs by date, by estimating the market volume growth in terms of units per quarter. In this way, the experience curve becomes the cost timetable. It shows the slope down which costs must be driven. Adding gross profit margins yields a price timetable—which must also be checked against the analysis of prices the market will be willing to pay.

However, to answer the price question for Speak and Spell, Frantz was faced with an immediate problem: how would he go to the marketplace to determine the need and affordable price of a product nobody had ever seen or conceived of? He was forced to rely on a conventional TI market research tool: the focus group. A focus group consists of a panel of consumers asked to assess what they would pay for a given product. Ordinarily, focus groups were given the actual product, or a reasonable facsimile, to evaluate. But, in the case of Speak and Spell, no such product nor reasonable facsimile existed. The need to convince potential consumers that the product was not another "pull-string" toy was crucial. As a substitute for the actual device, Frantz decided to use a combination of a tape recording, an artist's rendering, and an oral description for focus group presentations made during the summer of 1977. Preliminary analysis of the responses from consumers indicated that a successful retail price would lie somewhere between $40 and $50.

Turning to the second question, Frantz noted that he initially thought Speak and Spell would use a two-chip system: one chip for the controller and synthesizer, the other for storing the machine's vocabulary. (See Exhibit 5

EXHIBIT 5 Alternative Circuit Designs for Speak and Spell

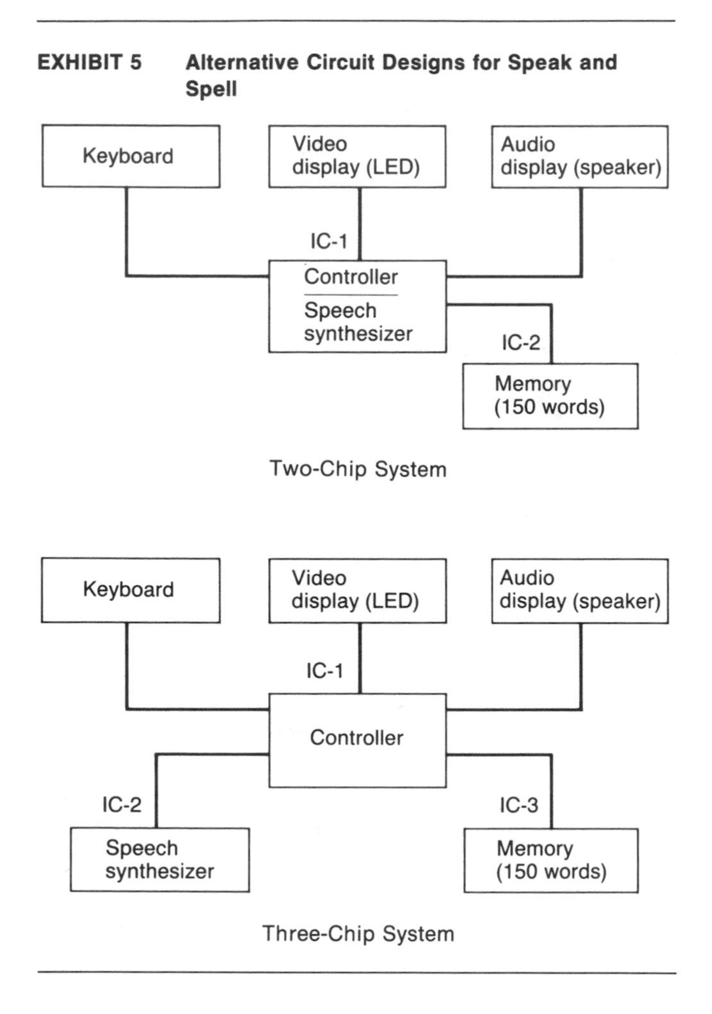

for an illustration of the possible internal designs of Speak and Spell.) Some of Frantz's engineers, however, argued for separating the controller and synthesizer into two separate chips. They argued that a separate synthesizer chip could be put to use in several other applications, e.g., warning devices for automobiles and aircraft. The existence of other uses for a speech synthesis integrated circuit would allow its R&D costs to be spread over a wider base of products. Moreover, other applications would mean increased profits for TI as a whole. The probability and extent of being able to spread R&D costs and of increased returns from expanded applications of a separate speech synthesis chip had to thus be factored into the decision between a two- or three-chip system.

The advocates of the three-chip system also argued that a three-chip system would involve significantly less risk in the manufacturing process. Their contention was based on the fact that a separate controller and separate synthesizer would be easier to fabricate than a more complicated controller/synthesizer.

The only way Frantz could evaluate the argument that there would exist other demands for a speech synthesis chip was through interdivisional communication. After several months of positive feedback from other TI divisions, Frantz was convinced of the need for flexibility: a three-chip system was adopted.

By October of 1977, Frantz had the Speak and Spell breadboard system operating. The next phase would be the design and fabrication of a chip set that would incorporate the circuitry worked out on the breadboard system. This last step took six months. By late April 1978, Frantz was being congratulated as the father of a new line of voice products. With the semiannual Consumer Electronics Show only a month away, he hurried to put the finishing touches on the prototype to be rolled out at the show.

SUPERVOICE

In developing Speak and Spell, a large part of Frantz's efforts was focused on converting TI's laboratory knowledge of speech synthesis into a marketable consumer product. Yet, in commenting on the hurdles that he and his colleagues had to surmount, Frantz noted that "the real problem was never designing the product's logic and eventually incorporating it in a two- or three-chip integrated circuit set. Rather, it was finding a voice that could satisfy three basic requirements: quality, clarity, and amenability to being digitalized."

Voice quality is simply the characteristic of a voice which makes it pleasing to listen to; voice clarity is the characteristic which renders a voice easily understood in terms of pronunciation; and, amenability to being digitalized is the characteristic which allows a voice to be electronically synthesized such that the least amount of information needs to be processed per unit of time. It was Frantz's hope that one voice could satisfy all three requirements in an optimal way.

The search for this optimal voice, i.e., "Supervoice," began very early in the Speak and Spell development program. The search consisted of taking hundreds of tape recordings of different voices and then subjecting them to the three requirements. The search stretched from professional speakers in Dallas to the Shakespearian actor whose voice was used for the computer "HAL" in the movie *2001: A Space Odyssey*. As intuitively expected, TI's researchers learned that a normal friendly voice was most readily accepted by a small child. More important was the finding that children were indifferent to the sex of a voice. This information was warmly received by Frantz, since synthesizing a male voice required less information processing per unit of time than synthesizing a female voice.

Through trial and error, Frantz finally decided to use the voice of a Dallas disc jockey for the Speak and Spell prototype to be introduced in June 1978. This voice was also used in the fall 1978 production runs of the product. Yet, by the start of 1979, Frantz was not entirely certain that the disc jockey's voice was as close to "Supervoice" as possible. Indeed, as of January 1979, TI was still looking for voices to be used in synthesizer circuits of the future.

EDUCATIONAL MERCHANDISING

As TI gained merchandising experience with its hand-held calculators, it learned that the written informational materials included with its products were extremely important. *Owners Manuals* and *Programmed Learning Guides* became important tools in preventing returns of calculators thought by their owners to be faulty, and in building consumer loyalty to TI. The importance of these materials was reflected in the establishment of an Education and Communication Center. Managed by Don Scharringhausen, who reported directly to Ron Ritchie, the ECC was directly responsible for preparing the consumer literature which accompanied TI's products.

In the first half of the 1970s, the consumer

literature accompanying a specific product, such as a simple four-function calculator, was largely prepared after the product itself was fully developed and ready for production. The move into more sophisticated programmable calculators and learning aids, such as Little Professor, Dataman, and Speak and Spell, however, brought a change in the process. As the importance of *Owners Manuals* and *Programmed Learning Guides* increased, the drafters of this literature became more involved in the actual design of the products. For the learning aids line, a separate Product Customer Center was set up in the Education and Communication Center. This PCC, Educational Merchandising, was managed by Ralph Oliva, a Ph.D. in solid-state physics from Rensselaer Polytechnic Institute.

Oliva was responsible for the "learning factors engineering" underlying the development of TI's learning aid line. "Learning factors engineering" was the process of bringing three points to bear in the development of each learning aid: first, the operation of each learning device had to be easily understood by its potential user; second, the product itself had to be able to telescope its value to prospective purchasers viewing the item in a retail outlet; and, third, the learning procedures carried out by the device had to be educationally sound.

To achieve these three goals, it was necessary for Oliva to assert some sort of balance to the design engineers responsible for the development of the actual hardware. For example, in the case of Speak and Spell, Oliva had to take charge of the list of words which would be programmed into the memory of the device. Although this assumption of power by Oliva was by no means a major matter, it marked the transition from the era in which all product design matters were handled exclusively by the semiconductor engineers.

To assure that TI's learning aids were each educationally sound, Oliva enlisted the support of outside authorities. Using these authorities raised an important question: viz, should the names of education authorities be displayed anywhere in the literature included with the product or in advertisements for the product? If this question was answered affirmatively, TI would have to worry about royalty payments as well as lump-sum consulting payments. For TI's first five learning aids (Little Professor, Dataman, Speak and Spell, Spelling B, and First Watch) it was decided that the authorities consulted in the development of the products would not be mentioned in product literature or advertisements.

In addition to worrying about the role of authorities in the development and merchandising of learning aids, Oliva was confronted with two broader questions. The first covered the area of market research; the second, the area of merchandising.

As noted above, TI used several focus groups during the development of Speak and Spell to acquire a feel for the market's interest in the product. Based on this research, TI decided the retail price for Speak and Spell would initially be set at $55. At this price, however, TI discovered that it could not make Speak and Spell fast enough to keep up with demand. This fact was consistent with reports in January 1979 from manufacturers of toys employing integrated circuits which were being sold during the 1978 Christmas buying season. These manufacturers, as well as the semiconductor manufacturers of the integrated circuits being used in the toys, felt that at least a million more products could have been sold during the holiday buying season if demand had been more accurately estimated and production appropriately adjusted. To avoid a recurrance of this problem, Oliva was concerned with developing a more reliable mechanism for predicting demand for future learning aids.

Another question confronting Oliva involved the marketing of learning aids. Some observers outside TI wondered if the company would try to use the route of school textbook publishers: viz, market the devices to school districts. TI's earlier experience with the mathematics packages, however, seemed to suggest that the school district route would not be a good one for TI's learning aids. To market to school districts, a network of sales agents would have to be set up. Oliva wondered if there was enough margin in the products to support such a system. An alternative was getting into a joint venture with a textbook publisher with a sales agent network already in place. At the start of 1979, Oliva viewed such a venture as possible, but in need of further analysis.

PRODUCTION PROBLEMS

As noted above, TI's Semiconductor Group supplied its end-user product divisions with integrated circuits and other semiconductor components. This sole source of supply presented the Speak and Spell Product Customer Center with a couple of problems. First, all integrated circuits developed by any PCC in the Consumer Product Group had to meet standards set forth by the Semiconductor Group. Moreover, since any chip design modifications would entail modifications in the chip manufacturing process, all such design changes had to be approved by the Semiconductor Group. The existence of a modification/approval bureaucracy meant some reduction in the speed with which products like Speak and Spell could be developed and improved.

To some degree, the inability of the Consumer Product Group to manufacture Speak and Spell fast enough was due to an inability to get the Semiconductor Group to manufacture the needed integrated circuits fast enough. As Larry Brantingham, the principal integrated circuit design engineer for the speech synthesizer chip, stated: "To the Semiconductor Group, we (the Speak and Spell PCC) are just another customer." Indeed, the Semiconductor Group serviced hundreds of customers—end-product PCCs at TI, as well as outside original equipment manufacturers.

One way of overcoming the sole source of supply problem presented by the Semiconductor Group was backward integration by the Consumer Product Group.

Quality control presented Gene Frantz with another production problem. During most of the 1970s, the consumer products manufactured by TI were inspected by performing visual tests of the products' displays. On a typical calculator assembly line, for example, 10 percent of the people on the line would be employed at checking product displays for expected results of pushing various buttons on the calculator keyboards. Through the experience of assembling millions of items, TI learned that the vision of inspectors viewing displays for several hours will gradually grow more critical. As the inspector's vision became more critical, he or she would throw increasing numbers of good products onto the scrap heap. To counteract this "false negative" phenomenon, TI learned how to calibrate its effects into the production economics at hand.

In the late 1970s, TI began a concerted effort to automate its consumer product assembly lines. As one line manager stated: "We want to go from making one calculator in x seconds to x calculators in one second." Part of the automation effort involved the use of computerized video test systems. The latter were made possible through technological improvements in electronic optical character reading devices. A computerized video test system consisted of a video camera which watched calculator displays respond to bursts of air trained on the keyboards. The camera would feed the resulting information on the displays to a computer which could determine if the calculators were functioning properly.

Speak and Spell, of course, required audio testing in conjunction with visual testing. Therefore, TI was precluded from using the computerized test systems on the Speak and Spell devices. Instead, human inspectors had to be employed. This, however, presented TI with a problem it had never before experienced. Whereas video testing by humans would result in increasing numbers of false negatives, audio testing by humans would result in increasing numbers of false positives. The reason for the false positives was due to the fact that the human ear will acclimate itself to the same sound heard repeatedly over a period of time. Thus, as a typical day wore on, the average inspector would approve increasingly more Speak and Spell devices which, if heard at the beginning of the day, would have been rejected. With Speak and Spell, therefore, the quality control process had to be entirely recalibrated.

FUTURE OF LEARNING AIDS AND SPEECH PRODUCTS

By the end of 1978, TI's line of learning aids consisted of five products: Little Professor, Dataman, Speak and Spell, Spelling B, and First Watch. The Spelling B product was a derivative of the Speak and Spell; it performed many of the learning games programmed into Speak and Spell, but did not include speech synthesis. First Watch, as its name suggests, was developed as a device to help children learn about time. Fortuitously, TI learned that small children enjoyed pushing the button on an LED digital watch and viewing the result. In contrast, adults very much disliked the need to have to push a button on a digital watch to learn the time. With the advent of liquid crystal displays (LCDs) for use in digital watches, the adult segment of the digital watch market rejected the LED digital watch. Instead of facing a mature, or declining, market for LED digital watches, TI was able to make use of them in the First Watch Product.

After three years of growing sales in its expanding learning products line, TI was determined to increase the range and capability of such products. One way of improving the capability of such products was to increase the amount of information contained in their memory chips. Rather than increasing the number of memory chips to be simultaneously attached to a learning device controller, TI decided to develop a line of memory chips which could be plugged into the learning products and bought separately. For example, each of the memory chips in the Speak and Spell contained about 150 words. With plug-in memory chips, to be used one at a time, the vocabulary could be expanded to thousands of words. The development of a line of plug-in memory chips presented more of a marketing problem than a technological one—TI had been selling plug-in modules containing programs for use in its line of programmable calculators as early as 1977.

One of the ways by which TI hoped to expand the breadth of its speech products line was through the development of devices that could understand voices as well as synthesize them. Speech recognition, however, required five times as many calculations per unit of time as did speech synthesis. As of 1979, further development in digital speech technology and integrated circuit technology would be needed before speech recognition could become a commercial reality.

Perhaps, most important, learning aids would help bridge the gap between an era of passive, hand-held digital devices and an era of very sophisticated home computer systems. In the late 1970s, predictions were made that sales of home computers would reach x million units by the early 1980s. Some observers, however, cautioned that many of these sales would be to small businesses who viewed such devices as relatively inexpensive substitutes for the minicomputers being marketed by traditional computer mainframe manufacturers. To sell home computers to households, a long period of consumer education and familiarization as to their uses would be necessary. Some people at TI thought that learning aids would help perform this function.

APPENDIX I

Excerpts from a paper presented by Dr. Morris Chang of Texas Instruments at a conference on "Tomorrow in World Electronics," London, 1974.

VERTICAL INTEGRATION: COMPONENTS TO SYSTEM

WHY DO SEMICONDUCTOR MANUFACTURERS INTEGRATE UPWARDS INTO THE SYSTEMS BUSINESS?

Increased Value Added

The most important motivation is the opportunity for higher growth resulting from increased value added for products where the principal function is already performed by semiconductors. Figure 3 shows the semiconductor content in some of the electronic equipment. Consumer calculators have a semiconductor content of 30 percent to 35 percent. This means, of course, that if the semiconductor manufacturer sells $15 of semiconductors to a calculator manufacturer, he gets only $15 of sales, but if he makes and sells the calculator himself, he gets $45 of sales. The same magnitude of increase in value added exists in minicomputers with semiconductor memories. For data terminals, point-of-sale systems, and electronic watches, the multiplying factor is even larger. For the first five types of equipment that I have listed (Figure 3), semiconductors in fact constitute the heart of the end-equipment. Other technologies are undoubtedly also necessary for those five types of equipment, but the semiconductor technology is the most important and perhaps the most difficult to master among all the technologies that are required to make each of those types of equipment. For that reason, those five types of equipment are very logical candidates for the vertical integration of semiconductor manufacturers.

Item 6 of Figure 3 is solid-state color TV. Here, while semiconductors perform the vital signal processing functions, the present picture tube technology is not within the present semiconductor technology. In order to be successful in color TV, a semiconductor manufacturer would have to either acquire the present picture tube technology, or develop a substitute technology for picture tubes. Neither is likely in the short term. Therefore, color TV is not an easy candidate for vertical integration of semiconductor manufacturers in the near term.

Item 7 of Figure 3 is mainframe computers. Semiconductors perform the key function of logic and, increasingly, the memory function. However, the expertise required in systems marketing, software, and technologies other than semiconductor constitute great barriers to a semiconductor manufacturer attempting to integrate upward.

This list (Figure 3) is, of course, not intended as an all-inclusive one for all the candidate areas that semiconductor manufacturers could integrate into. Rather, it serves to illustrate the motivation of upward integration: namely, increased value added. It also illustrates the criterion that determines whether an upward integration is likely, namely, whether semiconductor technology is the dominant technology in the end-equipment or not.

FIGURE 3 Semiconductor Content in End-Equipment (as percentage of factory selling price)

Product	Percentage Semiconductor Content
1. Consumer calculators	30–35
2. Mini and microcomputers (CPU with SC memory)	25–35
3. Data terminals	12–15
4. Point-of-sale systems	12–15
5. Electronic wristwatch (with SC display)	8–15
6. Color TV (solid state)	15
7. Mainframe computers	<10

Shrinking Product Life Demands Close Coupling with End-Market

The second reason for the semiconductor manufacturers to integrate vertically is that the rapid advance of semiconductor technology has shrunk the product lifetime of the end-equipment, and this phenomenon in turn demands a close coupling between the semiconductor manufacturer and the end-market.

. . . The rapid advance of semiconductor technology demands that the design of the semiconductor components, if it can still be called a component, proceed in parallel with the design of the end-equipment in which the component will be used. This parallel development can, of course, be done between a component vendor and a systems customer, but the trade secrets and proprietary inhibitions are such that parallel development between two companies is seldom as satisfactory an arrangement as parallel development within the same company. When component development and system development do not proceed in parallel, valuable time is lost. And when you think of the average product life cycle as two years, the loss of a few months would be a serious matter.

I also want to make an additional point. As semiconductor components become more complex, the investment that the semiconductor manufacturer makes in component design and development becomes more and more substantial. And as I pointed out earlier, the rapid product obsolescence allows only a short time window in which to recover the substantial design and development investment. It becomes essential for the semiconductor manufacturer to utilize this time window to the maximum. Therefore, it is essential to have equipment using your advanced components on the market as early as possible. The only way to do it is to make and market your own equipment. I said earlier that the first reason for semiconductor manufacturers to integrate vertically is to increase value added. But really, in order to recover his investment in component design, it may become mandatory in many instances for the semiconductor manufacturer to integrate vertically and therefore to get his own equipment on the market as soon as possible.

Benefits to System Manufacturers

What are the benefits of semiconductor makers' vertical integration to traditional system manufacturers? Are they the losers? The answer is no. I can see two very clear benefits to the traditional system manufacturers.

The most important benefit is lower component prices. As you know, semiconductor prices generally follow experience curves with a slope of 70 percent to 80 percent, which means that each time the cumulative unit volume doubles, the price declines 20 percent to 30 percent. When a semiconductor manufacturer first uses his components in his own equipment, the cost of the components is high because it is at the beginning of the experience curve. As the components are made available to other equipment manufacturers, the other equipment manufacturers share the lower component cost with the semiconductor manufacturer.

Another valuable benefit of vertical integration to the traditional system manufacturer is the broadening of markets resulting from the innovative end-product and the semiconductor cost reduction.

Let us look at the worldwide market growth of three of the seven end-products I discussed earlier. In Figure 4, the period 1970 to 1974 has been selected because this is the period of initial impact of higher density integrated circuits. You will recall from Figure 3 that these products have a high semiconductor content—in the 25 percent to 35 percent range—and their performance and price are greatly influenced by semiconductor technology. Note that their annual growth rate in the past three years is in the 40 percent to 50 percent range, which is about four to five times the growth rate of the

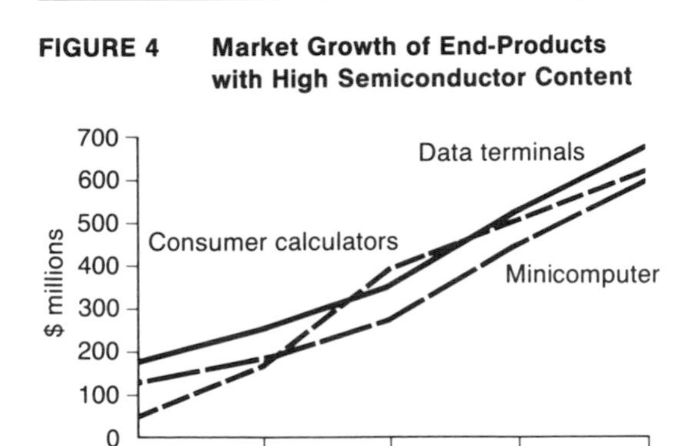

FIGURE 4 Market Growth of End-Products with High Semiconductor Content

total electronics market over the same period. Such explosive growth benefits all the manufacturers of those types of equipment since the market size becomes so large as to accommodate many participants.

BARRIERS TO DOWNWARD INTEGRATION INTO SEMICONDUCTORS

So far I have talked about semiconductor manufacturers integrating upward into the equipment business. What about the possibility of equipment manufacturers integrating downward into semiconductor components? This, of course, can happen. It has been done successfully in quite a few cases and has been attempted in even more cases. As I see it, there are two major barriers to downward integration. The first barrier is that a system manufacturer may not have semiconductor technology. Semiconductor technology has progressed to a degree of sophistication and complexity now that it will take a company not having it considerable time, investment, and talents to develop it. In this respect, a system manufacturer attempting to integrate downward into semiconductor components will have the same barrier as a semiconductor manufacturer integrating upward into such fields as TV picture tubes or large computers. The second barrier is that an equipment manufacturer must develop a large production base in semiconductors and sufficient research and development in order to remain competitive in semiconductors. The large production base is necessary because of the experience curve effect that I talked about earlier in semiconductors. The research and development is necessary because of the quick pace of advance in the semiconductor technology.

IMPLICATIONS OF VERTICAL INTEGRATION

Now I want to comment on some of the implications of vertical integration. The first and perhaps obvious implication of vertical integration is that the horizons of semiconductor companies have now broadened to extend their reach beyond the traditional component business, as a growing array of sophisticated end-equipment is now being built by what were hitherto known as semiconductor manufacturers. With vertical integration and the use of high-density components, the semiconductor producer is becoming an important practitioner of system design. Consequently, the boundary lines between component houses and system houses are becoming increasingly blurred.

Yet another important implication of vertical integration by semiconductor companies is the reduced labor content in end-equipment. Large-scale integration has simplified assembly by incorporating most of the complexity in the chip and drastically reducing the number of components needed. This tears down national or geographic barriers for the production of new equipment with a high semiconductor content. For example, a calculator can now be assembled in almost any country with relatively little difference in cost. In the near future, I can see that an electronic watch or a minicomputer can be assembled in any part of the world with relatively little difference in cost. New products with high semiconductor content could be produced indigenous to the market. This change has important ramifications for trade, industrial development, and for the growth of multinational companies.

Electronics is a dynamic industry which depends heavily on technological developments. Since semiconductor companies are at the leading edge of technology and have demonstrated their ability to introduce innovative, high-volume electronic products, they will come to play an increasingly important role in the total electronics industry. Vertical integration is not a newfangled craze but a viable way of growth for component manufacturers. Companies that develop and capitalize on the potential of new technology can achieve high rates of growth. Therefore, small companies today can become giants in just a few years. And giants today, unless they also continually develop and harness technology, can become noncompetitive and cease to grow in a short time. Vertical integration therefore has opened up possibilities for a reordering of hierarchy in the electronics industry.

APPENDIX II

Excerpts from: "Marketing in a Goal-Oriented Organization," by J. Fred Bucy, president and chief operating officer, Texas Instruments Incorporated; November 17, 1976, New York University Key Issues lecture series.

Marketing is but one leg of the "create," "make," and "market" functions of our business at Texas Instruments. Marketing strategy is a part of overall product strategy, and product strategy is derived from corporate strategy.

The "Marketing Concept" teaches that a company must focus on making what the customer wants, rather than on selling what the company makes. This is sound advice. But, like many simplistic statements, it obscures some major issues. First, there is a belief by the stronger proponents of the Marketing Concept that competitive products are essentially alike, and, therefore, success goes to the company that places the highest priority on marketing.

This belief in the supremacy of marketing over the "create" and "make" functions is based on the idea that equal technology quickly becomes available to all participants in a market through the mobility of the technical community and today's communications. If this were true, competition for market share would be won or lost based on the mechanics of bringing the product to the user, as the concept suggests. But this simply is not the case in high-technology businesses. Marketing is vitally important, but technology is still a prime determinant.

Technology encompasses the thousands of detailed steps that are necessary to develop and manufacture a product. Science gives us knowledge, but not concepts. Science may suggest what can be built, but only technology tells us how to build it. Frequently, technology alone permits us to invent new products to create markets, and new *science* is not always needed.

It's true that research findings are often made widely available—but the fruits of technology development rarely are. It's the lifeblood of competitive leadership, and successful companies guard it jealously.

In the United States, government-sponsored research accounts for 53 percent of all R&D. Except for a few areas crucial to defense, these research findings are available to all. Yet, certain companies consistently use this widely available research to produce superior products at lower prices—because they have developed their own superior technologies.

In addition to the free availability of government-sponsored research, private industry carries on active exchanges of research data through technical symposia, the publication of papers, and the like. In fact, our dissemination of this information is so free that Eastern Bloc countries have long been amazed at the ease with which they can acquire what has been so costly for us to learn. But data acquired in the course of research doesn't reveal much of the "how to" needed to design and manufacture a product.

This know-how is so crucial that, at TI, we develop and build most of our critical manufacturing systems—and we would like to build all of them. First, this permits us to keep the performance characteristics of these systems confidential. Second, advances we make in design and manufacturing technology give us an important competitive edge. When production equipment is bought on the open market, by definition your productivity is about the same as your competition's.

A company that enjoys life based on the assumption that there is no better way to design and produce a product than the currently available technology can be in for a rude awakening. It's happened time and again—in electronics, in business machines, in calculators, and now in watches, with electronic watches displacing many mechanical watches just as pin-lever movements displaced many Swiss movements a generation ago.

Technology is by no means a constant in the competitive equation. To the contrary, it is a most important variable.

The second point that the Marketing Concept tends to obscure is that create, make, and market functions must be tightly interwoven in a system of management, corporate philosophy, and the corporate purpose. None of these functions can operate effectively on a stand-alone basis.

At TI, we define a basic corporate philosophy and enunciate corporate goals that represent good citizenship in the broadest sense. These goals establish the basic purpose of the corporation, which neither we nor society should be allowed to forget. The basic purpose of corporations is to exercise wise stewardship in managing a large share of the physical assets of society. We strive to manage in a way that produces the maximum return to society—in new and better products, in creation and upgrading of jobs, in concern for the environment, and in community well-being. In short, TI's purpose is to provide a higher living standard, in both quality and quantity, at lowest cost, for TI's employees, its customers, and the community at large. Fulfilling this purpose requires that we make an adequate return on assets.

This basic philosophy is as old as TI. We formalized it in our corporate objective in 1961, and it has remained substantially unchanged. A fundamental element of this objective is a statement of corporate ethics—ethics that go further than just being within the law—and that cover situations where no law exists. We communicate these standards to all employees, and tolerate no deviation. Expedient compromises may promise short-term gains, but they can grow into a corporate cancer that will destroy the institution.

In addition to providing this ethical framework, TI's corporate objective defines the corporation's business intentions and goals; the kinds of business in which it will engage; the mix of business and geography; the company's posture toward employees, stockholders, customers, vendors, governments, and politicians; its specific profit goals and growth goals; and the methodology of growth. TI has always emphasized *internal* growth.

Within the corporate objective, we define a number of "business objectives" that specify the short- and long-range goals for major groupings of our business. These goals cover the worldwide markets to be served, the products and services for these markets and how to sell them, growth goals by product and market, technology requirements, and financial goals by product. This brings us to TI's Objectives, Strategies, and Tactics system.

This system, which we call "OST," provides an organization that overlays our decentralized product-customer center structure. It cuts across conventional organizational lines, so that an objective manager often has strategy managers reporting to him from various divisions of the company. A strategy manager, in turn, may have tactical action program (TAP) managers from many areas outside his conventional functional responsibility. As a result, a TI manager's responsibility usually exceeds his line authority. It also means that strategic organizations can quickly be formed or altered to meet rapid changes in our dynamic markets. Each business objective manager, strategy manager, and tactical action program manager is responsible for achieving agreed-upon product goals.

High volumes alone do not drive costs down; they merely provide the opportunity. At TI, capitalizing on this opportunity starts with "design-to-cost." This involves deciding today what the selling price and performance of a given product must be years in the future and designing the product and the equipment for producing it to meet both cost and performance goals. Stated another way, unit cost is a primary design parameter. It is a specification equal in importance to functional performance, quality, and service. This parameter takes the form of a timetable of steadily decreasing costs over the entire lifetime of the product.

In part, the cost timetable is determined by the price elasticity curve, which is a plot of the relationship of the price of a product to its potential sales volume. Unit price is the independent variable. Volume is the dependent variable. Typically, these curves have a knee where the market growth rate in units increases rapidly once a certain price level is penetrated.

This is the kind of result used to make the critical trade-off between cost and performance that sets the final design parameters. Constructing this curve is one of the most challenging of all marketing problems. Perhaps the most effective approach for both consumer and industrial products is to segment the market carefully and then to conduct broad customer surveys within the segments you expect to penetrate. The focal point of these surveys is what price the customer is willing to pay for specific product characteristics.

The cost timetable, in addition to being a function of the elasticity curve, is also a function of how quickly the knee of the curve will be approached, how large the start-up volume will be, and how quickly the volume increases on the experience curve.

The discipline created in an organization by setting pricing for the lifetime of a product is itself a powerful tool in the management of cost reduction. At the same time, the volume-dependent nature of the experience curve means that continual review and revision of cost goals are necessary as the future unfolds and volume requirements become more predictable. Thus, the design-to-cost approach becomes a forcing function for continuous productivity improvement throughout the entire lifetime of a product.

In addition to designing and manufacturing-to-cost, we must also distribute-to-cost. It's a mistake to devote thousands of engineering hours to wringing pennies out of manufacturing costs, only to have multi-tiered distribution add back dollars to the ultimate price paid by the customer. TI feels a responsibility to be innovative in using the right distribution channels to make sure that manufacturing economies are passed through distribution to the customer. The experience curve phenomenon, which requires us to produce at continually lower costs, also requires that we distribute at continually lower costs.

One example: five years ago, the small businessman who needed a mechanical four-function calculator had little choice but to call an office equipment dealer. Since a calculator then sold for perhaps $1,000, the dealer could justify several personal sales calls, as well as personal service after the sale. His gross margin on the sale of this product would be $400 to $450 a unit.

Today, an electronic calculator selling for less than $100 will outperform that old $1,000 machine. The margin available to the retailer who sells this product is about $35 per unit. This won't support the large sales and service organizations that previously existed. The result is that department stores now serve much of the small business market, and serve it at substantially less cost to the customer. The high level of personal service is no longer required because of the higher reliability of the electronic calculator. The business equipment dealers have reoriented their thrust toward the high end of the programmable calculator market.

Since the founding of TI, we have worked to create an organization and an institutional "culture" in which continuing productivity increases, cost reduction, and the design-to-cost approach are viewed as moral obligations to society.

There are times when companies have no choice but to raise prices. But it is morally wrong for institutions to believe that just because their labor and material costs have gone up, they are automatically entitled to raise their prices on an equivalent basis. It is just as wrong to expect wages to go up automatically, unless equal or greater gains can be made in productivity.

Price and wage increases will always be the first choice, though, unless design-to-cost and productivity improvement are implicit in the "culture" of the corporate institution. These elements must be so thoroughly built into the company's behavioral pattern that it actively, automatically, and continuously seeks to improve its mode of operation in order to give its customers more for less.

Reading II–4
How to Put Technology into Corporate Planning

Alan R. Fusfeld

Every executive knows of corporate successes in which technology has played a dominant role. Almost everyone in venture capital and entrepreneurship has a personal list of these successes to emulate. Dreams of technology turned to profit are nurtured by real-life success—Intel Corporation, Minnesota Mining and Manufacturing (3M), Polaroid, Hewlett-Packard, and Digital Equipment Corporation, to name a few of many.

Despite the obvious role of technology in superlatively successful enterprises, technological issues only occasionally are included explicitly in typical corporate strategy reviews, and only rarely are they among the regular inputs to corporate planning and development.

TECHNOLOGY: THE UNDERUTILIZED INPUT TO PLANNING

Most executives have limited management experience with technology. They see research and development as a black box: money and manpower resources are put in, but what should come out? How should these resources be directed and managed? And what should be the characteristic delays, success rates, and managerial control variables? General business management lacks an intuitive feel for strategically directing and positioning research and development investments as compared with similar investments in marketing, sales, and manufacturing. The result is that technology issues tend to be downgraded in overall importance to the business. Technology is addressed in strategic plans only implicitly, except in the case of special endeavors which are outside the main lines of production—new and joint business ventures, licensing, and acquisitions. In these, technology cannot be overlooked; it is often a major ingredient and even rationale in a purchase or joint venture plan.

In general, key management decision makers have inadequate background and ability to make judgments and forecasts in the area of technology. Without that ability, their options in utilizing technology in corporate strategy are severely limited.

There are many reasons for this blindness to technology and its management in our traditional administrative practices:

■ Most managers have been trained and have made their successful contributions in marketing, manufacturing, law, accounting, or some other corporate function. Their limited training in science or engineering is not enough to give them confidence in dealing with technological change.

For similar reasons, corporate economists fail to recognize the process of technological change in their economic forecasts. They either consider all products as homogeneous or see technological change as a wildcat input to their processes—something that comes from heaven or not at all.

Market research, too, has drawn very little on the technological field. Market researchers typically focus on short-term perspectives. Good, future-oriented market research should provide information that puts together a corporate strategy involving a realistic contribution from technology.

■ We know very little about the process of technological change; the knowledge we have is new (accumulated in the last 10 to 15 years) and has yet to be synthesized.

■ Partly due to limited experience, we lack adequate frameworks for viewing technological change. There is nothing comparable in this field to the simplifying frameworks for strategic business planning which have become prevalent in the last decade. The management of technology is, in fact, the only functional area which is not represented by a discipline within any management school.

■ Technological change proceeds slowly: significant change requires 5 to 10 years. This time span meshes poorly with the planning objectives of most American corporations. Although most corporations have five-year plans, 90 percent of their research and development activities are designed to be implemented within three years, and the remaining 10 percent within four years. Most corporations outline their strategic objectives on the short time horizon enforced by their need to manage short-term cash flow needs. That's not a time horizon appropriate for significant technological change.

Most research and development objectives are biased toward existing needs—such defensive goals as product improvement and cost reduction. This bias toward the use of technology in the support role to implement strategic objectives planned for three or four years in advance is the obvious result when managers lack an intuitive understanding of any larger goal for their research and development investments.

■ Most U.S. corporations are organized around the production process. They are not organized to recognize or to reward the uncertainties, risks, and time constraints of the technological innovation process. Not surprising, then, that most significant technological change originates outside of the firm—or even of the industry—that eventually uses it.

In only three areas of strategic corporate planning has technological change been widely—and, in general, wisely—considered in corporate planning. Acquisition has been a major activity of corporate development and diversification in the last half-century, and expected technological change and the acquisition of new technology has usually been an explicit consideration in this area. Technology has also been addressed explicitly in the licensing area, and it is an implicit part of new venture activities. In all these cases, technology is the essential element of the new opportunity.

PUTTING TECHNOLOGY IN ITS PLACE

Put yourself in the place of an executive assigned to set forth a corporate strategy. You must consider many elements—the broad characteristics of the industry, the qualifications of your firm's competitors in it, and your organization's corporate resources—managerial, financial, organizational, research and development, manufacturing, marketing, and distribution.

Technological issues enter as a result of activities both inside and outside the industry. They can affect the whole range of corporate activities: management, materials procurement, manufacturing, marketing, financial results, and future growth through new products and into new markets.

As you begin your analysis of corporate strategy, ask yourself such questions as these:

- How are technological issues recognized by your senior management? As a black box? As an input to long-range planning? For meeting short-term objectives? How explicit is the recognition of technology in each of these roles?
- How has management used technology to implement strategic objectives?
- How has technology been monitored? (One of the simplest and most conventional ways is by simply maintaining a research and development department to keep abreast of the state of the art. Other methods include outside technology boards and liaison activities to keep informed on areas where your own technical resources are limited.)
- How are activities relevant to technology recognized and organized in your enterprise? Where are they located, and how are they rewarded? (The typical corporate reward system is biased to short-term, cash flow performance; these criteria are simply not appropriate to the risks that must be taken in a viable technological development system.)

THE FUNDAMENTAL UNITS OF TECHNOLOGY

To improve your understanding of technology in your corporation, you will need first of all an adequate unit of analysis.

When we talk about technologies, we tend to speak of specific techniques and products—internal combustion engines, refrigeration and air conditioning, and machine tools, for examples. But technology flows in and out of such products as these, and they do not provide the fundamental basis by which to measure technological change. The analysis must be on the level of generic technologies. A carburetor, for example, is an application of the generic technology of vaporizing a liquid and mixing it with a gas. The same technology applied in the paint industry might become an automatic paint sprayer or in the aerospace industry a jet backpack. This way of focusing on generic technologies and the variety of technical applications of each is necessary if your planning is to be effective at capturing the implications of technological change that's going to affect a company's general product area. Consider, for examples, how Raychem and Hewlett-Packard have succeeded by concentrating on a single generic technology, developing and exploiting it in countless products for many different industries.

SEVEN DIMENSIONS OF PRODUCT ACCEPTABILITY

Having defined the unit of technology for analysis, you now need some basic parameters for explicit analysis of how a given technology is to be applied in your company's products and how effective they will be as a result.

After collecting information from many corporations on the characteristics of successful new products, I have found seven qualities which determine the success of any embodiment of any generic technology by any industry:

- *Functional performance*—an evaluation of the basic function that a device is supposed to perform. For example, the functional performance of a household refrigerator is to remove heat, and engineers evaluate a refrigerator's performance of this basic task in terms of what is called "pull-down" efficiency.

- *Acquisition cost*—in the example of the refrigerator, the price per cubic foot.
- *Ease-of-use characteristics*—the form of the user's interface with the device; in the example of the refrigerator, magnetic door latches and automatic defrosters contribute to the consumer's acceptance of the technology.
- *Operating cost*—in the case of the refrigerator, the number of kilowatt-hours used per unit of service performed.
- *Reliability*—the question of how often the device or process normally requires service, how free it is from abnormal service requirements, and—ultimately—what its expected useful lifetime is.
- *Serviceability*—the question of how long it takes and how expensive it is to restore a failed device to service.
- *Compatibility*—the way the device or product fits with other devices in the context of the larger system.

These are useful categories for analyzing applications of technologies because they are general, applying to everything from refrigerators to jet engines to medical services; they describe technology in a specific application very quickly and very adequately; they describe the goals of most research and development efforts; and they describe most of the emphasis in advertising and marketing strategies. Without such a set of dimensions, you will find yourself talking about the costs and benefits of potential technological change in haphazard, incomplete ways.

TECHNOLOGY DEMAND ELASTICITIES

Economists talk about price elasticity for a product, an indication of the role of price in determining demand. In the same way, each of the different dimensions in which technological change can affect the acceptability of a product is subject to evaluation in a fashion analogous to price elasticity. For example, you can analyze the change in demand for a product when its functional performance has been improved, when its ease of use has been increased or its service requirements lowered. In some cases elasticity will be low, in other cases high. Such data can be measured and used in the same way as the economists obtain and use price elasticity.

Two types of elasticity—absolute and relative—are very important in technology planning. Absolute elasticity represents the responsiveness of total market demand to improvements in function, ease of use, reliability, cost, etc. Relative elasticity is a similar measure of the tendency for shifts in market share to occur as competitors introduce new products with better performance in one or more of the various dimensions.

To see how these ideas enter into technological planning, consider a piece of medical equipment. In one case the product is destined for emergency-room use in a hospital; in another the same functions are to be performed in an individual doctor's office. Some characteristics will be more important in one market than in the other. Cost and ease of use will be relatively unimportant to the hospital; medical insurance will pay most of the bills, and the machine in the emergency room will be operated by a technician. The individual doctor, who must collect from individual patients and use the machine without a technician's help, will put a higher priority on low cost and ease of operation.

Or consider the example of Black and Decker, a company that once concentrated exclusively on commercial and industrial construction tools. The technological demands and price constraints of that market are different from those of the market for home use; and until Black and Decker recognized the differences and developed its technology accordingly, its penetration of the home tool market was very small.

In short, there are significant differences among customers' preference sets and hence different technological market elasticities. Calculation of technology elasticity results from analyzing statistically different market segments according to priorities in purchase decisions which can be established for each individual class of buyer.

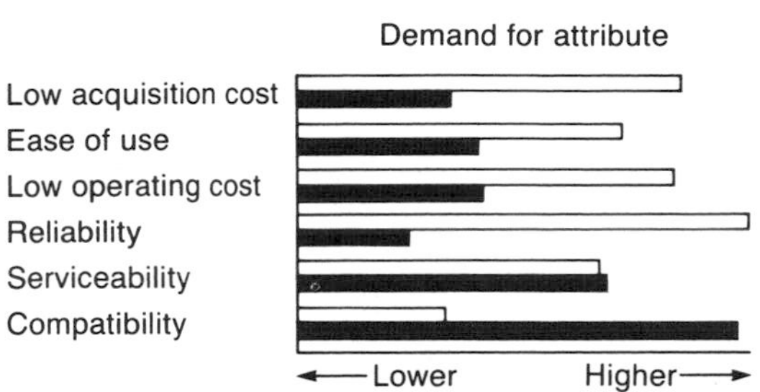

Market segment 1: private physicians
Market segment 2: health institutions

Matching the qualities which technology can impart to a product with the different needs of potential consumers. Cost is relatively unimportant to an institutional buyer of medical equipment—but it is a major consideration of the physician in private practice. Reliability is important to both; but the hospital may have a technician to effect repairs promptly. An analysis such as this demonstrates the different technological market elasticities for the same product in different markets—an important concept in any firm's planning for future investment in technology.

PROFILING TECHNOLOGY BY MARKET SEGMENTS

You now have determined a unit of technology on which to concentrate, the dimensions in which it is embodied in the market, and the relative demand for those dimensions. The next step is to apply these analyses to compare your company's technology with the needs by market segments, producing a competitive technological profile. Where do competitors' technologies stand in relation to yours in any particular market? Where did they stand a few years ago? And where are they going? Your goal, of course, will be to answer a question such as this: by the time my company's new technology-based product or service is in operation or on line, what will the competitive situation be?

The competitive profile helps answer that question by answering some simpler ones: what have the rates of change been in the past? And can one project a continuation of those rates into the future? How fast must one company move to gain ground on the others?

ASSESSING THE TECHNOLOGY AND PRODUCT PORTFOLIO

To develop an overall technology strategy around your company's existing technical and business strength, draw a chart (such as Exhibit 1) to show the generic technologies in which your company is engaged and the products in which they are applied. Such a chart provides a profile of the portfolio of technologies and products which may be illuminating to a management whose business has grown and developed in an opportunistic way. A similar chart for competitors in your company's product lines will help reveal what competitors are doing and what has been their strategy.

The company represented in Exhibit 2 may have difficulties in the future because it is trying to manage under the same roof different kinds of technology in which the manufacturer has different roles.

Exhibit 2 reveals that particle separation is a primary driving force of this company's corporate strategy, and there is at least the possibility that the company intends to be a leader in that field. The chart also shows that another part of

EXHIBIT 2

Principal product applications:

Generic technologies	Industrial pollution	Commercial filtration	Medical filtration	Construction	Wall/floor coverings	Automotive engines
Particle separation	○	○	○			
Metal fiber formation				○		
Molded material formation					○	
Noise control						○
Static electricity control					○	
Energy conservation					○	

A profile of the generic technologies in relation to the principal products of a hypothetical company. All six of this company's technological interests are good examples of what the author calls generic technologies, all sensible responsibilities of a central research and development laboratory. But only one of them is germane to more than one of the company's products; three of them result from the company's interest in the wall/floor coverings industry. Such a chart could be revealing to a management whose business has grown and developed in an opportunistic way. In this case it reveals that this company is actually two companies—one driving the technology of particle separation and one driven by the several technologies involved in floor coverings of several kinds.

this company is pursuing strategy that emphasizes a product area, picking up all kinds of technologies because of their common applications. This company may be said to have two parts—one technologically driven and one driving technology. Their technological strengths, their laboratories, their organizations, their pursuit of joint ventures and acquisitions programs, and their technology strategy in general are different. Indeed, these two parts of the company are so different that the company as a whole may be weakened by having to accommodate two such very different enterprises in its management structure.

A chart of this kind is the first stage in combining all the ideas previously discussed so that you may understand the role of technology in your company and weigh the investment and strategy options that are open to it. To select a particular strategy, begin by considering the generic technological strengths of your enterprise. You may find that you have no adequate strengths; you may find that your strengths in technology are not complemented by strengths in manufacturing or marketing, for example. Or you may find that your organization is fully prepared to drive a particular technology into many different product applications; or you may see that your best strategy is to capitalize on similar applications of different technologies. Depending on your analysis, you may want to add by merger a new generic technology in order to extend your applications area one step further; or you may want to offer your technology through merger to some firm which is equipped to capitalize on it through manufacturing and marketing.

In making these decisions, review the profile of your technologies by your market segments. Which technological dimensions of your products are important? Reliability? Function? Ease of use? Operating cost? How much emphasis will you place on reducing acquisition costs? On increasing ease of use? On reducing operating costs? On improving service?

This evaluation of technology dimensions in relation to market needs and competitive thrusts—the elasticity of technology demand—is the part that's missing from most research and development plans. But it provides answers to the crucial questions: what is your basic competitive advantage relative to other people? Why is your product or service going to sell? Why is it going to work in the marketplace?

Answering such questions succinctly and consistently will help many managements increase the strategic use of technology in their corporate planning.

Reading II–5
Patterns of Industrial Innovation

William J. Abernathy and
James M. Utterback

How does a company's innovation—and its response to innovative ideas—change as the company grows and matures?

Are there circumstances in which a pattern generally associated with successful innovation is in fact more likely to be associated with failure?

Under what circumstances will newly available technology, rather than the market, be the critical stimulus for change?

When is concentration on incremental innovation and productivity gains likely to be of maximum value to a firm? In what situations does this strategy instead cause instability and potential for crisis in an organization?

Intrigued by questions such as these, we have examined how the kinds of innovations attempted by productive units apparently change as these units evolve. Our goal was a model relating patterns of innovation within a unit to that unit's competitive strategy, production capabilities, and organizational characteristics.

This article summarizes our work and presents the basic characteristics of the model to which it has led us. We conclude that a productive unit's capacity for and methods of innovation depend critically on its stage of evolution from a small technology-based enterprise to a major high-volume producer. Many characteristics of innovation and the innovative process correlate with such an historical analysis; and on the basis of our model we can now attempt answers to questions such as those above.

A SPECTRUM OF INNOVATORS

Past studies of innovation imply that any innovating unit sees most of its innovations as new products. But that observation masks an essential difference: what is a product innovation by a small, technology-based unit is often the process equipment adopted by a large unit to improve its high-volume production of a standard product. We argue that these two units—the small, entrepreneurial organization and the larger unit producing standard products in high volume—are at opposite ends of a spectrum, in a sense forming boundary conditions in the evolution of a unit and in the character of its innovation of product and process technologies.

One distinctive pattern of technological innovation is evident in the case of established, high-volume products such as incandescent light bulbs, paper, steel, standard chemicals, and internal-combustion engines, for examples.

The markets for such goods are well defined; the product characteristics are well understood and often standardized; unit profit margins are typically low; production technology is efficient, equipment intensive, and specialized to a particular product; and competition is primarily on the basis of price. Change is costly in such highly integrated systems because an alteration in any one attribute or process has ramifications for many others.

In this environment innovation is typically incremental in nature, and it has a gradual, cumulative effect on productivity. For example, Samuel Hollander has shown that more than half of the reduction in the cost of producing rayon in plants of E. I. du Pont de Nemours and Company has been the result of gradual process improvements which could not be identified as formal projects or changes. A similar study by John Enos shows that accumulating incremental developments in petroleum refining processes resulted in productivity gains

EXHIBIT 1

The changing character of innovation, and its changing role in corporate advance. Seeking to understand the variables that determine successful strategies for innovation, the authors focus on three stages in the evolution of a successful enterprise: its period of flexibility, in which the enterprise seeks to capitalize on its advantages where they offer greatest advantages; its intermediate years, in which major products are used more widely; and its full maturity, when prosperity is assured by leadership in several principal products and technologies.

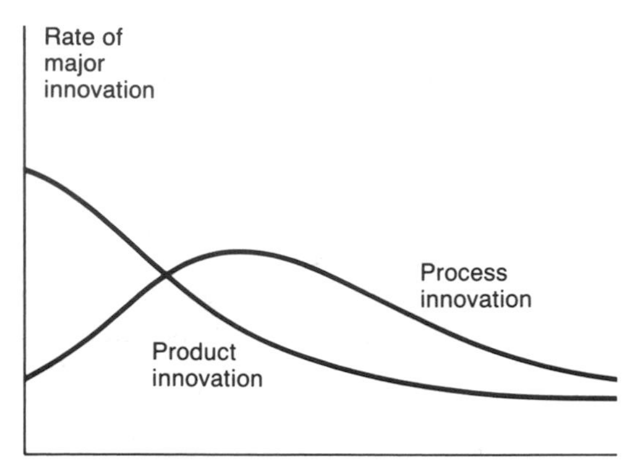

	Fluid Pattern	Transitional Pattern	Specific Pattern
Competitive emphasis on	Functional product performance	Product variation	Cost reduction
Innovation stimulated by	Information on users' needs and users' technical inputs	Opportunities created by expanding internal technical capability	Pressure to reduce cost and improve quality
Predominant type of innovation	Frequent major changes in products	Major process changes required by rising volume	Incremental for product and process, with cumulative improvement in productivity and quality
Product line	Diverse, often including custom designs	Includes at least one product design stable enough to have significant production volume	Mostly undifferentiated standard products
Production processes	Flexible and inefficient; major changes easily accommodated	Becoming more rigid, with changes occurring in major steps	Efficient, capital-intensive, and rigid; cost of change is high
Equipment	General-purpose, requiring highly skilled labor	Some subprocesses automated, creating "islands of automation"	Special-purpose, mostly automatic with labor tasks mainly monitoring and control
Materials	Inputs are limited to generally available materials	Specialized materials may be demanded from some suppliers	Specialized materials will be demanded; if not available, vertical integration will be extensive
Plant	Small-scale, located near user or source of technology	General-purpose with specialized sections	Large-scale, highly specific to particular products
Organizational control is	Informal and entrepreneurial	Through liaison relationships, project and task groups	Through emphasis on structure, goals, and rules

which often eclipsed the gain from the original innovation. Incremental innovations, such as the use of larger railroad cars and unit trains, have resulted in dramatic reductions in the cost of moving large quantities of materials by rail.

In all these examples, major systems innovations have been followed by countless minor product and systems improvements, and the latter account for more than half of the total ultimate economic gain due to their much greater number. While cost reduction seems to have been the major incentive for most of these innovations, major advances in performance have also resulted from such small engineering and production adjustments.

Such incremental innovation typically results in an increasingly specialized system in which economies of scale in production and the development of mass markets are extremely important. The productive unit loses its flexibility, becoming increasingly dependent on high-volume production to cover its fixed costs and increasingly vulnerable to changed demand and technical obsolescence.

Major new products do not seem to be consistent with this pattern of incremental change. New products which require reorientation of corporate goals or production facilities tend to originate outside organizations devoted to a "specific" production system; or, if originated within, to be rejected by them.

A more fluid pattern of product change is associated with the identification of an emerging need or a new way to meet an existing need; it is an entrepreneurial act. Many studies suggest that such new product innovations share common traits. They occur in disproportionate numbers in companies and units located in or near affluent markets with strong science-based universities or other research institutions and entrepreneurially oriented financial institutions. Their competitive advantage over predecessor products is based on superior functional performance rather than lower initial cost, and so these radical innovations tend to offer higher unit profit margins.

When a major product innovation first appears, performance criteria are typically vague and little understood. Because they have a more intimate understanding of performance requirements, users may play a major role in suggesting the ultimate form of the innovation as well as the need. For example, Kenneth Knight shows that three quarters of the computer models which emerged between 1944 and 1950, usually those produced as one or two of a kind, were developed by users.

It is reasonable that the diversity and uncertainty of performance requirements for new products give an advantage in their innovation to small, adaptable organizations with flexible technical approaches and good external communications, and historical evidence supports that hypothesis. For example, John Tilton argues that new enterprises led in the application of semiconductor technology, often transferring into practice technology from more established firms and laboratories. He argues that economies of scale have not been of prime importance because products have changed so rapidly that production technology designed for a particular product is rapidly made obsolete. And R. O. Schlaifer and S. D. Heron have argued that a diverse and responsive group of enterprises struggling against established units to enter the industry contributed greatly to the early advances in jet aircraft engines.

A TRANSITION FROM RADICAL TO EVOLUTIONARY INNOVATION

These two patterns of innovation may be taken to represent extreme types—in one case involving incremental change to a rigid, efficient production system specifically designed to produce a standardized product, and in the other case involving radical innovation with product characteristics in flux. They are not in fact rigid, independent categories. Several examples will make it clear that organizations currently considered in the "specific" category—where incremental innovation is now motivated by cost reduction—were at their origin small, "fluid" units intent on new product innovation.

John Tilton's study of developments in the semiconductor industry from 1950 through 1968 indicates that the rate of major innovation has decreased and that the type of innovation shifted. Eight of the 13 product innovations he considers to have been most important during that period occurred within the first seven years, while the industry was making less than 5 percent of its total 18-year sales. Two types of enterprise can be identified in this early period of the new industry—established units that came into semiconductors from vested positions in vacuum tube markets, and new entries such as Fairchild Semiconductor, IBM, and Texas Instruments, Inc. The established units responded to competition from the newcomers by emphasizing process innovations. Meanwhile, the latter sought entry and strength through product innovation. The three very successful new entrants just listed were responsible for half of the major product innovations and only one of the nine process innovations which Dr. Tilton identified in that 18-year period, while three principal established units (divisions of General Electric, Philco, and RCA) made only one quarter of the product innovations but three of the nine major process innovations in the same period. In this case, process innovation did not prove to be an effective competitive stance; by 1966, the three established units together held only 18 percent of the market while the three new units held 42 percent. Since 1968, however, the basis of competition in the industry has changed; as costs and productivity have become more important, the rate of major product innovation has decreased, and effective process innovation has become an important factor in competitive success. For example, by 1973 Texas Instruments, which had been a flexible, new entrant in the industry two decades earlier and had contributed no major process innovations prior to 1968, was planning a single machine that would produce 4 percent of world requirements for its integrated-circuit unit.

Like the transistor in the electronics industry, the DC-3 stands out as a major change in the aircraft and airlines industries. Almarin Phillips has shown that the DC-3 was in fact a cumulation of prior innovations. It was not the largest, or fastest, or longest-range aircraft; it was the most economical large, fast plane able to fly long distances. All the features which made this design so completely successful had been introduced and proven in prior aircraft. And the DC-3 was essentially the first commercial product of an entering firm (the C-1 and DC-2 were produced by Douglas only in small numbers).

Just as the transistor put the electronics industry on a new plateau, so the DC-3 changed the character of innovation in the aircraft industry for the next 15 years. No major innovations were introduced into commercial aircraft design from 1936 until new jet-powered aircraft appeared in the 1950s. Instead, there were simply many refinements to the DC-3 concept—stretching the design and adding appointments; and during the period of these incremental changes, airline operating cost per passenger-mile dropped an additional 50 percent.

The Unit of Analysis

As we show in this article, innovation within an established industry is often limited to incremental improvements of both products and processes. Major product change is often introduced from outside an established industry and is viewed as disruptive; its source is typically the start-up of a new, small firm, invasion of markets by leading firms in other industries, or government sponsorship of change either as an initial purchaser or through direct regulation.

These circumstances mean that the standard units of analysis of industry—firm and product type—are of little use in understanding innovation. Technological change causes these terms to change their meaning, and the very shape of the production process is altered.

Thus the questions raised in this article require that a product line and its associated production process be taken together as the unit of analysis. This we term a "productive unit." For a simple firm or a firm devoted to a single product, the productive unit and the firm would be one and the same. In the case of a diversified firm, a productive unit would usually report to a single operating manager and normally be a separate operating division. The extreme of a highly fragmented production process might mean that several separate firms taken together would be a productive unit.

For example, analysis of change in the textile industry requires that productive units in the chemical, plastics, paper, and equipment industries be included. Analysis involving the electronics industry requires a review of the changing role of component, circuit, and software producers as they become more crucial to change in the final assembled product. Major change at one level works its way up and down the chain, because of the interdependence of product and process change within and among productive units. Knowledge of the production process as a system of linked productive units is a prerequisite to understanding innovation in an industrial context.

The electric light bulb also has a history of a long series of evolutionary improvements which started with a few major innovations and ended in a highly standardized commodity-like product. By 1909, the initial tungsten filament and vacuum bulb innovations were in place; from then until 1955 there came a series of incremental changes—better metal alloys for the filament, the use of "getters" to assist in exhausting the bulb, coiling the filaments, "frosting" the glass, and many more. In the same period, the price of a 60-watt bulb decreased (even with no inflation adjustment) from $1.60 to 20 cents each, the lumens output increased by 175 percent, the direct labor content was reduced more than an order of magnitude, from 3 to 0.18 minutes per bulb, and the production process evolved from a flexible job-shop configuration, involving more than 11 separate operations and a heavy reliance on the skills of manual labor, to a single machine attended by a few workers.

Product and process evolved in a similar fashion in the automobile industry. During a four-year period before Henry Ford produced the renowned Model T, his company developed, produced, and sold five different engines, ranging from two to six cylinders. These were made in a factory that was flexibly organized much as a job shop, relying on trade craftsmen working with general-purpose machine tools not nearly so advanced as the best then available. Each engine tested a new concept. Out of this experience came a dominant design—the Model T; and within 15 years, 2 million engines of this single basic design were being produced each year (about 15 million all told) in a facility then recognized as the most efficient and highly integrated in the world. During that 15-year period, there were incremental—but no fundamental—innovations in the Ford product.

In yet another case, Robert Buzzell and Robert Nourse, tracing innovations in processed foods, show that new products such as soluble coffees, frozen vegetables, dry pet foods, cold breakfast cereals, canned foods, and precooked rice came first from individuals and small organizations where research was in progress or which relied heavily upon information from users. As each product won acceptance, its productive unit increased in size and concentrated its innovation on improving manufacturing, marketing, and distribution methods which extended rather than replaced the basic technologies. The major source of the latter ideas is now each firm's own research and development organization.

The shift from radical to evolutionary product innovation is a common thread in these examples. It is related to the development of a dominant product design, and it is accompanied by heightened price competition and increased emphasis on process innovation. Small-scale units that are flexible and highly reliant on manual labor and craft skills utilizing general-purpose equipment develop into units that rely on automated, equipment-intensive, high-volume processes. We conclude that changes in innovative pattern, production process, and scale and kind of production capacity all occur together in a consistent, predictable way.

Though many observers emphasize new-product innovation, process and incremental innovations may have equal or even greater commercial importance. A high rate of productivity improvement is associated with process improvement in every case we have studied. The cost of incandescent light bulbs, for example, has fallen more than 80 percent since their introduction. Airline operating costs were cut by half through the development and improvement of the DC-3. Semiconductor prices have been falling by 20 to 30 percent with each doubling of cumulative production. The introduction of the Model T Ford resulted in a price reduction from $3,000 to less than $1,000 (in 1958 dollars). Similar dramatic reductions have been achieved in the costs of computer core memory and television picture tubes.

MANAGING TECHNOLOGICAL INNOVATION

If it is true that the nature and goals of an industrial unit's innovations change as that unit matures from pioneering to large-scale producer, what does this imply for the management of technology?

We believe that some significant managerial concepts emerge from our analysis—or model, if you will—of the characteristics of innovation as production processes and primary competitive issues differ. As a unit moves toward large-scale production, the goals of its innovations change from ill-defined and uncertain targets to well-articulated design objectives. In the early stages, there is a proliferation of product performance requirements and design criteria which frequently cannot be stated quantitatively, and their relative importance or ranking may be quite unstable. It is precisely under such conditions, where performance requirements are ambiguous, that users are most likely to produce an innovation and where manufacturers are least likely to do so. One way of viewing regulatory constraints such as those governing auto emissions or safety is that they add new performance dimensions to be resolved by the engineer—and so may lead to more innovative design improvements. They are also likely to open market opportunities for innovative change of the kind characteristic of fluid enterprises in areas such as instrumentation, components, process equipment, and so on.

The stimulus for innovation changes as a unit matures. In the initial fluid stage, market needs are ill-defined and can be stated only with broad uncertainty; and the relevant technologies are as yet little explored. So there are two sources of ambiguity about the relevance of any particular program of research and development—target uncertainty and technical uncertainty. Confronted with both types of uncertainty, the decision maker has little incentive for major investments in formal research and development.

As the enterprise develops, however, uncertainty about markets and appropriate targets is reduced, and larger research and development investments are justified. At some point before the increasing specialization of the unit makes the cost of implementing technological innovations prohibitively high and before increasing cost competition erodes profits with which to fund large indirect expenses, the benefits of research and development efforts would reach a maximum. Technological opportunities for improvements and additions to existing product lines will then be clear, and a strong commitment to research and development will be characteristic of productive units in the middle stages of development. Such firms will be seen as "science based" because they invest heavily in formal research and engineering departments, with emphasis on process innovation and product differentiation through functional improvements.

Although data on research and development expenditures are not readily available on the basis of productive units, divisions, or lines of business, an informal review of the activities of corporations with large investments in research and development shows that they tend to support business lines that fall neither near the fluid nor the specific conditions but are in the technologically active middle range. Such productive units tend to be large, to be integrated, and to have a large share of their markets.

A small, fluid entrepreneurial unit requires general-purpose process equipment which is typically purchased. As it develops, such a unit is expected to originate some process-equipment innovations for its own use; and when it is fully matured, its entire processes are likely to be designed as integrated systems specific to particular products. Since the mature firm is now fully specialized, all its major process innovations are likely to originate outside the unit.

But note that the supplier companies will now see themselves as making product—not process—innovations. From a different perspective, George Stigler finds stages of development—similar to those we describe—in firms that supply production-process equipment. They differ in the market structure they face, in the specialization of their production processes, and in the responsibilities they must accept in innovating to satisfy their own needs for process technology and materials.

The organization's methods of coordination and control change with the increasing standardization of its products and production processes. As task uncertainty confronts a productive unit early in its development, the unit must emphasize its capacity to process information by investing in vertical and lateral information systems and in liaison and project groups. Later, these may be extended to the creation of formal planning groups, organizational manifestations of movement from a product-oriented to a transitional state; controls for regulating process functions and management controls such as job procedures, job descriptions, and systems analyses are also extended to become a more pervasive feature of the production network.

As a productive unit achieves standardized products and confronts only incremental change, one would expect it to deal with complexity by reducing the need for information processing. The level at which technological change takes place helps to determine the extent to which organizational dislocations take place. Each of these hypotheses helps to explain the firm's impetus to divide into homogeneous productive units as its products and process technology evolve.

The hypothesized changes in control and coordination imply that the structure of the organization will also change as it matures, becoming more formal and having a greater number of levels of authority. The evidence is strong that such structural change is a characteristic of many enterprises and of units within them.

FOSTERING INNOVATION BY UNDERSTANDING TRANSITION

Assuming the validity of this model for the development of the innovative capacities of a productive unit, how can it be applied to further our capacity for new products and to improve our productivity?

We predict that units in different stages of evolution will respond to differing stimuli and undertake different types of innovation. This idea can readily be extended to the question of barriers to innovation; and probably to patterns of success and failure in innovation for units in different situations. The unmet conditions for transition can be viewed as specific barriers which must be overcome if transition is to take place.

We would expect new, fluid units to view as barriers any factors that impede product standardization and market aggregation, while firms in the opposite category tend to rank uncertainty over government regulation or vulnerability of existing investments as more important disruptive factors. Those who would promote innovation and productivity in U.S. industry may find this suggestive.

We believe the most useful insights provided by the model apply to production processes in which features of the products can be varied. The most interesting applications are to situations where product innovation is competitively important and difficult to manage; the model helps to identify the full range of other issues with which the firm is simultaneously confronted in a period of growth and change.

CONSISTENCY OF MANAGEMENT ACTION

Many examples of unsuccessful innovations point to a common explanation of failure: certain conditions necessary to support a sought-after technical advance were not present. In such cases, our model may be helpful because it describes conditions that normally support advances at each stage of development; accordingly, if we can compare existing conditions with those prescribed by the model, we may discover how to increase innovative success. For example, we may ask of the model such questions as these about different, apparently independent, managerial actions:

- Can a firm increase the variety and diversity of its product line while simultaneously realizing the highest possible level of efficiency?
- Is a high rate of product innovation consistent with an effort to substantially reduce costs through extensive backward integration?
- Is government policy to maintain diversified markets for technologically active industries consistent with a policy that seeks a high rate of effective product innovation?
- Would a firm's action to restructure its work environment for employees so that tasks are more challenging and less repetitive be compatible with a policy of mechanization designed to reduce the need for labor?
- Can the government stimulate productivity by forcing a young industry to standardize its products before a dominant design has been realized?

The model prompts an answer of no to each of these questions; each question suggests actions which the model tells us are mutually inconsistent. We believe that as these ideas are further developed, they can be equally effective in helping to answer many far more subtle questions about the environment for innovation, productivity, and growth.

R&D Resource Allocation

Case II–5
Golden Gate Semiconductor

S. Koreisha and M. A. Maidique

In late October of 1975, Mr. Robert Kerr, senior vice president of research and development of Golden Gate Semiconductor (GGS), asked Dr. Martha Meyer, manager of the planning department, to help him formulate an R&D budget for the coming fiscal year (July 1, 1976). The plan, Kerr explained, should pay special attention to the possibility of GGS entering the field of magnetic bubble memories, a technology which has been reported to have the potential of revolutionizing the computer memory business.

GGS AND THE ELECTRONIC COMPONENTS INDUSTRY

GGS was founded by a group of scientists and engineers from the University of California, Berkeley, and Stanford shortly after the outbreak of World War II. In its early years, the company was engaged in the development, manufacturing, and marketing of electrical and electronic equipment for military purposes. With the end of the war, the company began to expand its line of business to the civilian sector of the economy. The company's growth paralleled the advancements of the electronic industry. In fact, much of the initial development in solid-state devices originated at GGS.

Today GGS's divisions and subsidiaries are engaged in the manufacturing and selling of what we shall term electronic components and systems. Solid-state components are the principal business of the firm. GGS is engaged in the development, manufacture, and sale of semiconductor solid-state devices such as transistors, diodes, and integrated circuits, such as memories and microprocessors, hybrid microcircuits, light sources, optoelectronic arrays, and microwave devices. These devices generate, control, store, amplify, and process electric signals. Solid-state devices are used in virtually every type of electronic circuit, including those in computers, radio, television, and other home entertainment units, aircraft, missiles, space vehicles, communications apparatus, automatic controls, and a variety of other industrial, commercial, and military equipment.

The solid-state or semiconductor industry is characterized by rapid technological advances and by severe price reductions. In such an environment, new product development is a way of life. A substantial amount of the company's sales of semiconductor devices in any year consists of products that were not sold in the preceding year.

New semiconductor devices initially sell at a high price. The price declines sharply as production efficiency and competition increase so that eventually they sell for only a small fraction of their original selling price. Further price declines may occur as a result of competitive pressures or the development of technologically superior devices. For these reasons, GGS has placed considerable emphasis on developing new products and on increasing the yields of its production runs by introducing new production techniques and more efficient assembly equipment.

GGS maintains an active research and development activity in the solid-state field which is designed to produce new products and process concepts, and to seek new uses for existing products in those fields of electronics in

Harvard Business School case 9-680-038

which the company believes it has particular technological competence. Among its current projects is further miniaturization of existing component parts as well as experimentation in microcircuitry. Although efforts are concentrated on semiconductors, other areas of development have included semiconductor test equipment, microwave generators, and memory systems.

GGS's semiconductor devices are sold by a sales force of salaried salesmen, of whom most are graduate engineers, and by a number of independent distributors in the United States. The company's sales force is supplemented by a product engineering group which works with the salesmen in developing and recommending specific semiconductor devices for customer applications. A substantial volume of sales to equipment manufacturers is attributable to this custom engineering service. GGS also employs application engineers who are engaged in developing new uses for the company's devices.

A significant portion of GGS's business consists of sales to a relatively small number of customers, primarily in the data processing and home entertainment industries. In fiscal year 1974, GGS's five largest customers accounted for approximately 35 percent of its domestic electronic components and systems sales.

In addition to solid-state devices, GGS also develops, manufactures, and sells electronic test equipment and special electronic tubes. The principal electronic test equipment products are solid state and are used for testing semiconductor devices. Special electronic tubes produced by GGS include cathode-ray tubes, which briefly display electronic impulses and are used in products such as oscilloscopes, direct-view display storage tubes for high-power transmission for radar and other applications. These products are sold primarily for industrial uses, although there are some military applications.

Although GGS is one of the largest producers of solid-state devices, it faces intense competition from both large and small manufacturers. The company also develops and manufactures products under government contract. These generally consist of complex devices designed for specialized uses, many of which are classified and, consequently, GGS does not have reliable statistics pertaining to its relative position with respect to these products. GGS's competition includes a large group of well-known, highly capitalized, technologically advanced producers such as Fairchild, Motorola, and Texas Instruments, as well as a greater number of smaller, more specialized companies, some of which have exceptional competence in certain product areas.

Price competition has historically been severe in semiconductor components, frequently resulting in periods of severe price reductions. Pricing policy within the industry generally follows reductions in manufacturing costs estimated on a learning curve basis. From time to time, price reductions for particular products may exceed cost reductions. Profitability of existing products is frequently reduced by new product developments which result in partial technological obsolescence of existing products and manufacturing technologies. Product life cycles were often three to five years. Some major customers of the semiconductor industry, such as the largest manufacturers of computers, telecommunications equipment, and consumer products, also have a captive semiconductor production capability. Other important factors in competing within the industry include manufacturing efficiency and reliability plus a reputation for dependability as a source of supply for particular products.

MANAGING THE SITUATION

Following her meeting with Kerr, Martha Meyer called the three planning analysts in her department to discuss their new assignments. All three, Jorge Serpa, Ian Madfes, and Joel Ratner, were recent business school graduates who had been with the company for at least two years. Meyer, in addition to her M.B.A., held a Ph.D. in electrical engineering (see Exhibit 1 for additional background).

After explaining some of the details of her conversation with Kerr, Dr. Meyer went on:

> We need to give some concrete plans to Kerr next week. The earlier the better. I think that the best thing we can do right now is to start delegating some of the data-gathering tasks and meet again in two days. I suppose we will need the obvious type of data such as GGS's sales, advertising, and R&D budget histories as well as similar data on the industry.

"This really shouldn't be too hard to come up with," replied Madfes. "I've been working on some projects with John Fletcher's people in marketing and I already have much of that data." (See Exhibits 2, 3, 4, and 5.)

"Good!" answered Meyer. "Since you already have much of the data, why don't you also prepare a short memo, based on your experience at the MIT Seminar on Innovation, on how research and development budgets are formulated in other technology-based companies. (See Exhibit 6.) It would be better if you didn't offer criticisms of each of the methods. Together, we will discuss the pros and cons of each approach at our next meeting."

"I can look into the new magnetic bubble memory technology," interrupted Ratner.

EXHIBIT 1 Biographical Sketches

	Position	Education	Comments
Bob Bottiglia	Sr. vice president, digital integrated circuits	Ph.D. Physics U.C., Berkeley	One of the founders of the company. Highly respected as manager and scientist.
John Fletcher	Manager, marketing	M.B.A., B.A. Political Science, Chicago	Ten years with company.
Robert Kerr	Sr. vice president, research and development	Ph.D. Electrical Engineering, Stanford	Fifteen years with the company. Considered to be an outstanding manager.
Ian Madfes	Sr. planning analyst	D.B.A. B.S. Economics, Harvard	Two years with the company. Considered to be a likely candidate for vice presidential position.
Martha Meyer	Manager, planning	M.B.A. Harvard Ph.D. Electrical Engineering, U.C., Berkeley	Four years with the company. Bright; fast track executive.
Joel Ratner	Planning analyst	M.B.A., Wharton BSEE, Illinois	Three years with company. Still getting exposure.
Jorge Serpa	Planning analyst	M.B.A., University of Florida BSEE, MIT	Three years with the company. Highly regarded by upper management.

EXHIBIT 2 Sales, Cost of Sales, and Profit Statistics for GGS (millions of dollars)

	58	59	60	61	62	63	64	65	66	67	68	69	70	71	72	73	74	75*
Sales	128	142	147	152	156	156	160	187	196	202	205	259	228	201	230	361	396	429
Cost of sales:																		
Materials and labor	80	88	95	104	104	108	109	120	129	130	145	178	170	138	156	296	272	278
Administrative and selling	12	16	19	18	18	22	24	25	33	32	31	35	25	34	27	24	27	38
Research and development	6	7	6	9	11	10	10	9	6	10	12	16	18	12	14	18	19	25
Engineering†	8	9	8	10	11	10	6	10	6	14	14	20	23	17	16	20	21	22
Other	10	8	7	2	5	8	9	8	12	8	12	8	10	6	4	3	5	6
Profits before taxes	12	14	12	9	7	(2)	2	15	10	8	(9)	2	(18)	(7)	13	50	52	60
Profits after taxes	6	8	7	5	4	(3)	1	8	4	5	(11)	1	(23)	(10)	7	27	28	32

* Budgeted.

† Engineering expenditures differ from R&D expenditures in that allocations to the engineering department are directed toward product or product line efficiency improvements, while R&D allocations are made for new product or technology development, including the first production line.

EXHIBIT 3 Dollar Change in GGS Sales, Cost of Sales, and Profits from Previous Year (millions of dollars)

	59	60	61	62	63	64	65	66	67	68	69	70	71	72	73	74	75
Sales	14	5	5	4	0	4	27	9	4	3	54	(31)	(27)	29	131	35	33
Cost of sales:																	
Materials and labor	4	7	9	0	4	1	11	9	1	15	33	(8)	(32)	18	90	26	6
Administrative and selling	4	3	(1)	0	4	2	1	8	(1)	(1)	4	(10)	9	(7)	(3)	3	11
Research and development and engineering	4	(2)	5	3	(2)	(4)	3	(7)	12	4	10	5	(12)	1	8	2	7
Other	(2)	(1)	(5)	3	3	1	(1)	4	(4)	4	(4)	2	(4)	(2)	(1)	2	1
Profits before taxes	2	(2)	(3)	(2)	(9)	4	13	(5)	(2)	(17)	11	(16)	11	20	37	2	8
Profits after taxes	2	(1)	(2)	(1)	(7)	4	7	(4)	1	16	12	(22)	1	(3)	20	1	4

EXHIBIT 4 Sales and R&D Figures for the Electronic Components Industry,* 1958 to 1977 (billions of dollars)

	58	59	60	61	62	63	64	65	66	67	68	69	70	71	72	73	74	75†	76‡	77‡
Sales of electronic components	2.4	2.9	3.1	3.4	3.6	3.7	3.9	4.5	5.5	5.4	5.3	5.7	5.1	4.8	5.5	7.8	8.3	6.7	8.8	9.9
Annual R&D funding	.22	.27	.31	.33	.32	.32	.31	.31	.34	.32	.31	.33	.34	.34	.32	.38	.45	.47	NA	NA
R&D as percentage of sales§	9.1	9.3	10.1	9.7	8.8	8.7	7.9	6.8	6.2	6.0	5.8	5.8	6.5	7.1	5.8	4.9	5.4	7.0	NA	NA

Notes: NA = not available.

* The electronic components industry includes a broad range of establishments engaged in the manufacture of electron tubes, solid-state devices, passive and mechanical parts.

† Estimated based on three quarters.

‡ Projected.

§ Includes engineering and government contracts. Government contracts accounted for about one third of all R&D expenditures.

SOURCE: National Science Foundation.

EXHIBIT 5 Percentage Increase in Sales and R&D Expenditures as a Percentage of Sales for the Leaders in the Electronic Components Industry

Competitors		70	71	72	73	74	75*
I	R&D (% sales)	NA	17.0	14.5	6.9	7.8	10.6
	% increase in sales	NA	149	183	103	2	NA
II	R&D (% sales)	NA	9.6	16.4	11.8	8.8	NA
	% increase in sales	81	(8)	53	68	116	NA
III	R&D (% sales)†	18.8	17.1	16.3	14.4	14.5	11.6
	% increase in sales	81	(8)	55	67	115	NA
IV	R&D (% sales)	12.6	10.7	8.3	11.5	9.6	NA
	% increase in sales	20.5	18.2	15.3	19.9	17.1	NA

* Estimated.

† Includes engineering.

SOURCE: Annual reports of various companies.

EXHIBIT 6 Golden Gate Semiconductors

MEMORANDUM

TO: Martha Meyer

FROM: Ian Madfes

SUBJECT: R&D Budgets

My conversation with several R&D managers at the MIT Seminar on Innovation confirmed my prior notion that gut feel and historical trends play a major role in the actual allocation of R&D funds. Furthermore, companies that consider themselves as leaders generally spend more on R&D than those which are primarily followers. Leaders generally spend on the neighborhood of 8 to 10 percent of sales revenues on R&D whereas followers tend to spend around 4 to 5 percent.

Although a great deal of similarities exist in the way innovators and followers formulate their budgets, there are sufficient differences in their approaches to allow for some categorization.

Innovators

Among the companies which can be classified as innovators, I have identified at least three different approaches to the formulation of R&D budgets. These are top-down with compromise, correlation analysis, and bottom-up. Companies in this group generally have deep commitments to R&D. Incorporated in their strategies is the fundamental premise that R&D is vital to their success, and that they have to be ahead of the competition. Below is a description of the above-mentioned methods.

Top-Down with Compromise: Corporate staff determines what level of R&D funding will be available for the following year by subtracting from expected revenue a desired profit level and all the non-R&D costs associated with sales, e.g., materials, labor and equipment, advertising, engineering, etc. The balance becomes the preliminary figure from which to judge the merits of potential projects. The derived figure is generally compared with the company's R&D budget history as well as with figures for other industry leaders to check for major discrepancies.*

Once the figure has been more or less worked over at corporate level, discussion with division managers takes place to evaluate the various projects being contemplated by the divisions. The stage of the particular product's life cycle, marketability, development, timing, type of technology required (new versus modified), are some of the factors considered in these discussions. The outcome of these discussions is elimination of some projects and a ranking of each division's projects. The recommendations from the divisions are then studied by senior management to see how they fit within the overall corporate strategy. The available R&D funds are allocated to the projects which management feels provide balanced portfolio for the company.

Correlation Analysis:

a. Many companies have noticed that peaks in their sales are generally associated with the introduction of new products or a new technology. Using specific examples from the past, they have been able to associate specific correlations between R&D and sales. For example, suppose that a specific percentage of sales revenues were spent on research and development for a particular product over a known period of time. With the introduction of the product in the market, suppose that sales increased by a certain percentage. An association, presumably, can be made between the amount spent on R&D and the increase in sales. Thus, by having several types of product categories, one could have a baseline for judging the potential of products with similar characteristics. Input from the marketing department plays an important role in the funding decision for companies which utilize this approach. Using these correlation coefficients for specific products or at times for particular divisions together with the desired sales growth rate, one can arrive at an R&D figure. For instance, assume that 1 percent of sales revenue maintained over the years in-

EXHIBIT 6 *(concluded)*

creases sales by 3 percent, and the desired annual growth rate for that division = 30 percent. Thus, 10 percent of the sales revenues associated with that division should be allocated to R&D. Using sales forecasts for the coming year, a direct monetary figure can be obtained.

b. Other companies use a similar approach, but rather than associating sales growth rates with R&D expenditures for their own divisions or products, they calculate the correlation coefficients for the leaders in the industry and then decide which coefficients should apply to themselves.

c. Some companies formulate econometric relationships between sales and factors such as R&D, advertising (Adv), capacity expansion, etc. Included in many regression equations are combinations of the above type of variables together with corresponding lagged values. A typical equation might be:

$$\begin{aligned} \text{Sales}_t = {} & a(\text{R\&D})_t + b(\text{R\&D})_{t-1} \\ & + \cdots + c(\text{R\&D})_{t-x} + \\ & d(\text{Adv})_t + e(\text{Adv})_{t-1} \\ & + \cdots + f(\text{Adv})_{t-y} + \\ & g(\text{Sales})_{t-1} + h(\text{Sales})_{t-2} \\ & + \cdots + i(\text{Sales})_{t-z} \end{aligned}$$

where the subscript t = time.

Bottom-Up: Projects are evaluated on their worth and potential without the confines of specific R&D budget figures. The costs associated with projects which passed through various screening levels are summed together with the total becoming the R&D budget for the coming year. Project considerations do not generally restrict the formulation of the R&D budget. Companies with such policies clearly regard R&D as investment rather than as expenditure. Often this policy characterizes firms with longer than usual development periods for their products such as the aircraft companies.† A great deal is generally invested on diverse types of projects because the consequences will not be known for quite some time.

Followers

Companies which can be classified as followers spend considerable time and effort on improvements of either existing products or of the manufacturing processes of products developed by competitors. It should be emphasized, however, that process improvements particularly in the semiconductor industry may require a commitment comparable to the development of the original process.

Generally, R&D budget formulations among these companies are based upon stringent and often rigid bounds. Many companies are not on a sufficiently sound financial situation to be able "to afford luxuries." One R&D executive said: "We find it very difficult to spend more than $1 million on any one new product."

The methods described in the previous section, with some modifications, such as less emphasis on what the competition and industry leaders are spending, are frequently used by many followers. For completeness, I have incorporated one more approach which typically characterizes companies in this group: fixed percentage:

> Fixed Percentage: Based on past company history, a fixed percentage of sales revenue—3 to 5 percent—is allocated to R&D. The same procedure is followed for the other expense categories. Provisions are generally made for drastic changes in the economic situation of the company in order to prevent violent fluctuations in the actual monetary allocations. Three to five-year rolling sales averages are frequently used as a way to smooth recessionary or inflationary patterns. Distribution of funds is based upon management's opinion of the contribution expected from each project. The derived upper ceiling on R&D expenditures is adhered to firmly. Postponements of development projects are frequent.

---

* A variation of this technique is to perform an analysis of the firm's recent history of R&D expenditures and simply ask management to stay within the historic levels.

† A large portion of the R&D function in the aircraft industry is subsidized by the federal government. The company's direct allocations, although large, are more closely tied to budget and profit considerations.

"Fine," interjected Meyer. "But let's get what you find on paper. Why don't you team up with Jorge and prepare a brief background paper on the subject. We will integrate what you find with our business strategy for digital integrated circuits."

"No problem. I suppose we should also consult with Dr. Bottiglia since after all he is the person who knows most about that business," added Ratner.

"Good. Let's get together in two days to discuss what we've found out."

MEETING WITH DR. BOTTIGLIA

"I've recently learned quite a bit about magnetic bubble memories," said Bottiglia as he talked to Serpa and Ratner. "I am very impressed with their potential. We've always been innovators in our field and I think we really should make a commitment to bubble memory technology."

"At this point I still have several reservations," rejoined Ratner. "We already have quite a few major projects going on like the high-density MOS chip[1] program which is central to our MOS memory business, and the fiber optic modulator product line, a promising new opportunity.

"Historically," he added, "at least over the past 10 years, our expenditures on R&D have been around 5.5 percent of sales. We already earmarked $20 million for the fast, higher density MOS logic circuits, further development on Isoplanar technology, and $8 million for several other ongoing R&D projects. I don't really know if we can spare too much for additional projects. Yet to really get into magnetic bubble memories, I would say that, based on our bipolar experience, we would need to make a $30 million commitment over the next two to three years."

"Initially, say over a two-year period," added Madfes, "we would need $8 to $10 million for development work on the new technology and a very small pilot line. A minimum increment of manufacturing capacity would cost around $20 million, and would take about 18 months to construct. The facilities for the new plant would be similar to some of our already existing plants. The new plant would be about 30 to 40 percent different from our MOS plants. This means that we have the option of building the plant while developmental work is still being done—getting in the market early. Alternatively, we could wait until that work is done—minimizing risk and avoiding a great deal of capital expenditures right away. But, let's not forget inflation. Availability of people to run the project should not be a problem. We already have some very competent scientists, and as you know, we've been in close contact with some of the original people at Bell Labs who developed magnetic bubble technology. I think that there are two or three of them that we could attract."

"There are always more projects to work on than we think we can handle or afford," replied Bottiglia. "But, what young guys like you have to remember is that when a major technological innovation comes around, you've got to be prepared for it. Your philosophy about the company has to change. You have to assume that we're back in the 50s, during our embryonic stage of development, and have some vision. The profits will come later. When we started out, our R&D expenditures were roughly 25 to 30 percent of sales for several years and our profits, well, they didn't exist for a while. After that, we grew by leaps and bounds. But we had to invest and be leaders in R&D to get techniques; it's a major new technology. And we have to get in early if we are going to continue to be leaders in the semiconductor market."

After some further discussion with Bottiglia, Serpa and Ratner returned to their offices to prepare for their meeting the next day with Meyer.

---

[1] The term *chip* was an industry term for integrated circuit.

THE PLANNING DEPARTMENT MEETING

After receiving and discussing all the reports, Meyer suggested that work be done on the actual generation of the budget. (See Exhibits 7 and 8.)

"What we should do," she said, "is test various approaches to formulating R&D budgets to see what results we obtain. Each of the approaches has its own inherent biases and limitations. Ultimately, however, we'll have to settle on a rationale for making our recommendations to Kerr."

EXHIBIT 7 Golden Gate Semiconductor

MEMORANDUM

TO: Martha Meyer

FROM: Jorge Serpa and Joel Ratner

SUBJECT: Computer Memories

In a recent article in *Electronics,* a senior scientist at the Electronic Research Division at Rockwell succinctly summarized the situation in the computer memory business when he said, "The memory business is great [because] no one wants to forget anything—they just keep adding capacity." In the past two decades, immense progress has been achieved in the capabilities of memory units: while storage capacities have been increasing at phenomenal rates, access time and costs have been commensurately decreasing. From tapes, disks, and transistors to bubbles and charge packets, the revolution in computer technology rushes on unabated.

One of the first casualties of the computer memory revolution will no doubt be magnetic tape units. In magnetic tape systems, digital information is stored as dots of magnetism in tapes similar to those used in tape recorders. Tape drives may have as many as 16 separate parallel tracks. The capacity of such a storage system is virtually without limit (up to 10 million words per reel). The main problems with such archival systems, however, are access time and design. Since tape storage is sequential, it may be necessary to traverse large portions of tape (which can take a minute or more) to obtain the desired information. Furthermore, the sophisticated equipment required to handle tape speeds of 200 inches per second with the capability of starting and stopping in a few thousandths of a second is susceptible to mechanical failure.

Magnetic tape units are one of the few major components of the whole computer system which require manual intervention. Someone must physically select a tape from a library containing hundreds, if not thousands, of tapes, and then manually insert the tape in the reading unit. *Science News* (September 13, 1975, p. 170) reports that "this manual intervention may cost as much as $2.50 for every tape change, and that because material at the end of a tape takes so long to reach, 95 percent of the users fill less than half a tape."

Data cartridges, which are very similar to cassette tapes, are starting to replace the bulky magnetic tapes. "A 10-foot-long cabinet can hold 2,000 of these cartridges, replacing some 4,000 to 16,000 tapes and holding the information available for purely automatic retrieval—with an average access time of three seconds." (*Science News,* September 13, 1975, p. 170.)

While tapes are suited for applications in which data has been prestored in a logical sequence, they leave a lot to be desired when data are needed at unpredictable intervals. Magnetic drum and disk storage systems, on the other hand, have the capability of accessing the data randomly, but are mechanically more complex than tapes. A drum, for our purposes, can be viewed as a magnetic tape, carrying up to 200 tracks, wrapped around a cylinder rotating at approximately 3,000 rpm. The maximum access time to the information is about 1/50 of a second, orders of magnitude faster than magnetic tapes, but storage capacity is limited to several tens of thousands of words.

EXHIBIT 7 *(continued)*

A disk is a round, flat surface coated with magnetic material which rotates on a shaft with a reading head positioned to read a desired track. The operation of disk systems is very similar to that of a second album on a turntable except that up to 20 disks may be rotated on a common shaft. The entire system looks very much like a jukebox. The heads in disk drives do not come in contact with the recording surfaces during the read and write operations. Instead, they are aerodynamically designed to fly over the rotating disk surfaces. The microscopic distances between the surfaces require that the surfaces be very smooth and clean. A contaminant on the surface of a disk may collide with the head, thus, possibly causing or cause the head to skip the reading of some data. Positioning of the heads is rather complicated: the difficulty increases with the density of the track. Access time for disks is, thus, somewhat longer than that of drums, but storage capacity is larger.

Memories that can be accessed at random and at high speeds are essential for the large, high-speed, modern computer operations. Such memories became possible with Jay Forrester's revolutionary invention of the magnetic core.* Magnetic cores are small doughnut-like cores (usually 1 to 2 millimeters in diameter) made up of hard magnetic material (ferrite) that can be magnetized in either of two directions by current in the wires passing through the hole. Advances in ferrite technology have produced faster materials. Reduced core size has made possible faster switching, shorter line length, and increased density. Electrical problems associated with resistance of the small wire required to string the core, and mechanical problems associated with the strength of the core, however, limit the development of larger memory capacities.

The development of semiconductor memories marked the genesis of a second computer memory revolution. Speed, density, and cost parameters improved significantly. At present, however, semiconductors' memories still have a volatility problem (content of memory is erased when power is shut off†), but they bring to memory technology the batch-processing capability that has in the past accounted for dramatic breakthroughs in prices.

The majority of commercially built semiconductor memories (logic devices) are comprised of bipolar and MOS devices. Some read-write memories require periodic rewriting of existing data to compensate for leakages that cause information to be erased.‡ These devices, known as dynamic memories, are very sensitive to power line variations. Other integrated circuit memories, known as static memories, maintain their data as long as the power is on.

Bipolar memories are generally static devices. In their circuitry, "one transfer is normally on and the other off. When the off transistor is forced into the on state by an external signal, the on transistor is turned off. It is possible to store information by defining the stable states as binary 1 and 0 with the normal plate identified with the no-data condition. Information is stored or retrieved by means of gating techniques." (*Control Engineering,* January 1972, p. 57.) These devices operate at relatively high power levels. Typical access times for bipolar memories range from 10 to 70 nanoseconds. The number of bits stored per chip varies from 70 to 1,000. Costs are still high: 2 to 10 cents per bit, but are dropping rapidly.

An important group of memory devices utilize the metal-oxide-semiconductor (MOS) transistor. All the electronic active areas of this transistor are on the same plane, and with the exception of a window opening to connect the terminals to different regions, the entire surface is covered with silicon dioxide. The major advantage of the silicon dioxide is the reduction of surface contaminants and impurities that cause surface leakage currents, which reduce the time that information can be stored without rewriting.

There are two types of MOS memories: static and dynamic. Static memories for small capacities (less than 256 bits)§ are more economical than dynamic ones because the cost required to refresh dynamic memories is proportionately more expensive for small memories. Dynamic memories have higher packing densities. They usually require external clocking pulses in addition to power supply voltages. In general, MOS memories have longer access times than bipolar memories, but MOS chips have greater storage densities. Typical access times range from 200 to

EXHIBIT 7 *(continued)*

800 nanoseconds and chip densities vary from 256 to 1,024 bits. The costs per bit are around 2 cents. Various hybrid versions of MOS memories exist.

Special semiconductor memories have been constructed to take advantage of situations in which the data being stored is more or less permanent and volatility eliminates read-write memories from consideration. These types of memories are called read-only memories (ROM). They have storage densities which are 3 to 4 times greater than the read-write memories because the write circuitry has been eliminated. There are two basic types of ROMs: fixed mask and field programmable. Fixed mask ROMs are programmed at the time of manufacture. "This type of ROM can be the least expensive for volume applications, but the program is permanently fixed and cannot be altered. Moreover, there is an overhead cost incurred in developing the special masks for each unique application. This cost in conjunction with the turn-around time required for mask changes has restricted this type of memory to applications where the data pattern is fixed for all time and the quantity is large enough to amortize the cost of the mask, typically $1,000 to $2,000." (*Control Engineering,* January 1972, p. 62.) As the name implies, field programmable ROMs are memories designed for on-site programming. The programmable capabilities are made by making or breaking aluminum interconnections in the chips. A typical ROM device has an access time of about 800 nanoseconds; a storage capacity of 2,048 to 4,096 bits/chip; and costs around 1 to 4 cents per bit.

Another product just coming on the market is the charge-coupled device (CCD) developed by Bell Labs. The CCD is an offspring of MOS technology. It operates "by shifting packets of charge trapped in its semiconductor substrate. As voltage changes along the line of metal 'gates' lying above the substrate, the packets of charge below move down the line correspondingly. Bits of data are stored according to whether a particular region contains a charge packet or not, and data are read out serially at the end of the line. Access time depends on how long the particular data line is; if short access time is required, more expensive read/write centers must be added." (*Science News,* September 13, 1975, p. 170.) The access time of CCD memories ranges from 25 microseconds to a millisecond. But data can be stored more densely than in the faster semiconductor devices. CCDs form the basis of several new products such as solid-state TV cameras and filters for discriminating certain radar signals. The main disadvantage associated with CCDs lies in the serial nature of the memories: they are slower than random access memories, and each time the packets of charge are shifted, a little charge gets lost. Compared with rotating magnetic memories, CCDs offer several advantages: access time of CCD memories is about one-fifth of that of a magnetic drum (for comparable capacity) and the power requirement is roughly one-sixth of that of a drum.

Next on the horizon of computer memories are magnetic bubbles. The technology is based on minute magnetic domains ("bubbles") that appear, disappear, and move around certain kinds of crystalline materials such as garnets under the control of magnetic and electric fields. "When a thin wafer of garnet is placed in a magnetic field, the naturally occurring, randomly spaced islands of magnetic anomaly (domains with field orientation opposite to that of the rest of the material) shrink into tiny round bubbles. These can then be attracted to the ends of tiny magnets deposited on the garnet surface. If the external magnetic field is rotated, the poles of the deposited magnets change and the bubbles will shift position. Through careful construction, a long string of these magnets can be created with bubbles running along its length as the external field rotates. By adding a bubble generator and detector, a serial memory is created." (*Science News,* September 13, 1975, p. 171.) Bubble memories are only about 1⁄100 as fast as CCDs, but are capable of storage densities which are one or two orders of magnitude greater than CCDs, and require considerably less power to run. According to some experts, bubble memories could "drastically change the architecture of future computers by shrinking the size of the total system considerably from half a floor to the size of a suitcase." John Douglas, an authority in computer memories, has noted that "the attractiveness of this apparently clumsy device is that the bubbles can be

EXHIBIT 7 *(concluded)*

made so small that theoretically a billion bits of data could be stored in a square inch of material—a density approximating the human brain." He has also observed that it is possible to construct bubble logic circuits which could combine storage and arithmetic operations on one tiny chip. Bubble memories also have the advantage of being "nonvolatile," i.e., information doesn't disappear during power shutoff. Costs of bubble memories are higher than some semiconductor memories, but experts predict that they will fall at a faster rate than semiconductors as more research is conducted. Many supporters of bubble memories are careful to point out that the manufacture of integrated circuits is more complex and takes more steps than bubble circuitry.

Another memory device presently being contemplated is the electron-beam-accessed memory (EBAM). "EBAM works much like a television picture tube, except that the screen is replaced by a set of MOS chips. Electrons from the cathode induce a packet of charge to form at the interface of the oxide and silicon layers. The presence or absence of charge at any given location can be determined by changing voltage on the chip and seeing whether a pulse is sensed when the beam passes over the desired location." (*Science News*, September 13, 1975, p. 171.) EBAMs require large numbers of power supplies, have access times of tens of microseconds, and can store up to 30 to 100 million bits at surprisingly low costs. Table 1 provides an estimate of the costs/bit between CCDs, bubbles, EBAMs, and disks. A comparison based upon storage capacity and access time between various memory systems is found in Figure 1.

FIGURE 1 Estimated Capacity of Various Memory Systems and the Access Times Needed in Each

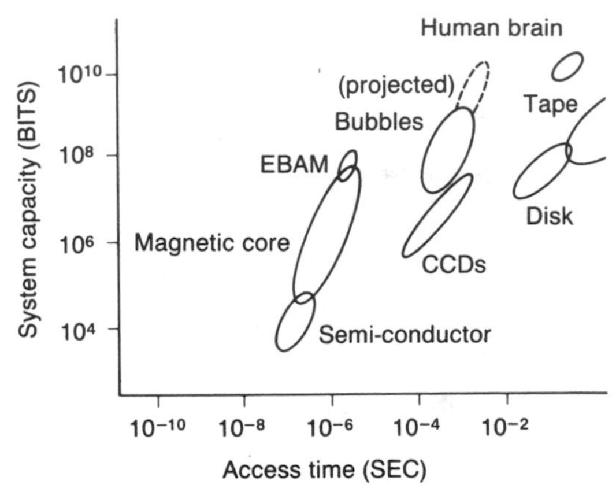

SOURCE: Adapted by casewriter from *Science News*, September 13, 1975, p. 173.

TABLE 1 Comparative Estimates of Projected Costs/Bit between CCDs, Bubbles, EBAMs, and Disks

Memory	*Cost/Bit*
CCDs	$.00015
Bubbles	$.00005
EBAMs	$.00002
Disk	$.00003

SOURCE: *Science News*, September 13, 1975.

---

* Arrays of magnetic cores are often called "core memories."

† Erasing time varies widely depending on which semiconductor devices are employed.

‡ In computer language, "write" means putting information into a storage register; "read" means taking information out of a register for use; and "erase" means completely removing the information from a register.

§ In 1975, the highest capacity commercial semiconductor memories stored approximately 4,000 bits.

EXHIBIT 8 Proposed 1976/1977 Budget Allocations and Expected Sales

Expected sales:	
Marketing forecast	$479 million
Planning department forecast	$458 million
Outside econometric forecast	$450 million
Cost of sales:	
Materials, labor*, manufacturing, engineering, distribution, and overhead	$320 million†
Expenses:	
Administrative and selling	$45 million
Research and development‡	$28 million
Other	$6–12 million
Profit before tax§	$65 million

* Does not reflect possible labor contract changes which may come about in January 1976 when union contract expires. It is expected that the settlement would require wage increases of 7 to 10 percent. Labor costs comprise roughly one fourth of the costs in this category.

† Estimate based on planning department forecast.

‡ Preliminary R&D department request exclusive of bubble memory project.

§ Management goal.

Reading II–6
Linking R&D to Strategy

Richard N. Foster

As a high-technology company grows more complex and decentralized, its procedures may fail to keep pace with change, weakening vital links between business operating and technical people, leaving key decisions to be made by default, and resulting in ultimate loss of competitive edge. As the author of this fictionalized case study shows, to ensure that research efforts are clearly geared to priority strategic projects, changes may be necessary not only in planning processes but also in organization structure and executive personnel development systems.

McKinsey Quarterly, Winter 1981. Reprinted with permission.

Harlow McLaughlin, the 58-year-old chief executive officer of United International, was just concluding his first annual report on the state of the corporation to the 15 members of his board: "Well, gentlemen, there have been some rough spots over the past year, but I think that you can see that United has emerged from them in sound condition to face the future. All in all, I am convinced that we are well positioned to meet our planned objectives—which, as I've pointed out, are probably the most ambitious we have ever had. Thank you."

Harlow looked around the room. Instead of the nods he had expected to see, most of the 15 board members wore dour expressions; some doodled nervously on their yellow pads.

The only question, however, came from Steele Blackall, managing partner of United International's chief legal counsel. "Harlow," he said, "I thought that was a first-rate presentation, especially considering that it's your first since you've been in office. But I wonder if you could review for us again why you think UI's historical performance patterns are about to be reversed? You've said that we can expect to see UI's profits increase in step with your sales instead of bumping along on the level the way they've done for five years now. What's actually going to make that happen? Your plans as you've described them seem perfectly sound, but I don't detect any radical departures from the sort of plans your predecessor used to present. We approved those, but as you know they didn't lead to much better performance. In short, I think all of us here applaud your objectives—but can you make them come true?"

"Yes, Steele, I believe we can," replied Harlow briskly. "We'll be able to achieve these results because of three factors: first, the conservative financial policies that we've put in over the past decade, which will just begin bearing fruit in the next several years; second, selling off those three unprofitable divisions; and third, the rationalization of our production processes that I was describing to you 10 minutes ago."

But even as he sat down, Harlow became aware of the same queasiness he felt every time he was asked how he was going to make it all happen. One did not finance, divest, and rationalize a company into greatness. How, indeed, could he be sure the actions he had cited would really cure some of United's underlying problems—specifically, its loss of clear competitive leadership in several product lines and its failure to come up with important product innovations over the past several years? How would conservative financial policies, factory rationalization, or corporate sell-offs stimulate innovation or improve the quality of the main divisions' products? But no one challenged his answer, and with it the meeting adjourned.

On the way out, however, Dick Fredericks, the recently retired chairman of a world-famous electrical equipment and electronics company, asked for a quiet word in Harlow's office.

"Harlow," said Fredericks, as they stood together by the window, "that was really a good job in there, but something makes me a little uneasy and I was trying to put my finger on it during the meeting. I think what bothered me was that something was missing in all the talk we had today about United's financial condition and long-term plans. You know, we didn't have any real discussion of your underlying competitive position. And that, unless I miss my guess, has slipped rather badly over the past four or five years."

Sharp fellow, Harlow thought.

"I know you guys at UI are considered one of the best managed companies in the business, and very definitely a technological leader—historically, anyway. What I'm wondering, frankly, is how long you can continue living up to that reputation. You know Caspar Wallingham, the CEO at Farflung Industries? He said something to me a year or two ago that I think may be pretty relevant to your present situation. The gist of it was that if a company is having trouble with its competitive product position, there's an awfully good chance that it's doing something wrong in its technical program. Specifically, Caspar said that it's probably because the R&D program isn't closely enough integrated with the business goals. In fact, he went on to argue that since R&D goals are part of the company's overall competitive strategy, you've got to know the technological opportunities *plus* the strategic objectives for the businesses or product lines to make sound technical program decisions.

"That may sound obvious, Harlow, but in my experience most companies stumble over it because of the enormous and increasing complexity of their business and their product lines and their technology. I get an uneasy feeling that here at United International the linking may not be taking place and, what's worse, people may not realize it. Of course, I could be way off base. What do you think?"

"Frankly, Dick, I can't give you a straight answer," Harlow said after a moment's pause. "I've always delegated the responsibility for the technical program to the division technical directors because it didn't seem to me that I had much to add from the corporate level. Also, you know, even though I was trained as an engineer, there's a lot of this new technology I'm not familiar with."

"Well, Harlow, I know that you believe strongly that the only way to manage this business is by being committed to decentralization and I don't disagree with you. But I think you've got to be careful that you're not inadvertently setting up a system within United International that perpetuates traditional patterns of investment. You've got to be alert to shifts in the strategic potential of your operating divisions and technical opportunities play a very big role in that. In a company the size of United International, you really can't afford to simply assume that the technical and business plans are linked. Because, if it isn't true, it could be disastrous for the company and for you. I can't give you very concrete suggestions for how to go about answering this question, but I think it's extremely important for you to get after it and, from what you say, I sense that you may think so, too."

"You're right, Dick, I *am* uneasy, and all the more so because I can't think of how to get hold of the problem. Perhaps what I should do is call together the operating committee and get their views on the subject. Maybe collectively we can fashion an approach to getting a hold on this question."

As Harlow reflected later on the conversation with Dick Fredericks, he understood why he had been uneasy at that day's meeting. Was something fundamentally wrong in the development process at UI? If so, he had a major problem on his hands, and no idea how to solve it.

THE OPERATING COMMITTEE MEETS

At the end of the next operating committee meeting, Harlow raised Dick Fredericks' concern.

"Gentlemen, I think you all know that, in addition to our regular agenda today, I'd like to get your views on a comment made to me by Dick Fredericks after the last board meeting. Dick made the point that while all of our financial and planning numbers were impressive, he would be more confident that we were going to be able to reach our objectives if he knew that our technology was on track. He meant that the strategic objectives of our businesses should be set with an eye toward the technological opportunities we see ahead, and that the technical plans that we have in place should accurately reflect the strategic business objectives. It was Dick's argument that unless these plans are linked, our R&D program will not be productive and that, of course, will lead to a deteriorating competitive position. Reading between the lines, I think Dick also meant that if we didn't get this linking right in the first place, no amount of project control could make any impact on our performance. Frankly, that's a point of view that I share. So the question I'm asking is: How well linked are our strategic and technical plans?"

"Well, Harlow, I'm glad to see somebody on the board has finally recognized how important technology is," said Doc Erlynmeyer, vice president for technology. "I agree with Dick Fredericks that if we don't have linking, little else makes any difference. However, I can assure you that our programs are closely linked to the strategic plans of our businesses."

"Oh, come off it, Doc!" growled Larry Bendix, division general manager for mechanical products. "You guys have been working on that graphite frame design for damn close to three years and all the time we've been telling you to use aluminum. You just won't listen. 'Not invented here!' "

"Larry, let's not start that all over again. We have to reserve the right in the technical department to make those kinds of technical decisions."

"It's not a question of departmental rights," said Larry. "It's a question of linking technology to our competitive strategy. You said the linkage was there and I'm saying it's pretty damn hard to find. And what about the new program in microprocessors we've been trying to get you started on for the last two years? We haven't even been able to get a proposal out of your guys on that."

"Well," said Doc, "we just can't free up time enough from our ongoing programs to start in right away on every new project you guys dream up. Plus our guys, who, I'd like to point out, have been designing integrated circuits for 10 years, tell me there's little advantage for microprocessors in our particular applications."

"How would they know?" put in Brad Pollini, head of engineering. "And even if they did, they'd be dumb to push a technology that would just put them out of business."

"Oh come on, Brad, you know our guys are honest."

"It's not a question of honesty, Doc, it's a question of knowledge. If somebody hasn't been working with a new technology, you can't expect them to be able to put it to work."

"Well, in any case, Harlow," said Doc, trying weakly to recover from the barrage, "it's my judgment as the technical head that the programs *are* linked. I'd go further and say that we're doing as well in this area as anyone could ask."

"Possibly so," said Harlow, "but somehow I'm not totally convinced. I see some other signs that we might have a linkage problem. Take our estimates of unit costs—how do you explain the fact that they routinely seem to increase by 25 percent after the start of production? If we're really linking our production requirements with our design efforts, we should be able to do better than that. And why do we miss so many of the most important innovations in the industry? We do about 35 percent of the total industry development work and we're not getting anything like that many of the significant innovations. Yet our marketing guys have told me time and again that they can see these developments coming months before the competition introduces them."

"Harlow, I hope you're not suggesting that all our R&D projects ought to be sure-fire hits. That's totally unrealistic. After all, R&D is a risky business. If it was just a question of engineering, Brad could do it all himself."

"No, Doc, I just want our fair share," said Harlow. He turned to Fred Broadhead, vice president for medical products. "Fred, you've had some experience in both development and marketing. What do you think?"

"Harlow, what I've heard here so far reinforces my view that we do have a problem in this area," Fred said slowly. "I can agree with Doc that a good many of our technical programs are linked to our strategic business objectives, but I believe that in quite a few cases that linkage is lacking. I do know we're continually getting outmaneuvered in the marketplace. Nothing major, mind you, but quite a number of small things. One day we're going to wake up and find that we're too far behind to easily recoup. So I think Dick Fredericks has a point, and I think we ought to take a look at it."

"Good idea, Fred. Why don't you put together a subgroup of this committee and figure out a way to do that. How much time do you think you'll need?"

"Hold on there, Harlow, I haven't even said I'd do it yet!" Fred said with a grin. "But I have a funny feeling that I'm going to anyhow. I can't say yet how long it'll take. Let me talk to some of the others, give it some thought, and get back to you a week from today with an approach for getting at the problem in a systematic way, and then we can talk about time."

THE NEW APPROACH

"Well, Fred, have you got a handle on this thing yet?" asked Harlow on the appointed day.

"I'm beginning to," Fred answered. "I've asked Doc, Brad, and Larry to work with me on it, and since it ties in so closely with our strategic plans I also asked Chris Thorn from strategic planning to join our team. Collectively, we think there are two questions that have to be addressed.

"The first is, are we putting our technical dollars into those areas that will give us the best possible returns? Specifically, are we putting our long-range, high-risk dollars into the areas of greatest opportunity? The same question goes for our exploratory R&D dollars. In short, is our program emphasis right from a corporate point of view? Our feeling is that there probably isn't much difference between the percentage distribution of funds we're putting into our product lines now and what we were putting in five years ago, despite changes in their potential. Accordingly, we'd like to rack up those areas that seem to have the greatest potential and see what kind of effort we're putting into them.

"Second, even if we *are* putting our technical dollars into the areas of greatest opportunity, are there still potential mismatches between the individual technical program and the business plans? For example, when we started out on this report, we thought that assessing opportunity would be a simple matter of looking at the strategic classification of each one of our businesses—you know, our harvest/divest or invest/grow classification system—and putting our most extensive technical efforts behind those in the invest/grow areas. But Doc Erlynmeyer pointed out that that was pretty silly. You've only got to look at truck leasing, one of our hotter businesses, but there isn't a thing to be done in it technically. At the other extreme, we've got communications equipment. That one's currently classified as a harvest/divest business, yet Doc tells us he's working on some laser technology that's got enormous potential for our communications division—in fact, it could revolutionize our whole view of the future of that product line.

"We found Doc's arguments pretty persuasive, so we modified our original approach to consider both the technical opportunities and the market opportunities of the business. It makes good sense, we think, to plot the market and technical opportunities for each business on a grid, like this *(Exhibit 1)*. You'd have a different technical program for each box in the grid. For example, where both technical potential and market opportunity are high, we'd like to support a full R&D program, basic research and all, because the market could use it and we could handle it technically. On the other hand, we'd like to cut way back in areas of low market and technical opportunity. The other corners of the grid (that is, those areas where we have either a high technical or market opportunity but not both) are areas where we should proceed with great caution, because we could waste a lot of time and effort. From this grid, we've classified products into three categories:

EXHIBIT 1 The R&D Effort Portfolio

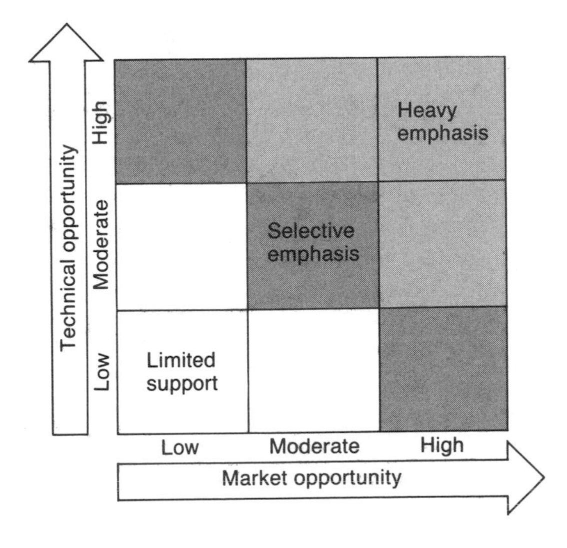

- *Heavy emphasis*—deserving full support, including basic R&D.
- *Selective opportunistic development*—may be good or may be bad; requires a careful approach and top management attention.
- *Limited defensive support*—merits only minimum support."

Harlow held up a hand. "Let me see if I follow you, Fred. You're saying that if we know our technical and market opportunities, we have a good start in thinking about what our technical program should be. Right?"

"Exactly, Harlow. And, given that understanding, we can compare what we should be doing to what we actually are doing, to find any weak spots. We've even gone one step further by laying out the kinds of programs we'd expect to see in each of those areas, so as to have a clear idea of what we're looking for when we start *(Exhibit 2)*. The technical opportunities concept was a little bit tough to get hold of. What we finally came up with was a rather neat

EXHIBIT 2 Implied Nature of R&D Effort

	R&D Program Elements					
R&D Emphasis	Level of Funding	Primary Focus of Work	Level of Basic Research	Technical Risk	Acceptable Time for Payoff	Projects to Exceed or Maintain Competitive Parity
Heavy	High	Balance between new and existing products	High	High	Long	Many
Selective	Medium	Mainly existing products	Low	Medium	Medium	Few
Limited	Low	Existing processes	Very low	Low	Short	Very few

concept we call the 'technology gap.' In a nutshell, that's simply the difference between the best anybody is able to do today, and the best that's theoretically possible, given the laws of nature.

"No product or process can be improved indefinitely. In electronics, for example, they've been getting more and more circuits on a chip. But they can't keep on doing it forever. Each of these little circuits generates some heat, you see, and if you pack them too closely together, the chip is going to overheat and not perform as designed.

"Now, there's an absolute limit to the rate at which heat can be removed from the chip. When the electronics boys hit that limit—and they probably will before the end of the decade—they're going to have to go through a dramatic change in strategy. No more 50 percent-a-year price decreases after that! So some big changes are in store. You can say the same thing for production processes. For example, in chemicals, the final process yield is determined by the laws of thermodynamics. You can get just so much and no more—Mother Nature won't let you."

"What you're saying, Fred," ventured Harlow, "is that just as in market potential, we've got to know where the areas of technical potential are."

"Exactly, Harlow. And that means we've got to take a look at the limits for each of our product lines and see how close to them we are, or how far off. Now, often we don't know where the limits are, and it would take a pretty clever research program to locate them precisely. Doc Erlynmeyer points out that this is just the sort of challenge that the boys in the corporate labs thrive on, so he thinks we should give the limits work to them. If this idea can be made to work, I must say it would be one of the few times I've seen the corporate labs pointed in a direction that's clearly relevant to the needs of our business. As a matter of fact, everyone I've talked to likes the idea."

Harlow nodded. "So do I."

MATCHING STRATEGY AND TECHNOLOGY

"OK," Fred went on, "now let's look at the second big issue, linkage. What we think we ought to do is ask each of the business managers to lay out his strategy and his technical plan on a single piece of paper, like this *(Ex-*

EXHIBIT 3 Testing Business Strategy/Technical Program Linkage

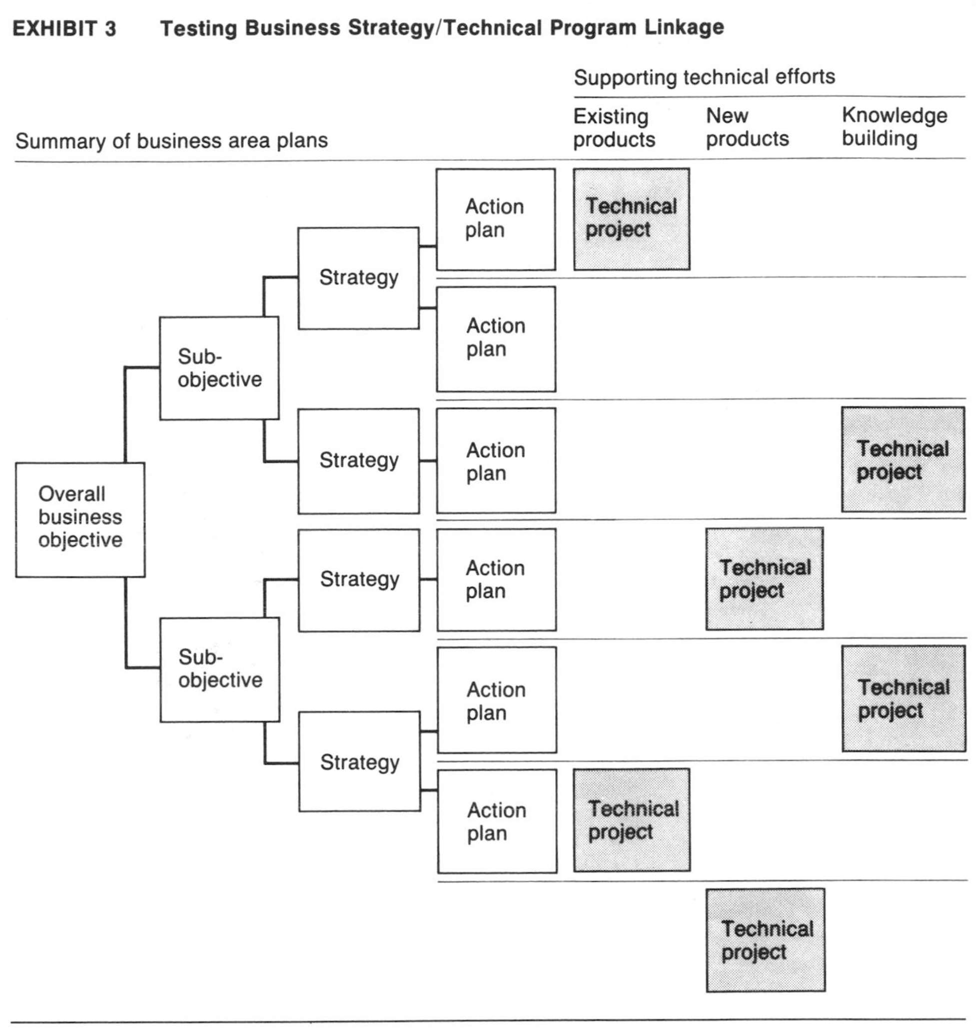

hibit 3). We've tested it in Toby Barry's communications division and I must say there were some surprises. Toby was convinced that his technical program was tightly linked to his business plans, but when he sat down and analyzed it he found that roughly 20 percent of his plans weren't getting anything like adequate technical support, and fully 60 percent of his technical program wasn't directly supporting the requirements of the business. Needless to say, Toby's busy right now reorienting his program! Anyway, these are the questions we think should be answered, Harlow. And we think we know how to do it."

"I must say you fellows have done an impressive piece of work here, Fred. You've given me a good framework for thinking about the strategic guidance of technology. I like your notion of the technical limits. It really gets to the heart of the question, and it's also something I think I can explain to the board. It seems to me that if we're not operating close to the limits of technology, probably neither are our competitors and, if there is a gap, *we* should be the ones to fill it. Well, Fred, how will you go about getting an analysis like this and how long do you think it'll take?"

"It looks like about four months," said Fred. "We have around 100 product lines and maybe 700 or so R&D projects, so we'll divide up the investigation into tasks that can be done without too much disruption of daily work."

"That's a little longer than I bargained for," said Harlow as he stood up. "But if you guys say it's going to take four months, why don't I just inform the board and tell them that we'll take up this subject again the meeting after next."

EXHIBIT 4 Actual R&D Emphasis versus Planned Income Generation

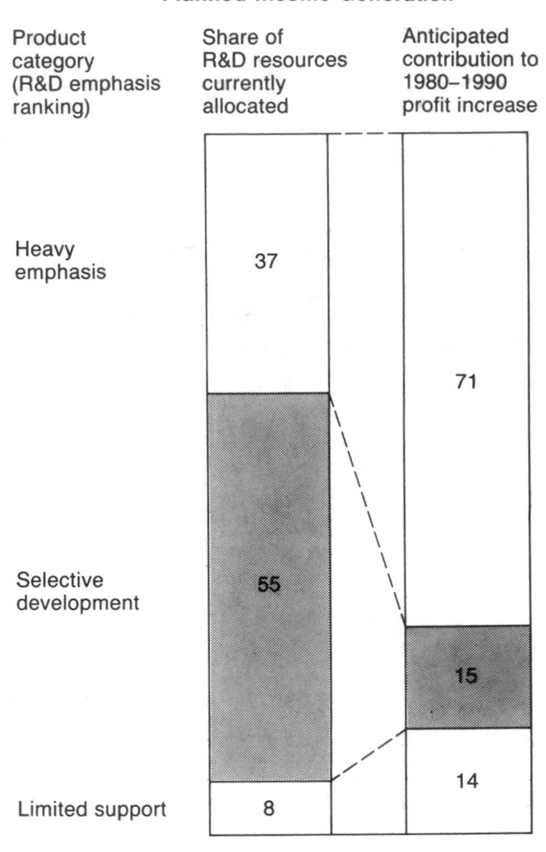

THE FOUR-MONTH REPORT

Fred began his report to Harlow by highlighting one of his main findings: UI had been underinvesting.

"First of all," he said, "you can see from this analysis *(Exhibit 4)* that while we expect over 70 percent of our future income growth to come from the product lines with heavy technological emphasis, we're only putting 37 percent of our funds into that area. Conversely, we're devoting 55 percent of our resources to the opportunistic development product area where we expect only 15 percent of our future growth. Well, of course, this all by itself doesn't mean that we're underinvesting in the heavy-emphasis area, but it does suggest that R&D for these opportunistic development product lines is a lot more expensive in terms of future contribution to earnings. There must be some underlying reason for this if it's to make any sense. However, when we looked into it, we could find nothing to account for the difference. For example, there are no economies of scale in the R&D for the heavy-emphasis product lines that we wouldn't also find in the opportunistic development product lines. So it must be something else.

"As we looked further into the situation, it became even more intriguing. In general, we found that UI just hasn't been differentiating between technical programs intended to support our heavy-emphasis, selective-development, and defensive-support technical opportunity areas. All three had roughly the same risk profile and offensive/defensive posture. The time required to get to commercialization was about the same for each group, too. In other words, we were taking no more risk in our heavy-emphasis, high-opportunity product lines than in our limited, defensive-support product lines. Nor were our programs any longer-range. In effect, we've been playing Russian roulette with our most important future product lines and engaging in technological overkill in our opportunistic development product lines.

"On the question of mismatching technical programs and business plans, our analysis turned up several kinds of problems. There are business programs that don't have any supporting technical efforts. There are technical programs that aren't properly geared to the relevant business objectives. There are dangling R&D projects that don't support any business program at all. Overall, only 50 percent of our business plans have the appropriate technical support, and only about half of our technical effort is supporting the business plans without a major problem.

"No wonder you were a little uneasy when Dick Fredericks raised his point, Harlow. We've been systematically underinvesting in our most important product lines and, even in those areas, less than half of our technical program has a chance of being truly productive. Fortunately, this exercise has not only pinpointed the problems, it has also shown a way for their solution. I'll come to that presently."

"Fred, that is an absolutely staggering report," Harlow broke in. "I had no idea that we were getting so far out of whack. Doc, what do you have to say about our linking now?"

"Candidly, Harlow, I'm a bit shaken. All this has come as quite a shock. I guess it's a salutary shock, though. It's really opened our eyes in R&D."

"Don't take too much of the blame, Doc," said Larry Bendix. "I think we on the business side bear a good deal of responsibility for this, too. I know when I went back to my guys and told them about the study, they said they thought things were in pretty good shape. So we've learned a lot from this exercise, too."

UNDERLYING CAUSES

Harlow leaned forward. "Fred, you said that your analysis helped pinpoint the problems and also suggested some solutions. Let's hear about that."

"Well, Harlow, the reasons why we got into this fix aren't really very surprising. Our problems are simply a consequence of the growing size and complexity of our business. Ten years ago, things were in pretty good shape, our emphasis was right, and our business and technical plans seem to have been closely matched. But over that decade, as you know, we've doubled our product lines and more than doubled our sales. We've opened up new plants both here and abroad. And the technologies central to about half our product lines have changed radically.

"All this is ancient history, of course—it's why we decentralized and set up profit centers four years ago. And while that pushed a lot of the decisions for marketing and manufacturing down to the level where they could best be dealt with, it tended to leave a lot of the technical decisions up for grabs. Typically the technical guy would find himself caught in the middle of a bunch of marketing, manufacturing, and planning guys from two or more divisions, trying to figure out what to do on a day-to-day basis to keep his programs straight, while at the same time he was supposed to be worrying about which areas of technology he ought to be pursuing for the long-term benefit of the corporation.

"Meanwhile, top management was assuming that the decentralization had solved most of the tough problems. We forgot that any decentralized organization builds up a certain inertia to maintain the status quo. There's a strong tendency to keep ratcheting the budget up at some percentage rate year after year, regardless. And that makes for a very slow reallocation of resources, even though the outlook of the business might change dramatically. That, plus the fact that R&D people need time to build up new skills, results in unusual stability of R&D—more stability, in fact, than may be warranted, given the changing outlook for our various businesses. Over time, then, things get out of whack.

"So it seems that in decentralizing, we created some new problems in the course of solving some old ones. That doesn't say it was all a mistake. What it does say is that we've got to come to grips with these more recent problems and overcome the handicaps they impose. Now, as I see it, there are three areas where we can really bring about some constructive change.

"The first is in management processes, planning in particular. We should incorporate the analysis of technical opportunity, and particularly technical limits, into our planning process so that we can give balanced consideration to both the technical and the marketing opportunities for each of our businesses. If this is done jointly by the marketing and technical people working together, I believe we'll have taken a long step toward building linkage right there. I also think that the one-page summary of business plans and technical plans, added to the annual planning process, would reinforce that linkage.

"Our second area of opportunity is in strengthening our organization. As you know, even in our decentralized divisions, we have R&D departments organized by function. What this means on a practical level is that the researcher in the lab is often working in support of half-a-dozen different objectives and setting his own priorities among them. If everyone's priorities happen to agree with those of the corporation, it's only by good luck. This problem can be avoided somewhat by organizing labs according to the organization of our business departments. Now the R&D boys say that we will lose efficiency this way and I must say they make a good point. But what we have to decide is whether we're prepared to sacrifice a certain amount of organizational efficiency for the sake of a real gain in the strategic effectiveness of our technical efforts. I think we'd all agree that it's a price worth paying.

PERSONNEL MANAGEMENT

"The third area is what you like to call people management, Harlow. One of the problems of decentralization is that, while we have a perfect system for controlling the short-term outlook of the business, we have an imperfect one for controlling the long-term outlook. That's because of both the short tenure of any manager in a given job and because the normal performance measure is contribution to this year's earnings. Probably this is why the time profile we found in the heavy-emphasis product lines wasn't significantly different from the one in the limited, defensive-support product lines.

"It's a problem that can be fixed, first of all, by working closely with our personnel people to set up a system which rewards people for their contributions to the long term as well as to the short term.

"Second, we've got to take a hard look at career planning. We ought to be moving technical people around, over time, in a carefully planned way, not only within the technical organization, but also marketing and manufacturing. So far, we have moved quite a few, but after they've gained their business exposure we seem to forget all about bringing them back into the technical organization at a higher level. So the flow is in only one direction—out of research. That leaves us with less than perfect linking.

"So, Harlow, while decentralization has created part of the problem that we're in, the solution isn't to recentralize anything but to carry the decentralization that one logical step further."

"Well, Fred, I must say I'm relieved to hear you say that," Harlow remarked wryly. "I was getting pretty nervous for a minute! Seriously, what you say makes a lot of sense. But it does strike me that you're proposing a fairly mechanical and systematic approach. We don't want to start rewarding our researchers for their ability to fill out forms, do we? Isn't this approach of yours going to dampen their creativity?"

"Harlow, let me answer that," said Doc Erlynmeyer. "As I see it, what Fred has laid out here is really a way of thinking a lot more clearly about our technical efforts, and seeing them in the light of our business objectives. I think what it'll do is focus our creativity on the most important problem areas. I'm already beginning to see a remarkable increase in the number of new ideas coming out of the labs. I think it's going to stimulate our creativity rather than the reverse."

"One other point, Harlow," interjected Larry Bendix. "I agree that creativity is key in R&D, but don't forget Edison's formula for innovation: 1 percent inspiration, 99 percent perspiration. The perspiration part needs plans and schedules. Fred has given us a way of putting those plans and schedules into strategic perspective."

"All this is very encouraging," Harlow said after a pause. "But I'm also concerned about the vast cost overruns we've had on our projects; the unpredictability, or the seeming unpredictability, of controlling them and the difficulty of getting the timing right. Aren't those important problems, too?"

"Sure, Harlow," said Fred, "but our fundamental premise was, and is, that unless you have the direction right nothing else—and that includes project control—is going to make any difference. Just to check this out, we looked into the causes of the delays on our past projects. More than three quarters of them could be traced to changes in the strategic objectives of their product lines rather than to problems in the technical programs per se. It seems that what we generally do is let the projects meander along until somebody decides what the real project objectives are. Then we have to do a costly major about-face in the development process. I think you'll find that project control processes will also work a lot better as a result of the program we're suggesting. Of course, we may have to strengthen our project control procedures too, but let's give this plan a try first.

"While I doubt that there's any process that will eliminate all the problems in R&D, I think it's important to make some progress in the short term, and it looks like this is our best option from that standpoint. Maybe we'll run into problems when we transfer our projects from the lab to the manufacturing line. If so, we can work on those as they come up."

"Okay, Fred," said Harlow. "I agree that we'll never get it perfect, and this approach seems like the best one we've had so far. So I'll take that material from you now, if you don't mind. I'll need it to prepare my speech to the board."

NEXT BOARD MEETING

Harlow commented on a proposed acquisition, accepted some questions from the members of the board, and moved on to the last topic. "Gentlemen, as I told you at the last board meeting, we've been undergoing a rather intensive review of our technical programs at the suggestion of Dick Fredericks. We discovered that Dick's concerns were amply justified. In fact, we were on the verge of having a major problem.

"In a nutshell, our decentralization had seriously weakened some vital links between our business operating and technical people. As a result, our technical efforts were getting out of synch with our business efforts—pretty seriously so, in fact. Well, I'm happy to say that all that is behind us now. We've taken effective short-term actions to redirect our technical program to the most important business objectives and to ensure that individual technical projects are well lined up with the specific business plans in each of our project lines. Perhaps more important, we've begun to modify our system of management to cope with problems that developed in the wake of decentralization. We're making a series of changes in our planning processes, our organization structure, and our career-path development and motivation systems to ensure that linkage problems don't recur. More accurately, if they do recur, we'll know it far enough in advance to take corrective action.

"In the areas where we've started this program, the results have been good. For example, our new blood dialyser is being redesigned to reflect business plans. It's been on the market a month now and already profits are double what they were because of reduced assembly costs on the new design. Now, do any of you have any questions?"

Steele Blackall raised his hand. "Harlow, six months ago I sensed something was wrong at UI but couldn't put my finger on it. I must say I'm gratified to hear the progress that you've already made in this area and I think all of us here would agree that you and your operations committee are to be congratulated on an outstanding job."

With that the meeting was adjourned. As they were walking out of the door Harlow felt Dick Fredericks' hand on his arm. "Well, Harlow, it really looks like you've made progress. Congratulations on getting through to the problem."

"I can only claim part of the credit, Dick," said Harlow. "A lot of it goes to Fred Broadhead and his team. But thanks to your own prompting at the beginning, we've learned a lot at UI."

CODA

That *Business Week* photographer took a damn good picture, thought Harlow a year later. And for the cover, no less. Good title, too: "McLaughlin Puts New Life into United International: A Venerable Competitor Shows How to Stay Ahead." The only trouble is, they make it sound easy!

Licensing and Marketing Technology

Case II–6
Biodel, Inc. (A)

J. M. Crowe and M. A. Maidique

Dr. Oscar Feldman, founder and president of Biodel, Inc., sat back for a moment and reflected. The year 1979 had recently ended. It had been a constructive 12 months for his small biotechnology company, yet Feldman knew that several difficult strategic choices loomed before him in 1980.

Biodel stood at an enviable crossroad. Feldman was confident that Biodel had distinct competences in its current biotechnologies. These competences currently provided the company with competitive advantages on which the president had resolved to capitalize. Should the company pursue the significant growth prospects in these current technologies—cell biology, molecular biology, and immunodiagnostics? Or should the company expand its technological focus to include genetic engineering, a field poised at the threshold of exciting advances? If Biodel were to pursue genetic engineering, how should it do so?

Finally, Feldman wondered if Biodel had sufficient personnel and funds to pursue an aggressive strategy.

COMPANY BACKGROUND

Oscar Feldman was originally from Scotland, where his father had been a successful businessman. After a McGill chemistry Ph.D. and post-doctoral work at Harvard Medical School, Dr. Feldman taught at Stanford University. His work there centered on applying chemistry to biological and medical problems, including cancer research. During his 11 years at Stanford, Dr. Feldman published almost 100 papers, enjoyed widespread popularity, and established a base of contacts in the academic community which he valued highly.

Biodel was founded in 1962, shortly after Dr. Feldman obtained a contract for research which he felt could be executed more effectively in a commercial setting. The contract required the combination of several disciplines, including chemistry, biochemistry, biology, and enzymology, in order to obtain the best results. In an effort to cover initial working capital and facilities requirements for the start-up, Dr. Feldman raised $50,000 from local businessmen. He considered seeking a larger sum, but decided that he would rather remain the principal shareholder of a smaller enterprise.

Dr. Feldman's initial business objective for Biodel was simply to establish a position of technological leadership in the biomedical industry. A key to his strategy was the leveraging of his academic contacts. Dr. Feldman relied on his contacts at Stanford, for example, to bring on scientists to commence work on Biodel's first contract. He also planned to obtain additional government research contracts. Such contracts would provide Biodel with the financial support necessary to build technological leadership through the development of high-quality facilities and staff. Finally, Dr. Feldman began to assemble a small group of leading academics to advise Biodel. He stated at the time: "It is important to associate with none but the very best minds in the field."

During the start-up phase, Dr. Feldman held several views about his company's long-term strategy. Once Biodel had established itself in the contract research marketplace, he expected that the company would expand its focus. Dr. Feldman's exposure to his father's lumber business had convinced him that earning contract revenues and royalties would not be a sufficient long-term business base for Biodel. He envisioned a time when his company would manufacture and market biomedical products developed through its own research. Dr. Feldman was also wary of Biodel depending only on a few government agencies for its revenues. Marketing a product, he believed, would endow his company with a broad base of customers.

Despite slow but steady growth throughout the 1960s, Dr. Feldman's fears about dependence on government contracts were eventually realized. (See Exhibit 1.) In the early 1970s, government cutbacks resulted in the loss of Biodel contracts with the Surgeon General, the Quartermaster Corps, and other agencies. At the time, the government had been responsible for 85 percent of Biodel's revenues. Biodel was forced into its first layoff, which troubled Dr. Feldman greatly. He considered the technological expertise of his employees to be one of Biodel's significant assets and regarded layoffs as damaging to the company's long-term potential.

This period of cutbacks and layoffs was a crucial point in the company's history. Biodel faced the threat of bankruptcy. Dr. Feldman later claimed that "It was a good thing I wasn't such a good businessman, otherwise I would have realized that Biodel was insolvent."

EXHIBIT 1 Financial History

Fiscal Year	Revenues	Net Income	PAT
1962	$ 250,000	$ 5,000	2%
1965	510,000	45,000	9
1968	619,000	31,000	5
1969	647,000	59,000	9
1970	352,000	(32,000)	−8
1971	289,000	(44,000)	−11
1972	394,000	26,000	7
1973	460,000	15,000	3
1974	583,000	49,000	8
1975	748,000	62,000	8
1976	1,011,000	88,000	8.7

Concerns with its long-term survival caused Biodel in the early 1970s to move into the business of scientific research products, an area which Dr. Feldman had not originally anticipated. Biodel's scientific research products, numbering approximately 500, were initially items which the company produced to utilize in its own research efforts. Biodel discovered, however, that biochemists and molecular biologists in other organizations also needed such research products, yet often lacked the technical expertise to make them. Biodel found a ready market for various reagents and synthetic nucleic acids which it had been using internally. Relying on word of mouth as its basic marketing tool, Biodel generated enough demand for its research products to reverse the sharp decline in revenues from lost contracts. The company was so successful in commercializing research products throughout the 1970s that research products constituted approximately 60 percent of the company's revenues by 1980.

In January 1980, Biodel conducted all of its research and development at a 14,000-square-foot leased location in Menlo Park, California. Due to the rapid growth of the late 1970s, quarters in the aging building were becoming cramped and Dr. Feldman knew that he had to start looking for additional space soon.

TECHNOLOGIES AND PRODUCTS

By the end of the 1970s, Biodel had developed special expertise in three areas of biotechnology: molecular biology, cell biology, and immunology. The company conducted research and sold research products to scientists working in each of the three areas.

All of Biodel's researchers were in some way studying cells, which are the basic biological units of life. Each cell possesses the biochemical machinery to grow and reproduce. An important component of this machinery is nucleic acid, a form of which is deoxyribonucleic acid (DNA). DNA is a relatively large molecule which consists of small building blocks called nucleotides. Specific arrangements of nucleotides, called genes, determine the production of specific proteins through a sequence of steps aided by biocatalysts called enzymes. Proteins provide the machinery by which cells utilize nutrients to grow and reproduce.

The techniques of *molecular biology* have played a leading role in determining the molecular structure and function of DNA and relating this structure to the production of proteins. In 1980, Biodel was using the techniques of molecular biology to isolate and prepare biologically active substances, such as nucleic acids and enzymes. The company then marketed the products to other researchers in molecular biology and genetic engineering.

Cell culture technology, a technique of *cell biology,* concerns itself with the growth of mammalian cells. Such cells have stringent nutrition requirements normally supplied by serum, the fluid portion of the blood. In this area, Biodel was primarily involved in manufacturing and marketing cell growth factors, products which could be used, either partially or completely, to replace serum in helping cells to grow. Adequate quantities of uniformly high-quality serum (usually derived from horses, pigs, and calves) were proving difficult for researchers to obtain. Thus, Biodel had enjoyed increasing success in the late 1970s in selling its cell growth factors to scientists who could not locate serum for use in their cell proliferation research. By 1980, Biodel's pioneering efforts had paid off. The company dominated the growth factor market with about 60 percent share.

Biodel's third area of special expertise was *immunodiagnostics.* Immunodiagnostics is one field within immunology, which is the study of how organisms protect themselves against infection. When foreign substances (antigens) are introduced into an organism, the organism responds by producing antibodies which bond themselves to the antigens. The presence of a specific antigen in a sample may be measured by adding to the sample a known level of antigen which has been radioactively tagged. The radioactive antigen competes with the sample's antigen for the antibodies in the sample. By measuring the residual radioactive antigen not attached to the antibodies, the level of antigen in the sample can be accurately estimated. Biodel had research expertise and a small product line in radioactive immunodiagnostic products.

CURRENT ORGANIZATION

Contract Research

During the 1970s, Biodel reported its revenues in two lines: contract research and research products. (See Exhibit 2.) The contract research activities were projected to generate $1 million in revenues in fiscal 1980. Seventy percent of those revenues related to industrial research, the two prime customers being a large pharmaceutical company and a large chemical company. The government accounted

EXHIBIT 2 Selected Financial Data (for fiscal years ending August 31)

	1977	1978	1979	Est. 1980
Revenue:				
Product sales	$ 598,941	$ 738,732	$1,153,749	$1,450,000
Contract revenue	754,207	836,385	730,942	1,000,000
Royalty and license income	—	—	—	50,000
Total revenue	1,353,148	1,575,117	1,884,691	2,500,000
Cost of revenue:				
Cost of product sales	271,225	324,781	489,091	750,000
Cost of contract revenue	550,652	659,480	667,548	800,000
Total cost of revenue	821,877	984,261	1,156,639	1,550,000
Gross profit	531,271	590,856	728,052	950,000
Operating expenses:				
Research and development	146,228	193,285	274,224	200,000
Selling, general, and administrative	205,592	245,475	436,057	650,000
Total operating expenses	351,820	438,760	710,281	850,000
Net interest income	—	—	2,000	—
Income before income taxes	179,451	152,096	19,771	110,000
Income taxes	81,400	56,200	2,000	10,000
Net income	98,051	95,896	17,771	100,000
Net income per common share	.08	.08	.01	.07
Common shares outstanding	1,351,875	1,351,875	1,351,875	1,351,875
Working capital	449,209	485,587	476,698	325,000
Total assets	803,238	875,063	965,559	1,400,000
Long-term debt, including capital lease obligations	127,095	108,414	114,732	30,000
Stockholders' investment	433,233	529,129	546,900	650,000

for the remaining 30 percent of the contract research.

The scope of Biodel's contract research included work in the company's three primary areas of expertise (molecular biology, cell biology, and immunology) as well as in fields such as cancer chemotherapy and enzymology. Within those areas, the company offered its customers high-quality technical advice, numerous links to the scientific community, and a highly sophisticated contract research and development service with a record of many successes.

Dr. Feldman marketed Biodel's contract research efforts. He personally secured the contracts through his relationships with scientists in government and industry. Dr. Feldman also supervised the ongoing contract research activities. He managed the activities informally, preferring not to set exceedingly detailed milestones and budgets. He commented: "I consider my researchers to be professionals. I see no need for me to continually monitor them. Scientists are motivated by new technical challenges, not by heavy-handed supervision."

Research Products

Dr. Feldman expected sales of research products to reach $1.5 million in fiscal 1980. Research products consisted of three interrelated product lines corresponding to the company's three areas of scientific expertise: molecular biological products, cell biological products, and immunodiagnostic products. The product lines were generally sold to researchers in universities, private laboratories, and industrial firms. Despite a limited marketing effort, sales had been growing at a 35 percent clip over the last several years.

In the area of molecular biology, Biodel prepared and stocked the largest commercially available selection of synthetic nucleotides. Researchers used nucleotides as substitutes and primers for nucleic acid enzymes, as reference compounds for sequence analysis in studies of nucleic acids, for the development of new separation techniques, and as tools in recombinant DNA research. Nucleotides accounted for 50 percent of the sales of all research products. (See Exhibit 3.)

Cell growth factors, Biodel's primary product offspring in the cell biology field, generated 40 percent of the research product revenues. Sales of cell growth factors had risen rapidly over the past several years, and Dr. Feldman believed that they represented a fertile area for future growth. There did exist disagreement within the company's management team, however, over the company's current competitive position in cell growth factors. Dr. Feldman considered Biodel to be the technological leader, yet several top employees believed that this assessment might be too optimistic. All did

EXHIBIT 3 Research Product Sales

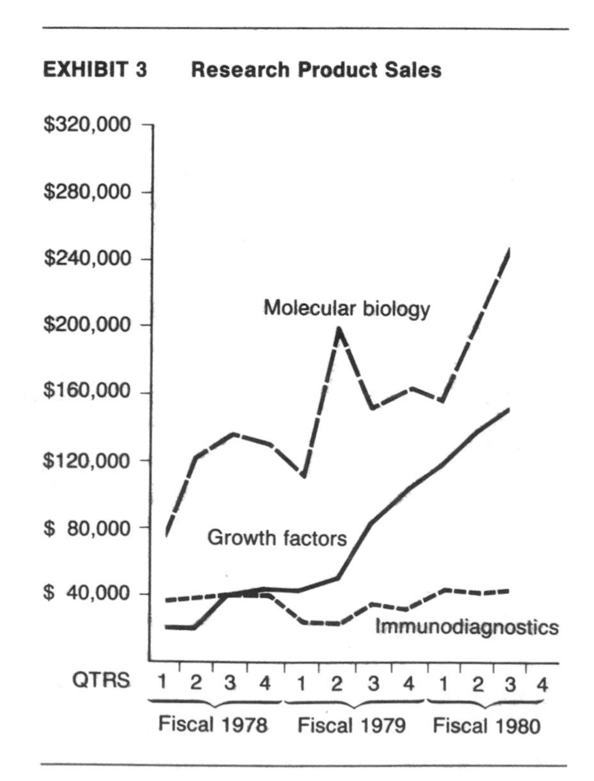

agree, however, that they lacked the necessary market research data to back their conclusions with confidence.

Biodel had been a major factor in the immunodiagnostics market for several years until several large firms aggressively entered the field and slashed the company's market share. The product line had not expanded since that time and constituted 10 percent of the sales of research products in 1980. Further significant growth in radioactive diagnostic products was not considered likely.

Profitability for these research products varied, depending upon the intensity of the product's research and development. Operating profit margins, after charges for product cost, research and development, and marketing, were estimated at 20 percent on the aggregate. Biodel's current accounting system, however, did not provide product-by-product profitability data and aggregate data was clouded by overhead allocations that in the opinion of some managers were "arbitrary."

Personnel

Biodel was organized along lines of scientific expertise. (See Exhibit 4.) The three operating groups—molecular biology, cell biology, and immunodiagnostics—were each under the control of a manager who reported directly to Dr. Feldman. All three managers were experienced scientists who were long-time employees of the company. Each manager supervised R&D and production and held some marketing responsibility. The organization within each operating group was highly fluid. Generally, those scientists who completed research on a particular product would then turn to manufacturing the product in small quantities and would also determine which scientists in other organizations would be likely customers for that product.

From the standpoint of staff, Dr. Feldman had kept Biodel a lean organization. Biodel had 55 employees, most of whom were scientists or technicians. The company employed neither a

EXHIBIT 4 Organization Chart

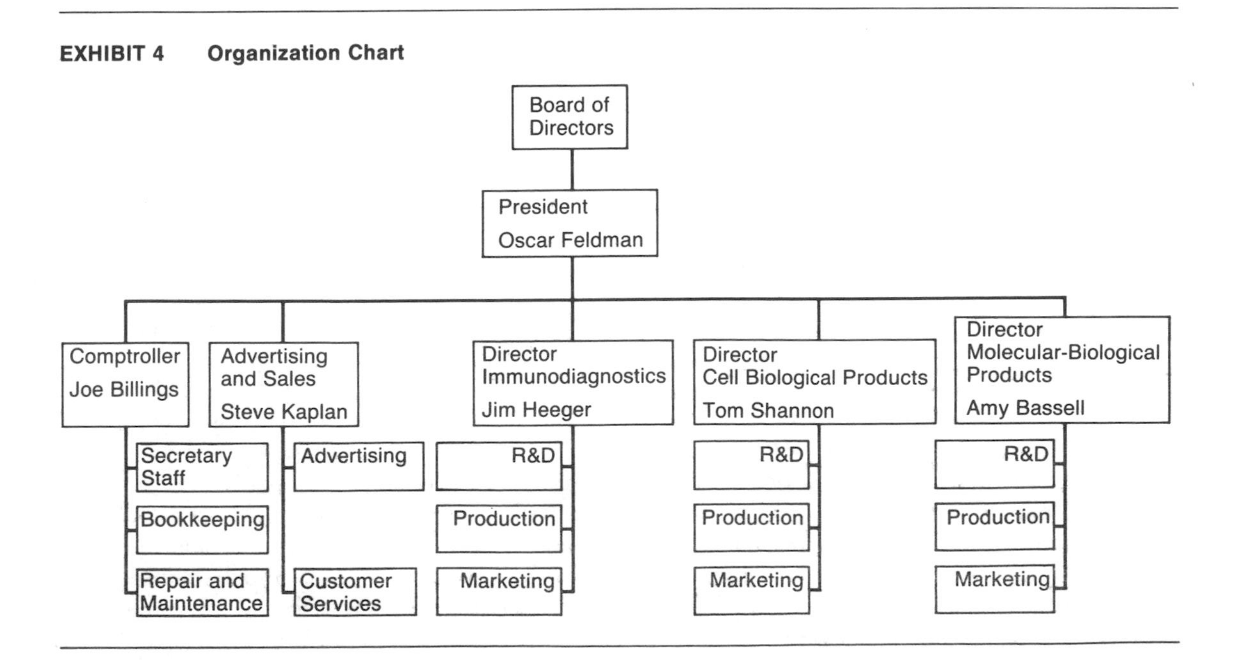

marketing manager nor a research director. Dr. Feldman filled both roles due to his widespread contacts and scientific expertise. All of Biodel's financial functions were handled by an accountant who was also a long-time employee. Insofar as Dr. Feldman perceived major strategic, financial, or administrative issues, he invariably made the decisions himself. He rarely convened staff meetings and did not require regular reports from his subordinates. Dr. Feldman considered his management style well-fitted to Biodel's organization and atmosphere:

> We are a paternalistic company. I believe in management by walking around and talking to people. That's the best way for me to stay on top of what is happening. I don't want our employees tied up in paper shuffling. And anyway, "professional management," goals, budgets, and meetings are not my "schtik."

One employee characterized Biodel's board of directors as "Dr. Feldman and his friends." The board consisted of four members, including Dr. Feldman. A corporate lawyer, age 79, sat on the board. He had been involved with the company since its founding. Dr. Feldman had also wanted scientific expertise on the board and had persuaded a gifted Stanford scientist, now retired, to join him. The fourth member was an associate of Dr. Feldman who was an officer of an investment banking firm. The board was convened infrequently, whenever Dr. Feldman felt a need to discuss the company's affairs.

From the standpoint of the researchers, Biodel was an exciting place to work. Dr. Feldman believed that he placed heavy pressure on his researchers and in return offered them projects at the cutting edge of technology. Although Biodel's equipment was not the most sophisticated and the company's quarters were spartan, the work itself was more challenging and fruitful than that of many commercial labs. In fact, employees likened the company's atmosphere to one of an academic facility. They considered the combination of informality and high challenge to be attractive. Turnover among employees, especially the senior technical people, was extremely low. The technical staff expressed pride in the company's work, and one referred to the firm's reputation as the "Cadillac of the industry." Dr. Feldman, who had distinct automotive preferences, preferred to refer to it as the "Mercedes of the industry."

A second reason for the low turnover was Oscar Feldman himself. Dr. Feldman was generally regarded as the hub of Biodel's universe. The senior employees were unanimous in their affection for the president. One of the senior scientists explained the phenomenon as follows:

> Simply put, Oscar is an attractive man. His warmth and effusiveness is infectious. He is so naturally witty and charming that you can't help but enjoy working with him. Oscar gets so enthusiastic about the work that it's contagious.

A senior manager added,

> Oscar is unique. Irrespective of the situation, he never fails to appear distinguished. He wears perfectly tailored suits, drives a stately old convertible, and has impeccable manners. He also looks remarkably trim and healthy for his age. There are not too many people still skiing avidly at age 65. True, sometimes he'll forget a fact or two, but he can also be extraordinarily articulate.

Marketing

Throughout the 1970s, Biodel's marketing effort was an informal mixture of different activities. Research products were sold by mail, with customers typically having heard of the company through word of mouth. Trade shows, advertising, direct mail, and phone solicitation were also employed from time to time. Order processing and shipping were handled informally, without much emphasis on control. Dr. Feldman cited Biodel's customer service as "almost laughable."

In 1979, Dr. Feldman determined that the company needed to market its research products more aggressively and systematically. He decided that he needed someone who had familiarity both with the sales function and with biotechnology. In May 1979, he hired Steve Kaplan, who had been marketing manager at a large pharmaceutical company.

Tension eventually developed between Dr. Feldman and his new manager. Dr. Feldman wanted Kaplan only to organize a sales effort and gather information on customers and competitors. He still felt that he should direct Biodel's marketing strategy himself. Kaplan, on the other hand, perceived a need for focus in the company's marketing strategy. In addition, he concluded that Biodel was understaffed and proceeded to hire additional salespeople, an administrative assistant, an order entry clerk, and a secretary.

The results were mixed. Sales of research products increased 65% in the first quarter of fiscal 1980, an achievement for which Kaplan took credit. In addition, the customer service function began to respond more systematically to shipment delays and other problems. On the other hand, marketing costs increased 500 percent, resulting in sharply reduced profits despite the jump in sales. Dr. Feldman began to wonder if his marketing group was too large for Biodel, given its size and stage of development. He also began to question Steve Kaplan's tendency to make solo decisions regarding the company's marketing direction.

GROWTH OPPORTUNITIES

While Dr. Feldman was satisfied with the course Biodel had taken over the past 10 years, he knew that important choices remained to be made. Several of his top scientists were excited about two of the company's new product developments in cell biology and immunodiagnostics. At the same time, interest was building rapidly in scientific and financial circles in genetic engineering. The new genetic engineering technology was closely related to Biodel's expertise in molecular biology and could be a natural extension of the company's scientific focus. Each of the growth opportunities looked attractive, and Dr. Feldman wondered how he should decide which path to pursue.

Cell Biology

Based on its experience and expertise in using cell growth factors as components of serum substitutes, Biodel had under development several synthetic sera which were formulated to satisfy the growth requirements of a variety of cell lines in tissue culture. The synthetic serum substitutes would replace natural fetal calf serum, which was currently the most widely used source of growth for cells. Horse serum was second in market importance. The price and quality of fetal calf serum had been unstable over time because the availability of the product depended upon the slaughter of cattle, which tended to be cyclical. Biodel's researchers projected the market for fetal calf serum at about $50 million domestically in 1980 and $80 million worldwide and growing at 15 percent a year. Biodel believed that the market for synthetic sera of uniformly high quality and reliable supply would be even larger. However, these numbers were somewhat speculative, for the firm had not conducted a systematic analysis of the serum market.

Dr. Feldman believed that the company would have a competitive edge in synthetic sera which would be difficult for other firms to overcome. This advantage would allow Biodel to achieve a market share of up to 20 percent of the current market. The sera would be produced by adding to distilled water certain combinations of cell growth factors which would not be easy to break down and analyze. Even if a competitor could break the combinations down, Dr. Feldman believed that developments in this scientific discipline could not be quickly duplicated. It might require several years between the time a firm initially studied an area and the time it commercialized a product. Dr. Feldman was not certain whether other firms were currently pursuing the same course as his company. Finally, Biodel planned to cement its advantage by applying for patent protection, although it was by no means certain that a patent could be obtained.

Tom Shannon, the cell biology manager, felt that Biodel could eventually produce the synthetic sera at costs which would allow it to price the products competitively with fetal calf serum. At this point, Shannon guessed that a $1 to $2 million investment would be needed in manufacturing facilities. Dr. Feldman thought that the company would also need additional management personnel to oversee the venture. Both he and Tom Shannon were unsure as to just how best to market the product and what product introduction and marketing costs would likely be.

Immunodiagnostics

Within immunodiagnostics lay another opportunity for Biodel to enter markets vastly larger than its current customer base in research organizations. The company had under development a new testing technology based on enzyme membranes rather than radioactivity. Jim Heeger, the immunodiagnostic manager, expected that the new product (called DEMA) would have many applications in clinical, medical, environmental, and industrial testing. The product could determine the presence and level of many substances, including hormones, enzymes, drugs, viruses, and bacteria. The tests could include, among others, those for pregnancy, syphilis, hepatitis, cancer, toxins in food, and carcinogens in the environment.

Heeger considered DEMA an alternative to tests based on radioactivity. It appeared to share the high sensitivity of radioactive tests without the drawbacks and hazards associated with radioactivity. In addition, Heeger believed that DEMA tests would be simpler, faster, and less expensive than radioactive tests. Other enzyme-linked immunodiagnostic technologies, such as EMIT and ELISA, were already in existence, but Heeger judged them to be less sensitive and applicable to fewer substances than DEMA. The company had filed for patent protection and had been encouraged to believe by its patent attorneys that a patent for the technology would be forthcoming.

As with synthetic sera, the market for DEMA tests appeared to be vast, perhaps in excess of $100 million. One analyst's estimate placed the potential home market at over $1 billion. Again, however, Dr. Feldman and his subordinates were unsure how to bring the product to the marketplace. Investments in the necessary manufacturing and marketing facilities and personnel could easily total in the millions. Further R&D costs would range from $1 to $3 million. On the other hand, several large drug companies had expressed an interest in exploring a joint venture or licensing agreement. Under such conditions, license percentages ranged from 4 percent to 7 percent, depending on the fraction of R&D costs funded by the sponsoring firm. Dr. Feldman wanted Biodel to have some manufacturing and marketing capability, but he did not know how much of the marketplace his company could feasibly pursue on its own. One possibility was to have Biodel target the clinical diagnostics market, which was limited and well defined. Medical clinics would be a logical place to introduce the new DEMA technology. On the other hand, drugstore sales of DEMA potentially could generate enormous revenues and would be best pursued in conjunction with a partner that had an established distribution system and brand name. Pharmacies greatly outnumbered clinics, and DEMA would easily have numerous applications in the vast consumer markets.

Genetic Engineering

Dr. Feldman saw genetic engineering as a third opportunity for Biodel's expansion. The company currently had no direct expertise in the field, although it was closely associated with genetic engineering laboratories by virtue of its work as a supplier of molecular biology products. The nucleotides and synthetic genes which Biodel produced and sold were used as support products by genetic engineers. In some cases, Biodel was the sole supplier.

The opportunity for Biodel to move into genetic engineering itself arose through Dr. Feldman's contacts. Dr. Daniel Ballantine, a Berkeley Nobel Laureate and a pioneer in genetic engineering, was a longtime friend of Dr. Feldman and a consultant to Biodel. He had risen to prominence in the 1970s, and for the last two years he had been suggesting to Dr. Feldman that genetic engineering offered Biodel explosive opportunities for growth. Dr. Feldman began to consider seriously his associate's recommendations when he noticed an intensive interest developing in financial circles in the concept of genetic engineering.

The technology of genetic engineering is not complex in theory. To "engineer" a cell to produce a specific product, DNA containing the desired sequence must be isolated. The desired gene is either obtained from a biological source or is synthesized chemically. The gene is then spliced into a carrier molecule, called a vector, to form a recombinant DNA molecule. Control sequences which program the cell to produce the product coded by the gene are introduced into the vector, which itself could be a virus or a plasmid. The vector carries the new gene into the host cell, thereby programming the host cell to manufacture the desired product. The most widely used host cell has been Escherichia coli, or E. coli. (See Appendix A for more detailed technical explanation of the genetic engineering.) More is known about E. coli than about any other bacteria.

While the theory of genetic engineering may be easy to understand, the techniques have been difficult to perform. Procedures for isolating DNA and for utilizing vectors were not discovered until the early 1970s. By the mid-1970s, however, the academic world realized the future of gene splicing and many major universities launched DNA research programs. A critical breakthrough came in 1973 when Stanley Cohen of Stanford and Herb Boyer of the nearby University of California in San Francisco first chemically translated DNA from one species to another by gene splicing. (See Appendix A.) In contrast, the commercialization of the technology began slowly. In 1971, Cetus was the only genetic research firm. Genentech and Bethesda Research Labs were founded in 1976, followed by Genex in 1977, and Biogen in 1978. (See Exhibit 5.) Venture capitalists and large pharmaceutical, chemical, and energy companies provided the financing for the start-up phases of the fledgling firms. The large corporate investors had two goals: (1) to establish a technological window in a potentially revolutionary technology; and (2) to make a profitable investment.

The early investments were already generating large capital gains by 1979. In 1976, Inco purchased $400,000 of Genentech's stock, only to sell it to Lubrizol four years later for $5.2 million. In early 1980, Genentech estimated that the market value of its privately held stock exceeded $100 million—one half of which was owned by the company's officers, directors, and employees. A frenzy was enveloping the whole field of genetic engineering. Investors seemed willing to stake significant sums of money on almost any company who employed well-known scientists with connections to gene splicing. Financial journals continually touted genetic engineering's revolutionary potential to impact the manufacturing processes of products in the chemical, pharmaceutical, and petrochemical industries. *The New York Times* editorialized, "Recombinant DNA technology seems poised at the threshold of advances as important as antibiotics or electronic semiconductors." (January 19, 1980.)

Despite the euphoria surrounding genetic engineering, no firm had yet sold a genetically engineered product in mass quantities. Inves-

EXHIBIT 5 The Four Pacesetting Genetic Engineering Firms

GENENTECH, INC.:

Headquarters: South San Francisco. Founded in 1976; 110 employees. Has announced more DNA-made products than competitors. Joint ventures with Eli Lilly for human insulin; with A. B. Kabi of Sweden for human growth hormone; with Hoffmann-La Roche for interferon. Half-owned by employees. Lubrizol, a lubricating oil company, holds 20 percent; venture capitalists own the rest.

CETUS CORPORATION:

Headquarters: Berkeley, California. Founded in 1971; 250 employees. Concentrates on industrial and agricultural chemicals, also interferon. Joint ventures with Standard of California for chemicals and fruit sugar; with National Distillers for fuel alcohol. Founders, employees, and private investors own almost 40 percent; Standard California, 24 percent; National Distillers, 16 percent; Standard Indiana, 21 percent.

GENEX CORPORATION:

Headquarters: Rockville, Maryland. Founded in 1977; 50 employees. Concentrates on industrial chemicals. Has interferon research contract with Bristol-Myers; another contract with Koppers, a mining and chemicals company. Management owns about 45 percent; Koppers, 30 percent; InnoVen, a venture capital company backed by Monsanto and Emerson Electric, about 25 percent.

BIOGEN S.A.:

Headquarters: Geneva, Switzerland. Founded in 1978; about a dozen employees, plus others under contract. First to make interferon. Schering-Plough, a New Jersey pharmaceutical company that owns 16 percent, plans to begin pilot production of the antiviral drug using Biogen's process. Inco, formerly International Nickel, owns 24 percent. Remainder held by management and various outside investors.

SOURCE: *The New York Times,* June 29, 1980, section 3, p. 1. Reprinted by permission.

tors were lured by the prospects of production of a host of recombinant products, including pharmaceuticals, biologicals, chemicals, and fuels. Several firms had announced product capability—Biogen was making interferon; Genentech, interferon and insulin—but observers believed that years would transpire before any genetically engineered product generated significant revenues. The major pharmaceutical and chemical companies had set up their own gene splicing departments, but they, as well as the small firms, had yet to understand the intricacies of production on a mass scale.

Amidst the mounting excitement over gene splicing's long-run potential, Dr. Feldman pondered Biodel's role. While the business of selling support products to the genetic engineering firms was expected to grow at a 30 percent to 50 percent clip over the next several years, it held neither the glamor nor the potential for explosive expansion associated with genetic engineering. One of Biodel's competitors in the molecular biology products industry was quoted as saying: "Our market won't ever compare to the markets for genetically engineered products. After all, it only takes one dollar of our stuff to make a thousand or a million dollars of their stuff." Cetus was the world's biggest user of enzymes in its genetic engineering research, yet it could have bought one year's supply for $12,000. The market for synthetic nucleotides, enzymes, and the like seemed limited.

What Dr. Ballantine offered Biodel was a route to expand into genetic engineering itself. He proposed a novel approach to the problem of growing cells by using yeast organisms as hosts in place of the E. coli predominantly used by other genetic engineering firms. For the past two years, Dr. Ballantine had been collaborating with three other renowned scientists on the development of yeast as the host cell in the genetic engineering process. The four men believed that yeast cells would ultimately prove more attractive than E. coli for industrial applications of genetic engineering. Yeast cells were easier and less costly to grow and it was believed by Biodel scientists that they could be grown to higher yields and thus with lower costs than E. coli. In addition, yeast cells contained biochemical machinery, absent in E. coli, which allowed for the possibility of programming the yeast cells to produce glycoproteins (proteins which contained carbohydrates).

Interferon and urokinase were two examples of glycoproteins which, although currently produced by conventional extraction processes, could potentially be manufactured by yeast cells through genetic engineering techniques. Interferon was a protein which performed a regulatory function in the body; it appeared to inhibit the multiplication of viruses and cancerous tissue cells. Because of the extraordinary difficulty of producing it in large quantities through conventional techniques, interferon was highly valued in medical circles. In 1980, its price exceeded $1 billion per pound. Urokinase was an enzyme produced in the human body as an agent to dissolve blood clots. Sales of urokinase up to 1980 had been limited due to the complication and high cost of conventional extraction processes. Biodel had already had some experience producing urokinase through a tissue culture process and knew that several drug companies were interested in securing a large, stable supply of the enzyme if it could be genetically engineered. In short, Biodel scientists believed that genetic engineers using yeast might have an advantage over researchers using E. coli in producing both interferon and urokinase.

Dr. Ballantine indicated his willingness to convene his three colleagues with Dr. Feldman in order to discuss a possible association with Biodel. Dr. Ballantine and his friends were all experts in yeast genetics. Full professors at the nation's most distinguished university laboratories, the four had been elected to the National Academy of Sciences and had jointly won all of the coveted biochemistry and molecular genetics prizes in American science. Dr. Feldman was excited at the possibility of attracting them to Biodel. He commented:

> As a group, they have talents unsurpassed in genetic engineering. James Finney, Columbia's leading biochemist, possesses one of the most penetrating intellects I've come across. He is a mature scientist with unimpeachable integrity. He's the type of person you'd want at your side when the going gets tough. Ralph Davidson is noted among the scientists at Cal Tech for his brilliant creativity. Despite his quiet nature, he could make invaluable intellectual contributions to our activities. Dennis Bernstein generates more ideas than 10 scientists combined. His work at University of Wisconsin has earned acclaim throughout scientific circles. And of course, we have my good friend Daniel Ballantine. He is simply the best there is. If the four of them worked with us, it would give our company a tremendous edge. If even one of them joined us, we'd have an advantage over Genentech, Cetus, and the rest. However, if we plan to land any or all of them, we will have to make an extraordinary offer.

Indeed, all four individuals were in high demand. They had offers from large chemical and pharmaceutical companies for positions as senior scientists with salaries ranging from $75,000 to $100,000. Smaller biotechnology firms were luring them with stock option packages which included 1 percent to 4 percent of the companies' outstanding stock. Leading universities were proposing prestigious endowed chairs with unparalleled academic freedom and clout. Even venture capital firms had approached them, exploring the possibilities of a start-up. One venture capitalist had asserted that he could raise $5 million on the strength of Dr. Ballantine's reputation alone.

Biodel, however, was not without its attractions. The company could offer the scientists both the freedom to start up their own gene-splicing R&D operation and the expertise in key related areas. Biodel had placed itself at the leading edge of technology and had earned a position of respect in science. The scientists' ideas could be further developed and enhanced through an association with the company. Despite the intangible benefits, however, Dr. Feldman knew that he would also have to structure a lucrative financial package to lure them away, even on a consulting basis, from their well-established academic environments.

Although there was much uncertainty surrounding genetic engineering—estimates varied widely on market sizes and on the time required for successful refinement of production processes—Dr. Feldman hoped that Biodel would be able to quickly generate revenue if it secured the services of the four scientists. He felt that the company could land a gene-splicing research contract from one of several large corporations for $5 to $10 million over a period of five years. From such an arrangement, Biodel could earn as much as a 25 percent margin after deduction for salaries, capital expenditures, and other associated costs. More important, the company would retain licensing rights at agreed-upon rates for potential products. In effect, Biodel would be conducting research at the expense of a commercial sponsor who sought to participate in genetic breakthroughs but who lacked the necessary technical capability.

ACTIONS TO BE TAKEN

Several routes lay open to Biodel at this point. Shannon was pushing to develop synthetic sera. He thought the company would lose any competitive edge it might have in cell biology if it did not bring the sera to market as soon as possible. Heeger, in contrast, pressed for more investment in DEMA. The immunodiagnostics manager argued that DEMA could reach the marketplace in a year if his group could obtain substantial additions in people and facilities. Both managers believed that Biodel could within reason meet whatever goals Dr. Feldman might set simply by pursuing the company's present product opportunities. They saw little reason to look elsewhere. Genetic engineering, on the other hand, represented a potentially lucrative expansion of Biodel's focus and a considerable boost to the company's prestige. Dr. Feldman was fascinated by the idea of having world-famous scientists officially and intimately associated with his firm.

No matter what course Biodel chose, tough financial decisions needed to be made. Development of synthetic sera was projected to cost more than $500,000; development of DEMA was estimated to be several times more expensive. On the advice of a finance professor at the Graduate School of Business at Stanford, Dr. Feldman held informal conversations with local bankers and venture capitalists. He discovered from the bankers that a loan above $500,000 would require his personal guarantee. One bank was willing to supply Biodel with as much as $1 million. In exchange, it wanted two points over the prime rate (currently 17 percent) and covenants restricting further debt, dividend payments, equity issues, and mergers and acquisitions. Venture capitalists generally expressed reluctance to invest in Biodel unless it strengthened its management team. One venture capital firm, however, tentatively offered Dr. Feldman $2 million for 40 percent of the company's equity. On January 1, 1980, Biodel had 1 million shares of common stock outstanding. Upon hearing the proposal, Dr. Feldman exclaimed that his company was worth many times more, prompting the investor to dryly remark that the financial community would not commit large sums of money simply for the potential of developing synthetic sera.

Another issue was the financial package that Biodel might offer to Dr. Ballantine and his cohorts. One alternative was to set up a separate subsidiary of Biodel, with all transactions between the parent and the subsidiary at arm's length. The four scientists would not work directly for the subsidiary, but they would act as an advisory board and recruit other top scientists to work for it. In return, they would receive a consulting fee and restricted stock in the subsidiary, which would vest over a four-year period at 25 percent per year. In this way, the four could maintain their affiliation with their respective universities. The finance professor guessed that a per diem of $500 to $1,000 and ownership of 2 percent to 10 percent of the subsidiary for each of the scientists would be a reasonable range within which Dr. Feldman could make an offer.[1]

A second possibility was to hire one or more of the geneticists as employees of the company itself. Dr. Feldman knew that to employ the scientists directly, he would have to match any salary and equity combination that another company might offer. This would pose a commitment larger than Dr. Feldman was used to for his own employees. Biodel's three managers currently earned salaries under $35,000, and their ownership of the company's stock jointly totaled less than 2 percent. Dr. Feldman owned over 80 percent of the outstanding stock. Friends, business associates, and relatives of Dr. Feldman owned another 10 percent. On the other hand, by hiring directly, Biodel would get at least one top scientist solely committed to the company's efforts and an immediate boost to its reputation. Dr. Feldman believed that one of the geneticists ought to be employed full time if Biodel planned to set up and operate a significant genetic engineering operation. He was not certain, however, of how difficult it would be to entice one of the scientists away from his academic research on a full-time basis.

A third alternative was to retain the scientists as technical consultants. Biodel would pay them a per diem fee of $800 to $1,200 in exchange for guidance on the company's current projects, proposals for new avenues of research, and recruitment of geneticists. In addition, the company would offer incentive agreements which would allot the scientists stock options on the basis of the company's performance. One possible measure of performance was revenue earned from genetic engineering contracts and products. The Stanford finance professor suggested the following proposal: Biodel would grant the scientists, as a group, options to purchase 50,000 shares at 10 cents per share for each incremental $1,000 in annual revenues related to genetic engineering. The grants would be made yearly for the next four years, based on Biodel's genetic engineering revenues in the particular year. Dr. Feldman estimated that over a four-year period, genetic engineering contract revenue could rise to about $5 to $10 million. The options would be exercisable starting one year from the date of grant if, and only if, the scientists were still consulting for the company. In this way, the Stanford professor noted, the scientists would have an incentive to remain with Biodel. They would also be motivated to help the company grow enough to go public, a development which would greatly increase the value of their options. A final advantage of this alternative was that Biodel would be able to avoid a drain on its cash flow stemming from large salaries.

Dr. Feldman felt that he had to make a move. Shannon and Heeger were pressing for more money, more people, and more facilities. Dr. Ballantine warned that his colleagues were being pressured to accept individual offers. It was a time for decisions.

[1] Maximum allowed consulting time at major universities was one day per week.

APPENDIX A PUTTING DESIGNER GENES TO WORK

Channing Robertson

Over a century ago, the Swiss biochemist Johann Friederich Miescher reported that the

Excerpt from Fall/Winter 1981–82 issue reprinted by permission of *The Stanford Engineer*.

contents of nuclei obtained from human cells were rich with organic acids containing nitrogen and phosphate. To this mixture he gave the name nucleic acids. In 1929, the geneticist, Muller, proposed that every living cell possessed "genes," and that the information required to perform metabolic functions and self-replication was in some way mapped onto genes. In the years that followed, genes and nucleic acids were shown to be one and the same, and in 1953, James Watson, Francis Crick, Rosalind Franklin, and others working in the Cavendish Laboratories at Cambridge University reported the structure of DNA (deoxyribonucleic acid). Of the several nucleic acids, it was this one that represented the data bank, common to all organisms, needed to remember the past and preserve the present. And of utmost importance to species propagation, it was this storehouse of information that must be transmitted to future generations with unfailing accuracy.

Since 1953, much of the molecular information code of DNA has been deciphered and more recently, with code in hand, molecular biologists have been attempting to create new forms of DNA. For billions of years, DNA has been altered in random ways by nature, an event referred to as evolution through mutation and natural selection. Eight years ago, DNA was altered, for the first time, by humans, in a predictable and controlled fashion. Such alterations are now commonplace activities in university, government, and industrial laboratories throughout the world.

Without a doubt, man's ability to reprogram and someday synthesize the master molecule of life, DNA, will affect our society in ways none of us can even imagine. At the very least, I believe it will lead to the development of an entirely new chemical process industry, one based on biological feedstocks that in time will supersede and ultimately replace the fossil materials-based chemicals industry now in existence. For this to happen, engineers and technologists will have to conceive, develop, and put into place processes that accommodate living organisms and biological catalysts in an optimally efficient and economic way for the synthesis of tomorrow's chemicals.

When nature is viewed at its most fundamental levels, one cannot help but be awed by the presence of the underlying symmetry and simplicity. For instance, all of the instructions to propagate every life form known to man reside within the chemical structure of a highly organized linear polymer known as DNA. How is this possible?

On a molecular level, each DNA molecule consists of parallel strands of polymerized

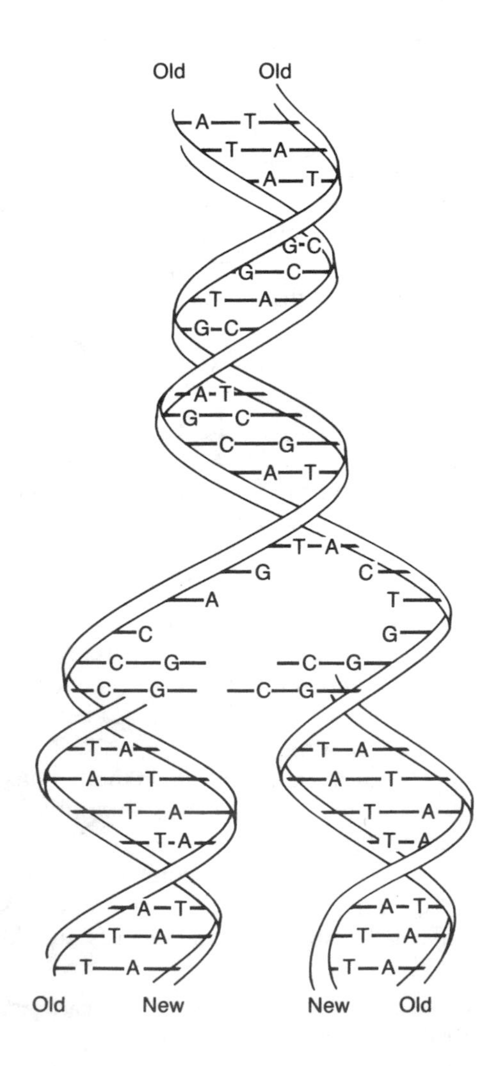

sugar residues. Every sugar residue is attached to one of four heterocyclic bases: adenine (A); thymine (T); cytosine (C); or guanine (G). These bases fill the space between the two polymer strands and are juxtaposed according to an inviolate base-pairing rule. Only a T may be associated with an A, and G with a C. As shown in the figure, the molecular order in one strand imposes a complementary molecular order in the opposite strand. The two strands are held together rather loosely by hydrogen bonds (dotted line). As discovered by Crick and Watson, the dual strands are twisted about one another like two intertwined spiral staircases (the double helix) with each turn incorporating 10 base pairs. In a simple bacterial cell, the DNA polymer contains approximately 2.5 million base pairs, whereas animal cell DNA has roughly 5 billion base pairs. To give an idea of the scale involved, imagine if each base (A, T, G, or C) were a model H-O scale railroad car (about 3 inches long). The two juxtaposed "trains" represented by a bacterial DNA molecule would then be a little over 100 miles in length. This "train" in turn would have to be packed in a space (a bacterium is about 2mm in diameter) approximately four football fields in diameter. In human cells, the equivalent DNA "train" is about 200,000 miles long, about eight times the earth's circumference. This train would have to be confined within a space approximately two thirds of a mile in diameter. In either case, one is impressed by the densities of information packing that nature is able to achieve.

In the vernacular of information processing, nature uses a three-bit per word vocabulary; that is to say, each sequence of three base pairs along the double helix of DNA (called a codon) specifies one of the 20 amino acids that are themselves joined into a linear polymer to form polypeptides and proteins (both are linear amino acid sequences). A sequence of base-pair triplets (or codons) together with start/stop codons is then sufficient to instruct for protein synthesis. This entire sequence of base pairs is called a gene. For instance, the protein insulin is composed of 51 amino acids. Therefore, the insulin gene must be 153 base pairs in length. A few additional base pairs are required to provide start/stop instructions so the organism can identify the insulin gene on its DNA and ascertain its length. Given the number of base pairs in bacteria ($\sim 2.5 \times 10^6$), several thousand protein molecules may be synthesized from the information encoded on bacterial DNA. In animal cells, instructions for several tens of thousands of proteins are contained in the DNA. One might expect this number to be higher in animal cells since there are about 2,000 times more base pairs available than there are in bacteria. A particular distinction between animal cells and bacterial cells is that in the former, only about 10 percent of the base pairs code for proteins. That is to say, 90 percent of the information data bank is blank—perhaps allowing for future programming either by natural or synthetic processes.

To this point, I have discussed only how information is stored by DNA. How is this information passed on to succeeding generations with such precision? Conceptually, when a cell divides to form two progeny, the DNA unzips and complementary base pairs are simultaneously retrieved from the cellular contents. As a result, two identical DNA molecules are synthesized, one being allocated to each of the progeny of cell division. This bifurcation process coupled with complementary base-pairing insures that all the genes are preserved for future generations.

In many ways, DNA recombination occurs all the time in nature. Sexual reproduction is a good example wherein the gene pool of the progeny is contributed to in equal amounts by each parent. Mistakes or errors in DNA replication and in DNA recombination are known to occur. Some may be traced to the presence of toxic chemicals, radiation, and the like, whereas others seem to be chance errors. The result of such mistakes are genetic mutants, and typically they are at an evolutionary disadvantage in competing with nonmutants of their species (recall Darwin's theories on the subject) and in due time most, although not all, are eliminated together with their erroneous genes.

DNA recombination performed as an orderly, conscious, and explicit act was first reported in 1973 by Stanley Cohen of Stanford and Herbert Boyer of the University of California at San Francisco and their colleagues. Using chemical scissors (enzymes known as restriction endonucleases), they cut open a small circular strand of extrachromosomal DNA (known as a plasmid) in a bacterium. Plasmids are small rings (they have roughly 100 times fewer base pairs than does the primary chromosomal DNA) of DNA found in bacteria that are not needed, in principle, for survival, but which impart useful properties to the bacterium, such as antibiotic resistance. They also used the same chemical scissors to cut up a DNA molecule from an unrelated bacterial species. A useful property of the enzyme employed is that it severs any DNA molecule at the same base-pair location (i.e., the G-A bond in a G-A-A-T-T-C sequence). Consequently, when the ruptured plasmid DNA and the fractured foreign DNA were mixed, some of them joined together according to the base-pairing rules discussed earlier. The plasmid chimera was reinserted into its original bacterial species and was replicated during cell division. As a result, the foreign DNA was "cloned" into the unrelated bacterium. As one might imagine, since 1973 numerous techniques have become available to chemically translate DNA from one species to another. The utility of doing so is best understood by example.

Consider the protein hormone insulin. Currently insulin is extracted from the pancreatic glands of cows and pigs. It is a costly and tedious process that results in a product (animal insulin) that is not quite the same as human insulin. Porcine insulin differs in only a single amino acid from that produced in humans, whereas bovine insulin has three nonoverlapping amino acids. Nevertheless, both bovine and porcine insulin work well, in most cases, in humans who suffer from insulin deficiency. There are, however, ways in which one might imagine obtaining human insulin. One approach would be to culture human pancreatic cells in vitro and collect the insulin they manufacture. Unfortunately, large-scale culture of mammalian cells is very difficult to achieve. Yet another approach would be to excise the gene (i.e., sequence of base pairs) that codes for human insulin and insert it into a bacterial plasmid. The bacteria could then be cultured en masse and, provided the human insulin gene was expressed (i.e., that it indeed caused insulin to be synthesized), we would have a bacterial insulin factory. Unfortunately, it is not enough merely to insert the human gene amid the myriad of bacterial genes. It must be placed in such a way that it does not interfere with any of the normal bacterial functions while at the same time it must be duplicated on cell division and also transcribed into its amino acid (protein) analog. These are all exceedingly nontrivial considerations.

In essence, recombinant DNA, molecular cloning, and DNA recombination all refer to the same activity, namely, the insertion of a DNA segment into an intact DNA molecule that is replicated faithfully by the descendants of the organism into which it is placed. This is not unlike coupling a new set of boxcars into the molecular "train" used earlier as an example. Furthermore, the instructions for polypeptide or protein synthesis must be recognized and obeyed by the bacterium.

Given the ability to augment the molecular instruction book for chemicals manufacture by cells, is it reasonable to ask what one might accomplish? Generally speaking, there are two approaches to be taken. Cells (bacterial, animal, or plant) can be made to express a chemical in far greater quantities than they normally would by multiple copying of the gene (or genes) that codes for the chemical. For instance, a particular microbe might produce a useful substance, a protein; however, the rate at which the protein is synthesized may be low. By "stitching" in multiple copies of the organism's gene that codes for the protein, then it would, in principle, be possible to amplify chemical productivity. This procedure would be akin to xeroxing the same page in the microbe's instruction manual many times, yet having them all read more or less simultaneously. The other approach is that of removing the instruction page (the gene) responsible for a particular chemical's synthesis and pasting it in someone else's manual (e.g., an unrelated organism). An example would be the insulin example referred to earlier. In either case, we are talking about synthesizing chemicals already made by cells. By virtue of DNA recombination, the cells will either make the material in more copious quantities (e.g., at higher rates), or else manufacture chemicals they normally (as a result of evolutionary pressures) did not but which instead were produced by some other cell line.

Since almost every chemical substance, or at least the precursor to same, is or can be synthesized by some type of cell, the possibilities for putting this new technology to work are indeed staggering. Already a number of biochemicals have been synthesized using molecular cloning techniques. Among them are: human insulin; human interferons (antiviral proteins); and somatostatin (a brain hormone). The current interest in interferons is particularly keen due in part, I presume, to their current retail value of $20 billion per pound.

In time we will see a host of recombinant products become available. Most of them will fall into one of the following classes: pharmaceuticals and biologicals; fine, intermediate, and bulk chemicals; and fuels. Examples of each include: pharmaceuticals and biologicals—human-growth hormone, antibiotics, nerve-growth hormone, other immune proteins besides the interferons, and vaccines; chemicals—epoxides, oxides, glycols, alcohols, acids, herbicides, pesticides, enzymes, lubricants, and sugars; fuels—alcohols, hydrogen, and methane. The impact of this new technology will also be felt by the farming and food industries. Fertilizers are normally produced from ammonia, which in turn is manufactured using methane, a dwindling resource. Some bacteria are able to form a symbiotic relationship with plants, the result of which is the direct fixation of atmospheric nitrogen to organic nitrogen, thereby obviating the need for externally applied fertilizer. Gene-splicing can, in principle, be employed to genetically alter plants to give them pest resistance and tolerance to drought and to saline soils, or to enhance their photosynthetic efficiency. Cells cloned to manufacture prodigious amounts of protein offer a new source of food and food supplements.

Reading II–7
Taking Technology to Market

David Ford and Chris Ryan

Corporate management of technology requires careful planning of the relationships among a company's technologies, its markets, and its development activities. It requires coordination of R&D activities to ensure an optimum research level—depending on a company's available resources, competitive pressures, and market requirements. And it requires systematic linkage between a company's product and process technologies: the products developed must also be produced efficiently.

Current management wisdom says that a company invests its skills and resources in developing products or services that are of value to its customers. However, we argue that to

---

maximize the rate of return on its technology investment, a company must plan for the fullest market exploitation of all its technologies. These technologies may, but need not necessarily, be incorporated into that company's own products or services. In fact, the growth of low-cost Third World producers will make it increasingly difficult for Western companies to exploit fully their technologies through their own production alone.

Thus, a company's marketing strategy may—and probably should—provide for the sale of technologies for a lump sum or a royalty. (By *sale,* we mean either the direct sale of a technology or the sale of a license to use it.)

The marketing literature, however, provides little help for the manager who wishes to exploit fully his or her company's technologies. Many critical questions remain unanswered, among them:

- What problems are involved in selling a technology?
- Is a company that sells a technology giving away its "seed corn" and thus prejudicing its future?
- How, to whom, and when should a technology be sold?
- What is the relationship between the sale of a technology and the sale of a product based on that technology?

In this article, we draw on our own research to examine these important questions within the conceptual framework of the technology life cycle (TLC), which traces the evolution of a technology from the idea stage through development to exploitation by direct sale. In particular, we examine the relationships between product and technology sales and describe the important choices and strategies open to management throughout the TLC. Specifically, we examine the value of shifting from managing a product portfolio to managing a portfolio of technologies. We also consider the impact of the TLC positions of technologies in the portfolio and the significance of a company's level of dependence on individual technologies.

WHY SELL TECHNOLOGIES?

Let us look at some of the reasons that even a successful company cannot fully exploit its technologies through product sales alone:

1. The ever-increasing costs and risks of R&D, especially of basic research, mean that companies must be certain to get the most out of the technologies they develop—including those that do not have immediate relevance to their own lines of business. For example, General Electric in the United States developed a microorganism that destroys spilled oil by digesting it. However, after winning a well-publicized patent case in the U.S. Supreme Court, GE is now offering this technology for sale because it does not fit into the company's major lines of business.

2. Some technologies may not fit into a company's overall strategy. Their application may be in markets that are too small or undesirable, or a company may have ethical objections to their use. Technologies may even cease to fit with a corporate strategy after they have been in use for some time. For example, GE sold off its mature Fluidics technology (which uses pressure changes in gas streams for measurement purposes and has wide application in textiles and metalworking) because it no longer fitted the company's major strengths or strategy.

3. Producers of products based on new technologies frequently rely on patents as a protection against competitive pressures, but patent rights offer only limited protection. After all, most technology can be copied by other producers, given time. These producers have cost advantages in not having to amortize the high R&D costs incurred by the originator. And Third World producers, in particular, have the benefits of lower labor costs. The recent battle between Eastman Kodak and Polaroid over in-

stant-picture technology is an extreme case of heavy reliance on patents that were unable to prevent competitive entry into a market.

4. Similarly, a company may refuse to sell a technology to an overseas producer because it fears competition at home or in other markets. If, however, a competitor agrees to sell that technology, then the first company will face the same competition without any compensating royalty payments or any control over the technology's diffusion.

5. A company may develop a new technology and then, because of financial woes or restrictions on its own production capabilities, may not be able to exploit it fully in the market. Sinclair Radionics, a small British company in the consumer electronics industry, claims to have made a breakthrough in the technology of flat-screen color TVs. The screens, measuring several feet in length and width and three quarters of an inch in thickness, would render obsolete hundreds of millions of dollars of investment in conventional cathode-ray tube plants. The first company to succeed in producing them at a competitive cost would have a decisive advantage in a market worth more than $20 billion a year. But Sinclair's past marketing problems have made it impossible for the company to find further risk capital. Thus, it has sought joint venture arrangements with large multinational electronics companies as a way of combining resources.

6. A company may not be able to capture through its own production all the world markets for a given technology. For one thing, sophisticated import restrictions on direct sales of manufactured products are common in Third World countries. For example, American Motors—which produces jeeps through subsidiaries in South Korea, India, and Australia and through license agreements in the Philippines, Pakistan, Sri Lanka, Thailand, and Bangladesh—is trying to penetrate the Chinese market. It is doing this through a license arrangement with the Beijing Automotive Industrial Corporation. Beijing has produced a line of four-wheel drive vehicles since 1964 and now wants to acquire the technology to produce certain jeep models in an existing plant.

7. Finally, a company may be restricted in its direct exploitation of technology by antitrust legislation. One well-documented example is the Federal Trade Commission's complaint against Xerox. The FTC charged Xerox with using its patent position to acquire a complete monopoly over the paper copier market. The proceedings were terminated in 1975 when the court entered a consent decree under which Xerox offered its competitors nonexclusive licenses on a number of its patents at either no royalty or a minimal royalty.

THE TECHNOLOGY LIFE CYCLE

Given these restrictions on the direct embodiment of technology in products, a company that wants to get the most out of its technology must plan carefully to realize the full market value of that technology at all stages of its TLC evolution.

In developing the concept of the TLC from the more familiar notion of the product life cycle,[1] we examined technology development, application, and degradation in such diverse industries as electronics components, consumer electronics, automobiles, shoe manufacturing, construction, mining equipment, and air conditioning.

Exhibit 1 shows the complete TLC for a major technology application. It is equivalent to the product life cycle for an entire industry for a generic product. Included are the proportion of the total use of the technology accounted for by the originator's technology sales and the

[1] Theodore Levitt, "Exploit the Product Life Cycle," *HBR,* November–December 1965, p. 81. See also Rolando Polli and Victor Cook, "Validity of the Product Life Cycle," *The Journal of Business,* vol. 42, no. 4, October 1969, p. 385.

product or, more accurately, the *production* life cycle of the original manufacturer.

We now turn to the particular issues in marketing and selling a technology that face an originating company in each of the stages of the TLC.

Stage 1: Technology Development

This stage begins long before any production, when research indicates a potentially valuable technology. The major issue the company faces here is whether further development of the technology should take place. Normally, development continues if:

- The technology has an obvious application in a readily identifiable market that fits with the company's overall strategy.
- The company has the financial resources to develop the technology, and the technology is compatible with that company's production and marketing skills.
- The projected rate of return on development is favorable when compared with alternative investments.

The situation is often less clear-cut. The technology may have several potential—but unclear and possibly unrelated—applications. Say a pharmaceutical company were to make a breakthrough in the technology surrounding the complex group of chemicals known as prostaglandins, which have wide applications in medicine, agriculture, and bioengineering. Early on, it might not be clear which application had the most potential or was most capable of realization. And even if the best application were readily identifiable, it might not match the company's strategy or resources.

Under these circumstances, the company would face more complex questions:

- Should it find a partner with whom to develop the technology further, especially if related technologies or additional financial resources were required?
- Should it try to sell the technology if it did not have immediate application within present markets or strategy?

Technology sales at GE

GE in the United States has formalized the latter approach by "packaging" for sale technologies that are too small or that lie outside its areas of interest. Since 1968, it has sold "surplus" technologies through a technology marketing operation, which is staffed by five technically trained people with experience in business planning, product sales, product plan-

EXHIBIT 1 The Technology Life Cycle

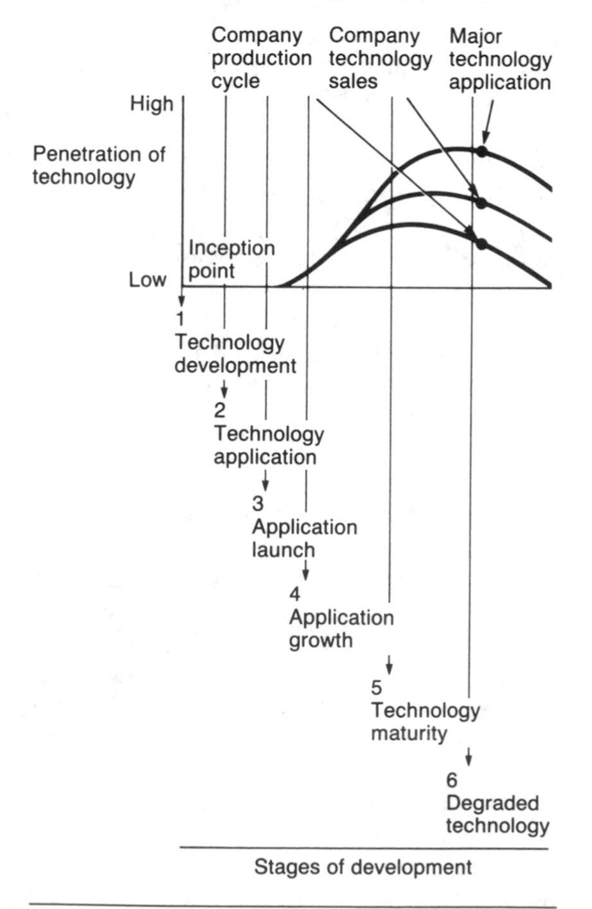

ning, or market research, as well as by a licensing counsel. In each case, the technology marketing operation, after being approached by the operating division involved, seeks potential buyers on the basis of the data file that it maintains on known customer needs, which it cross-references with GE technologies available for sale.

This operation is also responsible for *Selected Business Ventures,* a monthly publication based on potentially useful and salable technologies. Ninety-eight percent of the ideas in it do not emanate from GE but are collected by it for the 400 to 600 companies that subscribe to the publication. In addition, the unit updates and sells to subscribers technical manuals, such as those on heat transfer and fluid flow, developed by the company. It does not, however, involve itself in major licensing decisions, which remain the responsibility of the individual product divisions.

Benefits of hanging on

Even if a company wished to sell a technology this early in the TLC, it might have difficulty doing so; potential applications or the costs of developing marketable applications might be too uncertain. There is, however, an alternative at this stage that companies frequently ignore: their technology might be marketable but not in its present form. If developed further and incorporated into a product, it might then be salable to other companies. Frequently, technologies are ditched when they do not immediately fit into a company's product strategy, despite the fact that further development might lead to a marketable technology for someone else. Of course, this alternative also preserves a company's own option to produce products based on the technology.

Importance of staffing

Companies face still another major issue in this early stage of the TLC: they must have staff who are as sophisticated in the marketing of technologies as in the marketing of products. Unfortunately, this is rarely the case.

The people who make decisions early on about a technology's future are frequently nonmarketing staff. Decisions to kill development of a technology often take place before a company's marketing staff are involved. On the other hand, when marketers are involved, they will more likely see the future of a technology solely in terms of its translation into a company's own products. They are often unable either to think about selling a technology or to analyze the potential market for a technology. More important, the market for a new technology may lie among a company's product competitors, and marketing staff may often have difficulty in dealing with these competitors as potential customers.

Stage 2: Technology Application

After a company has decided to apply a technology to a new product—whether for its own production or for production by others—it incurs its first major costs. These costs are likely to be the primary factor in the company's decision either to continue with development or to sell the technology during this stage.

Few stockholders of a public corporation will tamely allow the development of technologies that even at the outset involve both high costs and high risks.[2] For instance, the Pilkington Company's successful development of the float glass process was brought to fruition only because of the support of the privately held company's board and because of the determination of the entrepreneur involved.

Further, when embodying technology in a product, a company is likely to face heavy costs in other areas—for example, in developing as-

---

[2] See L. Nabseth and G. F. Ray, *The Diffusion of New Industrial Processes* (Cambridge, England: Cambridge University Press, 1974), p. 200.

sociated process and product technologies. The company may find it necessary to buy into these associated technologies either by licensing or on a risk-bearing basis. Perhaps the largest recent example is the design for the European A300 Airbus, whose initial success led to the production consortium that included such companies as British Aerospace.

At this stage of the TLC, regardless of the high costs involved or the possibility of buying into related technologies, a company must not base its development decisions on the projected returns from product sales alone. Instead, it should consider potential returns from the technology as a whole—including product sales, license revenues, and perhaps turnkey deals.

Burroughs, for example, has put its computer technology to work in a network that allows some 700 of the world's largest banks to make transactions anywhere in the world in a matter of seconds. The banks run the network as a nonprofit organization, but it is staffed by Burroughs' people using Burroughs' equipment. A similar computer system, Ethernet, is being launched by Xerox in conjunction with Digital Equipment and Intel. This system allows participating companies to set up total communications networks within single buildings or closely built clusters of buildings. The three companies have made a special point of allowing other manufacturers to license their network system—in this case, as a way of encouraging use of their protocols and equipment.

Sell or license?

Possible applications of a technology become clearer and financing is more likely to be forthcoming when the technology has progressed beyond the idea stage to the point of practical demonstration. Similarly, potential licensees are likely to become more interested in a technology after it has been brought to the prototype or preproduction stage.

Nevertheless, decisions on the sale of technology at the end of stage 2 are primarily determined by the development costs incurred and by the projected initial revenues from licenses or product sales or both.

Dolby Laboratories provides an example of the problems created by high development costs and the need to ensure an adequate return. Ray Dolby initially saw his company as a research laboratory that would sell its noise reduction technology for tape equipment to the professional electronics industry. He wished to sell to this market first to gain a quality reputation among recording engineers. But, having invested $25,000 of personal savings and loans from friends, he quickly found that to survive he needed the income which could come only from product sales. Thus, he decided to make and sell noise reduction units initially for the professional music recording market.

The professional market in the electronics field is small, and any attempt to exploit it through the licensing of technology would not have provided sufficient revenue to justify Dolby's initial investment. On the other hand, the smallness of the market made it possible for a new company such as Dolby's to exploit the technology without major investment capital and without attracting overwhelming competition from large rivals. Once Dolby established the reputation of his technology, he had a strong marketing position when he later introduced the technology into the mass market for consumer tape-recording equipment.

Unfortunately, there are no universally applicable rules of thumb for making sound decisions about the sale or production of a technology at this stage. Too many variables enter into the equation: initial development costs, a company's cash position, its other development activities and their requirements, the technology's potential market, the possibilities of market segmentation, and—perhaps most important—the extent to which a technology is perceived as *essential* to a company's present or future activities.

Stage 3: Application Launch

The application launch stage of the TLC corresponds to the first, or "performance maximizing," phase of Abernathy and Utterback,[3] during which a company is likely to be developing its technology further—either through product modification or through application to different or perhaps wider product areas. But if a technology has been developed to the point of product launch without the involvement of potential buyers, decisions on its exploitation become more complex. The originating company, having faced both the high costs of development and of product launch, now confronts a number of issues that work *against* its recovery of some of those costs through license arrangements.

First, there may not be enough companies around with the skill to employ the new technology properly. The hasty sale of licenses could easily damage the reputation of the technology. This factor weighed heavily, for example, in the considerations of the Pilkington Company in exploiting the technology for glass-reinforced concrete.

Pilkington bought the license for this technology from the British Building Research Council with the aim of selling it to individual construction companies. Thus, Pilkington served as middleman between a research-oriented development company and the market because of its experience in license deals and its understanding of glass technology. However, the license process proved very lengthy, for Pilkington found it necessary to restrict the number of buyers by rigorous company appraisal and product inspection in order to maintain high product quality.

Sale of a technology during stage 3 may also be delayed by the long lead times involved in customer purchase of a relatively unproved technology. The purchase may depend on government backing for the buyer, which in turn may depend on a country's industrial policy. In addition, the sale of a technology may be held up or prevented by government restrictions on the seller, especially where the technology has strategic or military implications—as, for example, in such fields as computer networks, high-energy lasers, wide-bodied aircraft, and diffusion bonding. In the United States, the Technology Transfer Ban Act, updated in 1978, prohibits the sale to any communist country—or to any country that fails to impose restrictions on such a sale—of any "significant" or "critical" technology or product with a potential military or crime control application.

Consider, as well, how a sale of technology affects those technologists who are responsible for development within the purchasing company. They may see a purchase as an indication of failure and therefore may try to delay the decision to buy a technology while pressing for funds to develop their own. The "not invented here" syndrome is rife, although it may be more prevalent in some countries than others. In Japan, for example, the ratio of license revenues to license payments has remained fairly constant at around 1 to 8; in West Germany, the ratio is 1 to 2.5. By contrast, the ratio in the United Kingdom is approximately 1 to 1; in the United States, 10 to 1.

The final market factor working against technology sales at this stage is that customer purchase usually requires major changes in the purchaser's way of doing things. A company may be unwilling to undertake these changes until a technology is proved through more extensive product application or until its own technology is seen to be clearly inadequate.

On the other hand, the originating company itself may now wish to delay the sale of a technology, thinking that its potential value will increase with greater market acceptance. The company may also feel the need to recoup its development costs while taking advantage of the opportunity to skim the market as a mo-

[3] William J. Abernathy and J. M. Utterback, "A Dynamic Model of Process and Product Innovation," *Omega,* vol. 3, no. 6, 1975, p. 639.

nopoly supplier of a possible major technology. Further, it may wish to control the use of the technology in order to use its own production facilities to capacity.

The Swedish ASEA Company, for instance, provides electric locomotives for U.S. railroads. Although the technology involved in electric locomotive construction is relatively well known, the ASEA locomotives have a sophisticated electronic control system that makes them especially attractive to American railroads. Hence, the electric locomotive as a product is literally the vehicle for a technology sale. By tying the technology sale to its own product, ASEA hopes to maximize its revenues at this stage of the TLC.

Stage 4: Application Growth

Until this point in the TLC, the major issue restricting technology sales has been development costs. It is this fourth stage that Abernathy and Utterback call the stage of "sales maximization."[4] As an originating company begins to reap the rewards of increasing product sales, a number of strong reasons *for* technology sales begin to surface. The arguments made within the company, however, are likely to be in favor of delaying any sale. Thus, the crucial issue here is *timing.*

Growth in customer demand usually coincides with great interest in a technology by the developer's competitors. These competitors may well wish to avoid the high costs of developing their own alternative versions of products based on the technology. Therefore, the market value of the technology is probably now at its maximum. Nonetheless, the originating company's success in its own product sales, together with the discomfort of its competitors, is often a persuasive argument against selling the technology. In fact, a decision to sell during stage 4 is one of the most difficult that a company can make.

[4] Ibid., p. 643.

A technology sale is thus often delayed until later in the TLC when the value of the technology has decreased, both because of lessened customer interest and because of the development of alternative and perhaps improved technologies by competitors. This is often a mistake. A cold assessment of market potential could lead many companies to sell their technologies at the very moment that their own sales are increasing and before their markets are saturated. Such an assessment should include consideration of:

1. Market size. A decision to sell a technology through geographically selective license arrangements can lead to increased revenues based on wider application of the technology. More generally, a company faced with booming or novel market demand for an innovative technology may not be able to generate cash quickly enough to exploit the technology fully through its own production.

2. Technological leadership. The willingness of a company to share a technology much in demand can reduce its competitors' incentive to engage in their own technological development. The originator, by investing its additional revenues in further R&D, can better maintain its leadership position. Of course, such an approach must rest on a careful strategic assessment of whether the company's strengths are more in the creation of new ideas or in the reduction of old ideas to practical implementation.[5]

3. Standardization. The issue of government and industry standards is often vital in the growth phase of the TLC. The originator of a technology has a clear early advantage: the first product on the market *is* the standard. However, by stage 4 some of the company's compet-

[5] H. Igor Ansoff and John M. Stewart, "Strategies for a Technology-Based Business," HBR, November–December 1967, p. 71.

itors may develop alternative technologies and, if they have production advantages, may soon flood the market with their own products.

The active sale of licenses by the originating company will help ensure that its technology is incorporated into the production of as many companies as possible. Different technologies are often incompatible, and thus the first company to have its technology widely adopted may well set the technology standard for all. For example, Philips N.V. successfully achieved such standardization in the market for pocket dictating machine cassettes. Although Philips does not produce all the cassettes for all the machines in the world, most are produced according to its design and are subject to a royalty payment to Philips.

Another example is Ray Dolby's strategy for exploiting his noise reduction technology in the tape-recorder market. Dolby sold his technology to professional users in the form of equipment only (Dolby units) but has not allowed other professional equipment manufacturers to use his technology in their products. The market was small enough for him to do this without provoking competition.

Had he tried the same strategy in the consumer cassette market, which is vast in scale, he would immediately have invited rivalry. His decision there was to offer his technology to all manufacturers on a license basis and to require that licensees display the Dolby name and logo on the front of their equipment. In return for their license fees, manufacturers can submit their new products to Dolby for detailed criticism and advice, and improvements in the Dolby circuitry are made available to them without charge.[6] Even so, standardization has been difficult to achieve. Philips produced and promoted a rival technology system and only later was won over to become a licensee of Dolby.

Exhibit 2 summarizes the crucial factors in stage 4 of the TLC. In general, the best strategy is to seek both wide application and standardization of a technology while discouraging other companies from producing substitute technologies. Technology sales have an important part to play. Delay in sales here can mean that the value of the technology decreases, leaving the company to exploit the technology by other means after it has passed its peak value.

EXHIBIT 2 The Critical Timing Decision during Stage Four

Factors **for** Early Sale	Factors **against** Early Sale
Difficulty of developing new market alone	Low value of technology until proved
Lack of process or support technologies in company	High initial investment by developer
Cash shortage	Need to use production facilities
Importance of achieving standardization	High value added in production
Wide potential application for technology	

Stage 5: Technology Maturity

By the time a technology reaches maturity, it will have been modified and improved, not only by the originator but also by competing companies. No longer is timing of technology sales crucial. Instead, the originating company will be concerned with its production costs, the involvement with buyers that technology sales would now bring, and the relationship between those sales and its own production.

The originator's production will level off or decline as the overall market for products based on the technology stabilizes. The only fresh markets for the technology will now be found in less advanced countries, which are

[6] See the *Financial Times*, September 11, 1979.

eager to substitute their own production for imports.

Technology transfer to a Third World country often takes place on the basis of standard turnkey deals. However, a number of developing countries have tied the technology seller into ever more complicated arrangements. For example, the Algerians have increasingly sought to transfer technology through *clef en main, produit en main,* and *marché en main* purchases (literally, "key in hand," "product in hand," and "market in hand").

In the *clef en main* arrangement, the technology seller's involvement continues past the point of completing a production facility to training of staff. *Produit en main* transactions are not complete until the facility is fully on stream and has delivered products for an extended time. In the *marché en main* arrangement, the technology seller provides both *produit en main* services and a guaranteed market.

Dangers of technology transfer

During stage 5, which corresponds to Abernathy and Utterback's "cost reduction" phase,[7] the originating company must reduce costs to compete in its own markets. Hence, any decision to transfer technology to a low-cost producer must take into account the effects of that transfer on the company's own manufacturing plans. No producer wants to stumble by accident into the kind of competition Fiat now faces in its Western European car markets from its licensees in the Soviet Union and Poland.

In general, developing countries, like those of the Eastern bloc, wish to add value to their natural resources by buying sophisticated process technologies. Brazil, for example, wants to sell steel, not iron ore, and may be able to export steel relatively cheaply because it possesses key raw materials. This need to turn raw materials into finished or semifinished products is good news for companies like Davy International of the United Kingdom, which now has a big Brazilian steel plant under way. But the buy-back and barter arrangements on which such deals often rest make it essential that a technology seller consider the effects of those arrangements on its own plants, work force, and other product areas.

[7] Abernathy and Utterback, "Dynamic Model," p. 644.

Stage 6: Degraded Technology

The final stage of the TLC occurs when a technology has reached the point of virtually universal exploitation. By this time, license agreements will probably have expired, and the technology will be so well known as to be of little commercial value for direct sale.

However, many older technologies may still have market value in Third World countries. For example, some Middle Eastern countries had to import from Western Europe prefabricated ventilating ducts, which were essentially large boxes of air that were expensive to transport. But imports ceased when old-fashioned spot-welding technology was sold to the importing countries. Similarly, a small British company sold the technology for manufacturing simple wooden school furniture to another Middle Eastern country.

Cyril Hobbs, managing director of R&D at Laing Construction Group (a U.K. company), believes that what these "countries are looking for is basic standard technology. The question is how to identify what to us is old hat but may be just what other countries need. Here it [is] useful to have an outsider looking in. . . . I'm constantly explaining that I'm not selling Britain's seed corn just because technology is involved. We're selling yesterday's and today's know-how to the developing countries . . . it will take most of them 20 years to absorb what we're throwing at them now."[8]

[8] *London Sunday Times,* May 16, 1976.

Technology middlemen

Many transactions, including the two just described, are arranged by one of the growing number of technology middlemen, who are in business to bring together potential buyers and sellers of technology. The number of companies or individuals acting as intermediaries in technology transactions has grown considerably in recent years. They usually operate for a fee paid by the technology seller based on a percentage of royalty or lump-sum revenue.

A member company of the British Technology Transfer Group, which was set up as a nonprofit body by 10 companies that had a wide variety of skills of particular interest to less developed countries, confirms the value of these intermediaries:

"When we were trying to sell our expertise in the Middle East, we found out how difficult it was as an individual company, however well known we were in the U.K. You have two different levels of selling. First you have to convince the technical experts, then they have to persuade the nontechnical people in charge of the purse strings. If you can say you are part of an organization recognized by the British Overseas Trade Board, that is a reference straightaway."[9]

IN CONCLUSION

Fifteen years ago, Theodore Levitt's article on the product life cycle suggested that a company's basic technology should be embodied in a range of products.[10] We argue here that the full exploitation of a company's technology should include not only product applications but also technology sales. The complete marketing of a technology requires at least the following prerequisites:

[9] Ibid., January 30, 1977.

[10] Levitt, "Exploit," p. 81.

- *Development of a coherent strategy for a full portfolio of technologies.* Just as a company analyzes its product portfolio according to the position of its products in their life cycle, so it should pay attention to the TLC positions of its existing product and process technologies. Are its products, although selling well, based on a technology that is now widely available to other, perhaps lower cost, competitors? Is the company heavily dependent on a single main technology or on vulnerable sources of raw materials?

One British company, a market leader in specialized industrial pumps, reinforced its position with a series of new product introductions, each incorporating refinements on previous products. However, all the company's products were based on a single main technology that was increasingly available to other lower cost producers. Analysis of the situation led the company to rethink its overall strategy. First, it embarked on a program of license and buy-back arrangements to exploit more fully its existing technology; and second, it changed the direction of its R&D activity away from past over-reliance on a single, widely available technology.

- *Decisions on acquisition or divestment of individual technologies.* TLC planning involves at the outset clear *marketing* decisions about the whole course of a technology's development. The possibility of license or sale must be built into development plans, which need constant review both before and after application launch.
- *Awareness of the value of developing technology primarily for direct sale without incorporation into products.* This can occur in the case of technologies which do not fit into a company's main strategy or for which the company lacks the required production or marketing resources. It is likely that there will be a growth in the number of companies whose sole aim is the development of technologies to the application launch stage for subsequent sale to other companies.

■ *Clear understanding of the relationship between the sale of a technology through license and the sale of products based on that technology.* All too frequently, the licensing of a technology is delayed until the company's product sales and, thus, the market value of the technology start to decline. Full exploitation of a technology frequently involves *earlier* rather than *later* license or sale.

■ *Recognition that a technology buyer often has a better idea of its needs and opportunities than a technology seller.* Most companies find it easier to analyze an inadequacy in their own technology when they know of a product or process innovation elsewhere in use than to assess the value of their own potentially marketable technologies or the appropriate customers, prices, and overall strategies for them.

■ *Reliance on technology marketers.* All too frequently, technology sales are the part-time responsibility of top management. The marketing of a technology during all the stages of its TLC requires specialized decisions usually beyond the expertise of top corporate managers as well as conventional product marketers. Our research suggests that this marketing function be separated both from a company's overall strategic planners and from its regular marketing staff. Only after these specialists have carried out detailed analyses of a technology and its potential markets should their work be integrated with that of general strategic planners.

Integrating Business and Technological Strategies

Case II–7
Silicon Valley Specialists, Inc. (A)

M. A. Maidique, P. H. Thurston, and W. J. Abernathy

To be lastingly successful, a small semiconductor company must be able to introduce sophisticated new products while manufacturing competently its existing product lines. For a while, we slipped well below a competitive level in manufacturing—our yield in wafer fabrication plummeted—but we have since brought this under control. Now I am planning changes which will revitalize and organize our work in new product development. Such changes and the decisions we make about new projects like the "fuel control" chip now under consideration will greatly influence the company's future.

Harvard Business School case 9-677-053

The speaker was Jason Andrea,[1] president and general manager of Silicon Valley Specialists, Inc., a fast-growing California company in the electronics industry.

JASON ANDREA

At Berkeley, Dr. Andrea had been a teacher and researcher in the field of electronics, with some of his pioneering work in semiconductors leading to patents. Five years previously, in 1970, he and another man left the Berkeley campus to form Silicon Valley Specialists, Inc., financed by Stanton Caldwell. Jason Andrea was the company's principal technologist for three years. With the departure of his original partner to pursue other interests, Dr. Andrea's responsibilities had been broadened and he had assumed the presidency 14 months before.

[1] All names of persons and figures are disguised.

Dr. Andrea commented, "At age 33 I passed the threshold from technology management and engineering into general management—in this fast-paced, highly technical field it is a step I virtually cannot reverse."

Dr. Andrea personally owned 12 percent of the corporate stock, but the controlling interest, 70 percent, was owned by Stanton Caldwell, Inc., a financial firm. Mr. Caldwell took time from his other interests to serve as chairman of the board of Silicon Valley Specialists, Inc., and also as chairman of the executive committee and chairman of the finance committee. (See Exhibits 1 and 2 for financial statements.)

PRODUCT/MARKET INFORMATION

Silicon Valley Specialists defined its business as "development, manufacturing, and marketing analog[2] integrated circuit products for precision measurement and control." The total electronics industry was estimated at $40 billion, and of this, 5 percent ($2 billion) was in integrated circuits. The layman might describe these as "chips" containing solid-state circuits. Digital circuits were the largest segment ($1.8 billion), purchased mainly by the 10 major computer companies. The successful digital chip manufacturers were large and few (example: Texas Instruments) and the products highly standardized.

The analog market, in which Silicon Valley competed, was now $200 million and growing at 25 percent. It could be divided into two roughly equal segments, standardized and specialized. Products in the standardized segment

[2] "Analog" referring to proportional "analogous" measurement as distinct from a "digital" or symbolic representation of a number or measurement.

EXHIBIT 1

SILICON VALLEY SPECIALISTS, INC.
Income Statement
(in thousands)

	Year 1*	Year 2	Year 3	Year 4	Year 5
Net sales	$2,000	$5,000	$9,000	$15,000	$20,000
Cost of sales	1,320	3,900	6,650	10,200	12,600
Gross margin	680	1,100	2,350	4,800	7,400
Percent of sales	34%	22%	26.1%	32.0%	37.0%
Research and development	290	740	810	1,250	1,400
Percent of sales	14.5%	14.8%	9.0%	8.3%	7.0%
General, administrative, and marketing	190	760	1,540	2,150	2,400
Income before taxes	200	(400)	—	1,400	3,600
Provision for taxes	100	(100)	—	600	1,800
Net income	100	(300)	—	800	1,800
Percent of sales	5.0%	(6.0)%	—	5.3%	9.0%

* During the first year of its operations, the company operated its capital-intensive final test department as an incoming inspection service for purchasers of integrated circuits. This policy was discontinued during the second year of operation when SVS's own IC production began to absorb a substantial fraction of test department capacity.

EXHIBIT 2

SILICON VALLEY SPECIALISTS, INC.
Balance Sheet
(in thousands)

	Year 1	Year 2	Year 3	Year 4	Year 5
Assets					
Current assets:					
Cash and other	$ 50	$ 100	$ 200	$ 400	$ 600
Accounts receivable	250	600	1,100	2,200	3,700
Inventories	200	700	1,700	3,600	4,900
Total current assets	500	1,400	3,000	6,200	9,200
Property, plant, and equipment:					
Machinery and equipment	520	630	1,520	3,700	5,650
Office equipment	70	80	140	180	250
Leasehold improvements	260	300	660	1,110	1,700
	850	1,010	2,320	4,990	7,600
Less: Accumulated depreciation	150	260	470	1,190	1,800
Net property, plant, and equipment	700	750	1,850	3,800	5,800
Total assets	$1,200	$2,150	$4,850	$10,000	$15,000
Liabilities and Stockholders' Equity					
Current liabilities:					
Accounts payable	$ 500	$ 400	$1,110	$ 430	$ 770
Accrued liabilities	150	70	120	170	310
Total current liabilities	650	470	1,220	600	1,080
Long-term debt	250	680	1,630	4,100	6,820
Stockholders' equity:					
Paid-in capital	200	200	200	200	200
Add paid-in capital	—	1,000	2,000	4,500	4,500
Retained earnings	100	(200)	(200)	600	2,400
Total stockholders' equity	300	1,000	2,000	5,300	7,100
Total liabilities and stockholders' equity	$1,200	$2,150	$4,850	$10,000	$15,000
Total asset/sales	.60	.43	.54	.67	.75
Average asset/sales	.30	.34	.39	.50	.63
Debt/equity ratio	.833	.680	.815	.774	.961
Shares outstanding	20	195	370	720	720
Earnings per share	$5.00	—	—	$1.12	$2.50

were characterized by long runs and low margins. Performance specifications did not vary significantly from manufacturer to manufacturer. Price and delivery played a key role in a customer's choice of producer. Specialized products, while not customized for each customer, generally had more narrow applications than standardized. Short runs and high margins were the norm. Selling of specialized products was a more complex task than for "off the shelf" items since the customer was often not familiar with the performance characteristics of the products of the different manufacturers. The same customer often bought both standardized and specialized chips. Some users preferred to source all their analog business to just one or two companies when possible.

The three major firms in the analog industry, National Semiconductor, Fairchild, and Motorola, competed in both segments. National Semiconductor had annual analog sales of $40 million with both Fairchild and Motorola at the $30 million level. According to one observer, the three dominated the market both in terms of pricing and process technology advances. Silicon Valley, though concentrating principally in the specialized segment, captured 10 percent of the total analog market. A number of smaller firms split the remaining 40 percent.

A typical product for Silicon Valley might be a "chip" for an aircraft instrumentation manufacturer, where readings would be fed in from various aircraft controls; the chip would amplify, combine, or compare these signals, performing subtraction, addition, division, power functions, etc., as required, and produce an output signal to indicate whether the operating status of that element of the aircraft was satisfactory. In other instances, the signal might actually control an operating function automatically. Silicon Valley received orders for chips of specialized nature for use in industrial applications including medical electronics, automation of scientific instruments, process control of chemicals, and heat control in industrial furnaces. The possibility of developing a chip that would be a key component of a more efficient ignition control system, for one of the Big Three U.S. automobile producers, was under consideration.

Silicon Valley, with 400 employees, has some market position across the board, including less expensive chips, but an increasing share in the high-performance segment. The company's competitive strength lies in marketing and in the use of technical abilities to develop innovative products that precisely match the market's needs. The company has innovated to some extent in the manufacturing process, particularly in the use of factory adjustment of chips, but for the most part Silicon Valley has been a product leader, not a process technology leader.

The price that Silicon Valley quotes depends upon many factors: the amount of additional technical development required, the number of chips in the order, the complexity of the circuit, the risk in meeting specifications only at a low-yield rate, and the prospect of future business. For normal sales, the price per chip ranges from about $5 for the more common chips to more than $50. If the order is largely a custom development effort, involving only a few prototype chips, other pricing considerations are involved. Such orders were, however, unusual.

The real question is securing a fair return on the critical new-product development resources. Some managers at Silicon Valley Specialists question whether the pricing policy doesn't allow these scarce talents to be sold too cheaply. On the other hand, to seek business in high-volume chips at prices below the current pricing policy would involve a significant change in current strategy.

Silicon Valley, like other semiconductor manufacturers, employs two distinctly different manufacturing processes. One is wafer fabrication, which is predominantly a chemical process.[3] Carefully controlled impurities or "dopants" (boron, phosphorus, and arsenic are used frequently) are added to silicon to produce regions of different conductivity types which are in contact with other regions on an extremely small scale. Unintended impurities, even in minute amounts, usually give undesirable results. Accordingly, there is a pronounced emphasis on purity pervading the 40-plus steps in the process for making an integrated circuit wafer.

[3] Up to "Test, Scribe, and Dice" in the process, as shown in Exhibit 3.

EXHIBIT 3 Integrated Circuit Manufacturing Process

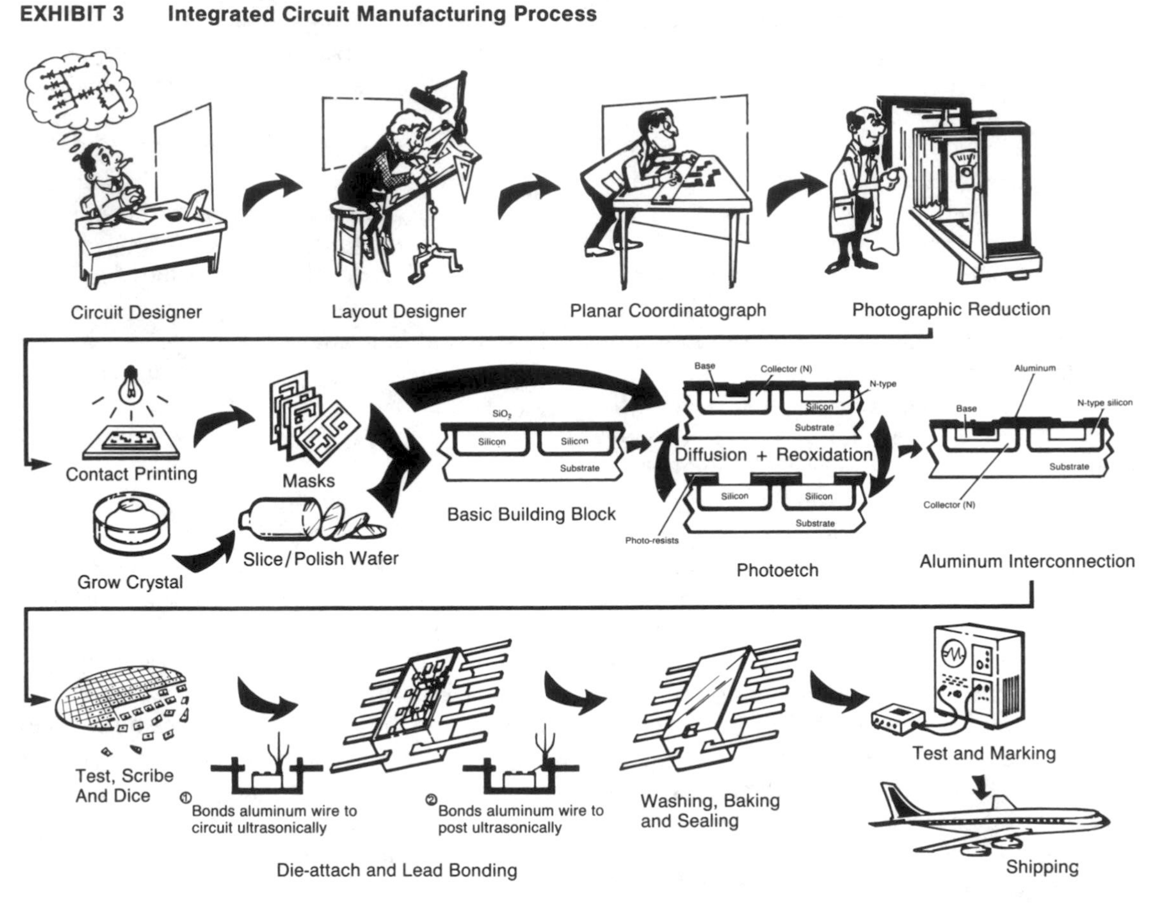

Exhibit provided through the courtesy of the Harris Corporation, who is not associated with the case, and reproduced with permission.

The second manufacturing process is the more conventional batch process of electrical assembly and test. Wafers are cut into individual chips, tested and graded, assembled in required configurations with leads, and retested before packaging.

Silicon Valley does not generally make investments which seek to push the capabilities of the company's manufacturing processes beyond the state-of-the-art as that is advanced by the industry leaders. To compete successfully on the basis of price or process innovation would mean a significant stream of investment. Dr. Andrea estimated that for Silicon Valley to pioneer a major process innovation, an investment of $2 million per year for two years would be necessary. Instituting a major innovation pioneered by another company would still cost half the amount. In either case, a dedicated line would have to be set up, new equipment purchased, and the flow for wafer fabrication redesigned.

YIELD IN WAFER FABRICATION

Speaking of the two manufacturing processes, Jason Andrea said:

> My observation is that all *successful* semiconductor companies, large and small, are the work of two men. One has mastery of the chemical process and through his organization can control the literally thousands of variables that determine the quality and yield in wafer making. The other man, typically an electrical engineer, has mastery of the much different electrical design, test, and marketing activities.
>
> When we were a small company, my partner and I had control of wafer yield. Our batch sizes and total volume were smaller. We were close to the process ourselves. But as we grew, we seemingly lost control of yield. We stepped back from the manufacturing process ourselves and substituted, as factory managers, engineers who had high technical competence but little skill in working through supervisors and operators. Further, we made organizational mistakes. When we first ran into yield problems, we thought to break the problem into component responsibilities. We assigned process engineering for wafer fabrication to one engineering manager and production responsibility to another. These and other engineering managers reported to the operations manager responsible for all manufacturing. Yield deteriorated further. Things came to a head when we changed from two-inch wafers to three-inch wafers and the yield plummeted.
>
> More than a year ago, when I became general manager, I changed the responsibility for wafer fabrication in two respects. I assigned all aspects of wafer manufacturing to one man—and of at least equal importance, I got the right man. He was an outsider, Roger Sorenson, a chemist by trade, who had grown up with integrated circuits and whose skill lay not in innovating but in manufacturing. We not only got Roger, but also a few of his key team members from his old company. Roger has the skill to work through people to ferret out little difficulties in every aspect of the process, and to achieve standards for performance by people. Roger, in working with his superior, Jim Lawrence—the operations manager—increased yield by almost 100 percent while cutting the labor force by 20 percent. Assembly and test had never been a major problem, so profitability responded sharply to rise in wafer yield.

THE FUEL MISER

The "fuel control" chip project presented a specific decision in which to test some of Dr. Andrea's ideas about the management of Silicon Valley Specialists. Three months earlier, Max Lehrer, a highly motivated Silicon Valley engineer, had been approached through the sales force by Ford to develop and produce the "fuel control" chip, to provide "computer-controlled" ignition for improved automobile fuel use and performance. As a result, Silicon Valley now had a very attractive offer to develop, manufacture a number of prototype "fuel control" chips, and be involved in testing them. Successful completion of this program would lead to an option to produce 40 percent of the auto company's requirement for chips, at market price, for five years. Under the terms of the contract, product technology developed by Silicon Valley for the fuel control chip would not be proprietary; Ford probably would use two other sources to meet its total needs for the product.

Such a contract would involve production levels of around 1 million fuel control chips a year. This equaled over a quarter of Silicon Valley Specialist's current unit volume. The price would be determined at a fair market value, based on actual future conditions, but had been projected by Ford along a learning curve approximately as follows:

Chip Price	Quantities per Year	
$20 to $25	Initial	50,000 to 100,000
$10 to $15		100,000 to 300,000
$ 4		1,000,000

Dr. Andrea considered that these were valid price projections, that the auto company was sincere and quite willing to allow a fair price so that the necessary investment could be recovered early. Furthermore, he believed the technical risk was quite low. The problem was that a major investment of scarce technical personnel and capital would be needed to perform under the contract. Based on the best estimates available, however, a new, more automated manufacturing process dedicated only to the fuel control chip would be required to reduce the cost per chip to the $4 level. This would involve an investment of from $3 to $6 million in furnaces, reactors, and other equipment. Perhaps more important, it would require the dedication of the company's critical people resources. Roger Sorenson, who would be a key man in the project, was very enthusiastic and believed he could manage both the new process and the existing one.

He argued that if Silicon Valley was to ever develop an advanced capability in process technology, this was the opportunity to do so. Automotive electronics was a large market of growing potential, the major auto producers were not expert in managing semiconductor electronics, and here was one willing to finance their own entry into this field. Even if this project only broke even, it would have allowed Silicon Valley to develop a cadre of talent where it was weak—high-volume, low-cost production technology.

Actually, the economies also looked attractive. Sorenson pointed out that the projected learning curve was based on industry experience in a yield improvement rate. He was confident that he could manage a much better yield improvement profile over time so that investment recovery and, therefore, profitability, would be higher than planned.

PRODUCT INNOVATION

Dr. Andrea acknowledged remaining problems that helped shape options for the future. One was a decreased rate of product innovation by his company in the fast-moving semiconductor industry. Silicon Valley had a burst of new products during its first two years, mainly through the efforts of the then vice president of engineering, Dr. Andrea. But the number of new products per year had changed as follows:

Year	Number of Basic New Product Families
1st	8
2nd	11
3rd	5
4th	3
5th (current)	7

There were those in the company who interpreted this trend as a signal that the time was right for a change in strategy. Others, however, pointed out that at least one of the recently introduced products was selling at a $2,000 rate, several times the sales rate of the early products.

Had the innovative spark of the young company been replaced by mature competence? To retain momentum, some felt the need to translate this early strength into a dominant position in a particular market area. They argued the fuel control chip project was an opportunity to do just that. Others were not so concerned. Though the absolute number of new products had decreased after the first two years, the level of sophistication had increased as Silicon Valley expanded into new applications. Also the pickup in the number of new products in the last year was encouraging. Dr. Andrea expressed high hopes for these new products though he acknowledged the high degree of risk associated with them.

Dr. Andrea commented in general:

> My responsibilities can be divided into two time frames. One is operating and controlling. The other is preparing for the future. The operating and controlling responsibility corresponds, in time frame, to the annual budget updated with rolling, four-quarter budgets. This responsibility concerns process control, output levels, quality maintenance, selling and customer ser-

vice, and personnel. Here I think we are achieving a measure of success.

The forward-looking responsibility has a time dimension influenced by: First, the 12 to 18 months it takes to develop and introduce a new product; and, second, a longer time to effectively add engineers with new technical competence or modify the technical focus of professionals employed. Somewhat arbitrarily, this suggests a strategic planning framework reaching out at least three years. The specific output is new products. Here we are not where we'd like to be.

I am considering some changes to raise our technological competence and increase the flow of new products. An immediate change could be organizational. At present, a minimum of six different engineering specialists plus prototype manufacturing are involved in bringing out any distinctly new chips. Time is lost and there is miscommunication as new product ideas move between specialties. More direction and consistency might also be provided in new product development decisions. Further, we need more focused responsibility when there are technical questions within one of the existing product lines. Could these needs be solved by a matrix organization?

I have never heard of a matrix organization in a $20 million company, but that's what I'm considering. We would have not only our present functional organization, but also an overlay of engineering responsibility for all aspects of each different product line. Think of it as a matrix with products on the vertical axis and engineering specialties on the horizontal. Each product line including new products would become the responsibility of an engineer (vertical axis) helped by the specialists (horizontal axis) assigned part time to each product line. We don't want to add people; some of our engineers would simply wear one more hat to become the product line engineer. Will it work?

A second change would involve the expenditure of $600,000 to set up a partial prototype line for developing new products. For this amount, we could duplicate many of the manufacturing processes and keep our trial manufacturing out of the slow-moving sequence of higher volume production flow. A few expensive, key pieces of equipment could not be duplicated short of nearly $2 million. For these few steps, the trial runs will enter the production sequence and return immediately to the much faster "hot-lot," "line within a line." New products presently need about "five fixes," that is, five cycles through wafer fabrication to eliminate any difficulties in design. One full production cycle requires eight weeks. Even if some of the five development "fixes" only need some of the production steps, it may take nearly a half year to complete the five fixes. That is too long. I expect the "hot-lot" line would cut this to four to six weeks. But we have not yet approved the $600,000 expenditure.

EXHIBIT 4 Typical Trapezoidal—New Product Sales Buildup

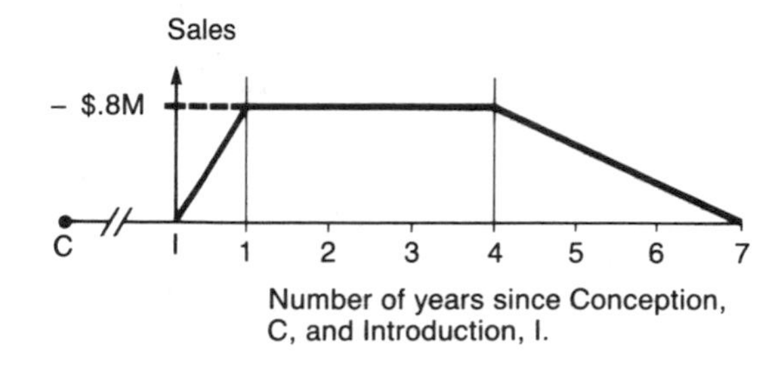

Dr. Andrea sketched out a trapezoidal sales pattern that he observed as typical for Silicon Valley's products (Exhibit 4). The initial phases, to full sales buildup, could be accelerated by the hot-lot concept. According to Silicon Valley's marketing director, earlier entry into the marketplace could also boost the average sales level of new products in the first year 25 percent beyond their present level.

THE AUTOMATIC PROJECTION ALIGNMENT SYSTEM

A third, but different, direction of change, Dr. Andrea went on to say, is suggested by a project that is of particular interest to Roger Sorenson, the production manager. This involves the purchase of a Kasper automatic projection alignment system. Mask alignment—a key step in IC manufacture—was performed at Silicon Valley Specialists by conventional Cobilt mask aligners which pressed the mask (see Exhibit 3) to the surface of the IC wafer to prevent diffraction of the exposure light. The continuous pressing of mask against circuit wafer resulted in rapid wearing of the contact printed masks (masks were internally manufactured at a cost of $2 each). Typically, one mask was worn out and discarded per IC wafer reaching wafer test. More important, topological imperfections on the surface of the IC prevented a perfect fit between mask and wafer causing imperfect component definition and lowered yield. Nevertheless, conventional contact alignment had been the industry standard alignment method for the past five years.

Several equipment manufacturers had recently developed advanced "projection alignment" systems which did not require contact between mask and wafer. These systems, now field proven in several IC plants, used collimated (parallel) light to expose the photographic material on the wafer and thus did not depend on the pressure of the mask to seal the exposed area. Using collimated light technology, mask life was limited primarily by obsolescence. Long-lived IC products were typically redesigned once a year, at least at Silicon Valley, after the initial preproduction cycle.

Silicon Valley was nearing capacity on its aligners. The company could either buy two more conventional aligners at a cost of $20,000 or buy the Kasper system. Roger Sorenson recommended purchasing the Kasper system. The system was fully automated and would handle twice the volume of four conventional aligners (four were now being used on the SVS line) and would cost $200,000; installation was estimated at an additional $50,000 plus three months of a senior engineer's time. Direct labor saving equal to four hours per week was projected. Maintenance would run about $20,000 per year. In addition, SVS technicians would have to learn how to service the equipment. Sorenson believed that the Kasper system would improve yields by 20 percent at the wafer test stage. It was also possible that there would be some reduction in breakage in the wafer manufacturing process. About 10 percent of the wafers were damaged or broken during the alignment process.

Silicon Valley was operating at a wafer start level of 2,000 three-inch wafers per week in 1975. Breakage and process variations in the wafer fabrication process caused 40 percent of the wafers started to be rejected before reaching wafer test. At wafer test, a computer-controlled tester dropped ink dots on chips that failed electrical testing; of those chips entering wafer test, 80 percent were rejected. These yields at Silicon Valley were somewhat lower than industry norms, largely it was thought because more challenging circuits were produced. Sorenson believed that once equipped with the new Kasper system, his line would run at yields somewhat higher than the rest of the industry. The standard cost for a wafer (entering wafer test) was $50.

MANAGING ENGINEERS

Dr. Andrea continued:

> Changing the organization and setting up a "hot-lot" line may help, but our long-range performance must rest on our selection and management of engineers. How competent are they? How motivated are they? How well are their different contributions focused? These are major concerns. In this industry, the most qualified people—the ones we need—are highly mobile. In this part of California, a good individual can easily change jobs overnight without changing the family home. With so many firms in the area, attracting and retaining top people is difficult.
>
> Are we doing a good job in managing our engineering personnel? This question is as big as the technological posture of the company. A major technical development will afford a wide profit margin into the future for a few operating quarters if the major developments are followed up

by several minor product improvements. Eventually margins shrink under competition and minor improvements are harder and more expensive to find. If this company seeks to be in the technological forefront, it must have engineers working at the lead technological edge at least with a minimum effectiveness to anticipate, judge, and secure for our company the important innovations of the industry. It is unrealistic to expect any one company to generate internally all the ideas that will give leadership. Rather an important function of top engineers is to have sufficient professional recognition and understanding to participate in and monitor the advance of their fields. Not just incidentally, such a technological forward look makes a company attractive as an employer for the best engineers and strengthens motivation.

We have tried to recognize skills and fit people in the organization accordingly. For a very strong individual contributor, we have created the "technical fellow" position. The function is to diffuse superior engineering knowledge in the organization. Even though, in a conventional sense, the "technical fellow" may manage only one person, the pay and prerequisites are comparable to those of the manager of people, the other advancement route. I like to think we have a place in the organization at any level for any outstanding engineer, whether as an individual contributor or as a manager.

One key to success is selecting good engineering managers. There are important characteristics of the engineer who will do well in a managerial job. It must be a truly committed person with a high degree of enthusiasm. Another matter is recognition of the importance of teamwork as opposed to individual effort. If a manager doesn't anticipate the effect of decisions on other parts of the organization and work with others to provide accordingly, there will be difficulty. We have learned that an engineer moving to supervisory responsibility who can be intimidated by the people below is the wrong person for the job.

We have engineers interested in becoming managers who didn't make the shift successfully. Even with strong interest, there is a big difference between an engineer being able to intellectually understand the concepts of administration and his being able to perform effectively in a managerial situation. It is a huge step to move from the responsibility for one specialty or for one function to the responsibility of coordinating the work of two or more functions. In our experience, the engineer who takes on managerial responsibility incrementally—goes from one functional responsibility to two and then adds more—has a better chance of success than the person who plunges into broad responsibility. People need time to develop.

In other situations we have learned that if you are going to change an engineer's responsibility because you are convinced he cannot surmount the administrative difficulties, make the change sooner rather than later. It is a mistake to wait too long. Bitterness can come in, not only with that individual but also with others if there are persistent difficulties in the organization. At worst, there can be infighting.

Within our organization we appoint the best people we can, with the director of R&D an excellent example. He has both immense talent and great sensitivity in handling people. He treats the 12 engineers who work for him as peers, and if you observed him casually, you might think of him as simply another member of the group. In fact, he has skills in motivation, in coordination, and in timing. He has grown in this job during a time in which we have been patient with him and he has been patient with us. I would emphasize that his commitment, particularly to new technologies, coincides with the company's direction.

Overall, I haven't gone wrong in giving big raises to able people and moving them into challenging jobs, and then not settling for second-best performance. Roger Sorenson's contribution to the company is a concrete example of the importance of bringing in the best people in manufacturing as well as product designs. Virtually no one has left of his own volition. The nice thing is that in a strong learning environment, the exceptionally able people pull the others along.

The question of challenge is very important. To provide challenge, we have shifted engineers who are doing very well on one job but ran the risk of going a little stale. Indeed, we have done this regularly in our short history. We try to hire individuals who reach for new challenges, and it follows naturally that if we keep them on the same jobs for three years their commitments can wane. Under the right circumstances, new challenges bring out the very best in people.

I should mention that of the kinds of engineering managers employed by Silicon Valley, the R&D-oriented product development specialists like Max Lehrer and the production-oriented manufacturing specialists like Roger Sorenson, it is the latter type that is most difficult to develop or recruit. In this part of the country, a group of bright, competent product development engineers emerge from the local universities each year; but the process manufacturing experience of a Roger Sorenson is a scarce resource and usually must be hired away from a competitor at a high cost.

Dr. Andrea continued:

> We have capable engineering managers and they should make the important, ongoing engineering decisions. My job is to point the direction and to create the proper environment. I refuse to take a hand in the easy decisions. No operating questions whatsoever are to wait for me if I am out of town. I do not sit on either the Operating Committee or the Product Development Committee, although I receive weekly written reports of their progress. I do, however, chair the Management Committee and the Product Planning Committee and sit on the Personnel Practices and Policies Committee.

INVESTMENT IN FUTURE

Jason Andrea and Stanton Caldwell had discussed the appropriateness of the company's strategy on technology. Had the company now become successful enough to begin to undertake a leadership position in manufacturing as well as product design? Could it afford the investment? On the other hand, was it wise to try to be among the technical leaders in process development? Current pretax profit was 18 percent of sales. Advanced engineering and efforts to develop new products consumed 7 percent of sales, and new-product marketing another 3 percent. A strategy directed only to existing products and technology would trim not only the 7 percent and the 3 percent, but also much of the approximately $2 million invested annually in new equipment.

If continued, dollars spent on technological advance could fall as a percent of sales as volume rose. But this was a chicken-and-egg proposition. Stanton Caldwell, Inc., had not taken any cash out of Silicon Valley (see funds flow analysis, Exhibit 5). When was it reasonable to expect positive cash flow? Indeed, to what extent could a fair return on capital be foreseen as Silicon Valley competed in its dynamic industry? Mr. Caldwell and Dr. Andrea hoped to develop a better framework in which to weigh different forces to evaluate the company's current and potential sizable investments.

At the specific level, Dr. Andrea learned that Ford had made its interest in having Silicon Valley participate in the "fuel control" project known in a tangible way. The auto company had separated the first phase—development and prototype production and testing—from the subsequent high-volume manufacturing contract. With success in the developmental phase, Silicon Valley would still have a first option on part of the long-run manufacturing contract. Since the costs of the development phase (estimated at $100,000) were to be reimbursed with profit, as a first phase contract, it appeared to some to be attractive to at least undertake this first step. Dr. Andrea was quite sure the sentiment in the Product Planning Committee was in favor of going ahead with this contract if he did not oppose it. He wondered if he should allow the matter to move ahead.

"It's easy to say we should stay where we are, but that's simplistic," Dr. Andrea said. "I have recruited talented engineers who just will not be satisfied to stand still. And, in addition, we are already feeling the competition of some small, new companies that are starting out just as we did, and a firm may eat into what has been our competitive advantage.

"I picture our strategic choice as having two possible branches for growth:

EXHIBIT 5

SILICON VALLEY SPECIALISTS, INC.
Funds Flow
(in thousands)

	Year 1	Year 2	Year 3	Year 4	Year 5
Sources:					
Net income	$ 100	$ (300)	$ —	$ 800	$1,800
Add back: Depreciation	150	110	210	720	610
Total funds from operations	250	(190)	210	1,520	2,410
Paid-in capital	200	1,000	1,000	2,500	—
Long-term debt (net)	250	430	950	2,470	2,720
Total sources	700	1,240	2,160	6,490	5,130
Uses:					
Additions to property, plant, and equipment	850	160	1,310	2,670	2,610
Total uses	850	160	1,310	2,670	2,610
Net changes in working capital	$(150)	$1,080	$ 850	$3,820	$2,520

1. One branch is the strategy of *better product performance* for the customer. This, in turn, has two possible subbranches:
   - *Product innovation,* which is primarily the route the company has taken so far. We have had generous customer support in product innovation.
   - *Product augmentation,* which essentially breaks down into technological and applications support for the customer in using the products; financial support for the customer; or other customer service.
2. The alternate branch has been the strategy of *selling primarily on price.* This, in turn, has the following subbranches:
   - *Process innovation* resting on process technology development.
   - *Process standardization* by limiting the choice of products and reducing the process options—or, in other words, 'any color as long as it's black.'"

Reading II–8
The Technological Dimension of Competitive Strategy

Michael E. Porter

Technological change ranks as one of the principal drivers of competition in industries. Schumpeter (1934) characterized technological change accurately as the source of "creative destruction" by which monopolies were destroyed and new industries created, and exam-

Research on Technological Innovation, Management, and Policy, vol. 1.

ples of the power of technology to change the boundaries or rules of competition are easy to find. Yet the study of technological innovation is too often decoupled from the study of competition and vice versa. Emphasis in the management of technology field is largely given to the internal conditions for success in R&D programs and to insuring the required linkage between innovation and customer needs. Conversely, research in industrial economics has taken a narrow view of technological competition, focusing on the relationship between technological inputs or outputs and firm size, diversification, industry seller concentration, and profitability,[1] and to a limited extent on technological spending as an entry barrier.[2] Missing in both fields is a comprehensive view of how technological change can affect the rules of competition, and the ways in which technology can be at the foundation of creating defensible competitive strategies for firms.

This chapter aims to establish a conceptual link between technological change and the choice of competitive strategy by the individual firm. It will begin by examining the potential impact of technology on industry structure that motivates technological activity. It will then link technological strategy to overall competitive strategy, and show how technological strategy can be a vehicle for pursuing generic competitive strategies aiming at fundamentally different types of competitive advantages. In this context, the structural determinants of the firm's choice of technological leadership or followership will be identified. Finally, I will build on the analysis of individual firm technological choices, in order to model the determinants of the pattern of technological change as industries evolve, and will develop some implications for the choice between pioneering and late entry and the choice of a late entrant's technological strategy.

I. TECHNOLOGY AND INDUSTRY STRUCTURE

The power of technology as a competitive variable lies in its ability to alter competition through changing industry structure. Viewing competition broadly, there are five fundamental competitive forces (shown in Figure 1) at work in any industry whose collective strength determines the ability of firms to earn rates of return on investment in excess of the opportunity cost of capital: the five forces are the threat of entry, substitution, bargaining power of suppliers, bargaining power of customers, and rivalry among incumbent competitors. Each of these forces has underlying economic, technical, and situational determinants that I have developed elsewhere.[3] These underlying determinants have been termed industry structure, and define the rules of competition in the industry. The same determinants can be examined at the level of the individual firm to assess its fundamental competitive position, using the concept of strategic groups.[4] The same factors that determine entry barriers also determine

[1] For a survey of this literature see M. Kamien and N. Schwartz, "Market Structure and Innovation: A Survey," *Journal of Economic Literature,* vol. 13 (1975), pp. 1–37; and F. M. Scherer, *Industrial Market Structure and Economic Performance,* 2d ed. (Chicago: Rand McNally, 1979). Attention has centered on the possibilities for mutual dependence recognition in R&D rivalry, particularly its relationship with industry concentration. Generally the direction of causality examined in the entire body of research is from size, diversification, or concentration to R&D, because the preoccupation is with how public policy might increase innovativeness.

[2] D. C. Mueller and J. E. Tilton, "Research and Development Costs as a Barrier to Entry," *Canadian Journal of Economics.* 2 (November 1969), pp. 570–79; and H. C. Grabowski and D. C. Mueller, "Industrial Research and Development, Intangible Capital Stocks, and Firm Profit Rates," *Bell Journal of Economics* 9 (Autumn 1978), pp. 328–43.

[3] This framework is discussed extensively in M. E. Porter, *Competitive Strategy: Techniques for Analyzing Industries and Competitors* (New York: Free Press, 1980).

[4] Ibid., chap. 7.

FIGURE 1 Forces Driving Industry Competition

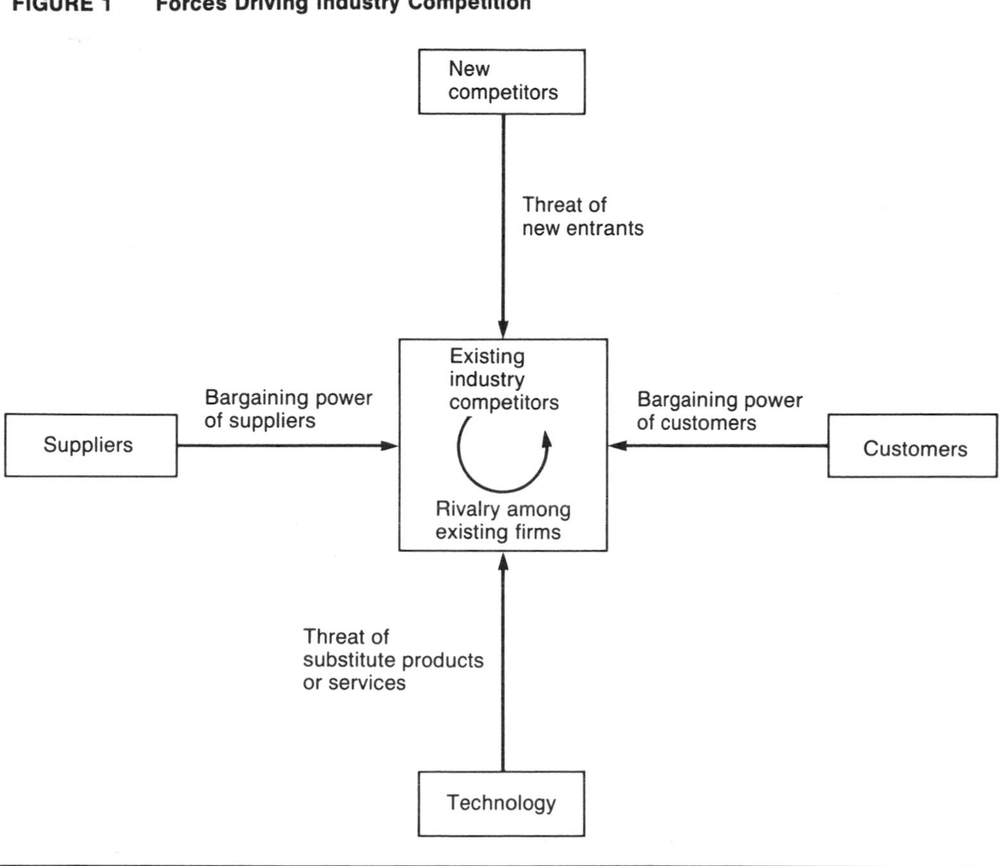

the mobility barriers that protect firms' strategic positions from attack by rivals, for example.

Industry structure defines the arena and rules of the game in which competitive strategy seeks to best position the firm. The aim of competitive strategy is to create a defensible position for the firm against the competitive forces. One approach to doing this is through finding industry positions least vulnerable to the forces, drawing on an understanding of their determinants; for example, one particular market segment might be less exposed to substitution than another. The other approach to competitive strategy formulation is to use strategy to *change* industry structure (the rules of competition) in the firm's favor. When the strategies of highly successful firms are examined, a striking number have succeeded in changing the competitive rules in their industries in ways that are defensible.

What makes technology unique as a strategic variable is its considerable power to change the competitive rules of the game. Technological change can be a great equalizer that nullifies the advantages of incumbents and creates opportunities for newcomers and followers. Technological change is perhaps the single most important source of major market share changes among competitors for this reason, and is probably the most frequent cause of the demise of entrenched dominant firms.

The competitive significance of a technological change depends neither on its scientific merit nor its effect on the ability of the firm to serve market needs per se, but on its impact on industry structure. Technological change can potentially affect a wide spectrum of determinants of industry structure. Where a firm's technological innovations are appropriable, these impacts of technological change on structure are the fundamental motivations underlying the firm's choice of technological strategies.

Technology and Entry Barriers

Technological change can affect industrywide entry barriers and the mobility barriers protecting individual firms' positions in a variety of ways. Technological change can raise or lower economies of scale in nearly every aspect of a firm's operations. While most discussion does not use this terminology, technological change is what underlies the learning curve: the learning curve creates an entry barrier if the learning can be kept proprietary. Proprietary technology can lead to absolute cost advantages in many other ways as well. Technological change can alter the capital requirements for competing in a business, both directly, through requiring firms to make R&D investment, or indirectly, through affecting the capital required for production, logistical, or other facilities. Technological change can enhance or eliminate opportunities for product differentiation, through proprietary product designs (enhancing differentiation), reducing the need for after-sale service (perhaps reducing differentiation), and the like. Technological change can also affect access to distribution through facilitating the circumvention of conventional distribution channels (like the low-cost Timex watch did), or conversely, through increasing industry dependence on distribution channels by requiring greater needs for product demonstration and after-sale service.

Technological change can also raise or lower switching costs, or buyers' fixed costs of changing suppliers. Technological choices by competitors determine the need for retraining users or reinvesting in ancillary equipment when changing from one supplier to another. Technological assistance by suppliers in designing products into a buyer's product can elevate switching costs, while standardizing technological interfaces between products can reduce them.

The R&D function itself also has potential implications for entry and mobility barriers. The essentially fixed costs of product development can lead to economies of scale favoring firms that can amortize these costs over large volumes. Maintaining an effective R&D program may itself be subject to economies of scale, though there is much disagreement about this in the literature.[5] It has also been argued that the need for substantial R&D creates capital barriers to entry.[6]

Whether technological change raises barriers industrywide or raises mobility barriers, protecting the strategic position of the innovating firm depends on the defensibility of technological changes from imitation by incumbent competitors. For example, Bausch and Lomb's low-cost spin casting production technique for making soft contact lenses has been defensible and a prime mobility barrier, while the automatic drip coffee maker was widely imitated and its effect on industry structure was industrywide. Defensible technological changes that raise mobility barriers usually carry the greatest competitive significance for a particular firm. However, a technological change that triggers industrywide imitation resulting in a favorable impact on industry structure may boost the profit potential of all industry partici-

---

[5] For a discussion, see Scherer, "Industrial Market Structure," chap. 15; Grabowski and Mueller, "Industrial Research"; and Mueller and Tilton, "Research and Development Costs."

[6] Grabowski and Mueller, "Industrial Research."

pants (including the initiator) enough to justify the cost to the innovator.

This analysis also suggests that technological change can *reduce* entry (or mobility) barriers and hence reduce the attractiveness of industry structure. Thus it should not be presumed that all technological change is beneficial from a strategic viewpoint, even if the firm can defend its innovation from imitation. A technological change that allows customer needs to be better met, for example, may reduce opportunities for product differentiation or lower the economies of scale in the business.

Another important implication of the discussion is that technology, though potentially creating entry/mobility barriers, can be the vehicle that allows new firms or industry followers to overcome entry/mobility barriers.[7] A technological breakthrough can provide the cost or differentiation advantage to allow the firm to fund the cost of overcoming other entry barriers. For example, its lead in radial tires gave Michelin an edge in product differentiation that allowed it to make significant inroads into the U.S. tire market, where distribution and advertising scale economy barriers are high. Rosenbloom's example of the penetration of the steam locomotive oligopoly by General Motors' diesel locomotive provides another potent example.[8]

Technology and Buyer Power

Technological change can shift the bargaining relationship between an industry and its buyers. As described above, technological change can change product differentiation or switching costs, which are both instrumental in determining buyer power. Technological change can also affect the ease of backward integration for the buyer, a key buyer bargaining lever. Technological change can also impact the relationship between the industry's product and the buyer's business and hence the basis of buyer choice. For example, technological change that allows the seller's product to favorably affect the performance of the buyer's product (for example, a better electric motor sold to an air conditioner manufacturer) may reduce the price sensitivity of the buyer.

Technology and Supplier Power

Technological change can also shift the bargaining relationship between an industry and its suppliers. Technological change can eliminate the need to purchase from a powerful supplier group or, conversely, can force an industry to purchase from a new, powerful supplier. Technological change can allow substitute inputs to be used in the firm's product, which creates bargaining leverage against suppliers. R&D investments by a firm to assure proprietary control over frontier technology for its raw materials or other inputs can facilitate the use of multiple suppliers, or allow the breakup of purchased systems into components that can be sourced individually, both reducing supplier leverage. These practices have been instrumental in enhancing the bargaining leverage of U.S. auto producers over component suppliers. Technology can also overcome switching costs that suppliers have put in place. Of course, technological change that locks the firm into particular inputs or has the effect of raising switching costs can worsen the bargaining relationship with suppliers.

Technology and Rivalry

Technology can alter the nature and basis of rivalry among existing competitors in an industry. Technological change can raise or lower fixed costs and hence the pressure for price

---

[7] The operation of technology and marketing as entry "gateways" has been analyzed in a doctoral dissertation by George S. Yip, "Barriers to Entry: A Corporate Strategy Perspective," Harvard Graduate School of Business Administration, 1980.

[8] R. S. Rosenbloom, "Technological Innovation in Firms and Industries: An Assessment of the State of the Art," in P. Kelley and M. Kransberg, eds., *Technological Innovation: A Critical Review of Current Knowledge* (San Francisco: San Francisco Press, 1978).

cutting. For example, the shift to continuous process technology in the corn wet milling industry has raised fixed cost and contributed to elevated warfare, as has the increasing size of oil tankers. Technology can affect product differentiation and switching costs, and hence the responsiveness of buyers to price cuts or other competitive moves and thereby the incentives for competitive warfare. Technology affects substitution which can enhance or reduce industry growth and hence rivalry. Technology can also raise or lower exit barriers that enhance rivalry by locking unsuccessful competitors into an industry. Finally, technological change can alter capacity utilization by raising or lowering effective capacity, or the ability to adjust capacity to demand. This affects the likelihood of repeated outbreaks of rivalry.

Technology and Substitution

Perhaps the most commonly discussed effect of technology on industry structure is its impact on product substitution. Substitution is a function of the relative price/performance of competing products and the switching costs of changing between them. Technological change creates entirely new products or product uses that substitute for others, such as fiberglass, personal computers, and microwave ovens. Technological change also can impact both relative price/performance and switching costs among existing products, as manifested in the two-decade-old struggle for supremacy between steel and aluminum beverage cans. The technological battle to improve relative price/performance between industries producing close substitutes is at the heart of the substitution process.

A technological change is potentially significant for competitive strategy if it significantly changes structure through one or more of the mechanisms described above. The impact of technological change on industry structure can be either positive or negative, as has been illustrated. Technological change is not always good for firms' and industries' profitability, as is sometimes supposed; it can raise supplier power, increase rivalry, or otherwise change structure in ways that lower long-run profit potential.

As was noted earlier, a technological change that improves industry structure is generally most significant for a firm's strategy position if it can be defended from imitation. Even a technological change that threatens industrywide structure can improve the strategic position of an individual firm if it is not imitatable. A technological change that is not imitatable can sometimes differentiate the firm's product or improve its cost position, whereas the same technological change would destroy industry structure if it were widely imitated. For example, a reduction in downtime or spare parts utilization can significantly improve a firm's strategic position if the firm can protect its advantage from imitation or stay ahead of imitation. However, if the improvement becomes widespread, the resulting decrease in the need for manufacturer service organizations and the reduction in the proportion of sales to the price-insensitive aftermarket may reduce differentiation possibilities, enhance buyer power, and destroy industry profitability. Thus the analysis of the consequences of a technological change for industry structure both with and without imitation is central to the choice among technological alternatives.

Technology and Industry Boundaries

One of the ways in which technological change affects industry structure, foreshadowed by my discussion of substitutes, is through its impact on industry boundaries. The boundary of an industry is often an imprecise concept, because of the often arbitrary distinctions that exist between the industry's product and substitutes, between incumbents and potential entrants, and between incumbents and suppliers or buyers who may be (or have the potential to be) partly vertically integrated. The industry analysis framework described above mitigates the importance of drawing precise industry boundaries for strategic analysis, because it focuses on competitive forces beyond just rivalry among existing competitors. Nevertheless, it is important to recognize that wherever one chooses to draw industry boundaries, technological change can widen or shrink them.

Technology can widen industry boundaries in a variety of ways. It can reduce transportation or other logistical costs and thereby enlarge the geographic scope of the market. Technological change that reduces the cost of responding to national market differences can make global industries out of domestic ones. Technology can also enhance the functions the product performs, thereby bringing new customers (and competitors) into the market. In industries such as bank cash dispensers, watches, and telecommunications, technological change is blurring industry boundaries and folding whole industries together.

Technology can narrow industry boundaries as well. Technological change may allow the tuning of product characteristics to a particular business segment or specializing the system for producing and delivering a product to a particular business segment. Thus segments can, in effect, become industries with their own product differentiation and distinct production processes.

Finally, it must be stressed that technological change that creates entirely new products establishes new industry boundaries where there were none previously. Recent developments in areas such as bioengineering, consumer electronics, and ceramic materials provide examples of entirely new industries triggered by technological change.

II. GENERIC TECHNOLOGICAL STRATEGIES

Having cataloged the potential impacts of technological change on industry structure and competitive position, the question becomes: How can the firm conceptualize alternative technological strategies? While there have been a number of attempts to taxonomize technological strategy alternatives, none is entirely satisfactory because of the lack of a clear linkage between technological strategy and an overall view of industry competition.

The starting point for a framework for analyzing technological strategies must be a broader concept of overall competitive strategy. Competitive strategy is an integrated set of policies in each functional activity of the firm that aims to create a sustainable competitive advantage. Technological strategy is but one element of an overall competitive strategy, and thus must be consistent with and reinforced by the actions of other functional departments. A technological strategy designed to achieve product performance leadership will lose much of its impact, for example, if a technically trained sales force is not employed to explain the performance advantages to the customer, and the manufacturing system does not build in adequate provisions for quality control.

Having positioned technological strategy as one element of competitive strategy, also pointed out by others, I will now consider the ways in which technological strategy can contribute to overall strategy.[9] To do this, we must look fundamentally at the mechanisms by which competitive strategy creates sustainable advantages over rivals.[10] Competitive strategy can lead to two broad types of competitive advantage: lower cost or differentiation (uniqueness).[11] If a firm can develop a lower delivered cost of its product to the customer and can protect the sources of this cost advantage from imitation, then the firm has defenses against each of the five competitive forces and will

[9] For an interesting survey that advocates the concept of strategy as an integrating framework for thinking about innovation, see Rosenbloom, "Technological Innovation."

[10] See Porter, *Competitive Strategy,* chap. 2.

[11] Each element of industry structure can be translated into either an effect on the firm's costs, its ability to command premium prices, or both.

earn above-average returns for its industry.[12] Similarly, if a firm can achieve differentiation in some aspect(s) of its product or the manner in which the product is sold or supported that can be protected against imitation, then the firm will have established defenses against the competitive forces and earn above-average returns for its industry. Usually achieving both low cost and differentiation is inconsistent because they imply different collections of functional policies, not to mention different organizational arrangements and company cultures. In some situations, however, the two are not inconsistent, and simultaneous achievement of the two competitive advantages is additive in determining the competitive advantage of the firm. Achieving lower cost or differentiation is the result of differentially impacting the determinants of industry structure described above relative to competitors.

As I have described elsewhere, the two fundamental sources of competitive advantage translate into three generic competitive strategies depending on the scope of the firm's target market within its industry.

Overall cost leadership: The firm seeks industry-wide cost leadership in serving a broad range of industry segments.

Overall differentiation: The firm seeks industry-wide differentiation[13] in serving a broad range of industry segments.

Focus: The firm directs its entire strategy at a narrow target business segment,[14] foregoing sales to other segments. By tuning the strategy exclusively to the target, it seeks to achieve cost leadership or differentiation or both in serving this narrow target even though it does not achieve these advantages overall.

The essential nature of the focus strategy is to exploit the inability of a broadly based competitor to address all business segments equally well. Note that the choice of a particular generic strategy relates to the competitive *advantage* sought by the firm and does not imply that the firm can ignore other aspects of its business. For example, the overall cost leader must still possess a product of acceptable quality and the overall differentiator must have a cost level that is in proximity to other firms.

There are many ways to actually go about implementing each of the three generic strategies, and the particular functional policies (including technology) that will support each of the three differ from industry to industry depending on industry structure. Industry structure also determines whether each of the three strategies is attainable and defensible. In some industries, no firm can significantly differentiate itself in ways that cannot be readily matched by competitors, for example.

Technological strategy is a potentially powerful vehicle with which the firm can pursue each of the three generic strategies. We can link technological strategy to overall competitive strategy, then, by identifying the fundamental source and scope of competitive advantage that the technological strategy is attempting to create or reinforce. Depending on which generic strategy is being followed, the character of the technological strategy will be different, as shown in Figure 2.

An important observation from Figure 2 is that both product and process technological change can have a role in supporting each of

---

[12] The firm's level of return on investment will be partly determined by overall industry structure. In an industry with an unfavorable overall structure, even the above-average returns may be modest in an absolute sense.

[13] Differentiation can be the result of product differentiation or switching costs. Both allow the firm to command premium prices.

[14] The target may be defined by a particular geographic market(s), part(s) of the product line, customer group(s), distribution channel(s), or combination of these.

FIGURE 2 Technological Policies and Generic Competitive Strategies

	GENERIC STRATEGY			
	Overall Cost Leadership	Overall Differentiation	Focus-Segment Cost Leadership	Focus-Segment Differentiation
	TECHNOLOGICAL POLICIES			
Product technological change	Product development to reduce product cost by lowering materials content, facilitating ease of manufacture, simplifying logistical requirements, etc.	Product development to enhance product quality, features, deliverability, or switching costs.	Product development to design in only enough performance for the segment's needs.	Product design to exactly meet the needs of the particular business segment application.
Process technological change	Learning curve process improvement. Process development to enhance economies of scale.	Process development to support high tolerances, greater quality control, more reliable scheduling, faster response time to orders, and other dimensions that improve the ability to perform.	Process development to tune production and delivery system to segment needs in order to lower cost.	Process development to tune the production and delivery system to segment need in order to improve performance.

the generic strategies. While it is sometimes assumed that process technological change is cost oriented and product technological change is differentiation (performance) oriented, the possibilities are much richer. Process technological change may be the key to product performance and hence differentiation (a favorite Japanese company tactic), while product technological change may be at the heart of achieving low cost, either overall or with respect to a particular business segment.

In industries where technological change is rapid or the level of technological sophistication is high, the technological dimension of competitive strategy can be the primary source of competitive advantage in the generic strategy being followed by the firm. Other functional policies support and reinforce the advantage due to technology. In industries where technological change is slow or the level of technology is low, the technological dimension of strategy supports or reinforces competitive advantages fundamentally deriving from other functional areas. However, technology can rapidly become primary to competitive strategy in the latter types of industries if a scientific breakthrough or exogenous innovation occurs that impacts industry structure.

Technological Leadership or Followership

The distinction between technological leadership and followership is a central one in the literature on the management of technology, because it bears so centrally on the required skills and way of managing the R&D organization that have preoccupied scholars in the field. However, technological leadership and followership are empty as alternative technological strategies unless they are related explicitly to their impact on industry structure and to the sources of competitive advantage the firm seeks to gain through pursuit of one generic strategy or the other.[15]

While the notion of technological leadership is relatively clear, there are a variety of possible definitions or types of technological followership reflecting alternative modes in which the follower behaves.[16] Here I consider technological followership as an active strategy aimed at developing particular strategic purposes, and do not discuss the passive follower who knowingly or unknowingly disregards technological change. In general terms, the technological follower examined here exploits (1) the irreversibilities afflicting the leader due to moving first and (2) the potentially lower costs of adapting rather than creating technological change.

Our discussion of the relationship between technological strategy and generic competitive strategies has laid the groundwork for establishing the relationship between technological leadership/followership and competitive advantage, because technological leadership/followership are different ways of implementing the generic strategies. However, the relationship between technological posture and competitive advantage is a complex one, because technological leadership or followership can each be the path to achieving very different strategic purposes.

Figure 3 gives some examples of how technological leadership and followership can be related to the generic competitive strategies described earlier:

The first observation to be made about Figure 3 is that, conceptually, technological leadership or followership can each be ways of achieving any of the generic strategies, and hence any of the fundamental types of competi-

[15] The analysis of the firm's choice between technological leadership and followership is related to the analysis of the determinants of R&D spending in an industry. This is because the proclivity of firms to attempt technological leadership versus followership will strongly influence overall R&D spending, assuming that leaders bear higher R&D costs.

[16] See, for example, C. Freeman, *The Economics of Industrial Innovation* (Baltimore: Penguin Books, 1974).

FIGURE 3 Illustrative Links between Technological Leadership/Followership and the Generic Strategies

	Technological Leadership	Technological Followership
Overall cost leadership	First mover on lowest cost product* or process technology.	Lower cost of product or process through learning from leader's experience.
Overall differentiation	First mover on unique product or process that enhances product performance or creates switching costs.	Adapt product or delivery system more closely to market needs (or raise switching costs) by learning from leader's experience.
Focus—lowest segment cost	First mover on lowest cost segment technology.	Alter leader's product or process to serve particular segment more efficiently.
Focus—segment differentiation	First mover on unique product or process tuned to segment performance needs, or creates segment switching costs.	Adapt leader's product or process to performance needs of particular segment or create segment switching costs.

* Product as used here is defined broadly to include service, delivery, etc.

tive advantage. Most treatments of technological leadership and followership have explicitly or implicitly posited technological leadership as primarily a vehicle for achieving overall differentiation, while followership is the approach to achieving overall cost leadership or focus. While this may be the case, Figure 3 indicates that the possibilities are broader. If the technological leader is the first to begin going down a new, lower learning curve as a result of a new process, for example, the leader can become the overall low-cost producer. Or if the follower can learn from the leader's mistakes and alter the technology of the product to better meet the needs of customers, the follower can achieve overall differentiation. Thus technological leadership and followership are routes to a variety of strategic purposes. Depending on the strategic purpose and the structure of the particular industry, the nature of the technological activities involved in leadership or followership change.

Since technological leadership or followership can support any of the generic strategies, the meaning of technological leadership and followership cannot be precisely defined without reference to the specific competitive advantage the technological activity aims to achieve. There is no one kind of technological leadership or followership from a competitive strategy viewpoint.

The Determinants of the Leadership/Followership Choice

At the heart of the firm's choice between technological leadership and followership is the match between the type of competitive advantage that leadership or followership aims to achieve and industry structure. Industry structure determines the expected payout to achieving the alternative generic strategies that technological effort supports. Having established this broad point, a number of industry structural characteristics are at the core of the choice between leadership and followership:[17,18] technological opportunity to influence cost or differentiation, the uniqueness of the firm's technological skills, first mover advantages, the continuity of technological change, the rate of change in process technology or customer purchasing behavior, irreversibility of investments, uncertainty, and leadership externalities. High technological opportunity, unique technology skills, first mover advantages, and relatively continuous technological change favor technological leadership, while rapid change in process technology or customer purchasing behavior, technological discontinuities, irreversibility of investments, uncertainty, and leadership externalities favor technological followership. Technological licensing plays a complex role in the leadership/followership choice and will be discussed separately.

The attractiveness of technological leadership is clearly a function of the extent of technological opportunity to improve the product or process in ways that lower cost or enhance differentiation either overall or in particular business segments. Technological opportunity affects the possibility for the leader to create a cost or differentiation gap between it and the follower, and the size of the potential gap. The persistence of the gap and its strategic value will be influenced by factors to be discussed below. Technological opportunity, though intuitively clear, is difficult to specify empirically. It is a function of the degree of technological complexity in an industry, influenced by intrin-

[17] I consider here only the link between industry and competitive characteristics and the leadership/followership choice, and ignore internal organizational considerations that may favor one strategy or another. In practice, the firm's particular initial position, internal management structure, and culture, among other things, will play a large part in the choice as well.

[18] This analysis of technological leadership and followership can be readily generalized to the choice of leadership or followership in other dimensions of strategy such as marketing.

sic product features and by the richness of scientific enquiry in related fields.[19]

The uniqueness of the firm's technological skills will be a second essential determinant of the leader/follower choice. Leadership is favored if the firm has unique technological abilities that allow it to open up a larger technological gap, while the benefits of leadership are dissipated if multiple firms can achieve the hoped-for technological change through parallel technological investments. Where there are multiple firms with equivalent technological skills, leadership is optimal only if it offers other first mover advantages in addition to the technological edge itself that allow an initial jump in technology to be preemptive. The uniqueness of a firm's technological skills is partly a function of its ability to share research costs with related businesses and its ability to access supplier or customer technological help, as well as its technological capabilities in the particular business unit facing the leader/follower choice.

In some industries, a significant fraction of technological change flows into the industry from exogenous sources, such as suppliers, customers, or even industries unrelated vertically but sharing basic technologies. The existence of significant flow of technology from exogenous sources generally increases the risk of technological leadership, because it decouples competitors' level of technology from their internal skill and investment in R&D.[20]

First mover advantages are factors that allow a leader to translate a technology gap into other competitive advantages, which persist even if the technology gap closes. First mover advantages available to leaders can be divided into two broad types: intertemporal dependence of demand and intertemporal dependence of costs. Intertemporal dependence of demand is the situation where the demand for the firm's product in period B is a function of its sales in period A. This could be true for a number of reasons. First, sales in period A could establish the reputation or brand identity of the firm as a premier producer, leading both to repeat purchases by old customers in period B and the attraction of new customers in period B who are aware of this reputation.[21] Through this mechanism, the perceived product differentiation of the firm can accumulate. Intertemporal dependence of demand also can occur because of the existence of switching costs.[22] If the firm makes a sale in period A, the existence of switching costs gives the firm an advantage in repeat sales during period B even if its product no longer is superior to competitors'.

Intertemporal dependence of demand will potentially be greatest where switching costs are high, where there is moderately frequent repeat buying, and where gathering information for purchase decisions is expensive. Moderate rather than frequent or infrequent repeat buying maximizes the first mover effect because infrequent repeat buying minimizes the first mover's advantages with repeat buyers, while frequent repeat buying implies that the

[19] F. M. Scherer, "Firm Size, Market Structure, Opportunity, and the Output of Patented Inventions," *American Economic Review,* December 1965; and "Market Structure and the Employment of Scientists and Engineers," *American Economic Review,* June 1967. Scherer has attempted to measure technological opportunity empirically in a number of studies.

[20] See R. E. Caves, "Market Structure and Embodied Technological Change," in R. E. Caves and M. J. Roberts, eds., *Regulating the Product: Quality and Variety* (Cambridge: Ballinger Press, 1975), pp. 125–41, for a discussion and example of how vertical relationships have affected technological change in the aircraft industry.

[21] One reason for this may simply be that high market share in period A establishes the firm as a leader in the mind of the customer, in a way that is unavailable to the follower coming in as the second, third, or subsequent brand.

[22] Switching costs are fixed costs faced by buyers in changing from one supplier to another, distinct from the relative price/performance of alternative suppliers' products. Switching costs are discussed extensively in Porter, *Competitive Strategy,* chaps. 1 and 6.

risk of sampling a new brand is likely to be low. The cost of gathering information is important because costly information gathering will lead purchasers who have had an acceptable purchase experience in period A to repeat purchase from the first mover rather than search for a new brand, and will imply that first-time customers in period B will make use of reputation or past market share in their purchase decisions rather than utilize more extensive information gathering on many competing brands.

Intertemporal dependence of costs is present where the firm's costs in period B are a function of its having produced the product in period A. Intertemporal dependence of costs will be present when there is a learning curve operating that can be kept proprietary. The technological leader will enjoy higher cumulative volume due to moving first that will lead to a fast rate of learning. This first mover advantage is present whether the learning curve is a function of cumulative volume or calendar time, though the strategic implications are somewhat different.[23]

Intertemporal dependence of demand interacts with the presence of economies of scale to produce another potential cost advantage for the first mover. If there is intertemporal dependence of demand, the firm with sales in period A will be advantaged in achieving sales in period B, which implies a higher sales volume for that firm in period B than for an otherwise equivalent firm that begins production in period B. If there are scale economies present, then, the first mover will enjoy lower costs by moving down the static scale curve first.

First mover advantages are enhanced by the existence of lead time in bringing capacity onstream or in product or process change. Lead time lengthens the first mover's learning advantage where costs are intertemporally dependent, and lengthens the period during which the first mover has uncontested sales, which benefits sales in subsequent periods through intertemporal dependence of demand.[24] Lead time also obviously lengthens the duration of the period in which the leader has a technological gap even if there are no first mover advantages.

If first mover advantages are great, the leader may only need to open up a onetime technological gap to enjoy a substantial long-term competitive advantage. The first mover advantages leverage the transient technological gap into mobility barriers of other kinds. Where first mover advantages are slight, however, the technological gap must be maintained over time through continued innovation to offset the penalties of leadership to be discussed below.

The ability to reap first mover advantages presupposes technological change characterized by continuity and relatively stable process technology and customer needs. Continuity refers to the movement of technology along a fundamentally similar path, while technological discontinuity is the shift in technology to an entirely new path such as the shift from electromechanical to electronics that has impacted many industries. Intertemporal dependence of costs is nullified by process change that invalidates past learning and creates a new learning curve or obsoletes existing facilities. On the other hand, if process change is slow enough to allow significant transference of learning to the new technology, then the first mover retains its cost advantage. Or if technological change is characterized by continuity, then chances are the leader can remain on the same scale or learning curve.

Intertemporal dependence of demand is

[23] For a discussion of some alternative formulations of the rate of learning and some of their implications, see M. E. Porter, "Competitor Selection and Optimal Market Configuration," mimeographed course note, Harvard Graduate School of Business Administration, 1981.

[24] For a model that relates lead time to market share, see the chapter by M. Therese Flaherty in *Research on Technological Innovation, Management, and Policy,* vol. 1 (Greenwich, Conn.: JAI Press, Inc., 1983).

also nullified by significant changes in customer needs or purchasing behavior that cause either repeat or first-time buyers to no longer value information gained from sales of the first mover in period A. If customer needs change, the first mover may actually be at a disadvantage because its brand reputation is identified with the old needs. Sometimes reputation carryover still occurs despite changing customer needs, however, and here the first mover may retain an advantage over a follower.

A rapid rate of product change without a change in customer needs is not in itself a threat to first mover advantages unless the first mover is somehow identified in the customer's mind with old generation products, assuming the first mover has access to the new technology. This is because the first mover (leader) can modify its product to reflect the technological change. However, even without such a negative image carryover, rapid product change can put the first mover at a disadvantage if product and process investments are specialized and irreversible.

The irreversibility and specialization of investment requirements in an industry interacts with changing process or product technology to work against the first mover advantages of technological leadership. If investments in plant, tooling, training, and specialized personnel are specific to the particular process and product configuration of the leader (first mover) and costly to modify, a process or product change requires reinvestment by the leader to maintain its position. This implies that the total investment required by the leader exceeds that of a follower who has not made the initial investment, and favors the follower strategy, other things being equal.[25] On the other hand, if product or process modifications to keep up with technological change are not costly, or product or process investments are not specialized in the first place, then the leadership strategy is favored provided there are advantages from the sources previously described.

The discussion so far serves to illustrate a misconception that is sometimes translated into implications for the choice of technological strategy. It is sometimes argued that attention to cost is only sensible once a dominant product design has emerged, usually late in industry development. This can be true, but whether it is true in a particular situation depends on the intertemporal dependence of costs, the specialization and irreversibility of investments, and the degree of process technological change. If process change is slow or incremental, or process investments are unspecialized to product configuration, intertemporal cost dependence may imply that a follower who waits to invest in process technology is doomed to a significant cost disadvantage to the technological leader.[26]

Another determinant of the leader/follower choice is the extent of uncertainty present about the appropriate technological direction and the ability to achieve a desired technological breakthrough. The effect of uncertainty on the choice between leadership and followership depends on the particular nature of the uncertainty present, and presupposes that investments in process or product technology are specialized and irreversible. Uncertainty over future product and process configuration favors the attractiveness of the follower strategy compared to the leadership strategy, because the leader has a higher probability of having to reinvest to meet changed conditions than the follower who gains information by moving late, *ceteris paribus*. Once again, though, the specialization and irreversibility of investment is pivotal. Uncertainty over achieving the desired technological breakthrough unambiguously raises the expected cost of leadership.

---

[25] This discussion refers to the ex ante choice between leadership and followership, and not the choice of the leader about investing or not investing in next generation technology.

[26] These issues will be treated further in Part III.

Uncertainty over the size of future demand creates a trade-off for but does not unambiguously favor or disadvantage the leadership strategy, *ceteris paribus*. Uncertainty over the size of future demand raises the probability that the leader can preempt the market through moving first with a new product or process before the uncertainty is resolved. This is because as the follower waits to gain more information about future demand, future demand may prove to be fully served by the leader, thereby denying the follower market access at all. The trade-off comes from the risk that the technological leader builds more capacity than the demand can ultimately absorb.

The final important factor influencing the choice between technological leadership and followership is the extent of leadership externalities. Leadership externalities stem from investments the leader makes that become partial or complete public goods which benefit followers as well. Thus followers get a free ride on the leader's investments, lowering their required investment.

Leadership externalities can be divided into two broad types: technological and market. Technological externalities stem from the public good character of the leader's intangible assets in technology. The leader must make investments in product or process technology for which the returns may not be appropriable.[27] If diffusion of technology is rapid[28] and the follower's cost of imitation is low relative to the leader, then the leader's period of technological superiority is short. Technological information disseminates through product inspection and reverse engineering, suppliers, customers, ex-employees, and other sources. If these mechanisms are effective in an industry, the follower strategy is favored.[29] Conversely, if the leader's technology can be protected through patents, secrecy, or other means, then leadership is favored.[30]

A leader may also face market-oriented externalities, which derive from the leader's required investment in gaining market access broadly defined. Some of the most common of such investments are as follows:[31]

Market Externalities in Product Innovation

- Gaining regulatory approvals.
- Achieving code compliance.
- Winning customers away from substitutes (e.g., marketing costs, penetration prices).
- Customer education on product usage.
- Investments in infrastructure such as training independent repair and service personnel.
- Investments to improve the performance price, or availability of, complementary goods.

Market Externalities in Process Innovation

- Gaining regulatory approvals.
- Costs of convincing customers of the reliability of the new process.
- Investments in infrastructure such as supply sources for new raw material inputs, machinery, etc.

---

[27] The concept of appropriability is central in economists' discussion of the incentives for innovation, and is used as it is here to refer to the extent of diffusion of technological investments. See, for example, R. C. Levin, "Technical Change and Optimal Scale: Some Evidence and Implications," *Southern Economic Journal,* October 1977; and P. Dasgupta and J. E. Stiglitz, "Uncertainty, Industrial Structure, and the Speed of R&D," *Bell Journal of Economics* 2 (Spring 1980), pp. 1–28.

[28] Partly a function of lead time, as discussed earlier.

[29] In some industries, suppliers or customers can aid a follower in keeping up technologically through investments of their own.

[30] One device sometimes available to a leader is to patent not only its own technology but also the alternative technological paths to a given result. For a discussion, see R. J. Gilbert, "Patents, Sleeping Patents, and Entry Deterrents," in S. Salop, ed., *Strategy, Predation, and Antitrust Analysis,* Federal Trade Commission, 1981, pp. 205–69.

[31] In the case of process technology leadership, the leader may have to gain regulatory approvals to operate the new process or convince customers that the new process will be reliable in order to secure business. The analysis of these costs is the same as with marketing costs.

The size of these investments is a function of the industry and the nature of the technological change that the leader has developed (e.g., training independent repair personnel is not an issue if the technological change does not involve the repairable parts of the product). If followers can have a partial or complete free ride on these investments, then followers will be at a cost advantage compared to leaders, unless there are significant first mover advantages that are offsetting. Just as with technology diffusion, however, followers must generally bear some incremental cost of "adapting" the market-oriented public good created by the leader. For example, the leader may have secured regulatory approval of a new product variety or production process, but the follower must still bear some cost of getting its specific product or facility approved. Further, some of the benefits of investments in opening up the market for a new product technology may be appropriated by the leader because they are made under its brand name. The size of required investments in adaptation relative to the leader's investment influence the attractiveness of the follower strategy.

Technology Licensing and Leadership/Followership

Technology licensing is a device for leaders to potentially appropriate additional profits from their technological investments, or for followers to gain access to leaders' technology without investments in imitation. The optimal pattern of licensing is partly endogenous based on the factors described above, and partly a function of exogenous factors.

Where the leader can license technology to firms with which the leader does not actually or potentially compete, licensing unambiguously raises the expected profits of leadership and favors the leadership strategy. Licensing to competitors (or potential competitors such as foreign firms), however, involves a trade-off between licensing fees and strategic considerations. Licensing a follower reduces or eliminates the technology gap between the leader and the follower, which is a cost of licensing. However, in some circumstances the leader may benefit from the presence of a competitor with its technology, and here licensing may improve the leader's position and provide royalties.[32]

For followers, licensing may be a low-cost source of technology depending on licensing fees and the cost of imitation. Licensing fees will more likely be low relative to imitation costs if the source of technology is outside the market directly contested by the follower, because this implies that the licensee is less of a threat to the licensor and will view royalties as incremental profits. The attractiveness of licensing for a follower is a function of the cost of licensing and the particular follower's strategy. Licensing may lock the follower to the leader's technology, for example, in ways that will impede adaptation or focusing.

III. TECHNOLOGICAL CHANGE AND THE STATE OF INDUSTRY MATURITY

In view of the powerful role of technology in changing industry structure discussed in Part I, it is of interest to examine the ways in which technological change varies with the state of development of an industry. Are there characteristic patterns of technological innovation during different states of industry maturity that would allow predictions about the way in which industry structure might evolve, and do these have implications for the direction of a firm's technological activity and its choice of technological strategy?

Most research on the relationship between technological change and industry develop-

---

[32] For a discussion, see M. E. Porter, "Strategic Interaction: Some Lessons from Industry Histories for Theory and Antitrust Policy," in S. Salop, ed., ***Strategy, Predation, and Antitrust Analysis,*** Federal Trade Commission, 1981.

ment has grown out of the product life cycle concept. Many treatments of the product life cycle include an examination of how technological change varies as an industry moves from the emerging state through growth, maturity, and decline.[33] According to the life cycle model, early technological emphasis is focused on product innovations, while the production process is flexible and job shop in nature. As the industry matures, product design changes slow down and mass production techniques are introduced. Process innovation takes over from product innovation as the primary form of technological activity to reduce costs of the increasingly standardized product. Finally, all innovation slows down in late maturity and decline.

This view of how technological innovation varies with industry maturity has been deepened and refined considerably by the work of Abernathy and Utterback.[34] Abernathy and Utterback posit a transition over time from a "fluid" state of technology to a "specific" state. The spirit of their model is consistent with the life cycle view but considerably more developed. Initially, according to Abernathy and Utterback, product design is fluid, and substantial product variety is present. Product innovation is the dominant mode of innovation and aims primarily at improving product performance. Successive product innovations ultimately yield a "dominant design" where the optimal product configuration is reached. Process innovation is initially minor in significance, and early production processes are characterized by small scale, flexibility, and high labor skill levels. As product design stabilizes, increasingly automated production methods are employed and process innovation to lower costs takes over as the dominant innovative mode. Ultimately, innovation of both types begins to slow down.

These hypotheses about the variation of technological innovation with industry development are an accurate portrayal of technological change in a variety of industries, as the examples cited by Abernathy and Utterback attest. However, as Abernathy and Utterback themselves recognize, the hypotheses are not general and do not apply to every industry. For example, in industries with undifferentiated products (e.g., minerals, refined sugar, many chemicals), the sequence of product innovations culminating in a dominant design does not take place or takes place very quickly. In other industries (e.g., military and commercial aircraft, large turbine generators), automated mass production is never achieved and most innovation is product oriented. And these are just two examples of the wide variety of patterns of technological change that are observed in practice.

This suggests that a general theory of how innovation varies with industry development must have as its aim not a single pattern but rather the identification of underlying variables that interact to yield the pattern observed in a particular circumstance. It is also clear that a theory of how innovation varies with industry maturity must be embedded in a more general view of how industry structure evolves, because innovation is both a response to the incentives created by the overall structural configuration of an industry and a shaper of structure. While we cannot present a general theory of how innovation varies with industry maturity here, we can sketch some of its essential parts.[35]

It is useful to divide the determinants of the pattern of technological innovation as indus-

[33] For a generalized summary of the predictions of the life cycle literature, see Porter, *Competitive Strategy,* chap. 8. An example of the predictions for technology appears in S. Hirsch, "The United States Electronics Industry in International Trade," *National Institute of Economic Review,* no. 34 (1965).

[34] See W. J. Abernathy, *The Productivity Dilemma: Roadblock to Innovation in the Automobile Industry* (Baltimore: The Johns Hopkins Press, 1978), chaps. 4 and 7.

[35] Some beginning points for this theory are described in Porter, *Competitive Strategy,* chaps. 8 and 9.

tries evolve into two types: dynamic processes and underlying structural parameters that influence the extent and speed with which these processes occur. Since the pattern of technological change will be the outcome of the technological investments made by leaders and followers, the determinants of the pattern of technological change will be related to the factors discussed above that determine the leadership/followership choice.

A number of fundamental dynamic processes underlie the shifting pattern of innovative activity in industries over time, of which the following seem particularly important:

1. Scale change.
2. Learning curve in product design and process.
3. Uncertainty reduction and imitation.
4. Technological diffusion.
5. Diminishing returns to technological innovation in product and process.

The manner in which these processes operate and interact is best demonstrated by describing their effect on innovation in particular types of industry situations. Inherent in the patterns of innovative activity posited by Abernathy and Utterback is one set of implicit assumptions about these processes that will be illustrative. Abernathy's and Utterback's model assumes that as an industry develops from initially small scale to some scale in maturity, the unit volume to support mass production methods is forthcoming. Through successive product innovation (coupled with diminishing returns), the uncertainty about appropriate product characteristics is reduced and a dominant design emerges, reinforcing the growth in available scale by introducing product standardization. Technological diffusion must be present in Abernathy's and Utterback's world to eliminate product differences, and compels process innovation to push down the scale and learning curve to lower costs of the standardized product. Continued diffusion motivates continuous process innovations to lower costs of the standardized product even further. Ultimately, diminishing returns to process innovation set it to reduce innovative activity altogether.

The degree to which the Abernathy and Utterback pattern occurs depends on the speed and extent to which these dynamic processes proceed. This will be a function of some underlying industry structural parameters such as the following:

1. Intrinsic physical differentiability of the product.
2. Intrinsic segmentation of buyer needs.
3. Unit volume (scale) at maturity.
4. Potential scale economies and learning effects.
5. Linkage between product design and process.
6. Motivation for substitution.
7. Technological opportunity.

The first four parameters are intrinsic characteristics of the product and process, and can be discussed together. The Abernathy and Utterback pattern of innovation implies that the product is physically differentiable so that many varieties exist early in the evolutionary process, but that the intrinsic segmentation of buyer needs is low because a dominant design emerges that satisfies many buyers. Further, the dominant design must emerge (and buyers willing to compromise their particular needs), because standardization allows major cost reduction due to the implied presence of significant economies of scale. Abernathy's and Utterback's model also assumes high unit volume at maturity to support a highly automated production process.

The assumptions about these four parameters implicit in Abernathy's and Utterback's model are only one of many possible combinations, each with differing implications for the pattern of technological change. If the product is not physically differentiable, then product innovation largely stops with initial product introduction, and innovative effort centers around the process. Buyer's product needs are definitionally common, so that the extent and nature of process innovation depends on potential scale economies, learning effects, and the unit volume achievable in industry maturity. Such a pattern might be characteristic of industries such as high-fructose corn syrup[36] and copper mining.

If the product is physically differentiable, buyer needs are diverse, and potential scale economies are not too great or small unit volumes at maturity limit their achievement, the pattern of technological change will likely take yet another course. Here product innovation will be the predominant/dominant mode of technological activity, and the search for new varieties and new segments will be continuous and no dominant design will emerge. Process innovation will occur but only in a subordinate mode and to support product variety.

Commercial aircraft illustrates another possible pattern, where small unit volumes even for the largest selling model preclude mass production, and the production process is thus inherently flexible. The product is highly differentiable and buyer needs are potentially diverse, but the scale economies in product design are great enough to limit the number of product varieties developed for the sake of unit cost. The flexibility of production means that the production process places no barrier to continuous and long-lasting efforts at product innovation with secondary efforts on process innovation. Many other combinations of the first four parameters can also be observed in practice with resulting implications for the pattern of innovation.

At least three other structural parameters as previously noted are central to determining the nature of technological innovation as an industry evolves. The first is the linkage between product design and the production process. The Abernathy and Utterback pattern of innovation implies a strong linkage, where the process cannot be stabilized and automated without the product becoming standardized. With undifferentiated products the linkage from process to product is definitionally broken (though not vice versa), and it can also be broken in such industries as plastic injection molding and book publishing where highly automated production processes can be inexpensively set up for different product varieties. The existence of a product-process linkage creates pressures to standardize the product and then improve the process in that sequence, while the absence of the linkage allows product and process innovation to be independent and to reflect the respective technological opportunities and market needs represented in the other parameters.

A second additional parameter is the motivation for substitution that underlay the creation of the industry. Abernathy and Utterback (and most life cycle theorists) appear to assume that product performance is the dominant motivation for substitution that creates new industries, and hence the key objective of early innovative efforts. In some industries, however, cost saving can be the dominant or a primary motivation for substitution, implying a great deal of early innovative activity directed at cost reduction.[37] For example, disposable diapers did not succeed in the U.S. market until production process breakthroughs by Procter and Gamble brought the cost of a disposable diaper into the ballpark compared to cloth diapers and diaper services. Similarly, aluminum was substituted for copper in many types of electrical wiring almost exclusively for cost reasons. The motivation for substitution varies among industries, and with it the incentive for process versus product innovation in the early phases of industry development and the type of such inno-

[36] Here there was a second-generation product, but most technological effort has been directed at the production process.

[37] See Porter, *Competitive Strategy,* chap. 9.

vation (e.g., product innovation to lower cost or process innovation to improve tolerances or product performance).

A final important parameter influencing the pattern of innovation as an industry develops is the technological opportunity for improvement in the product and process. As discussed earlier, technological opportunity is a difficult concept to define precisely, though it is intuitively clear that not all products and processes offer equal opportunities for continued technological improvement. In a product like commercial aircraft, diminishing technological returns to efforts at product innovation come relatively slowly, while in others such as table salt they come quickly. The same could be said about process technology. Depending on technological opportunity, the relative effort devoted to product and process change and their duration as an industry evolves will vary.

Continuity versus Discontinuity in Technological Change

The pattern of technological change and its impact on industry structure is likely to differ greatly depending on whether technological change is characterized by continuity or discontinuity, just as this distinction influenced the choice between leadership and followership. Where there is technological continuity, the process of technological change is likely to be endogenously determined by actions of participants or spinoffs of industry participants. Exogenous sources of technology in this world tend to center primarily around existing suppliers to the industry, whether they be equipment suppliers or raw material producers. As described earlier, technological continuity favors first mover advantages which tend to stabilize competitive positions and market shares compared to the situation where discontinuity allows followers to leapfrog early entrants.

Technological discontinuity is likely to be associated with an entirely different process of technological change. The source of technology is much more likely to be exogenous, the result of entirely new competitors or of new suppliers to the industry who promote the new technology to existing competitors. Technological discontinuities pioneered by existing industry participants usually reflect great pressures from substitutes or an unusual commitment to basic research. Technological discontinuity tends to nullify many first mover advantages and mobility barriers built on the old technology. Hence, a period of technological discontinuity offers a window in which market positions become fluid and market shares can fluctuate greatly.[38]

Technological discontinuity tends to decouple the pattern of technological innovation from the state of industry maturity, because exogenous sources of technology are less responsive to industry structural circumstances than are the technological activities of existing industry participants. Hence technological discontinuities can impact mature industries (e.g., the radial tire, coffee percolators) or still developing industries (e.g., color TV) alike.

IV. INDUSTRY EVOLUTION AND TECHNOLOGICAL STRATEGY

The choice of technological strategy in evolving industries is, following the earlier discussion, a function of the generic competitive strategy that yields the greatest sustainable competitive advantage, given industry structure, and its implications for the desired nature of technological effort by the firm. Some parameters that influence the pattern of technological change and hence the ability to achieve competitive advantages of various sorts have been sketched in Part III. To carry this analysis further, two central strategic issues in evolving

[38] This effect was supported statistically in a cross-section of industries in R. E. Caves and M. E. Porter, "Market Structure, Oligopoly, and the Stability of Market Shares," *Journal of Industrial Economics,* June 1978, pp. 289–313.

industries that relate to technological strategy will be briefly examined here: timing of entry and the relationship of timing of entry to the choice of technological leadership or followership.

Framed most sharply, the choice of timing the entry is the choice between pioneering and being a later entrant. The choice of pioneering versus later entry is related to the choice between technological leadership and followership, though not exactly colinear, as shown in Figure 4.

Technological leadership and followership have an implicit time dimension, because followership can only exist relative to a technological standard established by a leader. Hence a pioneering strategy will imply technological leadership, at least initially. However, a later entrant is not necessarily a technological follower. A later entrant may instead attempt to leapfrog the incumbents' technology in product, process, or both. It should also be noted that the timing of entry choice is a once and for all decision, while the choice of technological posture can be changed over time. Thus, mixed technological strategies are possible which involve shifting between technological leadership and followership as an industry evolves.

The choice of timing of entry depends on many of the same considerations that bear on the choice between technological leadership and followership. Pioneering involves a broader form of leadership than technological leadership because the pioneer must move first on all elements of strategy, but the analysis of leadership is generalizable as noted earlier. The pioneering strategy rests fundamentally on confidence in the ability of the firm to maintain a technological gap or on the presence of first mover advantages. If first mover advantages are significant enough, the pioneer need not maintain technological leadership in order to enjoy a strategic advantage because the cost or differentiation advantages of moving first will create mobility barriers not dependent on a technology gap.

FIGURE 4 The Relationship of Technological Posture to Timing of Entry

Timing of entry	Technological posture: Technological leadership	Technological posture: Technological followership
Pioneer	X	
Later entrant	X	X

Working against the pioneering strategy are specificity and irreversibility of required investments, product, process and buyer behavioral change, technological discontinuities, uncertainty about product and process, and leadership externalities. The later entrant exploits these problems facing the pioneer. The discussion in Part III about the determinants of the rate and direction of technological change as an industry matures bears directly on the likely extent of these problems. Since the rate of change of product, process, and buyer behavior can often be great in the early stages of an industry's development, the need for first mover advantages to justify the pioneer strategy are accentuated.

The later entrant has a choice of technological strategies for entry. The choice of leadership or followership by the later entrant is a function of whether industry evolution subsequent to entry by pioneers has created *new* advantages of technological leadership. Some of the determinants of the choice of technological strategy by late entrants are given in Figure 5.

Late entry with a technological followership posture is primarily an approach to achieving overall cost leadership or cost leadership through focus. Late entry with a technological leadership strategy may potentially achieve any one of the generic strategies, depending on industry structure.

FIGURE 5 Determinants of Technological Strategy of Late Entrants

Late Entry with Technological Leadership	Late Entry with Technological Followership
Product, process, or buyer behavior change that allows technological leapfrogging.	Few first mover advantages of second-generation technological leadership.
Irreversible and specific pioneer investments in first-generation technology.	Irreversible and specific required investments.
Significant first mover advantages to second-generation technological leadership.	Large market-oriented externalities of leadership that persist in the second-technological generation.
Relatively small externalities of second-generation technological leadership.	Rapid, low-cost technological diffusion.
	Relatively stable product, process, and buyer behavior that does not support leapfrogging.

V. CONCLUSIONS

This article has developed a theoretical framework for determining the optimal choice of technological strategy by the firm. Technological change was shown to have a variety of potential impacts on industry structure which motivated technological strategies. Technological strategy was framed as a vehicle for implementing the three generic competitive strategies that seek fundamentally different types of competitive advantages. The choice between technological leadership and followership was modeled as a function of a number of industry structural characteristics, recognizing that the aims of technological leadership and followership differed depending on the generic strategy followed. The determinants of the pattern of technological change as an industry evolves were identified, and followed closely from the factors previously discussed that influenced the technological choices of individual firms. Finally, some implications of the pattern of technological change in evolving industries were derived for the firm's choice between pioneering and late entry, and the late entrant's choice of technological posture.

The theory presented here has numerous implications for empirical research which remain to be developed. Promising areas for empirical tests are explaining the mix and relative success of technological leaders and followers in a sample of industries, and relating the empirically observed pattern of technological change with industry development to the structural variables identified. If this essay will stimulate such empirical testing, as well as more theoretical work that examines the link between technological strategy and competition, then it will have achieved its desired technological effect.

ACKNOWLEDGMENT

The author has benefited from comments by Richard Rosenbloom, Richard Caves, and William Abernathy. September 1981.

REFERENCES

Abernathy, W. J. *The Productivity Dilemma: Roadblock to Innovation in the Automobile Industry.* Baltimore: The Johns Hopkins Press, 1978.

Caves, R. E. "Market Structure and Embodied Technological Change." In R. E. Caves and M. J. Roberts, eds., *Regulating the Product: Quality and Variety.* Cambridge: Ballinger Press, 1975, pp. 125–41.

Caves, R. E., and M. E. Porter. "Market Structure, Oligopoly, and the Stability of Market Shares." *Journal of Industrial Economics,* June 1978, pp. 289–313.

Dasgupta, P., and J. E. Stiglitz. "Uncertainty, Industrial Structure, and the Speed of R&D." *Bell Journal of Economics* 2, Spring 1980, pp. 1–28.

Freeman, C. *The Economics of Industrial Innovation.* Harmondsworth, England; Baltimore: Penguin Books, 1974.

Gilbert, R. J. "Patents, Sleeping Patents, and Entry Deterrence." In S. Salop, ed., *Strategy, Predation, and Antitrust Analysis.* Federal Trade Commission, 1981, pp. 205–69.

Grabowski, H. C., and D. C. Mueller. "Industrial Research and Development, Intangible Capital Stocks, and Firm Profit Rates." *Bell Journal of Economics* 9, Autumn 1978, pp. 328–43.

Hirsch, S. "The United States Electronics Industry in International Trade." *National Institute of Economic Review,* no. 34, 1965.

Kamien, M., and N. Schwartz. "Market Structure and Innovation: A Survey." *Journal of Economic Literature* 13, 1975, pp. 1–37.

Levin, R. C. "Technical Change and Optimal Scale: Some Evidence and Implications." *Southern Economic Journal,* October 1977.

Mueller, D. C., and J. E. Tilton. "Research and Development Costs as a Barrier to Entry." *Canadian Journal of Economics* 2, November 1969, pp. 570–9.

Porter, M. E. *Competitive Strategy: Techniques for Analyzing Industries and Competitors.* New York: Free Press, 1980.

_________. "Competitor Selection and Optimal Market Configuration." Mimeographed course note. Harvard Graduate School of Business Administration, 1981.

_________. "Strategic Interaction: Some Lessons from Industry Histories for Theory and Antitrust Policy." In S. Salop, ed., *Strategy, Predation, and Antitrust Analysis.* Federal Trade Commission, 1981.

Rosenbloom, R. S. "Technological Innovation in Firms and Industries: An Assessment of the State of the Art." In P. Kelley and M. Kransberg, eds., *Technological Innovation: A Critical Review of Current Knowledge.* San Francisco Press, 1978.

Scherer, F. M. "Firm Size, Market Structure, Opportunity, and the Output of Patented Inventions." *American Economic Review,* December 1965.

_________. "Market Structure and the Employment of Scientists and Engineers." *American Economic Review,* June 1967.

_________. *Industrial Market Structure and Economic Performance.* 2d ed. Chicago: Rand McNally, 1979.

Schumpeter, J. A. *The Theory of Economic Development.* Cambridge: Harvard University Press, 1934.

Yip, G. S. "Barriers to Entry: A Corporate Strategy Perspective." Unpublished doctoral dissertation. Harvard Graduate School of Business Administration, 1980.

Reading II–9
Technological Strategy

M. A. Maidique and A. L. Frevola, Jr.

Technology, a vital force in the competitive environment of the modern firm, is a "missing link" in corporate strategy. The majority of firms in technology-intensive industries such as aerospace, computers, chemicals, electronics, and pharmaceuticals do not generally include technology explicitly as part of their corporate strategy.

Technology, generally viewed as either "high" or "low" or often not at all, and technological decision making are given very little emphasis in the traditional management literature and in American business education in general. Management, more often than not, relies on corporate technical experts to make the proper technological choices. A top R&D executive, when asked to describe how he participated in technological decisions, explained, "That's simple. I don't."

It would be foolish to argue that senior executives should concern themselves in the minutiae of technological decisions. Yet many executives fail to recognize that all technological choices, when made in the corporate context, are also business decisions. Design choices, for example, involve trade-offs between cost, performance, reliability, and ease of use. And all of these generally have an important impact on competitiveness.

Both a forecast of expected technological directions and an environment that emphasizes the fundamental role of technological policies and decisions in the firm are required to make effective technological choices. The pattern of choices that the firm makes regarding technology becomes the technological strategy of the firm. By integrating technology with business policy in such a framework, firms can develop a well-defined, coherent posture toward technology and facilitate and encourage executive decisions in technology-related areas.

Because technological strategy addresses a distinct set of decisions, it should, however, be differentiated from manufacturing strategy, even though the two are closely related elements of business strategy. Manufacturing strategy, decisions regarding the location, scale, and organization of productive resources, is usually developed within the scope of a given technology. Technological strategy, on the other hand, is concerned with choices between alternative new technologies, the manner in which they are implemented into new products and processes, and the utilization of resources that will allow their successful implementation.

Properly defined, technological strategy cuts across such functional policies as manufacturing, marketing, finance, R&D, as well as corporatewide policies regarding product-market focus, personnel resource allocation, and control. As such, this note attempts to examine both the array of choices and plans that enables the firm to respond effectively to technological threats and opportunities and the relationship of technological strategy to some of the functional elements of corporate strategy.

In the development of its technological strategy, a firm must make decisions in the areas of technology selection and embodiment, technology sources, competitive timing, level of R&D investment, organization and policies concerning R&D, and competence levels. Within these areas, the firm should answer the following questions: (See Exhibit 1.)

- *Selection*—What technologies to invest in are promising from the perspective of existing, new, or related product lines? How

EXHIBIT 1 Technological Policy Framework

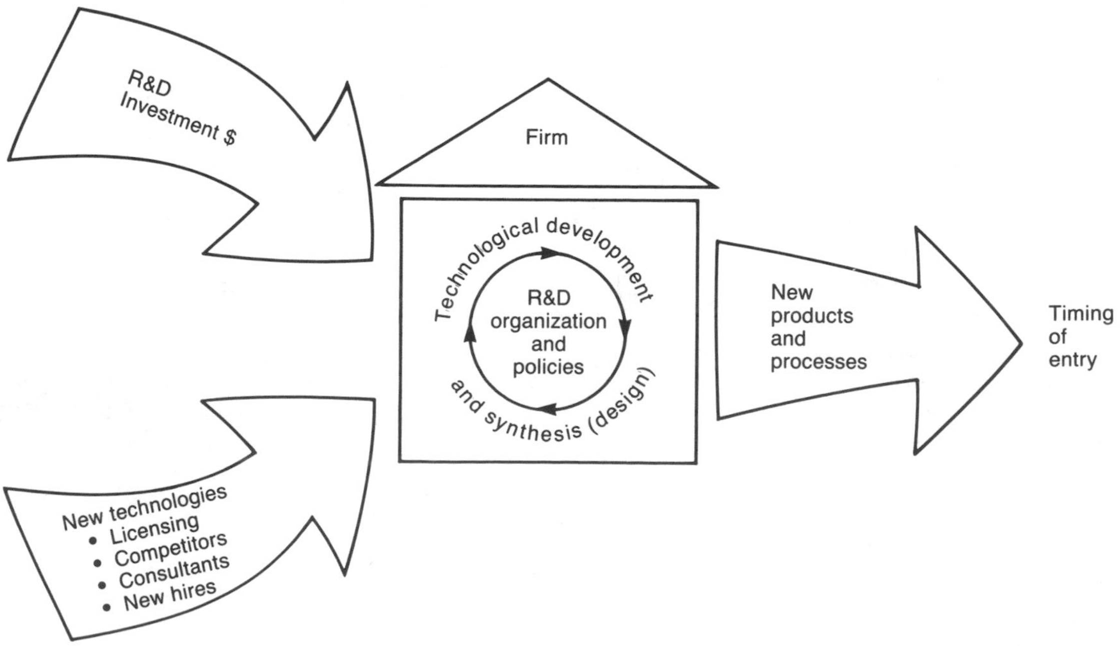

R&D Investment $
Firm
Technological development and synthesis (design)
R&D organization and policies
New products and processes
Timing of entry
New technologies
• Licensing
• Competitors
• Consultants
• New hires

should proposals for new technologies/products be evaluated? What technologies provide opportunities for improved product performance or lower cost?

- *Embodiment*—How should these technologies be utilized in new products? What performance parameters should dominate?
- *Technology sources*—To what extent should a firm rely on internal development? To what extent should external sources, such as contract research and licensing from individual inventors, research and engineering firms, and/or competitors, be relied upon?
- *Competitive timing*—Whether to lead or lag new product introduction. Consider the benefits of leading versus the risk of uncertain market acceptance of a new product. Are there benefits in developing an improved product after allowing a competitor to go first, then evaluating market acceptance?
- *Level of R&D investment*—How much to invest in technologies and internal staffing versus external staffing? Should the firm let R&D investment or profit oscillate?
- *Organization and policies for R&D*—Should there be a central R&D facility? How should it be structured? Is a separate career track needed for scientists with compensation compatible with or leading the industry? Project teams versus matrix organization for the sharing of resources? How closely will top management be involved in technological decisions? How to allocate funds for R&D projects? What should policies be concerning patents, publications, and protecting technological know-how?
- *Competence levels*—Given the competitive environment, how close to the state of the art should a firm be in a technology to achieve its objectives? How proficient to become in understanding and applying the technology? Should the firm emphasize straightforward applications of the technology through product engineering, or emphasize advancing knowledge of technology through basic or applied research?

Case II–8
Software Architects (A)

T. J. Kosnik and R. A. Burgelman

Harvey Mayerowicz, the president and founder of Software Architects, was in the process of reviewing the business strategy for 1982–83. Software Architects (SA) was a data processing consulting firm which provided customized computer programming services and technical seminars on various topics to companies in the Chicago area. A small, entrepreneurial enterprise, SA had enjoyed modest growth and profitable performance in the two-and-a-half years since its founding. Harvey's concern was to develop a clear understanding of the factors which had contributed to SA's past success, and to position the company for continued success in an industry that was experiencing rapid growth, increasing competition, and technological change.

COMPANY BACKGROUND

Software Architects was founded by six individuals in late 1979. Harvey Mayerowicz was the president, elected by his colleagues on the board of directors. The other five directors were Gloria Petersen, Gene Petrie, Edward Wroble, Bruce Parrello, and Fritz Wolf. All of the directors were experienced systems programmers, who performed technical consulting tasks in addition to their duties of managing the company.

Decision making at SA was a consensual process, with each of the six directors contributing ideas and opinions at the meetings held every one or two weeks. An idea which was of interest to a particular director was often explored outside of the meetings so that he or she was able to make recommendations and advocate a particular position to the others.

Responsibility for business functions was shared among the directors in the following way. Harvey and Gloria were responsible for marketing and new client development. Bruce, who had developed the automated accounting systems used by SA, was concerned with the financial side of the business. Fritz, Gene, and Ed focused their attentions on the conduct of consulting projects themselves, including technical work, supervision of SA employees, and managing the delicate relationships with SA clients.

In addition to the six founders/directors of SA, there were five other technical consultants who performed project work. Harvey anticipated that these employees might be groomed to take on additional responsibilities in marketing or project management as SA grew. They were encouraged to take initiative and to accelerate the timetable for their development, in an effort to provide challenge and stimulation on the job.

Financial Performance

The directors all took pride in the fact that SA had been profitable since the time the firm was established. Income statements for the 1980 and 1981 fiscal years indicated that retained earnings for 1981 had almost tripled the 1980 figure. The results for the first six months of the 1982 fiscal year (July through December of 1981) showed revenues of over $400,000 and before-tax income of over $140,000.

The accountants who reviewed SA's financial reports had given Harvey a set of financial performance ratios for firms in the computer programming and software services industry. The results of high-, average-, and low-performing companies are provided in Exhibit 1. It appeared to Harvey that SA was doing reasonably well, especially in view of the fact that it was a new venture in a fairly competitive marketplace.

Services Offered

SA provided services to clients in three main areas:

1. *Systems programming:* The design and programming of systems software to aug-

EXHIBIT 1 Key Financial Ratios for Firms in the Computer Programming and Software Service Industry Compared with Software Architects' Results

	Industry Performance in 1981			Software Architects' Results for 1981
Financial Performance Measures	High Performers	Average Performers	Low Performers	
Percent profit before tax/Sales	N/A	8.4%	N/A	13.6%
Percent profit before tax/Net worth	73.5%	45.2%	20.8%	59.3%
Percent profit before tax/Assets	22.1%	14.2%	5.7%	19.7%
Sales/Receivables	7.7	5.7	3.8	2.6
Sales/Assets	3.0	2.2	1.5	1.9
Debt/Worth	1.1	2.0	3.6	2.0
Current ratio	1.8	1.4	1.0	1.6

SOURCE: Robert Morris Associates 1981 and Software Architects internal records.

ment that provided by the hardware manufacturer. Examples included programs for measuring systems efficiency, compilers, utilities, and report generators.

2. *Application programming:* The design and programming of software to perform specific applications for the client, such as accounting, inventory management, and planning systems.
3. *Technical education:* The development and presentation of training seminars for employees in client organizations on a variety of technical subjects in which SA had special expertise.

Additional descriptions of systems and application software are provided in Appendix A.

SA had also done a project on a microcomputer network in the last year. It was possible that there might be other opportunities for SA related to microcomputers in the future, but it was not clear what form those opportunities might take.

SA's costs and revenues for the services it provided were largely based on the time spent by SA consultants to complete design and programming tasks on each project. The custom development of software was usually broken down into several phases. Although the tasks performed in each phase varied depending upon the methodology that was being used, most of SA's projects consisted of:

1. *General design:* In which the overall framework for the new system was developed and the needs of the users which were to be satisfied by the system were identified.
2. *Detailed design:* During which the general guidelines from the previous phase were elaborated to provide a clear, explicit blueprint for subsequent programming.
3. *Implementation:* During which programs were written, tested individually and as an overall system, and the final system was introduced into a "live" environment for use in day-to-day operations.

SA spent more time than many of its competitors in the detailed design phase. However, they were convinced that extra effort there allowed even greater time savings during implementation. In essence, their detailed planning allowed a smoother, faster execution and much less time in testing and debugging programs.

Keys to Success

In discussions with other members of the board, Harvey and Gloria expressed their feelings that in a consulting relationship, technical competence was a necessary but not sufficient condition for success. There were other essential ingredients as well. Harvey believed that a programming consultant had to have three qualities to satisfy most clients and earn an invitation for follow-on work. The individual had to be honest, competent, and personable. Gloria felt that the most important characteristic was the ability of the consultant to be empathetic with the client, understanding and listening to the client's business and personal concerns.

Their continuing discussions had yielded a list of factors which they believed were important to a prospective client in deciding whether or not to accept a proposal from a particular software consulting firm:

1. *Availability:* The ability of the consultant to meet staffing levels, start dates, and completion dates required by the client.
2. *Cost:* Measured both in terms of an hourly billing rate and an estimate of overall project costs.
3. *Honesty:* The willingness of the consultant to admit he/she either does not know the answer or had made a mistake. In addition, the unwillingness of the consultant to make promises in a proposal that

the firm may not be able to keep; e.g., the promise to meet a deadline desired by the client.

4. *Professionalism:* The consultant's respect for the client's work environment rules, punctuality, neatness, and personal appearance.
5. *Quality of past work:* As evidenced either by the recommendations of other satisfied clients or the quality of sample programming documentation, articles, technical papers, etc., provided by the consultant for review by the client. The quality of the analysis supporting the proposal was also important.
6. *Rapport:* A combination of what Harvey had identified as the personable quality and Gloria had called empathy. The sense of shared understanding, values, and personal friendship between the client and the consultant.
7. *Technical fit:* The match between the consultant's areas of competence and the client's needs for specialized expertise. It was possible for a highly technical individual to be undesirable to a client if the expertise was out-of-date or was not relevant in the client's problem situation.

Harvey and Gloria wondered whether their clients saw the keys to success in software consulting the same way that they did. They were also curious about how potential clients perceived SA in each of the areas above. In what respects was SA's position strong relative to its competitors? What were the major areas needing improvement? Were there other considerations in the decision to hire a firm like SA that the two of them had overlooked?

It seemed that getting answers to the questions above were crucial to understanding what SA had done right in the first two years. They were also critical to keeping SA well positioned for the future.

SOFTWARE ARCHITECTS' EXTERNAL ENVIRONMENT

Prospective Clients

As suppliers of customized programming services and technical education, SA had a wide variety of potential clients. However, the realities of the marketplace and the values of the SA principals both served to focus attention on a reasonably small group of companies.

The cost of custom programming was prohibitive for many organizations. SA believed that the size of the data processing budget was related to the annual revenues of a company, and its main targets were therefore the largest industrial firms. Because SA had only one office and the professional staff preferred not to travel out of town, Harvey generally restricted his clients to those in the Chicago metropolitan area. SA did not attempt to specialize in a particular application area (such as general ledger systems) or industry group (such as banking or forest products). However, it did try to concentrate on its areas of primary technical expertise. As a result, potential clients were screened based upon the hardware and systems software that they had in their data processing centers. SA preferred to work with IBM hardware and software, and rarely competed for projects in other hardware/software environments.

SA's potential clients, then, included large businesses in the Chicago area with IBM computer equipment. Most of the 19 clients for which they had worked in the first two years had all three of these characteristics. There were almost 50 prospective clients which met the three criteria and showed high potential for future SA work. Market research had identified approximately 50 other firms which were large enough and were based in Chicago. It was not yet determined what hardware and software was in place in each of the second group of companies.

The Chicago area was not a center of high-technology manufacturing like the areas outside Boston, Massachusetts, and San Jose's Silicon Valley. As a result, few of SA's potential clients made computer-related products. Most used computers for their internal administrative, accounting, inventory, and planning systems. The data processing departments varied in size from a few to several hundred people. Nearly every organization suffered from a shortage of skilled programming personnel. In fact, the shortage of people and technical skills was what provided the principal *raison d'être* for the hundreds of systems-consulting firms which operated in the Chicago area.

The data processing departments of most organizations had at least three separate groups. One was responsible for systems programming, and maintained the technical environment in which the other two groups worked. The technical skills required to work as a systems programmer were greater than those of other data processing personnel. The second group was responsible for application development. This group designed and tested programs for accounting, planning, or other functions which had been requested by nontechnical users in the organization. The third group was responsible for operations. They "ran" the hardware and software in order to process information and produce reports desired by management of the business for planning and control purposes. Computer operators were typically the least technically skilled of the three groups.

The decision to acquire consulting services from a firm like SA was sometimes made by the manager in charge of the systems programming group. At other times, the person in charge of the entire data processing operation was the final decision maker. Occasionally, and especially when a manager from another part of the potential client company had requested that an application system be developed to meet his or her needs, nontechnical management were involved in the process. If the contract was for a large dollar amount (greater than $50,000), or was for a project of critical importance to the organization, senior management approval was usually required.

No matter who was involved in the process of choosing the consultant, SA's "client" was almost always a member of management in the data processing organization. This individual was often under a great deal of stress to develop a new system under tight budget and deadline constraints. If the project timetable "slipped" or there were cost overruns, the manager's job could be in jeopardy. With billing rates between $30 and $120 an hour for outside programming support, a small error in estimating the scope of the project resulted in a large increase in cost.

The stories of projects that had been placed in the hands of data processing consultants and had subsequently gone awry were many. There were also frequent accounts of outsiders who had installed a system and left when the initial contract had expired. When the client later found "bugs" during the day-to-day operations of the system, there was often no one in-house who knew enough about the system to make the necessary repairs. The consultants were seldom available for assistance. The programmers were immersed in other projects with new clients. Frequently, the people who had actually worked on the system had moved on to other employers. Turnover among software programming houses was notoriously high.

Thus, the potential clients of SA shared two major concerns about the use of consultants: the fear of project delays and budget overruns in the short run, and the worry about being left "high and dry" when the project was over. Many companies had policies against the use of outsiders for system development. Unfortunately, that rarely offered protection from the two key risks. Inability to meet deadlines was at least as much a problem for internal programmers as it was for firms like SA. So was turnover. The competition for technical people was fierce enough that programmers were often lured away by other employers.

Harvey believed that if SA were able to demonstrate its technical competence and its ability to meet deadlines to a client, the chances for follow-on business were good. Firms who could deliver on promises made at the time of the initial proposal were rare. Repeat business also served to reduce the client's second area of risk. If the SA programmer were working on another project in the same firm, he or she was accessible to answer questions or provide a "quick fix" if problems arose.

The problem for Harvey was how subtly to address what he thought were the two areas of concern for most clients. He was unsure of how SA might demonstrate its competence and its willingness to provide continuity in a way that sounded like an honest promise rather than a sales pitch.

The clients for SA's technical seminars were sometimes the same individuals as for the programming services. At other times, people in the organization's professional training and development staff got involved. Although the risk of a poorly conceived and executed seminar did not appear to be as great as those for a computer system, the two groups shared something in common. They had to commit money and their reputations in advance to an outside supplier of a product which did not yet exist. This often made the criteria for selection difficult to articulate. The decision was rarely clear cut. Usually, the client had to rely on "gut feel" and hope for the best.

Competitors

Harvey and his colleagues were not certain who their closest competitors were in the systems and applications programming area. One study (see Appendix A) had estimated that there were thousands of firms providing custom design and programming services in the United States. Of these, there were only about 2,000 competitors nationwide who generated more than $250,000 each in annual revenues. There was no market research data available on the firms in the Chicago area. Harvey had found over 200 names of data processing consultants in the Chicago Yellow Pages. However, neither SA nor several of the firms against which they had bid on past projects were listed in the Yellow Pages. He was not sure how many firms were actually providing a similar type of service to the same target clients.

Harvey and Gloria had developed a list of potential competitors which included those against whom SA had prepared bids for work in the past, as well as those about whom SA had heard from clients and other contacts. Exhibit 2 contains information about billing rates of SA's major competitors on past projects.

SA's known competitors included several

EXHIBIT 2 Estimated Hourly Billing Rates of Software Architects' Competitors

	Estimated Hourly Rates for			
Competitor Name	Programmer/Analyst	System Designers	Project Leaders	Partners/Principals
Arthur Andersen	$38–$40*	$45–$55	$50–$85	$100–$150
Consumer Systems	$27–$31	N/A	$38	N/A
Farlow Associates	$25–$28	N/A	N/A	N/A
Giles Associates	$24–$32	$26–$36	$35 and up	N/A
McAuto	$46–$58	$66–$83	$109–$136	$127–$159

* For every five programmer/analysts, Arthur Andersen provided an experienced consultant as supervisor at no charge to the client.

very large, national firms, such as Arthur Andersen and McAuto. There were also a number of smaller companies. It was unclear what the best strategy might be for SA's positioning relative to this wide variety of opponents. How did potential clients perceive SA relative to a company with the resources and reputation of Arthur Andersen? Were there needs that SA might fill better than a larger, better-established firm? Where were the client's sensitive spots in dealing with an entrepreneurial company that Harvey had to address? Further, did SA appear to be special in the eyes of the client, or was it one of a myriad of small, nondescript "body shops" that abounded in the market place?

Learning where SA was from the client's point of view was an important first step. Creating an image for the firm as a high-quality, high-price source of services was the next item on the agenda.

Industry Facilitators

Although software consultants typically marketed their services by direct sales calls on potential clients, there were other institutions in the marketplace which could aid in the spread of reputation and the generation of leads for new business. Harvey considered at least three types of "industry facilitators":

1. *Hardware vendors:* Who often provided purchasers of equipment with the names of several software consulting firms if there were a need for help in setting up new applications.
2. *User groups:* Who met to discuss new ideas and common problems with hardware or software that they all had at their respective sites.
3. *University MIS professors:* Who provided consulting firms with programmers, and who also referred technical consulting business that they could not handle because of time commitments to software houses that they considered "top notch."

It seemed that SA should have a coherent strategy for their dealings with each of these industry groups. Harvey also wondered whether there were other "facilitators" that he might have overlooked.

Growth and Technological Change

Appendix A provides growth rates for different segments of the software industry. Professional services for custom software development showed a 20 percent annual growth forecast for the 1981–86 period. Harvey had no data for the growth in demand for software services in the Chicago area.

Technological change was an important and troublesome issue for SA's future planning. The manufacturers of hardware and systems software improved their products and developed new technologies at a breakneck pace. As a result, technical expertise became obsolete almost as fast as the products did. There was always a lag in the decline in demand for services, because clients who already owned older hardware did not discard it as rapidly as the technology changed.

Harvey had developed a list of the technological changes which offered the greatest risk of obsolescence (and opportunity for new business) for SA:

1. Changes in systems software technology.
2. Development of application generators.
3. Proliferation of microcomputers.

There were two impending changes in systems software technology that threatened SA: (1) IBM's promise of operating systems that eliminated the need for systems programmers, and (2) new data-base management software. SA's special expertise was in systems programming for IBM mainframe computers. That firm had recently announced a new operating system called the SSX system, which it claimed made it unnecessary to hire systems programmers. The software was supposed to allow a nontechnical user to start and operate the system without programming commands in Job Control Language (JCL). Harvey was a bit skeptical about IBM's claims, but felt that in the future, systems software might become more "user friendly." In fact, some of the minicomputer manufacturers had made great strides in that area.

SA's experience with data-base management systems was with a pair of competing technologies known as hierarchical and network architectures. Most of the data-base packages that had been sold in the last 10 years had been one of those varieties. While extremely powerful, these systems were known for their technical complexity, and data-base design and programming skills were in short supply. SA had a wealth of experience in several of these products, including IBM's flagship IMS, Cullinane's IDMS, and Software AG's ADABAS.

There had been a great deal of interest in the trade press in a new, simpler technology called relational data-base management systems. There were two or three new products that were actually on the market, and IBM had been working for several years to develop a relational DBMS of its own. If and when the new software came into widespread use, the threats to SA were twofold. First, it was a technology in which they did not have expertise to differentiate them from competitors. Second, the systems were supposed to be much simpler and "user friendly," reducing the need for specialized systems programming experience.

The shortage of skilled programmers and rising salary costs for technical personnel had provided the incentive to reduce the labor intensity of systems development. Many suppliers had developed application software packages which a client could buy "off the shelf" and use with little or no additional programming. A more radical solution to the problem was the concept of a product known as an application generator. This was an off-the-shelf package that a nontechnical person could use to translate English-like commands into machine-readable code. In essence, the idea of the application generator was to eliminate the "middle man" between the nontechnical user and the machine. If such a product were perfected, the need for application programmers would all but disappear.

Thus far, no one had succeeded in developing the concept of the application generator to its full potential. Nevertheless, products which vastly improved programmer productivity had been introduced. For example, Cullinane Corporation had recently advertised a package which it claimed reduced the time required for programming by 90 percent.

Most of SA's past experience had been with large IBM mainframes. IBM's share of the data processing industry's total revenues was almost 40 percent in 1980. Its nearest competitor, NCR, had 5 percent share of market in 1980. The large installed base of IBM mainframes made it the obvious choice of hardware in which to develop specialized expertise.

However, there were considerable differences in the growth trends for different segments of the hardware industry, as shown in Exhibit 3. In particular, sale of microcomputers had grown 85 percent from 1979 to 1980. Many industry observers believed that microcomputer sales might grow at a faster rate in the 1981–86 period, as new suppliers entered the marketplace, the technology improved, and the selling price per unit declined.

The massive influx of microcomputers was both a threat and an opportunity for SA. If users moved away from dependence on large mainframes and centralized data processing departments to meet their needs, SA's traditional clients might be faced with less work, lower data processing budgets, and less power in the decision to bring in consultants. On the other hand, many of the users of microcomputers were nontechnical people who preferred not to write their own programs. SA might be able to develop application software for micros and sell it to a large number of users.

Harvey was confident that SA had the technical skill to develop such software. In fact, SA

EXHIBIT 3 Selected Data Processing Industry Growth Rates, 1979–1980

TABLE I

TOP 10 DP REVENUES
(in millions of dollars)

	1980	1979	% Growth Rate
1 IBM	21,367	18,338	16.5
2 NCR	2,840	2,528	12.3
3 Control Data	2,791	2,273	22.8
4 DEC	2,743	2,032	35.0
5 Sperry	2,552	2,270	12.4
6 Burroughs	2,478	2,442	1.5
7 Honeywell	1,634	1,453	12.5
8 Hewlett-Packard	1,577	1,147	37.5
9 Xerox	770	570	35.1
10 Memorex	686	658	4.3
Total top 10	39,438	33,710	17.0
Total	55,626	46,220	20.4
Top 10 as a percent of total	70.9%	72.9%	

TABLE III

TOP 20 REVENUE GROWTH RATE
(in millions of dollars)

	Total Dp % Growth Rate	U.S. Dp % Growth Rate	Foreign Dp % Growth Rate	1980 Dp Rev.	1980 Earnings
1 Sanders Assoc.	208.5	91.5	NM	145.0	49
2 Apple	175.1	163.4	224.7	165.2	47
3 Philips Information Sys.	100.0	100.0	NM	50.0	98
4 Tandem	93.9	58.7	179.4	128.8	53
5 Intergraph	91.3	80.1	153.7	56.5	90
6 Dysan	86.1	79.9	127.7	62.9	85
7 Computervision	85.5	72.7	108.3	191.1	41
8 Paradyne	83.2	74.0	108.9	75.9	77
9 Prime	75.0	59.9	95.8	267.6	27
10 Teletype	72.4	62.1	NM	250.0	29
11 CPT	68.9	44.7	157.7	76.4	76
12 Wang Labs	66.1	68.8	61.5	681.8	11
13 Lanier	64.1	60.6	129.7	128.0	54
14 Triad Systems	61.0	61.0	NM	60.2	87
15 Anacomp	60.1	60.1	60.0	57.0	89
16 Commodore International	54.1	−13.2	105.2	98.7	66
17 Applicon	51.4	35.2	136.6	68.5	82
18 Auto-trol Technology	51.3	58.5	27.1	50.8	97
19 AM International	49.0	49.0	49.0	98.8	65
20 Printronix	48.8	37.7	93.0	48.9	99

NM—Not meaningful.

EXHIBIT 3 *(concluded)*

TABLE IX

DP REVENUES BY PRODUCT SEGMENT
(in millions of dollars)

	1980		1979		%
	$	%	$	%	Growth Rate
Systems:					
Mainframes	15,148	27.2	13,312	29.0	13.8
Minicomputers	8,840	15.9	6,916	15.0	27.8
Microcomputers	769	1.4	416	0.9	84.9
Word processing	881	1.6	538	1.2	63.8
Total systems	25,638	46.1	21,182	46.1	21.0
Ocm peripherals	3,968	7.1	3,128	6.8	26.9
End user peripherals	6,910	12.4	5,943	12.9	16.3
Data communications	1,141	2.1	927	2.0	23.1
Software products	1,738	3.1	1,347	2.9	29.0
Maintenance	3,388	16.0	7,372	16.0	20.6
Service	6,432	11.6	5,329	11.6	20.7
All other	911	1.6	772	1.7	18.0
Total	55,626	100.0	46,000	100.0	20.9

SOURCE: P. Wright, "The Datamation 100," *Datamation*, June 1981.

had a TRS-80 microcomputer, and Bruce Parello had written programs to do the company's project accounting, accounts payable and receivable, and general ledger in his spare time. However, SA had no idea how to market microcomputer software once it was developed. They had deferred discussion on diversification into that area until they learned more about the marketing channels for microcomputer software.

SA directors knew that the company could not expect to be writing systems and applications programs for IBM mainframes in 20 years. But how fast would the technology change? What could SA do in 1982 to anticipate these changes and to build a viable niche in the future data processing arena? With limited time and financial resources, he wondered where he should place his bets in the next two years. The alternatives were almost too numerous to list, much less scrutinize in detail. And it seemed imperative that SA move quickly to establish its position for the future.

SOFTWARE ARCHITECTS' BUSINESS OBJECTIVES

The business objectives which Harvey and the board established for Software Architects were:

1. Preserve the quality of life of the SA employees and directors.
2. Recruit additional consultants to support the growth in revenues and earnings without sacrificing SA's requirements for technical competence and the potential ability to deal effectively with clients and others in the firm.
3. Maintain exceptional quality of SA work product.

EXHIBIT 4 Projected Billings, Income before Tax,* and Staffing Levels for Software Architects

	June 1981 (Actual)	Period Ending June 1982 (Projected)	Period Ending June 1983 (Projected)
Billings by line of business:			
Systems software	$203	$245	$ 240
Custom application software	75	335	600
Education/training programs	60	105	360
Microcomputer systems	5	15	0
Total billings	$343	$700	$1,200
Income before tax	$ 47	$210	$ 360
Staffing levels:			
Number of Software Architect consultants	7	15	25
Number of support staff	0	1	2
Number of subcontractors	½	1½	2½

* Billings and income are in thousands of dollars.

4. Differentiate SA from competitors by demonstrating a perceptibly higher quality of work to existing and potential clients.
5. Charge a premium price based on the value of higher quality work to the client.
6. Increase annual billings to $700,000 in 1982 and $1.2 million in 1983.
7. Achieve a target percentage of before-tax income to total billings of 30 percent in both 1982 and 1983.
8. Explore new products and services for SA to introduce in order to meet its targets for growth and profitability, and to position the firm favorably in a rapidly changing technological environment.

The projected billings by category of service and staffing levels for SA are included in Exhibit 4. SA performance for the first five months of the 1982 fiscal year made the targets for billings and income seem reasonable. Year-to-date billings were $411,000, with before-tax income of over $147,000. This performance had been achieved with minimal requirement for overtime work.

SOFTWARE ARCHITECTS' BUSINESS STRATEGY

A strategy was under development to address each of SA's eight business objectives. While not yet completed and approved by the board of directors, most of the elements of the plan had fallen into place.

Quality of Life

The six founders of SA had all worked previously for other employers. They had left those organizations because they were dissatisfied with the stress, red tape, and sluggishness they had encountered. Each of them wanted to learn more in the technical area. They also shared the desire to develop their interests outside of their careers, and wanted to avoid the frenetic pace of tight deadlines and 60-hour weeks that were the rule in many software consulting firms. Therefore, the quality of life for SA personnel was of tremendous concern to Harvey, Gloria, and the other directors. They had adopted the following policies to demonstrate their commitment in this area:

1. The billable hours targeted for each employee were 1,680 a year, considerably lower than targets in other firms. This helped to prevent excessive overtime on projects.
2. Each employee was allowed 12 days of vacation, 12 days of sick leave, two weeks for training/development, and 10 holidays a year.
3. To prevent the stress of out-of-town travel, SA limited its marketing activities to the Chicago area. The directors had occasionally turned down lucrative contracts that required prolonged travel to other locations.

Recruiting and Staff Development

Harvey, with his board's approval, devised a strategy for preserving quality of service to customers that was closely tied to his staffing plans. Every professional at SA was a graduate of a rigorous program in computer science at Northern Illinois University. The school had achieved a prominent reputation in the Midwest for producing graduates of unquestioned technical competence who also had an appreciation for the needs for practical data processing solutions to business problems. Harvey planned to recruit Northern Illinois University students on campus, as well as graduates who had been in industry for a year or two. He also provided funds for continuing development of the technical and interpersonal skills of current SA employees. They were encouraged to take seminars in state-of-the-art technical subjects, as well as those which might refine their skills in marketing, managing subordinates, and interacting with clients.

SA had planned to add several members to the staff in late 1981. They had not yet been brought on board. Several people who had been made offers accepted jobs elsewhere. One indicated that he did not feel technically qualified to "measure up" to the current SA staff. Although there was some concern that SA might be too selective in its recruiting criteria, the directors were convinced that maintaining high standards of technical competence was critical to the preservation of the quality of the work on which SA's reputation was based.

Quality of Service

It appeared that the major causes of failure in systems development projects were the imposition of unrealistic deadlines, the use of programmers without relevant experience, lack of adequate planning in the design phases, and poor documentation in the implementation phase. Harvey and Gloria felt that their strategy for quality of service addressed each of those problems.

SA refused to commit to completion dates when submitting proposals for new business. They insisted on being able to work on the problem a few days or weeks in order to gain information to make realistic estimates of the elapsed time required to finish the job. This prevented the premature establishment of a timetable that was doomed to failure.

Many of SA's competitors staffed projects with one or two seasoned programmers and a large number of "green troops." SA used smaller project teams and experienced programmers almost exclusively. When a new person joined the firm, he or she was put on a project with the several SA veterans to learn the ropes and ensure that the quality of the new programmer's work was consistent with the rest of the firm's.

SA spent more time and energy during detailed design than many of its competitors. Planning and the preparation of detailed specifications reduced the risk of problems and delays during implementation. SA also documented their programs more extensively than did most other firms. Although this was somewhat time consuming during the first writing of the program, it made it much easier to test the program later on. Also, modifications were simpler once the program was installed. There was less dependence on a particular programmer, because good documentation was easily decipherable by someone who had not written the original code.

SA's directors agreed that all of these things made the company's work product better than that of most of their rivals. But they were not sure whether their clients perceived a difference in quality. Nor were they confident that such subtle gradations in quality of service could be effectively communicated to a prospect who was unfamiliar with SA.

Differentiation from Competitors

The strategy for differentiating SA from its competitors was still under review. The firm was considering a policy of guaranteeing their custom software work for a period after contract expiration. This practice was rare in the software industry, where the costs of time spent on "free" warranty work rather than billable projects were high and extremely visible. Harvey was not sure how effective a warranty might be in convincing potential clients that SA provided a quality advantage. It was not clear how to assess the potential benefits of new business versus the costs of honoring the warranty if the client encountered a problem.

Harvey and Gloria, in making their sales pitch, stressed the fact that SA programmers were all Northern Illinois University graduates in the belief that the school's reputation for data processing instruction might help to distinguish SA from other firms. But they needed to establish other unique advantages for SA which could be communicated to prospective clients. Harvey was trying to generate creative ideas for doing this.

Pricing Strategy

Harvey was not sure whether SA's billing rates were set at a "premium price." This was mainly due to uncertainty over the pricing strategies of competitors. It was difficult to obtain good estimates of what each firm charged for its services, since there were no industry guidelines for billing rates. Moreover, rates quoted by consultants in response to general inquiries were generally higher than those submitted during competitive bidding. Price cutting in order to win new business occurred frequently. The amount of the reductions varied and was impossible to predict.

It was evident that the rates customers would accept varied, depending upon the service performed. System software development skills were in short supply in the industry, and SA had no trouble charging $50 an hour for such projects. SA hourly rates for application software development, which was less complex and less specialized, ranged from $33 to $37. Data-base design and application programming in special environments commanded rates between $34 and $44. The development of technical seminars brought $50 to $65 an hour, while giving the seminars to clients often paid $75 to $100 per instructor-hour. However, clients typically wanted a more experienced individual to prepare the technical training than they did for programming tasks, so the higher billing rates were offset by higher salaries and opportunity costs for SA.

Marketing Strategy to Achieve Target Growth and Profits

It appeared to Harvey that the keys to meeting the goals for growth in revenues, profit, and staffing levels were effective recruiting and marketing programs. The recruiting strategy has already been discussed. The marketing strategy consisted of several elements:

1. Concentrate on the largest private sector enterprises in the Chicago area.
2. Where possible, use the growing network of key decision makers familiar with SA to obtain referrals to new prospective clients.
3. Where referrals are not possible, initiate "cold calls" on key technical managers in target organizations to expose them to SA and discuss current and future needs for services provided by the firm.
4. Concentrate on organizations using IBM hardware and system software, because of the specialized expertise of SA consultants in those systems.
5. Wherever possible, adopt the premium quality and price approach. In competitive bidding situations, if the potential for repeat business is high, reduce the hourly rate to not less than $33 an hour for the first contract. Increase the hourly rate for follow-on work, once the client has seen the high quality of SA consultants' performance.
6. Stress the links between SA and Northern Illinois University in order to develop the image of SA as a source of updated technical talent and high-quality work.

Development of New Products and Services

While the first seven objectives were addressed by the strategy Harvey had developed, he was uncertain about the best approach to meet the objective of exploring new products and services. There was no shortage of ideas from within the firm about potential areas of new business. Some of the suggestions included:

1. Development of application software for microcomputers, including a project accounting and general ledger system for a company in a service business like SA's.
2. Development of software to collect information on telephone calls made in a company using Bell equipment. The information would subsequently be used to produce accounting and resource usage reports for management.
3. Development of systems software packages for users of IBM computers in many organizations.
4. Preparation and presentation of courses on programming using techniques that SA had perfected.
5. Joint presentation of technical seminars in cooperation with several of the faculty members from Northern Illinois University.
6. Establishment of an institute in which nontechnical personnel provided by SA's clients would be given an accelerated course in programming. Harvey estimated that a bright liberal arts major could be transformed into a crack application programmer in four to six months.

These were but a few of SA's potential alternatives. What was missing was a method by which to review these and other alternatives from the strategic perspective. How might one assess the potential of alternative projects to contribute to SA's growth and profitability? How might one discern whether one project or another provided a better hedge against the risk of technological obsolescence? Developing a strategy and a system for the last objective appeared to be the most difficult task of all.

CONCLUSION

As Harvey reviewed the elements of the strategy he had assembled, he had three major concerns:

1. Was the strategy realistic, given SA's resources and the realities of the external environment?
2. Were the elements of the strategy internally consistent?

3. How might the short-term strategy be modified to better equip SA to deal with its long-term prospects?

The issues were important enough that he wanted an outside opinion. He decided to bring in a consultant that Gene Petrie had recommended to help him think them through.

APPENDIX A DATAGUIDE REPORT ON THE FUTURE OF THE SOFTWARE INDUSTRY

The software industry is booming! By 1986, it will be a $38 billion industry. Market growth will average 30 percent per year for the next five years. Today companies spend over $10 billion on external software expenditures.

This may seem an extraordinary growth from other market-size estimates of $3.5 billion today and $10 billion at mid-decade. However, it is important to realize that most estimates ignore the two largest segments of the software marketplace: professional services and turnkey software, each of which is as lucrative as the combined market for applications and systems software packages, as explained below.

Software is provided to users in two ways: standard off-the-shelf packages and customized products. In addition, software may be packaged with computer hardware and called a turnkey system. The software industry can be graphically depicted as shown in Figure 1.

Turnkey systems can use either standard off-the-shelf software products or can be customized for a client (professional services). It is estimated that between 50 percent and 60 percent of all turnkey systems use off-the-shelf software products while the other 10 percent to 50 percent use software created specially for the client through a professional services arrangement.

Warren N. Sargent, Jr. and Paul Colen, Palo Alto Management Group, "Software," *Dataguide,* Fall 1981.

FIGURE 1 Software Industry Composition in 1981

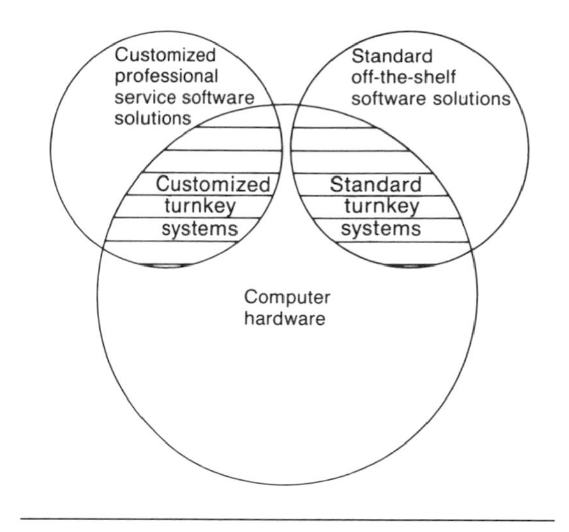

Turnkey systems are generally perceived to be a relatively recent product innovation, although the concept is at least 15 years old. Historically, software vendors sold software products and/or professional services, but rarely involved computer hardware in the sale. Today, computer manufacturers offer quantity discounts to turnkey system vendors, who can then add software and sell the system to a user, theoretically generating profit from the software as well as the hardware (marked up to the list price for the user). More will be said about this theoretical relationship later.

Software can be separated into two categories. Systems software enables the computer communications system to perform basic operations. Applications software provides solutions to specific user requirements. It is the application software that is most visible to the user, while the systems software supports the functions called for by the application. Figure 2 depicts a computer system showing the main functions of systems and applications software.

Each of the four elements of the software industry will grow rapidly (see Figure 3). Turn-

FIGURE 2 Computer System Architecture

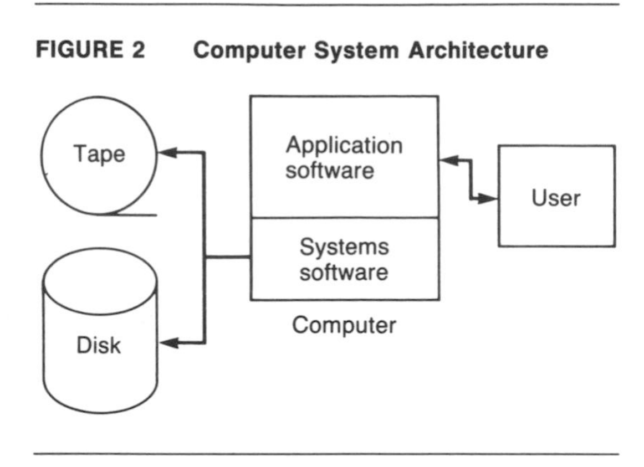

key systems user expenditures will grow the fastest: 35 percent per year for the next five years. The practice of computer manufacturers charging separately for systems software products will continue to increase over the next five years and will therefore drive the growth forecast for these products to over 30 percent per year.

FIGURE 3 User External Software Expenditures

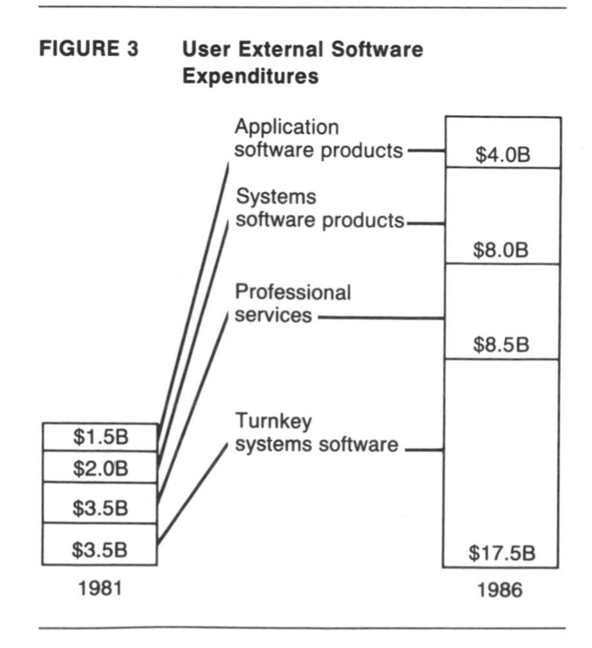

The rate of growth in user expenditures for application software products, 25 percent per year, is deceptive. Turnkey systems use application software products in at least 50 percent of the cases, and turnkey systems user expenditures are growing at 35 percent per year. Therefore, the true growth for all uses of application software (products and turnkey systems) exceeds 30 percent per year. Professional services user expenditures, excluding turnkey systems, will grow at 20 percent per year for the next five years. However, the true growth rate for all types of professional services (including turnkey systems) will be 25 percent per year.

SYSTEMS SOFTWARE

Systems software, which enables the computer communications system to perform basic functions, is classified by three types—system operation, system utilization, and system implementation.

System operation products manage computer communication system resources during program execution. Examples of such products are operating systems, data-base management systems, and telecommunication monitors.

System utilization products help manage the computer system operation more efficiently. Performance measurement systems, job accounting systems, and system utilities are examples.

System implementation products prepare applications for execution by assisting application design, programming, testing, and related functions. Examples of such products are assemblers, compilers, software design productivity aids, report writers, and program library systems.

There are probably 4,000 or more systems software products, marketed by nearly 1,000 vendors today. It is extremely difficult to count vendors and products because of the changing nature of the industry. Some vendors announce a new product and subsequently drop it because they can't generate sales, or they decide to sell the product and licensing rights for the product to another vendor. Vendors change the name of some products to create a new image even though the product is not actually new. There has been an explosion of new products for personal and small computers. Since many of the vendors of these systems software products are still operating regionally, it is difficult to identify their market presence.

At one time, over 80 percent of user expenditures for systems software products were for IBM and IBM-compatible large computers. That percent is decreasing as users of small computers become more sophisticated and realize the value of the systems software products available in the marketplace.

The industry trends of increased use of telecommunications and automation in the office have a positive impact on the market for systems software products. Systems software is the "glue" that joins the computer hardware and application software together. Changing functionality in the computer hardware creates a need for new systems software. Enterprising organizations have recognized this need and have been responding with new and improved products.

Computer manufacturers have historically fostered this industry by their attitude toward independent software vendors in that they have not "locked out" the independent vendor's software from operating on the manufacturer's computer. Manufacturers have tacitly encouraged independents to develop user-friendly systems that facilitate computer use: the easier computers are to use, the more will be sold.

The future of the systems software market is unclear in the late 1980s and beyond. On the one hand, the computer manufacturer must consider its users and cannot lock out independent systems software products. On the other hand, computer manufacturers could develop similar product capabilities to the independent vendors' products. The investment required to do this is probably prohibitive. The computer manufacturer is left in a precarious position: i.e., how can it maintain control over its customers?

One solution to this dilemma is to combine the systems software and the computer hardware (firmware) together in future computer generations. As users buy these new computers, they will not have a need or perhaps the ability to add additional systems software products to their computers. Will this happen? Intel has stated that software (including applications) will be combined with the hardware in the 1980s. It is unlikely that all systems or application software will be combined with hardware in this decade, but it appears clear that a substantial part of the systems software *could* be combined with the computer hardware in the next 10 years.

APPLICATION SOFTWARE

Application software products perform specific application functions for end-user organizations. These products solve specific user problems.

Application software products can be divided into two types: cross-industry products and industry-specialized products. Cross-industry products perform applications common to many industries, such as payroll, general ledger, fixed asset accounting, accounts receivable, or inventory management. Industry-specialized products perform applications specific to industries such as banking, medical, or insurance. Examples of industry-specialized applications are demand deposit accounting (banking), shop scheduling (manufacturing), and policy administration (insurance).

Users of application software products are people involved in day-to-day company business. Thus, typical users of an accounts receivable product include clerical personnel who perform data entry, credit managers who use CRT terminals for online inquiry, and the vice president of finance who uses summary reports. All of the functions described are performed by the application software product. Generally, the people involved with computer operations are not the users of application software products. Computer operations personnel typically use systems software products. However, both groups of people help evaluate application software products for company use.

Nearly 10,000 applications software products are marketed by approximately 2,000 vendors. These numbers grow daily due to the tremendous growth in use of small computers. No vendor, including computer manufacturers, has captured more than 10 percent of this market. At one time, the majority of application software products was written for IBM and IBM-compatible mainframes, but this is no longer true today. Over one half of the application software products and vendors now serve the personal and small computer used by many user markets.

Application software product vendors have penetrated the banking, insurance, and discrete manufacturing sectors with industry-specialized application products to the extent that nearly 60 percent of user expenditures for application software products occur in these three industries. Discrete manufacturing application software expenditures is the fastest growing section due largely to the need for distributed applications for plant operations in diverse locations. However, other industries such as retail, process manufacturing, and transportation will also experience rapid growth in user expenditures for application software products, as these industries have not yet been heavily penetrated by product solutions.

PROFESSIONAL SERVICES

Professional services are used for the design and programming services performed for users who require software tailored to their specifications. In some cases, users want software capabilities that are not available in standard products. In other cases, the user wants very specific functions performed according to set procedures that rule out product solutions. Most professional services are for application software although some systems software is written for mainframes or unique distributed processing projects.

There are thousands of professional services vendors. Every independent consultant who designs software or programs on a custom basis provides professional services. However, probably 2,000 vendors each generate more than $250,000 revenue in professional services per year. The amount of client contact required means these vendors generally serve geographical markets. There are only several hundred firms that offer professional services in more than one geographical area.

The use of professional services is increasing primarily due to the backlog of programming work in data processing departments, the shortage of software designers and programmers, and the lack of specific skills in data processing departments needed to develop certain applications. The backlog of application development in most data processing departments is 24 months and growing. The only way to reduce this backlog without hiring permanent employees (assuming that they could be found) is to use outside professional services. The shortage of software designers and programmers tends to raise the compensation level for people with these key skills. In order to attract and retain these people, professional services firms offer higher salaries and more diverse project experience. This creates a greater personnel shortage in other types of firms, reinforcing the need for outside professional services.

Professional services firms have developed technical skills to handle new applications in distributed processing, telecommunications, data-base systems, and office automation. Most companies don't have the diversity of needs to have developed these skills in-house. Now that these skills are required, it may be more economical to obtain them from a professional services vendor.

Many professional services vendors have begun offering turnkey systems to their clients. Although it is too early to measure the success of most of these offerings, the professional services firm is extremely well positioned to offer unique turnkey systems because of its established base of technical expertise.

TURNKEY SYSTEMS

Turnkey systems solve an application problem for a user. The key features are that the computer hardware and software are sold at the same time, and that the problem addressed is an application (as opposed to systems software). Application software not sold at the same time as the computer hardware is considered a product (or professional service) and would be included in the application software product (or professional service) category. User expenditure data shown in Figure 3 includes only the software portion of the turnkey system sale. The computer hardware, training, supplies, and computer hardware maintenance components are independent of and not included in Figure 3.

There are approximately 5,000 turnkey system vendors. Some turnkey system vendors also sell software products and/or professional services. The largest turnkey system suppliers today provide systems for computer-aided design and manufacturing (CAD/CAM). Approximately 15 percent to 20 percent of all turnkey system user expenditures today are for CAD/CAM. The smallest turnkey system suppliers typically serve the general business accounting environment. The supplier sells payroll, accounts receivable, accounts payable, and general ledger software along with the computer hardware. Most of these suppliers are marginally profitable due to intense competition in this market.

The key to successful turnkey systems is offering specialized systems to a narrow audience. Industry-specialized turnkey systems have been successfully sold and serviced and are easily sold by customer referrals. Vendors who have specialized systems have less competition, can command higher prices, and have higher profits. Vendors that offer generalized systems have intense competition (from computer processing services firms and computer manufacturers, as well as other turnkey system suppliers) and generally have to discount list prices of hardware. Thus, it is a myth that the generalized turnkey system vendor will make profit on *both* the computer hardware and software.

The future for independent turnkey system vendors should involve the packaging of small computers with highly specialized software. The more unique the system, and consequently the smaller the market niche, the better for the vendor. The successful vendor will become the leader in serving these market niches and hence move down the experience curve and ultimately raise the barriers of entry to the competition.

Computer manufacturers will concentrate on offering generalized solutions that will appeal to many buyers. However, these solutions will not have the specialization needed to serve small market niches.

DRIVING FORCES

The past five years have shown a dramatic increase in user demand for software. Substantially improved computer hardware cost-performance has lowered the threshold and opened new application areas. A new generation of management is (by education and training) increasingly more aware of the types of applications that can be addressed with computer software. Increased scarcity of software designers and programmers, coupled with the impact of inflation on personnel costs, has forced managers to explore alternatives to in-house software development.

Computer manufacturers have devoted most of their resources to development and marketing of computer hardware and systems software, leaving voids to be filled, particularly in application software. Computer manufacturers such as Atari and IBM are encouraging software developers to write software for the manufacturer's hardware. Government regulation in such industries as banking and insurance, and in such functional areas as human resources and taxes, has had a substantial impact on demand for external software. The net result has been the entry of many small companies, with a corresponding proliferation of software.

New forces are coming into play that will add to the factors influencing the software marketplace in the next five years. Computer manufacturers such as Honeywell and Burroughs and other large companies such as Xerox and Exxon are entering the marketplace. The accompanying buying process—better coordinated, more sophisticated, demanding improved support service—will force greater competition among vendors of software solutions. Advances in technology in areas such as data base, microprocessors, image processing, telecommunication networks, and distributed processing will furnish new opportunities for new software solutions. Future success will require vendors to excel in software development, sales, marketing, and in customer maintenance and support. Computer manufacturers and selected large companies, together with large independent computer software solution firms, will aggressively acquire smaller vendors, resulting in a market consolidation in which a relatively small number of firms will dominate the profitable market sectors.

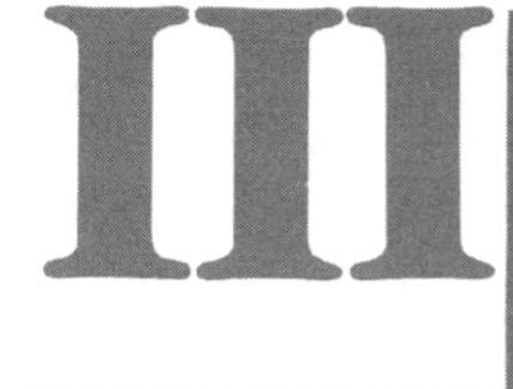

Developing New Products and New Businesses

New product and new business development are fundamental to the success of the technology-intensive firm. For these firms, product life cycles are short. A large portion of current sales is usually generated by products developed during the past five years. Successful new product development in the technology-based firm requires the integration of technical and human systems through effective management of complex functional interfaces at every stage of new product development. Some of the key skills required in managing new product development are:

- Selecting, developing, and managing project teams.
- Balancing the concerns of different professional orientations.
- Keeping various functional groups responsive to each other.
- Linking user needs to internal technical capabilities in product design.
- Keeping project development consistent with business strategy.
- Managing the product champion/venture manager types.

New product development can be grouped into two types: *incremental* and *radical*. Incremental new product development refers to enhancements of existing models and/or the development of second-generation models, and does not involve significant changes in the underlying technology. Radical new product development requires fundamental changes in the design parameters of the product concept, and often will involve technology that is significantly new. Radical product development usually destroys much of the capital base associated with the current line of products and may be a catalyst to major organizational change.

Successful new business development requires all of the skills listed above as well as some others. For a new business to succeed, a whole new class or "family" of new products must be developed. In addition to new knowledge, skills, equipment, and people, new business development usually also requires the development of a new unit to house the new business in the corporate structure.

The new product and new business development process can be usefully conceptualized in terms of major *stages*. Each of these stages involves different *key activities/problems* which are performed/solved by people or groups with different functional skills and capabilities (these include R&D, product design, process design, manufacturing, marketing, etc.). These groups have different professional, emotional, and time orientations. How to manage the *interfaces* between these people and groups within different stages and in the transition from one stage to another raises some important but difficult challenges. Each of the stages has a different output (going from product concept through prototype design to volume production) which is an important building block for the next stage. As the stages of the development unfold, the *commitment of resources* increases exponentially, raising some important issues concerning project milestones and associated "go/no go" decisions.

Central in the study of new product and new business development are the roles of the *product champion* and the *project manager*. The project manager role is usually well defined in the organization but is nevertheless a demanding one because the responsibilities associated with it require subtle skills over and beyond the exertion of "authority" in order to get things done. The product champion role typically emerges in an unplanned fashion. Not infrequently product champions are necessary in order to overcome the resistance to change encountered by new product development efforts in both the external and internal (to the firm) environments.

Studies of success and failure in new product development have consistently found that good communication and cooperation between functional groups is critical in the new product development process. Achieving such communication and cooperation is not easy because of the differences that exist between these groups.

Most recent research suggests that the success of the new product development process is to a large extent dependent on a *learning process*. This in turn suggests that a "family" of products may be a more useful unit of analysis for studying new product development than a single product development. Another important recent finding in the research literature is that there are not only things to be learned but also to be *unlearned,* so as to make room for new techniques and approaches or simply to do away with routines that turn out to be ineffective.

We noted at the outset that it is useful to distinguish new business development from new product development. One important additional dimension has to do with the involvement of different levels of management in the different stages of the development. Even though different management levels are important for new product development, usually the new product's position in the firm's operating structure is well defined, and the channels for strategic decision making are well established and clear. In the new business development process, the situation is different. Here, the strategic decisions fall outside of the regular channels, and hence it is important to understand the emerging patterns of involvement of different levels of management in the development process. Instead of just folding into an already existing organizational unit, new businesses require the development of a new unit to house it.

Whereas in new product development product championing often plays an important role, the network of activities (and roles) in new business development is more complex. Here, the role of middle-level and top-level managers in the strategic management process can be quite different from the traditional view (direction from the top). To conceptualize the new business development process, it is therefore useful to construct what will be called a "process" model which combines sequential (stages related) as well as simultaneous (multilevel related) activities.

The cases and readings in Part III reflect the various themes and issues discussed above. Together they provide the opportunity to become familiar with the issues and problems that new product developers, corporate entrepreneurs ("intrapreneurs"), and their managers will generate and encounter in the development process in established organizations. These issues and problems are more of an organizational and implementational nature, but they build on the ones concerning strategic positioning that were already addressed in Part II of this book.

New product and new business development processes within established firms constitute one major mode of innovation. Equally important is the process of external new venture development through the venture capital mode. Recently, a number of studies have shed light on the factors associated with success and failure of hi-tech start-ups and on the process of collaboration between venture capitalists and independent entrepreneurs. Part III of this book would not be complete without providing some materials highlighting the venture capital approach. The concluding discussion of Part III could raise the issue of similarities and dissimilarities between internal and external new business development. Such discussion will set the stage for Part IV of the book.

Interfaces: Marketing—Engineering—Manufacturing

Case III–1 Apple Computer (A)

J. S. Gable, S. Tylka, and M. A. Maidique

As Wil Houde, vice president and general manager of the Apple Computer Personal Computer Systems Division (PCSD), turned to look at the clock, his eyes fell upon a white teddy bear sitting on his shelf. For months that bear had been symbolically rotated between individuals responsible for solving the problems associated with Apple's newest product, the Apple III. The bear was his for now.

As part of his new responsibilities, Houde was about to present a major proposal at an Apple executive committee meeting in February 1981. He was not especially concerned that his proposal would be accepted. After all, it had been shaped in consultation with nearly everyone who would be there. However, he couldn't help wondering if this approach to the problem was correct. Probably, he reflected, it was.

APPLE COMPUTER HISTORY

In 1975, the Apple Computer Company was founded as a partnership. Microcomputers—computers based on the recently developed microprocessor chip—were new in 1975. The chips on which these small computers were based had an active area smaller than a fingernail and computing power greater than a room filled with vacuum tube computers.

The year the Apple partnership was founded, Steve Wozniak, 24, was working at Hewlett-Packard designing hand-held calculators, while Steve Jobs, 22, was designing video games at Atari. The two computer buffs were interested in having their own microcomputer, but they really couldn't afford their cheapest option, a $600 kit. Jobs recalls:

> We bought one microprocessor chip (between the two of us) and designed a microcomputer. Woz designed about 75 percent of it, and with some Hewlett-Packard and Atari parts, we built one. Our first computer, the Apple I, looked like a mess, but it worked.

Jobs and Wozniak belonged to the Home Computer Club in Palo Alto. As was the group custom, they showed off their Apple I. It wasn't much to look at, being little more than a single printed circuit board with wire streaming out from it, but all their friends wanted one. They were soon spending all their spare time building computers. Jobs explained:

> It took about 40 hours to get a working model: 20 hours to build one, and 20 hours to troubleshoot it. We decided that we could cut our time in half if we made a printed circuit board. So Woz sold his HP calculator and I sold my VW van for $1,325. This was the beginning of Apple Computer.

In 1976, a friend of Jobs laid out a printed circuit board (PCB) for the computer. The PCB made the computer production process much easier. Users could now insert the parts with a higher probability that the board would work. The PCB design cost $2,500. Jobs and Wozniak planned to make 100 boards and to sell them for $50 each, leaving them with $2,500 in profit. Maybe, they thought, they could recoup their initial investment of a calculator and a van.

Initially the partners sold the boards to hobbyists through trade magazines. But Jobs was interested in expanding business through other distribution channels. "I went to the original computer store in Mountain View (California) called the Byte Shop. The owner was really excited and said he'd take 50 of them. Dollar signs!"

But the owner added the condition that the boards be fully assembled. At that time all computers were kits, so Jobs and Wozniak went to local parts distributors and convinced them to sell $25,000 worth of parts on net-30-days terms. Jobs explained, "We bought the parts on sheer enthusiasm. We had no collateral—our VW and calculator were long gone. We didn't even know what net-30-days meant."

First, Jobs and Wozniak built 100 boards in their garage. They took 50 down to the Byte Shop and were paid cash. "We paid off the suppliers in 29 days, and we've run our business on cash flow ever since." They farmed out the assembly of the printed circuit boards and then plugged in the integrated circuits. About 200 Apple I computers were sold.

APPLE II

By the fall of 1976, Jobs and Wozniak had designed a new home computer, the Apple II. The Apple II computer mainframe was similar to Apple I, but it had additional circuitry, a keyboard, and was packaged in a plastic housing. "We had learned a lot from the Apple I. We learned that 8K bytes of memory really weren't enough. We were using the new 4K RAMS (Read and Write Memories). At that time no one else was using RAMS. Going out with a product using only dynamic memories was a risky thing." On the other hand, the manufacturing process in the Apple II was essentially the same process originally worked out in the garage for the Apple I.

Wozniak designed most of the internal working of the new computer while Jobs was primarily responsible for its overall concept and appearance. Jobs wanted to strip the computer of its mysterious and sometimes threatening reputation. This led to a low profile, a plastic case, and no blinking lights or confusing knobs and dials. In many ways this design philosophy made the Apple II the first truly successful "personal computer." Apple also made available every piece of technical data relating to the machine, a highly unusual move in an industry where secrecy had always been tightly maintained. This open policy allowed sophisticated users to design circuit boards that could plug into the computer and expand its capabilities. In fact, several empty slots were built into the Apple II for just this purpose. Independent vendors began marketing hardware and software enhancements.

At this point, the two Steves were hesitant about going into business by themselves. They showed their prototype to others at the computer club. Commodore showed a special interest in the Apple II. (The Apple II almost became the Commodore I.) The partners also took their Apple II prototype to both Hewlett-Packard and Atari. Jobs stated:

> Atari rightfully said, "We have so much going on we have to stay focused on games," which was true. Hewlett-Packard's response was that Woz and I didn't have degrees and couldn't possibly know what we were doing. They decided to do it themselves and have since then. (Back in 1976, Hewlett-Packard was working on the HP-85.)

Each step along the way, the partners were virtually forced into the business, from design to manufacturing to marketing. But as Jobs explained, in time they found themselves more and more excited:

> You don't realize that there's a feeling developing, the feeling of being able to make something (especially an artifact) and see it get out in the world. It's up there with the "best" of feelings. It is an underlying passion that entrepreneurs feel that stems from the idea that you really can impact the world. You really can change how this world is being run.

During 1976, Jobs realized that the market was growing faster than they could internally finance.

> We needed some more money, so I called up a venture capitalist and he tooled over in his Mercedes. He said that he basically invested in companies right before they were about to take off, and we still had a few years. (He later ended up investing, at a substantially higher price.) But he gave me the name of a friend, Mike Markkula, who invested in riskier operations. So I called up Mike, and he tooled over in his Corvette.

Markkula was 32 when he went to discuss the business with Jobs. He had been in marketing at Fairchild Semiconductor, and later became a marketing manager at Intel Corporation. Then he "retired." Markkula advised Jobs that he needed a business plan and together they worked on the project.

At this time the company developed its first business strategy. Their concept was to use microprocessor technology to create a personal computer for individuals to use in a wide variety of applications. They also needed a name for the company. Jobs explains how they arrived at that name:

> I had traveled some before I worked for Atari. In fact, I went to India, and—like most tourists—ended up with dysentery. So when I returned to the States, I was interested in diet. I believe that man is architected as a fruitarian, and basically, that is the way we should eat. And the apple is the staple of the fruits. So we needed a name, and people were suggesting all of these funny, technical-sounding names like XXX or Matrix Manipulation. Just to spur creativity, I deemed that we would call it Apple Computer unless someone had a better idea by 5:00 P.M. Well, it stuck and it got us ahead of Atari in the phone book.

Most of the initial cash invested in Apple Computer came from Markkula, who invested $91,000 in January 1977. Jobs and Wozniak put up $2,654.48 each. Another $517,500 was raised from various venture capitalists in early 1978. This made Apple a rather inexpensive start-up. As Jobs commented, "It was clear that we didn't necessarily want Markkula's money—we wanted Markkula. I really admire Mike. He's a really good coordinator." They then hired Mike Scott, who had been an office mate of Markkula at Fairchild, to be the president. Markkula became chairman of the board and vice president of marketing, and Jobs, vice president of product development.

Meanwhile, other firms had also become interested in the personal computer market. In early 1977, Commodore announced the Pet, the first personal computer, priced at $495. Interest in the product was intense, but no one could get one. In hopes of purchasing a Pet, people mobbed the Commodore booth in April 1977, at the first West Coast Computer Fair in San Francisco. However, Commodore had problems. The price of Pet computers had been raised to $595, $100 over their advertised price. More importantly, they weren't actually available until August—or September—or October.

But Apple was right across the hall introducing the Apple II. Moreover, the Apple II could do some things that the Pet could not, such as color graphics and sound effects. Although the Apple II was somewhat more expensive than the Pet, priced between $1,195 and $1,395, it *was* available.

The Apple II was a product for which the customer had very low expectations, in part because it had little competition. The early personal computers were purchased by hobbyists. Customers were 98 percent male, typically 25 to 45 years old, earned a salary over $20,000 a year, and had the knowledge to program their own routines. (There was little standard software available.) Customers, for instance, commonly programmed games such as Star Trek and Adventure.

THE APPLE COMPUTER COMPANY

Apple's first full fiscal year of operations was 1978. For financial data, see Exhibits 6 to 9. During that year, the company organized a distribution network through independent distributors. This expanded distribution system contributed greatly to the increase in sales. Sales of Apple products to the ComputerLand retail chain accounted for 14 percent of the net sales for 1980 ($118,000). No other retail chain or store accounted for more than 3 percent of net sales.

Products were sold through 950 independent retail computer stores in the United States and Canada, and internationally through 30 independent distributors who supplied 1,300 retail stores. In 1980, however, Apple terminated independent distribution arrangements and established its own sales organization to serve retail computer stores. Apple absorbed the costs to repurchase inventory from the former distributors. Over 100,000 units had been sold by September 1980. (See Exhibit 1.)

In the late 1970s, many people still considered personal computers as something of a fad. By 1980, however, the outline of an industry had begun to take shape. Apple, Radio Shack, and Commodore were the principal manufac-

EXHIBIT 1 Apple Product Shipments

APPLE II
Annual Shipments

	Year Ending				
	September 1977	September 1978	September 1979	September 1980	March 1981 (6 months)
Number sold	570	7,600	35,100	78,100	75,000

APPLE III
Monthly Shipments: Plan versus Actual

	Month		
	December 1980	January 1981	February 1981
Shipments planned	300	2,500	7,500
Shipments realized	125	500	2,500

turers of systems retailing for less than $5,000. Apple had the second largest installed base for such systems in the United States, next to Radio Shack (a subsidiary of Tandy Corporation) which enjoyed strong market penetration due to its large number of company-owned retail stores (over 8,100 worldwide). Commodore, however, benefited from broader international retail distribution. (For an overview of the personal computer industry's products and markets, see Exhibit 2.)

The independent retail dealers who sold Apple computers were trained to replace and exchange most computer system components at the store. Apple required them to enter into dealer service agreements under which they could purchase service kits containing spare parts, components, manuals, and diagnostic programs. Apple typically offered a 90-day full parts and labor warranty for its products and, since January 1980, offered an extended limited warranty on the Apple II at a price of $225 for each year of coverage. Approximately 5 percent of the purchasers of Apple IIs entered into extended service agreements. About 90 percent of all repair work and diagnostic testing on the Apple II systems was provided by dealers.

MANUFACTURING AND PERSONNEL POLICIES

Manufacturing at Apple was essentially buy, assemble, and test. Most components in Apple products were purchased from outside vendors who built chips, boards, cases, and other parts to Apple specifications. Some of Apple's

EXHIBIT 2 Personal Computer Industry

Market Size (in millions of dollars)

	Year	
	1980	1985 (est.)
Home	120	475
School	35	145
Small business	590	2,700
Office	90	1,450
Scientific	220	1,020

Competition—An Overview of Key Competitors

Company	Product	Price*	Memory
Apple	Apple II, Apple III	$1,330 to $5,000	16K to 256K
Atari	400, 800	$400 to $1,000	16K to 48K
Commodore	VIC-20, PET	$300 to $3,000	5K to 96K
Hewlett-Packard	HP-83, HP-85	$2,250 to $3,250	16K
IBM	Personal computer	$1,600 to $6,000	16K to 256K
Radio Shack	Color computer, TRS-80	$400 to $8,000	4K to 64K
Texas Instruments	99/4	$525	16K
Xerox	SAM	$3,000	64K
Zenith	Z-89	$2,900	48K

* Suggested retail prices, varying with memory size and attachments.
SOURCE: *New York Times,* August 23, 1981.

competitors, in contrast, manufactured many of their own components. Manufacturing facilities were maintained in Cupertino, San Jose, and Los Angeles, California; Dallas, Texas; and County Cork, Ireland.

Apple had developed production modules which could be duplicated within each plant. A module for the Apple II filled 30,000 square feet, required a crew of 70, and produced between 450 and 500 units per day. Apple utilized 90 percent of its manufacturing space capacity, operating one work shift a day, five days a week. The Cupertino plant had a special role in that it maintained at least one production module for each product. This gave the company a single location for developing new production techniques and a manufacturing location near top management.

Quality control and final system testing and inspection were performed at production facilities. In the testing process, Apple computers with special software performed diagnostic tests to isolate and identify problems. As a part of the final testing process, all systems were "burned-in" to provide assurance of electronic and mechanical functions.

Management carefully monitored sales-per-employee, which was $176,000 in 1980. In March 1981, the company employed 1,530 full-time employees, including 230 in marketing and sales, 250 in research, product development, and related engineering, 950 in manufacturing, and 100 in general management and administration. (See Exhibit 3 for organization chart.) The company never had a work stoppage and no domestic employees were represented by labor unions.

Apple had liberal employee benefits, such as profit sharing. Virtually all professional employees owned stock, and there was an employee purchase plan in which all employees could participate. Under this plan, employees

EXHIBIT 3 Corporate Structure

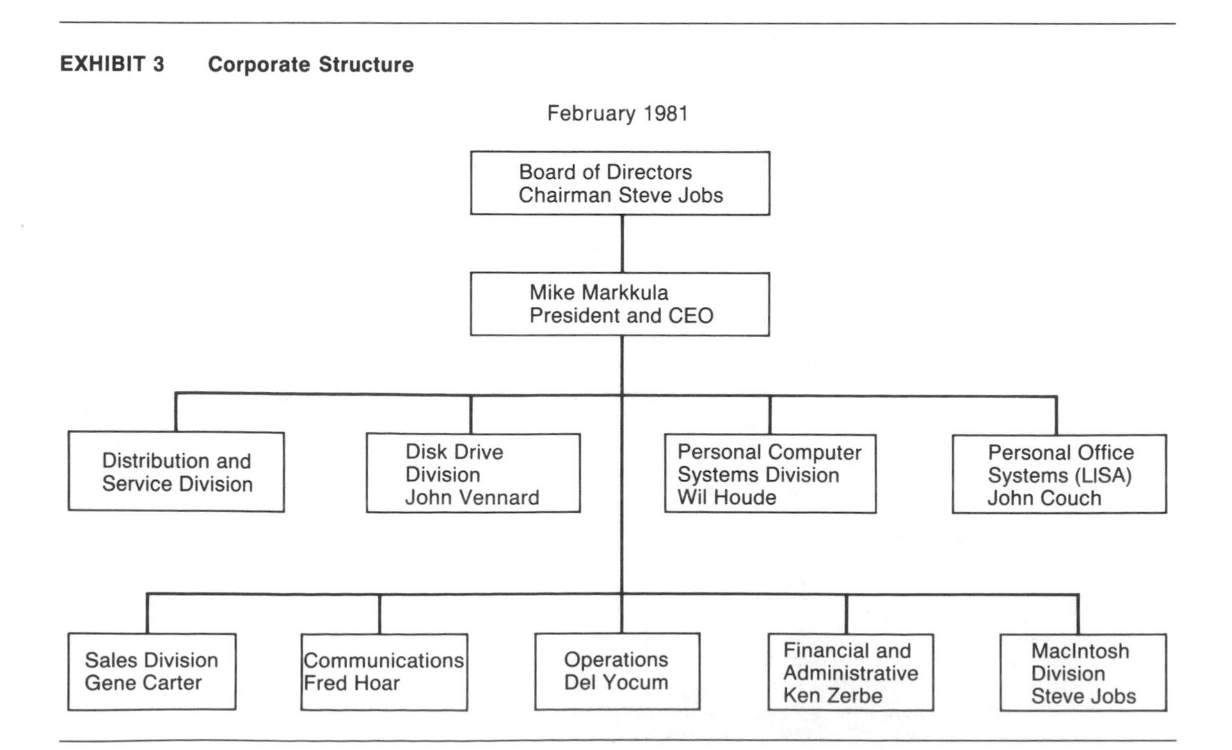

could purchase stock up to 10 percent of their salary every six months at 85 percent of the price at the beginning or end of the six months, whichever was lower. It was not unusual to find workers on the assembly floor who owned Apple stock.

In December 1980, Apple Computers made its first public offering, selling 4.6 million shares at $22. The reasons for going public, as with many other things at Apple, were unconventional. Jobs recalled:

> When a company exceeds a number of shareholders (something like 500), it has to start reporting to the Securities Exchange Commission just as if it were public. Therefore, we had the choice either to stop giving stock options to new employees or essentially to go public. It was a binary choice. Since it is Apple's philosophy that we should be an employee-owned company, the choice was simple. We went public. Every professional at Apple has a major stock option. Over 200 Apple employees have become millionaires in the past two years.

Apple also had a "loan-to-own" program through which, after one year of employment, the company gave the employee a free computer if he could demonstrate minimal computer skills. About 60 percent of the employees had an Apple computer at home. The company also paid for one half of employee lunches. There were also occasional special benefits. When Apple had its first $100,000 quarter, everyone was given an extra week of vacation that year, for a total of four weeks.

The company published a set of beliefs known as "Apple Values." (See Exhibit 4.) These values helped guide decisions, policies, and procedures throughout the organization. They were designed to provide continuity and a sense of tradition as the company grew. In short, Apple Values was an attempt to delineate that which was unique to Apple. Jobs described it in this way:

> Our greatest joy is hiring people who are better at doing things than we were. Our managerial style is to hire really great people and create an environment where people can make mistakes and grow: the Apple culture. The Apple Values were, in effect, originally contained in our first business plan.

EXHIBIT 4 Apple Bulletin

Apple Values
September 23, 1981

Apple Values are the qualities, customs, standards, and principles that the company as a whole regards as desirable. They are the basis for what we do and how we do it. Taken together, they identify Apple as a unique company.

Values are important because they determine our culture, our style. They guide our decisions, our policies, and our procedures. When they are identified and supported through management practices, they foster trust and unity, thus reducing the need for cumbersome rules and unnecessary supervision. They help Apple employees to understand the company's philosophy and objectives . . . and how they themselves contribute as individuals. Values are a compass for all. As the company grows, they provide continuity and the thread of tradition that marks Apple as different from others.

The attached list of Apple Values was created by an employee task force, based upon input from the Executive Staff. Both the Apple Values Task Force and the Executive Staff recognize that these values are GOALS. We don't always live up to them. For example, although we value teamwork, there is clearly room for improvement in the ways we work together. We should strive to live

EXHIBIT 4 *(continued)*

up to these values as the standard, and be looking for ways to do better. The values are guides to help us make decisions.

Identifying our values is important, but fostering them in practice is even more important. The draft of Apple Values will always be a draft, i.e., open for redefinition or affirmation. Please let us know what you think of the draft, and give us your ideas on what we can do to help Apple live up to these goals. And most of all, consider what you can do personally to support these values—to make them as alive as possible within each of us at Apple.

Thanks!!

The Apple Values Task Force

Task Force Members	*Executive Staff Members*	
Ron Boring	S. Bowers	S. Jobs
Trip Hawkins	C. Carlson	T. Lawrence
Pat Marriott	G. Carter	A. C. Markkula
Bob Montgomery	J. Couch	A. Sousan
Phil Roybal	A. Eisenstat	J. Vennard
Pat Sharp	F. Hoar	S. Wozniak
Ken Victor	R. Holt	K. Zerbe
Roy Weaver	W. Houde	

Draft: Apple Values

EMPATHY FOR OUR CUSTOMERS/USERS

One person, one computer.

We provide our dealers and users with reliable products of lasting value and dependable service. We expect our products to respond to the user's needs: To be seen as friendly, natural tools that can extend each person's analytical and imagining abilities. Most of all, we want our customers to feel that they receive more benefit from Apple than they paid for.

ACHIEVEMENT/AGGRESSIVENESS

We are going for it and we will set aggressive goals.

We are all on an adventure together.

We will continue to set high goals and high standards of performance because we realize that through meeting such challenges we will advance as a company and as individuals. We do not value risk taking for its own sake. Rather, we regard taking calculated risks as part of the adventure and the challenge of accomplishing something significant.

POSITIVE SOCIAL CONTRIBUTION

We build products we believe in.

We are here to make a positive difference in society, as well as make a profit.

Apple contributes to society by providing the power and usefulness of the computer to individual people. With this tool, people are improving the way they work, think, learn, communicate, and spend their leisure hours. As a corporate citizen of this world, we will respect our social and ethical obligations. Our profits are the result and an important measure of how well we succeed in making this contribution.

INDIVIDUAL PERFORMANCE

Each person is important; each has the opportunity and the obligation to make a difference.

The individual worth of each employee as a person is highly valued. We recognize that each member of Apple is important, that each can contribute directly to customer satisfaction. Our results come through the creativity, craftsmanship, initiative, and good work of each person as a part of a team.

TEAM SPIRIT

We are all in it together, win or lose.

We are enthusiastic!

We strive for a cooperative, friendly work environment that supports individual contribution as well as team effort. As a company, we know that we are all working together for a common goal. Accordingly, we want to keep our organization simple and flexible so that ideas and information can pass freely among those who need it. Indeed, our work environment serves to reflect and support all of our values.

INNOVATION/VISION

We are creative; we set the pace.

Apple was founded on the conviction that the computer could be, and should be, a personal

EXHIBIT 4 *(concluded)*

tool. From simple beginnings, the company has accomplished in a short time what many others thought impossible. Innovation, aggressiveness, and responsible risk taking, combined with a vision of the future in which the personal computer will serve us all, will continue to motivate us.

INDIVIDUAL GROWTH/REWARD

We want everyone to enjoy the adventure we are on together.

We are committed to providing a work environment based on mutual respect and will strive to support our employees in achieving their personal objectives in line with their contribution at Apple. We encourage the growth of each individual. Sharing the tangible rewards of Apple's success as well as the challenges and satisfactions that come with it is part of our management philosophy.

QUALITY/EXCELLENCE

We care about what we do.

We take pride in our integrity and the quality of our products. We strive for absolute fairness in our dealings with customers, vendors, and competitors as well as among ourselves. In cases of dispute, we are willing to go the extra mile. The quality of our work stems directly from the care we express in all we do. It is an attitude that unites us.

GOOD MANAGEMENT

We want to create an environment in which Apple Values flourish.

We want to foster the best environment for individual initiative while maintaining the highest standards in business ethics, personal and professional integrity, and achievement.

PERSONAL COMPUTING

Apple Computer had been an innovator in personal computers. Aside from the computers themselves, Apple's most significant innovation was the low-cost, 5¼-inch, flexible or "floppy" disk drive for the Apple II—introduced in mid-1978. The floppy disk storage replaced the less efficient cassette tape storage. It provided file memory capacity of up to 143K bytes of data, vastly increased data retrieval speed, and provided random access to stored data. These hardware enhancements increased the power and speed of the Apple II and also sped up the development of software. Over 100 independent vendors were developing Apple II software and peripheral equipment for text editing, small business accounting, and teaching.

Some of these programs were extraordinarily successful. The best seller was a financial modeling program called Visicalc, developed by two Harvard Business School students. Over 250,000 copies of Visicalc were sold at about $250 a copy. *Forbes* magazine (August 2, 1982) called it "the infant personal computer software industry's first gorilla hit."

By 1980, Apple had begun to introduce its own software, but Apple expected to write no more than 1 percent of the software that would eventually be available for their computers. Even 1 percent of these programs would be a large number, since over 100,000 were then available for the Apple II.

Apple also had introduced new peripheral devices to expand the computer's applications. Peripheral accessories manufactured by Apple included a graphics tablet, the floppy disk, a thermal printer, and interface circuit boards. Apple computers incorporated standard interfaces, permitting the use of peripherals designed and manufactured by others as well as those offered by Apple. These peripherals included medium-speed printers for home or business applications requiring letter-quality output; modems which provided a data communication link over telephone lines to access time-sharing services, computerized bulletin boards, or other computers; music synthesizers; portable power units; and more. The development of this software and equipment helped the growth of the personal computer business segment by increasing the variety of applications for which the computer could be used.

THE APPLE III

The project to build the successor to the Apple II began in late 1978 under the code name Sara, after the daughter of Wendell Sander, Apple's 16th employee, who was the designer of the new computer. Sander held a Ph.D. in electrical engineering, and had worked for several years in the defense and semiconductor industries. In general, Apple was feeling a need to build a product which would eliminate the Apple II's shortcomings. "We had listened very carefully to our customers," remembered Jobs. In fact, a great deal of marketing research had been conducted for the Apple III. Not only did this work help detail technical specifications, but it also called for a sense of urgency. "We were told that sales of the Apple II would peak around 10,000 units per month," said Houde.

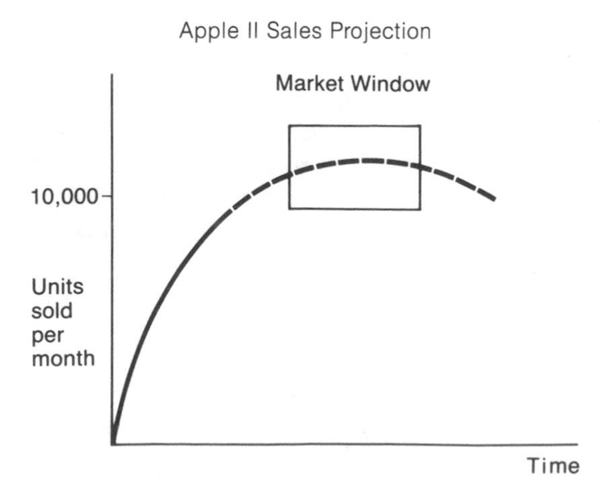

In order to sustain the company's growth, the Apple III had to be introduced during the "market window." As it turned out, Apple II sales charged through the 10,000 mark without a pause. Sander recalled, "We had no idea back then that the Apple II's popularity would last so long."

Heading the Personal Computer Division at this time was Tom Whitney, a Ph.D. formerly with Hewlett-Packard. His background in electrical hardware included work with the HP-35 calculator and the HP-3000 computer. The initial outline of the new product concept was largely his work.

From the very beginning, the new computer was envisioned to have an 80-character wide screen (instead of 40), upper- and lowercase capability, and a built-in disk drive. These made the new machine much closer to a full word processor—an application for which Apple II owners had been buying special plug-in cards. A built-in drive was only logical, since microcomputers were virtually useless without the programs on the floppy disks, or "diskettes." At that time, the price of RAMS was dropping so the new machine gained additional internal memory. Computer memory was measured in thousand bytes, and the Apple II typically had 8 or 16K (8 or 16 thousand bytes). Whenever a program's memory requirements exceeded this internal quota, it had to utilize the diskettes, which had a longer access time. The common purchase of special cards by users to expand the II's internal memory pointed to this shortcoming also.

This new product, which was to become the Apple III, was envisioned to be ready in approximately one year—a short turnaround time. It would consolidate Apple's leadership role in personal computers. Although it would be far ahead of any competition, the Apple III did not presume to establish radically new standards for the industry. Indeed, some of Apple's best engineers embarked at this same time on a separate project named Lisa. Lisa would have a longer time horizon and was expected to be the company's next revolutionary product.

During the next few months, the design evolved to a product quite different from the Apple II. The new computer was to have a numeric keypad in addition to a standard typewriter keyboard. The memory was expanded considerably to 96K. An entirely new operating system, the Sophisticated Operating System (SOS, pronounced "sauce"), was created. The aluminum case was designed and cast. The hardware had been given the Apple look and feel.

SOS made the Apple III a much more versatile machine. A computer needed an operating system to create the framework in which application programs could operate. The Apple II had a comparatively inflexible system that would often call for minute changes in programs if a peripheral, say a printer, had to be changed. SOS introduced a "driver" system that allowed programs to address a "PRINTER" routine. This way, a new printer required only a new driver, not a change in every program that used it. Additionally, SOS could save files in a hierarchical framework, which was important for saving large quantities of data. SOS was also more "user friendly." The operating system was considered so advanced that Apple engineers insisted on keeping its specifications secret. However, this technical information was released a few months after the initial product introduction.

The system had made a transition from a follow-on product for the Apple II to an entirely new product. Instead of replacing the Apple II, the Apple III would coexist. (See Exhibit 5 for Apple III product illustration and data sheet.)

Most significantly, the decision was made to incorporate an emulation mode so the new machine could simulate an Apple II and run software written for the II. "Essentially, emulation performs a frontal lobotomy on the machine and makes it forget that it ever was an Apple III," is how one person at Apple described it. This decision, which was hotly debated, may have added as much as 25 percent to development costs, although the additional cost per unit in hardware and labor was small. It was felt that one of the elements in Apple's success was the "cottage industry" that had sprung up to write software for the Apple II. Emulation assured the new machine access to about 90 percent of this vast resource.

Marketing

"The Apple III gave us a classic product positioning problem," said Barry Yarkoni, the product manager of the Apple III. "It originally set out to be a replacement unit for the Apple II but became much more powerful and much more expensive. We couldn't target for the same markets without losing Apple II sales and being overpriced." The new computer cost more than two times to make as did a comparable configuration of the Apple II.

It was important for maintaining sales of the II to assure the public that the III was considerably different and not a replacement. Part of Apple's solution to this problem was to suggest prices for the new computer as high as possible so different buyers would be attracted to the two computers. While the Apple II now ranged from $1,330 to $1,530, the Apple III was announced at prices ranging from $4,700 to $7,800 (depending on the configuration).

"Our strategy was to roll out the III to selected markets in a rifle shot approach, as opposed to the shotgun approach of the II," explained Yarkoni. The first market targeted was the professional user. This market had not been penetrated deeply by the Apple II, and this plan would protect the II's position in other segments, such as education, small business, and hobbyists.

This strategy led to the development of the Information Analyst package. It dictated a software "bundle" that would include a word processor, business basic language, and Visicalc, the popular program for financial planning—about $700 of software. The Apple III was now

EXHIBIT 5

The Apple III Personal Computer System
The Most Powerful Professional Computer System in Its Class

The Apple III is a powerful desktop computer system available as part of custom-tailored packages designed to solve your complex application needs. For managers, financial planners, analysts, and others who need to organize facts and figures, there's the Apple III Information Analyst System. It offers special features that make it the most powerful, easy-to-use timesaver available.

A wide variety of Apple III system configurations can be tailored to meet your specific needs. Consult your dealer for information.

Powerful Solutions for Complex Applications

The Apple III Computer System has been designed to tackle the tasks that keep you from being as productive as you'd like to be. With an Apple III, you can . . .

- Plan budgets, compare actuals with forecasts, and modify projections.
- Calculate rate-of-return, pro formas, and financial statements.
- Develop highly accurate forecasting models and pricing strategies.
- Create scientific and engineering models, and study causes, effects, and trade-offs.
- Compose, revise, and print all kinds of documents—from memos and brochures to form letters and book-length manuscripts—quickly and easily.
- Maintain and update comprehensive mailing lists, sort them by name, ZIP code, or special key, and selectively print mailing labels and phone lists.
- Write complex computer programs in a variety of languages, including Apple Business BASIC and Pascal.

The casewriters with the Apple III in a typical configuration.

Professional Features for Professional Needs

In addition to its outstanding applications software, the Apple III offers a powerful operating system and all the hardware features professionals look for.

Apple III's Sophisticated Operating System (SOS)

Designed to control all of the Apple III's hardware for you. SOS handles interrupts, manages the system's memory and periph-

EXHIBIT 5 *(continued)*

erals, provides the foundation for graphics, and performs comprehensive file management.

Apple III's Keyboard

The typewriter-style Apple III keyboard has been sculptured for maximum typing speed and accuracy. It contains 61 alpha keys and a separate 13-key numeric pad.

Four dedicated cursor control keys provide single-keystroke cursor movement; each key also fast repeats when held down, so that you can move quickly from point to point in the text. The alpha-lock key shifts only the alphabetic keys into upper case, leaving the number row unaltered. To speed numeric data entry, the layout of the numeric keypad is identical to that of a standard calculator.

Apple III's Disk Drive

A built-in, 140K-byte, flexible disk drive makes the Apple III a compact, space-saving unit. System expansion is cost effective, too, because you can add up to three external disk drives without the need for additional control hardware or software.

Apple III's Back Panel

The system's back panel (as well as most of its case) is formed of diecast aluminum. The aluminum fins on the back of the unit keep the system cool and eliminate the need for a fan.

Most peripheral devices plug directly into the Apple III's back panel connectors. Additionally, there are four large slots in the back panel for input/output connectors mounted on optional peripheral cards.

As many as three expansion floppy disk drives can be used with the Apple III. The first additional drive plugs into the "floppy disk" connector on the system's back panel; then, in "daisy chain" fashion, the second drive plugs into the first, and the third plugs into the second.

Back panel connectors are also provided for two joysticks. Application programs can be designed to take advantage of a joystick (for example, to move the cursor around the screen, or to point to displayed items). Also, one of the joystick connectors can alternately be used to connect a Silentype thermal printer to the Apple III.

The Apple III allows you to use a wide variety of video display devices. The high-resolution Monitor III connects to the system by a shielded cable, which plugs into a jack on the Apple III's back panel. Color video monitors—including NTSC (standard) color, and RGB (for exceptional color purity and resolution)—require commercially available video adaptors, which attach to the Apple III by means of a 15-pin connector that provides all the power and signals necessary.

Built into the Apple III is a two-inch speaker, which produces sound of such high quality that it can even be used to generate voice. An audio output jack located on the back panel of the Apple III also allows for connection of a separate earphone or external speaker (plugging into this jack silences the Apple III's built-in speaker).

An RS-232C connector, also located on the Apple III's back panel, provides for direct attachment of many types of "serial" input/output devices. Using a modem, for example, your Apple III can connect to other computers and data banks by phone line. Or you can quickly add a variety of high-speed or letter-quality printers—even other terminals or computers—to your system, simply by plugging them into the RS-232C connector.

Inside the Apple III

Removing the top cover reveals the peripheral card expansion section of the Apple III.

EXHIBIT 5 (*continued*)

Up to four different peripheral cards can be used at one time to supplement the Apple III's built-in peripheral interfaces.

For optimal computing speed, the Apple III's CPU can be "interrupted" by your system's peripheral devices whenever they require CPU control. Alternately, the CPU can also poll the devices to determine which need attention—thereby minimizing the software required for peripheral control.

The Apple III also contains many features that enhance its utility, including a number of powerful text, graphics, and color capabilities. In all text modes, for instance, the character set that appears on your monitor can be chosen from several available fonts. Special characters, graphics symbols, and even foreign language character sets can be selected quickly and easily from a diskette, and used by any program on the computer.

Through the Apple III's professional display, you view 80 characters by 24 lines of easy-to-read characters. The system can also display 40 characters by 24 lines of color text on color background to add dramatic emphasis to programs.

Along with various text modes, several graphic modes are also available with the Apple III, including an ultrahigh-resolution, monochromatic mode. Up to 16 different colors are available in the high-resolution color mode, and even higher resolution color can be generated by restricting color changes slightly. On a monochromatic monitor, color is displayed as 16 different shades, making it easy for you to use shading and highlighting to enhance and emphasize your displays.

The Apple III also has an Apple II emulation mode for those users who already have an investment in Apple II software. This mode enables you to run most Integer BASIC and Applesoft programs on your Apple III. (Minor modifications may be required for Apple II programs which use the game paddles or other peripherals, however.) Because the Apple III in emulation mode behaves exactly like an Apple II Plus, the screen will display 40-character by 24-line text.

Technical Specifications

Physical dimensions:
Height: 4.8 in. (12.20 cm.)
Depth: 18.2 in. (46.22 cm.)
Width: 17.5 in. (44.45 cm.)
Weight: 26 lb. (11.8 kg.)
Cast aluminum base with molded plastic.

Processor:
Apple-designed processor utilizes 6502A as one of its major components. Other circuitry provides extended addressing capability, relocatable stack, zero page, and memory mapping.

Emulation mode:
Provides hardware emulation of 48K-byte Apple II Plus. Allows Apple II programs, with the exception of Pascal and FORTRAN, to run without modification.

Clock speed:
1.8 MHz with video off. 1.4 MHz average; 1.0 MHz in emulation mode.

Main memory:
128K (131.072) bytes of dynamic RAM memory.

ROM memory:
4K (4096) bytes used for self-test diagnostics.

Power supply:
High-voltage switching type +5, −5, −12, −12 volts.

Mass storage:
One 5.25-inch floppy disk drive built-in; 140K (143.360) bytes per diskette. Up to three additional drives can be connected by daisy-chain cable (560K bytes on-line storage).

Keyboard:
74 keys (61 on main keyboard, 13 on numeric pad);
Full 128-character, ASCII encoded;
All keys have automatic repeat;
Three special keys: SHIFT, CONTROL, ALPHA LOCK;
Two user-definable "Apple" keys;

EXHIBIT 5 *(continued)*

Four directional arrow keys with two-speed repeat;
Four other special keys: TAB, ESCAPE, RETURN, ENTER.

Screen:
Three upper/lower case text modes:
80-column, 24-line, monochromatic:
40-column, 24-line, 16-color foreground and background;
40-column, 24-line, monochromatic;
All text modes have a software-definable, 128-character set (includes upper and lower case) with normal or inverse display.
Three graphic modes:
280 × 192, 16 colors (with some limitations);
140 × 192, monochromatic;
560 × 192, monochromatic;
plus Apple II modes.

Video output:
RCA phono connector for NTSC monochromatic composite video;
DB-15 type connector for:
NTSC monochromatic composite video;
NTSC color composite video;
−5, −5, −12, −12 volt power supplies;
Four TTL outputs for generating RGB color;
Composite sync signal;
Color signals appear as 16-level grey scale on monochromatic outputs.

Audio output:
Built-in two-inch speaker;
Miniature phono jack on back panel;
Driven by six-bit digital analog converter or fixed-frequency "beep" generator.

Serial I O:
RS-232C compatible. DB-25 female connector;
Software-selectable baud rate and duplex mode.

Joysticks:
Two DB-9 connectors for two joysticks with push-buttons and switches.

Printer:
One DB-9 connector (shared with second joystick) for Apple Silentype printer.

Expansion:
Four 50-pin expansion slots inside the cabinet.

SOS:
Sophisticated Operating System handles all system I O;
SOS can be configured to handle standard or custom I O devices and peripherals by adding or deleting "device drivers";
All languages and application programs access data through the SOS file system.

Languages:
Apple Business BASIC and Pascal.

The Apple III Package

With your order for any Apple III configuration, you will receive:

- Apple III Professional Computer System with built-in disk drive, keyboard, serial (RS-232C) and Silentype printer interfaces, and 128K bytes RAM.
- Apple's Sophisticated Operating System (SOS) package, with:
  - System Owner's Guide.
  - DOS 3.3 diskettes.
  - DOS 3.3 instruction manual.
  - Standard drivers manual.
- Apple's Business BASIC programming software package with:
  - Business BASIC diskettes.
  - Instruction manual.

Plus the following (based on the configuration ordered):

Apple III Information Analyst

In addition to the basic hardware and software, with your Apple III Information Analyst order you will also receive:

EXHIBIT 5 *(concluded)*

- VisiCalc™ III software package with
  - VisiCalc III diskette.
  - VisiCalc III manual.
  - Toolkit sampler diskette (with prewritten VisiCalc III worksheets to help you get started.
  - Toolkit sampler manual.
- Monitor III.
- Second disk drive (Disk III—with options B and C).
- Apple III Silentype thermal printer (with option C).
- All necessary cabling, accessories, and blank diskettes to put your system to work immediately.

Apple III Information Analyst
Order No. A3P0001 (Option A)
Order No. A3P0002 (Option B)
Order No. A3P0003 (Option C).

™ VisiCalc is a trademark of Personal Software, Inc.

SOURCE: Company literature.

positioned as a "discretionary purchase by professional people to improve their productivity."

Another marketing manager added, "In those days, it was THE computer. There were no other products competitive with it." The Apple offered far more memory, higher resolution, and a more advanced operating system than anything else on the market. The machines that would later compete, like the IBM, were not due out until mid-1981.

Hardware and Software

The Apple III was a self-contained microcomputer with a built-in disk drive. The unit's flexibility was expanded by the inclusion of several peripheral ports in the back. These ports were simply jacks for plugs to printers, speakers, monitors, and several other "peripheral" devices. Ports were provided also for additional disk drives and, of course, joysticks for games.

The main unit of construction was an all-aluminum casing under which were attached the power supply and circuit board. This casing was the result of early concerns regarding shielding requirements that the FCC was preparing. Because the project could not wait for the new regulations, the casing had been designed to pass the most stringent of tests.

Another design problem was the printed circuit board itself. Essentially, the space allotted was insufficient for the circuitry needed, and Apple preferred not to abandon its "one-board" philosophy established with the Apple II. Nonetheless, these space limitations forced Apple to place the memory chips on a separate board which rested on top of the larger logic board like a bunk bed.

Overall styling followed the Apple II's lead with a similar color scheme and smooth lines. A matching monitor fit neatly on top of the computing unit.

User expectations were quite different for the two machines. While the hobbyist might buy an Apple II and write his own programs, an Apple III purchaser would expect a large selection of tools. The Information Analyst package was an initial approach to this demand.

The development of this software, however, was far behind hardware development. For marketing, this was a major block to selling. During the summer of 1980, nothing was available except for emulation mode and promises for the future.

Software development was understaffed and overly optimistic. "Really, no one understands how to schedule software," said Yarkoni. A shortage of reliable hardware slowed program writing still more.

Product Introduction

The National Computer Conference (NCC) was upcoming in May of 1980. The NCC was a key forum for the computer industry, and Apple decided to use the event to introduce the Apple III. This put tight time constraints on the development project. Also, word of the new computer was becoming the talk of the industry, not against Apple's wishes. "Apple leaked like a sieve. Our distributors thought they knew all about the Apple III," said one executive.

Apple treated participants of the 1980 NCC in Anaheim, California, to a great deal of fanfare, and to the new Apple III. Forced into a separate building for the amount of space they wanted, Apple provided a double-decker bus, appropriately decorated with Apple colors, to ferry people to their exhibits. In addition, the company rented Disneyland for an evening to celebrate its success. Anyone stopping by their booth could get a free ticket to the park for that evening—on Apple.

The public stock offering was also stirring up even more attention. Top management was becoming increasingly sensitive to criticism of Apple being "a single-product company." To refute this view, Apple had announced the Apple III in the prospectus. By that claim, the computer would be available by the end of 1980.

Internally, the people at Apple had an extremely difficult time getting the machine and related material ready for the show on time. Mike Scott, then Apple's president, had set the NCC deadline about eight weeks earlier, and everyone involved was racing for that milestone. With great effort, Apple engineers had produced several prototypes of the Apple III for the show. Marketing had worked to the last moment to have literature ready for the Information Analyst package. "The ink wasn't dry." Although everyone at Apple knew that a few bugs still remained to be worked out, the overall attitude at the firm towards the new product was positive.

And customers were putting down $500 deposits.

Volume Production

The transition to volume production, however, became an endless stream of problems. Engineering and production for the Apple III would later be merged into the Personal Computing System Division (PCSD), but at this point they were still separate departments. The entire production system had to be designed, and every aspect had its challenges. None of the first five or six prototypes worked.

A conflict over the internal cables was a case in point. Engineers in manufacturing had the attitude that, "The cables won't fit into the machine, which is an unmanufacturable pile of junk anyway," while their design engineers would complain back, "If you would listen to us, we would show you how to roll the cables up and slip them in." One production manager explained:

> Normally one should go through four steps in design: first, design; second, design validation; third, process design; fourth, process validation. In the case of the Apple III, we skipped both validation steps. The machine's development called for an overly ambitious schedule.

The units were burned-in for three days each. Burning-in was merely plugging in a machine (or part of a machine) and waiting to see what happened. Often units would be switched on and off or subjected to environmental tests during the burn-in to simulate actual operating conditions. Later, one participant claimed, "We needed 500 machines on (burn-in) racks for six months."

Another problem Apple had throughout this period was dealing with its tremendous growth. During 1980, sales rose from $48 million to $118 million. The number of employees increased from 240 to 670. "It seemed that everytime I wanted someone, he had a new phone number," said one engineer. Production moved in September to a new facility 10 miles from the corporate offices. This 45,000-square-foot pilot plant had 100 employees and was to work out volume production methods for plants in Dallas and Ireland. Although most computers would ultimately be manufactured in these other plants, the pilot plant would always make a small volume to keep production information and experience close to the main office.

"The Apple III was the first new product Apple really had introduced. Initial volume was to be 1,000 a month, instead of 10 or 20, as with the Apple II," remembered Sander.

Reliability Problems

It was becoming painfully clear that the Apple III was fraught with problems. Design engineering had been attributing reliability failures to thermal stress induced by the completely enclosed aluminum case. Three months of temperature tests were made, cooling fins were added, holes were drilled, and the computers were still breaking down. "Weird, intermittent failures kept happening. At times, we were grasping at straws," one engineer reported.

This concern with thermal issues delayed the discovery of other sources of failure. The space constraint on the printed board had several unexpected and unwanted implications. First, the trace lines were excessively thin and sometimes didn't work. Trace lines were the copper strips on a printed circuit board that carried signals between the semiconductor devices. With great difficulty, the board was completely redesigned with wider traces. The memory board, supported over the logic board by a pair of vertical connectors, would sag in the middle—traces would break and chips would pop out of their sockets. Stiffeners were added to the board, which also helped dissipate heat. Furthermore, the first vertical connectors used were not especially good; the supplier had substituted the plating and they often lost their conductivity. This took some time to discover. They were replaced with better quality parts. One design engineer remembered with wry amusement:

> We had what we called the "in-house standard operations fix" for the Apple III. It was to pick up the front of the unit about four inches and drop it. This makes some sense because it would jolt those connectors back into action, but you couldn't very well tell a dealer to do that.

The connectors highlighted another characteristic of the Apple III's troubles—the company had several problems with suppliers. Various sources sold unreliable equipment to Apple and Apple didn't have sufficient inspection or burn-in tests to discover many of them. "We were still a small company and totally lacked things like a component evaluation group. You need a fairly sizable firm to do all that. So we just went for it."

One exciting element of the computer, for example, was the clock chip, which would run even when the machine was unplugged and track time for various programs. Unfortunately, the supplier just couldn't produce reliable chips as promised. And no second source existed because this chip was uniquely suitable for the crammed circuit board. Reliability was so low that this clock was quietly dropped from the product.

Shipments

For a few months following the NCC show, only a few Apple IIIs trickled out to dealers. Quantity shipments started in November. Mike Scott had established the 15th as the due date for 100 units to be shipped. Again, it was a deadline not easily met. Although Scott personally oversaw the production of the hardware, the software was still limited. Furthermore, packaging for the Information Analyst was squeezed out at the last moment. "You cannot believe the documentation necessary to produce six or seven booklets and a box to hold them. The specifications for the Information Analyst package made a stack several inches high," Yarkoni recalled.

Demand initially was extremely strong; however, customer reaction to the new product was not good. Because there was a certain lag time in shipping from inventory, it took a month or two for the message to become clear: The Apple III was in deep trouble. "The response was very strong," said Yarkoni. "Either they got one that worked and loved it, or they got one that didn't and hated us." Some customers were returning machines four or five times for repairs. One marketing manager recalled, "The market reputation of the III was extremely low. It didn't do any good to advertise for it because customers would go to the store only to be talked out of an Apple III by the dealer who felt he didn't need the headache."

In early 1981, Apple was considering halting the production and shipment of the Apple III. In the field, approximately 20 percent of the machines were dead-on-arrival, another 20 percent had major problems, and still more had minor bugs, such as a keyboard that would periodically "freeze." Furthermore, morale was extremely low throughout the company. Forty employees had been simultaneously let go a few weeks earlier. "Considering that we had hired something like 1,500 people in the last year," said Wil Houde, "it is actually surprising that we had only 40 'bad apples.' But it was very shocking to drop them all at once." Around this time both Mike Scott and Tom Whitney had also parted company with Apple, partly due to the Apple III. Houde, in fact, had assumed Whitney's position.

The situation prompted a *Forbes* article (April 13, 1981) entitled "Apple Loses Its Polish." *Forbes* went on to say, "Apple still has to prove that it can put together the kind of innovative products, manufacturing skills, and long-range growth that competitors like Tandy, Texas Instruments, and Hewlett-Packard have been demonstrating for years."

Explanations

Reasons given for the new product's problems were as numerous and varied as the respondents:

Design engineers

> Marketing had sold the machine long before it was ready. If we could have had another six months, none of this would have happened.
>
> The guy in charge was an "ivory tower" kind of guy who couldn't get something built.
>
> If we had known that the Apple II was going to continue to be as successful as it has been, we would not have speeded up so much on the Apple III.

Manufacturing engineers

> Design engineering just dumped the design on manufacturing and said, "Build it." They even went ahead and ordered parts for 10,000 machines.
>
> The company's experience with the Apple II, a much simpler machine, left us totally unprepared for the Apple III.
>
> As we were developing the III, we were also racing to keep up with Apple II demand.

Management

> The Apple III was a machine that an organization had to build, not a small group of individuals.
>
> The democratic system at Apple led to overexcitement and the premature introduction.
>
> The Apple II had taken the world by storm. We felt we could do most anything. We got cocky.
>
> The project was plagued by "creeping elegance," that is, never-ending improvements to the design. Once you have the idea, you should

lock the visionary in the closet and build the product.

The people who put the Apple III together weren't here to learn the lessons of the Apple II. They didn't appreciate the details put into Apple II.

No one in this valley had built computers in high volume. There was nowhere to find the experience.

Alternatives

In early February 1981, Wil Houde and the rest of the PCSD management team were reas-

EXHIBIT 6 Consolidated Balance Sheet

Assets

	September 30, 1979	September 26, 1980	March 27, 1981 (Unaudited)
Current assets:			
Cash and temporary cash investments	$ 562,800	$ 362,819	$ 74,086,000
U.S. government securities, at cost which approximates market in 1980 and at market March 27, 1981	—	2,110,710	1,800,000
Accounts receivable, net of allowance for doubtful accounts of $1,225,000 ($617,763 in 1980 and $400,000 in 1979)	9,126,010	15,814,000	34,538,000
Other receivables	52,301	1,557,000	7,887,000
Inventories:			
Raw materials and purchased parts	5,607,596	13,857,943	24,924,000
Work in process	3,727,008	9,625,234	14,838,000
Finished goods	768,113	10,708,443	13,090,000
	10,102,717	34,191,620	52,852,000
Prepaid expenses		70,066	1,989,000
Total current assets	19,843,828	54,106,215	173,152,000
Property, plant, and equipment, at cost:			
Land and buildings	—	242,851	2,084,000
Machinery, equipment, and tooling	404,127	2,688,787	6,017,000
Leasehold improvements	384,186	710,556	1,678,000
Office furniture and equipment	321,718	1,673,225	3,317,000
	1,110,031	5,315,419	13,096,000
Less: Accumulated depreciation and amortization	(209,824)	(1,311,256)	(2,808,000)
Net property, plant, and equipment	900,207	4,004,163	10,288,000
Leased equipment under capital leases, net of accumulated amortization of $572,000 ($205,358 in 1980 and $32,627 in 1979)	195,764	774,988	1,584,000
Cost in excess of net assets of purchased business, net of accumulated amortization of $39,785 ($13,193 in 1980)	—	514,592	488,000
Reacquired distribution rights, net of accumulated amortization of $782,326 ($90,022 in 1980)	—	5,311,304	4,619,000
Other assets	231,180	639,079	608,000
	$21,170,979	$65,350,341	$190,739,000

SOURCE: Company records.

EXHIBIT 7 Consolidated Balance Sheet

Liabilities and Shareholders' Equity

	September 30, 1979	September 26, 1980	March 27, 1981 (Unaudited)
Current liabilities:			
Note payable to bank	$ —	$ 7,850,000	$ —
Note payable	—	1,250,000	1,460,000
Accounts payable	5,410,879	14,495,143	23,458,000
Accrued liabilities	1,264,459	5,241,945	10,013,000
Accrued key employee bonus	456,000	554,000	2,174,000
Income taxes payable	1,879,432	7,474,170	—
Deferred taxes on income	2,051,000	661,000	1,446,000
Current obligations under capital leases	21,823	253,870	877,000
Total current liabilities	11,083,593	37,780,128	39,428,000
Noncurrent obligations under capital leases	203,036	670,673	1,179,000
Deferred taxes on income	204,000	951,000	3,086,000
Commitments and contingencies	—	—	—
Shareholders' equity:			
Common stock, no par value:			
160 million shares authorized, 55,237,000 shares issued and outstanding (48,396,928 in 1980 and 43,305,632 in 1979)	4,297,729	11,428,438	116,800,000
Common stock to be issued in business combination	—	920,210	—
Retained earnings	5,907,884	17,605,867	34,202,000
	10,205,613	29,954,515	151,002,000
Less: Notes receivable from shareholders	(525,263)	(4,005,975)	(3,956,000)
Total shareholders' equity	9,680,350	25,948,540	147,046,000
	$21,170,979	$65,350,341	$190,739,000

SOURCE: Company records.

sessing the Apple III. A variety of alternatives had been discussed which could be broadly encompassed by three categories: to abandon, replace, or correct the Apple III.

To abandon the machine would be quick and inexpensive, allowing reallocation of internal resources to other projects. There were approximately 6,000 units out the door, selling for about $4,000 each to the final customers. Apple could show character by publicly admitting its mistake, and vowing never to repeat it. Or the Apple III could die a quiet, unpublicized death with no promotion from the company.

Replacing the machine with a new one begged many questions. What would be new about the Apple IV, or whatever? Furthermore, millions in development costs would have been wasted. The casting for the case itself had cost over $200,000.

Correcting the machine would entail a continuation of the "Band-Aid" approach that had been followed inconclusively for months. Every major subassembly required some

EXHIBIT 8 Consolidated Statement of Income

	January 3, 1977 (Inception of Corporation) to September 30, 1977	Fiscal Year Ended			Six Months Ended	
		September 30, 1978	September 30, 1979	September 26, 1980	March 28, 1980 (Unaudited)	March 27, 1981 (Unaudited)
Revenues:						
Net sales	$ 773,977	$ 7,856,360	$47,867,186	$117,125,746	$43,090,000	$146,386,000
Interest income	—	27,126	71,795	775,797	546,000	4,178,000
	773,977	7,883,486	47,938,981	117,901,543	43,636,000	150,564,000
Costs and expenses:						
Cost of sales	403,282	3,959,959	27,450,412	67,328,954	23,813,000	82,617,000
Research and development	75,520	597,369	3,601,090	7,282,359	3,173,000	9,111,000
Marketing	162,419	1,290,562	4,097,081	12,109,498	3,478,000	16,378,000
General and administrative	76,176	485,922	2,616,365	6,819,352	1,939,000	8,519,000
Interest	5,405	2,177	69,221	209,397	13,000	737,000
	722,802	6,335,989	37,834,169	93,749,560	32,416,000	117,362,000
Income before taxes on income	51,175	1,547,497	10,104,812	24,151,983	11,220,000	33,202,000
Provision for taxes on income	9,600	754,000	5,032,000	12,454,000	5,785,000	16,606,000
Net income	$ 41,575	$ 793,497	$ 5,072,812	$ 11,697,983	$ 5,435,000	$ 16,596,000
Earnings per common and common equivalent share	$*	$.03	$.12	$.24	$.11	$.30
Common and common equivalent shares used in calculation of earnings per share	16,640,000	31,544,000	43,620,000	48,412,000	47,394,000	55,051,000

* Less than $.01.

EXHIBIT 9

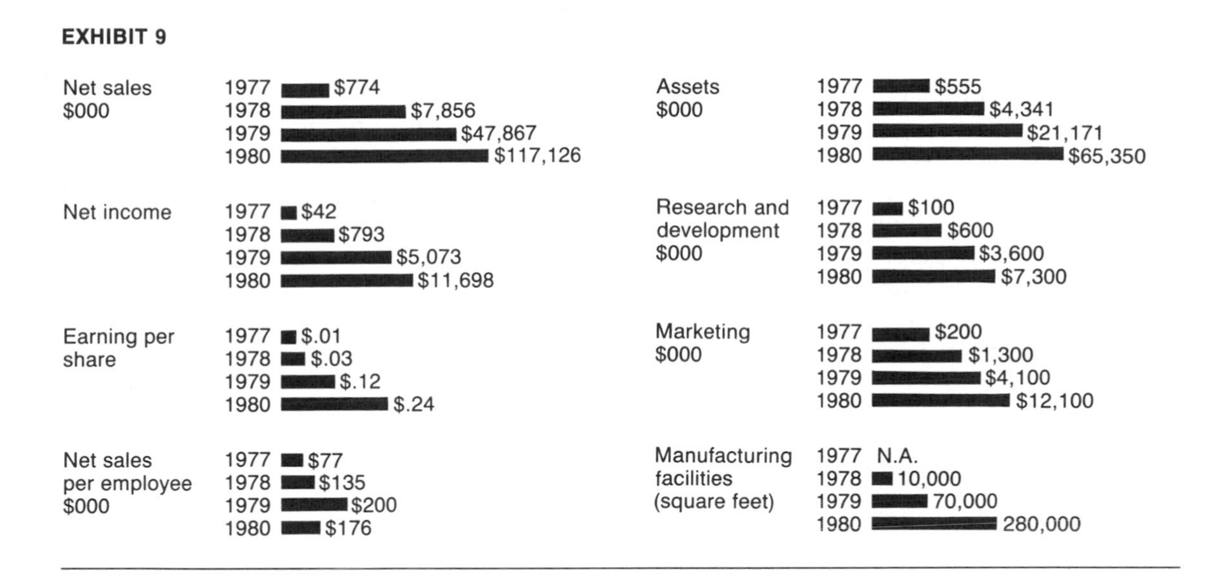

change, and no one knew yet how many changes were to come.

Even if Apple could correct the machine, they still had to convince the dealers and customers. Marketing estimated the cost of a reintroduction plan to be $5 million. Development of the machine itself had been only about $3 million. How should this be done? Should Apple repair old machines? Replace them? Or give dealers special upgrade kits?

Wil Houde hoped that the decision he and his colleagues were about to make at the executive committee meeting would be the right one.

Reading III–1
Communication between Engineering and Production: A Critical Factor

H. E. Riggs

Extensive communication between engineering and production is critical to implementing the firm's technical policy. Communication must be both formal and informal. Informal communication should be encouraged in all the ways that have become common in high-technology companies, from "beer busts" to technical symposia to off-site, multiday discussion sessions.

Regardless of the amount of informal communication, formal communication is also essential. The formal communication system between engineering and production must deal with three important, related, but distinct, challenges:

1. Introducing new products from the development laboratory to the production floor.
2. Providing the optimum—neither maximum nor minimum—level of documentation on existing products.
3. Facilitating orderly and cost-effective changes to products now in production.

INTRODUCING NEW PRODUCTS TO MANUFACTURING

Handing over the new product from engineering to manufacturing tests the cooperation and communication between engineering and production personnel as does no other activity. The high-technology company that manages this transition well stands to gain timing, cost, and quality advantages that can have substantial payoffs in the marketplace.

Where departmental barriers are high and engineers are encouraged or permitted to be myopic, design engineers will attempt to maximize product performance and manufacturing engineers will try to redesign the product to reduce its cost. Such a two-step process is highly inefficient and very time consuming. Optimizing across the conflicting priorities of cost and performance—often called value engineering—must be the responsibility of everybody engaged in new product creation, including marketing managers who have a hand in setting the new product's target specifications.

Extensive communication must both precede and follow the formal transition from engineering to production. Periodic product design meetings that involve design engineers, manufacturing engineers, and material planners (and sometimes product managers from marketing and others as well) should be held monthly during the early design stage and perhaps weekly just prior to and following the "hand over" of the new product from engineering to production. In these meetings, the design staff consults with manufacturing personnel regarding design alternatives under consideration and gains insight into the issues of producibility as the product or system is being designed. Realistic tolerances are specified, and engineers are encouraged not to tighten tolerances to expedite design or to gain an additional margin of safety. (Excessively tight tolerances almost always increase manufacturing costs.) Manufacturing should share with the design team its experience with present vendors and subcontractors as engineering is selecting sources for parts or processing for the new product or system.

Both manufacturing and engineering personnel must be aware that the design process is generally not complete when manufacturing commences. Design errors that need attention may be uncovered; change requests initiated by manufacturing to facilitate fabrication or assembly must be evaluated; and operating performance that met specifications in the laboratory but cannot be replicated on the manufacturing floor must be reassessed. Proper reliance on prototype units and pilot production runs before full-scale production is attempted can reduce costly errors.

Prototype, Pilot, and Production Runs

Engineering typically produces prototypes (the first one or two units of a new product or system). The engineers and design technicians construct them at considerable cost, frequently building and rebuilding them. They use techniques appropriate to the lab but inappropriate to full-scale production—"breadboards" instead of printed circuit boards and fabricated instead of cast metal or injection-molded parts—in order to facilitate design changes and minimize tooling costs and lead times. These prototypes, which are often necessarily quite different physically from the units ultimately supplied to customers, should be both thoroughly tested in the laboratory and subjected to some field testing. The purpose of prototypes is to prove design concepts and confirm product specifications.

Once the basic design concepts have been proven in prototype and satisfactory operating specifications have been met, a pilot production run (the production of a limited quantity) should be initiated. The design used for the pilot production run should be the one that is expected to be used in full-scale production—for example, breadboards are now replaced with printed circuit boards, and substantially more investment is made in tooling. The purpose of the pilot production run is to test product producibility and to work out any bugs in the final design before the company scales up to full production. (When total anticipated volume of the product or system is small, this pilot production step can be eliminated.)

In some companies, pilot production runs are undertaken by the engineering department and, in others, by the manufacturing department. The exact reporting relationship is not particularly significant. What is important is to recognize that pilot production runs are inevitably the joint responsibility of engineering and production.

Freezing Designs

Before full-scale production is undertaken, the design must be frozen, after which time formal engineering change notices are the only mechanism for effecting changes. In the absence of a pilot production run (and sometimes even with it), the point at which design becomes final, or frozen, is often unclear. At the prototype stage, the design must be allowed to remain fluid, permitting design changes at minimum cost and documentation. But it is human nature to seek almost endlessly for small improvements and refinements. This propensity is as true for the design engineer who is a parent of the new product as it is for the artist in her sculpture or the writer in his manuscript. Just as editorial changes are expensive to effect once the manuscript has been set in type, so product design changes are expensive to effect once manufacturing has commenced.

Thus, at some point, the design of the new product must be frozen, and both manufacturing and engineering must agree upon that point. Subsequent design changes can no longer be made unilaterally by the design engineer, as they could during the prototype phase.

You can often gain important timing and cost advantages by freezing certain portions of the design before other portions. For example, in a complex computer-based system, the selection of the system's minicomputer can and should be frozen long before other portions of the system are designed, in order to provide sufficient time for the programmers to develop the necessary software and for purchasing to negotiate OEM contracts with the minicomputer supplier. Sequential freezing is appropriate: Freeze parts or components known to have long procurement lead times early; leave standard components and those parts requiring little or no tooling unfrozen until late in the design cycle. A complex design project can usefully be subjected to PERT (program evaluation and review technique) analysis to reveal the critical design concepts or components that need early freezing. This process of sequential freezing of portions of the design implies close working relationships and much communication throughout the engineering organization and between engineering and production.

Top managers of high-technology companies should see to it that procedures for freezing designs are both established and adhered to.

Using a "Skunk Works"

Some technology-based companies have successfully used an unusual organizational technique to expedite new product design and introduction. When a new product requires (1) a number of engineering disciplines, (2) careful attention to manufacturability and cost, and (3) a telescoping of the design and introduction stages, a separate task force may be created, drawing personnel from a number of functional departments in the company. When this task force is assigned separate facilities, sometimes with extra security against industrial espionage, these facilities are often referred to as the "skunk works."

The objective is to recapture the advantage of the small company: high motivation, focused purpose on a single product, system, or process, and intensive and informal communication with a minimum of organizational barriers. The task force is accorded (or assumes) high prestige in the organization, and assignment to it is eagerly sought. Extra resources are typically made available to the task force.

Arguing against the establishment of a skunk works is the fact that creating one or a series of these task forces can be disruptive to the organization. Other development projects may be interrupted and key technical personnel assigned to the task force may be unavailable for informal counsel and advice on projects to which they are not formally assigned. Acceleration of the design can also cause some loss of efficiency.

This organizational device should be used only when competitive conditions demand fast action, either to protect an existing market position or to gain a jump on anticipated competition. Although the device has proved highly effective in a number of instances, resulting in a dramatic product unveiling that left both customers and competitors in awe, its overuse reduces the opportunities for specialization, economies of scale (experience curve economies), and routinizing of procedures.

A variation on this organizational device is to assign a group of engineers to "follow" a new design through the laboratory and onto the production floor. That is, rather than turning over its design (and prototypes) to manufacturing engineering, a portion or all of the engineering design team is assigned the responsibility for moving with the product from the design engineering organization to the manufacturing engineering organization. The tradeoff is that the "following" engineers will know the new product in detail, thus eliminating the need for manufacturing engineers to learn the new product, but will be less experienced and probably less capable in attacking the problems of producibility and tooling. However, a design engineer who has spent some time wrestling with new products from a manufacturing engineering viewpoint will be a more effective design engineer when he or she returns to the laboratory and another new product. Again, a type of job rotation has occurred.

Moving from Single to Multiple Products

Many emerging technical companies—that is, small but rapidly growing companies—encounter real turmoil as they move from relying on a single product to offering multiple products to the market. A small technology-based business focusing on designing and manufacturing a single product is often wonderfully efficient. It minimizes conflicting priorities because all hands are devoted to the single product. As the business grows and more product lines are added to the company's portfolio, choices must be made. The general management task suddenly becomes much more complex.

In engineering, the task of product maintenance engineering on the older products competes for attention with new product development. The need for standardization of components and subassemblies across product lines becomes evident. Compromises between standardization and optimum price/performance suddenly become necessary.

In manufacturing, quality problems that remained under control because of the undivided attention of manufacturing and engineering on the single product line now drift out of control as technical attention is diffused across many products. The existence of multiple products on the manufacturing floor complicates production scheduling. These products require both unique and common skills and often incorporate common parts or subassemblies that ought to be produced in larger lots.

The interaction between engineering and manufacturing was extensive on the company's first product line. This intimate, one-on-one communication needs to be continued on the newer products, but coordination on older products needs to be more routinized.

A key test for an emerging high-technology company is its ability to move successfully from engineering and producing a single product to engineering and producing a portfolio of product lines. The transition requires that a manufacturing engineering function be established, as well as a data base and reference system to aid in standardizing components. Task assignments in engineering must clearly recognize the dual responsibility of product maintenance and product development. A formal documentation and engineering change request system must replace the informal communication that sufficed when the company was small and produced only one product.

ENGINEERING DOCUMENTATION

Formal communication between engineering and production demands product and process documentation: drawings, bills of material (parts lists), schematics, assembly prints, software listings, and many other elements of paperwork (and now, increasingly, microfiche, computer data bases, videotapes, and other media). Most of these communication media are created by engineering and represent the detailed specification of the product to be produced or the process to be operated.

Level of Detail

A persistent dilemma facing management in high-technology companies is the decision of just how much detail to incorporate into the documentation of particular products and processes. Detailed documentation, taking the form of prints, parts lists, assembly drawing, process and assembly instructions, and sometimes audio, video, and other nonprinted media, is expensive to create, control, and update. However, skimpy documentation may be risky, allowing design changes to be effected without thorough review. Such incomplete documentation may also inhibit accurate and complete communication among the functional departments of the business.

The dilemma is resolved primarily on the basis of the relative importance the high-technology company places on manufacturing flexibility and product costs. Very detailed documentation is required when (1) production volumes are high, (2) automation and tooling are relied upon to reduce costs, and (3) less-skilled manufacturing labor is to be utilized. More elaborate documentation is a prerequisite to the aggressive pursuit of learning curve economies. Such elaborate documentation is not justified, however, when volumes are small, a skilled work force can be relied upon to operate with limited instructions, and design changes are implemented at a rapid rate. As a general rule, more documentation is appropriate, justified, and necessary as one moves along the continuum from custom to standard products.

High-technology companies most frequently err on the side of too little documentation. This tendency is not surprising. In the early stages of the life of products and technologies, a minimum of documentation is appropriate. As the company, products, and technologies mature, there is a reluctance to invest engineering time and attention in paperwork on existing products rather than in designing new products. Companies that neglect documentation, however, find they are forever running to catch up with the required documentation.

General managers must strike the proper balance between too much documentation and too little. Despite protests to the contrary from most manufacturing managers, more complete and thorough documentation is not always appropriate. The right balance is a function of the overall business strategy and of the position of the particular product or product family within the product-process matrix. When the strategy is geared to a succession of new, high-technology products, skimpy documentation is both appropriate and cost effective. When the strategy depends upon achieving learning curve economies—the company is operating down and to the right on the product-process matrix—complete, up-to-date, and reliable documentation is essential.

Effects on Inventory Control

Effective inventory planning and control requires very accurate bills of material (that is, listings of individual parts, components, and subassemblies that go into a finished product). Inaccurate or incomplete bills of material preclude using sophisticated planning techniques, such as MRP. The result is that excessive raw material inventories are held in order to guard against shortages. Moreover, the omission of one or more parts from a bill of materials can cause a halt in the assembly process while the missing part is located. The result is that in-process inventories also balloon. Thus, improved inventory control in high-technology companies, an objective stressed repeatedly throughout this book, requires the active participation of engineering, as well as of the production and finance departments.

PROCESSING ENGINEERING CHANGES ON EXISTING PRODUCTS

Life in a technology-based business would be substantially simplified if all documentation, once created, could be relied upon to be both accurate and stable. Neither condition is easily achieved when both technology and product change is an ever-present fact of life. All engineering changes, whether to improve performance or to reduce costs, must be reflected in changed documentation. In addition, design errors uncovered by engineering or manufacturing personnel (and sometimes by field service personnel) must be corrected and the corrections incorporated into the documentation system.

Thus, requests for changes to existing products can—and should—emanate from all corners of the organization:

1. From engineering to take advantage of new technology or to incorporate a new product feature.
2. From the field service organization to improve reliability or to facilitate field repair.
3. From purchasing to take advantage of a new supplier or a lower price of a substitute component.
4. From marketing to improve the competitive posture of the product.
5. From manufacturing engineering to permit the use of more sophisticated tooling.
6. From production and inventory planning to permit standardization of components across product lines.
7. From production supervisors to reduce tolerances, and thereby costs, or to facilitate processing or assembly in some other way.

Just as requests for changes can emanate from all corners of the organization, so implemented changes affect all corners of the organization, including particularly purchasing, inventory control, marketing, field service, production supervision, and cost accounting. Because these organizational units will be affected by the change, they must have a hand in deciding whether the requested change should be adopted (and when), and they must be notified in a timely fashion of approved changes.

The number of change requests may be very high—in the tens for a simple product, the hundreds for a complex instrument, and the thousands for a comprehensive system. Each change is likely to have a ripple or domino effect on documentation. For example, the change of a single component may require a change in the drawing on which it first appears, on one or more bills of materials, on drawings of parts or assemblies farther up the product tree, on assembly instruction sheets, and so forth. Each change may have both obvious and not-so-obvious consequences; these need to be anticipated, evaluated, and, if appropriate, tested.

Technical companies should develop and institute formal procedures and paper flow systems to be certain that all necessary documents are changed as required, that changes do not become incorporated into the documentation before they are appropriately authorized, and that all affected individuals and groups within the organization are aware of the nature and effective date of the change in sufficient time to adapt accordingly. The process must be both rapid and thorough, but it also must be routine, if production and engineering activities are not to grind to a halt either as a result of a preoccupation with processing changes or a lack of coordination among the changes themselves.

Discipline must be built into the engineering change request system so that procedures are not short-circuited. If control of documentation is lost, the following conditions can occur:

1. Quality problems multiply as exact specifications of components become impossible to trace and unanticipated consequences arise from unauthorized design changes.
2. Inventory investments and write-offs increase as parts are made obsolete without notice and the production cycle lengthens because newly specified parts are not planned and acquired in a timely manner.
3. Manufacturing labor costs escalate as expediting, troubleshooting, and additional setups consume both direct and indirect labor-hours.

Proper handling of engineering changes is the bugaboo of documentation methods in many high-technology companies.

COST AND BENEFIT TRADE-OFFS

All changes have both benefits and costs, even those that simply correct drawing errors. The challenge is to make the proper trade-off. Engineering changes that alter the physical specifications of particular components may render obsolete present components now in inventory and necessitate rescheduling of manufacturing work orders or purchase orders with vendors. Such obsolescence and rescheduling costs must be weighed against the advantages to be achieved from the change to decide both if and when the change should be effected. The optimum decision is often to delay the change until present inventories are depleted, until new vendors can be brought on stream, or until other conditions occur that will minimize disruption.

Some changes—for example, in computer software—must be expedited to fix a bug in a particular program, with notification rushed to various parts of the organization and to customers. Other changes in the software—changes designed to enhance capabilities or improve execution efficiencies—should be saved up and incorporated with other alterations in periodic rereleases of entirely new generations of software. Changing software documentation is expensive, and such changes typically require changes in operating, training, and service manuals as well. Batching changes may be efficient, but this advantage must be weighed against the disadvantage of delaying the introduction of an improved product to the marketplace.

The initiator of a change request may be unaware of the full ramifications of the proposed change. A change in part M may require an adaptation of part P or assembly T, expensive reworking of tooling, or a change in maintenance procedure that must be communicated to customers and the field service force. The possibility that the benefit sought from the engineering change request could be more expeditiously accomplished by an alternative change must be evaluated. For example, a problem that could be corrected by a hardware change might also be correctable by a less-expensive software change.

Evaluate all changes on the basis of costs and benefits. Making the trade-offs between the costs and the benefits of change is complicated within most high-technology companies by the fact that the relevant data on both costs and benefits are not readily available to the decision maker. Manufacturing cost penalties or savings may be ascertainable (although even here most cost accounting systems do not reveal the incremental costs), but the tangible and intangible benefits or costs associated with changes in competitive position, in ease of field maintenance, or in vendor relationships are often uncertain. The costs of effecting the change—engineering time, clerical effort on documentation, renegotiation by purchasing, and possible scrapping of inventory—must be factored into the decision.

Engineering change requests must be routed for approval through each affected department: design engineering, manufacturing engineering, material planning, quality assurance, and field service. (In some companies still other departments should formally approve changes.) Each of the evaluators must be alert to the possible need to solicit input from other functions, such as marketing or finance. Checklists and rules of thumb may help streamline the process. An engineering change committee, which is responsible for making the final cost-benefit trade-off when disputes arise, should be constituted.

NEW MODELS VERSUS INCREMENTAL CHANGES

I spoke earlier of the importance of freezing new product design, and now I have suggested that engineering changes may occur in large numbers. What factors should management consider in deciding how much product evolution to permit through the engineering change request mechanism?

First, saving up (or batching) engineering change requests in order to effect many changes at one time can have distinct advantages in reducing implementation costs. Disruptions in both production and engineering are minimized.

More important advantages often attend the introduction of a brand new model or line of a product. First, the company's image in the marketplace may be enhanced when it introduces a new product or model that delivers significantly improved performance. The opportunity may exist to leapfrog the competition. A series of incremental changes may not have the same marketing impact on customers as the introduction of a new product, and competitors may be better able to react to, and sell against, a series of small improvements. When these conditions are present—as they usually are—management should restrict product evolution through small, incremental changes, even when such changes would result in some improvement in performance.

The engineering staff may benefit from an opportunity to start over on a product line, to incorporate new technology or new design concepts that cannot be utilized given the constraints of the present product. Such starting over is, of course, expensive, but new competitors entering the market are not constrained by present products. Thus, if the removal of such constraints represents an important design advantage, management should be certain that its own design engineers are not denied that advantage. For example, the full benefits of a new software language probably cannot be realized without starting over, and the maximum benefit of VLSI circuits is not realized by designing incrementally from present products.

Relatedly, new models or product lines, rather than incrementally improved present products, often permit adoption of manufacturing techniques that provide the company with significant cost and quality advantages. The use of robots in fabrication and assembly typically requires some product redesign to make optimum use of the robots' capabilities. The opportunities for automation may not be evident or, if evident, may not be economically justified if product design is accepted as a given. The concept of the product-process matrix presumes that both product design and process design are subject to changes and that the changes can and should be related.

The case should not be overstated, however. A market leader, such as IBM in mainframe computers, may need to pay particular attention to thwarting competitors' attempts to copy (often referred to as "reverse engineer") its products. A continuing series of well-planned design changes can severely complicate the process of reverse engineering and permit the leader to sustain a technological and performance edge over its competitors.

ALLOCATION OF ENGINEERING RESOURCES

Related to this problem of new products versus incremental changes is the inherent risk in high-technology companies that excessive engineering resources will be diverted from the truly new product to service the existing products. New products are the lifeblood of such companies; the more the company relies on technology to differentiate itself from competitors, the more this statement holds. Two sources of diversion are prevalent: product line maintenance and customer "specials."

In this chapter, I have been emphasizing that continuing engineering of existing products is not unimportant, particularly as production seeks to improve product manufacturability and reduce its costs and as the need for improved documentation is realized. But such maintenance must not be permitted to consume all engineering resources.

Customers' requests for product modifications to meet their particular requirements consume precious engineering resources. The more the company accommodates such requests for specials, the more the company takes on the aspects of an engineering consulting firm rather than a manufacturing company. When important customers make such requests, they may have to be accommodated. But too often technology-based companies drift into producing increasing numbers of specials when such activity is clearly not consistent with their overall strategy.

The balanced allocation of engineering resources, assuring adequate attention to the development of truly new products, is an important challenge to general managers. When the dominant view in the councils of management is production, excessive investment in product maintenance engineering will result. When the dominant view is marketing, excessive pressure for accommodating customers' requests for specials may result—to be followed soon by dissatisfaction at the slow pace of new product development. When the dominant view in management councils is development, essential product maintenance engineering may be shortchanged and very attractive opportunities for specials may be overlooked. No such myopic views can be permitted to dominate.

■ Communication between production and engineering is particularly intense, and often necessarily nonroutine, in connection with introducing new products onto the production floor from the engineering laboratory. Prototype and pilot production runs can assist in the transfer, as can mutual agreement on timely freezing of the design. The more dependent the company is on process technology, rather than state-of-the-art product technology, the more thorough must be the product and process documentation. Because documentation is both expensive and difficult to control, high-technology companies typically underemphasize it. To maintain careful control of products, processes, quality, costs, and inventory investments, you must subject suggested changes in existing products to strict and well-defined procedures to be certain that the myriad potential ramifications of the change are fully evaluated. In formulating its engineering change policy, the high-technology company should consider the trade-off between introducing a new model and permitting product evolution by means of a series of incremental changes. The policy must also assure that engineering resources are not so committed to product maintenance and customer specials that new product development is shortchanged.

The New Product Development Learning Cycle

Case III–2
Apple Computer: The First 10 Years

C. C. Swanger and M. A. Maidique

Apple Computer is a vintage story of American technological entrepreneurship. From the now-famous story of Steve Jobs and Steve Wozniak selling their calculator and van in 1975 to build themselves a microcomputer, to the $1.5 billion in revenues Apple expected to realize during its 1984 fiscal year, Apple's extraordinary success could hardly be questioned. (See Exhibit 1 for selected company financial data.)

The evolution of the company makes a fascinating story. One particularly interesting area was the evolution of the Apple product line and the related waves of renewal that each of Apple's new products had prompted. There had been several successes, but some interwoven failures as well. This posed a complex dilemma, i.e., how should Apple continue to develop as a company and increase the likelihood of product successes?

As Apple matured, so had the personal computer industry. In 1984, the competitive situation was much more complex and the stakes much higher than they had been in 1975. Market growth continued to be strong, but the competitive environment was drastically different.

Numerous firms had entered the personal computer industry during the last decade, including such giants as IBM, Digital Equipment Corporation, and Hewlett-Packard. Most notably, in 1983, IBM had surpassed Apple as the leading personal computer company in terms

EXHIBIT 1 Selected Financial Data

(Fiscal Year-End September 30)

	1977	1978	1979	1980	1981	1982	1983	1984
Net sales ($000)	774	7,856	47,867	117,126	334,783	583,061	982,769	1,515,876
Net income ($000)	42	793	5,073	11,698	39,420	61,306	76,714	64,055
Earnings per share ($)	.01	.03	.12	.24	.70	1.06	1.28	1.05
Assets ($000)	555	4,341	21,171	65,350	254,838	357,787	556,579	788,786
Research and development ($000)	100	600	3,601	7,282	20,956	37,979	60,040	71,136
Marketing ($000)	200	1,300	4,097	12,619	55,369	119,945	229,961	392,866
Cash and temporary cash investments ($000)	24	775	563	363	72,834	153,056	143,284	114,888
Current liabilities ($000)	N/A	N/A	N/A	37,780	70,280	85,756	128,786	255,184
Noncurrent liabilities ($000)*	N/A	N/A	N/A	1,622	7,171	14,939	49,892	69,037
Ratios:								
Gross margin (%)	N/A	N/A	42.7	42.5	46.1	50.6	48.5	42.0
PAT (%)	5.4	10.1	10.6	10.0	11.8	10.5	7.8	4.2
R&D/Sales (%)	12.9	7.6	7.5	6.2	6.3	6.5	6.1	4.7
Marketing and distribution/ Sales (%)	25.8	16.5	8.6	10.8	16.5	20.6	23.4	25.9
Sales/Employee ($000)	77	135	200	176	190	200	212	282

* Noncurrent liabilities consisted of obligations under capital leases plus deferred income taxes. Apple Computer had no long-term debt.

of sales dollars (though Apple still led in units shipped). Numerous bankruptcies had occurred, particularly among start-up firms (Osborne, Franklin, Victor, and others) and many other firms were in trouble financially.

Many questions remained to be resolved regarding Apple's future direction. Developing a new plan required dealing with several complex interconnected issues such as: (1) what to do with the Apple II family; (2) where to go with the Macintosh family; (3) product family integration issues; (4) the role of accessory products; (5) meeting the IBM challenge; (6) managing Apple's growth while maintaining the distinctive culture and working environment; (7) the pursuit of new areas, such as lap-size computers, educational computers, systems, etc.; and (8) evaluating possible diversification alternatives.

COMPUTER INDUSTRY HISTORY

Mainframes

The origins of computers date back to the 1800s when Charles Babbage, George Boole, and Charles Sanders Pierce individually worked on switching circuits and advanced algebraic techniques. As the switching circuits progressed from mechanical to electromechanical during the 1920s and 1930s, to vacuum tubes during the 1940s, the possibility of building commercially useful computers rapidly increased.

Following World War II, the modern computer industry emerged, initially led by the Remington Typewriter Company, which later became Sperry Univac. When International Business Machines (IBM) realized computers

EXHIBIT 2 Mainframe and Minicomputer Manufacturers Market Shares

A. Mainframe Computer Manufacturers (percent of market)

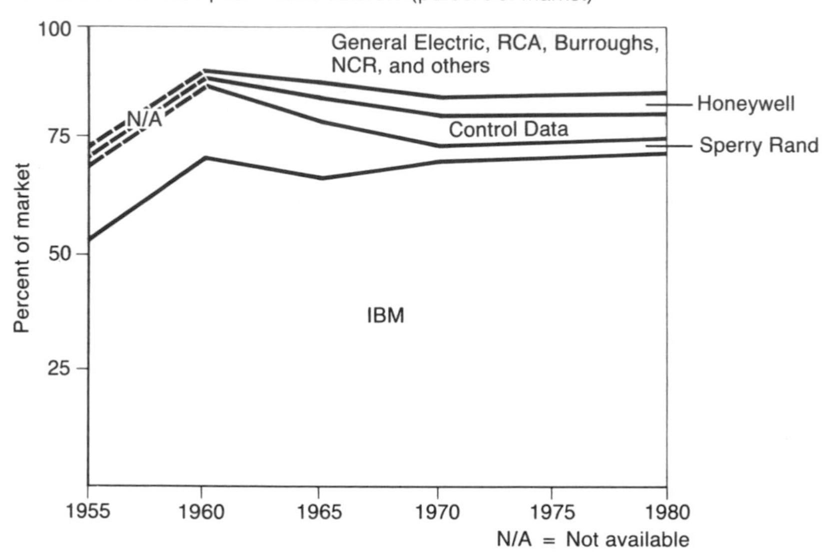

SOURCE:
1. 1955–1970: Gerald W. Brock, *The U.S. Computer Industry: A Study of Market Power* (Cambridge, Mass.: Ballinger Publishing Company, 1975), pp. 21–22.
2. 1980: Casewriters' estimate.

B. Minicomputer Manufacturers (percent of market)—General-Purpose Microcomputers

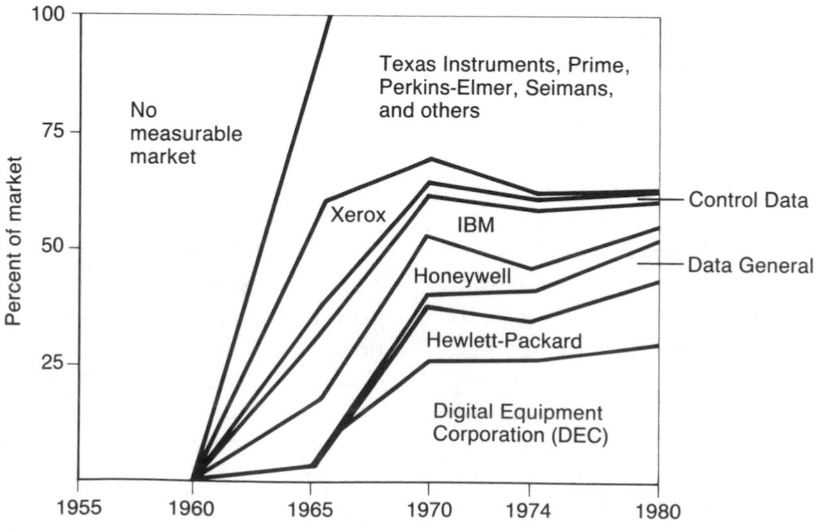

SOURCE:
1. 1960–1974: Montgomery Phister, *Data Processing Technology and Economics*, 1976, p. 291.
2. 1980: Dataquest, Inc., San Jose, California.

were going to take away some of their customers for their large-scale adding machines (such as the U.S. Census Bureau), IBM entered the business. By the late 1950s, Control Data Corporation, an IBM spin-off, was established, and Honeywell, Burroughs, General Electric, RCA, and National Cash Register (NCR) were also producing computers.

IBM quickly came to dominate the industry with a market share that had risen to roughly 70 percent by the early 60s. (See Exhibit 2.) The computer industry was often referred to as "Snow White" (IBM) and the "Seven Dwarfs" (the other competitors).[1]

[1] Gerald W. Brock, *The U.S. Computer Industry: A Study of Market Power* (Cambridge, Mass.: Ballinger Publishing Company, 1975), pp. 3–24; and Katherine Davis Fishman, *The Computer Establishment* (New York: Harper & Row, 1981), pp. 19–47.

In addition to IBM dominance, the computer industry was characterized by continuing and dramatic technological advances, which resulted in increasing computer speed, power, and memory while decreasing costs. (See Exhibit 3.)

Minicomputers

In the late 1940s and early 1950s, a sequence of technological breakthroughs resulted in the diffused transistor, a microscopic arrangement of crystal layers with electrical properties similar to a vacuum tube amplifier. This development led ultimately to the invention of the integrated circuit around 1960. IBM and most of the other existing computer manufacturers used these advances to increase the speed and power of their computers while reducing the

EXHIBIT 3 Trends in Microelectronics

Trend in density:
Maximum number of components per electronic circuit:
1959 = 1; 1969 = 1,024; 1979 = 1 million; 1985 = over 50 million?
Maximum number of binary digits (bits) per memory chip:
1970 = 1,024; 1980 = 65,536; 1985 = over 500,000?

Trend in speed:
Speed of an electronic logic circuit:
Mid-1950s (vacuum tube circuit) = one microsecond.
Early 1960s (transistorized printed circuit) = 100 nanoseconds.
Late 1970s (integrated circuit chip) = 5 nanoseconds.
Mid-1980s (integrated circuit chip) = 1 nanosecond?

Trend in cost:
Cost per integrated circuit chips:
1964 = $16; 1972 = 75¢; 1977 = 15¢; 1985 = 1¢?
Cost per bit of integrated circuit memory chip:
1973 = 0.5¢; 1977 = 0.1¢; 1985 = 0.005¢?

Trend in reliability:
Reliability of electronic circuits:
Vacuum tube = One failure every few hours.
Transistor = 1,000 times more reliable than vacuum tube.
Integrated circuit = 1,000 times more reliable than transistor.

SOURCE: James A. O'Brien, *Computers and Information Processing* (Homewood, Ill.: Richard D. Irwin, 1983), p. 25.

power consumption. Other firms concentrated on using the new technology to design smaller, cheaper computers.

The first firm to commercialize a smaller, cheaper computer was Digital Equipment Corporation (DEC). In 1967, DEC introduced the first "minicomputer," so named after the popular miniskirts of that time. DEC's machine sold in the tens of thousands of dollars and took up much less space than the mainframe computers. Soon thereafter, Hewlett-Packard became the other major minicomputer manufacturer. IBM, however, did not enter the market until much later.[2]

Personal Computers

In the late 1960s and early 1970s, the trend to continually smaller and less expensive computers was clear. The development of the microprocessor, a fingernail-sized computer central processor, in 1971 provided a key catalyst to the industry.

A number of engineers in the established mainframe and minicomputer firms approached top management with proposals to develop what would have become personal computers. They knew smaller computers could be developed, and their experiences of having to wait to use the larger computers and not be able to work at home provided the real motivation. For a variety of reasons, the major firms chose not to pursue the proposals during the early 1970s. DEC, for example, was one of the most likely candidates to develop personal computers because they already made the smallest computers available, i.e., computers small enough to fit inside the trunk of a car. DEC, however, was interested in industrial markets and had not mastered the art of selling to individual consumers.

Eventually, some engineers and programmers struck off on their own to develop smaller computers. Throughout most of the 1970s, and still to a large extent during the early 1980s, personal computers were developed by entrepreneurs. However, by the mid-1980s, many industry observers believed the industry had entered an era where size and economy of scale were key to survival, thus sharply reducing the possibilities of new Apple-like garage start-ups.

From its modest beginnings as a recognizable industry in the mid-1970s, the personal computer industry reached estimated annual sales of 1.5 million units during 1982, representing revenues in excess of $3 billion. Those sales came from four primary markets: business (975,000 units or 65 percent), science (240,000 units or 16 percent), home (225,000 units or 15 percent), and education (60,000 units or 4 percent).[3] The 1982 factory shipments or wholesale revenues consisted of $.7 billion from units generally priced under $1,500 and used primarily in the home, and $2.6 billion from units generally selling from $1,500 to $5,000 and used in offices. *Business Week* forecasts projected $2 billion and $7 billion revenue, respectively, in 1984 from the two categories.[4] (See Exhibit *4a*.) Thus, the personal computer industry was forecast to grow at about 65 percent per year from 1982 to 1984.

Three of the initial personal computer developers (MITS, IMSAI, and Process Technology) participated in the business primarily in 1976. From 1978 to 1981, Apple Computer, Inc., Radio Shack (owned by Tandy Corporation), and Commodore Business Machines were the primary manufacturers of personal computers. Radio Shack had the largest market share due primarily to its extensive worldwide network of 8,100 company-owned retail stores. However, Radio Shack's market share declined

[2] Ibid.

[3] "Personal Computers," *Scientific American,* December 1982, pp. 96, 99.

[4] *Business Week,* June 25, 1984, p. 106; and August 15, 1983, p. 89.

EXHIBIT 4*a* **U.S. Factory Shipments of Personal Computers**

Units generally priced *under $1,500* and used primarily in the home:

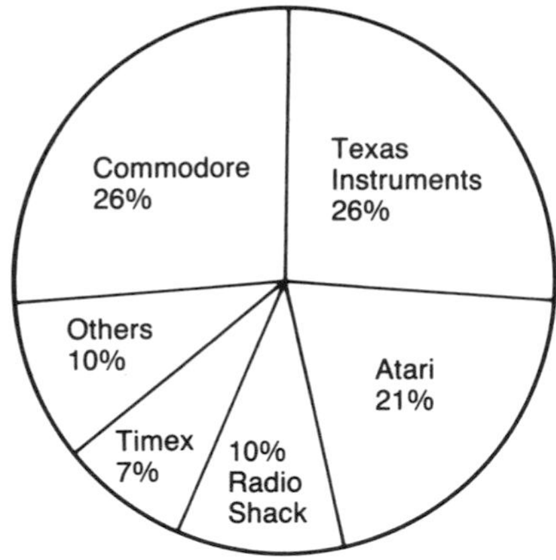

1982 total: $.7 billion

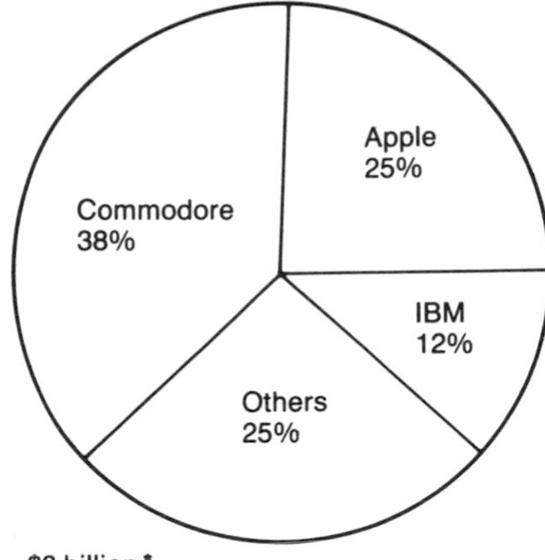

1984 total: $2 billion*

Units generally selling from *$1,500 to $5,000* and used in offices:

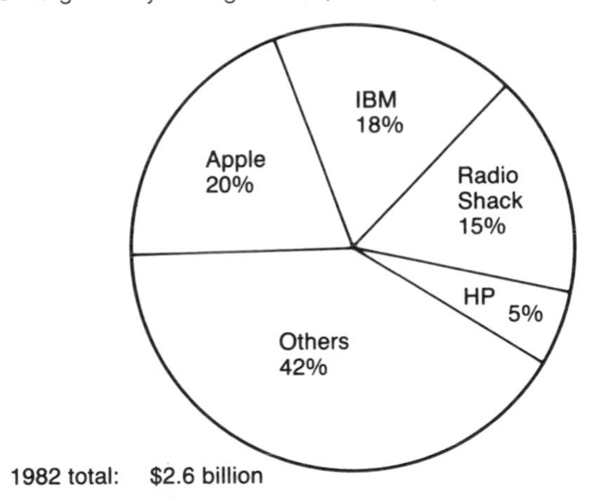

1982 total: $2.6 billion

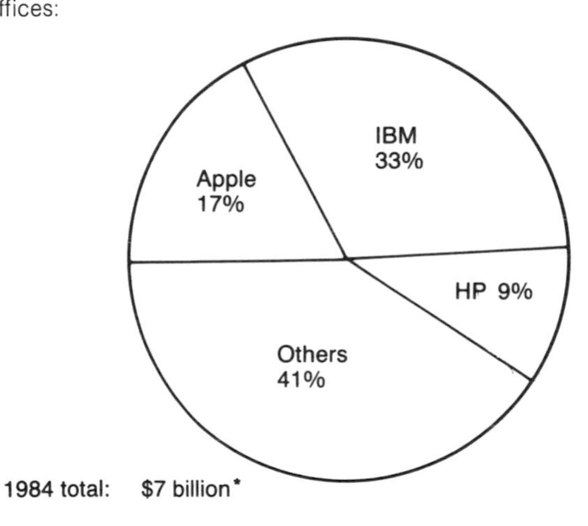

1984 total: $7 billion*

SOURCE: *Business Week,* June 25, 1984, p. 106
* Forecasts.

steadily, and in 1981 Apple surpassed them in number of units shipped. In addition, while Commodore continued to enjoy broad retail distribution internationally, its U.S. share dropped also. (See Exhibit 4*b*.)

IBM entered the personal computer industry in August 1981, and precipitated dramatic changes in market share and the nature of the industry. As evidence of IBM's commitment and objectives, the company named its computer the IBM Personal Computer, or PC for short. IBM also broke with historic company tradition in several key areas, utilizing outside suppliers and software developers, standard parts, and the so-called "Woz principle" (the principle espoused so strongly by Apple co-

EXHIBIT 4*b* **U.S. Factory Shipments of Personal Computers Excluding Peripherals and Software**

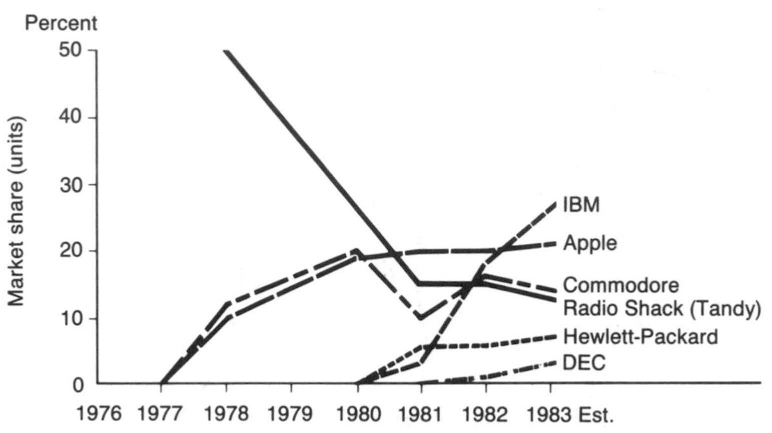

SOURCES: *Scientific American*, December 1982, p. 99.
Business Week, October 3, 1983, p. 77.
Casewriter's estimates.

EXHIBIT 4*c*

U.S. Market Share of Personal Computer Systems Sold (based on dollar retail value)

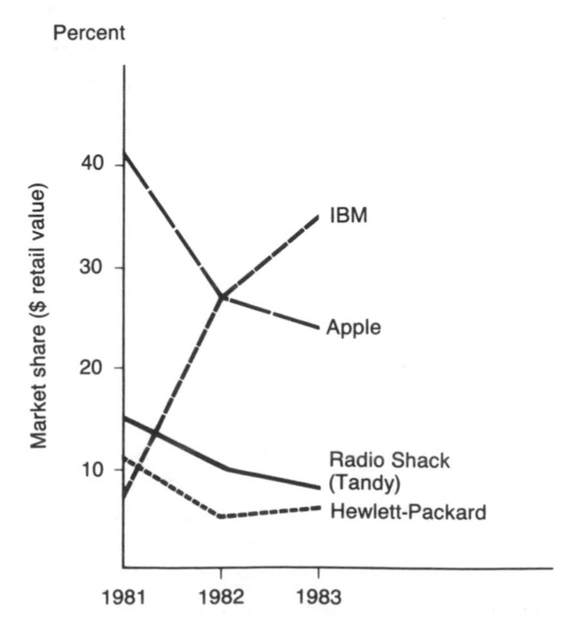

SOURCE: *Business Week*, January 16, 1984, p. 79.

U.S. Personal Computer Production

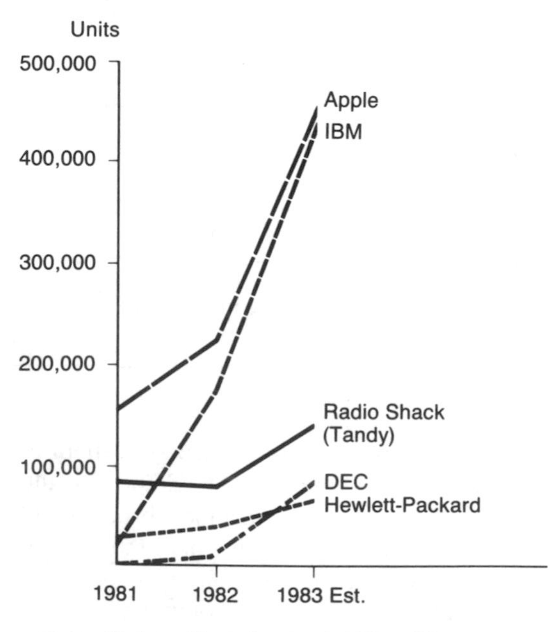

SOURCE: *Business Week*, October 3, 1983, p. 76.

founder Steve Wozniak of open computer architecture that allowed independent software developers to write programs for the machine).[5] By the end of 1983, IBM had surpassed Apple in dollar volume, though Apple led slightly in units shipped. (See Exhibit 4*c*.)

With projected growth rates of 100 percent-plus a year, the personal computer market was a major attraction to a number of firms in the early 80s. In fact, at the beginning of 1983, over 100 companies were each vying for 10 percent to 20 percent of the market. The shakeout that had been predicted for 1985 began in 1983. It was particularly severe in the home computer segment of the personal computer industry.

Apple was the only major competitor whose entire business was built around this single industry. (Even with a quarter of the market in 1983, IBM's personal computer division would account for no more than 5 percent of IBM corporate sales.) Thus, for Apple, its new product activities were not only a response to additional opportunity but a defense of the core of its business.

ORIGINS OF APPLE COMPUTER

Apple Computer Company had been founded in 1975 as a partnership by Steve Wozniak and Steve Jobs. Wozniak, 24, had been working at Hewlett-Packard, designing handheld calculators, while Jobs, 20, had been designing video games at Atari. Their first product, the Apple I, evolved from their interest in having their own microcomputer. It cost about $10,000 to develop, which they raised from their own savings and by selling a calculator and a van. Once they had built one for each of themselves, they showed it to other members of the Home Brew Computer Club in Palo Alto, and several people wanted one. At first, they built units in their garage but in 1976 decided to farm out the printed circuit board (PCB) operations and to do only the final integration and testing of the finished product.

THE APPLE II

By the fall of 1976, the Apple II had been developed with Wozniak designing the majority of the internal workings and Jobs defining the overall concept and appearance.

Jobs wanted to strip the computer of its mysterious and sometimes threatening reputation. This led to a low profile, a plastic case, and no blinking lights or confusing knobs and dials. In many ways, this design philosophy made the Apple II the first truly successful "personal computer."

While similar to the Apple I, the Apple II included additional circuitry, a built-in, Teletype-style keyboard, BASIC in ROM (Read-Only Memory), high-resolution color graphics, and a 16K memory, all enclosed in a plastic housing. (See Exhibit 5 for product specifications.) On the other hand, the manufacturing process in the Apple II was essentially the same process originally worked out in the garage for the Apple I. Jobs recalled:

> We had learned a lot about product design from the Apple I. We learned that 8K bytes of memory really weren't enough. We were using the new 4K RAMS (Read and Write Memories). At that time, no one else was using RAMS. Going out with a product using only dynamic memories was a risky thing.[6]

With the design and a prototype in hand, Jobs and Wozniak approached Hewlett-Packard and Atari to see if they might be interested in involvement in the project. For a variety of reasons, neither of those firms chose to pursue it, although Commodore did come very close to reaching an agreement with the two partners.

---

[5] Paul Freiberger and Michael Swaine, *Fire in the Valley: The Making of the Personal Computer* (Berkeley, Calif.: Osborne/McGraw-Hill, 1984), p. 280.

[6] "Apple Computer (A)," Stanford University Graduate School of Business, Case S-BP-229A, revised June 1983, p. 2.

EXHIBIT 5 Product Specifications

	Base List Price	Microprocessor	RAM	ROM	Operating System
Apple I	N/A		8K		
Apple II	N/A	8-bit	16K	Including BASIC	DOS 3.3
Apple III and Apple III plus	N/A $2,995	8-bit 8-bit	96K 128K	4K, for self-test diagnostics	DOS 3.3; "Sophisticated Operating System" (SOS)
Apple IIe	$1,224	8-bit, 6502 micro-processor	64K	16K bytes including Apple-soft BASIC, self-test, and 80-column routines	DOS 3.3 Pro DOS
Lisa 2 Lisa 2/5 Lisa 2/10	$3,495 $4,495 $5,495	32-bit, Motorola 68000	512K 512K and 5MB hdd* 512K and 10MB hdd*	64K bytes	No name given: Apple wants user to think of operating system as "invisible"
Macintosh	$2,195 $3,195	32-bit, Motorola MC 68000	128K bytes or 512K bytes	64K bytes	No name given: same reasons as for Lisa
Apple IIc	$1,295	65CO2 CMOS version of 6502	128K bytes	16K bytes including Apple-soft BASIC and Mousetext	Pro DOS

	Graphics	Disk Drive	Other	Documentation
Apple I			Single printed circuit board; no keyboard; no plastic housing	None
Apple II	Color graphics		Teletype-style keyboard; upper case only; 40-column display	
Apple III and Apple III plus	280 × 192, 16 colors (limited); 140 × 192, 16 colors; 560 × 192 monochromatic; Apple II modes	One 5¼-inch floppy built in 140K bytes per diskette	Aluminum case; 60-key keyboard plus 13-key numeric pad; 80 or 40 character wide columns; upper and lower case letters; Audio-Video output	Instruction manual

Apple IIe	40 × 48, 16 colors; 280 × 192 bit-mapped array (can provide, e.g., 560 × 192 monochromatic limited; 280 × 192 monochromatic; 140 × 192, 4 color)	One 5¼-inch floppy	63-key keyboard; upper and lower case; 7 expansion slots; video interface	Manuals; tutorials
Lisa 2		One 3½-inch floppy w/400K bytes	Mouse; 2 serial ports; video cable	
Lisa 2/5	512 × 342 pixel bit-mapped display; free-hand drawing	Same as Lisa 2, plus 5MB hdd*		Manual; interactive tutorial diskettes
Lisa 2/10		Same as Lisa 2, plus 10MB hdd*		
Macintosh	512 × 342 pixel bit-mapped display; free-hand drawing	One 3½-inch disk drive with 400K bytes	Mouse; 2 serial ports; video cable; numeric pad	Manual; cassette plus tutorial diskettes and tape cassettes
Apple IIc	6-color graphics with 280 × 192 resolution; 16-colors with 560 × 192 resolution	One 5¼-inch low height floppy disk drive	Two keyboard layouts (standard QWERTY, or Dvorak); video cable; 2 serial ports	Manual and six interactive tutorial diskettes

	Options	Software
Apple I	—	—
Apple II	Plug-in cards for word processing; 64K maximum memory	
Apple III and Apple III plus	Emulator for All 3 floppy disk drives	Can run majority of Apple IIe software. Included is Information Analyst: word processor, business BASIC, Visicalc
Apple IIe	128K maximum memory; up to 6 140K byte; 5¼″ floppy disk drives	Runs most Apple II software; Applewriter IIe word processor; Quickfile IIe database system
Lisa 2	1 megabyte memory	
Lisa 2/15		Can run most Macintosh software
Lisa 2/10	All Lisas can emulate Macintosh	
Macintosh	Second disk drive; modems	Included is MacWrite, MacPaint, MacFile
Apple IIc	Mouse; flat-panel LCD display; second disk drive; modems	Can run most Apple IIe software

* hdd = hard disk drive.

Finally, Jobs decided that a formal corporate organization would need to be developed and that outside financing would be required. In 1977, Mike Markkula, who had previously been employed by Fairchild Semiconductor and Intel but was now "retired," put up $91,000 to supplement the $2,600 each put up by Jobs and Wozniak. Even more important than the money, Markkula agreed to work with Jobs and Wozniak to formalize a strategy and to draw up a business plan. Their concept was to use microprocessor technology to create a personal computer for individuals to use in a wide variety of applications. Another key contribution Markkula made was to identify and bring in Mike Scott as president.

The Apple II was introduced during the summer of 1977 at the West Coast Computer Fair. Its development and introduction had cost approximately $2 million, $1.7 million of which went to development costs.[7]

[7] Casewriter's estimate.

Apple made available every piece of technical data relating to the machine, a highly unusual move in an industry where secrecy had always been tightly maintained. This open policy allowed sophisticated users to design circuit boards that could plug into the computer and expand its capabilities. In fact, several empty slots were built into the Apple II just for this purpose. Independent vendors soon began marketing hardware and software enhancements.

Because the market was embryonic, the Apple II was a product for which the customer had very low expectations, in part because it had little competition. The early personal computers were purchased by hobbyists. Customers were 98 percent male, typically 25 to 45 years old, earned a salary over $20,000 a year, and had the knowledge to program their own routines. (There was little standard software available.) Customers, for instance, commonly programmed games such as Star Trek and Adventure. (See Exhibit 6 for yearly sales.)

EXHIBIT 6 Yearly Sales of Apple Products (thousands of units)

	1977	1978	1979	1980	1981	1982	1983	1984 est.
Apple I	.2	—	—	—	—	—	—	—
Apple II/IIe*	.6	7.6	35(1)	78(1)	180(2)	279(3)	420(4)	500+(5)
Apple III	—	—	—	—	6	20	35(6)	40(6)(8)
Lisa	—	—	—	—	—	—	20(7)	100(8)
Macintosh	—	—	—	—	—	—	—	250(9)–500(10)
Apple IIc	—	—	—	—	—	—	—	150(8)–400(12)
Total	.8	7.6	35	78	186	299	475	1,200(11)

* Beginning in 1983, all these units were Apple IIe.

SOURCES: (1) Paul Freiberger and Michael Swaine, *Fire in the Valley,* 1984, pp. 231, 234.
(2) Apple Computer, Inc., 1981 annual report, p. 6.
(3) Apple Computer, Inc., 1982 annual report, p. 1.
(4) Apple Computer, Inc., 1983 annual report, p. 2.
(5) *InfoWorld,* April 9, 1984, p. 54.
(6) *InfoWorld,* April 9, 1984, p. 55.
(7) *Forbes,* February 20, 1984, p. 88.
(8) Casewriter's estimate.
(9) *Silicon Valley Tech News,* May 21, 1984, p. 4.
(10) *Fortune,* February 20, 1984, p. 100.
(11) *Forbes,* February 13, 1984, p. 40, states that 100,000 Apples are bought each month.
(12) *The Wall Street Journal,* April 24, 1984, p. 23.

The Apple II sold steadily throughout its product life cycle. Demand for the Apple II was enhanced by Visicalc, a financial modeling program developed by two Harvard Business School students. (See the Accessory Products section.) In the summer of 1982, it appeared that sales were slowing, so Apple bundled the II with a monitor and software for $2,000. The results were impressive, with 30,000 units per month being sold during the 1982 Christmas season. The $2,000 price point seemed critical, according to Dave Larson, later the IIe marketing director.[8]

[8] Interview with Dave Larson, October 1984.

THE APPLE COMPUTER COMPANY

During Apple's first full fiscal year (1978), the company raised $517,000 from outside sources, organized a network of independent distributors, and greatly expanded production of the Apple II. Much of the financing came from Venrock Associates, a New York–based venture capital firm founded by the Rockefeller family to invest in high-technology enterprises. Arthur Rock became a member of Apple's board of directors. (See Exhibit 7 for major company milestones, and Exhibit 8 for organization charts.)

Apple II manufacturing was essentially buy, assemble, and test. Under these manufacturing

EXHIBIT 7 Selected Milestones

Date	Event	
Jan. 1975	Development of Apple I.	
	Apple Computer Company partnership formed.	
Jan. 1976		
	Fall.	Apple II development completed.
Jan. 1977	Jan.	A. C. "Mike" Markkula joined. Mike Scott named president.
	Apr.	Apple II announced and introduction.
Jan. 1978	First full fiscal year.	
	Beginning of Apple III development.	
Jan. 1979	Mar.	Initial ideas for Lisa.
Jan. 1980	May.	Apple III announced.
	Jul.	Lisa development underway.
	Nov.	Apple III quantity shipments began.
	Dec.	Initial public offering.
Jan. 1981	Feb.	Apple III production and sales halted.
	Mar.	Macintosh project initiated.
	Mar.	Mike Markkula named president and CEO. Steven Jobs, chairman of the board.
	Dec.	Apple III hardware and software reintroduction.
Jan. 1982	May.	Apple IIc work begun.
Jan. 1983	Feb.	Apple IIe introduction.
	Mar.	Lisa introduction. John Sculley named president. Markkula became consultant.
	Sept.	Lisa 2 introduction.
Jan. 1984	Jan.	Macintosh introduction.
	Apr.	Apple IIc introduction.
	Jun.	Lisa second-generation software available, and networking and data communications.

EXHIBIT 8 Company Organization, 1978

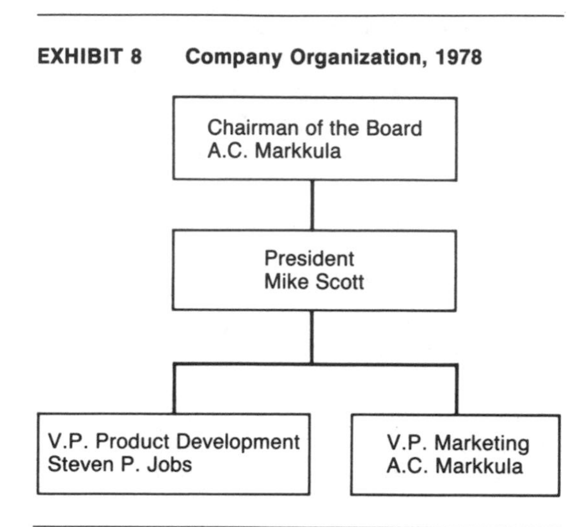

policies, which later came to be called the "conventional" approach, most components in the Apple II were purchased from outside suppliers who built chips, boards, cases, and other parts to Apple specifications. Only final assembly and test were done in-house.

In the testing process, Apple computers with special software performed diagnostic tests to isolate and identify problems. As a part of the final testing procedures, all systems were "burned-in" for two days to provide assurance of electronic and mechanical functions. In addition, the processes used were primarily manual, labor-intensive ones.

With regard to facilities, Apple operated out of one plant in California until June 1980. At that time, Apple instituted the "module" concept. This included the "parent plant" for the Apple II, located near corporate headquarters in California and in close contact with the engineering staff. Satellite production facilities (called "modules") that replicated the assembly and test portion of the parent plant were then developed as additional capacity was required. The satellite facilities for the Apple II were in Dallas, Texas (June 1980), and Cork County, Ireland (November 1980). Apple opened the facility in Ireland because of tax and transportation considerations for selling into the European market, and after coming to believe it would be more effective to sell to the European countries from a nearby location.

The module scheme allowed management and engineers at headquarters to maintain close contact and control over design and manufacturing operations, while spreading some of the risks associated with a single site or an excessively large facility. A module for the Apple II filled 30,000 square feet, required a crew of 70, and produced between 450 and 500 units per day.

By the end of 1980, sales were moving along at a $100 million-plus annual rate through a network of 950 independent retail computer stores in the United States and Canada (serviced by Apple's own sales organization) and 1,300 retail stores elsewhere in the world (serviced by 30 independent distributors). Virtually all of those sales were coming from the Apple II. While the company had introduced some of its own software for that product, it relied on independent developers (often users) to develop the vast majority of the needed software. Apple's policy was to try to price its products so that the retail price equaled approximately four times the manufacturing cost.

In December of 1980, Apple Computer went public, selling 4.6 million shares at $22 per share, resulting in an overall market value in excess of $1 billion. As a consequence of the offering, over 200 Apple employees became millionaires.

THE APPLE III

In 1981, Apple expanded its product offerings through the development and introduction of the Apple III. This product operated much as the Apple II but included additional features specifically aimed at an office environment.

The project to build the successor to the Apple II began in late 1978. In general, Apple was feeling a need to build a product that would eliminate the Apple II's shortcomings. "We listened very carefully to our customers," remembered Jobs. A great deal of market research had been conducted for the Apple III. Not only did this work help detail technical specifications, it also called for a sense of urgency. "We were told that sales of the Apple II would peak around 10,000 units per month," said Wil Houde, vice president and general manager of the Personal Computer Systems Division (PCSD). In order to sustain the company's growth, the Apple III had to be introduced during the "market window." That window was projected to be in approximately one year, so Apple planned to introduce the III at that time. In fact, the decision to introduce the III by the end of 1980 was announced in the prospectus for Apple's initial public stock offering. As it turned out, however, Apple II sales charged through the 10,000 mark without a pause. Wendell Sander, the Apple III's designer, recalled, "We had no idea back then that the Apple II's popularity would last so long."

The Apple III was envisioned to consolidate Apple's leadership role in personal computers. Although it would be far ahead of any competition, it did not presume to establish radically new standards for the industry. Indeed, some of Apple's best engineers embarked at this time on a separate project named Lisa. Lisa would have a longer time horizon and was expected to be the company's next revolutionary product.

From the beginning, the III was envisioned to have an 80-character wide screen (instead of 40), upper- and lowercase capability, and a built-in disk drive. These made the new machine much closer to a full word processor—an application for which Apple II owners had been buying special plug-in cards. The III was also to have additional memory, since memory limitations appeared to be a second main shortcoming of the II.

Eventually, the Apple III made a transition from an Apple II follow-on product to an entirely new product. The III was to have a numeric keypad in addition to a standard typewriter keyboard, and 96K memory. The hardware was given the Apple look and feel, and enclosed in an aluminum case. (See Exhibit 5.)

The III was to have an entirely new operating system called the "Sophisticated Operating System," or SOS. SOS made the III more versatile and more user friendly than the II. Indeed, the operating system was considered so advanced that Apple engineers insisted on keeping its specifications secret. However, this technical information was released a few months after the initial product introduction. Apple also decided to incorporate an emulation mode so that the III could simulate a II and run software written for the II. This decision, which was hotly debated, may have added as much as 25 percent to the development costs, which totaled $2.5 million.

Since the Apple III had evolved into a more powerful and expensive machine than the Apple II, the company needed to target a different market segment in order to protect the II's position in the education, small business, and hobbyists segments. It became positioned as a "discretionary purchase by professional people to improve their productivity."

This marketing strategy led to the development of the Information Analyst package of bundled software that included a word processor, business BASIC language, and Visicalc (the popular program for financial planning). User expectations were predicted to be quite different for the III than for the II. While a hobbyist might buy an Apple II and write his or her own programs, an Apple III purchaser would expect a large selection of tools. Unfortunately, software development lagged significantly behind the hardware development and the introduction timetable.

Eight weeks before the National Computer Conference in May 1980, Apple president Mike

Scott decided to unveil the Apple III at that conference. At the time, the engineers had numerous design difficulties, most notably with the spacing of components on the printed circuit boards. However, with great effort, the engineers produced several prototypes of the Apple III for the show. Marketing also worked until the last moment to have literature ready for the Information Analyst package. Although everyone at Apple knew that certain bugs still remained to be worked out, the overall attitude at the firm towards the new product was positive.

Amid great publicity and fanfare, the Apple III was introduced as planned at the May 1980 National Computer Conference in Anaheim, California. The product elicited great interest at the conference and Apple rented Disneyland for an evening to celebrate its success, and gave free tickets to anyone who had stopped by the Apple display at the conference. Introduction costs plus later Apple III product launch costs were estimated at $.5 million.

During the fall of 1980, when the transition to volume production was to occur, an endless stream of problems surfaced. Some could be attributed to Apple's rapid growth and the difficulty the company was having managing that growth. Others had to do with supplier problems. Many, however, were related directly to the Apple III design. Apparently, there was unanticipated thermal stress induced by the completely enclosed aluminum case that warped the circuit boards. In addition, the space constraint on the printed circuit board could cause several major failures in the memory board and the copper trace lines that carried the signals between the semiconductor devices.

Nevertheless, the Apple III was shipped and made available to customers in December 1980 as stated in Apple's stock offering prospectus. Of the machines going out the door, 20 percent did not function at all, another 20 percent had major problems, and many more had minor bugs. Some customers brought them back to the dealers four or five times for repairs.

Demand for the Apple III was strong initially, but it declined rapidly as news of the reliability problems spread and as the dealers encouraged potential customers to purchase other machines so they wouldn't have to cope with the extensive repair work the Apple IIIs required.

Because of the severe problems with the Apple III, company executives considered abandoning, replacing, or correcting the machine. In February 1981, the Apple Executive Committee decided to halt production of the III and to redesign it as necessary. The Executive Committee had concluded that the credibility of the entire firm, not just a single product, rested with the future of the Apple III. Mike Markkula announced to the public that Apple would not "abandon" the III. Wil Houde, vice president and general manager of the Personal Computer Systems Division (PCSD), stated: "We felt that we were protecting the reputation of all our future products, not just this one."

Internally, the process of correcting the machine's problems and redesigning the manufacturing process was characterized by one executive as "infinite attention to finite detail." Every major subassembly except the power supply was modified before the computer was ready for reintroduction. Apple IIIs were then subjected to stringent in-house testing, consisting of six phases. Aside from the more rigorous testing, Apple III manufacturing emulated the Apple II processes described earlier in the case. The software bottleneck was overcome by farming out a good deal of the task.

In March, on what came to be called Black Wednesday, Mike Scott fired 40 employees and terminated several hardware development projects he believed were taking too long. The company was stunned by the firings, a substantial number of which many employees believed were unjustified. Shortly thereafter, Mike Scott left Apple, Mike Markkula took over as president, and Steve Jobs became chairman of the board. (See Exhibit 9 for organization chart.)

EXHIBIT 9 Company Organization, 1981

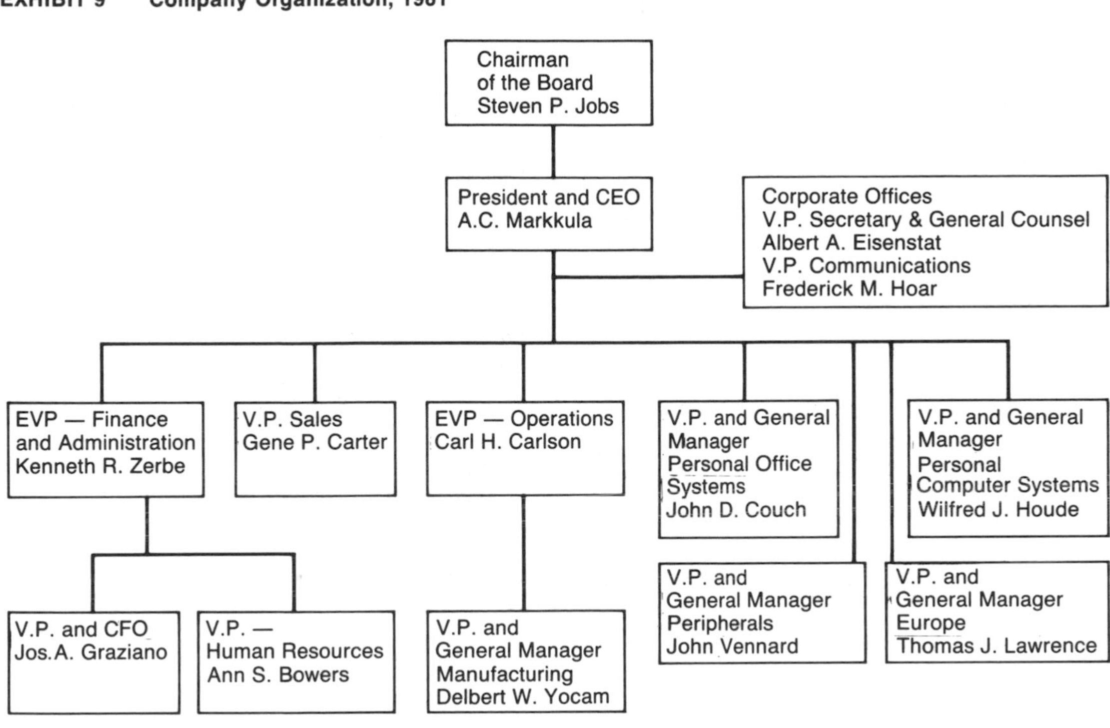

In December 1981, Apple announced a massive hardware/software reintroduction program for the Apple III. Code-named the THUD project, the company aimed to catch the country's attention with the reintroduction. Apple intended to replace every Apple III that had been sold. The owners would receive completely new machines, including updated software and expanded memory (from 96K to 128K), free of charge. The redesign and reintroduction cost Apple $5 million, bringing the total cost for Apple III development to $8 million. Of the $5 million spent on the redesign and reintroduction, a larger portion went into the reintroduction campaign than had gone to the initial Apple III introduction.[9]

[9] Casewriter's estimate.

Following the Apple III reintroduction, numerous Apple managers reflected back:[10]

Design engineers:

Marketing had sold the machine long before it was ready.

If we could have had another six months, none of this would have happened.

Manufacturing engineer:

Design engineering just dumped the design on manufacturing and said, "Build it."

Management:

The Apple II had taken the world by storm. We felt we could do anything. We got cocky.

[10] "Apple Computer (B)," Stanford University Graduate School of Business, Case S-BP-229B, 1983, p. 2.

The Apple III was a machine that an organization had to build, not a small group of individuals.

The people who put the Apple III together weren't here to learn the lessons of the Apple II. They didn't appreciate the details put into the Apple II.

Now we're much more formal. We write things down on paper. But we're still quite unstructured and informal relative to many established firms.

Steve Jobs, chairman of the board, added to these comments:

There is no question that the Apple III was our most maturing experience. Luckily, it happened when we were years ahead of the competition—it was the perfect time to learn.

Despite the reintroduction, sales of the Apple III never accounted for more than a small percentage of Apple II sales. Estimates indicated that approximately 75,000 Apple IIIs had been sold by April 1984,[11] and that at most, 100,000 cumulative would be sold by the end of Apple's fiscal year 1984. (See Exhibit 6.)

APPLE IIe

In February 1983, Apple Computer introduced the IIe, or the so-called "enhanced" Apple II. The IIe had been under development since early 1981 and was initially called the "LCA," for Low Cost Apple. According to Dave Larson, the Apple IIe director of marketing:

The choice was between developing the IIe or what later became the IIc. There were resources available only for one. It was an emotional time at the company. With VLSI (very large scale integrated circuits), it was possible to develop a computer with similar capabilities to the II and a 30 percent reduction in the number of chips from the II. Eventually, we decided to develop the IIe—it was a smaller jump . . . a new model for existing markets.

The attitude at the time was that you couldn't make a mistake, like you can now. Anything you could do would be successful. You just had to choose. It was difficult to focus.[12]

The Apple IIe group began with 7 people and eventually grew to 30, and came from a variety of places. These included the Apple II group, the Apple III area, and outside firms. For example, Peter Quinn, who was responsible for designing the integrated circuit, had previously worked for Xilog and Fairchild. Dave Larson, who became the marketing director, came from Measurex via the Apple III group, and other key technical people, including Steve Wozniak, came from the Apple II group.

According to *Byte Publications, Inc.:*

It had been obvious for a while at Apple Computer that a replacement for the Apple II was needed. The Teletype-style keyboard—uppercase-only, 40-column display—and the maximum of 64K bytes of memory were becoming limitations as the marketplace changed and software became more sophisticated. The design was getting old and technology had changed enough to allow a redesign with significantly fewer parts. A new design could also address foreign requirements for special keyboards, displays, and video signals better than the Apple II. Although the Apple II was a tremendous success, it was clearly time to design a successor.[13]

Remarkably, however, less clear was the breadth of support for the project within Apple. Hardware designer Peter Quinn commented:

The IIe was almost a closet project. Of course, we had support from Steve Jobs and Mike Markkula, but everyone pretty much left us alone. Many of the hotshots were working on Macintosh and Lisa at the time, and that left Wozniak free to pursue his interest of customizing the Apple II. Initially, we had 7 people working on the project though that grew to 30. And for most of the year, there was no leader.

---

[11] *InfoWorld,* April 9, 1984, p. 55.

[12] Interview with Dave Larson, October 1984.

[13] *Byte Publications,* February 1983.

We developed the product specs from our gut. We didn't do market research. We had a respected customer base of hobbyists and hackers, and, you know, that's what we were too. So we built the product like we thought it should be built.[14]

The Apple IIe included 64K bytes of memory (expandable to 128K bytes), Applesoft BASIC in ROM, a 63-key keyboard with upper- and lowercase letters, seven expansion slots for I/O (input/output) devices, and a video interface. The keyboard was essentially an improved version of the Apple III's keyboard (but without a numeric pad) and was the most visible difference between the II and the IIe.

Internally, though, the IIe was significantly different from the II. The keyboard was completely new, and the main printed circuit board was totally redesigned. The IIe used just 31 integrated circuits compared to approximately 120 in the II to achieve similar capabilities. A substantial part of the reduction resulted from the use of 64K-bit rather than 16K-bit dynamic memories.

Market surveys had indicated that the volume of software available was a key consideration among personal computer purchasers. Therefore, Apple decided the IIe had to be compatible with existing Apple II hardware and software products. This proved to be quite a challenge since the internal workings of the IIe had been designed from scratch. The first two major software products that were designed to use all the II's features (Applewriter IIe and Quickfile IIe) were enhanced versions of the same programs for the Apple III, and were reported to be very user friendly, providing clear prompts, multiple "menus," and numerous "help" screens.

Apple also developed a keyboard tutorial and wrote a precise technical manual to aid third-party software development efforts.

Apple spent about $8 million to develop the IIe and $3 million to launch it. According to Dave Larson, the advertising consisted primarily of "corporate umbrella messages," and little advertising in computer magazines. He added that the customer base was not differentiated at this time.[15]

The Apple IIe was manufactured using essentially the same production processes as were used for the II. In fact, at the end of 1982, Apple stopped making IIs completely and started making only IIes. Considering that the II accounted for almost all of the company's revenue stream at the time, this was a bold move.

The printed circuit boards (PCBs) for the IIe were made in Singapore, as they had been with the II since the Singapore facility opened in July 1981. Previously, Apple had been having quality control problems with its eight domestic PCB producers. The first pass yield rate of the PCBs assembled by the eight U.S. firms (i.e., the percentage of PCBs that had zero defects following initial assembly) ranged from 60 to 70 percent. At Apple's Singapore facility, the first pass yield rate ranged from 90 to 98 percent. By 1984, the Singapore facility also handled all the purchasing, board testing, and subcontracting with other Asian countries, in addition to board stuffing. Thirty to 40 percent of the board was stuffed using automated processes, as compared to 0 percent in 1981.

Regarding the evolution of final assembly and test, Apple began using a series of "linear manufacturing lines" early in 1983. The Dallas facility was responsible for the IIe, III, and upcoming Lisa product, and consequently needed some individual product line flexibility. The design of the lines allowed changeover to another product in the space of a couple of days. The burn-in, or testing time for the IIe, took eight hours, compared to two days when the II was first manufactured.

The manufacturing goal for fiscal years 1984 and 1985 was to develop a "computer integrated flexible manufacturing system." The first

---

[14] Interview with Peter Quinn, October 1984.

[15] Interview with Dave Larson, October 1984.

step would be to computerize all the work in process inventory tracking and the management information system. The second step would include some degree of workstation automation, automated storage and retrieval, and automated handling. Most of the initiative for automating the IIe derived from the development of the Macintosh factory. (See Macintosh section.)

In the years following the IIe introduction, Apple continued to offer additional enhancements for the product. These included a hard disk, Appleworks integrated software package, a mouse, and MS–DOS add-on capability. Many of these enhancements made the IIe easier to use.

In addition, third-party software developers and users themselves continued to write applications programs for the IIe. Even in 1984, the limits of the machine's capabilities were just beginning to be stretched. With over 10,000 programs written for it, the IIe had had more software developed for it than had any other personal computer.

Marketing at Apple developed through continual refinement and improvement. The retail distribution channel, based on national distribution through manufacturers' representatives, provided the mainstay. The largest retailers, such as Computerland and Businessland, for example, were serviced through a Central Buy organization. A Value-Added Resale channel existed for customers wanting systems integration or customized specific applications. Direct channel representatives sold to both the education—kindergarten through 12th grade—and the adult/university markets. In 1983, the University Consortium was established to sell to selected universities.

The Apple IIe customer mix also changed over the years: from predominantly business and corporate to more small business, school, and some home.

The Apple IIe sold extremely well and captured 55 to 60 percent of the U.S. education market for personal computers. By 1984, roughly 1.5 million Apple IIs and IIes had been sold. (See Exhibit 6.) Thus, the IIe, combined with its direct predecessor, the II, represented one of the most successful products in the history of U.S. business.

However, in late 1983 and early 1984, it appeared that IIe sales were slowing. In response, Apple cut the price from $1,800 to $995 in April 1984.

Peter Quinn reflected on the success of the IIe and the situation in early 1984:

> When the IIe came out, it got a corner of the cover of *Popular Computer* magazine, and Lisa got the rest. Yet Apple might not have made it through 1983 if the IIe hadn't existed. Lisa wasn't selling. Mac wasn't available. Sculley arrived in June 1983 and gave a year to Apple. The company was a mess when he arrived. There were shareholders calling all the time.[16]

APPLE IIc

Work on the Apple IIc ("c" for compact) began in mid-1982 when:

> Apple chairman Steve Jobs walked into engineer Peter Quinn's office. He slapped an Apple IIe main circuit board on Quinn's desk and plunked a low-profile keyboard at one end of it and a disk drive at the other. Jobs pointed to the assemblage of parts and declared, "*That* is a great product. Do you want to do it?"
>
> Quinn quickly replied, "Sure, Steve, we're half done."
>
> Actually, they weren't even half started. Apple had just introduced the IIe, an enhanced version of the Apple II. But Quinn realized that the work on that computer could be easily applied to the machine Jobs envisioned: a portable Apple II.[17]

By 1982, portable computers had begun to appear in the marketplace. Most notable among these was the Osborne I, which was

---

[16] Interview with Peter Quinn, October 1984.

[17] *Popular Computing,* June 1984.

introduced in April 1981. While the Osborne was not a spectacular machine, it was adequate, it was available, and it proved that one of the next directions for personal computers was towards portability.

Most of the people who developed the IIc came from within Apple, and most notably from the IIe group. Peter Quinn (designer), Dave Larson (marketing), and Dave Patterson (engineering) came from the IIe group, while Wil Houde had been involved with development of the III.

One key issue at the time, according to Dave Larson (marketing manager), was how to get over the hurdle to the next layer of customers. Concurrently, the market for personal computers had become much more complicated. Some Apple managers claimed that more market research went into the development of the IIc than into any other Apple product.

Designed specifically to fit inside any briefcase, the IIc measured 11.5 × 12 × 2.25 inches. The IIc weighed 7.5 pounds, had a single 5.25-inch built-in, low-height disk drive, and 128K bytes of RAM. It had a slightly flatter keyboard than the IIe, utilized the new Motorola 65C02 microprocessor, and needed 21 integrated circuits (compared to the IIe's 31). Importantly, Apple chose to replace many of the IIe expansion slots with fixed functions in the IIc, thus creating a "closed" architecture machine rather than its traditional open architecture approach. This meant that the IIc had limited expansion capabilities and could not run CP/M software. Also, importantly from a technological viewpoint, the IIc had an optional flat-panel LCD (liquid crystal display) display that was reported to be more readable than other flat panel displays available at the time, and to support a full 24 × 80 columns of text and the high-resolution graphics. However, the flat panel was not available at the introduction. The IIc also could be switched from traditional QWERTY keyboard layout (found on virtually all typewriters) to a Dvorak keyboard layout, with the lettered keys arranged differently, which many claimed was easier to learn and use and would become more frequently used than QWERTY. Finally, the IIc included a number of details: a 12-foot video cable, a "splash pad" to protect the circuitry from liquids, auditory-tactile response on the keys. (See Exhibit 5.)

The IIc's software resulted primarily from market research studies Apple conducted to try to understand "popular taste in computers." Some of the managers claimed that the IIc was the first Apple product targeted to a specific customer group. The market research indicated that consumers were spending two to three hours setting up their computers, which was too long. So, Apple built in all the protocols so that the user just needed to plug in the cables in order to begin. The research also showed that consumers became frustrated wading through large manuals. Therefore, Apple decided to bundle an introductory guide and six interactive tutorials with the IIc. Even more significantly, project manager Peter Quinn claimed that 90 to 95 percent of IIe software was compatible with the IIc. Additional third-party software development was encouraged. Finally, the IIc's ROM included mouse-controlled pictorial icons similar to those used with Lisa and Macintosh. (See Macintosh and Lisa sections.)

Apple targeted the IIc for a broad consumer market, Apple's first product aimed at this market. Specifically, the IIc was aimed at "serious home users," i.e., college-educated professionals with school-age children. The company also used consumer-oriented advertising, such as extensive use of TV spots, and more focused, events-oriented promotional activities. Industry observers speculated that the IIc market positioning and strategy represented the influence of Apple's new president, John Sculley. Mr. Sculley, who joined Apple in April 1983, was previously a leading Pepsi-Cola executive and was skilled in mass marketing techniques. He replaced Mike Markkula, who was now worth $100 million and wanted to retire.

Many industry observers claimed that the IIc marked a dramatic change in the Apple II family, both in marketing and product appearance. But also indicated was that Apple had decided to continue the II family. The IIc clearly was positioned against the IBM PCjr, which had been a disappointment for IBM due in large part to its "chiclet" keyboard and its limited memory. In fact, IBM had decided to redesign and reintroduce the machine. Some expected the IIc to cannibalize IIe sales.

The IIc was manufactured according to Apple's conventional manufacturing techniques. The printed circuit boards were assembled in Singapore, most components purchased from subcontractors, and then the final assembly and test performed at Apple's California, Texas, and Ireland facilities.

By the time the Apple IIc was introduced at a computer dealer exposition in San Francisco in April 1984, it had been developed into a fully portable computer with sophisticated software. The IIc development and introduction cost Apple $50 million.[18] Apple spent $15 million on the IIc launch alone. Initial sales reports indicated that the IIc was selling well. One article stated that Apple had forecast 400,000 IIc sales during calendar 1984 and that it would be outselling the IIe by year-end.[19]

ACCESSORY PRODUCTS DIVISION

About 15 percent of Apple's sales were derived from peripheral devices to expand the computer's applications. These hardware devices and software programs did as much as anything to stimulate sales of the Apple II. Apple's most significant innovation, aside from the computers themselves, was the low-cost, 5.25-inch, flexible or "floppy" disk drive for the Apple II, which was introduced in mid-1978. The floppy disk storage replaced the less efficient cassette tape storage. It provided file memory capacity of up to 143K bytes of data, vastly increased data retrieval speed, and provided random access to stored data.

This hardware enhancement increased the power and the speed of the Apple II and also sped up the development of software. Over 100 independent vendors were developing Apple II software and peripheral equipment for text editing, small business accounting, and teaching. Some of these programs were extraordinarily successful. The first best-seller was a financial modeling program called Visicalc that two Harvard Business School students had developed. Over 250,000 copies had been sold by 1982.

By 1980, Apple had begun to introduce its own software, but they expected to write no more than 1 percent of the software that would eventually be available for their computers. Even 1 percent of the programs would be a large number, since over 10,000 were then available for the Apple II. This "open system" of software development proved crucial to Apple's success.

Apple had also introduced new peripheral devices to expand the computer's applications. In addition to the floppy disk, other peripheral accessories manufactured by Apple included a graphics tablet, a thermal printer, and interface circuit boards. Apple computers incorporated standard interfaces that permitted the use of peripherals designed and manufactured by others as well as those offered by Apple. These included medium-speed letter-quality printers, modems that provided a data communication link over telephone lines, computerized bulletin boards, other computers, music synthesizers, portable power units, and others.

In 1981, Apple introduced the Profile hard disk mass storage unit, initially for the Apple III. It held 5 million bytes and, as such, had 35 times the capacity and 10 times the speed of the standard floppy disk drives available at the time. Apple believed the Profile would be an important building block for future computer

[18] Interview with Peter Quinn, October 1984.

[19] *The Wall Street Journal,* April 24, 1984, p. 23.

products, potentially as important as the Apple II's Disk II was in the 1970s.

In 1982, an Accessory Products Division was formed. The company introduced two brand-name printers, a dot matrix and a letter quality, for the Apple II and III. Nearly 1,000 third-party software suppliers were developing applications for Apples.

Apple established even tighter, stronger links to independent software developers in 1983. The company also reduced efforts in disk drive manufacturing, shifting to outside sources. This decision followed the difficulties and abandonment of development efforts for the "Twiggy" disk drive for the Macintosh.

The Accessory Products Division was located in Garden Grove, California (outside Los Angeles) in 1984. Some keyboards were assembled there, and development of additional add-ons for the IIe continued. As part of the 1984 reorganization, it became part of the Apple II Division.

LISA

During the 1970s, Apple had conducted market research in an effort to understand the Apple II customer base and how the product was being used. The results indicated that by and large, people were using the Apple IIs for basic spreadsheet and word processing activities (text editing and filing). They also needed graphics and communications capabilities. More importantly, however, the research indicated that the average consumer spent 30 to 40 hours learning how to use their new personal computer. The top managers and engineers at Apple, represented most vocally by Steve Jobs, concluded in 1981 that the key to expanding the personal computer market was in making a machine that was significantly more user-friendly. Only by doing so, the argument went, would large numbers of new users be attracted to personal computers. In fact, Jobs had held this view since at least 1976 when he designed the Apple II to be less threatening than traditional mainframe computers.

The ideas for the Lisa originated in 1979 with Steve Wozniak and Steve Jobs. Wozniak wanted to design a computer using a bit slice central processor, i.e., a computer that would include a purchased central part of the processor and an Apple-designed control for the processor. Jobs, then chairman of the board and vice president of product marketing, had an initial concept to produce an advanced, easy-to-use office computer. Jobs envisioned a computer with five integrated software applications, i.e., a computer with which data could be moved between the software applications packages relatively easily. The original target market was to be the "productivity marketplace," primarily executives.

Later in 1979, Jobs and some of the engineers visited Xerox's Palo Alto Research Center (PARC). The visit proved particularly stimulating. Some of the Xerox people, including noted developer Alan Kay, were also focusing on increasing personal computers' user-friendliness. They had developed a language called Smalltalk, which used a "mouse" or hand-held pointing device that greatly reduced the number of complex commands a user needed to type into the computer. Also, advances had been made with bit map displays where every pixel or point on the computer screen could be controlled, thus greatly expanding graphics capabilities.

Work on the Lisa project began in earnest in mid-1979 and accelerated during 1980. Some engineers from within Apple joined the group, notably Bill Atkinson, who rejoined Apple at this time and who played a critical role in the software development. Atkinson had previously developed the Pascal programming language for use on the Apple II, and more recently had studied neurology at the University of California at San Diego and the University of Seattle. Others from outside Apple joined the project also, including about 15 engineers from

Xerox's PARC facility and several people from DEC.[20]

Mike Scott, Apple's president, named John Couch as head of the Lisa project, insisting that Steve Jobs was too inexperienced to manage such a large project. Jobs remained as vice president of product marketing. The Lisa project staff began with a core of approximately 20 Apple people. In addition, 40 outsiders were subsequently hired, including some from Hewlett-Packard and a handful from Xerox. Ultimately, about 100 people worked on Lisa. The total development cost approached $40 million.

During 1980, hardware development breakthroughs occurred after three quarters of a year of false starts. The Motorola 68000 microprocessor replaced the Intel 8086 because it was more powerful for supporting graphics than the 8086.

Also during 1980, the software engineers began to solve the problems of integrating the software. A key to Lisa was the "windows" software, which allowed a user to view on one screen parts of several different "screens" of information at once. The Lisa also used a mouse, or hand-held pointing device, which facilitated movement around the screen. Instead of hitting cursor keys, the mouse would be moved over the desktop surface, and its movement translated into cursor movements on the screen.

By September 1980, three or four simple electronics prototypes were running, and demonstration software was available. Essentially, the hardware never changed after the end of 1980. The plan was to ship beginning in May 1981.

Wayne Rosing, general manager, recalled his perspectives from this time:

> With only six months until the projected product release date, we were in a "Lisa late" mode. No one had a total vision of the product, and the vision people did have was out of sync with the marketplace. The engineering group had no clear leader at the top, no single technical visionary, no intellectual. Instead, there were four or five warring camps. The product concept decision led to a political atmosphere resembling guerrilla camps. In addition, no one in the marketing group (which was more loyal to Jobs than Couch) had ever done a major new product.[21]

As development on Lisa continued, it became clear that a hard disk would be required to handle the massive software programs that would be run on the machine. But this was hard to swallow: it made Lisa expensive. In early 1981, Steve Jobs stated that Lisa had lost it—that it was just too big, that what he wanted was a "personal productivity appliance." Shortly thereafter, Jobs initiated work on what became the Macintosh.[22]

During 1982, an outside consultant brought a PERT program to Apple for the Lisa, and work began on a terminal emulator. Apple began introducing potential corporate customers to Lisa in June 1982 through the Lisa Sneak Program. The vast majority of the potential customers thought the product was great, except that it didn't have data communications, programmability, or networking capabilities.

The Wall Street Journal reported in January 1983 that:

> One reason the Lisa has taken so long getting to market is that Apple was undertaking an enormous development job for such a young company. It is understood to have written all the basic programs itself, even though most personal computer manufacturers use independent authors, even for the most fundamental software. Apple is also said to have devised its own video tube, disk drive, and mouse, which required considerable time and capital.
>
> Apple also had little incentive to move faster until recently, when competition increased so dramatically. The Apple II was selling well, and other companies were coming out with products

[20] *Newsweek*, January 31, 1984, p. 74.

[21] Interview with Wayne Rosing, July 1984.

[22] Ibid.

> that could be added to the Apple II to increase its capacity and keep it competitive.
>
> In addition, three years ago, when Apple introduced its last entry, the Apple III, it stumbled badly. A lot of the bugs hadn't been worked out, and the machine had to be redesigned.[23]

Apple targeted the Lisa at large corporate customers, a market traditionally dominated by IBM and new to Apple. Apple initially supplied only 150 dealers with Lisa, out of 1,800 total dealers, and expanded the national sales force to 100.

Beginning in the latter part of 1982, Lisa received worldwide news coverage, allegedly more than any other product except the IBM 370, according to Apple executives. The advanced technology captured people's imaginations. Apple spent approximately $10 million introducing and launching the Lisa.

Apple introduced the Lisa in March 1983, at a price of $9,995. The product was manufactured in California using conventional Apple manufacturing techniques, including overseas parts suppliers and assembly and only final assembly and test performed in-house. The Lisa hardware that had been developed in 1980 remained virtually unchanged.

Though Lisa had sparked tremendous interest within the computer industry, the fanfare did not materialize into high sales volume. As President John Sculley said in October 1984, "Lisa captured people's imaginations with its technology, but it did not capture the desktops of America."[24]

The 16,000-machine backlog that existed in September 1983 evaporated 1,200 machines later,[25] and only 20,000 were sold during all of calendar 1983.[26] Apple had difficulty closing sales to large corporations and was unaccustomed to operating in the corporate environment where purchase decisions often took a year. In addition, Apple sold the Lisa from stock in distribution (rather than through classic bookings), so didn't know the demand at any particular point in time. The competitive environment had toughened, also, due in large part to IBM's introduction of its Personal Computer and Vision's "Windows" programs in August 1983. Some Apple executives claimed in retrospect that Lisa sales would have been lower if it hadn't been for several popular software programs.

Apple's 1983 annual report said:

> Sales of the Lisa system, while on track with our early, modest expectations, were generally disappointing in fiscal 1983. Partly because of lengthy evaluation and purchasing cycles, we did not receive the volume orders we would have liked from the *Fortune* 1000 market, upon which we focused to the exclusion of other segments.
>
> Lisa should be viewed as much as a beginning—the threshold of new technology that will give rise to a family of new products from Apple—as it is an end in itself. Lisa technology is what Apple Computer is all about—in terms of the elegance of the product, ease of learning how to use it, and what its development says about Apple's commitment to innovation.[27]

In the fall of 1983, with less than 20,000 Lisas sold in total, Apple announced that Lisa manufacturing would be consolidated in the Texas and Ireland high-volume plants earlier than expected. Apple also increased the number of authorized Lisa dealers, unbundled the software, and introduced three modified versions of the Lisa. Each had different memory configurations and lower prices than the original version:

	Memory	Price
Lisa 2	512K	$3,495
Lisa 5	512K, 5MB hard disk drive	$4,495
Lisa 10	512K, 10MB hard disk drive	$5,495

---

[23] *The Wall Street Journal,* January 1983.

[24] Speech by John Sculley, Stanford Graduate School of Business, October 1984.

[25] Personal communication, Apple manager.

[26] *Forbes,* February 20, 1984, p. 88.

[27] Apple Computer, Inc., 1983 annual report, p. 6.

By mid-1984, second-generation software was available, and networking and data communications capabilities had been added. Company executives claimed demand was brisk because of the improved price/performance characteristics and the fact that some potential Macintosh customers (see following section) were deciding they preferred a machine with larger memory (such as the Lisa). Discussion about the future of the Lisa continued within Apple.

Business Week reported that Lisa had missed its goal because of several marketing mistakes, notably lack of ability to communicate with other computers (which was later remedied), and too-slow dealer and sales-force training. Tom Peters added:

> Apple suffered from technological hubris. It took the native Silicon Valley view that Lisa would walk into the boardrooms of corporate America and sell itself.[28]

MACINTOSH[29]

When Steve Jobs chose to pursue a different product from the Lisa, he was adamant about certain features for the new computer: price around $2,000, user-friendliness, and short initial learning time. He also wanted to automate manufacturing.

In order to focus on these objectives, a separate team of people was assembled to work on the project. Some key Macintosh technical people came from other parts of Apple, including the Apple II group and Lisa (Bill Atkinson and the Lisa graphics code). Many others were hired from outside Apple from firms including Xerox PARC, Texas Instruments, Hewlett-Packard, and Four-Phase. (See Exhibit 10.)

In the words of Joseph Graziano, vice president of finance and chief financial officer, "The Macintosh was intended to be like a car. We don't think much anymore of driving our car—we just get in and do it. We want to make this product equally as comfortable and natural." An important aspect of making the Macintosh "natural" or, in computer terms, "user-friendly," was the development of software that would make the human interaction with the Macintosh much easier than with traditional computers. This had been accomplished in the product design through the development of pictorial or visually oriented commands and a mouse that permitted the user to move a pointer (cursor) on the video screen. In developing the software, the Macintosh people benefited greatly from Lisa's graphics systems and development systems (i.e., the systems that programs are developed on). Also, the Lisa programs provided the Macintosh team with a base of experience and a standard against which to measure themselves and their improvement.

A second feature of the Macintosh, intended to increase user-friendliness, was a substantial decrease in the initial learning time required for the new user. This was achieved through the development of a more conversational language and ancillary support materials that did not require a computer background for understanding. When combined with a price under $2,500, top management expected the "every person's computer" to reach a large and primarily untapped number of customers, eventually realizing volumes of up to 1 million units of Macintosh per year. (Final Macintosh specifications are summarized in Exhibit 5.)

In terms of product positioning and target market segments, Mike Murray, director of marketing for the Macintosh, described the product as "a desktop appliance for the knowledge worker." The idea was to provide knowledge workers, that is, people whose major resource was information and ideas (and who in their jobs would take information and ideas

---

[28] *Business Week,* January 16, 1984, p. 78.

[29] The majority of this section comes from "Apple Computer, Inc.—Macintosh (A)," Stanford University Graduate School of Business Case S-BP-234, pp. 5–8, 11.

EXHIBIT 10 Company and Macintosh Organization (February 1983)

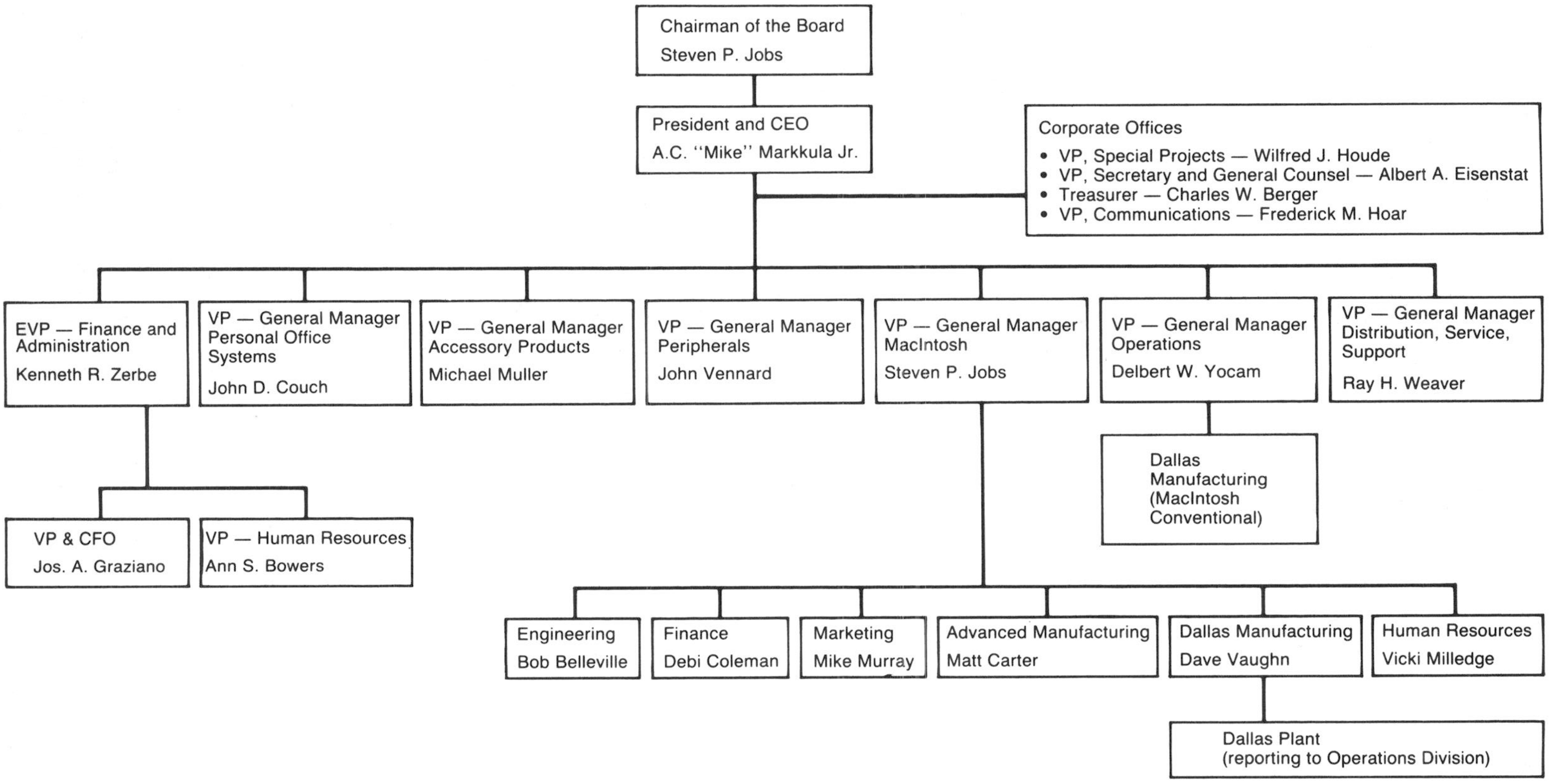

Chairman of the Board
Steven P. Jobs

President and CEO
A.C. "Mike" Markkula Jr.

Corporate Offices
- VP, Special Projects — Wilfred J. Houde
- VP, Secretary and General Counsel — Albert A. Eisenstat
- Treasurer — Charles W. Berger
- VP, Communications — Frederick M. Hoar

EVP — Finance and Administration
Kenneth R. Zerbe

VP — General Manager Personal Office Systems
John D. Couch

VP — General Manager Accessory Products
Michael Muller

VP — General Manager Peripherals
John Vennard

VP — General Manager MacIntosh
Steven P. Jobs

VP — General Manager Operations
Delbert W. Yocam

VP — General Manager Distribution, Service, Support
Ray H. Weaver

VP & CFO
Jos. A. Graziano

VP — Human Resources
Ann S. Bowers

Dallas Manufacturing (MacIntosh Conventional)

Engineering
Bob Belleville

Finance
Debi Coleman

Marketing
Mike Murray

Advanced Manufacturing
Matt Carter

Dallas Manufacturing
Dave Vaughn

Human Resources
Vicki Milledge

Dallas Plant
(reporting to Operations Division)

SOURCE: "Apple Computer, Inc.—Macintosh (A)," Stanford University, Graduate School of Business Case #S-BP-234, p. 25, revised 3/12/84.

and turn them into memos, reports, policies, and plans), with an "appliance" that would enhance their productivity and creativity. (The type of audience at whom the Macintosh was aimed could be described graphically as shown in Exhibit 11.) When initially introduced, it would focus primarily on the office segment and have only small penetration in the fields of education, home, and small business. However, by 1985, the Macintosh was envisioned as the heart of Apple's product offerings in sales volume. As suggested in Exhibit 11, a replacement for the IIe would complement the Macintosh and Lisa products.

EXHIBIT 11 Product Positioning and Target Segments for Apple Products

A. 1981

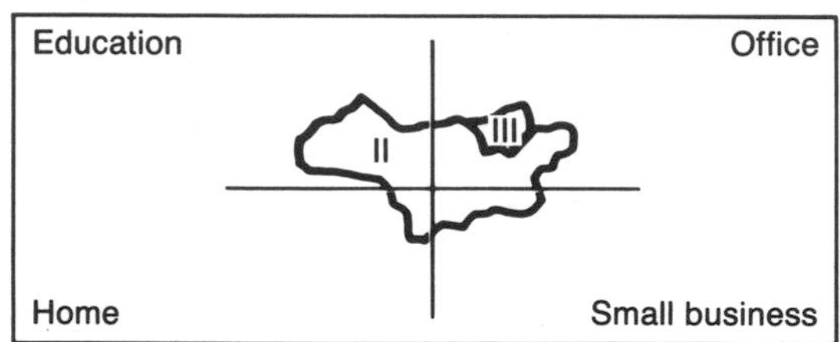

B. 1983-1984 Macintosh introduction

Target markets for Apple IIe		Target markets for Macintosh	
Business	42%		
Family	30%	Business	70%
Education	20%	College	20%
Science/Industry	8%	Home	10%

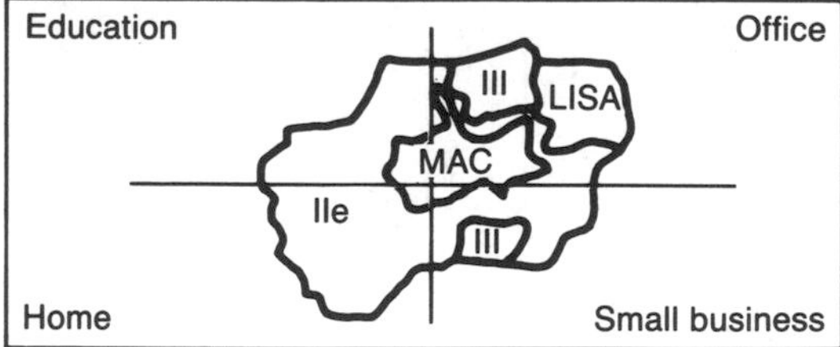

C. 1985-1986

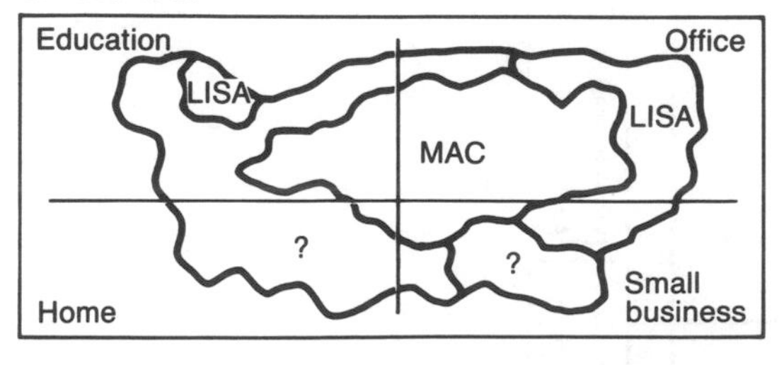

SOURCE: "Apple Computer, Inc.—Macintosh (A)," Stanford University, Graduate School of Business Case #S-BP-234, revised 3/12/84, p. 21.

From the point of view of all of those working closely with the Macintosh project, the product design represented only half the challenge and half the opportunity associated with Macintosh. The other half involved the transformation of the manufacturing function itself. Here Macintosh not only differed from the Apple II, IIe, and III, but also differed significantly from the Lisa. All of these products were manufactured using the company's "conventional" manufacturing approach (described earlier in the case).

The ideas regarding manufacturing for the Macintosh originated with Steve Jobs and Rod Holt. Rod Holt was an Apple Fellow, one of about half a dozen leaders in the field who were employed by Apple to pursue whatever they wanted. Holt came to Apple in 1977 following work at both National Semiconductor and Kodak. Steve Jobs had read about Japanese manufacturing techniques and became convinced Apple needed to learn and adopt those techniques in order to compete versus IBM and Japan in the long run. Rod Holt had been introduced to automated manufacturing at Kodak and was just as much in favor of it as was Steve.

The manufacturing approach proposed for the Macintosh involved several basic changes from Apple's conventional one. Two of these were particularly significant—the involvement of manufacturing in the product design stage and the incorporation of new technology in the final production process.

In the conventional design process, the product development team essentially completed its design of the product before getting involved with manufacturing as to how the product would be produced. Since Apple (and most other electronics equipment firms, for that matter) tended to use simple, readily available, worker dependent manufacturing process technologies, products often could be designed without much consideration of tight tolerances or new capabilities that would require uniqueness in the product's manufacturing process. In fact, in some companies this product design-manufacturing handoff sequence was referred to as "throwing it (the design) over the wall."

The Macintosh was to be designed for manufacturability using a parallel, interactive process between product development and manufacturing process development. In the short term, it was hoped that this parallel, interactive approach would lead to a smoother start-up of production. In the intermediate term, it was intended to result in a more manufacturable product design, providing more stable design, thereby giving lower costs and higher quality. In the long term, an additional potential benefit was that more (and) better features could be built into next-generation products because better processes would have been developed to provide such features. (That is, the product designers would be able to take better advantage of enhanced process capabilities.) In addition, further automation could take place in manufacturing because the initial design would have incorporated the tighter tolerances needed for that automation.

Unfortunately, in the near term, this approach had at least two significant drawbacks: product designers could feel a loss of freedom and, they would argue, a resulting decline in creativity because they were constrained by manufacturing; and, the development process initially might take longer because the systems for the interaction would need to be developed and installed.

A second particularly important aspect of the proposed Macintosh manufacturing approach was that of automation. Simply stated, the concept was to provide better—more efficient, higher quality, and more reliable—production processes. The Macintosh team referred to this as "an enhanced manufacturing capability." One aspect of this enhanced capability was that it required broadening the scope of activities performed on site: that is, with the automated process, more of the manufacturing tasks would be done by Apple rather than being subcontracted or done at Apple's Singapore site. Another was that it implied putting in place a form of technology with the expectation that it would be continually upgraded and enhanced (rather than being put in place once and for all and left unchanged until completely replaced). Two other aspects of this enhanced capability were the development of a focused facility that used a single process technology to produce a single product, and development of a set of control systems uniquely designed to operate that facility.

A third part of the plan called for closer ties to parts vendors and subcontractors. Apple wanted to establish stronger relationships in order to work with the vendors to decrease the percentage of defective parts.

Finally, the Macintosh team planned to automate in two stages. Phase I, to be part of the initial plant construction, would include automated testing, parts documentation, storage and loading, automatic insertion of components into printed circuit boards, and automated final test. Phase I would not impact future product design flexibility and provided somewhat of a hedge against the automation since all of the individual parts of the process had been done somewhere else before. What was different was that no one had put them all together as yet. Phase II would encompass additional automation but would not be developed until 1985, when the Mac team would have more hands-on automated manufacturing experience.

Debi Coleman, controller for the Macintosh, had summarized some of the team's views with regard to the automated production option:

> The automation issue is really one of out-Japanesing the Japanese. Can we? Should we? If we were to do Mac the old way, would we be giving the business away to the Japanese? . . . The type of equipment we'll be using to manufacture Mac has been used in Japan for three to five years. What's new is to put it all together in one factory.[30]

Great debate occurred within Apple regarding the Macintosh manufacturing process. Many people believed the Macintosh team was wildly optimistic about the time frame required to build and ramp up an automated manufacturing facility. There was considerable risk associated with an automated plant: the schedule couldn't slip too much without significant competitive implications regarding the appropriate time to introduce the Macintosh, yet everyone remembered the Apple III problems and so wanted to be absolutely sure the product was debugged, inventory in place, and production ramp-up well under way before the Macintosh was announced. Also, questions arose about the reporting relationships. Traditionally, the product and engineering groups turned the plant over to the Manufacturing Division once it was operational. But the Manufacturing Division had little experience with automated production, and the Macintosh team did not want to turn it over to manufacturing.

For quite some time, the plan was to set up a conventional manufacturing line that would be phased out within about six months, after the automated facility was up and running. However, in the spring of 1984, Steve Jobs decided to cancel the conventional line. At about the same time, the plant location was moved from Dallas, Texas, to Fremont, California, in order for it to be near corporate engineering and headquarters. Apple also decided to do Phase I of the automation at start-up, and Phase II a year later in 1985. Steve Jobs summarized some of his views about the Macintosh and its significance to the entire corporation this way:

> Mac is the Apple II of the 80s. Apple needs to become a manufacturing company and the Macintosh strategy is an offensive manufacturing strategy. With regard to some of the manufacturing options being discussed, I don't believe in alternatives. I believe in putting all my eggs in one basket and then watching the basket carefully. Then I either succeed or fail colossally. We have three to four years before the Japanese figure out about computers. We made the decision to do Mac essentially because we wanted to do it; it's a long-term decision.[31]

The development of the Macintosh cost Apple approximately $35 million. Apple introduced Macintosh on January 24, 1984, with a $15 million launch campaign. On May 21, the company reported it had shipped more than 70,000 Macintosh computers and that it expected to sell 250,000 by the end of 1984. In order to meet the demand, Apple announced that Macintosh manufacturing capacity would be doubled by the end of 1984.[32]

Sales was a key criterion by which Apple had chosen to measure the success of the Macintosh during the first 100 days. Other criteria included response from independent software developers and dealers. Apple reported it received over 5,000 inquiries from independent software developers during the first 100 days, sold 1,000 machines to these firms, and anticipated 150 software programs by the end of 1984 (50 by the end of the summer). Steve Jobs, chairman, commented:

> Apple set very aggressive goals and we believe that we have exceeded those goals. Based on the response we've had so far, we feel that the Macintosh is on its way towards becoming the third industry standard product in the personal computer business.[33]

---

[30] "Apple Computer, Inc.—Macintosh (A)," Stanford University Graduate School of Business, Case S-BP-234, revised March 1984, p. 11.

[31] Ibid., p. 1.

[32] *Silicon Valley Tech News,* May 21, 1984.

[33] Ibid.

THE SITUATION IN LATE 1983 AND EARLY 1984

The second half of calendar 1983 saw IIe sales weaken, Lisa and III sales continue low, depressed earnings, and the stock price fall from a high of $63 to $17 per share. Apple management presented a plan to securities analysts which stated that Apple would be a marketing innovator and an equal leader with IBM in the personal computer industry—and that they would do it in a year through new product introductions, a $100 million advertising budget, and depressed fiscal year 1984 earnings. As President John Sculley later said, "There's nothing like a life-and-death situation to focus your priorities."[34]

[34] Speech by John Sculley, Stanford Graduate School of Business, October 1984.

In January 1984, Apple announced a major company reorganization into two divisions: the Apple II Division and the Apple 32 Division (referring to the 32-bit microprocessor used in the Lisa and Macintosh products). The divisions were essentially parallel in structure and included all the major functional areas.

The Apple II family would include the IIe, III, and IIc products. It would focus on the home, education, kindergarten-through-12th-grade, and small business markets. It represented older technologies, a consumer goods orientation, and computer ease of use that was more likely to be added on than built in. Del Yocam, who had been vice president of manufacturing, was named executive vice president of the Apple II Division.

The Apple 32 Division would house Lisa and Macintosh and focus more on the college education and corporate environment markets. It

EXHIBIT 12 Company Organization, September 1984

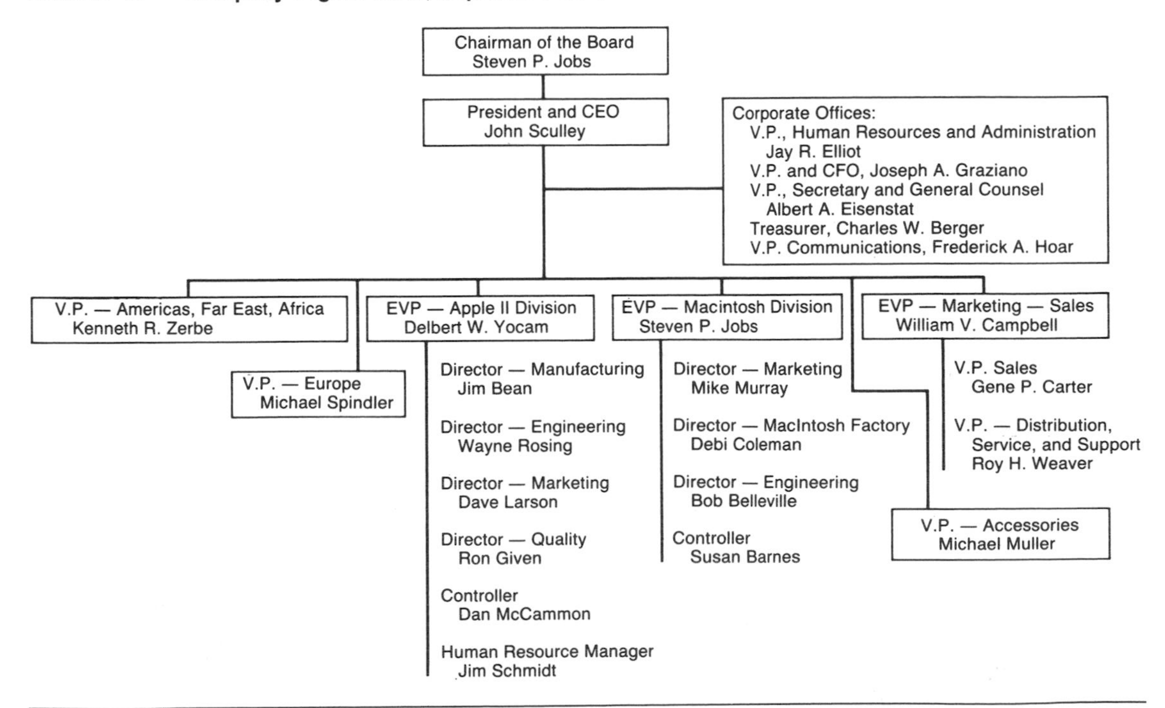

had newer technology, and the ease-of-use characteristics tended to be built into the computers rather than added on through enhancements. About a month later, the new division was renamed the Macintosh Division. Steve Jobs, 29, became general manager and retained the role of chairman of the board. As Steve explained, "He (Sculley) is my boss, but in a sense I'm his boss."[35] The majority of the other key management positions in the new division went to Macintosh people. (See Exhibit 12 for mid-1984 company organization chart.)

The marketing function also underwent change. In the summer of 1984, Apple announced that it would terminate its manufacturing representatives and replace them with Apple-trained and -employed reps. Bill Campbell took over as head of the sales area.

Future products allegedly under development included a "Fat Mac" with expanded memory, a terminal emulator, a lap-sized Mac coordinated by Alan Kay, and possibly an educational product.

THE COMPETITIVE ENVIRONMENT, 1984

By 1984, the competitive situation in the personal computer industry was ferocious. IBM continued to gain market share but had been hampered by poor sales of its PCjr. Bill Gates, chairman of the board of Microsoft and the developer of the IBM PC operating system MS–DOS, stated, "IBM is committed to the PCjr. It just has to get the formula right."[36] IBM announced it would modify the PCjr, and in August 1984 announced a much improved machine that included a keyboard similar to the PC, 256K memory (which meant virtually all the PC programs could now be run on the PCjr, including the best-selling integrated spreadsheet program 1-2-3® from Lotus®), and memory expandable to 512K. A few weeks later, IBM announced two PC networking products. (See Exhibit 13 for Apple and IBM product and price comparisons.) In many ways, IBM had broken with their corporate tradition to do things the way other personal computer firms did (e.g., open hardware architecture, use of outside suppliers and software developers, distribution through retail channels). However, the question remained whether the personal computer market would ever resemble the mainframe market of the 1960s when a popular saying was that "IBM was not the Competition, it was the Environment."[37]

Hewlett-Packard was also clearly committed to the personal computer market, although H-P got a slow start, having begun development of a personal computer in 1976 but not having introduced one until 1981. H-P's 150 was reputed to be selling well to corporate accounts (their historic strengths), though not nearly as well through retail distribution channels.

The demise of four other personal computer manufacturers had occurred in the previous 12 months: Victor, Franklin, Atari, and Osborne. Gavilan Corporation, which had planned to introduce a lap-sized computer in mid-1983, still couldn't ship product due to manufacturing difficulties. As a result, Gavilan had missed its window of opportunity in the marketplace, and many industry observers believed the company would not survive. Grid Systems, one of the first firms to develop a portable computer, was now focusing primarily on military applications and customers since its high quality and high price were out of most business and consumer price/performance ranges.

There also continued to be concern about the Japanese. One key concern was potential aggressive entry into the U.S. personal computer market by Japanese firms. Another was the projection that by the end of 1984, the Japanese would have overtaken the United States in *every* major consumer electronics industry.

---

[35] "Apple Bites Back," *Fortune*, February 20, 1984, p. 100.

[36] Freiberger and Swaine, *Fire in the Valley*.

[37] Ibid., p. 276.

EXHIBIT 13 Selected Apple and IBM Product Offerings*
(Retail Prices, October 1984)

Price	Apple	IBM
$0–$1,000		▪ Notebook (anticipated) ($800–900) – 16-bit, 128K – No monitor
$1,000–$2,000	▪ IIe ($1,224) – 8-bit, 64K, 1dd† ▪ IIc portable ($1,295) – 8-bit, 128K ▪ IIe ($1,795) – 8-bit, 128K, 2dd	
$2,000–$3,000	▪ Macintosh ($2,195) – 32-bit, 128K, 1dd, mouse ▪ III + ($2,995) – 8-bit, 256K, 1dd	▪ PCjr expanded ($2,035) – 16-bit, 128K, 1dd ▪ PC ($2,920) – 16-bit 256K, 2dd – Color monitor
$3,000–$4,000	▪ Macintosh ($3,195) – 32-bit, 512K, 1dd, mouse ▪ Lisa 2 ($3,495) – 32-bit, 512K, 1dd, mouse	▪ PC ($3,920) – 16-bit, 512K, 2dd – Color monitor
$4,000–$5,000	▪ Lisa 2/5 ($4,495) – Same as Lisa 2, plus 5MB‡ hdd	▪ PC XT ($4,800) – 16-bit, 256K, 1dd, 10MB hdd
$5,000–$6,000	▪ Lisa 2/10 ($5,495) – Same as Lisa 2, plus 10MB hdd	▪ PC AT ($5,220) – 16-bit, 256K, 1.2MB dd
$6,000–$7,000		▪ PC AT enhanced ($6,320) – 16-bit, 512K, 1.2MB dd, 20MB hdd

* All product prices include a monochrome monitor and adapter (if needed), unless otherwise stated.
† 8-bit microprocessor, 64 Kilobytes memory, 1 floppy disk drive.
‡ MB = megabyte, hdd = hard disk drive.
SOURCES: Businessland sales representative, Los Altos, California.
Computerland sales representative, Los Altos, California.
The Wall Street Journal, August 15, 1984, p. 3.
Fortune, February 20, 1984, p. 88.

THE SITUATION IN 1984, AND A LOOK TOWARDS THE FUTURE

As computer industry experts observed the personal computer industry in mid-1984 and speculated on Apple Computer, Inc.'s, future, the key question regarding Apple was "Where to go from here?"

There was a whole range of strategic questions for Apple to confront:

1. What will their long-term position be vis-à-vis IBM in 1987 or 1990, when the personal computer market reaches $25 billion in size? Will Apple be able to maintain a large market share?
2. What should be done with the Apple II family? Especially given the fact that the IIe was back selling well above company expectations, and that initially, at least, it appeared that Macintosh sales were not cannibalizing IIe sales to the degree that had been anticipated.
3. Where to go with the Macintosh?
4. What to do about the Accessory Products Division?
5. How to handle issues related to systems integration, and integrating communications capabilities (many other firms were entering partnerships in these areas)?
6. How to continue managing growth, at approximately 25 percent per year, while maintaining the Apple culture?
7. How to improve the product development process and related questions of organization, learning, etc.?

John Sculley, president and CEO of Apple since June 1983, summarized his views on Apple's past, present, and future:

> In the past, people at Apple were free to work on any neat idea. There was no strategic plan, no strategic discipline, little thought of where products fit in the marketplace.
>
> From day 1 of my arrival through month 10, I spent lots of time delineating a product line strategy. Now, all products fit into our two families of products: the Apple II family for education, small business, and serious home users and the Macintosh Division for users in an office environment.
>
> I want to cautiously avoid market research. The industry is too new, market research doesn't have a good track record; intuition and insight are still required. I'm against market research designing products. If research did it, there'd be other great products on the market. I think research designed the IBM PCjr. However, I firmly believe you still have to do lots of basic homework before you design a product: such things as technology, competition, distribution. Then I prefer a small team of people.
>
> Apple is a great product company. It has a group of people who love great products, who have driving passion. But I want it to be a great product marketing company also.[38]

Reading III–2
The New Product Learning Cycle

M. A. Maidique and B. J. Zirger

This paper summarizes our extensive study (n = 158) of new product success and failure in the electronics industry. Conventional "external factor" explanations of commercial product failure based on the state of the economy, foreign competition, and lack of funding, were found not to be major contributors to product failure in this industry. On the other hand, factors that can be strongly influenced by management such as coordination of the create, make, and market functions, the quality and frequency of customers' communications, value of the product to the customer, and the quality and efficiency of technical management explained the majority of the variance between successful and unsuccessful products. From

---

[38] Interview with John Sculley, September 18, 1984.

Research Policy, December 1985.

these findings a framework for understanding and managing the new product development process that places learning and communication in the center stage was developed.

Successes and failures in our sample were strongly interrelated. The knowledge gained from failures was often instrumental in achieving subsequent successes, while success in turn often resulted in unlearning the very process that led to the original success. This observation has led us to postulate a new product "learning cycle model" in which commercial successes and failures alternate in an irregular pattern of learning and unlearning.

1. INTRODUCTION

Many factors influence product success. That much is generally agreed upon by researchers in the field. The product, the firm's organizational linkages, the competitive environment, and the market can all play important roles. On the other hand, the results of research on new product success and failure[1] is reminiscent of George Orwell's *Animal Farm* in that some factors seem to be "more equal than others." But, exactly which set of factors predominates seems to be, at least in part, a function of both the methodology and the specific population studied by the researcher.[2]

The Stanford Innovation Project (SINPRO)

In a survey of 158 products in the electronics industry, half successes and half failures, we developed our own list of major determinants of new product success.[3] The eight principal factors we identified are listed below roughly in the order of their statistical significance. Products are likely to be successful if:

1. The developing organization, through in-depth understanding of the customers and the marketplace, introduces a product with a high performance to cost ratio.
2. The create, make, and market functions are well coordinated and interfaced.
3. The product provides a high contribution margin to the firm.
4. The new product benefits significantly from the existing technological and marketing strengths of the developing business units.
5. The developing organization is proficient in marketing and commits a significant amount of its resources to selling and promoting the product.
6. The R & D process is well planned and coordinated.
7. There is a high level of management support for the product from the product conception stage to its launch into the market.
8. The product is an early market entrant.

The study that led to these conclusions consisted of two exploratory surveys described in detail elsewhere.[4] The first survey was open ended and was divided into two sections. In the first part we asked the respondent to select a pair of innovations, one success and one failure. Successes and failures were differentiated by financial criteria. The second section of the original survey asked each respondent to list in his own words the factors which he believed contributed to the product's outcome. Seventy-nine senior managers of high-technology companies completed this questionnaire.

---

[1] R. C. Cooper, "A Process Model for Industrial New Product Development," *IEEE Transactions on Engineering Management,* EM-30, no. 1 (1983), pp. 2–11.

[2] M. A. Maidique and B. J. Zirger, "A Study of Success and Failure in Product Innovation: The Case of the U.S. Electronics Industry," *IEEE Transactions on Engineering Management,* EM-31, no. 4 (1984), pp. 192–203.

[3] Ibid.

[4] Ibid.

The follow-up survey was structured into 60 variables derived from three sources: (1) analysis of the results of the first survey, (2) review of the open literature, and (3) the authors' own extensive experience in high-technology product development. Each respondent, on the basis of the original two innovations identified in survey 1, was asked to determine for each variable whether it impacted the outcome of the success, failure, neither, or both. Survey 2 was completed by 59 of the original 79 managers.

The results from these two initial surveys were reported earlier.[5] To summarize, we conducted several statistical analyses for each variable and innovation type including determination of means and standard deviations, binomial significance, and clustering. Table 1 shows the binomial significance for the 37 variables which differentiated between success and failure. Combining our statistical results with the content analysis of the initial survey, we derived the eight propositions listed earlier.

[5] Ibid.

FIGURE 1 Diagram of the Critical Elements of the New Product Development Process

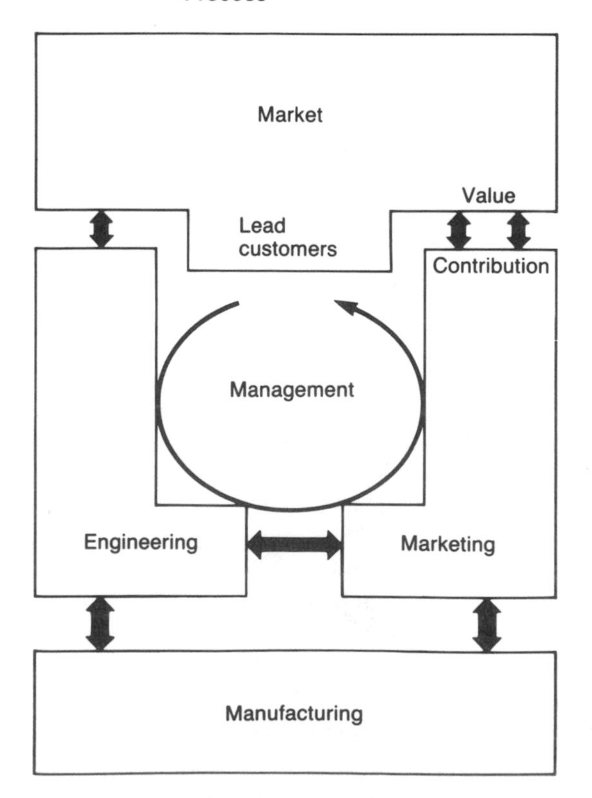

Using these eight factors as a starting point, we then developed a block diagram of the new product development process that focuses on the product characteristics and the functional interrelationships and competences that are most influential in determining new product success or failure (Figure 1). In our view, the innovation process is a constant struggle between the forces of change and the status quo. Differences in perceptions between the innovator and the customer and also between the groups that make the building blocks of the innovation process—engineering, marketing, and manufacturing—all conspire to shunt new product development or to deflect it from the path of success. Effective management attempts to integrate these constituencies and to allocate resources in a way that makes the new possible. These ideas are the basis of a model of the new product development process that we describe more fully and validate empirically in a forthcoming paper.[6]

The eight propositions resulting from our analysis were the objective "truths" that resulted from statistical analysis of our large sample of new product successes and failures. Though coincident in their salient aspects with the work of others,[7] these results, however, did not fully satisfy us. Had we missed important variables in our structured surveys? Had our respondents understood our questions? Had we failed to detect significant relations be-

[6] B. J. Zirger and M. A. Maidique, (forthcoming), "Empirical Testing of a Conceptual Model of Successful New Product Development, to be submitted to *Management Science*.

[7] Cooper, "Process Model."

TABLE 1 Significant Variables from Survey 2 Grouped by Index Variable

Successful Innovations Were:	Number of Observations	Cumulative Binomial	Significance Rating
1. Better matched with user needs.			
Better matched to customer needs.	44	8.53 E-09	+ + +
Developed by teams which more fully understood user needs.	44	1.27 E-05	+ + +
Accepted more quickly by users.	49	7.01 E-04	− − −
2. Planned more effectively and efficiently.			
Forecast more accurately (market).	43	1.25 E-07	+ + +
Developed with a clearer market strategy.	45	1.24 E-04	+ + +
Formalized on paper sooner.	45	3.30 E-03	+ + +
Developed with less variance between actual and budgeted expenses.	46	2.70 E-02	− −
Expected initially to be more commercially successful.			
3. Higher in benefit-to-cost.	42	8.21 E-02	+
Priced with higher profit margins.	51	6.06 E-08	+ + +
Allowed greater pricing flexibility.	52	1.02 E-06	+ + +
More significant with respect to benefit-to-cost ratio			
4. Developed by better-coupled organizations.	43	6.86 E-03	+ + +
Developed by better-coupled functional divisions.	39	1.68 E-07	+ + +
5. More efficiently developed.			
Less plagued by after-sales problems.	35	5.84 E-05	− − −
Developed with fewer personnel changes on the project team.	28	6.27 E-03	− − −
Impacted by fewer changes during production.	41	1.38 E-02	− −
Developed with a more experienced project team.	39	2.66 E-02	+ +
Changed less after production commenced.	47	7.19 E-02	−
Developed on a more compressed time schedule.	39	9.98 E-02	+
6. More actively marketed and sold.			
More actively publicized and advertised.	39	4.74 E-03	+ + +
Promoted by a larger sales force.	28	6.27 E-03	+ + +
Coupled with a marketing effort to educate users.	37	1.00 E-02	+ +
7. Closer to the firm's areas of expertise.			
Aided more by in-house basic research.	25	7.32 E-03	+ + +
Required fewer new marketing channels.	25	7.32 E-03	− − −
Closer to the main business area of firm.	30	8.06 E-03	+ + +
More influenced by corporate reputation.	29	3.07 E-02	+ +
Less dependent on existing products in the market.	36	6.62 E-02	−
Required less diversification from traditional markets.	24	7.58 E-02	−
8. Introduced to the market earlier than competition.			
In the market longer before competing products introduced.	44	1.13 E-02	+ +
First-to-the-market type products.	39	1.19 E-02	+ +
More offensive innovations.	46	5.19 E-02	+
Generally not second-to-the-market.	36	6.62 E-02	−
9. Supported more by management.			
Supported more by senior management.			
Potentially more impactful on the careers of	31	1.66 E-03	+ + +
the project team members.	32	5.51 E-02	+
Developed with a more senior project leader.	39	9.98 E-02	+
10. Technically superior.			
Closer to the state-of-the-art technology.	36	3.26 E-02	+ +
More difficult for competition to copy.	45	3.62 E-02	+ +
More radical with respect to world technology.	42	8.21 E-02	+

tween some of the variables we identified—or between these and some yet undiscovered factors? How valid were our final generalizations? And most important, what were the underlying conceptual messages in this list of factors? In short, we were concerned that perhaps our statistical analysis might have blurred important ideas.

Reflecting on his research on the individual psyche, Carl Jung once put it this way:[8]

> The statistical method shows the facts in the light of the average, but does not give a picture of their empirical reality. While reflecting an indisputable aspect of reality, it can falsify the actual truth in a most misleading way. . . . The distinctive thing about real facts, however, is their individuality. Not to put too fine a point on it, one could say that the real picture consists of nothing but exceptions to the rule, and that, in consequence, reality has predominately the characteristic of irregularity.

Such irregularities have caused one of the most experienced researchers in the field to wonder out loud if any fundamental commonalities exist at all in new product successes. "Perhaps," Cooper observed, "the problem is so complex, and each case so unique, that attempts to develop generalized solutions are in vain."[9]

2. METHODOLOGY

To address the concerns noted above, we prepared individual in-depth case studies for 40 of the original 158 products to search for methodological flaws or significant irregularities that might challenge the results of our statistical analysis (Table 1). The case studies were prepared under the supervision of the authors by 45 graduate assistants.[10] Seventeen West Coast electronics firms which had participated in the 1982 Stanford-AEA Executive Institute and in our original two surveys served as sites for the 20 case studies. This subset of the original product pairs served as the subject of analysis for the case studies. Two or more project assistants interviewed managers and technologists and prepared written reports that included interview transcripts or summaries, background information on the firm, the competitive environment, the product development process, the characteristics of each of the two products, validation of the original survey 2, and a critical review of the factors that contributed to success or failure in each case. Overall, 101 managers and technologists were interviewed in 148 hours of interviews.

Most of the companies supplied the research teams with detailed financial, marketing, and design information regarding each one of the products, including in some cases internal memoranda that traced the products' development histories. Because of the confidentiality of this data, we must not identify any of the firms, much as we would like to thank them for their contributions to the project. In some cases, to illustrate a point, we have chosen to use examples from the public domain, or from published cases we or others have written about, and we may mention a company by name; however, the companies that collaborated with the project are either left anonymous or given ficti-

---

[8] C. G. Jung, *The Undiscovered Self* (New American Library, 1957), p. 17.

[9] R. C. Cooper, "The Dimensions of Industrial New Product Success and Failure," *Journal of Marketing* 43 (1979), p. 102.

[10] The authors wish to express their appreciation to the following graduate students and doctoral candidates, who assisted in preparation of the individual case studies: P. Achi, G. Ananthasubramanianium, R. Angangco, C. Badger, B. Billerbeck, R. Cannon, D. Chinn, L. Christian, B. Connor, A. Dahlen, S. Demetrescu, B. Drobenko, R. Farros, H. Finger, H. Jagadish, L. Girault-Cuevas, R. Guior, T. Hardison, Y. Honda, J. Jover, C. Koo, T. Kuneida, S. Kurasaki, M. Lacayo, D. Lampaya, D. Ledakis, L. Lei, R. Ling, S. Makmuri, P. Matlock, C. Mungale, R. Ortiz, B. Raschle, R. Reis, B. Russ, E. Saenger, J. Sanghani, V. Sanvido, F. Sasselli, R. Simon, P. Stamats, R. Stauffer, L. Taurel, B. Walsh, F. Zustak.

tious names which, when first introduced, are placed in quotation marks.

This paper reports how these case studies and the associated interview transcripts enriched our earlier conclusions. In section 3, we begin to clarify the terms that we had employed in our survey, specifically "user needs" and "product value." In section 4, we explore the meaning of success. The case studies led us to expand our concept of success and failure beyond the one-dimensional confines of financial return. Indeed, success and failure often appear to be close partners, not adversaries, in organizational and business development. Finally, in section 5, we postulate an evolutionary model of new product development, which we believe leads to a better understanding of the relationship between success and failure. For many of the propositions we present here, we lack the analytical support that underlies the eight factors identified in our original research. Nonetheless, we feel that these findings, which we hope will help to illuminate further research—including our own—are as important as our statistical results.

3. DEFINING "USER NEEDS" AND "PRODUCT VALUE"

The detailed case studies largely reinforced the principal findings of the overall study.[11] But the case studies also enriched some of the findings from structured questionnaires by providing fresh insights on several of the key variables. In this paper, we focus on the most important and perhaps the least specific variable, "understanding of the market" and "user needs" which is believed to result in products with "high value."

One of the principal findings of our large sample survey was that "user needs" and "customer and market understanding" are of central importance in predicting new product success or failure, a result that parallels the findings of the pioneering SAPPHO pairwise comparison study.[12] This result, however, does little to illuminate how a firm goes about achieving such understanding. What's more, citing user needs ex post facto as a key explicatory variable in product success can be simply disregarded as tautological. Of course, it can be argued the company "understood" user needs if the product was successful. Expanding on such criticisms, Mowery and Rosenberg have pointed out that the term "user need" is in any event vague and lacks the precision with which economists define related market variables such as demand.[13] What seems to be important, however, is to determine whether there are identifiable ex post ante actions that organizations take that develop and refine the firm's understanding of the customer's needs.

In most of the instances in which interviewees indicated a product had succeeded because of "better understanding of customer needs," they were able to support this view by citing specific actions or events. Both the experiential background of the management and developing team as well as actions taken during the development and launch process were viewed as important.

One line of argument went thus: we understood customer needs because the managers, engineers, and marketing people associated with the product were people with long-term experience in the technology and/or market. In such a situation, some executives argue, very little market research is required because the company's management has been close to the

[11] Maidique and Zirger, "A Study of Success."

[12] R. Rothwell, C. Freeman, A. Horley, V. I. P. Jervis, Z. B Robertson, and J. Townsend, "SAPPHO Updated—Project SAPPHO, Phase II," *Research Policy* **3** (1974), pp. 258–91. See also C. Freeman, *The Economics of Industrial Innovation* (Harmondsworth: Penguin Books, 1974), pp. 161–97.

[13] D. Mowery and N. Rosenberg, "The Influence of Market Demand upon Innovation: A Critical Review of Some Recent Empirical Studies," *Research Policy* 8 (1979), pp. 101–53.

customer and to the dynamics of his changing requirements all along. As the group vice president of a major instrument manufacturer explained, "We were able to set the right design objectives, particularly cost goals, because we knew the business, *we could manage by the gut* (authors' emphasis)."

This approach was evident in other firms also. When "Perfecto," a leading U.S. process equipment manufacturer, induced by a request from one of its European customers, commissioned a domestic market survey to assess potential demand for a new product that combined the functions of two of its existing products, the result was almost unanimously negative. Because of a quirk in the process flow in U.S. plants (which differed from European plants), domestic customers did not immediately see significant value in the integrated product. Notwithstanding the market survey data, Perfecto executives continued to believe that the product would prove to be highly cost effective for their worldwide customers. Buoyed by enthusiasm in Europe and a feeling of deep understanding of his customers that was the result of 13 years of experience in the numerically controlled process equipment market, Perfecto's president gave the project the go ahead. His experience and self-confidence paid off. There was ultimately a significant demand for the new machine on both sides of the Atlantic.

These experience-based explanations, however, are only partially useful blueprints for action. The argument simply says that experienced people do better at new product development than the inexperienced, a hypothesis confirmed by our earlier research and that of others.[14]

Most of our informants, however, characterized the capture of "user needs" in action-oriented terms. For the successful product in the dyad, they described the company as having more openly, frequently, carefully, and continuously solicited and obtained customer reaction before, after, and during the initiation of the development and launch process. In some cases, the attempt to get customer reaction went to an extreme. "Electrotest," a test equipment manufacturer, conducted design reviews for a successful new product at their lead customers' plants. In general, the successful products were the result of ideas which originated with the customers, filtered by experienced managers. In one case, customers were reported to have "demanded" that an instrument manufacturer develop a new logic tester. As a rule, the development process for the successful products was characterized by frequent and in-depth customer interaction at all levels and throughout the development and launch process. While we did not find (and did not look for) what Von Hippel discovered in his careful research on electronic instruments, that users had in many cases already developed the company's next product, it was clear that, more so than any other constituency, they could point out the ideas that would result in future product successes.[15]

But when listening to customers, it's not enough to simply put in time. It is of paramount importance to listen to potential users without preconceptions or hidden agendas. Some companies become enamored with a new product concept and fail to test the idea against the reality of the marketplace. Not surprisingly, they find later that either the benefit to the customer was more obvious to the firm than to the customer himself or the product benefits were so specific that the market was limited to the original customer. For these reasons, the president of an automated test equipment manufacturer provided the following admonition, "When listening to customers, clear

[14] A. C. Cooper and A. V. Bruno, "Success among High-Technology Firms," *Business Horizons* 20, no. 2 (1977), pp. 16–22.

[15] E. A. von Hippel, "Users as Innovators," *Technology Review,* no. 5 (1976), pp. 212–39.

your mind of what you'd like to hear—Zen listening."

Unless this careful listening cascades throughout the company's organization and is continually "market"-checked, new products will not have the value to the customer that results in a significant commercial success. A predominant characteristic of the 20 successful industrial products that we examined in our case studies is that they resulted in almost immediate economic benefits to their users, not simply in terms of reduced direct manufacturing or operating costs. The successful products seemed to respond to the utility function of potential customers, which included such considerations as quality, service, reliability, ease of use, and compatibility.

Low cost or extraordinary technical performance, per se, did not result in commercial success. Unsuccessful products were often technological marvels that received technical excellence awards and were written up in prestigious journals. But typically such extraordinary technical performance comes at a high price and is often not necessary. "Very high performance, at a very high price. This is the story behind virtually every one of our new product failures," is how a general manager at "International Instruments," an instrument manufacturer with a reputation for technical excellence, described the majority of his new product disappointments.

In contrast to this phenotype, new product successes tend to have a dramatic impact on the customer's profit-and-loss account directly or indirectly. "Miltec," an electronic systems manufacturer, reported that its successful electronic counter saved their users 70 percent in labor costs and downtime. "Informatics," a computer peripheral manufacturer, developed a very successful magnetic head that was not only IBM compatible but 20 percent cheaper and it offered a three times greater performance advantage. An integrated satellite navigation receiver developed by "Marine Technology," a communications firm, so drastically reduced on-board downtime in merchant marine ships in comparison to the older modular models that the company was overwhelmed by orders. The first 300 units paid for the $2.5 million R & D investment; overall 7,000 units were sold. By comparison, the unsuccessful products provided little economic benefit. Not only were they usually high priced but often they were plagued by quality and reliability problems, both of which translate into additional costs for the user.

4. HOW SHOULD PRODUCT SUCCESS BE MEASURED?

Our original surveys used a unidimensional success taxonomy. Success was defined along a simple financial axis. Successful products produced a high return while unsuccessful products resulted in less than break-even returns. Using this measure, our population of successes and failures combined to form a clearly bimodal distribution (Figure 2) that reinforced our assumption that we were dealing with two distinct classes of phenomena. While obtaining and plotting this type of data went beyond what most prior success-failure researchers had deemed necessary to provide, our detailed case studies lead us to conclude that this may not have been enough.

Success is defined as the achievement of something desired, planned, or attempted. While financial return is one of the most easily quantifiable industrial performance yardsticks, it is far from the only important one. New product "failures" can result in other important by-products: organizational, technological, and market development. Some of the new product failures that we studied led to dead ends and resulted in very limited organizational growth. On the other hand, many others—the majority—were important milestones in the development of the innovating firm. Some were the clear basis for major successes that followed shortly thereafter.

FIGURE 2 Distribution of Successes and Failures by Degree of Success/Failure for Case Studies

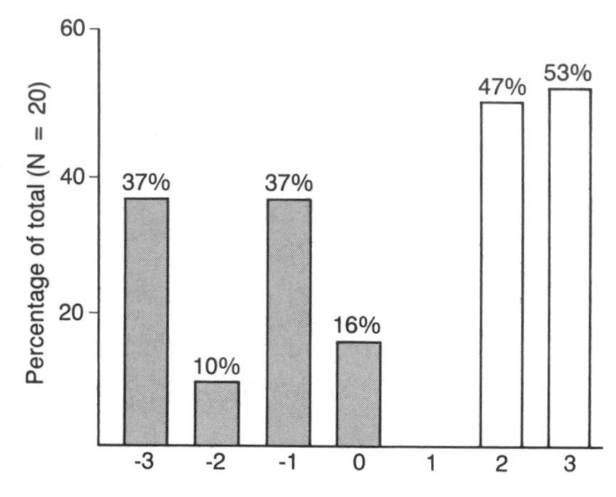

Less successful innovations
More successful innovations

International Instruments, a large electronics firm, developed a new instrument based on a new semiconductor technology (diode arrays) that the firm had not yet used in one of its commercial products. The instrument, though technically excellent, was developed for a new market where the company did not have its traditionally keen sense of what value meant to the customer. Few units were sold and the product was classified as a failure. On the other hand, the experience gained with the diode array technology became the basis for enhancement of other product families based on this newly gained technical knowledge. Secondly, the organization learned about the characteristics of the new market through the diode array product, and, armed with new insights, a redesigned product was developed which was a commercial success. Was the diode array product really a failure, its developers asked?

In this and other cases we observed, the failure contributed naturally to the subsequent successes by augmenting the organization's knowledge of new markets or technologies or by building the strength of the organization itself. An example from the public domain illustrates this point. After Apple Computer had been buffeted by the manufacturing and reliability problems that plagued the Apple III launch, which caused Apple to lose its lead in the personal computer market and to yield a large slice of the market to IBM, Apple's chairman summed up the experience thus, "There is no question that the Apple III was our most maturing experience. Luckily, it happened when we were years ahead of the competition. It was a perfect time to learn."[16] As demonstrated by the manufacturing quality of the Apple IIe and IIc machines, Apple, that is the Apple II division, learned a great deal from the Apple III mishaps. Indeed, Sahal has pointed out that success in the development of new technologies is a matter of learning.[17] "There are few innovations," he points out, "without a history of lost labor. What eventually makes most techniques possible is the object lesson learned from past failures." In his classic study of technological failures, Whyte argues that most advances in engineering have been accomplished by turning failure into success.[18] To Whyte, engineering development is a process of learning from past failures.

Few would think of the Boeing Company and its suppliers as a good illustration of Sahal's and Whyte's arguments. Rosenberg, however, has pointed out that early 707s, for many years considered the safest of airplanes,

---

[16] M. A. Maidique, J. S. Gable, and S. Tylka, "*Apple Computer (A) and (B),*" Case #S-BP-229(B) (Stanford Business School Central Services, Graduate School of Business, Rm. 1, Stanford, CA, 1983). See also M. A. Maidique and C. C. Swanger, "*Apple Computer: The First Ten Years,*" Case #PS-BP-245 (Stanford Business School Central Services, Graduate School of Business, Rm. 1, Stanford, CA, 1985).

[17] D. Sahal, *Patterns of Technological Innovation* (Reading, Mass.: Addison-Wesley Publishing, 1981), p. 306.

[18] R. R. Whyte, *Engineering Progress through Trouble* (London: Institution of Mechanical Engineers, 1975).

went into unexplainable dives from high-altitude flights. The fan-jet turbine blades used in the jumbo jet par excellence, Boeing's famed 747, failed frequently under stress in the 1969–70 period.[19] Despite these object lessons, or perhaps because of them, Boeing makes more than half of the jet-powered commercial airliners sold outside of the Soviet bloc. According to the executive vice president of the Boeing Commercial Airplane Company, himself a preeminent jet aircraft designer, "We are good partly because we build so many airplanes. We learn from our mistakes, and each of our airplanes absorbs everything we have learned from earlier models and from other airplanes."[20]

Learning by Doing, Using, and Failing

It has long been recognized that there is a strong learning curve associated with manufacturing activity. Arrow characterized the learning that comes from developing increasing skill in manufacturing as "learning by doing."[21] Learning by doing results in lower labor costs. The concept of improvement by learning from experience has been subsequently elaborated by the Boston Consulting Group and others to include improvements in production process, management systems, distribution, sales, advertising, worker training, and motivation. This enhanced learning process, which has been shown for many products to reduce full costs by a predictable percentage every time volume doubles, is called the experience curve.[22]

Rosenberg, based on his study of the aircraft industry, has proposed a different kind of learning process, "learning by using."[23] Rosenberg distinguishes between learning that is "internal" and "external" to the production process. Internal learning results from experience with manufacturing the product, "learning by doing"; external learning is the result of what happens when users have the opportunity to use the product for extended periods of time. Under such circumstances, two types of useful knowledge may be derived by the developing organization. One kind of learning (embodied) results in design modifications that improve performance, usability, or reliability; a second kind of learning (disembodied) results in improved operation of the original or the subsequently modified product.

In our study, we found another type of external learning, a "learning by failing," which resulted in the development of new market approaches, new product concepts, and new technological alternatives based on the failure of one or more earlier attempts (Figure 3). When a product succeeds, user experience acts as a feedback signal to the alert manufacturer that can be converted into design or operating improvements (learning by using). For products that generate negligible sales volume, little learning by using takes place. On the other hand, products that fail act as important probes into user space that can capture important information about what it would take to make a brand new effort successful, which sometimes makes them the catalyst for major reorientations. In this sense, a new product is the ultimate market study. For truly new products it may be the only effective means of sensing market attitudes. According to one of our respondents, a vice president of engineering of "California Computer," a computer peripherals manufacturer, "No one really knows if a truly new product is worth anything until it has been

---

[19] N. Rosenberg, *Inside the Black Box, Technology and Economics* (Cambridge: Cambridge University Press, 1982), pp. 124–6.

[20] J. Newhouse, *The Sporty Game* (New York: Alfred A. Knopf, 1982), p. 7.

[21] K. Arrow, "The Economic Implications of Learning by Doing," *Review of Economic Studies,* June 1962.

[22] B. Henderson, *Perspectives on Experience* (Boston Consulting Group, 1968) (third printing, 1972).

[23] Arrow, "Economic Implications," pp. 120–40.

in the market and its potential has been assessed."

Another dimension of "learning by failing" relates to organizational development. A failure helps to identify weak links in the organization and to inocculate strong parts of the organization against the same failure pattern. The aftermath of the Apple III resulted in numerous terminations at Apple Computer, from the president to the project manager of the Apple III project. Those remaining, aided by new personnel, accounted for the well-implemented Apple III redesign and reintroduction program and the highly successful Apple IIe follow-on product.[24]

When the carryover of learning from one product to another is recognized, it becomes clear that the full measure of a product's impact can only be determined by viewing it in the context of both the products that preceded it and those that followed. While useful information can be obtained by focusing on individual products or pairs of products, *the product family is a far superior unit of analysis from which to derive prescriptions for practicing managers.* The product family incorporates the interrelationship between products, the learning from failures as well as from successes. Thus, it is to product families, including false starts, not to individual products, that financial measures of success should be more appropriately applied.

Consider a triplet of communications products developed over a 10-year period by an electronics system manufacturer. For several years, Marine Technology had developed and marketed commercial and military navigation systems. These systems were composed of separate components manufactured by others, such as a receiver, teletype, and a minicomputer, none of which was specifically designed for the harsh marine environment. Additionally, this multicomponent approach, though technically satisfactory, took up a great deal of space, which is at a premium on the bridge of a ship. Each of Marine Technology's new product generations attempted to further reduce the number of components in the system. By 1975, a bulky HP minicomputer was the only outboard component.

FIGURE 3 A Model of Internal and External Learning

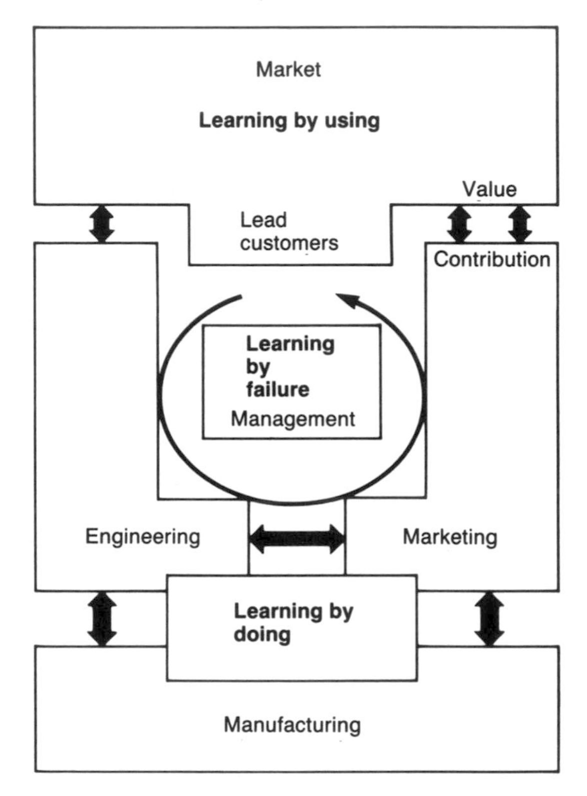

The need for a compact, rugged, integrated navigation system had thus been abundantly clear to Marine Technology engineers and sales people. Therefore, when microprocessors became available in the early 1970s, it was not surprising that Marine Technology's general manager initiated a program to develop a new lightweight integrated navigation system

[24] Maidique, Gable, and Tylka, *"Apple Computer (A) and (B)."* Also, Maidique and Swanger, *"Apple Computer: The First Ten Years."*

specially designed for the marine environment. The product was developed by a closely knit design team that spent six to eight months working with potential customers, and later market testing prototypes. Two years later, the company introduced the MT-1, the world's first microprocessor-based integrated navigation system. The product was an instant success. Over 7,000 units were sold at a price of $25,000 per unit. At this price, margins exceeded 50 percent.

Shortly after the success of the MT-1 was established, engineering proposed a new product (the MT-2) to Marine Technology's newly appointed president. The MT-2 was to be about one-sixth the volume of the MT-1 and substantially cheaper in price. The president was so impressed with a model of the proposed product that he directed a team, staffed in part by the original MT-1 design team members, to proceed in a top secret effort to develop the MT-2. The team worked in isolation; only a handful of upper management and marketing people were aware of the project. Three and a half years and $3.5 million later, the team had been able—by sacrificing some features—to shrink the product as promised to one-sixth the size of the MT-1 and to reduce the price to about $10,000. But almost simultaneously with the completion of the MT-2's development, a competitor had introduced an equivalent product for $6,000. Furthermore, the product's small size was not considered a major advantage. Key customers indicated the previous product was "compact enough." The company attempted to eliminate some additional features to tailor the product to the consumer navigation market, but it found that it was far too expensive for this market, yet performance and quality were too low for its traditional commercial and military markets. The product was an abject economic failure. Most of the inventory had to be sold below the cost.

A third product in the line, the MT-3, however, capitalized on the lessons of the MT-2 failure. The new MT-3 was directed specifically at the consumer market. Price, not size, was the key goal. Within two years, the MT-3 was introduced at a price of $3,000. Like the MT-1, it was a major commercial success for the company. Over 1,500 have been sold and the company had a backlog of 600 orders in 1982 when the case histories were completed.

At the outset of this abbreviated product family vignette, we said the family consisted of three products. In a strict sense, this is correct, but in reality there were four products, starting with what we will call the MT-0, the archaic modular system. The MT-0 was instrumental to the success of the MT-1. Through the experience with customers that it provided, it served to communicate to the company that size and reliability improvements would be highly valued by customers in the commercial and military markets in which the company operated. With the appearance of microprocessor technology, what remained was a technical challenge, usually a smaller barrier to success than deciphering how to tailor a new technology to the wants of the relevant set of customers, as the company found out through the MT-2.

The success of the MT-1 was misread by the company to mean, "the smaller, the more successful," rather than, "the better we understand what is important to the customer the more successful." The company had implicitly made an inappropriate trade-off between performance, size, and cost. They acted as if they had the secret to success—compact size—and by shutting off its design team from its new as well as its old customers ensured that they would not learn from them the real secrets to success in the continually evolving market environment. It remained for the failure of the MT-2 to bring home to management that, by virtue of its new design, the company was now appealing to a new customer group that had different values from its traditional commercial and military customer. Equipped with this new learning, the company was now able to develop the successful MT-3.

5. THE NEW PRODUCT LEARNING CYCLE

There are several lessons to be learned from the history of Marine Technology's interrelated succession of products. First, their experience clearly illustrates the importance of precursors and follow-on products in assessing product success. To what extent, for instance, was the M-2 truly a failure, or, alternatively, how necessary was such a product to pave the way for the successful MT-3? With hindsight one can always argue that the company should have been able to go directly to the MT-3, but wasn't to some extent the learning experience of the MT-2 necessary? Secondly, the story illustrates once again the importance of in-depth customer understanding as well as continuous interaction with potential customers throughout the development process even at the risk of revealing some proprietary information. Whatever learning might have been possible before entering the consumer market was shunted aside by the company's secretive practices.

The product evolution pattern of "Computronics," a start-up computer systems manufacturer, reinforces these findings. One of Computronics' founders had developed a new product idea for turnkey computer inventory control systems for jobbers (small distributors) in one of the basic industries. From his experience as a jobber in this industry, he knew that it was virgin territory for a well-conceived and supported computerized system. During the development process, the company enlisted the support of the relevant industry association. Association members offered product suggestions, criticized product development, and ultimately the association endorsed the product for use by its members. The first Compu-100 system was shipped in 1973. Ten years later, largely on the strength of this product and its accessories, corporate sales had doubled several times and reached nearly $100 million, and the company dominated the jobber market.

As the company's market share increased, however, management recognized that new markets would have to be addressed if rapid growth was to continue. In early 1977, the company decided to take what seemed a very logical step to develop a system that would address the needs of the large wholesale distributors in the industry. Based on its earlier successes, the company planned to take this closely related market by storm. After a few visits to warehousing distributors, the product specifications were established and development began under the leadership of a new division established to serve the high end of the business. Since no one at Computronics had firsthand experience with this higher level segment of the distribution system, a software package was purchased from a small software company, but it took a crash program and several programmers a year to rewrite the package so that it was compatible with Computronics hardware. After testing the Compu-200 at two sites, the company hired a team of additional sales representatives and prepared for a national roll-out. Ten million dollars of sales were projected for the first year.

First-year revenues, however, were minimal. Even three years after the launch, the product had yet to achieve the first year's target revenue. What had happened? The new market would appear at first glance to be a perfect fit with Computronics' skills and experience, yet a closer examination revealed considerable differences in the new customer environment which, nonetheless, were brushed away in a cavalier manner by Computronics' management, who were basking in the glow of the Compu-100's success.

As organizations in such a euphoric state often do, Computronics grossly underestimated the task at hand. The large market for warehousing distributor inventory systems was attractive to major competitors such as IBM and DEC. But only a cursory study of the new customers and their buying habits was carried out. The tacit assumption was that large warehousing distributors were simply grown-up jobbers. Yet these new customers were now much more sophisticated, had used data processing equipment for other functions, and generally required and developed their own specialized software. Increased competition and radically greater customer sophistication combined to require that Computronics be represented by a highly experienced and competent sales force. But because of the hurry to launch the product, Computronics skipped the customary training for sales representatives and launched the field sales force into a new area for the firm: the complex long-term business of selling large items ($480,000 each) to a technically knowledgeable customer. By believing that repeating past practices would reproduce past successes, Computronics had turned success into failure.

"Every victory," Carl Jung once wrote, "contains the germ of a future defeat."[25] Starbuck and his colleagues have observed that successful organizations accumulate surplus resources that allow them to loosen their connections to their environments and to achieve greater autonomy, but they explain, "this autonomy reduces the sensitivity of organizations to changing environmental conditions . . . organizations become less able to perceive what is happening so they fantasize about their environments and it may happen that reality intrudes only occasionally and marginally on these fantasies."[26] Fantasies create a myth of invincibility, yet an old Chinese proverb says, "There is no greater disaster than taking on an enemy lightly."[27]

Marine Technology's management fantasized that they had the secret of success: smaller is better. Computronics had a fantasy that similarly extrapolated their past victories: new markets will be like old markets. This is a pattern that repeats itself over and over in business. We have already alluded to one of the best publicized contemporary examples of this phenomenon: Apple Computer's Apple III on the heels of its colossally successful Apple II precursor. Even IBM is not exempt from this cycle. After taking one third of the market with its PC (personal computer) despite its late entry, a senior executive at the IBM PC Division stated, "We can do anything."[28] "Anything" did not, as it happens, include the follow-on product to the PC, the PCjr, which—in contrast to the original PC—was not adequately test marketed to determine how consumers would react to its design features and ultimately had to be dropped from the IBM product line. IBM and Computronics, however, have both, at least temporarily, become more humble. Both are soliciting customer inputs so that they can redesign their disappointments.

The flow from success to failure, and back to success again, at Marine Technology illustrated a rhythm that we were to encounter repeatedly in our investigations. In the simplest terms, failure is the ultimate teacher. From its lessons the persistent build their successes. Success, on the other hand, often breeds complacency. Moreover, success seems to create a tendency to ignore the basics, to believe that heroics are a substitute for sound business practice. As the general manager of "Automatrix," a test equipment manufacturer, pointed out, "It's hard, very hard, to learn from your successes." Ironically, success can breed failure for firms that continue to view the future through the prism of present victories, especially in a dynamic industry environment.

These observations have led us to propose a model of new product success and failure in which successes and failures alternate with an irregular rhythm. This is not to say that for

---

[25] C. G. Jung, *Psychological Reflections* (New York, Princeton University Press, 1970), p. 188.

[26] W. Starbuck, A. Greve, and B. L. Hedberg, "Responding to Crisis," *Journal of Business Administration,* no. 9 (1978), pp. 111–37.

[27] Lao Tzu, *Tao Te Ching* (New York: Penguin Books, 1983), p. 131.

[28] D. Le Grande, as quoted in "How IBM Made Junior an Underachiever," *Business Week,* June 25, 1984, p. 106.

every success there must be a complementing failure. Most industrial products—about three out of five—succeed despite popular myths to the contrary.[29] For some highly successful companies, as many as three out of four new products may be commercially successful. Most companies continuously learn by using, through their successful new products, and—as in the case of Marine Technology—they continuously develop improved designs. This is what most new product efforts are about—minor variations on existing themes.

But continued variations on a theme do not always lead to major successes. In time, further variations are no longer profitable, and the company usually decides to depart from the original theme by adopting a new technology—microprocessors, lasers, optics—or to attack a new market—consumer, industrial, government—or, alternatively, organizational changes, defections, or promotions destroy part of the memory of the organization so that the old now seems like the new. Changes in any of these three dimensions can result in an economic failure, or, in our terms, new learning about a technology, a market, or about the strengths and weaknesses of a newly formed group as shown in Figure 4, an extension of the familiar product-customer matrix originally proposed by Ansoff.[30] This recurrent cycle of success and failure is shown graphically in Figure 5.

In the model, a sequence of successes is followed by either a major organizational change, changes in product design, technology, or market directions that prompt an economic failure, which in turn spurs a new learning pattern. The model assumes a competitive marketplace, however, and is less likely to be applicable to a monopolistic situation in which a single firm dictates the relationship between customer and supplier. A second caveat is that while the pattern is roughly depicted as regular, in general it will be irregular, but the cycle of oscillation between economic success and failure, we believe, will still hold.

[29] R. G. Cooper, "Most Products *Do* Succeed," *Research Management,* November–December 1983, pp. 20–25.

[30] H. I. Ansoff, *Corporate Strategy* (New York: McGraw-Hill, 1965), pp. 131–33.

FIGURE 4 Learning by Moving Away from Home Base

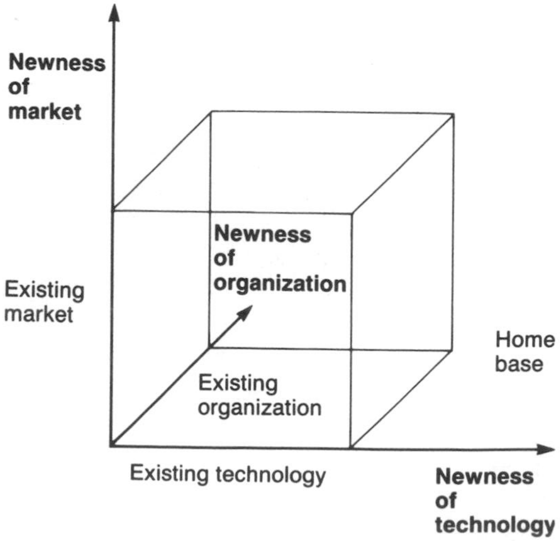

Newness of market
Existing market
Newness of organization
Existing organization
Home base
Existing technology
Newness of technology

FIGURE 5 A Typical New Product Evolution Pattern

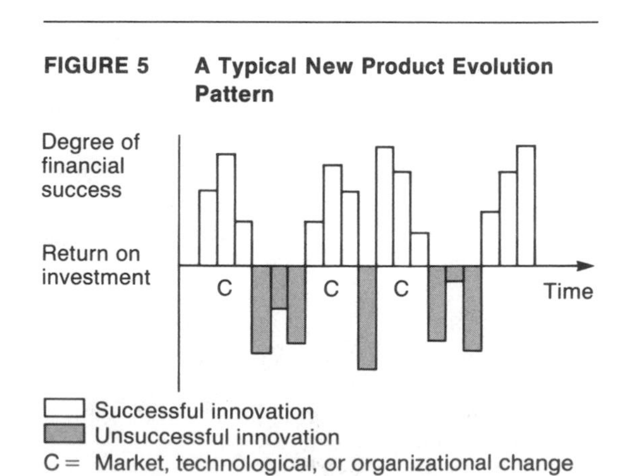

5.1 Success as a Stochastic Process

Success in new products is never assured. Too many uncontrolled external variables influence the outcome. Occasional or even frequent failure is a way of life for product innovations. As Addison reminds us in his Cato, "Tis not in mortals to command success."[31] But while it is not possible to assure the outcome of any one product trial, it is possible to increase the likelihood that a product or group of products will be successful. Addison goes on to add, "We'll do more, Sempronius, *we'll deserve it.*" (Authors' emphasis.) The eight factors that we identified at the outset of this article, and the cyclic model we propose, are an attempt to help managers conceptualize the new product development process so as to improve the proportion of economic successes. But on the other hand, it would be a mistake to attempt to increase the number of successes by reducing new product risk to zero by cautious, deliberate management. In the process, rewards may be also reduced to the same level.

Failure, as we have tried to argue, is part and parcel of the learning process that ultimately results in success. Sahal sums up the process thus, "What eventually makes the development of new techniques possible is the object lesson learned from past failures . . . profit by example."[32] The important thing is to have a balance between successes and failures that results in attractive returns. Here a lesson from experienced venture capitalists, masters at the success forecasting game, is useful. As part of another research project, the authors have interviewed some of the nation's most successful and experienced venture capital investors whose portfolios have generally shown gains of 25 to 35 percent over the past 10 years. Given a large pot of opportunities, experienced venture capitalists believe they can select a group that will on the average yield an excellent return, yet few professionals are so sanguine that they believe that they can with certainty foretell any one success; they've seen too many of their dreams fail to meet expectations.

A new product developed by "Electrosystems," a military electronics company, seemed to fit a venture capitalist's dream. The company's new product, a phase-locked loop, had its origin with one of their key customers, the requisite technology was within their area of expertise, and a powerful executive championed the product throughout its development. A large market was anticipated. Thus far, a good bet, one might conclude. What's more, the resulting product was a high-quality instrument. Yet the product brought little in the way of revenues to the firm for an alternative technology that solved the same problem in a cheaper way was simultaneously developed and introduced by a competitor.

Five years later, again spurred by a customer requirement, Electrosystems developed an electronic counter as part of a well-funded and visibly championed development program. This time, however, the product saved the customer 70 percent of his labor costs, considerably reduced his downtime, and there was no alternative technology on the horizon. This product was very successful. The point here is not that the product that ultimately produced a cost advantage to the customer was more successful. That much is self-evident. The point is that at the outset both products looked like they would provide important advantages to the customer. After all, they both originated with customers. Both projects were well-managed and funded and technologically successful. Yet one met with unpredictable external competition that blunted its potential contribution. The company did not simply fail and then succeed. It succeeded because it pursued both seemingly attractive opportunities. In other words, success generally requires not one but several, sometimes numerous, well-managed trials. This realization prompted one of our

---

[31] J. Bartlett, *Bartlett's Familiar Quotations,* 14th ed. (New York: Little, Brown, 1968), p. 393.

[32] Sahal, *Patterns.*

wisest interviewees, the chief engineer of "Metalex," an instrument manufacturer, to sit back and say, "I've found the more diligent you are, the more luck you have."

This is the way both venture capitalists and many experienced high-technology product developers view the new product process. Venture capitalists who have compiled statistics on the process have found that only 60 percent of new ventures result in commercial success, the rest are a partial or complete loss. (Not surprisingly, this is about the same batting average that Cooper found in his study of industrial products.) About 40 to 50 percent of new venture-capital backed ventures produce reasonable returns, and only 10 to 15 percent result in outstanding investments. But it can be easily computed that such a combination of investments can produce a 25 to 30 percent return or more as a portfolio.

5.2 New Research Directions

Our research on new product success and failure has led us to reconsider our unit of analysis. Choosing the new product as the basic unit of analysis has many advantages. New products are clearly identifiable entities. This facilitates gathering research data. New products have individualized sales forecasts and return on investment criteria, and managements generally know whether these criteria are satisfied. "Successes" can be culled from "failures."

Our results, however, indicate that if financial measures of success are to be applied as criteria, a more appropriate unit of analysis is the product family. Before an individual product is classified as a failure, its contribution to organizational growth, market development, or technological advance must be gauged. New products strongly influence the performance of their successors, and in turn are a function of the victories and defeats of their predecessors. Before the laurels are handed over to a winning team, an examination should be made of the market, technological, and organizational base from which the team launched its victory (Figure 4).

One of IBM's most notable product disasters was the Stretch computer. IBM set out to develop the world's most advanced computer, and, after spending $20 million in the 1960s for development, only a few units were sold.

On the heels of the Stretch fiasco came one of the most successful products of all time, the IBM 360 series. But when IBM set out to distribute kudos, it recognized that much of the technology in the IBM 360 was derived from work done on the Stretch computer by Stephen Dunwell, once the scapegoat for the Stretch "setback." Subsequently Mr. Dunwell was made an IBM fellow, a very prestigious position at IBM that carries many unique perks.[33] As Newton once said, "If I have seen far it is because I stood on the shoulders of giants."[34]

We were able to gain insight into this familial product interrelationship because our success-failure dyads were often members of the same product family. But even though they were interrelated, they represented only a truncated segment of a product family. Nonetheless, in some sites, for example Marine Technology, we were able to collect data on three or four members of a product family. On the other hand, our efforts, to date, fall far short of a systematic study of product families. This is the central task of the next stage of our research.

Our limited results, however, bring into question research that focuses on the product as the unit of analysis, including our own. Consider one of our principal research findings, which is also buttressed by the findings of several prior investigators: successful products benefit from existing strengths of the developing business unit. The implication of this find-

[33] T. Wise, "IBM's $5B Gamble," *Fortune,* September 1966; "A Rocky Road to the Marketplace," *Fortune,* October 1966. Also, Bob Evans, personal communication (Mr. Evans was program manager for the IBM-360 system).

[34] J. Bartlett, *Bartlett's Familiar Quotations,* 13th ed. (New York: Little, Brown, 1968), p. 379.

ing is that organizations should be wary of exploring new territories. In contrast to this result, our observations would lead us to argue just the opposite, that firms should continuously explore new territories even if the risk of failure is magnified.[35] The payoff is the learning that will come from the "failures" which will pave the way for future successes.

Careful validation of the cyclic model of product development proposed here could have other important consequences for our understanding of technology-based firms. If indeed the pattern proposed in Figure 5 is generalizable to firms that are continuously attempting to adapt to new markets and technologies, then there are important implications for management practice.

First, the model implies that new product development success pivots on the effectiveness of intra and inter company learning. This conclusion puts a premium on devising a managerial style and structure that serves to catalyze internal and external communication. Second, by implicitly taking a long-term view of the product development process, the model emphasizes the importance of long-term relationships with employees, customers, and suppliers. Out of such a view comes a high level of understanding, and therefore of tolerance for failure to achieve commercial success at any one given point in the product line trajectory. Firms need to learn that product development is a journey, not a destination. These preliminary findings are compatible with an exploratory study of new product development in five large successful Japanese companies completed by Imai and his colleagues.[36] One of the principal findings of their research was that the firms studied were characterized by an almost "fanatical devotion towards learning—both within organizational membership (sic) and with outside members of the interorganizational network." This learning, according to the authors, played a key role in facilitating successful new product development. It appears that when successful at new product development, small and large U.S. companies operate in a very similar manner to the best-managed Japanese firms.

Many key questions, however, remain to be settled. Is there an optimal balance between successes and failures? Are Japanese firms susceptible to the same oscillating pattern between success and failure as American firms? How does this balance change across industries? How can tolerance for failure be communicated without distorting the ultimate need for economic success? How can a firm learn from the failures of others? Are there characteristic success-failure patterns for a group of firms competing in the same industry? These and other related questions will occupy us in the next phase of our research.

---

[35] In an exploratory study of the relationship between the degree of "newness" of a firm's portfolio of products and its economic performance, the authors concluded that some "newness" results in better economic performance than "no newness." M. H. Meyer and E. B. Roberts, *New Product Strategy in Small High-Technology Firms,* WP #1428-1-84 (Sloan School of Management, Massachusetts Institute of Technology, May 1984).

[36] K. Imai, I. Nonaka, and H. Takeuchi, *Managing the New Product Development Process: How Japanese Companies Learn and Unlearn,* Institute of Business Research, Hitotsubashi University, Kunitachi (Tokyo, Japan, 1982), pp. 1–60. See also P. R. Lawrence and D. Dyer, *Renewing American Industry* (New York: The Free Press, 1983), p. 8.

Corporate Venturing Process

Case III–3
Medical Equipment (A)

R. A. Burgelman and T. J. Kosnik

INTRODUCTION

In January 1977, Daniel Burns was confronted with a thorny situation as he assumed his new responsibilities as director of Gamma Corporation's New Venture Division (NVD). Burns had been asked to review the performance of each of the projects underway in the NVD and to decide what actions, if any, were needed. One venture, the Medical Equipment Project, had experienced explosive growth in the early 1970s. However, the drastic increase in revenues had not been accomplished without problems. During the course of his evaluation, Burns discussed the history of Medical Equipment with Dr. Stephen Sherwood, the venture manager. He also consulted with Dr. Franz Korbin, who had managed Sherwood's efforts for several years before Burns had arrived. In the course of his discussions with other individuals within and outside Medical Equipment, Burns gained additional observations.

Burns was reviewing those conversations as he prepared for his meeting with Dr. Korbin. He had asked Korbin to come to the meeting with recommendations about what should be done with Medical Equipment and with Dr. Sherwood. Burns wondered what Korbin might recommend. He was also thinking through what his options were in response to Korbin's potential suggestions.

GAMMA CORPORATION AND THE NEW VENTURE DIVISION

Gamma Corporation is a multibillion dollar, U.S.-based industrial company. It started out as a holding company and operated for many years as a group of relatively autonomous components with limited direction from top management. The change from a holding company to a more integrated diversified firm occurred gradually through the 1960s. This integration was formalized through the adoption in the early 1970s of a comprehensive long-term program to improve all elements of human resources and business management. In recent years, a formal, unified management system has been instituted throughout the corporation. In the context of this new system, the operating components of Gamma develop interlocking objectives that are consistent with overall corporate objectives and goals. Gamma's Management Committee, with the support of the executive staff, oversees this process. It is deeply involved in setting overall direction, and monitors the evolution of business strategies and the allocation of capital.

The implementation of a uniform corporate management system was paralleled by an effort to introduce a uniform, integrated strategic planning approach for the management of Gamma's businesses. Gamma's businesses are now organized into strategic business units (SBUs). More than 150 SBUs are divided into six categories according to the criteria of present and expected future performance in terms of profitability, growth rate, and competitive position. The first four categories represent ongoing businesses offering successively lower gradations of business strength and potential. The fifth category represents business areas where withdrawal is indicated. The sixth is made up of emerging new businesses.

This large agglomeration of widely diversified but partially related businesses is grouped into a dozen major operational divisions under the administrative umbrella of corporate management and its executive staff.

Gamma's Unrelated Diversification Efforts

In view of its strategy of relying on the development of a "sustainable technical advantage" to compete in a particular market, Gamma has consistently spent vast amounts of resources on R&D activities both at the divisional and corporate levels. As a result, the firm has one of the highest records of obtaining annual awards for significant technological developments. The bulk of Gamma's business, however, traditionally consists of selling "commodities" and "intermediates" to major firms in the steel, auto, construction, appliance, textile, pharmaceuticals, and food industries, among others. Gamma's direct contacts with final consumers or users, though substantial in an absolute sense, have constituted a relatively minor part of its business.

During the late 60s, corporate management wished to dampen the effects of the business cycle on Gamma's activities, to protect itself from the long-range threats to its basic commodities and intermediate businesses, and to catch a greater part of the value added in the chain from raw materials to finished products. As a result, Gamma became interested in developing new business opportunities in unrelated areas, especially in the direction toward reaching the final consumer or user.

Before the creation of the New Venture Division (NVD) in 1970, such unrelated diversification efforts were carried out in the context of the operating divisions, with the help of the corporate R&D department. But, by the end of the 1960s, corporate management had become increasingly concerned about the relevance of the corporate R&D efforts for future growth and diversification in unrelated areas, and about its ability to turn new ideas into new businesses through the divisions.

The New Venture Division

By 1970, a separate group was formed for the corporate development efforts. At the outset, this group was loosely organized. It encompassed New Business Development (NBD) and New Business Research (NBR) departments. In 1971, the corporate R&D (CR) department was integrated into this group. The heads of all these departments reported directly to the corporate vice president for technology. Initially, the idea was to generate new technical projects in the corporate R&D department, manage them through a development phase, and then transfer them as new departments to existing divisions, or grow them into new freestanding divisions.

The new NBD department comprised the Gascoal, Firefit, Environmental Systems, Farming Systems, Sea Products, and Medical Equipment Projects. These six projects were managed as new ventures with their own organizations under the supervision of the NBD manager.

During 1973, the concept of a New Venture Division (NVD) took clearer shape. The appointment of a first NVD director, to whom the NBD, NBR, and CR departments now reported, resulted in a more integrated management structure for the NVD. Don Sharp, the newly appointed NVD director, had been the NBD department head before taking up his new responsibilities. He was replaced as NBD manager by Dr. Franz Korbin. Exhibit 1 illustrates this management structure as it existed in January of 1973.

Don Sharp and Franz Korbin were generally considered among the top-notch managers in the entire corporation. They were consistently referred to as brilliant conceptualizers and strategists, as the typical high-level entrepre-

EXHIBIT 1 Gamma Corporation: New Venture Division Organization Chart

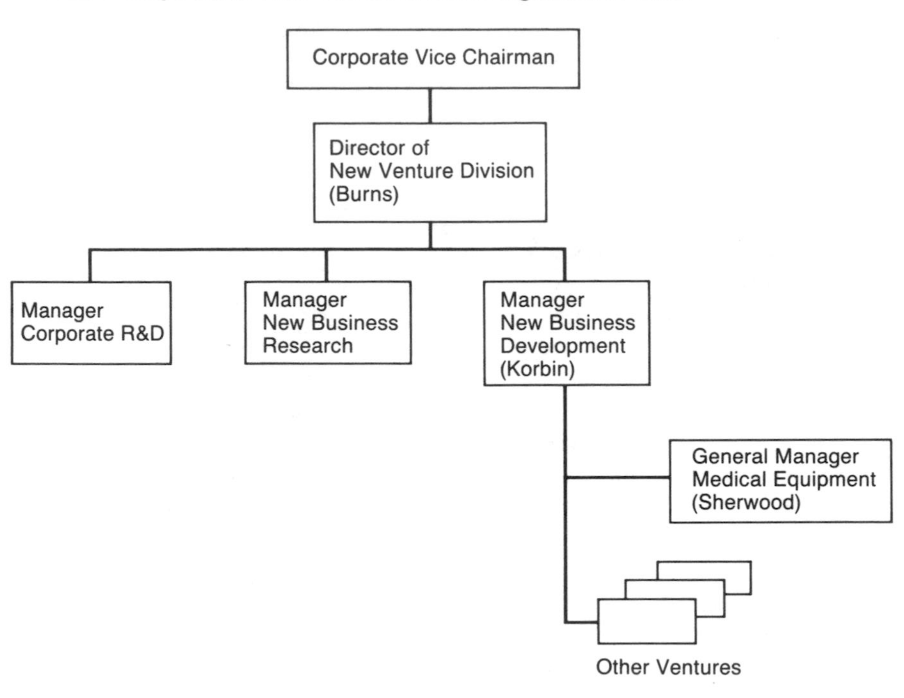

neurial managers. One observer articulated their philosophy:

> Sharp's and Korbin's approach was for you to learn while you are doing. Once you have a strategic concept you should start something, push it on its feet, and let it run and grow fast and thereby secure continued support from top management.

In this philosophy, the role of the entrepreneurial venture manager was very important. It encouraged strategic initiatives at lower levels, and allowed people to run hard within the as-yet-vaguely delineated boundaries of new business arenas.

However, by the end of 1976, the point had been reached where the number of projects and the total amount of required resources surpassed the corporation's readiness to absorb and fund. In view of a perceived need for consolidating the unrelated diversification efforts, corporate management promoted Don Sharp to the presidency of one of the operating divisions and brought Dan Burns in as the new NVD director in January of 1977. Burns was generally known in the corporation as the solid business type: realistic, cautious, and oriented towards "the numbers." He was expected to pull things together more, to establish a managerial rather than an entrepreneurial style throughout the NVD.

THE MEDICAL EQUIPMENT (ME) VENTURE

Burns was aware that Gamma Corporation had been interested in the health care field for many years. As early as the 1960s, the company had acquired a pharmaceuticals firm. Although that firm was subsequently divested in the 1968–69 recession, it provided part of the seeds from which the Medical Equipment Venture emerged. Other seeds included the company's experience in producing chemical reagents, nuclear medicine equipment, and electronic gear.

Evolution of the Project

From his conversation with Korbin, Sherwood, and others, and from a detailed study of the written long-range plans for the venture in the period 1973–76, Burns had obtained a fairly clear chronology of the venture's development. Exhibit 2 provides an overview of the evolution of the venture's sales volume and net income in the period 1970–77.

In 1968, Dr. Stephen Sherwood and several associates developed a business plan for a venture to take advantage of various technologies available within Gamma for measurement instruments. An instruments department was formed in the Electronics Division, and they began to explore potential high-growth areas, including the medical instrumentation business. At the same time, although the pharmaceutical company was divested, Gamma kept the medical instruments part of that firm.

The corporate R&D group of Gamma was interested in exploring the medical instruments area. When in 1969 the recession forced a reduction in the resource allocation to instruments in the Electronics Division, Sherwood was transferred to corporate R&D where he continued to explore potential applications in the medical technology area.

At a think tank session in R&D, another scientist mentioned a novel idea for a computerized blood analyzer. The idea was for a clinical centrifugal analyzer. In Sherwood's words,

> [It was]. . . an entirely new concept—not at all a "me too" type of product. So I made some proposals that were accepted. The other scientist had very fixed ideas about how the product should look as a commercial product. But I had experience in commercial computers and knew the need to match a technology with a market configuration. . . that you must develop a "needs oriented" approach. As the project manager, I insisted on doing market research . . . . We ended with a radical departure from the original approach. We used only the nucleus of the physical concepts.

For the first three years, the Medical Equipment Project was essentially focused on the development of one product: the blood analyzer, which was named the ANABASE system. Later, other products and projects were added. However, the ANABASE system was still the mainstay of the business in 1976.

In 1972, Sherwood proposed to sell the ANABASE system via a sales force specifically for the Medical Equipment Project. He met with organizational resistance, partly because management of Gamma was unsure of its ability to market to medical institutions.

Sherwood had told Burns:

> When we proposed to sell the ANABASE product by our own selling force, there was a lot of

EXHIBIT 2 Evolution of Sales and Net Income, 1970–1977 (in thousands of dollars)

	1970	1971	1972	1973	1974	1975	1976	1977	1978
Sales	700	1,200	2,000	5,000	10,000	18,000	32,000	40,000*	60,000*
Net income	(470)†	(720)†	(760)†	(720)†	(1,000)†	(860)†	330†	350*	3,000*

* Estimated.
† Exclusive related corporate R&D expenditures.

resistance out of ignorance. Management did numerous studies, had outside consultants on which they spent tens of thousands of dollars; they looked at XYZ Company for a possible partnership.

Management was just very unsure about its marketing capability. I proposed to have a test marketing phase with 20 to 25 installations in the field. We built our own service group; we pulled ourselves up by the "bootstraps." I guess we had more guts than sense.

By 1973, corporate management began to realize the substantial potential of the ANABASE business, and a new competitive arena was defined by the management of the New Venture Division (NVD) and endorsed by corporate management. According to the corporate long-range plan of 1973, the corporation wanted to

> Become a leading supplier of coordinated mechanical systems and related products in a market currently dependent on manpower, and limited annual operations for clinical testing in hospitals and private laboratories.

In 1974, the project obtained official venture status, and moved to the jurisdiction of the New Business Development department (NBD) in the NVD. Dr. Sherwood reported there to Dr. Korbin. During 1974, the ME venture continued to grow very rapidly because of the market success of the ANABASE system domestically and internationally, leading to a doubling of the 1973 $5 million sales volume. The newly formed ME venture encompassed, in addition to the ANABASE project and the selling of reagents, a radio-diagnostic product line and immunoassay technology. The backbone of the venture in terms of sales volume, however, was formed by the ANABASE system.

During 1975, the business strategy for the ME arena continued to be based primarily on market penetration with the ANABASE system which had been further improved and expanded to include computerized data processing components. Sales volume reached over $15 million for the year. The venture's development efforts were primarily absorbed in the straightening out of product quality problems with the ANABASE product.

The further development of the biomedical technology (hardware and reagents) remained the responsibility of the corporate R&D department. Korbin and Sherwood recognized that these continued efforts would be very important for the venture to be able to reach the corporate objectives in terms of size for new ventures.

New product development for the ME arena, started in 1973, had resulted in the introduction of the ANALYZ automated immunoassay system for tests involving hormones, steroids, and therapeutic drugs, and in a radio-diagnostic product line for the detection of pathology of organs. These new products for the ME arena differed from the ANABASE product in that they attempted to conform more closely to an "integrated systems" approach, which had become the keystone of the arena strategy:

> Whereas ANABASE was designed to utilize reagents available from multiple sources in relatively low dollar volumes, subsequent products have been designed to utilize consumables primarily and in much higher dollar volumes, relative to those of the hardware. The use of sole-source consumables is traditionally repugnant to most customers, so it was necessary to demonstrate strong credibility in the marketplace before moving too far in that direction. Present indications suggest the achievement of that credibility.

By 1975, the ME venture had identified the key characteristics of the new business field in which it was operating: high technology; pull-through demand engendered by public desire for health care; cost and efficiency pressures; strong regulatory pressures; high competition from largely fragmented and specialized suppliers. These characteristics reinforced the importance of the "systems approach" articulated by venture management for the ME field:

> The development of automated methods with consumables, service, and education to maximize reliability, efficiency, and economy for the users.

Korbin and Sherwood showed a strong awareness of the continued importance of the marketing function, and the dangers of becoming technology and/or facilities oriented. They were explicit in stating what they perceived to be the key to success for the venture in the 1975 long-range plan:

> Success is dependent on the ability to identify needs sufficiently in advance of their realization to permit the long lead times associated with high technology and cumbersome regulations.

In response to concerns about too limited a product range for the ME venture, there had been efforts to look around in the company for other, health-related projects which might be agglomerated with the venture. A major new arena concerned "blood banking," for which applicable technology was being developed in two different divisions.

In 1975, it also became clear that the future growth of the ME arena would be accomplished through products acquired from outside the corporation rather than exclusively from internal development. The venture was expected to be able to make such acquisitions because of its capability to offer private inventors or small companies strong marketing organization and product engineering functions. Imaging instrumentation and microbiological applications were envisaged to offer good opportunities in this respect. The choice of the imaging arena highlights the growth in understanding the nature of the opportunities in the ME field. The long-range plan stated:

> The "imaging" arena breaks down the classical separation between radiology, nuclear medicine, and other disciplines, just as ANALYZ breaks down the dichotomy between nuclear medicine and pathology; and, our planned activities in microbiology are likely to cross into the hematology segment. Such states of market flux, or discontinuities, are prime targets for rapid growth in the ME business strategy.

Still another new arena concerning support devices for patients—ranging from artificial hip joints to artificial organs—was identified in 1975 as one where a systems approach would be feasible. The exact content of this arena had not yet been defined, but it was already clear that the opportunities in the low end of the technology spectrum would be marginal. As with the ANABASE product and with a particular product in the blood banking arena, management envisioned that here, too, a first product based on entirely new technology would serve as the nucleus around which a variety of other products could evolve.

During 1976, the acquisition of a company with imaging technology was accomplished. Initial reaction to ANALYZ seemed favorable. A new radio-diagnostic product line had also been favorably received by customers, even though slower penetration than for ANALYZ was expected. For the patient support devices arena, the strategic design was more clearly articulated during 1976. The potential for value-upgrading of traditional raw materials was perceived to be great.

The ME venture continued to grow very rapidly, with sales of over $30 million for the year. During 1976, the ANABASE product line was reaching a stage of maturity. Sales for this product alone were more than $20 million.

The outlook for the future was quite optimistic in 1976. It was expected that the new government administration would mark the advent of "national health insurance" which would lead to an expected increase in diagnostic testing expenditures of 150 percent in 1980 over the 1975 forecast, and would result in substantial changes in health care delivery patterns. This anticipated change led to a further refinement of the original arena's new product development strategy:

> Product development strategies are setting objectives for small, freestanding, fast, discretion-

ary, low-skilled labor systems which can be combined to allow higher capacity systems for the hospital setting with high reliability through redundancy.

For 1977, the new blood bank product line was expected to begin penetrating the market. Still another product, for rapid cardiac diagnosis, was scheduled to be introduced during 1977. The latter product was oriented toward a first penetration of the physician's office and the clinical, rather than therapeutic, segment of the market. Sales volume was expected to reach $40 million. Even if the venture would absorb all related corporate R&D expenditures, management expected to show a small profit for the year.

Roles of Key Players in the Development Process

Burns also had tried to develop a clear understanding of the specific contribution of Korbin and Sherwood in the development of the Medical Equipment Venture. In an earlier conversation, Sherwood had explained to Burns that there had been three phases in the development from a prototype product to commercially viable large-scale production of the ANABASE system. First was the technology development stage, in which the various new technologies were integrated, tested, and refined to produce a system that was reliable and easy to use. That phase took nearly three years. To some extent, technology development continued throughout the product life cycle as problems were discovered and enhancements were made. For example, the number of samples that the system was able to analyze simultaneously was increased in later models. In addition, the data processing capabilities of the system for the handling of patient records throughout the laboratory test process had been enhanced.

The second phase was market development. During that period, the contacts were made in the medical community which were necessary for the introduction and testing of the new product. Announcements about the product and its capabilities were made to the public. In the case of ANABASE, the first product announcements were made before Sherwood and his colleagues were certain about how the commercial product would actually function. As part of the market development phase, versions of the system were installed and tested in the field. These "beta test sites" provided the Medical Equipment Project with valuable information about how to improve the system that was to be offered during the subsequent phase.

The third phase of the process was commercialization. During that period, the manufacturing capability to produce the product in volume was established, and thoroughly tested systems were shipped to customer sites.

Dr. Sherwood had shared his views about that part of the process:

> Commercial development is a continuum, stretching over a three- to five-year period. That must be an integrated process, a continuous path. When you have different groups with different organizational barriers, you have problems. For instance, between production engineers and the people who look for a product adequate for the market. I want to generate an organization that would make that continuous process possible.

Daniel Burns realized that a large amount of the credit for the development of the Medical Products Venture to a $30 million business belonged to Dr. Sherwood. At the same time, the business was experiencing difficulties that seemed to be getting more acute as it grew larger.

Sherwood's role had been similar in some ways to that of an entrepreneur starting a new business. He had had to find the resources from within and outside Gamma Corporation to propel the ANABASE project through the technological development and market development phases to commercialization. He worked long hours throughout the process, driving the people who worked for him to do the same. He was, in the eyes of everyone with whom Burns met, the champion of the ANABASE product line. His single-minded commitment to the success of the project had forced the team over some difficult hurdles.

However, Sherwood was, like many entrepreneurs, not oriented toward long-term strategic thinking. In the words of one of his subordinates:

> Some of the problems probably relate to being a new venture and reaching a certain size. But another part is probably due to Dr. Sherwood's efforts to organize the venture in a way that is orienting us toward solving particular problems of an immediate nature, without taking into account some of the broader implications of the design.

There was also a concern that Sherwood was unable or unwilling to delegate responsibility. One of his subordinates had intimated that:

> Sherwood can't cut the rope. Once you pass the $15 million mark you must start delegating or you just get overwhelmed. He can't do that; he has to be in on all decisions. He even wants to decide the colors of the rooms. He'd run down the hall to have a say in that, too.

The people Sherwood had brought in under him on the Medical Equipment Venture were usually generalists rather than technical or functional specialists. Sherwood also had hired a number of people from outside Gamma Corporation to staff the new venture. Thus, both the organizational form and the people who filled the key roles contributed to the entrepreneurial atmosphere in which a manager was responsible for a variety of activities in different functional areas. For instance, Sherwood's "right-hand man" was an M.B.A. from a well-known western business school. He had been in charge of both R&D and manufacturing of the reagents for ANABASE. He had interviewed and hired research engineers for the project, and his lack of technical background led to several serious mistakes. Individuals not competent to carry out the work on new products were brought on board, and the development timetable suffered.

Sherwood's practice of hiring for the Medical Equipment Venture from outside Gamma had created some resentment among longtime Gamma personnel. The Gamma old-timers thought that Sherwood did not trust their competence. He was alleged to have said that "You can't hire good people from within Gamma Corporation."

Some people felt that many of the outside hires did not have the best interests of Gamma at heart. One longtime Gamma employee, now a manager in the venture, had told Burns:

> These tend to be opportunistic people who are often more interested in making a name for themselves, so that they can jump to a next, even better position. They will tend to push the things in their area, so that they can put on their resumes specific things that they have accomplished, but which in the overall framework of the new business may not be the best things. We have quite a bit of this "hot shot" type of routine.

As the Medical Equipment Venture grew, there was a need for strong functional managers to develop the systems and procedures to run the business more efficiently. In order to deal with the operating problems and create a more continuous flow for new product development, some of the generalists were replaced by functional specialists. These managers, in turn, put pressure on Sherwood to pay more attention to the development of internal systems. The manufacturing systems, in particular, had been neglected in the rush to build sales volume rapidly. The manufacturing function in the venture had been haphazard, confusing, and poorly planned. The new manufacturing manager had told Burns:

> Sherwood and Korbin made a very good strategic move by going from R&D directly to the marketplace. They minimized investment, bought time, and tried to maximize the cash flow. But coming from an R&D environment, they

missed in the selection of key manufacturing people. That's still an ongoing problem. There is still some normalization to be done.

As a result of some of these pressures toward more professional management, conflicts had arisen between Sherwood and the new managers. In mid-1976, a recently hired, very capable marketing manager had left for the competition.

Sherwood had been preoccupied with rapid growth for the Medical Equipment Venture. He had felt that in order for the venture to reach maturity as a new business organization within the Gamma corporate structure, it was necessary to reach sales volumes of $50 to $100 million within 5 to 10 years after the project began.

Sherwood had focused his efforts on the development and marketing of the ANABASE system. He became so caught up in that process that the new product development activities suffered. Relationships with corporate R&D were neglected. As a result, the flow of new product development was not yet brought under control. In the opinion of one of Sherwood's associates:

> Every ounce of effort with Dr. Sherwood is spent on the short run. There is no strategizing. New product development has been delayed, has been put to corporate R&D. . . . Every year we have doubled in size, but things never get any simpler.

There were serious communications problems within the Medical Equipment Venture and between the venture and the rest of Gamma Corporation. The product managers were often involved in disputes with the manufacturing and accounting functions. The venture had all but severed relations with corporate R&D, despite the fact that R&D was developing new medical products for them. One transferee from corporate R&D to the venture had told Burns:

> We were basically separated from the R&D group in the venture. The people in that group wanted to identify themselves. They did it to such an extent that they put a wall between themselves and us. In a way it was ironic. We were funded by the venture, and the technology that we developed was not accepted by them. Now, however, we are trying to reestablish the working relationship.

After the transfer of the ANABASE project as a new venture to the NBD department, Sherwood reported to Dr. Korbin. Burns knew that Dr. Korbin had also played a crucial role in the evolution of Medical Equipment. In the words of one of the people from Gamma's R&D group:

> You need a champion at the technical level to start it. But then you also need an "organizational champion" who, ideally, should come from outside R&D. . . . For Medical Equipment, Dr. Sherwood was the "product champion" here in the lab, and Dr. Korbin was the guy who was running around in the corporation and whispering in the right people's ears.

Korbin had earned the respect of peers and subordinates by the way he had handled his relationship with Medical Equipment as well as a number of other ventures before it. He was willing to take risks and to allow the venture manager a great deal of latitude in running the business. He sought out and worked with a wide range of product champions like Sherwood. He had an uncanny ability to cut deep into the heart of a problem, despite its complexity. One researcher said of Korbin:

> He doesn't need the in-depth familiarity with the technical aspects. He has the intellectual ability to transcend jargon, to thread through the logic in spite of the jargon.

Korbin had described his management approach to Burns as follows:

> First, I look for demonstrated performance on an arbitrarily—sometimes not even the right one—chosen tactic. For instance, doing "X" may not be the right move, but it can be done and one can gain credibility by doing it. So, what I am really looking for is the ability to predict and plan adequately. I want to verify the venture man-

> ager's claim that he knows how to predict and plan, so he needs a "demonstration project," even if it is only an experiment. The second thing that I look for is the strategy of the business. That is most important. The strategy should be attractive and workable. It should answer the question where he wants to be in the future and how he is going to get there. It is important for two reasons. First, once he has a strategy and can answer those questions, and I can verify a piece of it through a demonstration project, I can give strategic guidelines; I can let people make more independent decisions. That builds momentum, and I don't have to go to top management every day; so, it's also very important from an operational viewpoint. Second, it allows planning. And that, in turn, allows me to go to the corporation and stick my neck out.

Korbin had been the architect of the broader strategic design which guided the venture's development after 1973. This strategic design included:

1. Internal growth through new product development.
2. Agglomeration of medical-related projects from other parts of the company under the administrative umbrella of Medical Equipment.
3. External growth through the acquisition of related small companies.

As a consequence of this design, Dr. Korbin played an active role in attracting other projects from around the company to the Medical Equipment Venture. One manager described this approach:

> It was the genius of Dr. Korbin to attempt to get a better focus by bringing all of the capabilities relevant to the medical equipment field existing in the corporation under one management. That is to say, the refrigeration know-how of the Industrial Gases Division, the plastics know-how of Packaging Products, and the electronics know-how of Medical Equipment.

The product development and agglomeration process resulted in three lines of business within the Medical Equipment Venture: clinical chemistry (the original blood analyzer), nuclear medicine and imaging, and blood banking. The products for nuclear medicine were also used in laboratories to perform tests involving nuclear materials. While the technology was different, the marketing requirements meshed closely with the blood analyzer business.

The blood banking business was conceptually different from the original laboratory diagnostic products. It consisted of equipment which froze blood for subsequent use in transfusions to patients. Several managers speculated that it might be necessary to create separate organizations for the different product lines if they grew rapidly to a sufficient size. However, for the time being, they had agreed with Korbin that a critical mass was needed in order for a venture to succeed, and that combining the different products under one management was the best approach.

In 1976, only one outside company had been acquired and folded into the Medical Equipment Venture. However, there was no problem in taking that approach if an opportunity presented itself. Dr. Korbin was somewhat concerned that Dr. Sherwood might have overlooked such opportunities in the turmoil of fighting fires that sprang up in the day-to-day operations of the business.

CONCLUSION

As Burns reviewed the jumble of facts and opinions that had been uncovered in his discussions, he wondered what could have been done better in the case of the Medical Equipment Venture. What might the director of the NVD and senior management of Gamma have done to take advantage of Sherwood's entrepreneurial talents without letting things get out of hand? Were there things that Dr. Korbin might have done differently? He had championed the project to his superiors and brought related initiatives under one roof, but what might he have done for Sherwood? As for Sherwood himself, how might he have avoided some of the difficulties? The lessons learned from Medical Equipment were important, as other new ventures were likely to encounter similar situations in the future. Burns recalled that he had not been able to resist telling Sherwood at one point in his conversations with him that "You guys are so different. The corporation doesn't understand you. You must involve us more in the business."

Burns also wondered what should be done next. It was not clear what approach Dr. Korbin might propose when he arrived for the meeting. Burns wanted to have clear in his mind the advantages and the risks associated with each alternative. He wondered what potential solutions to the problem he might have overlooked. He felt he should have a tentative position about which action to take in case he was asked during the discussion. Finally, Burns was troubled that so much of his energies since he had taken over the NVD had been devoted to fire fighting, operating decisions like the one facing him with regard to Sherwood. He knew that some careful strategic thinking was necessary to determine what steps to take next with regard to the Medical Equipment Venture once the decision about the venture manager was made.

Reading III–3
Managing the Internal Corporate Venturing Process: Some Recommendations for Practice

R. A. Burgelman

Internal Corporate Venturing (ICV) is an important avenue for corporate growth and diversification.[1] Systematic research, however, suggests that developing entirely new businesses in the context of established firms is very difficult even when a separate new venture division is created for this purpose.[2]

Reprinted from *Sloan Management Review,* Winter 1984, vol. 25, no. 2, pp. 33–48, by permission of the publisher.

STAGE MODELS OF ICV

Typical conceptualizations of ICV use a "stage model" approach.[3] Stage models provide a framework for discussing many important problems concerning the sequential development of new ventures. They focus on within stage problems as well as on issues pertaining to the transition between stages. They emphasize the different requirements of different stages in terms of key tasks, people, structural arrangements, leadership styles, etc. The problems most naturally addressed in a stage model are the ones that are important in growing any new business. But, many of the more difficult problems generated and encountered by ICV result from growing a new business in the con-

[1] An overview of different forms of corporate venturing is provided in E. B. Roberts, "New Ventures for Corporate Growth," *Harvard Business Review* 58 (July–August 1980), pp. 132–42.

[2] An overview of early studies on new ventures is provided in E. von Hippel, "Successful and Failing Internal Corporate Ventures: An Empirical Analysis," *Industrial Marketing Management* 6, pp. 163–74. Von Hippel has noted the great diversity of new venture practices. Some of this diversity, however, may be due to a somewhat unclear distinction between new product development and new business development. Von Hippel also identifies some key factors associated with success and failure of new ventures but does not document how the *process* takes shape.

Concerning the use of the new venture division (NVD) design, see N. D. Fast, "The Future of Industrial New Venture Departments," *Industrial Marketing Management* 8, pp. 264–79. Fast has observed the precarious, unstable function of NVD in many firms. He explains the evolution in terms of shifts in corporate strategy and/or in the political position of the NVD.

[3] For a recent example of a stage model, see J. R. Galbraith, "The Stages of Growth," *Journal of Business Strategy,* 1983.

text of an established organization. They result from the fact that strategic activities related to ICV take place at different levels of management simultaneously. Such problems are not easily incorporated in a stage model, and tend to be discussed only in somewhat cursory fashion.

A PROCESS MODEL OF ICV

Recently, I have proposed a new model based on the findings of an exploratory study of the complete process through which new ventures take shape in the context of the new venture division (NVD) in large, diversified firms.[4] Figure 1 shows the "process model" of ICV. The methodology of the study is briefly described in Appendix 1.

Figure 1 shows the *core* processes of ICV as well as the *overlaying* processes (the corporate context) in which the core processes take shape. The core processes of ICV comprise the activities through which a new business becomes defined and its development gains impetus in the corporation. The core processes subsume the managerial problems and issues that are typically addressed in stage models of ICV. The overlaying processes comprise the activities through which the strategic and structural contexts are determined. Structural context refers to the various organizational and administrative mechanisms put in place by corporate management to implement the current corporate strategy. Strategic context refers to the process through which the current corporate strategy is extended to accommodate the new business resulting from ICV efforts. Both the core and overlaying processes involve key activities (the shaded area) and more peripheral activities (the nonshaded area), situated at different levels of the organization.

[4] R. A. Burgelman, "A Process Model of Internal Corporate Venturing in the Diversified Major Firm," *Administrative Science Quarterly* 28 (1983), pp. 223–44.

FIGURE 1 Key and Peripheral Activities in a Process Model of ICV

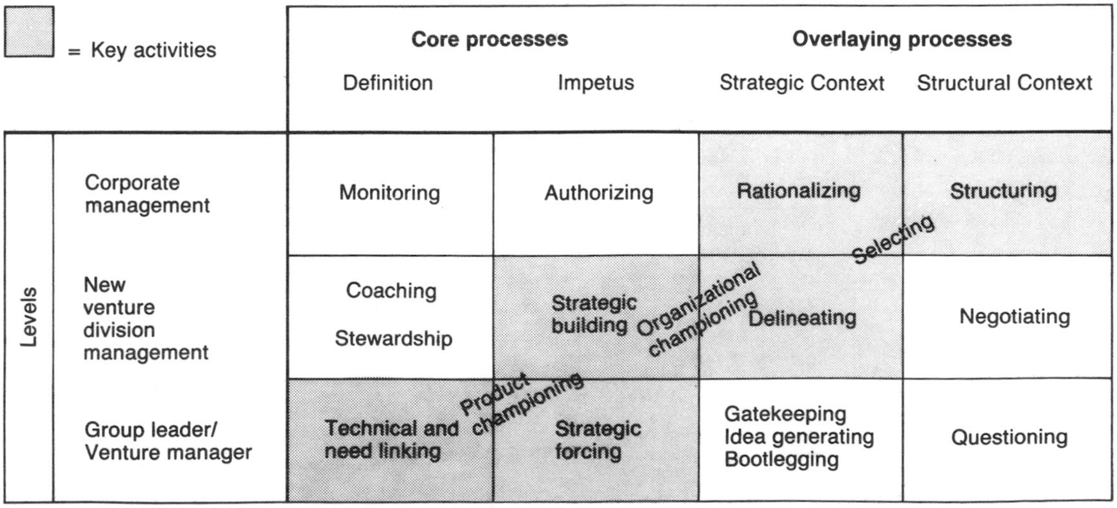

[shaded] = Key activities

		Core processes		Overlaying processes	
		Definition	Impetus	Strategic Context	Structural Context
Levels	Corporate management	Monitoring	Authorizing	Rationalizing	Structuring
	New venture division management	Coaching Stewardship	Strategic building	Delineating	Negotiating
	Group leader/ Venture manager	Technical and need linking	Strategic forcing	Gatekeeping Idea generating Bootlegging	Questioning

Selecting

Organizational championing

Product championing

MAJOR PROBLEMS IN THE ICV PROCESS

The process model indicates that ICV involves the interlocking strategic activities of managers at different levels in the organization. These strategic activities are enacted without an existing master plan in which they all neatly fit together. Figure 2 provides an overview of some of the problematic aspects of the strategic situation at each level in each of the processes that constitute ICV.

Vicious Circles in the Definition Process

At the corporate level, managers tend to have a highly reliable frame of reference to evaluate business strategies and resource allocation proposals pertaining to the main lines of business of the corporation. By the same token,

FIGURE 2 Major Problems in the ICV Process

Levels		Core Processes		Overlaying Processes	
		Definition	Impetus	Strategic Context	Structural Context
	Corporate management	Top management lacks the capacity to assess the merits of specific new venture proposals for corporate development.	Top management relies on purely quantitative growth results to continue support for a new venture.	Top management considers ICV as insurance against mainstream business going bad. ICV objectives are ambiguous and shifting erratically.	Top management relies on reactive structural changes to deal with problems related to ICV.
	NVD management	Middle-level managers in corporate R&D are not capable of coaching ICV project initiators.	Middle-level managers in new business development find it difficult to balance strategic building efforts with efforts to coach the venture managers.	Middle-level managers struggle to delineate the boundaries of a new business field. They spend significant amounts of time on political activities to maintain corporate support.	Middle-level managers struggle with unanticipated structural impediments to new venture activities. No incentive for star performers to engage in ICV activities.
	Group leader Venture leader	Project initiators cannot convincingly demonstrate in advance that resources will be used effectively. They need to engage in scavenging to get resources.	Venture managers find it difficult to balance strategic forcing efforts with efforts to develop the administrative framework of the emerging ventures.	Project initiators do not have a clear idea which kind of ICV projects will be viable in the corporate context. Bootlegging is necessary to get a new idea tested.	Venture managers do not have a clear idea what type of performance will be rewarded except fast growth.

their capacity to deal with substantive issues of new business opportunities is limited, and their expectations concerning what can be accomplished in a short time framework is often somewhat unrealistic. Also, ICV proposals compete for scarce top management time. Their relatively small size combined with the relative difficulty in assessing their merit make it at the outset seem uneconomical for top management to allocate much of their time to them. Not surprisingly, top managers tend to *monitor* ICV activities from a distance.

Middle-level managers in corporate R&D (where new ventures usually originate) experience a tension between their resource *stewardship* and *coaching* responsibilities. Such managers tend to be most concerned about maintaining the integrity of the R&D work environment which is quite different from a business-oriented work environment.[5] They are comfortable with managing relatively slow-moving exploratory research projects and well-defined development projects. But they are reluctant to commit significant amounts of resources (especially people) to suddenly fast-moving areas of new development activity which fall outside of the scope of their current plans and which have not yet demonstrated technical and commercial feasibility.

Operational-level managers typically struggle to conceptualize their still somewhat nebulous (at least to outsiders) business ideas, which makes communication with management difficult. The results of their *technical* and *need-linking* efforts often go against conventional corporate wisdom. They cannot clearly specify the development path of their projects, and cannot demonstrate in advance that the resources they need will be used effectively in uncharted domains.

[5] These differences are discussed in greater depth in R. A. Burgelman, "Managing Innovating Systems: A Study of the Process of Internal Corporate Venturing," unpublished doctoral dissertation, Columbia University, 1980.

Demonstrating technical feasibility

The lack of articulation between different levels of management results in a vicious circle in resource procurement. Resources can be obtained if technical feasibility is demonstrated, but demonstration itself requires resources. *Product championing* activities serve to break through this vicious circle. Using bootlegging and scavenging tactics, the successful project champion is able to provide positive information reassuring middle-level management and providing them with a basis for claiming support for ICV projects in their formal plans. This dynamic explains the somewhat surprising finding that middle-level managers often encourage and do not just tolerate such sub-rosa activities.

Demonstrating commercial feasibility

Even when a technically demonstrated product, process, or system exists, corporate management is often reluctant to start commercialization efforts because they are unsure about the firm's capabilities to effectively do so. To overcome such hesitancies, product champions engage in corner cutting: activities which deviate from the official corporate ways and means (e.g., contacting customers from other divisions for tryouts, hiring sales people in disguise, etc.). Or, they may choose an approach that is more acceptable in the light of corporate management's concerns but which may not be optimal from the long-term strategic point of view (e.g., propose a joint venture with another firm).

Managerial Dilemmas in the Impetus Process

Product championing resulting in preliminary demonstration of technical and commercial viability of a new product, process, or system sets the stage for the impetus process. In the course of the impetus process, an ICV project receives "venture status," i.e., it becomes a quasi-independent new business with its own budget and general manager. Often the product champion becomes the venture manager. Even though there are misgivings expressed about it, this happens naturally. First, for the product champion this constitutes the big, but also the only, reward. Second, there is usually just nobody else around who could take over and continue the momentum of the development process.

Continued impetus depends on the *strategic forcing* efforts of the venture manager level: attaining a significant sales volume and market share position within a limited time horizon.[6] Strategic forcing efforts center around the original product, process, or system. To implement a strategy of fast growth, the venture manager attracts generalist helpers who can cover a number of different functional areas reasonably well. With the growth of the venture organization and under competitive pressures due to product maturation, efficiency considerations become increasingly important. New functional managers are brought in to replace the generalists. They tend to emphasize the development of routines, standard operating procedures, and the establishment of an administrative framework for the venture. This, however, is time consuming and detracts from the all-out efforts to grow fast. Thus, the venture manager is increasingly faced with a dilemmatic situation: continuing to force growth versus building the organization of the venture. Growth concerns tend to win out, and organization building is more or less purposefully neglected.

Whilst the venture manager creates a "beachhead" for the new business, the middle level engages in *strategic building* efforts to sustain the impetus process. Such efforts involve the conceptualization of a master strategy for the broader new field within which the venture can fit. They also involve the integration of projects existing elsewhere in the corporation and/or of small firms that can be acquired with the burgeoning venture. These efforts become increasingly important as the strategic forcing activities of the venture manager reach their limit. At the same time, the administrative problems created by the strategic forcing efforts require increasingly the attention of the venture manager's manager. Hence, like the venture manager, the middle-level manager is also confronted with a serious dilemma: focusing on expanding the scope of the new business versus spending time coaching the (often recalcitrant) venture manager and building the organization. Given the overwhelming importance of growth, the coaching activities and organization building tend to be more or less purposefully neglected.

Corporate-level management's decision to *authorize* further resource allocations to a new venture are to a large extent dependent on the credibility of the managers involved. Credibility, in turn, depends primarily on the quantitative results produced. Corporate management tends to develop somewhat unrealistic expectations about new ventures. They send strong signals concerning the importance of making an impact on the overall corporate position soon. This, not surprisingly, reinforces the emphasis of the middle and operational levels of management on achieving growth.

New product development lags behind

The lack of attention to building the administrative framework of the new venture prevents it from developing a continuous flow of new products. Lacking carefully designed relationships between R&D, engineering, marketing, and manufacturing, new product schedules are delayed and completed new products often show serious flaws.

[6] The need for strategic forcing is also consistent with the findings that suggest that attaining large market share fast at the cost of early profitability is critical for venture survival. R. Biggadike, "The Risky Business of Diversification," *Harvard Business Review* 57 (May–June 1979), pp. 103–11.

The demise of the venture manager

Major discontinuities in new product development put more stress on the middle-level manager to find supplementary products elsewhere to help maintain the growth rate. This, in turn, leads to even less emphasis on coaching the venture manager. The new product development problems also tend to exacerbate the tensions between the venture manager and the functional managers. Eventually, the need to stabilize the venture organization is likely to lead to the demise of the venture manager.

The Indeterminate Strategic Context of ICV

The problems encountered in the core processes of ICV are more readily understood when examining the overlaying processes within which ICV development takes shape. Corporate management's objectives concerning ICV tend to be ambiguous. Top management does not really know which specific new businesses they want until the latter have taken some concrete form and size, and decisions must be made whether to integrate them or not in the corporate portfolio through a process of *retroactive rationalizing*.

Middle-level managers struggle with *delineating* the boundaries of a new business. They are aware that corporate management is interested in broadly defined areas like the "health field" or "energy." But it is only through the middle-level manager's strategic building and the concomitant articulation of a master strategy for the ongoing venture initiatives that the new business fields become concretely delineated and the possible new strategic directions determined.

At the operational level, managers engage in *gatekeeping, idea generating,* and preliminary *bootlegging* activities which may lead to the definition of ICV projects in new areas and/or in new business fields which are already emerging as a result of the ongoing ICV activities. These activities are autonomous because they are basically independent of the current strategy of the firm. Managers at this level have no clear idea at the outset which kinds of ICV projects will be viable in the corporation, but they seem to have a sense for avoiding those that have no chance of receiving support (e.g., because there have been some earlier failures in the area, or there are some potential legal liabilities associated with it).

Determining the strategic context

The indeterminateness of the strategic context of ICV requires middle-level managers to engage in *organizational championing* activities.[7] Such activities are of a political nature, and time consuming. They require an upward orientation (as one venture manager put it) which is very different from the venture manager level's substantive and downward (hands-on) orientation. The middle-level manager must also spend time to work out the frictions with the operating system that may exist when the strategies of the venture and of mainstream businesses interfere with each other. The need for these activities tends to reduce further the amount of time and effort spent by the middle level on coaching the venture manager.

Oscillations of corporate strategy

New ventures take between 8 and 12 years on the average to become mature, profitable new businesses.[8] Top management's time horizon, however, is usually limited to three to five

---

[7] The importance of the middle-level manager in ICV was already recognized by Von Hippel, "Successful and Failing Internal Corporate Ventures." I. Kusiatin, "The Process and Capacity for Diversification through Internal Development," unpublished doctoral dissertation, Harvard University, 1976; and M. A. Maidique, "Entrepreneurs, Champions, and Technological Innovations," *Sloan Management Review* 21, pp. 59–76, also have discussed the role of a "manager champion" or "executive champion."

[8] Biggadike, "The Risky Business."

years. Corporate management's objectives tend to be shifting: new ventures are viewed by top management as insurance against mainstream business going bad rather than as a corporate objective per se.[9] Middle managers are aware that there are short-term windows for corporate acceptance which must be taken advantage of. This also puts pressure on them to grow new ventures as fast as possible.

The Selective Pressures of the Structural Context

Top management establishes a structural context to support the corporate strategy. The structural context provides strategic actors at operational and middle levels of management with signals concerning the types of projects that are likely to be supported and rewarded. It operates as a selection mechanism on the strategic behavior in the organization. ICV projects, by definition, fall outside the scope of the current corporate strategy and must overcome the selective pressures of the structural context.

The incompleteness of the structural context

Establishing a separate new venture division facilitates the definition and early impetus processes of ICV projects. But, by itself, the new venture division constitutes an incomplete structural context. In the absence of measurement and reward systems tailored specifically to the requirements of new venture activities, venture managers do not have a clear idea what performance is expected from them except in terms of reaching a large size for their new business fast. Middle-level managers of the new venture division experience resistance from managers in the operating divisions when activities overlap. Ad hoc negotiations and reliance on political savvy substitute for long-term based, joint optimization arrangements. This leads, eventually, to severe frictions between ICV and mainstream business activities.[10]

Reactive changes in the structural context

When ICV activities expand beyond a level that corporate management finds opportune to support in light of their assessment of the prospects of mainstream business activities, or when some highly visible failures occur, changes are effected in the structural context to "consolidate" ICV activities. These changes seem reactive and indicative of the lack of a clear strategy for diversification in the firm.[11]

MANAGING THE CORPORATE CONTEXT OF ICV

The process model suggests that, at the corporate level of analysis, ICV is based on *experimentation and selection,* not a strategic planning process.[12] It is characterized by ambiguity, discontinuity, even an element of anarchy. Having identified some of the major problems in the ICV process, recommendations for improving the strategic management of ICV can be proposed. These can serve to alleviate, if not eliminate, these problems by making the corporate context more hospitable to ICV. Improvement of the overlaying process will, presumably, allow management to focus more on

[9] R. A. Peterson and D. G. Berger, "Entrepreneurship in Organizations: Evidence from the Popular Music Industry," *Administrative Science Quarterly* 16 (1971), pp. 97–106.

[10] These frictions are discussed in more detail in R. A. Burgelman, "Managing the New Venture Division: Research Findings and Implications for Strategic Management," *Strategic Management Journal* (forthcoming).

[11] Ibid.

[12] This argument is further developed in R. A. Burgelman, "Corporate Entrepreneurship and Strategic Management: A Review and Conceptual Integration," *Management Science* (forthcoming).

the problems that are inherent in the core processes and less on those that result from having to "fight the system." Figure 3 summarizes the recommendations.

Elaborating the Strategic Context of ICV

Top management should recognize that ICV is an important source of strategic renewal for the firm, and not just insurance against poor mainstream business prospects. ICV should therefore be considered an integral and continuous part of the strategy-making process.

The need for a corporate development strategy

To dampen the oscillations in corporate support for ICV, top management should create a process for developing an explicit and substantive long-term (10 to 12 years) strategy for corporate development, supported by a resource generation and allocation strategy. Both should be based on ongoing efforts to determine the remaining growth opportunities in the current mainstream businesses and the resource levels necessary to exploit them. Given the corporate objectives for growth and profitability, a resource pool should be reserved for activities outside the mainstream business. This pool should not be affected by short-term fluctuations in current mainstream activities. ICV as well as other types of activities (e.g., acquisitions) should be funded out of this pool. The existence of this pool of slack resources would allow top management to affect the rate at which new venture initiatives will emerge, if not their particular content.[13] This approach reflects a broader concept of strategy-making than maintaining corporate R&D at a certain percentage of sales.

[13] Ibid.

Substantive assessment of venture strategies

To more effectively determine the strategic context of ICV, and to reduce the political emphasis in organizational championing activities, top management should increase its capacity to make substantive assessments of the merits of new ventures for corporate development. Top management should learn to assess better the strategic importance to corporate development and the degree of relatedness to core corporate capabilities of ICV projects.[14]

One way to achieve this capability is for top management to include members with significant experience in new business development in the top management team. In addition, top management should require middle-level organizational champions to explain how a new field of business would further the corporate development objectives in *substantive* rather than purely numerical terms. Middle-level managers should have to explain how they create value from the *corporate point of view* with the new ventures they sponsor. Operational-level managers would then have a better chance to find out early which of the possible directions their envisaged projects could take will be more likely to receive corporate support.

Such increased emphasis on the part of top management would not necessarily mean having greater input in the specific directions of exploratory corporate R&D. Rather, it would increase their influence on the *business directions* that grow out of the exploratory R&D substratum.

[14] These two dimensions would seem to be important in deciding what type of arrangements to use. For instance, high strategic importance and high degree of relatedness might suggest the need to integrate the new project directly into the mainstream businesses (even if there is resistance). Very low strategic importance and very low degree of relatedness might suggest complete spin-off as the best approach. The NVD would seem to be most adequate for more ambiguous situations on both dimensions.

FIGURE 3 Recommendations for Making ICV Strategy Work Better

		Core Processes		Overlaying Processes	
		Definition	Impetus	Strategic Context	Structural Context
Levels	Corporate management	ICV proposals are evaluated in light of corporate development strategy. Conscious efforts are done to avoid subjection to conventional corporate wisdom.	New venture progress is evaluated in substantive terms by top managers who have experience in organizational championing.	A process is in place for developing a long-term corporate development strategy. This strategy takes shape as result of ongoing interactive learning process involving top and middle levels of management.	Managers with successful ICV experience are appointed to top management. Top management is rewarded financially and symbolically for long-term corporate development success.
	NVD management	Middle-level managers in corporate R&D are selected who have both technical depth and business knowledge necessary to determine minimum amount of resources for project, and who can coach star players.	Middle-level managers are responsible for the use and development of venture managers as scarce resources of the corporation, and facilitate intrafirm project transfers if the new business strategy warrants it.	Substantive interaction between corporate- and middle-level management leads to clarifying the merits of a new business field in light of the corporate development strategy.	Star performers at middle level are attracted to ICV activities. Collaboration of mainstream middle level with ICV activities is rewarded. Integrating mechanisms can easily be mobilized.
	Group leader Venture leader	Project initiators are encouraged to integrate technical and business perspectives. They are provided access to resources. Project initiators can be rewarded other than by becoming venture managers.	Venture managers are responsible for developing the functional capabilities of emerging venture organizations, and for codification of what has been learned in terms of required functional capabilities while pursuing the new business opportunity.	Slack resources determine the level of the emergence of mutant ideas. Existence of substantive corporate development strategy provides preliminary self-selection of mutant ideas.	A wide array of venture structures and supporting measurement and reward systems clarifies expected performance for ICV personnel.

Refining the Structural Context

Top management also needs to fine-tune the structural context and make it more compatible with the requirements of ICV.

More deliberate use of the new venture division (NVD)

Often, the NVD becomes the recipient of "misfit" and "orphan" projects existing in the operating system, and serves as the trial ground for possibly ill-conceived business ideas of the corporate R&D department. In some instances, greater efforts would seem to be in order to assess the possibilities of accommodating new venture initiatives in the mainstream businesses rather than transferring them to the NVD. In other instances, projects should be developed using external venture arrangements or be spun off. Such decisions should be based on an examination of where a project fits in the strategic context. They should be easily implementable by having a wide range of structures for venture-corporation relationships available.

Also, the NVD is a mechanism for *decoupling* the activities of new ventures and those of mainstream businesses. But, this decoupling usually cannot be perfect. Hence, integrative mechanisms (e.g., "steering committees") should be established to deal constructively with conflicts that will unavoidably and unpredictably arise.

Finally, top management should facilitate greater acceptance of differences between the management processes of the NVD and the mainstream businesses. This may lead, for instance, to more careful personnel assignment policies and to greater flexibility in hiring and firing policies in the NVD to reflect the special needs of emerging businesses.

Such measures to use the NVD more deliberately (and selectively) will reduce the likelihood of reactive changes in the structural context.

Measurement and reward systems in support of ICV

Perhaps the most difficult aspect of the structural context concerns how to provide incentives for top management to seriously and continuously support ICV as part of corporate strategy making. Corporate history writing might be an effective mechanism to achieve this. This would involve the careful tracing and periodical publication (e.g., a special section in annual reports) of decisions whose positive or negative results may become clear only 10 or more years after the fact. Corporate leaders (like political ones) would, presumably, make efforts to preserve their position in corporate history.[15]

To reduce the destructive emphasis on fast growth at middle and operational levels of management, the measurement and reward system must be tailored to the special nature of the managerial work involved in ICV. This would mean, for instance, greater emphasis on accomplishments in the areas of problem finding, problem solving, and know-how development than on volume of dollars managed. Efforts to develop the venture organization and the venture manager should also be included. These measures will alleviate the problems resulting from the pressures to grow fast, and more emphasis will be given by middle managers to coaching their venture managers, and to the administrative development problems of new ventures which otherwise tend to be neglected. More flexible systems for measuring and rewarding performance should accompany the greater flexibility in structuring the venture-corporate relations mentioned earlier.

---

[15] Some firms seem to have developed the position of corporate historian. Without underestimating the difficulties such a position is likely to encounter, one can imagine the possibility of structuring it in such a way that the relevant data would be recorded. Another instance, possibly a board-appointed committee, could periodically interpret this data along the lines suggested.

In general, the higher the degree of relatedness (the more dependent the new venture is on the firm's resources) and the lower the expected strategic importance for corporate development, the lower the rewards the internal entrepreneurs will be able to negotiate. Milestone points could be agreed upon to revise the negotiations as the venture evolves. To make such processes symmetrical (and more acceptable to the nonentrepreneurial participants in the organization), the internal entrepreneurs should be required to substitute negotiated for regular membership awards and benefits.

Furthermore, to attract "top performers" in the mainstream businesses of the corporation to ICV activities, at least a few spots on the top management team should always be filled with managers who have had significant experience in new business development. This will facilitate the determination of the strategic context and will eliminate the perception that NVD participants are not part of the real world and thus have not much chance to advance in the corporation as a result of ICV experience.[16]

At the operational level, where some managerial failures are virtually unavoidable given the experimentation and selection nature of the ICV process, top management should create a reasonably foolproof safety net. Product champions at this level should not have to feel that running the business is the only possible reward for getting it started. Systematic search for and screening of potential venture managers should make it easier to provide a successor for the product champion *in time*. Avenues for recycling product champions/venture managers should be developed and/or their reentry into the mainstream businesses facilitated.

[16] As some people in my study pointed out, there is no need to take the risks of new business development if you are identified as a star performer. Such performers are put in charge of the large, established businesses where their capabilities will presumably have maximum leverage.

MANAGING THE CORE PROCESSES OF ICV

Increasing top management's capacity to manage the corporate context of ICV will, in turn, facilitate the management of the core processes of ICV development. To alleviate some of the specific problems mentioned earlier, some further recommendations can be proposed.

Managing the Definition Process

The ICV projects in my study typically started with an initiative at the group leader level (first-level supervisor) in the corporate R&D department. Such initiatives were rooted in the periphery of the corporate technological capabilities, reflected the creative insight and entrepreneurial drive of the initiator, and were influenced by the latter's perception of the chances of getting the venture eventually accepted by top management as a major new area for the firm.

Of the many ICV projects that start the definition process, only a few reach "venture" status. Some of the ones that do not make it to that transition may find a home for further development as a new product line in one of the operating divisions; others may just be stopped, and result only in an extension of the corporation's knowledge base. Some of the ones that do not make it, however, could possibly have succeeded, and some of the ones that do obtain venture status should not have. Timely assessment of the true potential of an ICV project remains a difficult problem. This follows from the very nature of such projects: the many uncertainties around the technical and marketing aspects of the new business, and the fact that each case is significantly different from all others. These factors make it quite difficult to develop standardized evaluation procedures and development programs without screening to death truly innovative projects.

Managing the definition process effectively

thus poses serious challenges for middle-level managers in the corporate R&D department. They must facilitate the integration of technical and business perspectives, and must maintain a lifeline to the technology developed in corporate R&D as the project takes off. As stated earlier, the need for product championing efforts, if excessive, may cut that lifeline early on and lead to severe discontinuities in new product development after the project has reached the venture stage. The middle-level manager's efforts must facilitate both the product championing efforts and the continued development of the technology base, by putting the former in perspective and by making sure that the interface between R&D and business people works smoothly.

Facilitating the integration of R&D–business perspectives

To facilitate the integration of technical and business perspectives, the middle manager must understand the operating logic of both groups and must avoid getting bogged down in technical details, yet have sufficient technical depth to be respected by the R&D people. Such managers must be able to motivate the R&D people to collaborate with the business people toward the formulation of business objectives against which progress can be measured. The articulation of business objectives is especially important for the venture's relations with corporate management if the latter become more actively involved in ICV and develop a greater capacity to evaluate the fit of new projects with the corporate development strategy.

Middle-level managers in R&D must be capable to make the two groups give and take in a process of mutual adjustment toward the common goal of advancing the progress of the new business project. One of the key things here is creating mutual respect between technical and business people. Example setting by the R&D manager of showing respect for the business people's contribution is likely to have a carryover effect on the attitudes of the other R&D people. Regular meetings between the two groups to evaluate, as peers, the contribution of the different members of the team is likely to lead to much better integrated efforts.

The middle manager as coach

Such meetings also provide a vehicle for better coaching the product champion. The latter is really the motor of the ICV project in this stage of development. There are some similarities between this role and that of the star player in a sports team. Often, the situation with respect to product champions as star players is viewed in either-or terms: either he or she can do their thing, and then chances are that we will succeed, but there will be discontinuities, not fully exploited ancillary opportunities, etc.; or, we harness him or her, and they won't play.

A more balanced approach is possible if the R&D manager uses a process in which the product champion is recognized as the star player, but is, at times, challenged to maintain breadth by having to respond to queries like:

- How is the team benefiting more from this particular action than from others that the team may think to be important?
- How will the continuity of the efforts of the team be preserved?
- What will be the next step?

To back up this approach, the middle manager should have a say in how to reward the members on the team differently. This, of course, refers back to the determination of the structural context, and reemphasizes the importance at the corporate level to recognize that different reward systems are necessary for different types of business activities.

Managing the Impetus Process

Pursuing fast growth and administrative development of the venture simultaneously is a major challenge during the impetus process.

This challenge, which exists for any start-up business, is especially difficult for one in the context of an established firm. This is so because managers in ICV typically have less control over the selection of key venture personnel yet, at the same time, have more ready access to a wide array of corporate resources.[17] Thus, there is much less pressure on the venture manager and the middle-level manager to show progress in building the organization than there is to show growth.

The venture manager as organization builder

A more adequate measurement and reward system should force the venture manager to balance the two concerns better. The venture manager should have leeway in hiring and firing decisions but should also be held responsible for the development of new functional capabilities and the administrative framework of the venture. This would reduce the probability of major discontinuities in new product development mentioned earlier. In addition, it will provide the corporation with codified know-how and information which can be transferred to other parts of the firm, or to other new ventures even if the one from which it is derived ultimately fails as a business. Know-how and information thus become important outputs of the ICV process, in addition to sales and profit dollars.

Often, the product champion will not have the required capabilities to achieve these additional objectives. In such cases, the product champion should be warned that the business will probably have to be taken away from him or her at some later date unless the capability to handle the growing complexity of the new business organization is demonstrated. The availability of compensatory rewards and of avenues for recycling the venture manager would make it possible for management to tackle deteriorating managerial conditions in the new business organization with greater fortitude. Furthermore, the availability of a competent replacement (as a result of systematic corporate search) may induce the product champion to relinquish the venture rather than see it go under.

The middle-level manager as corporate strategist

Increasing the capacity of top management to assess new venture projects will reduce the need for organizational championing and free up time for the middle-level manager for more intensive coaching of the venture manager in his or her efforts to build the venture organization. This aspect of the middle-level manager's job should also be more explicitly considered in the measurement and reward system at this level.

The encouragement of star performers at the middle level to get involved in new business development will also enhance their key role in the development of corporate strategy. By getting deeply involved in new business development, such middle-level managers will not only learn to manage the ICV process, they will also get to know the new businesses which may be part of the mainstream by the time they reach top-level positions. Because new ventures often intersect with multiple parts of mainstream businesses, they will learn what the corporate capabilities and skills—and the shortcomings in them—are, and learn to articulate new strategies and build new businesses based on new combinations of corporate capabilities and skills. This, in turn, may also enhance the realization of the possibilities for new operational synergies existing in the firm. Middle-level managers thus become crucial

---

[17] Often, new ventures seem to be used for assigning personnel who do not fit well in the mainstream businesses (or are out of a job there).

linking and technology transfer mechanisms in the corporation.

CONCLUSION: NO PANACEAS

The recommendations presented in this paper may make the ICV development a better managed process, i.e., one less completely governed by the process of "natural selection." Yet, the implication is not that this process can or should become a planned one; or, that the fundamentally discontinuous nature of entrepreneurial activity can be avoided. Ultimately, ICV remains an uncomfortable process for the large, complex organization as it upsets its carefully evolved routines and planning mechanisms, threatens its internal equilibrium of interests, and requires a revision of the very image it has of itself.

The motor of corporate innovation consists of the strategic behaviors of individuals at different levels, willing to put their reputation and their career on the line in the pursuit of the big opportunity. As individuals, they act against and/or in spite of the system; not because they value corner cutting and risk taking per se, but because they must respond to the logic of the situation. And for radical innovation to take place, that logic entails—as Schumpeter posited some 70 years ago—the escape from routine; i.e., from the very stuff of which large, complex organizations traditionally are made. Yet, the success of such innovations is ultimately dependent on whether they can become institutionalized. This may pose the most important challenge for managers of large, established firms in the 80s.

This paper has proposed that managers can make a strategy involving radical innovations work better if they increase their capacity to conceptualize the organization's innovation efforts in process model terms. The recommendations, based on this point of view, should result in a somewhat better use of the individual entrepreneurial resources of the corporation, and thereby in an improvement of the corporate entrepreneurial capability.

APPENDIX 1: A FIELD STUDY OF ICV

A qualitative method was chosen as the best way to arrive at an encompassing view of the ICV process.

Research Setting

The research was carried out in one large, U.S.-based, high-technology firm of the diversified major type which I shall refer to as GAMMA. GAMMA had traditionally produced and sold various commodities in large volume, but it had also tried to diversify through the internal development of new products, processes, and systems so as to get closer to the final user or consumer and to catch a greater portion of the total value added in the chain from raw materials to end products. During the 60s, diversification efforts were carried out within existing operating divisions, but in the early 70s, the company established a separate new venture division (NVD).

Data were obtained on the functioning of the NVD. The charters of its various departments, the job descriptions of the major positions in the division, the reporting relationships and mechanisms of coordination, and the reward system were studied. Data were also obtained on the relationships of the NVD with the rest of the corporation. In particular, the collaboration between the corporate R&D department and divisional R&D groups was studied. Finally, data were also obtained on the role of the NVD in the implementation of the corporate strategy of unrelated diversification. These data describe the historical evolution of the structural context of ICV development at GAMMA before and during the research period.

The bulk of the data was collected in studying the six major ICV projects in progress at GAMMA at the time of the research. These ranged from a case where the business objectives were still being defined to one where the venture had reached a sales volume of $35 million.

Data Collection

In addition to the participants in the six ICV projects, I interviewed NVD administrators, people from several operating divisions, and one person from corporate management. All in all, 61 people were interviewed. The interviews were unstructured and took from one and a half to four and a half hours. Tape recordings were not made, but the interviewer took notes in shorthand. The interviewer usually began with an open-ended invitation to tell about work-related activities, then directed the discussion toward three major aspects of the ICV development process: (1) the evolution over time of a project, (2) the involvement of different functional groups in the development process, and (3) the involvement of different hierarchical levels in the development process. Respondents were asked to link particular statements they made to statements of other respondents on the same issues or problems and to give examples, where appropriate. After completing an interview, the interviewer made a typewritten copy of the conversation. All in all, about 435 legal-size pages of typewritten field notes resulted from these interviews.

The research also involved the study of documents. As could be expected, the ICV project participants relied little on written procedures in their day-to-day working relationships with other participants. One key set of documents, however, was the set of written corporate long-range plans concerning the NVD and each of the ICV projects. These official descriptions of the evolution of each project between 1973 and 1977 were compared with the interview data.

Finally, occasional behavioral observations were made, for example, when other people would call or stop by during an interview or in informal discussions during lunch at the research site. These observations, though not systematic, led to the formulation of new questions for further interviews.

Choosing a Venture Strategy

Case III–4
Medical Equipment (C)

M. van den Poel and R. A. Burgelman

Many things had happened since Burns had become manager of Gamma's New Venture Division in January 1977. First, Dr. Sherwood, the general manager of the Medical Equipment venture, had been transferred back to the corporate R&D department.[1] Second, Dr. Korbin, Sherwood's superior in the New Venture Division, replaced Sherwood as general manager of the Medical Equipment venture, which had grown to a $35 million business by mid-1977. Korbin, in turn, had hired Tom Westin from the outside as an assistant-general manager of Medical Equipment. Korbin focused primarily on the possibilities for acquiring small companies and new technologies for integration

[1] See the "Medical Equipment (B)" case for Sherwood's point of view.

within the venture. Westin was responsible for cleaning up the administrative problems in the venture, bringing the Medical Equipment project from a band of "entrepreneurial gunslingers" to a well-organized and smoothly functioning team.

Burns had overseen Korbin's and Westin's efforts, maintaining his distance at the beginning so as not to be drawn into the fire-fighting activities that were a natural part of managing a new venture.

By September 1977, Burns felt that things were well enough in hand to allow both him and venture management to take a long-range view of the problems and opportunities confronting the Medical Equipment venture. They needed to understand the key issues concerning the long-term strategic positioning of the venture in the health care industry and in the corporate context as well. A retrospective look at what had actually happened in 1977 compared with the plans and forecasts a year earlier was in order. Finally, Burns wanted to identify what their strategic alternatives were, so that they might select an appropriate course of action and recommend it to Gamma's senior management.

MEDICAL EQUIPMENT'S STRATEGIC PLANNING

For each of the last several years, the Medical Equipment venture had prepared a strategic plan which laid out the approach for generating new business and a forecast of the revenues expected in the next four years. Korbin, Westin, and Burns had the 1976 plan in hand as they prepared for the 1977 strategic planning process. Exhibit 1 shows the estimated revenues that Sherwood (the previous venture manager) had forecast as part of the 1976 plan.

In 1976, three major product areas constituted the Medical Equipment venture. Diagnostic equipment represented over 70 percent of ME's business due to the blood analyzer and

EXHIBIT 1 1976 Forecast (in millions)

	1977	1978	1979	1980
Net sales	40	60	78	105
Net income	.35	3.0	7.0	13.0

the radioimmunoassay system. Radiology and nuclear imaging was at this point limited to radiochemicals and automated preparation of radiodiagnostic scanning agents and represented some 25 percent of total sales, although some development work was done on imaging hardware systems. The remaining product area, patient support devices, was mainly concentrated on blood bank products.

Markets

Overall, the U.S. market for medical equipment and supplies had been growing steadily over the last 10 years and was estimated at $6.5 billion in 1977 (Exhibit 2). Different product groups addressed different markets with different competitors.

Clinical diagnostics was part of the clinical laboratory supplies and equipment market, an estimated $1 billion market in 1977, of which the clinical diagnostic equipment represented about one third. The market was expected to grow 11 percent a year for the next five years. Customers for clinical diagnostic equipment were either hospitals, clinical testing labs, and blood banks, buying sophisticated diagnostic testing systems for large tests; or doctors who were looking for fast, less expensive equipment for common blood tests performed in-house. While the market had been dominated in the 60s and early 70s by a few relatively small companies (Technicon, Hycel . . .), a number of large firms had now entered the picture (see Exhibits 3 and 4), bringing extensive corporate financial strength to develop competitive clinical diagnostic testing equipment.

EXHIBIT 2 **Growth Continues in Shipments of Medical Equipment and Supplies**

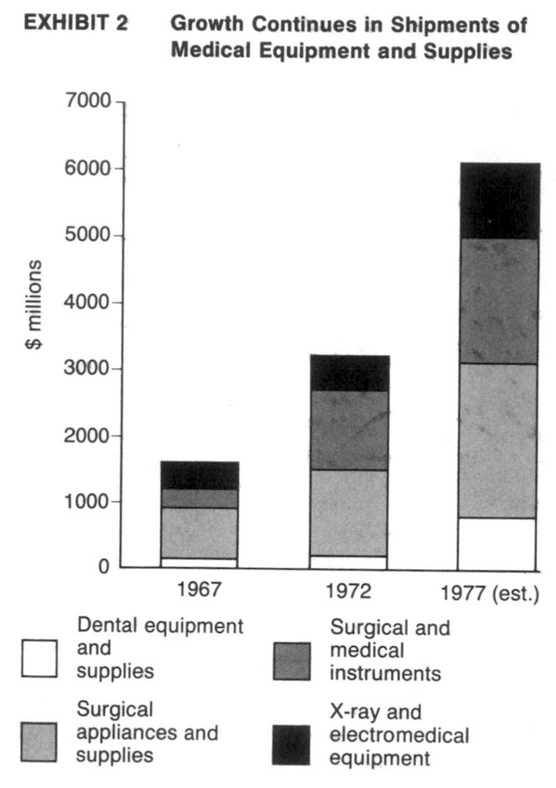

SOURCE: Bureau of the Census and BDC. Taken from *U.S. Industrial Outlook,* 1977.

Technicon had pioneered the automated blood analyzer in the mid-60s and dominated the market until the early 70s. By 1976, over 60 percent of Technicon's $225 million sales originated from automated blood and serum analyzers, the remainder from diagnostic reagents and medical information systems. A new blood analyzer, STAC (Single Test Analyzer and Computer), was introduced as a desktop microcomputer-controlled chemistry system, designed to offer laboratory results 24 hours a day. Technicon's new blood analyzer constituted an answer to Hycel's new product introduction. Originally only producing diagnostic reagents, Hycel developed blood analyzers in the late 60s. The Hycel Super 17, a computerized blood analyzer developed in 1972 and introduced in 1975, was very successful. By 1976, the Hycel 17 accounted for 65 percent of instrument's sales, which represented half of Hycel's $36 million sales.

Among the large firms, both SmithKline and Corning Glass Works introduced their first automated white blood cell analyzers in 1975. In Abbott Labs, the hospital product group was growing rapidly and Abbott brought their ABA-100 blood analyzer to the market. Union Carbide had also captured a part of the automated blood analyzer market with its Centrifichem analyzer, which was developed in the early 70s. In the area of radioimmunoassay systems, Rohm and Haas introduced in 1976 a new system, which measured hormone and drug levels. In addition, a few large computer firms such as Hewlett-Packard entered the radioimmunoassay systems market, bringing in their computer expertise and posing a threat to traditional medical equipment firms, who relied mainly on chemical analysis expertise.

Imaging represented a fast-growing market: U.S. industry shipments of X-ray and electromedical equipment had tripled between 1967 and 1975 (Exhibit 5). Business prospects seemed uncertain, however, mainly because four different technologies were in use: X ray, Ultrasound, Nuclear, and the recent CT (Computerized Tomography) scanners. This market represented in 1975 over $700 million with the X-ray systems taking 70 percent, but the CT scanners were slowly gaining a significant share, although the cost of a CT system was very high (Exhibit 6 explains the different technologies and gives average prices for the various systems). Competitors in this area included Technicare's Ohio-Nuclear, who was moving from nuclear imaging to CT scanners and was entering Ultrasound; Pfizer, using X ray and CT; General Electric with CT scanners; and EMI, the British company that developed the first head scanner in 1972 (Exhibit 7). Technology developments were changing rapidly in CT scanners, going from 4.5 minutes scan time with the first

EXHIBIT 3 Market Presence for Selected Companies

	Medical Supplies				Equipment and Associated Supplies			
	Disposable	Medical Care	Clinical Laboratory	Renal Dialysis	Diagnostic Imaging	Human Blood	Airways Management	Cardiovascular
Abbott Laboratories	X	X	X			X		X
American Home Products			X					X
American Hospital Supply	X	X	X				X	X
American Sterilizer Co.		X	X					
C. R. Bard, Inc.	X	X		X			X	X
Baxter-Travenol Labs., Inc.	X	X	X	X		X		X
Becton Dickinson & Co.	X	X	X	X				X
Bristol-Myers Co.		X						
Brunswick Corp.	X	X	X					
Colgate-Palmolive Co.	X	X					X	X
Cordis Corp.			X	X			X	
Dart Industries	X	X						
Du Pont			X		X			
Eastman Kodak Co.			X		X			
General Electric Co.					X			X
General Medical Corp.	X	X						
Hewlett-Packard			X		X		X	X
Hycel, Inc.			X					
Instrumentation Lab			X					X
IPCO Hospital Supply	X	X	X		X	X		
Johnson & Johnson	X	X	X	X	X	X	X	
3M Co.	X	X	X		X		X	
Pfizer, Inc.	X	X	X		X			
Puritan-Bennett Corp.	X						X	X
Richardson-Merrell, Inc.		X	X					
G. D. Searle & Co.	X	X			X			
SmithKline Corp.			X		X			
Squibb Corp.		X	X		X	X		
Sybron Corp.		X						
Technicare Corp.					X			
Technicon Corp.			X					
U.S. Surgical		X						

Market presence for selected SED companies. Data compiled by SRI International, September 1979.
SOURCE: *Medical Devices and Diagnostics Industry,* January 1981.

EXHIBIT 4

Technicon

1976 sales: $225 million
Net income: $ 23.4 million

Groups (percent of sales):	
Clinical systems	62%
Diagnostic reagents	22%
Medical information systems	7%
Other	9%

SmithKline

1976 sales: $674 million
Net income: $ 72 million

Groups (percent of sales):	
Pharmaceutical products	49%
Consumer products	24%
Ultrasonic products	10%
Medical diagnostics*	9%
Animal health products	8%

* Including Hematrak, automated differential white cell analyzer.

Union Carbide

1976 sales: $ 6.3 billion
Net income: $441 million

Groups (percent of sales):	
Chemicals and plastics	44%
Gases and related products	35%
Consumer and related products*	23%

* Blood analyzer business is part of consumer and related products.

Hewlett-Packard

1976 sales: $ 1.1 billion
Net income: $91 million

Groups (percent of sales):	
Test, measuring items	44%
Electronic data products	40%
Medical electronic equipment*	11%
Analytical instrumentation	5%

* Growing in importance.

Hycel

1976 sales: $36 million
Net income: $ 1.5 million

Groups (percent of sales):	
Clinical instruments	55%
Diagnostic reagents	45%

Corning Glass

1976 sales: $ 1 billion
Net income: $81 million

Groups (percent of sales):	
Electrical and electronic products	32%
Consumer products	30%
Technical* and other products	38%

* E.g., clinical chemistry instruments (LARC white blood cell analyzer and model 175 blood gas analyzer) and diagnostic testing systems.

Rohm & Haas

1976 sales: $ 1 billion
Net income: $11 million

Groups (percent of sales):	
Polymers, resins, and monomers	42%
Plastics	22%
Industrial chemicals	18%
Agricultural chemicals	13%
Health products*	5%

* Among which Concept 4, new radioimmunoassay system.

SOURCE: Annual reports.

EXHIBIT 5 **X-Ray and Electromedical Equipment: Trends and Projections 1967 to 1977 (in millions of current dollars except as noted)**

	1967	1972	1973	1974	1975[1]	1976[1]	Percent Change 1975–1976	1977[1]	Percent Change 1976–1977
Industry:[2]									
Value of shipments	233	444	476	650	685	890	30	1,110	25
Total employment (000)	8	12	13	16	15	16	7	17	6
Production workers (000)	4	7	8	10	9	10	11	11	10
Value added	136	311	315	433	456	593	30	—	—
Value added per production worker-hour ($)	15.60	22.08	21.13	22.55	25.33	—	—	—	—
Product:[3]									
Value of shipments	180	383	423	593	630	850	35	1,105	30
Value of imports[4]	22	66	92	121	173	192	11	221	15
Value of exports	39	81	107	152	196	235	20	282	20

[1] Estimated by Bureau of Domestic Commerce (BDC).
[2] Value of all products and services sold by the X-ray and Electromedical Equipment Industry (SIC 3693).
[3] Value of shipments of X-ray and electromedical equipment made by all industries.
[4] Import data revised from previous years due to change in import reporting categories.
SOURCE: Bureau of the Census, BDC.

U.S. Industrial Outlook 1977

World Market
CT Scanner, Nuclear
Medicine, Ultrasound

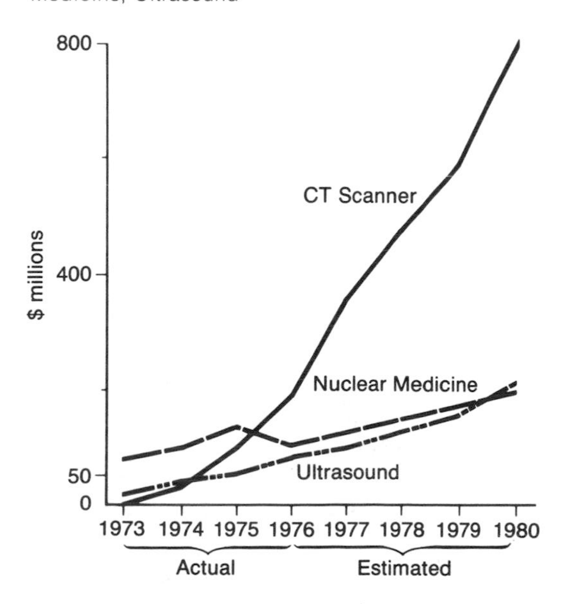

SOURCE: Technicare annual report, 1976.

EXHIBIT 6

I. U.S. Diagnostic Imaging Market (1975)

	U.S. Shipments (millions of dollars)	Average System Cost (thousands of dollars)
X ray	450–550	150
Ultrasound	35–45	35
Nuclear	80	80
CT scanners	85–100	550

SOURCE: Frost & Sullivan Report 252. 1976 American Hospital Association study.

II. Different Technologies

X rays, the oldest imaging technique, produce a two-dimensional picture of the three-dimensional body being photographed. The images of internal organs being superimposed; interpretation of these images requires considerable expertise and the interpretation is limited.

Ultrasound techniques started after World War II and consist of sound waves that are reflected back from the body tissues to form images. Limited to a tissue depth of 5 centimeters; applications are limited mainly to ophthalmology, obstetrics, urology, and certain types of heart diseases.

Nuclear imaging originated in the late 1960s, and involves introducing isotypes with short radioactive life into the body, providing useful information about the brain, liver, lung, and thyroid gland.

Computerized tomography (CT) constitutes the newest technique, where a fine beam of X rays is used to scan a thin section of the body. X rays taken from many different angles are used which are subsequently reconstructed by a computer, thus providing a cross-sectional, three-dimensional image of the body scanned.

generation of scanners in 1973 to 20 seconds scan time with the second generation in 1975. GE spent an estimated $15 million to develop a third-generation scanner with a 4.8-second scan time, to be marketed in 1977.

The *Patient* Support Devices area was part of the hospital and medical supplies market, a large, competitive market, estimated at $2.4 billion in 1976. Four top firms (Becton Dickinson, Johnson & Johnson, American Hospital Supply, and Abbott Labs—Exhibit 8) controlled over a third of this market. Several trends were occurring in the hospital supplies market: one was the growing importance of hospital group buying, which in 1976 accounted already for 15 percent of the total volume of hospital supplies. This trend had intensified price competition, and firms having a complete range of products had developed solid positions in the hospitals. Another trend was the growing use of "disposables," reducing the need for costly labor and lessening the possibility of cross-contamination. Gamma had entered this market with its disposable blood bag. The key marketing factor was to get the purchasing departments of blood banks to consider Gamma, when so many large competitors were well established and carried a broad range of products.

Strategic Directions

The strategic thrust of Medical Equipment's 1976 plan was summarized in the following excerpts.

EXHIBIT 7

General Electric

1976 sales: $ 15.7 billion
Net income: $931 million

Groups (percent of sales):	
Aerospace	11%
Consumer	18%
Industrial components and systems*	26%
Industrial power equipment	17%
International	22%
Natural resources	6%

* E.g., specialized medical equipment represented $400 million.

EMI

1976 sales: £671 million
Net income: £ 28 million

Groups (percent of sales):	
Music	51%
Leisure	12%
Television	6%
Electronics	25%
Medical electronics	6%

Pfizer

1976 sales: $ 1.3 billion
Net income: $160 million

Groups (percent of sales):	
Pharmaceutical and hospital products*	49%
Chemical products	16%
Agricultural products	15%
Consumer products	12%
Materials science products	8%

* E.g., ACTA scanner.

Technicare

1976 sales: $103 million
Net income: $ 7.7 million

Groups (percent of sales):	
Reactive metals	16%
Precision bearings	9%
Diagnostic imaging equipment*	51%
Dental equipment	8%
Wheelchairs and patient aids	16%

* Run by Ohio-Nuclear, Technicare subsidiary; product line Delta-Scan; Ohio-Nuclear represented in 1975 only 26 percent of Technicare's sales.

SOURCE: Annual reports.

For diagnostic equipment:

Major changes are envisaged: the advent of national health insurance in the next (Carter) administration will lead to an expected increase in diagnostic testing expenditures of 150 percent in 1980 over the previous forecast and will result in substantial changes in health care delivery patterns; a trend toward very fast diagnostic testing in outpatient clinics. Product development strategies are therefore setting objectives for small, freestanding, fast, discretionary, low-skilled labor systems which can be combined to allow higher capacity systems for the hospital setting with high reliability through redundancy.

For radiology and nuclear imaging:

Because of major advances in technology and changes in internal hospital administration, we forecast that the maintenance of separate customer functions based upon the particular technology involved (e.g., radiology and nuclear imaging) will break down and will coalesce into an "Imaging Department." . . . No financial provisions for Imaging are included in this long-range plan. It is intended that Imaging represents a quasi-separate activity in the Medical Equipment venture. Substantial synergism between ME and Imaging is possible.

Factors of resource and availability, timing, and competitive strength rule out an in-house development approach to this opportunity. Consequently an acquisition was sought.

For patient support devices:

Long Federal Drug Administration clearance and R&D lead times make acquisition desirable, particularly for blood bank products, since a

EXHIBIT 8

Johnson & Johnson

1976 sales: $ 2.5 billion
Net income: $205 million

Groups (percent of sales):	
Health care*	52%
Industrial	8%
International	40%

* Hospital and laboratory supplies have $500 million sales in United States.

Abbott Labs

1976 sales: $ 1 billion
Net income: $92 million

Groups (percent of sales):	
Hospital products*	41%
Pharmaceuticals	27%
Pediatric nutritionals	21%
Consumer agriculture	12%

* Mainly IV, diagnostic tests.

American Hospital Supply

1976 sales: $ 1.3 billion
Net income: $65 million

Groups (percent of sales):	
Science	31%
Hospital market*	29%
Pharmaceutical	10%
Medical specialties	6%
International	7%
Other	17%

* Product line of 44,000 items, e.g., surgical drapes, needles, diagnostic kits.

Becton Dickinson

1976 sales: $529 million
Net income: $ 43 million

Groups (percent of sales):	
Medical products*	61%
Laboratory products	24%
Industrial safety	15%

* Needles and syringes, blood collection systems, and diabetes care products.

SOURCE: Annual reports.

more rapid than anticipated FDA approval of the ME blood collection set (using nonleachable plastics) could lead to a severe bottleneck, which might well be relieved by such an acquisition. Building a direct marketing organization is likely to be required since most available acquisition candidates sell through dealers, whereas our emphasis on high-technology products would require direct selling.

By 1977, some forecasts had changed and some decisions had been implemented. For one, the expected increase in expenditures on medical equipment under the new Carter administration had not materialized; instead, Carter led off his administration proclaiming an absolute ceiling for hospital costs. Also, within Medical Equipment, the search for acquiring a company in the "imaging" area had led in early 1977 to the acquisition of a $10 million company that manufactured a medical diagnostic camera and to a distribution arrangement with another company that manufactured low-radiation breast X-ray machines. Gamma had set up a new subsidiary to handle those new products.

During the planning session with Burns and Korbin, Westin openly discussed some of his concerns relating to the ME strategy:

> We don't have a good strategy. We are operating in three different markets and we actually have a weak product strategy. This is somewhat understandable, because people were initially most concerned with reaching a critical mass for the business and they did not take enough care of the underpinnings. . . . We are constantly in need of organizational changes because things in the environment change so often. For instance, government intervention and regulations regarding capital acquisitions for hospitals, the competition in the computer business, which is a threat at

> the moment, etc. . . . Part of our strategy is the orientation toward the production and sales of consumable products to offset the capital crunch in the health care field, but this part of the strategy has not been implemented well.

Korbin and Burns agreed that they needed to draft a more coherent statement of the ME strategy. They were also concerned with the interplay between the strategy and the organizational structure of the venture.

MEDICAL EQUIPMENT'S ORGANIZATION

The ME venture had evolved from an organization devoted to a single product to one involved in three distinct lines of business. The original product, an automated blood analyzer called ANABASE, had been followed by improved models in the clinical chemistry product line. Several products were in development in the nuclear chemistry and imaging line. The patient support devices line was an umbrella for a variety of products. Some work had been done on artificial organs and artificial joints. The mainstay of the patient support devices line, however, was a blood banking system, which consisted of equipment which froze blood for storage and subsequent use in transfusion to patients, as well as blood storage containers.

Since each of the product lines involved a different set of technologies and markets, an organization was established with business managers for each of the three lines of products, as well as functional managers (accounting, marketing, manufacturing, R&D, personnel) from whom the business managers obtained support.

The business managers had four responsibilities:

- Business development, including strategy formulation, new product development, and resource allocation for each product.
- Day-to-day operating problems, including setting priorities and taking corrective actions.
- The integration across functions.
- The handling of business details.

Burns, Korbin, and Westin discussed whether the current organization made sense for Medical Equipment. In Westin's view:

> We are halfway between a matrix and a nonmatrix organization. The business managers are supposed to run their respective businesses, and they have what is called "coordinating responsibilities." In principle, the business manager sets the objectives and the line (functional) manager implements them, but it is never that clear cut. We need the business managers first to build credibility, and we still have too many line problems in the functional areas. . . . I want the functional managers to be in control at this point.

At one stage in their discussion, Burns, Korbin, and Westin had called in Jesse Grant, the business manager for the patient support devices, and he remarked:

> At some point, you need to reassert the management control of the business so that the functions do something *profitable,* not just "something."

MEDICAL EQUIPMENT'S MANAGEMENT CONTROL SYSTEMS

Westin felt that the systems and procedures that had been established in the personnel function were adequate for the venture. The Management Information System was somewhat behind, but in Westin's view it was progressing satisfactorily so it was not discussed further. In contrast, the systems to manage the marketing, manufacturing, planning, and financial accounting functions were woefully inadequate. According to Westin:

> The control system needs to be improved both in quantitative and qualitative terms. Quantita-

tively speaking, we should have clear objectives, numbers, and we don't! For instance, our sales records are handled manually! You just cannot get the information required. Three different people will come up with three different numbers, and you cannot decide which is the correct one. Qualitatively speaking, you must be able to set certain priorities for new product development, and you must be able to evaluate planning against intermediate objectives. And that doesn't happen yet.

We definitely need better financial control. It's partly a staffing problem. For instance, we know that we must clean up the accounts receivables mess on the West Coast, because we know that something is wrong there. Now, it's not so much a problem of developing procedures as one of having people to follow them. That's because our people are running all the time.

It seemed to Westin that the priorities for allocating resources and energies of people working in the venture were sometimes out of kilter. During one meeting he mused:

If you look back on it, we have been doing some things that are far too much ahead of our needs, and for others we are far too much behind. For instance, we have this tremendous market research library with all kinds of costly subscriptions, but nobody knows what questions to ask that can be answered by the information available! So I am trying to cut those expenses somewhat down. Also, we have developed product line P&L statements, but we haven't yet gotten the *total* under control. Different persons will give different estimates of the overall number, so we are focusing on the wrong thing. On the other hand, we have no systematic information on what we have actually sold!

Interfaces between Functional Areas

Westin and his superiors were aware that the hybrid organizational structure had had its share of communications problems. The business managers had occasionally mentioned the tension that existed between themselves and the functional managers. For example, Jesse Grant had stated:

The foremost problems are people and communications across the board. The major interface is, of course, with marketing. There are some positive and negative aspects. The major difficulty is with the development of new products: the question of who decides ultimately what is needed. The same problem exists with R&D. . . . I am somewhat concerned about the potential isolation of business management from marketing. I do not find this to be the case with Operations and R&D people so much, but that's because of my technical background. The background of the business manager is quite important for his interaction with the functional groups. If you have one that has a sales background, he will be able to interface easily with the sales people. If you have one with an R&D background, it is much more difficult.

There were also problems in the interactions between the functional areas. One of the R&D project managers, Dr. Edmunds, remarked to Westin that:

The geographic distance between corporate R&D and venture R&D was a big problem for new product development. There was also the problem of the organizational distance between venture R&D and manufacturing. The major problems are not research problems, but interactive ones with production, with marketing. . . . But you learn to adapt to the circumstances. For example, we needed a competent group in production, so I hired people in my group and then transferred them to production. There is no intermediate between R&D and manufacturing, and that creates a great problem. . . . There are no procedures, no ground rules, no provisions for relating the hardware to the reagents in the diagnostics area. . . . We need much more of a systems approach.

With regard to the relationship between R&D and marketing, Dr. Edmunds had said:

We need a more scientific approach in marketing. The marketing people don't get down to details. We need better guidance, indications of what to do.

One of the marketing people in Medical Equipment presented a different perspective:

> Marketing does not get involved in the conception phase. It is the technology area that determines what we sell! I feel we need more input from marketing in the new development of products . . . so we are working on some schemes to put our foot down without hurting people.

In Tom Westin's view:

> The impetus for new business development must come from R&D. In Gamma that happens. They provide an umbrella for a time, and have a commitment to review performance against plans. Furthermore, you know that there are many options when you work on something. It can be sold off, spun off. It can be dissolved and you can start with something new, or they can help it grow. Gamma is reasonably patient, even though sometimes you get unexpected reactions . . . like in Medical Equipment, where we have been growing at a rate of 70 percent and where top management visited us and said, "You are doing fine, but when are you going to have some impact on the overall Gamma position?"

The intensity of the time pressure on individuals working in all groups contributed to the problems in cross-group interaction. One of Westin's subordinates had observed:

> Because of the growth rate in new business development, there is no slack time, so communication has a tendency to stop and hence it is even more important to use the little time available to tell each other the truth.
>
> Also, a lot of people are a bit paranoid because of the continued change and uncertainty in a new business development environment. As a consequence, they want to protect their turf and position, and communication becomes very difficult.

CONCLUSION

At times during their conversations, Burns became frustrated by the preponderance of details that made it difficult to know where to start. He knew that decisions had to be made and corrective actions taken. But which were the critical decisions? Which were the areas that, if improved, would provide the greatest leverage to propel the venture to the revenue and profit objectives which were necessary for it to remain an attractive venture from the perspective of Gamma's senior management? Burns reflected that it was time to rise above the issues that seemed critical in day-to-day operations and take a strategic point of view.

Despite all the problems, the Medical Equipment venture had grown to a $40 million business in 1977. Burns wondered what the key factors were within the project and in the environment that had contributed to this success. Given the changing environment, he also wondered what changes would make it likely that the venture would be able to replicate its success in the future.

What were the future prospects for Medical Equipment? Given each scenario, did it make sense to infuse the venture with capital and talent necessary to grow into an operating division of Gamma Corporation? Or was it better to divest the venture, selling it while it was clearly worth something to potential acquiring companies?

Looking at Gamma overall, Burns realized that although one third—about $50 million—of Gamma's R&D expenditures were directed towards new business opportunities, the Medical Equipment venture represented only one of various new ventures, the others being in the specialty chemicals area and agricultural products development. Those ventures were somewhat closer to Gamma's mainstream businesses.

While Gamma's net income in 1976 had been more than adequate (8 percent of sales), earnings were expected to decline more than 10 percent for 1977, and Gamma started to review all its operations, seeking ways to concentrate efforts on businesses with the best long-term prospects and to identify and divest those businesses that did not fit Gamma's future plans.

Burns returned to his office to think through the issues that confronted him. The management of Gamma Corporation expected his recommendation on Medical Equipment at their annual planning meeting in a few weeks.

Reading III–4
The Middle Manager as Innovator

Rosabeth Moss Kanter

■ When Steve Talbot, an operations manager, began a staff job reporting to the general manager of a product group, he had no line responsibility, no subordinates or budget of his own, and only a vague mandate to "explore options to improve performance."

To do this, Talbot set about collecting resources by bargaining with product-line managers and sales managers. By promising the product-line managers that he would save them having to negotiate with sales to get top priority for their products, he got a budget from them. Then, because he had the money in hand, Talbot got the sales managers to agree to hire one salesperson per product line, with Talbot permitted to do the hiring.

The next area he tackled was field services. Because the people in this area were conservative and tightfisted, Talbot went to his boss to get support for his recommendations about this area.

With the sales and service functions increasing their market share, it was easy for Talbot to get the product-line managers' backing when he pushed for selling a major new product that he had devised. And, to keep his action team functioning and behind him, Talbot made sure that "everyone became a hero" when the senior vice president of engineering asked him to explain his success to corporate officers.

■ Arthur Drumm, a technical department head of two sections, wanted to develop a new measuring instrument that could dramatically improve the company's product quality. But only Drumm thought this approach would work; those around him were not convinced it was needed or would pay off. After spending months developing data to show that the company needed the instrument, Drumm convinced several of his bosses two levels up to contribute $300,000 to its development. He put together a task force made up of representatives from all the manufacturing sites to advise on the development process and to ensure that the instrument would fit in with operations.

When, early on, one high-level manager opposed the project, Drumm coached two others in preparation for an officer-level meeting at which they were going to present his proposal. And when executives argued about which budget line the money would come from, R&D or engineering, Drumm tried to ease the tension. His persistence netted the company an extremely valuable new technique.

■ When Doris Randall became the head of a backwater purchasing department, one of three departments in her area, she expected the assignment to advance her career. Understandably, she was disappointed at the poor state of the function she had inherited and looked around for ways to make improvements. She first sought information from users of the department's services and, with this information, got her boss to agree to a first wave of changes. No one in her position had ever had such close contacts with users before, and Randall employed her knowledge to reorganize the unit into a cluster of user-oriented specialties (with each staff member concentrating on a particular need).

Once she had the reorganization in place and her function acknowledged as the best purchasing department in the region, Randall wanted to reorganize the other two purchasing departments. Her boss, perhaps out of concern that he would lose his position to Randall if the proposed changes took place, discouraged her. But her credibility was so strong that her boss's boss—who viewed her changes as a model for

---

Reprinted with permission of *Harvard Business Review,* July–August 1982, vol. 60, no. 4.

improvements in other areas—gave Randall the go-ahead to merge the three purchasing departments into one. Greater efficiency, cost savings, and increased user satisfaction resulted.

These three managers are enterprising, innovative, and entrepreneurial middle managers who are part of a group that can play a key role in the United States' return to economic leadership.

If that seems like an overly grand statement, consider the basis for U.S. companies' success in the past: innovation in products and advances in management techniques. Then consider the pivotal contribution middle managers make to innovation and change in large organizations. Top leaders' general directives to open a new market, improve quality, or cut costs mean nothing without efficient middle managers just below officer level able to design the systems, carry them out, and redirect their staffs' activities accordingly. Furthermore, because middle managers have their fingers on the pulse of operations, they can also conceive, suggest, and set in motion new ideas that top managers may not have thought of.

The middle managers described here are not extraordinary individuals. They do, however, share a number of characteristics:

Comfort with change. They are confident that uncertainties will be clarified. They also have foresight and see unmet needs as opportunities.

Clarity of direction. They select projects carefully and, with their long time horizons, view setbacks as temporary blips in an otherwise straight path to a goal.

Thoroughness. They prepare well for meetings and are professional in making their presentations. They have insight into organizational politics and a sense of whose support can help them at various junctures.

Participative management style. They encourage subordinates to put in maximum effort and to be part of the team, promise them a share of the rewards, and deliver on their promises.

Persuasiveness, persistence, and discretion. They understand that they cannot achieve their ends overnight, so they persevere—using tact—until they do.

What makes it possible for managers to use such skills for the company's benefit? They work in organizations where the culture fosters collaboration and teamwork and where structures encourage people to "do what needs to be done." Moreover, they usually work under top managers who consciously incorporate conditions facilitating innovation and achievement into their companies' structures and operations.

These conclusions come from a study of the major accomplishments of 165 effective middle managers in five leading American corporations. I undertook this study to determine managers' contributions to a company's overall success as well as the conditions that stimulate innovation and thus push a business beyond a short-term emphasis and allow it to secure a successful future.

Each of the 165 managers studied—all of whom were deemed "effective" by their companies—told the research team about a particular accomplishment; these covered a wide range. Some of the successes, though impressive, clearly were achieved within the boundaries of established company practice. Others, however, involved innovation: introduction of new methods, structures, or products that increased the company's capacity. All in all, 99 of the 165 accomplishments fall within the definition of an innovative effort.

The Research Project

After a pilot study in which it interviewed 26 effective middle managers from 18 companies, the research team interviewed, in depth, 165 middle managers from five major corporations located across the United States. The 165 were chosen by their companies to participate because of their reputations for effectiveness. We did not want a random sample: we were looking for "the best and the brightest" who could serve as models for others. It turned out, however, that every major function was represented, and roughly in proportion to its importance in the company's success. (For example, there were more innovative sales and marketing managers representing the "market-driven" company and more technical, R&D, and manufacturing managers from the "product-driven" companies.)

During the two-hour interviews, the managers talked about all aspects of a single significant accomplishment, from the glimmering of an idea to the results. We asked the managers to focus on the most significant of a set of four or five of their accomplishments over the previous two years. We also elicited a chronology of the project as well as responses to a set of open-ended questions about the acquisition of power, the handling of roadblocks, and the doling out of rewards. We supplemented the interviews with discussions about current issues in the five companies with our contacts in each company.

The five companies represent a range of types and industries: from rather traditional, slow-moving, mature companies to fast-changing, newer, high-technology companies. We included both service and manufacturing companies that are from different parts of the country and are at different stages in their development. The one thing that all five have in common is an intense interest in the topic of the study. Facing highly competitive markets (for the manufacturing companies a constant since their founding; for the service companies a newer phenomenon), all of these corporations wanted to encourage their middle managers to be more enterprising and innovative.

Our pseudonyms for the companies emphasize a central feature of each:

CHIPCO: manufacturer of computer products

FINCO: insurance and related financial services

MEDCO: manufacturer of large medical equipment

RADCO (for "R&D"): manufacturer of optical products

UTICO: communications utility

Basic accomplishments differ from innovative ones not only in scope and long-run impact but also in what it takes to achieve them. They are part of the assigned job and require only routine and readily available means to carry them out. Managers reporting this kind of accomplishment said they were just doing their jobs. Little was problematic—they had an assignment to tackle; they were told, or they already knew, how to go about it; they used existing budget or staff; they didn't need to gather or share much information outside of their units; and they encountered little or no opposition. Managers performing such activities don't generate innovations for their companies; they merely accomplish things faster or better than they already know how to do.

In contrast, innovative accomplishments are strikingly entrepreneurial. Moreover, they are sometimes highly problematic and generally involve acquiring and using power and influence. (See the ruled insert on page 379 for more details on the study's definitions of *basic* and *innovative* accomplishments.)

In this article, I first explore how managers influence their organizations to achieve goals throughout the various stages of a project's life. Next I discuss the managerial styles of the persons studied and the kinds of innovation they brought about. I look finally at the types of companies these entrepreneurial managers worked in and explore what top officers can do to foster a creative environment.

THE ROLE OF POWER IN ENTERPRISE

Because most innovative achievements cut across organizational lines and threaten to disrupt existing arrangements, enterprising managers need tools beyond those that come with the job. Innovations have implications for other functions and areas, and they require data, agreements, and resources of wider scope than routine operations demand. Even R&D managers, who are expected to produce innovations, need more information, support, and resources for major projects than those built into regular R&D functions. They too may need additional data, more money, or agreement from extrafunctional officials that the project is necessary. Only hindsight shows that an innovative project was bound to be successful.

Because of the extra resources they require, entrepreneurial managers need to go beyond the limits of their formal positions. For this, they need power. In large organizations at least, I have observed that powerlessness "corrupts."[1] That is, lack of power (the capacity to mobilize resources and people to get things done) tends to create managers who are more concerned about guarding their territories than about collaborating with others to benefit the organization. At the same time, when managers hoard potential power and don't invest it in productive action, it atrophies and eventually blocks achievements.

Furthermore, when some people have too much unused power and others too little, problems occur. To produce results, power—like money—needs to circulate. To come up with innovations, managers have to be in areas where power circulates, where it can be grabbed and invested. In this sense, organizational power is transactional: it exists as potential until someone makes a bid for it, invests it, and produces results with it.

The overarching condition required for managers to produce innovative achievements is this: they must envision an accomplishment beyond the scope of the job. They cannot alone possess the power to carry their idea out but they must be able to acquire the power they need easily. Thus, creative managers are not empowered simply by a boss or their job; on their own they seek and find the additional strength it takes to carry out major new initiatives. They are the corporate entrepreneurs.

Three commodities are necessary for accumulating productive power—information, resources, and support. Managers might find a portion of these within their purview and pour them into a project; managers with something they believe in will eagerly leverage their own staff and budget and even bootleg resources from their subordinates' budgets. But innovations usually require a manager to search for additional supplies elsewhere in the organization. Depending on how easy the organization makes it to tap sources of power and on how technical the project is, acquiring power can be the most time-consuming and difficult part of the process.

Phases of the Accomplishment

A prototypical innovation goes through three phases: project definition (acquisition and application of information to shape a manageable, salable project), coalition building (development of a network of backers who agree to provide resources and support), and action (application of the resources, information, and support to the project and mobilization of an action team). Let us examine each of these steps in more detail.

Defining the project

Before defining a project, managers need to identify the problem. People in an organization may hold many conflicting views about the best method of reaching a goal, and discovering the basis of these conflicting perspectives (while gathering hard data) is critical to a manager's success.

In one case, information circulating freely about the original design of a part was inaccurate. The manager needed to acquire new data

[1] See my book *Men and Women of the Corporation* (New York: Basic Books, 1977); also see my article, "Power Failure in Management Circuits," *Harvard Business Review*, July–August 1979, p. 65.

to prove that the problem he was about to tackle was not a manufacturing shortcoming but a design flaw. But, as often happens, some people had a stake in the popular view. Even hard-nosed engineers in our study acknowledged that, in the early stages of an entrepreneurial project, managers need political information as much as they do technical data. Without political savvy, say these engineers, no one can get a project beyond the proposal stage.

The culmination of the project definition phase comes when managers sift through the fragments of information from each source and focus on a particular target. Then, despite the fact that managers may initially have been handed a certain area as an assignment, they still have to "sell" the project that evolves. In the innovative efforts I observed, the managers' assignments involved no promises of resources or support required to do anything more than routine activities.

Furthermore, to implement the innovation, a manager has to call on the cooperation of many others besides the boss who assigned the task. Many of these others may be independent actors who are not compelled to cooperate simply because the manager has carved a project out of a general assignment. Even subordinates may not be automatically on board. If they are professionals or managers, they have a number of other tasks and the right to set some of their own priorities; and if they are in a matrix, they may be responsible to other bosses as well.

For example, in her new job as head of a manufacturing planning unit, Heidi Wilson's assignment was to improve the cost efficiency of operations and thereby boost the company's price competitiveness. Her boss told her she could spend six months "saying nothing and just observing, getting to know what's really going on." One of the first things she noticed was that the flow of goods through the company was organized in an overly complicated, time-consuming, and expensive fashion.

The assignment gave Wilson the mandate to seek information but not to carry out any particular activities. Wilson set about to gather organizational, technical, and political information in order to translate her ambiguous task into a concrete project. She followed goods through the company to determine what the process was and how it could be changed. She sought ideas and impressions from manufacturing line managers, at the same time learning the location of vested interests and where other patches of organizational quicksand lurked. She compiled data, refined her approach, and packaged and repackaged her ideas until she believed she could "prove to people that I knew more about the company than they did."

Wilson's next step was "to do a number of punchy presentations with pictures and graphs and charts." At the presentations, she got two kinds of response: "Gee, we thought there was a problem but we never saw it outlined like this before" and "Aren't there better things to worry about?" To handle the critics, she "simply came back over and over again with information, more information than anyone else had." When she had gathered the data and received the feedback, Wilson was ready to formulate a project and sell it to her boss. Ultimately, her project was approved, and it netted impressive cost savings.

What Is an Innovative Accomplishment?

We categorized the 165 managers' accomplishments according to their primary impact on the company. Many accomplishments had multiple results or multiple components, but it was the breadth of scope of the accomplishment and its future utility for the company that defined its category. Immediate dollar results were *not* the central issue; rather, organizational "learning" or increased future capacity was the key. Thus, improving revenues by cutting costs while changing nothing else would be categorized differently from improving revenues by designing a new production method; only the latter leaves a lasting trace.

The Accomplishments Fall into Two Clusters:

Basic. Done solely within the existing framework and not affecting the company's longer term capacity; 66 of the 165 fall into this category.

Innovative. A new way for the company to use or expand its resources that raises long-term capacity; 99 of the 165 are such achievements.

Basic Accomplishments Include:

Doing the basic job—simply carrying out adequately a defined assignment within the bounds of one's job (e.g., "fulfilled sales objectives during a reorganization").

Affecting individuals' performance—having an impact on individuals (e.g., "found employee a job in original department after failing to retrain him").

Advancing incrementally—achieving a higher level of performance within the basic job (e.g., "met more production schedules in plant than in past").

Innovative Accomplishments Include:

Effecting a new policy—creating a change of orientation or direction (e.g., "changed price-setting policy in product line with new model showing cost-quality trade-offs").

Finding a new opportunity—developing an entirely new product or opening a new market (e.g., "sold new product program to higher management and developed staffing for it").

Devising a fresh method—introducing a new process, procedure, or technology for continued use (e.g., "designed and implemented new information system for financial results by business sectors").

Designing a new structure—changing the formal structure, reorganizing or introducing a new structure, or forging a different link among units (e.g., "consolidated three offices into one").

While members of the research team occasionally argued about the placement of accomplishments in the subcategories, we were almost unanimous as to whether an accomplishment rated as basic or innovative. Even bringing off a financially significant or flashy increase in performance was considered basic if the accomplishment was well within the manager's assignment and territory, involved no new methods that could be used to repeat the feat elsewhere, opened no opportunities, or had no impact on corporate structure—in other words, reflected little inventiveness. The manager who achieved such a result might have been an excellent manager, but he or she was not an innovative one.

Thus, although innovation may begin with an assignment, it is usually one—like Wilson's—that is couched in general statements of results with the means largely unspecified. Occasionally, managers initiate projects themselves; however, initiation seldom occurs in a vacuum. Creative managers listen to a stream of information from superiors and peers and then identify a perceived need. In the early stages of defining a project, managers may spend more time talking with people outside their own functions than with subordinates or bosses inside.

One R&D manager said he had "hung out" with product designers while trying to get a handle on the best way to formulate a new process-development project. Another R&D manager in our survey got the idea for a new production method from a conversation about problems he had with the head of production. He then convinced his boss to let him determine whether a corrective project could be developed.

Building a coalition

Next, entrepreneurial managers need to pull in the resources and support to make the project work. For creative accomplishments, these power-related tools do not come through the vertical chain of command but rather from many areas of the organization.

George Putnam's innovation is typical. Putnam was an assistant department manager for product testing in a company that was about to demonstrate a product at a site that attracted a large number of potential buyers. Putnam heard through the grapevine that a decision was imminent about which model to display. The product managers were each lobbying for their own, and the marketing people also had a favorite. Putnam, who was close to the products, thought that the first-choice model had grave defects and so decided to demonstrate to the marketing staff both what the problems with the first one were and the superiority of another model.

Building on a long-term relationship with the people in corporate quality control and a good alliance with his boss, Putnam sought the tools he needed: the blessing of the vice president of engineering (his boss's boss), special materials for testing from the materials division, a budget from corporate quality control, and staff from his own units to carry out the tests. As Putnam put it, this was all done through one-on-one "horse trading"—showing each manager how much the others were chipping in. Then Putnam met informally with the key marketing staffer to learn what it would take to convince him.

As the test results emerged, Putnam took them to his peers in marketing, engineering, and quality control so they could feed them to their superiors. The accumulated support persuaded the decision makers to adopt Putnam's choice of a model; it later became a strong money-maker. In sum, Putnam had completely stepped out of his usual role to build a consensus that shaped a major policy decision.

Thus, the most successful innovations derive from situations where a number of people from a number of areas make contributions. They provide a kind of checks-and-balances system to an activity that is otherwise nonroutine and, therefore, is not subject to the usual controls. By building a coalition before extensive project activity gets under way, the manager also ensures the availability of enough support to keep momentum going and to guarantee implementation.

In one company, the process of lining up peers and stakeholders as early supporters is called "making cheerleaders"; in another, "preselling." Sometimes managers ask peers for "pledges" of money or staff to be collected later if higher management approves the project and provides overall resources.

After garnering peer support, usually managers next seek support at much higher levels. While we found surprisingly few instances of top management directly sponsoring or championing a project, we did find that a general blessing from the top is clearly necessary to convert potential supporters into a solid team. In one case, top officers simply showed up at a meeting where the proposal was being discussed; their presence ensured that other people couldn't use the "pocket veto" power of headquarters as an excuse to table the issue. Also, the very presence of a key executive at such a meeting is often a signal of the proposal's importance to the rest of the organization.

Enterprising managers learn who at the top-executive level has the power to affect their projects (including material resources or vital initial approval power). Then they negotiate for these executives' support, using polished formal presentations. Whereas managers can often sell the project to peers and stakeholders by appealing to these people's self-interests and assuring them they know what they're talking about, managers need to offer top executives more guarantees about both the technical and the political adequacies of projects.

Key executives tend to evaluate a proposal in terms of its salability to *their* constituencies. Sometimes entrepreneurial managers arm top executives with materials or rehearse them for their own presentations to other people (such as members of an executive committee or the board) who have to approve the project.

Most often, since many of the projects that originate at the middle of a company can be supported at that level and will not tap corporate funds, those at high levels in the organization simply provide a general expression of support. However, the attention top management confers on this activity, many of our interviewees told us, makes it possible to sell their own staffs as well as others.

But once in a while, a presentation to top-level officers results in help in obtaining supplies. Sometimes enterprising managers walk away with the promise of a large capital expenditure or assistance getting staff or space. Sometimes a promise of resources is contingent on getting others on board. "If you can raise the money, go ahead with this," is a frequent directive to an enterprising manager.

In one situation, a service manager approached his boss and his boss's boss for a budget for a college recruitment and training program that he had been supporting on his own with funds bootlegged from his staff. The top executives told him they would grant a large budget if he could get his four peers to support the project. Somewhat to their surprise, he came back with this support. He had taken his peers away from the office for three days for a round of negotiation and planning. In cases like this, top management is not so much hedging its bets as using its ability to secure peer support for what might otherwise be risky projects.

With promises of resources and support in hand, enterprising managers can go back to the immediate boss or bosses to make plans for moving ahead. Usually the bosses are simply waiting for this tangible sign of power to continue authorizing the project. But in other cases, the bosses are not fully involved and won't be sold until the manager has higher level support.

Of course, during the coalition-building phase, the network of supporters does not play a passive role; their comments, criticisms, and objectives help shape the project into one that is more likely to succeed. Another result of the coalition-building phase is, then, a set of reality checks that ensures that projects unlikely to succeed will go no farther.

Moving into action

The innovating manager's next step is to mobilize key players to carry out the project. Whether the players are nominal subordinates or a special project group such as a task force, managers forge them into a team. Enterprising managers bring the people involved in the project together, give them briefings and assignments, pump them up for the extra effort needed, seek their ideas and suggestions (both as a way to involve them and to further refine the project), and promise them a share of the rewards. As one manager put it, "It takes more selling than telling." In most of the innovations we observed, the manager couldn't just order subordinates to get involved. Doing something beyond routine work that involves creativity and cooperation requires the full commitment of subordinates; otherwise the project will not succeed.

During the action phase, managers have four central organizational tasks. The technical details of the project and the actual work directed toward project goals are now in the hands of the action team. Managers may contribute ideas or even get involved in hands-on experimentation, but their primary functions are still largely external and organizational, centered around maintaining the boundaries and integrity of the project.

The manager's first task is to **handle interference** or opposition that may jeopardize the project. Entrepreneurial managers encounter strikingly little overt opposition—perhaps because their success at coalition-building determines whether a project gets started in the first place. Resistance takes a more passive form: criticism of the plan's details, foot-dragging, late responses to requests, or arguments over allocation of time and resources among projects.

Managers are sometimes surprised that critics keep so quiet up to this point. One manufacturing manager who was gearing up for production of a new item had approached many executives in other areas while making cost estimates, and these executives had appeared positive about his efforts. But later, when he began organizing the manufacturing process itself, he heard objections from these very people.

During this phase, therefore, innovative managers may have to spend as much time in meetings, both formal and one-to-one, as they did to get the project launched. Managers need to prepare thoroughly for these meetings so they can counter skepticism and objections with clear facts, persuasion, and reminders of the benefits that can accrue to managers meeting the project's objectives. In most cases, a clear presentation of facts is enough. But not always: one of our respondents, a high-level champion, had to tell an opponent to back down, that the project was going ahead anyway, and that his carping was annoying.

Whereas managers need to directly counter open challenges and criticism that might result in the flow of power or supplies being cut off, they simply keep other interference outside the boundaries of the project. In effect, the manager defines a protected area for the group's work. He or she goes outside this area to head off critics and to keep people or rules imposed by higher management from disrupting project tasks.

While the team itself is sometimes unaware of the manager's contribution, the manager—like Tom West (head of the now-famous computer-design group at Data General)—patrols the boundaries.[2] Acting as interference filters, managers in my study protected innovative projects by bending rules, transferring funds "illicitly" from one budget line to another, developing special reward or incentive systems that offered bonuses above company pay rates, and ensuring that superiors stayed away unless needed.

The second action-phase task is **maintaining momentum** and continuity. Here interference comes from internal rather than external sources. Foot-dragging or inactivity is a constant danger, especially if the creative effort adds to work loads. In our study, enterprising managers as well as team members complained continually about the tendency for routine activities to take precedence over special projects and to consume limited time.

In addition, it is easier for managers to whip up excitement over a vision at start-up than to keep the goal in people's minds when they face the tedium of the work. Thus, managers' team-building skills are essential. So the project doesn't lose momentum, managers must sustain the enthusiasm of all—from supporters to suppliers—by being persistent and keeping the team aware of supportive authorities who are clearly waiting for results.

One manager, who was involved in a full-time project to develop new and more efficient methods of producing a certain ingredient, maintained momentum by holding daily meetings with the core team, getting together often with operations managers and members of a task force he had formed, putting on weekly status reports, and making frequent presentations to top management. When foot-dragging occurs, many entrepreneurial managers pull in high-level supporters—without compromising the autonomy of the project—to get the team

[2] Tracy Kidder, *The Soul of a New Machine* (Boston: Little, Brown, 1981).

back on board. A letter or a visit from the big boss can remind everyone just how important the project is.

A third task of middle managers in the action phase is to engage in whatever **secondary redesign**—other changes made to support the key change—is necessary to keep the project going. For example, a manager whose team was setting up a computerized information bank held weekly team meetings to define tactics. A fallout of these meetings was a set of new awards and a fresh performance appraisal system for team members and their subordinates.

As necessary, managers introduce new arrangements to conjoin with the core tasks. When it seems that a project is bogging down—that is, when everything possible has been done and no more results are on the horizon—managers often change the structure or approach. Such alterations can cause a redoubling of effort and a renewed attack on the problem. They can also bring the company additional unplanned innovations as a side benefit from the main project.

The fourth task of the action phase, **external communication,** brings the accomplishment full circle. The project begins with gathering information; now it is important to send information out. It is vital to (as several managers put it) "manage the press" so that peers and key supporters have an up-to-date impression of the project and its success. Delivering on promises is also important. As much as possible, innovative managers meet deadlines, deliver early benefits to others, and keep supporters supplied with information. Doing so establishes the credibility of both the project and the manager, even before concrete results can be shown.

Information must be shared with the team and the coalition as well. Good managers periodically remind the team of what they stand to gain from the accomplishment, hold meetings to give feedback and to stimulate pride in the project, and make a point of congratulating each staff member individually. After all, as Steve Talbot (of my first example) said, many people gave this middle manager power because of a promise that everyone would be a hero.

A MANAGEMENT STYLE FOR INNOVATION . . .

Clearly there is a strong association between carrying out an innovative accomplishment and employing a participative-collaborative management style. The managers observed reached success by:

- Persuading more than ordering, though managers sometimes use pressure as a last resort.
- Building a team, which entails among other things frequent staff meetings and considerable sharing of information.
- Seeking inputs from others—that is, asking for ideas about users' needs, soliciting suggestions from subordinates, welcoming peer review, and so forth.
- Acknowledging others' stake or potential stake in the project—in other words, being politically sensitive.
- Sharing rewards and recognition willingly.

A collaborative style is also useful when carrying out basic accomplishments; however, in such endeavors it is not required. Managers can bring off many basic accomplishments using a traditional, more autocratic style. Because they're doing what is assigned, they don't need external support; because they have all the tools to do it, they don't need to get anyone else involved (they simply direct subordinates to do what is required). But for innovative accomplishments—seeking funds, staff, or information (political as well as technical) from outside the work unit; attending long meetings and presentations; and requiring "above and beyond" effort from staff—a style that revolves around participation, collaboration, and persuasion is essential.

The participative-collaborative style also helps creative managers reduce risk because it encourages completion of the assignment. Furthermore, others' involvement serves as a check-and-balance on the project, reshaping it to make it more of a sure thing and putting pressure on people to follow through. The few projects in my study that disintegrated did so because the manager failed to build a coalition of supporters and collaborators.

. . . AND CORPORATE CONDITIONS THAT ENCOURAGE ENTERPRISE

Just as the manager's strategies to develop and implement innovations followed many different patterns, so also the level of enterprise managers achieved varied strongly across the five companies we studied (see the Exhibit). Managers in newer, high-technology companies have a much higher proportion of innovative accomplishments than managers in other industries. At "CHIPCO," a computer parts manufacturer, 71 percent of all the things effective managers did were innovative; for "UTICO," a communications utility, the number is 33 percent; for "FINCO," an insurance company, it is 47 percent.

This difference in levels of innovative achievement correlates with the extent to which these companies' structures and cultures support middle managers' creativity. Companies producing the most entrepreneurs have cultures that encourage collaboration and teamwork. Moreover, they have complex structures that link people in multiple ways and help them go beyond the confines of their defined jobs to do "what needs to be done."

CHIPCO, which showed the most entrepreneurial activity of any company in our study, is a rapidly growing electronics company with abundant resources. That its culture favors independent action and team effort is communicated quickly and clearly to the newcomer. Sources of support and money are constantly shifting and, as growth occurs, managers rapidly move on to other positions. But even though people frequently express frustration about the shifting approval process, slippage of schedules, and continual entry of new players onto the stage, they don't complain about lost opportunities. For one thing, because coalitions support the various projects, new project managers feel bound to honor their predecessors' financial commitments.

CHIPCO managers have broad job charters to "do the right thing" in a manner of their own choosing. Lateral relationships are more important than vertical ones. Most functions are in a matrix, and some managers have up to four "bosses." Top management expects ideas to bubble up from lower levels. Senior executives then select solutions rather than issue confining directives. In fact, people generally rely on informal face-to-face communication across units to build a consensus. Managers spend a lot of time in meetings; information flows freely, and reputation among peers—instead of formal authority or title—conveys credibility and garners support. Career mobility at CHIPCO is rapid, and people have pride in the company's success.

RADCO, the company with the strongest R&D orientation in the study, has many of CHIPCO's qualities but bears the burden of recent changes. RADCO's once-strong culture and its image as a research institute are in flux and may be eroding. A new top management with new ways of thinking is shifting the orientation of the company, and some people express concern about the lack of clear direction and long-range planning. People's faith in RADCO's strategy of technical superiority has weakened and its traditional orientation toward innovation is giving way to a concern for routinization and production efficiency. This shift is resulting in conflict and uncertainty. Where once access to the top was easy, now the decentralized matrix structure—with fewer central services—makes it difficult.

EXHIBIT Characteristics of the Five Companies in Order of Most to Least "Entrepreneurial"

	CHIPCO	RADCO	MEDCO	FINCO	UTICO
Percent of effective managers with entrepreneurial accomplishments	71%	69%	67%	47%	33%
Current economic trend	Steadily up	Trend up but currently down	Up	Mixed	Down
Current "change issues"	Change "normal"; constant change in product generations; proliferating staff and units	Change "normal" in products, technologies; recent changeover to second management generation with new focus	Reorganized about 3–4 years ago to install matrix; "normal" product technology changes	Change a "shock"; new top management group from outside reorganizing and trying to add competitive market posture	Change a "shock"; undergoing reorganization to install matrix and add competitive market posture while reducing staff
Organization structure	Matrix	Matrix in some areas; product lines act as quasidivisions	Matrix in some areas	Divisional; unitary hierarchy within divisions, some central services	Functional organization; currently overlaying a matrix of regions and markets
	Decentralized	Mixed	Mixed	Centralized	Centralized
Information flow	Free	Free	Moderately free	Constricted	Constricted
Communication emphasis	Horizontal	Horizontal	Horizontal	Vertical	Vertical
Culture	Clear, consistent; favors individual initiative	Clear, though in transition from emphasis on invention to emphasis on routinization and systems	Clear; pride in company, belief that talent will be rewarded	Idiosyncratic; depends on boss and area	Clear, but top management would like to change it; favors security, maintenance, protection
Current "emotional" climate	Pride in company, team feeling, some "burn-out"	Uncertainty about changes	Pride in company, team feeling	Low trust, high uncertainty	High certainty, confusion
Rewards	Abundant. Include visibility, chance to do more challenging work in the future and get bigger budget for projects	Abundant. Include visibility, chance to do more challenging work in future and get bigger budget for projects	Moderately abundant; conventional	Scarce; primarily monetary	Scarce; promotion, salary freeze; recognition by peers grudging

As at CHIPCO, lateral relationships are important, though top management's presence is felt more. In the partial matrix, some managers have as many as four "bosses." A middle manager's boss or someone in higher management is likely to give general support to projects as long as peers within and across functions get on board. And peers often work decisions up the organization through their own hierarchies.

Procedures at RADCO are both informal and formal: much happens at meetings and presentations and through persuasion, plus the company's long-term employment and well-established working relationships encourage lateral communication. But managers also use task forces and steering committees. Projects often last for years, sustained by the company's image as a leader in treating employees well.

MEDCO manufactures and sells advanced medical equipment, often applying ideas developed elsewhere. Although MEDCO produces a high proportion of innovative accomplishments, it has a greater degree of central planning and routinization than either CHIPCO or RADCO. Despite headquarters' strong role, heads of functions and product managers can vary their approaches. Employers believe that MEDCO's complex matrix system allows autonomy and creates opportunities but is also time wasting because clear accountability is lacking.

Teamwork and competition coexist at MEDCO. Although top management officially encourages teamwork and the matrix produces a tendency for trades and selling to go on within the organization, interdepartmental and interproduct rivalries sometimes get in the way. Rewards, especially promotions, are available, but they often come late and even then are not always clear or consistent. Because many employees have been with MEDCO for a long time, both job mobility and job security are high. Finally, managers see the company as a leader in its approach to management and as a technological follower in all areas but one.

The last two companies in the study, FINCO (insurance) and UTICO (communications), show the lowest proportion of innovative achievements. Many of the completed projects seemed to be successful *despite* the system.

Currently FINCO has an idiosyncratic and inconsistent culture: employees don't have a clear image of the company, its style, or its direction. How managers are treated depends very much on one's boss—one-to-one relationships and private deals carry a great deal of weight. Though the atmosphere of uncertainty creates opportunities for a few, it generally limits risk taking. Moreover, reorganizations, a top-management shake-up, and shuffling of personnel have fostered insecurity and suspicion. It is difficult for managers to get commitment from their subordinates because they question the manager's tenure. Managers spend much time and energy coping with change, reassuring subordinates, and orienting new staff instead of developing future-oriented projects. Still, because the uncertainty creates a vacuum, a few managers in powerful positions (many of whom were brought in to initiate change) do benefit.

Unlike the innovation-producing companies, FINCO features vertical relationships. With little encouragement to collaborate, managers seldom make contact across functions or work in teams. Managers often see formal structures and systems as constraints rather than as supports. Rewards are scarce, and occasionally a manager will break a promise about them. Seeing the company as a follower, not a leader, the managers at FINCO sometimes make unfavorable comparisons between it and other companies in the industry. Furthermore, they resent the fact that FINCO's top management brings in so many executives from outside; they see it as an insult.

UTICO is a very good company in many ways; it is well regarded by its employees and is considered progressive for its industry. However, despite the strong need for UTICO to be more creative and thus more competitive and despite movement toward a matrix structure, UTICO's middle ranks aren't very innovative. UTICO's culture is changing—from being based on security and maintenance to being based on flexibility and competition—and the atmosphere of uncertainty frustrates achievers. Moreover, UTICO remains very centralized. Top management largely directs searches for new systems and methods through formal mechanisms whose ponderousness sometimes discourages innovation. Tight budgetary constraints make it difficult for middle managers to tap funds; carefully measured duties discourage risk takers; and a lockstep chain of command makes it dangerous for managers to bypass their bosses.

Information flows vertically and sluggishly. Because of limited cooperation among work units, even technical data can be hard to get. Weak-spot management means that problems, not successes, get attention. Jealousy and competition over turf kill praise from peers and sometimes from bosses. Managers' image of the company is mixed: they see it as leading its type of business but behind more modern companies in rate of change.

ORGANIZATIONAL SUPPORTS FOR CREATIVITY

Examination of the differences in organization, culture, and practices in these five companies makes clear the circumstances under which enterprise can flourish. To tackle and solve tricky problems, people need both the opportunities and the incentives to reach beyond their formal jobs and combine organizational resources in new ways.[3] The following create these opportunities:

- Multiple reporting relationships and overlapping territories. These force middle managers to carve out their own ideas about appropriate action and to sell peers in neighboring areas or more than one boss.
- A free and somewhat random flow of information. Data flow of this kind prods executives to find ideas in unexpected places and pushes them to combine fragments of information.
- Many centers of power with some budgetary flexibility. If such centers are easily accessible to middle managers, they will be encouraged to make proposals and acquire resources.
- A high proportion of managers in loosely defined positions or with ambiguous assignments. Those without subordinates or line responsibilities who are told to "solve problems" must argue for a budget or develop their own constituency.
- Frequent and smooth cross-functional contact, a tradition of working in teams and sharing credit widely, and emphasis on lateral rather than vertical relationships as a source of resources, information, and support. These circumstances require managers to get peer support for their projects before top officers approve.
- A reward system that emphasizes investment in people and projects rather than payment for past services. Such a system encourages executives to move into challenging jobs, gives them budgets to tackle projects, and rewards them after their accomplishments with the chance to take on even bigger projects in the future.

Some of these conditions seem to go hand in hand with new companies in not-yet-mature markets. But top decision makers in older, traditional companies can design these conditions into their organizations. They would be wise to do so because, if empowered, innovative mid-

[3] My findings about conditions stimulating managerial innovations are generally consistent with those on technical (R&D) innovation. See James Utterback, "Innovation in Industry," *Science*, February 1974, pp. 620–26; John Kimberly, "Managerial Innovation," *Handbook of Organizational Design,* ed. W. H. Starbuck (New York: Oxford, 1981); and Goodmeasure, Inc., "99 Propositions on Innovation from the Research Literature," *Stimulating Innovation in Middle Management* (Cambridge, Mass., 1982).

dle managers can be one of America's most potent weapons in its battle against foreign competition.

Building a Team

There was, it appeared, a mysterious rite of initiation through which, in one way or another, almost every member of the team passed. The term that the old hands used for this rite—West invented the term, not the practice—was "signing up." By signing up for the project, you agreed to do whatever was necessary for success. You agreed to forsake, if necessary, family, hobbies, and friends—if you had any of these left (and you might not if you had signed up too many times before). From a manager's point of view, the practical virtues of the ritual were manifold. Labor was no longer coerced. Labor volunteered. When you signed up, you in effect declared, "I want to do this job and I'll give it my heart and soul." It cut another way. The vice president of engineering, Carl Carman, who knew the term, said much later on: "Sometimes I worry that I pushed too hard. I tried not to push any harder than I would on myself. That's why, by the way, you have to go through the sign-up. To be sure you're not conning anybody."

The rite was not accomplished with formal declarations, as a rule. Among the old hands, a statement such as "Yeah, I'll do that" could constitute the act of signing up, and often it was done tacitly—as when, without being ordered to do so, Alsing took on the role of chief recruiter.

The old hands knew the game and what they were getting into. The new recruits, however, presented some problems in this regard.

From Tracy Kidder, *The Soul of a New Machine* (Boston: Little, Brown, 1981). Reprinted with permission from the publisher.

The Venture Capital Start-Up

Case III–5
Data Net

A. L. Frevola, Jr., and M. A. Maidique

It was a beautiful Sunday afternoon. The skies were clear, and there was only a slight breeze: a perfect day for flying. Richard Fretwell loved days like this because he could gather up his equipment, hit the open fields, and fly—model airplanes, that is. That is exactly what he decided to do this Sunday afternoon.

As his plane took off and proceeded to float above him, Richard's mind wandered. He was thinking that it had been almost six months now since he left his family in Arizona and moved to Miami, Florida, to become president of a start-up company named Data Net. It was a tough decision, he remembered. His family had stayed behind so his daughter could finish her last year of high school. Now, six months later, she was almost ready to graduate. His wife would soon be selling the house and moving to Miami.

The last six months had been quite productive, he thought. Without the distractions of a family—he had to be honest with himself—he and his colleagues had taken Data Net from an idea to a prototype ready for the AMS (Advanced Manufacturing Systems) show in Chicago.

"That's when the work really begins," he thought. "It's been a good six months, we're right on schedule, and there have been no major pitfalls. All we need now is to complete our first major round of venture capital financing."

THE FOUNDING

Data Net Corporation was founded to develop a proprietary, user-friendly Factory Data Collection (FDC) Terminal that would take advantage of the proliferation of the IBM–PC in the industrial workplace. Data Net's DNT 1000 (TM) terminals communicate to a host computer through proprietary communication cards resident in an IBM–PC, giving industrial customers quick and accurate collection of data on manpower utilization, work in process, inventory levels, and quality control.

Data Net was founded with a $500,000 seed capital investment from Hambrecht and Quist Venture Partners (H&Q), Southeast Venture Capital (SVC), and a small investment from a personal partnership managed by the sponsoring partner at H&Q. The origins of Data Net could be traced to discussions at Innovative Electronics, a small high-tech firm that had developed a small, unsophisticated office data collection terminal. Bob Williamson, Data Net's current vice president–finance, had been employed at Innovative Electronics in a similar capacity. Innovative aspired to expand into FDC, a market already served by companies such as Burr-Brown and NCR. One of Bob's principal responsibilities was dealing with potential venture capital investors. The venture capital community, however, had doubts about Innovative's leadership. After the company failed in several attempts to raise venture capital, Bob was unexpectedly fired on December 6, 1984. Bob, stunned, asked, "What do I do now?"

Ignoring offers from other firms, Bob contacted Mike Martin of H&Q with whom he had previously dealt in trying to raise money for Innovative. Soon thereafter, Bob contacted Richard Fretwell, whom he had met in the process of raising capital for Innovative.

The other two founders who soon joined Data Net's management team were Bruce Pelkey, vice president–engineering, and Carl Johnson, future vice president–operations. Bruce had gained extensive experience in FDC with NCR before going to Innovative. Soon after Bob was fired, Bruce resigned from Innovative to join the Data Net team. A future member of the team, Carl Johnson, retained his position with a local high-tech firm but planned to join Data Net as soon as production began.

When Richard said yes to the opportunity to head up the management team of Data Net, Jim Fitzsimons of SVC and Mike Martin of H&Q met with the four-man team. At the initial meeting, Jim Fitzsimons asked attorneys Vance Salter and Marc Watson to evaluate the employment contracts of each team member. The first question to be answered was, "Can these men start a new venture in the FDC industry without violating the terms of their current or previous employment agreements?" When Jim and Mike were persuaded that neither the Innovative nor Burr-Brown (Richard's company) agreements would restrict the group from competing in the Factory Data Collection market, they agreed to consider the possibility of funding the group.

The original business plan was written on Christmas Eve (see Exhibit A). On December 26, 20 days after he was fired, Data Net was incorporated by Bob Williamson. The following week, copies of a preliminary five-page plan were given to Jim and Mike. They agreed to get together in early January to discuss terms of the deal.

In mid-January, they hammered out an agreement. It was a one-page, handwritten equity agreement that specified ownership, capital commitments, management positions, board of directors structure, and conditions for the repurchase of stock should someone decide to leave (see Exhibit B). At that point, Mike Martin committed $225,000 from H&Q in return for 18 percent ownership, as well as $50,000 from ATF, a personal partnership, for a 4 percent share of the stock. Jim took about 10

EXHIBIT A Data Net Corporation

Data Net Corporation (DNC) is seeking $500,000 of seed capital to begin developing and marketing a state-of-the-art data collection system. Our products will enable factories, warehouses, and institutions to more quickly and accurately collect and use data generated by their operations. DNC will be successful because it is being founded by an experienced management team which has built similar successful businesses in the past and because it will sell to a large, $200 million market expanding at the rate of 35 percent per year.

DNC Product

DNC will design, build, and market Manufacturing Information Systems to be used for information collection and display in factories, warehouses, and institutions such as hospitals. A typical system will have numerous Application-Unique Terminals connected by simple multidrop wiring to a microprocessor-based controller/concentrator which, in turn, will communicate with the customer's host computer.

The Application-Unique Terminal is similar electronically to the standard CRT but generally has a smaller display, fewer keys on the keyboard, and a smaller footprint (50 to 70 square inches versus 300 square inches for a CRT). Most A-U Terminals are also designed for bar code, magnetic card, or other nonmanual inputs as well.

[International Data Corporation (IDC) coined the term *Application-Unique Terminal.* IDC uses this term to cover such devices as point-of-sale terminals (electronic cash registers), portable terminals, and fixed special applications terminals used in the factory. We use the term to mean DNC's DaNCer 1000 (TM) series of tethered and portable terminals.]

The controller/concentrator is similar to an IBM–PC, and some system integrators in our industry already use PCs in the factory environment. While some controllers do little more than concentrate the wires from the terminals, the trend is to smarter controllers with significant data processing and system control responsibilities.

Management Team

The four-man founding team has nearly 80 years of experience building similar businesses. Members of the team have:

1. Started up and run a successful manufacturer and marketer of factory data collection systems.
2. Developed and produced new lines of data collection terminals and controllers.
3. Set up and managed new manufacturing facilities including the most modern Surface Mounted Technology production plant in Florida.
4. Set up and managed all financial and administrative systems for numerous small, growing companies.

Richard Fretwell, President and CEO

Fretwell, marketing manager of Burr-Brown's factory data collection division has, in the last five years, built his company into a recognized factor in the industry. Upon taking over marketing responsibility three years ago, he quickly built Burr-Brown up to $7 million in systems sales this year.

Prior to joining Burr-Brown, Fretwell was the vice president–engineering of a modem manufacturer. He has a B.S.E.E. and M.S.E.E. from the University of Iowa.

Bruce Pelkey, Vice President–Engineering

Pelkey, who is vice president–engineering of Innovative Electronics, had the bulk of his experience at NCR's Data Pathing Division. He was in charge of data collection terminal design for NCR and during his tenure, he developed over 30 different types of A-U Terminals. At Innovative, Pelkey has been primarily involved in manufac-

EXHIBIT A *(continued)*

turing, as engineering development has been sharply curtailed.

Pelkey has a B.Eng. (Electrical) from McGill and an M.B.A. from Rollins.

Carl Johnson, Vice President–Operations

Johnson is director of manufacturing for Computer Products, Inc. At present, he is converting his $30 million sales division's products entirely to Surface Mounted Technology (SMT). When completed in April 1985, this will be the most modern SMT facility in Florida and enable Computer Products to enjoy a significant cost and size advantage over its competition.

Johnson has a B.S.E.E. from the University of Miami.

Robert Williamson, Vice President–Finance

Williamson had been vice president–finance of Innovative Electronics prior to founding DNC. He has arranged numerous bank and institutional financings for small, rapidly growing companies. He has broad experience setting up and managing accounting, auditing, and control functions.

Williamson has a B.S. (industrial design) with distinction and an M.B.A. from Stanford.

Market Analysis

IDC calls our overall market the Manufacturing Information Systems market and says that it was a $2.4 billion market in 1983. IDC expects the market to grow to $20.1 billion in 1990, a 35 percent annual rate of growth. Contained within that market is software packages such as MRP systems, minicomputer-based factory data collection systems, and products like ours based on A-U Terminals and microcomputer controllers.

We believe that our type of system accounted for $150 million of hardware sales in 1983 and over $200 million this year. Add to the hardware the various required or optional software packages purchased and installation costs and we believe total sales were over $500 million in 1984.

Our primary competitors include NCR's Data Pathing Division, Burr-Brown, Epic Data, IBM, and some of the bar code manufacturers such as Welch-Allyn and Intermec. No competitor is dominant and systems integrators are an important factor in the market.

Product Development

DNC will immediately begin development of advanced hardware and software for its market.

DaNCer 1000 (TM) Terminals

The most visible element of the system, a group of A-U Terminals, will be designed around a single printed circuit (pc) board. This single pc board will serve in all terminals from the simplest single function unit to the most complex, multifunction terminal. It will also be designed to serve in both tethered terminals and portable, hand-held units.

Our immediate commitment to SMT makes such a design possible. The pc board will be engineered to run the most complex terminal yet be small enough to fit in a portable, hand-held package. In general, SMT allows pc boards to be designed to fit in one half of the "real estate" of a comparable standard pc board.

Industrial design is an important element of DaNCer terminal development. While many competitive terminals have been factory hardened, they generally are both difficult to use and unattractive. We feel that ease of operation and appearance both play an important part in the buying decision and have budgeted for an experienced industrial design consultant during initial product development.

Data Net Controllers

We expect to use any one of a number of standard microcomputers in our system, such as the IBM–PC. With limited resources, it does not pay at present to develop our own controller when

EXHIBIT A *(continued)*

off-the-shelf products will give us 90 percent of the performance we need.

Data Net Software

Software development will occur on two fronts—a terminal user package and a controller-host communications package. Terminal user software will be significantly different from that presently offered. Because SMT allows for significantly higher data processing capabilities in the terminal, the terminals will not only be active data gatherers but also active data givers. Menu-driven software will allow the end user to configure terminals to interact with the manufacturing process, correcting errors and prompting next steps.

Manufacturing Facilities

DNC expects to contract out all production during the first 18 months. Management has identified a contract manufacturer in California which will be able to get us into immediate production. Initial capital expenditure allowances call for a CAD/CAM package for a PC to aid pc board layout. Initial calculations in the financial statements indicate that our own production facility will be economically feasible in June 1986. This decision, however, will be subject to extensive analysis before any commitment is made.

Marketing Rollout

The company is committed to a direct sales force as the most cost-effective means of selling. Initially Fretwell and a national sales manager, with Pelkey and Williamson's help, will call upon end users and systems integrators. By the end of the second fiscal year, we expect to have four regional offices staffed with direct salesmen and possibly a service technician. Initial leads will be developed through advertising, direct mail, and trade shows. The company expects to introduce itself at the AMS show in June 1985 and to display working prototypes.

Financial Projections

Financial projections are broken into two parts—the development period and the first two years of sales and operations.

Development Period

The development period is expected to be six months through June 1985. We expect no sales in this period and believe the $500,000 seed capital will be substantially used up by June.

This period will yield a working prototype of the terminals, some production tooling, marketing literature, and the initial sales leads that will help us close four system sales in the third quarter of 1985.

The loss for the first six months will be $389,000. Cash flow in this period will be:

	$000
Personnel	(228)
Marketing	(73)
Other	(74)
Capital equipment	(200)
Payables and debt	98
Total cash flow	(477)

Approximately $100,000 of the equipment to be purchased is standard engineering development tools. We believe we can lease a substantial portion of this equipment.

First Two Years

During the first six months, we expect to complete a $1.5 million equity placement. Assuming this is in place June 30, 1985, our projections are as follows:

EXHIBIT A *(concluded)*

	Year Ending June 30	
	1986	*1987*
Sales	2,417	8,123
Gross margin	937	4,318
Percentage gross margin	39%	53%
Operating income	(340)	1,818
Net before taxes	(274)	1,638
Percentage net before taxes	—	20%

We will be on a fiscal year ending June 30 and expect to move into our own production facility in mid-1986. We expect to negotiate a substantial line of credit soon after the $1.5 million funding.

Detailed financial projections are attached.

Proposed Capitalization

The $500,000 investment will be evidenced by preferred shares convertible into 40 percent of the common stock of the company. Common stock will be distributed to the founding team as follows:

Fretwell	21%
Pelkey	15%
Johnson	12%
Williamson	12%

days longer to convince his associates at SVC to invest $225,000 for another 18 percent share.

In total, the three investors purchased 400,000 shares of Data Net's series A convertible preferred stock for $1.25 per share. Another 600,000 shares of common stock were distributed among the management team. Richard received 210,000 shares (21 percent), Bruce received 150,000 shares (15 percent), and Carl and Bob each received 120,000 shares (12 percent). In addition, the parties agreed that 100,000 shares of common stock would be authorized and reserved for the purpose of establishing an incentive stock option plan for company employees other than the four-man management team. Additionally, there are reserved for issuance as of the closing date 400,000 shares of common stock for issuance upon conversion of the series A preferred stock.

During the next few weeks, the lawyers prepared the closing documents which were signed on January 24, 1985. Three days later, the funds were transferred to the company and Data Net was in business.

EXECUTIVES' BACKGROUNDS

Richard Fretwell received his bachelor's and master's degrees in electrical engineering from the University of Iowa from 1960 to 1967. After two years with the U.S. Army Security Agency, in 1969 he cofounded MI² Corporation with the hope of penetrating the data communications market developing due to the Carterfone Decision.

He left MI² in 1979, after designing 22 new products and receiving 12 patents, to join Burr-Brown Corporation's Data Acquisitions and Systems Division. Burr-Brown is a $100 million multidivision company that designs, manufactures, and markets monolithic and hybrid analog components and data acquisition products and systems.

"At MI², there was no budgeting, scheduling, operations, or strategic planning. At Burr-Brown, I had the opportunity to do that. Also, prior to Burr-Brown I was principally an engineer. Burr-Brown exposed me directly to sales and marketing, nationally and internationally." During his five years at Burr-Brown, Richard

EXHIBIT B

Points of Agreement 1/4/85

1. Data Net Corporation (Florida)

2. Ownership

	%	$
– VMA	18%	225,00
– H&Q	18%	225,000
– Mike/Personal	4%	50,000
TOTAL PREF.	40%	500,000
1,000,000 shares		
– Fretwell	21%	
– Pellay	15%	
– Williamson	12%	
– Johnson	12%	
TOTAL COMM.	60%	(OPEN)
TOTAL PREF. & COMM.	100%	

3. Board of Directors
Mitch Maidique
Jim Fitzsimons
Richard Fretwell
(chairman open)
+ 2 open

4. Mgt. stock can be repurchased if they leave – vesting schedule:

on joining	20%
after 1 yr	40%
2 yrs	60%
3 yrs	80%
4 yrs	100%

SOURCE: Data Net company records. Copy of original hand-written agreement.

EXHIBIT B *(concluded)*

DATA NET Corporation
Shareholders List*

Name of Shareholder	Number of Shares	Preferred, Common, or Option	Average Price	Percent Ownership
Richard D. Fretwell	210,000	Common	$0.025	19.09%
Bruce Pelkey	150,000	Common	$0.025	13.64%
Carl Johnson	120,000	Common	$0.025	10.91%
Robert Williamson	120,000	Common	$0.025	10.91%
Hambrecht & Quist	180,000	Preferred	$1.250	16.36%
Southeast Venture	180,000	Preferred	$1.250	16.36%
Mike Martin Partnership	40,000	Preferred	$1.250	3.64%
Steve Zarzecki	2,500	Option	$0.400	0.23%
Wendy Cerco	250	Option	$0.400	0.02%
Remaining options	97,250	Option	open	8.84%
	1,100,000			100.00%
New investment	$2,000,000			
Ownership Analysis	Percent			
Owned by founders	54.55%			
Owned by investors	36.36%			

* This a printed and updated version of the handwritten shareholder agreement.
SOURCE: Data Net company records.

held the position of engineering manager of a new start-up division formed to vertically integrate the company into systems markets; international sales manager—a new position to stimulate sales of divisions' products by international sales subsidiaries; and most recently, marketing manager.

Richard came to Data Net because of the opportunity to run his own show. "In a large company you're never really secure. What better way to be secure than to control your own destiny."

Bob Williamson, vice president–finance, received a B.S. in industrial design and an M.B.A. from Stanford University between 1962–68. By 1973, he was director of finance with the SSI Navigation Division of ITEL Corporation. Between 1977 and 1982, Bob served as vice president–finance for the Bernuth Corporation where he installed an accounting and control system for four divisions and obtained $12 million in financings.

In 1982, he began working for Innovative Electronics, Inc., as vice president–finance. Innovative develops and manufactures protocol converters and has sold some office and factory data acquisition systems. During his tenure at Innovative, Bob also became familiar with the venture capital community while attempting to raise capital for Innovative's new FDC division.

Bruce Pelkey, vice president–engineering, is also a cofounder of Data Net Corporation. He received his electrical engineering degree from McGill University in 1976 and his M.B.A. in finance from Rollins College in 1982. Bruce gained most of his experience in FDC terminal design with NCR Corporation's Data Pathing Division, a major supplier to the FDC marketplace. "No one has designed more terminals for this market than Bruce," boasts Bob.

Bruce has designed and managed the development of the NCR 2841 Time Clock and 2842 Data Collection terminals, developed the NCR 2825 Multifunction Industrial Data Collection terminal, and implemented numerous design improvements to existing terminal families. It was this experience he took to Innovative in 1984, endeavoring to compete in the FDC market. Unfortunately, due to Innovative's limited financial capability, he was forced to concentrate on previously developed communications products.

Carl Johnson, vice president–operations, is the fourth cofounder of Data Net. As of June 1985, he was with another company but had plans to join Data Net as soon as possible. Carl holds a degree in electrical engineering from the University of Miami and an M.B.A. from Florida Atlantic University.

In 1971, he became operations director at Coulter Electronics. With Coulter, he established a 100-employee plant in Puerto Rico and managed production growth from $20 million to $60 million in sales with a staff of 650. Between 1975 and 1980, he worked for SEL Gould, Hycel Corporation, and Cordis Corporation. During this period, he brought three new products from engineering to production and established and monitored manufacturing budgets.

In 1980, he took the position of director of advanced technology in the Real Time Division at Computer Products, Inc. CPI designs and manufactures industrial measurement and control systems for processing plants and utilities. While with CPI, Carl installed a surface mounted product line for a $30 million division and reorganized the manufacturing division.

THE MARKET

After the initial seed capital was secured, Richard moved to Miami to lead the management team. During the first few months, they spent a lot of time developing the business plan. "There are two reasons why we spent so much time on the business plan; first, it forces us to define exactly what we want to do, and second, it helps us raise money. But it's a real plan, we're not just trying to get money," explains Richard.

One of the main purposes of the business plan is to define a niche for Data Net in the FDC market. According to projections made by International Data Corporation (IDC), the FDC market was over $200 million in 1983 and is expected to reach sales of $522 million in 1985. Data Net's business plan projects a 20 percent per annum growth rate to $1.2 billion in 1990. Data Net's internal estimate is more conservative than IDC's projected 35 percent per annum growth rate and subsequent $2.0 billion+ market in 1990.

IDC found that $8.9 billion was spent on industrial data collection and control hardware and software in 1983. In 1985, FDC systems will generate $522 million of revenues as manufacturers look for the best way of directly collecting and using data from the manufacturing floor to increase efficiency. In 1983, closed loop or process control hardware and software made up 73 percent of the total FDC market

The Market—1983

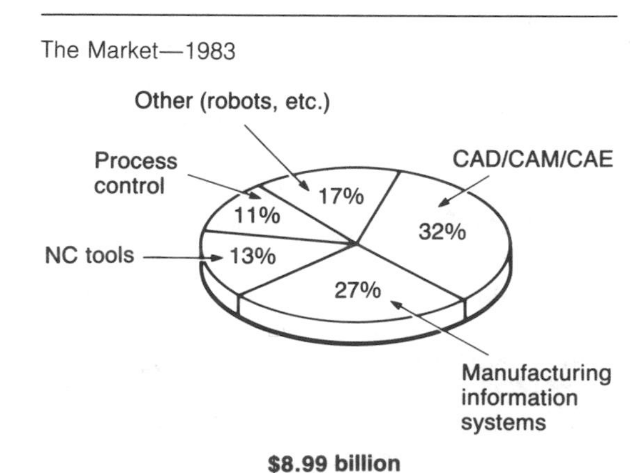

$8.99 billion

SOURCE: International Data Corporation.

while open loop systems (labeled "Manufacturing Information Systems" on the following chart) made up the remaining 27 percent.

The closed loop or process control segment is the more mature sector of the market and includes numerical control, robotics, CAD/CAM/CAE, and related systems. In a closed loop system, data gathering and processing causes an electromechanical response such as a valve to be closed or a conveyor belt to be speeded up. Competition in this segment is intense.

Manufacturing Information Systems (MIS), Data Net's target market, according to IDC, was a $2.4 billion market in 1983. In an open loop system, data is collected at the source and processed by a host, and reports are issued to management or workers for further action. At present, about 75 percent of each sale is hardware and 25 percent is software. Data Net expects this ratio to shift to a parity between hardware and software in the next five years.

The standard MIS installation has consisted of a paper-based recording system (travel cards, time cards, and the like), industrial hardened CRTs on the shop floor but not at the workstation, "keypunch" data entry in the office, and a mainframe/minicomputer host for data processing. Data Net calls these "Big Terminal" systems. An FDC system substitutes small terminals such as Data Net's DNT 1000 (TM) (to be described later) for the paper-based data records and CRTs. An FDC system is a "Little Terminal" system.

Data Net sees a number of reasons why the use of FDC systems will grow rapidly through the 1990s. Currently, less than 5 percent of all manufacturers have data-gathering systems, including those manufacturers doing direct reading of time card information. Error rates on manual systems are greater than 5 percent, and due to the time lag between data gathering and data entry/verification, reports are generated from days to weeks after the actual event. In a drive to become more efficient, manufacturers are adding Manufacturing Resource Planning (MRP) and other types of software (Just-in Time Inventories, etc.). These systems cannot be fully effective without direct, at-the-source data gathering. For this reason, Frost and Sullivan found:

1. Fifty-seven percent of factory data gathering will be done at the source by 1990, up significantly from less than 5 percent now.
2. Sixty-two percent of the MIS managers polled will be considering at-the-source, factory floor data collection in the next three years.

There are over 300,000 manufacturing entities in the United States. Data Net's primary target market is the 90,000 plants in SIC codes 34–39. Major sectors within these codes are electronics, automobiles, machinery, and medical equipment. Each of these segments can be characterized as an assembly operation with numerous parts and a high value added. Their complexity makes them prime candidates for an FDC system. Data Net will emphasize plants with more than 20 employees—about 30,000 facilities nationwide. Value-Added Resellers (VARs) selling to these small plants will be Data Net's direct customers.

THE COMPETITION

With an understanding of customer needs and market trends, Data Net's ambition was to pool the resources of the management team and design a product to serve the growing FDC market. However, the Data Net management team could not, in a short time frame, develop a superior product and capture a substantial market share, without a careful assessment of the strengths and weaknesses of their competition.

Major competitors with FDC divisions such as NCR, Hewlett-Packard, and Burr-Brown have not emphasized the products in these divisions. NCR's Data Pathing Division represents less than 1 percent of NCR sales. IBM does not make FDC products. Small competitors emphasize specialty items, like bar code wands, or are undercapitalized. Data Net knows of no FDC start-up with venture capital backing.

The Manufacturing Information Systems market is divided into "Big Terminal" systems and "Little Terminal" (FDC) systems like Data Net's. The Big Terminal systems predominate, being an extension of the office data processing system to the factory floor.

Big Terminal systems are predominately paper systems. Due to cost and size, few factory-hardened terminals are used on the factory floor. Major hardware vendors of "Big Terminal" systems include IBM, DEC, and Burroughs. Major software vendors include ASK (an H-P mainframe VAR) and McAuto (an IBM VAR). These systems are often integrated into a financial or accounting package and require a major money and time commitment. An ASK system can cost more than $100,000.

Since the bulk of the data is collected on paper and cannot be verified until data entry input is checked against existing jobs and parts files, error rates are high. Because Data Net's system communicates to a host such as a VAX, vendors in this segment are possible Original Equipment Manufacturers (OEMs).

In the FDC market, the primary competitors are NCR, Hewlett-Packard, Burr-Brown, Epic Data Systems, Intermec; and Data Net's most recent competitive analysis reveals significant cost advantages over the competition for both small FDC systems, a 5-terminal pilot system interfaced to a host computer with standard MRP software, and large FDC systems, a 128-terminal system interfaced to a host computer with standard MRP software.

Data Net can save its customers from 50 percent to 70 percent on the cost of hardware (terminals and concentrators), software (data collection, operating system, and data conversion), and installation (power supply and 4,000 feet of wiring) for a small FDC system. At a cost of $4,906 per terminal and a total system cost of $24,530, Data Net's system is nearly half the price of Burr-Brown's $47,900 system, the least expensive of the competition, and over two-thirds less than Intermec's $85,500 system, the most expensive system (see Exhibit C).

Data Net estimates its large FDC systems (128 terminals) will cost $213,555, with terminals priced at $1,668 each. For the same hardware, software, and installation fixtures, Burr-Brown, still the least expensive competitor, can sell the system for $252,600, 20 percent more than Data Net. The most expensive competitor is Epic, offering its system at a 40 percent greater cost of $369,200 (see Exhibit C).

Data Net attributes these savings to its PROPRIETARY CO–LOG (TM) Data Collection Software and low installation costs. The software runs on a standard MSDOS (TM) operating system and requires no programming to configure the data collection system to customer needs. Data Net also offers CO–LOG, data conversion software to interface to MRP during configuration. Installation costs are also reduced because the Data Net system utilizes multidrop communications and twisted pair wiring for up to 4,000 feet instead of expensive coax; also the installation does not require special tools or installation personnel.

NCR, through its Data Pathing Division, provides a wide range of terminals and controllers. NCR's primary strengths are its established sales force and service organization. Because of NCR's strategy to sell primarily to Fortune 100 companies, they are unable to address the smaller plant with 20 to 200 workers. The sales organization is unstable, having been assigned to three different vice presidents in the last two years. They also sell outdated hardware and software designs that require complex custom software.

Hewlett-Packard could become a dominant force in the FDC market because of its corporate commitment to product innovation and reliability. Its primary weakness, however, is that its system is only compatible with an HP host computer, and only 32 terminals fit on the sys-

EXHIBIT C

PRICE LIST

6/24/85
RDF

Prices effective 6/1/85.
Prices subject to change.

TERMINALS		PRICE SCHEDULE ($) Quantity		
Model	*Description*	*1–49*	*50–99*	*100 up*
DNT 1000	RS-422 Communications Interface	1,535	1,375	1,225
DNT 1100	RS-232-C Communications Interface	1,685	1,525	1,350
DNT 1200	20 ma Current Loop Communications Interface	1,650	1,485	1,325
TERMINAL OPTIONS				
F01	Carrying Strap	6	5	4
F10	Mag Swipe Reader	250	225	200
F20	Low Order Serial Port	150	135	120
F30	Relay Outputs (4)	100	90	80
F40	Bar Code Wand, Medium Density	360	325	290
F41	Bar Code Wand, High Density	380	340	300
F42	Bar Code Wand, IR, Medium Density	360	325	290
F43	Bar Code Wand, IR, High Density	380	340	300
F44	Bar Code Swipe Reader	450	400	360
F50	Power Supply, Terminal	30	27	24
F51	Spare Batteries, Terminal	40	36	32
F60	Wall Mount Cradle	250	225	200
F70	Junction Box, 16-Terminal	475	425	380
TERMINAL CONTROLLER				
DNC 5000	Controller Board, 128 Terminals for IBM PC, -XT, -AT	2,500	2,250	2,000
SOFTWARE				
CO–LOG (TM) (V 3.0)	Configuration and Control Software for IBM PC, -XT, -AT	5,000	4,500	4,000

SOURCE: Data Net.

tem. HP systems also require complex applications software. It is known that HP's focus is on establishing the company as a mainframe supplier to factories with minimal investment in the FDC market.

Burr-Brown is the largest terminal manufacturer and has over 70,000 terminals installed. The majority of these terminals are used as "front panel" controllers for stand-alone equipment such as test equipment or programmable controllers. Burr-Brown's primary strength lies in its established worldwide direct sales organization and its reputation for producing reliable products. Its primary weakness is that it

EXHIBIT C1

Data Net Corporation
RDF
Revised 8/15/85

Competitive Analysis—Small System Cost Comparison

ASSUME: 5 terminal pilot system interfaced to host computer with standard MRP software.

	Data Net	Epic	Burr-Brown	NCR	Intermec	HP
Hardware:						
Terminals	9,500	15,000	7,500	10,000	10,000	5,500
Concentrators	5,500	5,000	15,180	20,000	3,500	25,000
Software:						
Data collection	5,000	7,500	5,000	10,000	20,000	15,000
Operating system	55	7,500	10,000	10,000	30,000	15,000
Data conversion	0	10,000	5,000	10,000	10,000	10,000
Installation:						
Power supply	475	0	1,250	0	0	500
Wiring (4,000 ft.)	4,000	12,000	4,000	4,000	12,000	12,000
Total cost	24,530	57,000	47,900	64,000	85,500	83,000
Per terminal	4,906	11,400	9,600	12,800	17,100	16,600

The Data Net Advantage

50 percent to 70 percent lower cost.

Why?

1. PROPRIETARY CO–LOG (TM) data collection software:
   - No programming required to configure the data collection system.
   - Standard MSDOS (TM) operating system.
   - Data conversion software to interface to MRP system done by CO–LOG during configuration.
2. Low installation cost:
   - Twisted pair wiring for up to 4,000 feet.
   - Expensive coax not required.
   - Multidrop communications.
   - No special tools or installation personnel required.

SOURCE: Data Net.

EXHIBIT C2

Data Net Corporation
RDF
Revised 8/15/85

Competitive Analysis—Large System Cost Comparison

ASSUME: 128 terminal system interfaced to host computer with standard MRP software.

	Data Net	Epic	Burr-Brown	NCR	Intermec	HP
Hardware:						
Terminals	195,200	307,200	153,600	204,800	204,800	112,640
Concentrators	5,500	10,000	18,000	20,000	28,000	100,000
Software:						
Data collection	5,000	15,000	20,000	10,000	40,000	15,000
Operating system	55	15,000	20,000	10,000	30,000	15,000
Data conversion	0	10,000	5,000	10,000	10,000	10,000
Installation:						
Power supply	3,800	0	32,000	0	0	12,800
Wiring (4,000 ft.)	4,000	12,000	4,000	4,000	16,000	50,000
Total cost	213,555	369,200	252,600	258,800	328,800	315,440
Per terminal	1,668	2,885	1,973	2,022	2,569	2,465

Data Net Has the Advantage

20 percent to 40 percent lower cost.

Why?

1. PROPRIETARY CO–LOG (TM) data collection software:
   - No programming required to configure the data collection system.
   - Standard MSDOS (TM) operating system.
   - Data conversion software to interface to MRP system done by CO–LOG during configuration.
2. Low installation cost:
   - 128 terminals per concentrator.
   - Multidrop communication line wiring.
   - Twisted pair wiring for up to 4,000 feet.
   - Expensive coax not required.
   - Low-cost 24 VAC power required.

SOURCE: Data Net.

has simply adapted its "front panel" controller terminal and has not completely integrated the panel into a user-friendly system. Burr-Brown systems are not designed for a rugged environment, provide no tactile feel on small keyboards, have limited software capability, and require the purchase of an expensive concentrator.

Epic Data is a privately held Canadian company with estimated sales of $20 million that provides a complete line of FDC products. Its primary strengths are the versatility of its product line, an aggressive sales organization, and good systems software capability. Its primary weaknesses are the complicated programming required to make the system operational (can only be done by Epic engineers), mostly obsolete designs, and a nonintelligent controller. Epic is the smallest competitor and is undercapitalized.

Intermec manufactures a variety of products including point-of-sale terminals and bar code equipment. They offer a reliable, reasonably priced product. Because of a 32 percent annual sales growth over the last five years, they have an established reputation and a large installed base. However, Intermec products have limited systems capability and a nonintelligent data concentrator. They also rely on old, outdated designs that are no longer cost effective to manufacture.

Data Net feels it has significant overall advantages in the FDC market. One is focus. Except for Epic, Data Net is the only company where FDC is the primary business, not a sideline. Data Net's terminals have significant technical advantages over their competitors:

1. They can be used in either a fixed or portable mode.
2. Power is distributed along with the communications line, saving the use of an electrical wall socket at each workstation.
3. The two line display allows more detailed information and query response displays.
4. Average sell price will be less per terminal than competitors (see Exhibit C).

Also, Data Net's controller is resident in an IBM–PC and can handle 128 terminals. The Independent Data Collection System option requires no host computer, using low-cost standard data-base management software. Only Data Net, Epic, and Burr-Brown offer redundant operation, and no competitor offers battery backup on terminals.

Data Net does have some major disadvantages that will be addressed by its sales strategy. First, its competitors, especially NCR, Epic, and Burr-Brown, have good, direct sales forces in place. And second, its competition has an installed base, a real asset in the conservative manufacturing environment.

PRODUCT DEVELOPMENT

Richard Fretwell and Bruce Pelkey, vice president–engineering, came to Data Net because they were extremely frustrated in their previous positions. Both of them wanted to develop a totally new product, attractively packaged in a different type of case, and utilizing the most advanced design and production technologies.

Based on its experience and supplemented by a report of International Data Corporation (IDC), the Data Net management team found that the FDC market requires terminals specially designed for use on the factory floor. The standard CRT terminal is not suitable for factory use due to its cost, complexity, size, and vulnerability to damage in a harsh environment. With this in mind, Bruce and the design team immediately began developing a terminal for use on the factory floor.

February

Early in the month, Bruce and an independent hardware design consultant began the specifications work and subsequent electronic layout for the terminal's printed circuit (PC) board. Data Net also contracted William Smith, Associates, an industrial design consulting firm, to develop a logo for the company.

Pleased with their work on the logo design, Data Net retained the services of William Smith, Associates, to assist in the terminal's outer case design. Data Net's design team presented the consultants with their ideas for the product. Data Net wanted to use an LCD display, but was faced with the problems caused by the poor resolution outside a limited viewing angle. They also wanted a rugged case that could be easily mounted on a wall or sit horizontally on a work table, but was also small enough to be hand-held as a portable unit. No other manufacturer had ever developed a unit that could be used in either a portable or fixed mode.

March

In mid-March, William Smith, Associates, returned to Data Net with several variations of two different designs for the terminal. One was a terminal that was tilted on a cradle for use on a flat work area. The other design was a long, flat unit with an adjustable display. After a short review of both models, the two parties agreed that the terminal would use the best of both designs. The result was a design very similar to the end product, using the cradle as the handle for wall mount and portable use, and the adjustable screen to solve viewing problems.

With the design decision made, Data Net called in another consultant to assist in the layout for the power board. Work on the keyboard also began. Keyboard design was a special effort because it provides a full-travel keyboard in a sealed case, a feature unique to the industry. In late March, Bruce visited Hart Keyboard in Ohio to finalize design specifications and confirm Hart's ability to deliver a product on time.

Data Net also believed that to be cost effective to the end user, the FDC system must be designed to minimize the amount of custom software that has to be developed for each application. The system must be "configurable" rather than "programmable," enabling the user to set up the system without programming knowledge. An independent programmer was hired in March to begin the development of Data Net's software package. Bruce, the hardware consultant, and the software consultant all worked very closely during the rest of the development period.

April

In early April, Bruce visited MidWest Surface Mount to contract for the manufacture of the PC boards. After the design specifications were worked out, Bruce returned to Miami, but he was concerned that MidWest might not work quickly enough to meet the deadline required to make the AMS show in June.

Data Net also contracted with a reputable local firm, Boca Boards, Inc., for the production of the power boards. By contracting with more than one company, Data Net hoped to cover all their needs without relying heavily on any one firm.

May

During May, the hand tooling for the first few cases began, work continued on software and hardware design, and the management team began to prepare for the AMS show less than six weeks away. Within that time, all contracted work had to be complete so the product could be assembled and the final bugs worked out.

In late May, MidWest Surface Mount delivered the PC boards, on time and working. With the most critical element of the product functioning properly, it appeared that Data Net's concerns were behind them. Irony struck, however, as it began to appear as though Boca Boards, the more reliable firm, would not be able to deliver a workable board in time for the show.

June

Boca Boards had failed. The product they delivered did not work. With the AMS show less than two weeks away, Data Net immediately called in the consultant that had originally helped lay out the power board in early April. The original consulting designer came through and hand-built the five power boards needed.

The result was Data Net's DNT 1000 (TM) terminal (see Exhibit D), designed to meet the market needs for factory floor use. The DNT 1000 has an area less than 72 square inches and can be easily mounted both horizontally on the manufacturing workspace or vertically on a post, column, or wall, or the DNT 1000 can be used as a portable unit. The terminal is rugged enough to resist liquid spills, particle contamination, and hard physical use. Its environmentally sealed keyboard provides "Function Keys," tactile feedback, and a simplified keyboard layout for the unskilled factory worker. To be quickly and easily adapted by the user, it can be interfaced with a bar code reader and has full discrimination of major bar code families. The DNT 1000 has a large, easily read display, uses low-cost standard cables for communication and power distribution, and is easily installed.

Data Net's resulting software package is extremely user friendly. The entire user's manual is integrated into the configuration package so that the user simply asks the program for help when needed. To configure the program for a manufacturer's specific needs, the user needs only to sit at the terminal and answer a series of questions, referring to the help function when necessary.

Data Net also understood that host computers available today cannot efficiently communicate with a large number of terminals on a real time basis. They must be supported by intelligent Concentrators/Controllers, the second component of Data Net's system. The Concentrator is an intelligent multiplexer that funnels the many communications lines from the terminals down to a single communications line to the host computer. The Concentrator also provides local data processing and unburdens the host computer by handling all interactive communications with the terminals.

On June 15, only 4½ months after development began, the Data Net management team left for Chicago with a fully operational FDC system consisting of the DNT 1000 terminals, a controller/concentrator resident in an IBM–PC (or true compatible), and a family of software to run the terminals and provide host communications. By introducing their product in such a short time frame, Data Net gained instant credibility with venture capitalists and potential customers at the AMS show.

As stated earlier, Data Net received five handmade prototype terminals to be used in the AMS show in Chicago. The company also plans to fill the first few purchase orders with handmade terminals, while they continue to negotiate the details of an assembly contract. Data Net expects to contract out the assembly of its printed circuit boards through early 1986. Five firms have been identified that have the capacity for advanced surface mounted production. The first choice, MidWest Surface Mount, built the first five prototype boards used in the demonstration units.

System Configuration

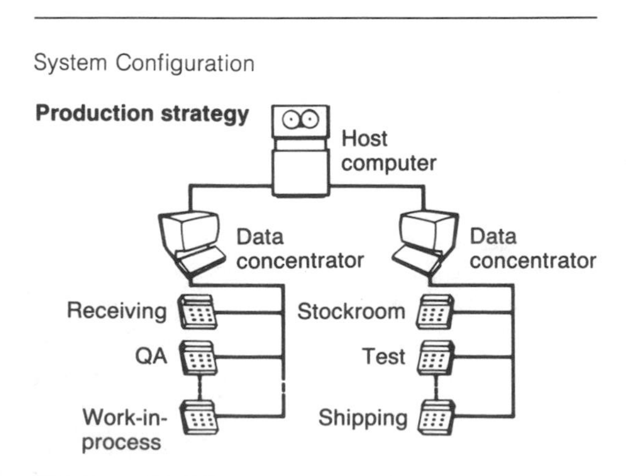

EXHIBIT D

Introducing the DNT 1000™

Factory
Data Collection
Terminal

The Next Generation™ by:

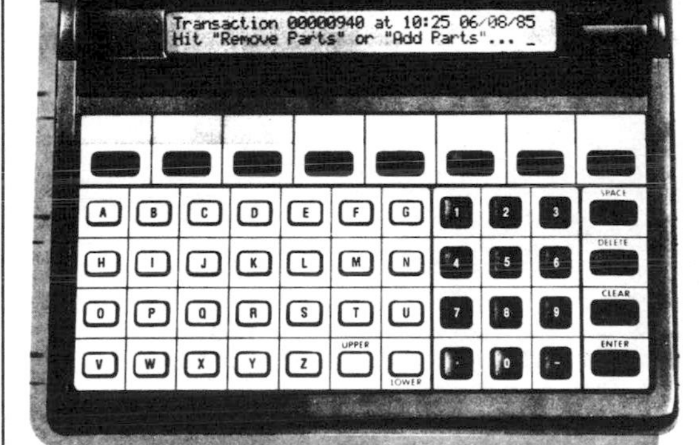

- 2 Line x 40 Character Display
- Adjustable Viewing Angle
- Display Backlighting
- Function Key Labels
- Wide Key Spacing
- User Programmable Function Keys
- Modified Full Travel Keys
- Separate Numeric Keypad
- Keyboard Overlay
- Sealed Keyboard
- Integral Bar Code Wand Holder

The DNT 1000 is specifically designed to meet the demanding requirements of FACTORY DATA COLLECTION systems in both PORTABLE and FIXED STATION applications. The compact, rugged DNT 1000 is the result of many years of data collection experience by the DATA NET design team and provides the exact combination of features required to solve your FACTORY DATA COLLECTION problems. These outstanding features include;

- Work-In-Process
- Inventory Control
- Quality Control
- Time/Attendance
- "3 in 1" Package
  Portable - Bench Mount - Wall Mount
- Human Engineered Package Design
- Works in Harsh Factory Environments
- Multidropped up to 4000 Feet From PC
- Easy to Read Display
- Rugged but Friendly Keyboard
- Bar Code Reader Option
  Code 3 of 9 - 2 of 5 -
  Interleaved 2 of 5 - UPC - Codabar
- Selective Code Read Control
- Mag Swipe Reader Option
- Distributed 24 VAC or
  Internal Battery Power
- Low Order RS-232-C Port
- Relay Contact Outputs -
  2 AMPS @ 120 VAC

DATA NET CORPORATION • 7255 CORPORATE CENTER DRIVE • MIAMI, FLORIDA 33126 • PHONE: (305) 593-1900

EXHIBIT D (*concluded*)

DNT 1000™ Terminal

SPECIFICATIONS

DISPLAY

- 2 Line, 40 Characters Per Line
- Liquid Crystal with Electroluminescent Backlight
- Adjustable Viewing Angle

KEYBOARD

- Modified Full Travel, Dust and Liquid Spill Resistant
- 52 Key Alpha Numeric, Function, Control and Special Symbols
- Separate Numeric Pad
- 8 User Definable Function Keys (Customer Changeable Labels)
- Changeable keyboard overlay provides custom key labels
- Optimized wide key spacing (0.75 inches)
- 10,000,000 actuation operating life

PACKAGING

- Rugged Injection Molded Polycarbonate
- Handheld, Benchtop, Wall, or Post Mounting
- Integral Bar Code Wand Holder
- Screw Terminal Wiring Connections
- Carrying Strap Holder for Portable Use
- Non-Skid Feet

COMMUNICATIONS

- 128 Multi-Drop Addresses (RS-422)
- RS-422, RS-232-C, or 20 Ma CL Communications Interface
- 300 - 19200 Baud
- Odd, Even, None Parity

MICROPROCESSOR

- 80C88, 16 Bit

MEMORY Maximum Configuration

- 64K RAM
- 32K EPROM

LOCAL I/O

- 4 Relay Contact Outputs, 2 AMP at 24 VDC/120 VAC, Double Pole
- RS-232-C Low Order Port

POWER

- Low Power CMOS Electronics
- Built-in High Efficiency Switching Power System
- 24 VDC Input
- Integral Battery Pack
- 8 Hour Operation on Batteries

SIZE

- 7.5″ x 9.5″ Footprint

OPTIONS

WALL/POST MOUNT CRADLE

- Screw Terminals for Communications, Power and Relay Outputs
- DB-25 Connector for Low Order Port
- Solenoid Actuated Security Lock that Releases Terminal for Portable Use upon Authorization by Controller

BAR CODE WAND

MAG SWIPE READER

CARRYING STRAP

The DNT 1000 terminal is directly compatible with the DATA NET model DNTCC controller board and CO-LOG™ software which converts an IBM® PC into a communications processor that controls up to 128 terminals in the FACTORY DATA COLLECTION system. The combination of the DNT 1000 terminal, IBM® PC, plug-in DNTCC controller board and CO-LOG™ software provides many features that significantly simplify the design and installation of the data collection system. These features include;

- IBM PC Based Data Concentrator
- Menu Driven Configuration
- 128 Terminals Per PC
- Multidropped Terminal Communications
- Distributed Terminal Power
- Fast Response Time
- Local Data Validation
- Redundant Configuration
- Host Computer Communications
- Intelligent Local Processing

DATA NET CORPORATION • 7255 CORPORATE CENTER DRIVE • MIAMI, FLORIDA 33126 • PHONE: (305) 593-1900

The cost of setting up a surface mounted production facility has continued to drop rapidly. Data Net expects to set up its own facility at a cost of less than $200,000 in 1986. Carl Johnson is just completing installation of one of the most advanced SMT facilities in the country. Data Net will analyze that experience to determine the exact nature of its own facility.

Data Net will move in July 1985 to a 7,500 square foot facility within its present industrial park. Carl Johnson has determined that this space will support a full manufacturing facility for projected sales volumes into early 1987. At that time, a larger space will be needed for production. Such space is available within Data Net's present industrial park and in the immediate area.

Data Net is currently looking to secure a contract for the tooling of a mold for the DNT 1000's external case. Tooling for the mold alone can cost as much as $300,000, and most companies require half of that in advance. Until Data Net obtains its first major equity backing, it cannot begin tooling, and production is delayed. As stated earlier, the first few sales orders will be filled with costly hand-made terminals.

MARKETING STRATEGY

Data Net's goal is quickly to sell four to six installations to verify the utility of the system to the conservative manufacturing market. This sales goal will be met by a direct sales effort on the part of Data Net's officers who will, at the same time, be building a direct sales organization. A direct, regionally based sales organization will sell directly to the customer and will direct the activities of Original Equipment Manufacturers (OEMs), Value Added Resellers (VARs), and select sales representatives. Data Net desires to have about 75 percent of sales direct and 25 percent through OEMs, VARs, and representatives. Since over 90 percent of revenues will come from off-the-shelf products and software, direct sales provide the best margins while avoiding the custom software trap. OEMs and VARs will be useful in specialized markets requiring more "hand-holding."

Richard Fretwell is responsible for Data Net's Marketing Strategy and will also head the sales organization until a vice president–sales is hired. Richard's experience with Burr-Brown led to the decision to develop a direct sales force. He feels there are four main reasons for a direct sales force.

1. *Time*—It takes 6 to 18 months to sell the first section of a system.
2. *Technical nature*—Because of the technical nature of the sale, the salesperson must be extremely knowledgeable and be able to analyze a problem and present Data Net's solution. Most customers are unsophisticated in FDC. All they know is that they have a problem.
3. *Order size*—An average order is between $30,000 to $50,000, with the potential to reach $100,000 to $300,000 because a factory orders three to four times as the FDC system is expanded to cover all areas of the production facility.
4. *Market feedback*—Data Net feels to be successful they have to be very "customer oriented." The customer tells you what you need to do for the future if you give them the right service.

Data Net's initial focus will be on the U.S. market, but will be expanded in 1987 to foreign sales. Both Burr-Brown and NCR sell over 50 percent overseas. Data Net's direct sales force will call directly on customers and coordinate the activities of factory support staff. System quotations, including product, software, and service, will be authorized only by the vice president–sales. In addition, the direct salesman will work with VARs, and, where applicable, sales representatives. Currently very little is being done in direct sales recruiting. "We can't take orders until we get a reasonable delivery date for the product," adds Richard.

Data Net is, however, presently looking for a vice president–sales. Richard would rather have a vice president–sales build his or her own organization than come in to head an organization already under development. The first responsibility of the new vice president–sales will be to establish regional sales offices, the first in Boston. Data Net understands high-tech business and wants to expand into Boston first because of the high concentration of electronic assembly customers there. With a customer base of 35,000, Data Net hopes to penetrate Boston by October 1985. Additional sales offices will be established in Los Angeles (November 1985—customer base 30,000), Chicago (January 1986—customer base 35,000), and Dallas (April 1986—customer base 13,000). Within a few years, regional offices will be comprised of a regional sales manager, two to four sales engineers, one to two service representatives, and staff.

Data Net's direct sales strategy calls for lead gathering to occur on two fronts, trade shows and targeted advertising. They introduced a prototype of their system at the Advanced Manufacturing Systems (AMS) show in Chicago on June 18, 1985, five months after seed funding. Product development of this speed is unprecedented in high-tech industry. The show yielded 200 leads.

Full-page advertisements will be placed in key journals, including *Modern Material Handling, P&IM Review, Bar Code News,* and *Industrial Engineering.* A direct-mail campaign will also be used at targeted mailing lists.

As stated earlier, Data Net hopes to have about 25 percent of sales through Original Equipment Manufacturers (OEMs) and Value Added Resellers (VARs). An OEM distributor would take Data Net's hardware, and perhaps some software, combine it with his own hardware and/or software, and then sell the entire product to customers under his company name. IBM buys their disk drives from other companies as an OEM, and other companies such as NCR and Burroughs are OEMs. "It is not unlikely that our first major contract will be an OEM," says Richard. "I don't like OEMs that much because then you have only a few customers controlling your volume. You have more control and the margin is greater with direct sales." Data Net has identified IBM and NCR as two primary OEM targets. They have not directly discussed Data Net's position but are currently identifying the people in those organizations who make the decisions regarding future FDC strategies.

VARs generally do not have hardware. They have no design or manufacturing capabilities. VARs may buy hardware from several different vendors and put it together, generally applying specialized software to the hardware. According to Richard, "they can be used to sell to a set of customers that Data Net really can't address (small companies), and generally those customers, because of size, will be specialized and unique."

Management will have a large selling responsibility. Richard will spend a substantial portion of his time in direct customer contact as part of the company's strategy of quickly gaining market acceptance. Both Richard and Bruce Pelkey are known to customers, and it is expected that their record in installing successful systems will help overcome the customers reluctance to use an unknown and untested company.

In accordance with the initial goals of the sales strategy, Data Net has identified two companies interested in the system and willing to become test sites. The first company is Cordis Corporation, a maker of pacemakers and other high-tech, precision electronics, medical devices. Because of government regulations, Cordis has a unique accountability problem. They need to be able to trace each of their pacemakers and other products. Cordis is advantageous because it is a local company and it represents an extremely sophisticated production environment.

Sagaz Industries is the other firm interested in Data Net. Sagaz is also a local company, but they represent the other side of the production spectrum. Sagaz manufactures seat covers in a large-volume, harsh production environment, and because of their high growth rate, they cannot keep track of production.

"We're selling them the systems at list price with a 100 percent money-back guarantee. If we give it to them for less, we lose respect. Also, the key is ownership and investment. The customer will do a better job evaluating the product if he's at risk. They get the advantage of modifications to the system that most others won't. I'm not worried about the money-back guarantee, the equipment will work; if it doesn't there won't be a company."

FINANCIAL PROJECTIONS AND FUNDING

The initial $500,000 seed investment made by H&Q and SVC was expected to take Data Net through the six-month development period ending July 1985. As Mike Martin of H&Q points out, "The $500,000 seed investment represents an unusual venture capital deal. Seed funding is generally raised by family and friends; venture capital backing is usually sought after development, and in larger amounts."

No sales were expected during the development period, and the initial seed capital was to be substantially used up. This period was expected to yield a working prototype of the terminals, some production tooling, marketing literature, and the initial sales leads that will help close four systems orders in the third quarter of 1985. The loss through July is expected to be $389,000 and was $247,000 as of May 31, 1985. In Bob's opinion, "We're in good shape financially. We'll have about $100,000 in cash left by the end of June, but we need the venture capital backing for two reasons; first to commit for the tooling for the mold of the terminal's external case, and second for beginning the sales rollout."

From 1985–87, sales projections track Richard's experience of the last three years at Burr-Brown. Data Net feels it is building a product superior to Burr-Brown's, and they plan to put substantially more effort into marketing than did Burr-Brown. "When I was with Burr-Brown, we received hundreds of leads from advertising every month and they were thrown away," Richard remembers. "Data Net will follow up every qualified lead." According to Data Net's projections, company sales could reach $100 million by 1991, or 10 percent to 15 percent of the market projected by Data Net (see Exhibit E).

In accordance with their business plan, Data Net is currently seeking their first major round of venture capital financing. They are planning to secure a $1.5 million equity placement by July 17, 1985. According to Bob Williamson, the $1.5 million figure was a compromise, determined after assessing Data Net's financial projections and learning what would be considered reasonable in the venture capital community.

The $1.5 million is consistent with Data Net's 1986 financial plan (see Exhibit F). If they were to seek only $1 million, more money would be needed by December 1985. "The $1 million would barely cover the $960,000 projected decrease in retained earnings incurred by December. Also, sales will have only just begun and it would be too soon to get a good bargain for a subsequent equity agreement," explains Bob.

A $2 million equity agreement would take Data Net well past June of next year if necessary. By that time, sales would be strong and the sales force would be established; therefore, they could command a premium price if additional financing was necessary. Since $2 million would cause greater dilution and would likely involve a longer fund-raising cycle, the management team set the $1.5 million figure as a target, but it remained open to raising somewhat more funds depending on the price and who the players were. Data Net also expects to negotiate a substantial line of credit soon after

EXHIBIT E

DATA NET CORPORATION
Summary Income Statement
(in thousands of dollars)

	1985	1986	1987	1988	1989
Net sales	75	3,810	8,102	14,452	34,203
Cost of goods	82	1,941	3,363	6,310	14,097
Gross margin	−7	1,869	4,739	8,142	20,106
Selling, General, and administration:					
Marketing	219	1,464	1,803	2,956	5,945
Engineering	223	539	967	1,609	3,183
Administration	159	473	734	979	1,805
Total selling, general and administration	601	2,476	3,504	5,544	10,933
Operating income	−608	−607	1,235	2,598	9,173
Other	26	49	20	−42	−424
Net before taxes	−582	−558	1,255	2,556	8,749
Taxes	0	0	0	1,022	3,500
Net after taxes	−582	−558	1,255	1,534	5,249
Statistical Analysis:					
Gross margin %	0.0%	49.1%	58.5%	56.3%	58.8%
Marketing % sales	291.8%	38.4%	22.3%	20.5%	17.4%
Engineering % sales	269.9%	14.1%	11.9%	11.1%	9.3%
Admin. % sales	212.5%	12.4%	9.1%	6.8%	5.3%
NBT % sales	−776.0%	−14.6%	15.5%	17.7%	25.6%
NAT % sales	−776.0%	−14.6%	15.5%	10.6%	15.3%

SOURCE: Data Net Business Plan.

the $1.5 million funding. In addition, Data Net projects the need for another round of financing when the $1.5 million runs out in the middle of 1986.

In a recent board meeting, the venture capital funding was the main topic of discussion. Richard reported to the other board members, Mike Martin of H&Q, Jim Fitzsimons of SVC, Tom Vogel of InstaRead, and Bob Williamson, that Data Net's development period was coming to a close. The prototype terminal was operational and the system would be ready for the AMS show the following week. The remainder of the discussion centered around stock prices, interested investors, and the current investors' position. Talk of raising as much as $2 million was entertained, but the main question was, "Where do we go from here?"

Both H&Q and SVC have the right of first refusal, and since they already have a considerable interest in Data Net, they asked to be penciled in at $1 million. Both Mike and Jim were quick to point out, however, that this move was intended to display their support of Data Net to

EXHIBIT F

DATA NET CORPORATION
1986 Financial Plan

Balance Sheet

	Sept 1985	← Oct	1st. Nov	→ Dec	← Jan	2nd. Feb	→ Mar	← Apr	3rd. May	→ June	← July	4th. Aug	→ Sep
Assets													
Current assets:													
Cash	10	10	10	10	10	10	10	10	10	10	10	10	10
Investments	980	926	753	576	479	323	2,170	2,057	1,954	1,865	1,828	1,802	1,813
Inventory	35	23	23	139	139	209	302	232	253	274	295	338	380
Receivables	82	64	64	381	381	572	826	699	762	826	889	1,016	1,143
Prepaid and other	1	1	1	1	1	1	1	1	1	1	1	1	1
	1,108	1,024	851	1,108	1,011	1,114	3,308	2,998	2,980	2,976	3,024	3,166	3,347
Plant and equipment:													
Plant	0	0	0	0	0	0	100	100	100	100	100	100	100
Equipment	378	453	483	513	574	584	594	694	754	814	834	844	854
	378	453	483	513	574	584	694	794	854	914	934	944	954
Depreciation	17	22	28	35	43	51	60	73	85	99	113	127	141
Total plant	361	431	455	478	531	533	634	721	769	815	821	817	813
Other assets	0	0	0	0	0	0	0	0	0	0	0	0	0
Total assets	1,469	1,455	1,305	1,586	1,542	1,648	3,942	3,720	3,749	3,792	3,845	3,984	4,160

EXHIBIT F *(concluded)*

DATA NET CORPORATION
1986 Financial Plan

Balance Sheet

	Sept 1985	← Oct	1st. Nov	→ Dec	← Jan	2nd. Feb	→ Mar	← Apr	3rd. May	→ June	← July	4th. Aug	→ Sep
Liabilities													
Current liabilities:													
Bank/lease debt	0	44	44	267	267	400	578	489	533	578	622	711	800
Accounts payable	20	17	17	104	104	157	226	174	190	206	222	253	285
Accrued	16	9	9	9	8	8	8						
	36	71	71	380	379	565	812	663	723	784	844	964	1,085
Long-term debt:													
Bank debt	0	0	0	0	0	0	0	0	0	0	0	0	0
Equipment leases	0	100	125	150	200	200	200	200	200	200	200	200	200
Total long-term	0	100	125	150	200	200	200	200	200	200	200	200	200
Total liabilities	36	171	196	530	579	765	1,012	863	923	984	1,044	1,164	1,285
Shareholders' Equity													
Preferred stock	2,000	2,000	2,000	2,000	2,000	2,000	4,000	4,000	4,000	4,000	4,000	4,000	4,000
Common stock	15	15	15	15	15	15	15	15	15	15	15	15	15
Retained earnings	−582	−731	−906	−960	−1,052	−1,132	−1,085	−1,158	−1,189	−1,207	−1,214	−1,196	−1,140
Total equity	1,433	1,284	1,109	1,055	963	883	2,930	2,857	2,826	2,808	2,801	2,819	2,875
Total liabilities and equity	1,469	1,455	1,305	1,586	1,542	1,648	3,942	3,720	3,749	3,792	3,845	3,984	4,160

SOURCE: Data Net Business Plan.

interested investors, and that they were willing to be flexible. Mike Martin told Richard that he should be seeking one or two additional investors to round out the deal.

Two weeks later, the Data Net team returned from a successful presentation at the AMS show in Chicago. Following his return, Richard updated his list of prospective venture capital firms to indicate those that had indicated interest at the show (see Exhibit G). Data Net is

EXHIBIT G Venture Capital Status Report

June 28, 1985

FIRM/Contact	Interest Level	Size	Status/Next Step
ADVANCED TECH. DEV. FUND Danny Ross	80%	RS*	Attended AMS, Richard visit in Atlanta
ADVANCED TECH VENTURES Bob Ammerman/David Walker	20%	NM	Richard contact
ALLIED CAPITAL CORP. George Parker	10%	NM	Richard contact Parker
BROVENTURE CAPITAL Bill Gust	60%	NM	Attended AMS, Branch to visit 7/2/85
BURR, EGAN, DELEAGE Bill Egan/John Flint	60%	NL	Attended AMS
CARDINAL DEVELOPMENT CAPITAL Richard Focht	?	RM	Richard follow-up
F. EBERTADT & CO. Rick Stone	?	NM	Visited AMS, Richard to follow up
FOSTIN CAPITAL CORP. Bill Woods/Thomas Levine	?	RS	Plan sent
HICKORY VENTURE CAPITAL Thomas Noogin	20%	RS	Richard follow-up after AMS
KITTY HAWK CAPITAL Walter Wilkinson	95%	RS	Attended AMS, pencil in for $200–300K
MERRILL LYNCH VENTURE Frederick Ruvkin	?	NL	Richard contact
MORGAN, HOLLAND VENTURES Robert Rosbe	?	NL	Richard contact
NCNB VENTURE Epes Robinson/Mike Elliot	50%	RS	Richard contact after AMS
SECURITY PACIFIC CAPITAL Dan Dye	30%	NL	Richard contact
SUMMIT VENTURES Thomas Avery/Larry Lepard	20%	NL	Visited at AMS, Richard call today
VenGROWTH CAPITAL FUNDS Andy Gutman/Mark Leonard	30%	RM	Visit here after AMS

* The first letter stands for either National(N) or Regional(R).
The second letter represents size, Small(S), Medium(M), or Large(L).
Therefore, the first firm is a regionally small firm.

currently especially interested in two firms which have also expressed a high interest in them. Richard hopes to include both of these investors in the deal with H&Q and SVC. "The more investors we have, the less control anyone obtains, and we get a broader perspective. We need people who can help us in the future with their contacts, influence, and management expertise. Also, H&Q and SVC cannot make an objective valuation of the company." If all three firms are interested, they make an independent, objective valuation of the company and the price of the stock.

The two key potential investors are Danny Ross of Advanced Technology Development Fund and Broventure Capital. Although Danny Ross's firm is not large, he can be very helpful because of his expertise and contacts. He is past president of Timex Computers and has many contacts within IBM. He is also highly involved with an Atlanta-based firm which is the MRP expert for IBM systems. He offers management expertise and is very knowledgeable in FDC. Broventure is a larger firm with deeper pockets if additional funding is necessary. Also, because of their size, they are very influential and can open many doors. "They can provide a strong backing and help us establish credibility," explains Richard.

A day after their return, at least two venture capital firms made appointments to visit Data Net the following week. One of the firms, Kitty Hawk Capital, a small venture capital firm interested in a small portion of the deal, attended the AMS show at the request of Jim Fitzsimons of SVC. Impressed with what they saw during their visit in Miami, Kitty Hawk asked to be "penciled in" for $200,000 to $300,000 before they left.

The interest in Data Net revives the questions raised in the board meeting two weeks earlier. How much stock should be sold, at what price, who should be the major investors, and how will these decisions affect future venture capital backings if needed?

Reading III–5
Linking Prefunding Factors and High-Technology Venture Success: An Exploratory Study

J. B. Roure and M. A. Maidique

ABSTRACT

This paper reports on the exploratory phase of a large-scale study of West Coast high-technology (principally electronics) start-up firms. The study was done in collaboration with a major West Coast venture capital firm that allowed the authors full access to the due diligence files, investment proposals, and closing documents associated with each venture. Half of the eight ventures studied are currently public companies with sales that range from $65 million to $500 million, with profitability of 10 percent or better. The other half have either been dissolved or did not reach $3 million in sales within the five years after they were founded.

Using the venture capital partnership's records, we obtained information on those prefunding factors that were available for investor review prior to funding, such as the individual founders' track records, the structure of the founding team, the nature of the target market, the technological strategy of the firm, the proposed composition of the board, and the deal structure.

Our research findings reflected significant differences in many of these variables between the successful and unsuccessful firms, notwithstanding the small sample size. The founders of the successful firms had previously worked together for a significant amount of time, tended

---

to form larger more complete teams, had senior-level experience in rapid growth high-technology firms, and had previously competed in the same industry as the start-up. The successful ventures targeted focused product market segments with high buyer concentrations where, through technological advantage, their products could attain and sustain a competitive edge. Often this advantage was achieved by careful management of the product development process which resulted in early market entry and its corollary, reduced competition.

On the other hand, some factors that we had anticipated would be important differentiators were not found to be significant. Both successful and unsuccessful ventures targeted high-growth markets, anticipated high gross margins, had founders with over five years of relevant experience, had experienced venture capitalists on their boards, and were characterized by a wide range of founder equity share.

INTRODUCTION

Most new companies fail soon after they are launched. High-technology start-ups fare better than the average new business, but their failures are usually more devastating because of the capital, time, and number of people involved.[1] Bruno and Cooper, who have followed a group of 250 Silicon Valley technology firms that were founded in the 60s, report that as of 1980, 36.8 percent had been discontinued, 32.4 percent had merged or had been acquired, and only 30.8 percent had survived as independent companies.[2] By 1980, on average each of these remaining independent firms did approximately $8.6 million in annual sales, employed 135 people, and had been in business for approximately 14 years.

Despite these substantial initial failure rates, the amount of venture capital committed to high-technology start-ups increased almost tenfold during the past decade.[3] In 1983 and 1984, venture capitalists, the principal financiers of early stage high-tech firms, invested over $6 billion in such businesses and started thousands of new high-tech firms. With so much capital invested at the preliminary stages of high-tech start-ups, it is important to correlate these investments with those factors that influence early stage performance of technological companies.

High-tech start-ups have been the subject of numerous case studies illustrating the many pitfalls to be avoided by entrepreneurs.[4] However, systematic research on the factors that explain the success or failure of new high-technology ventures is scarce. There are three studies that relate different factors of any given venture to certain measures of success. The first is Cooper and Bruno's longitudinal study of 250 technological firms.[5] That study concludes that the more successful companies (i.e., those that survived longest) were started by two or more founders, and had one or more founders with prior experience in the markets or technologies that they addressed. Furthermore, the authors found that entrepreneurs with prior experience in "large" organizations (those with over 500 employees) were even more likely to be successful.

---

[1] "Patterns for Success in Managing a Business" (New York: Dun & Bradstreet, 1967).

[2] A. V. Bruno and A. C. Cooper, "Patterns of Development and Acquisitions for Silicon Valley Start-Ups," *Technovation* 1 (Amsterdam, 1982).

[3] *Venture Capital Journal,* January 1985, p. 1.

[4] See for instances M. Maidique, course outline, GBM 698, "Managing Technological Enterprises," School of Business, University of Miami, Coral Gables. Copy available on request.

[5] A. C. Cooper and A. V. Bruno, "Success among High-Technology Firms," *Business Horizons,* April 1977, pp. 16–23.

In another study on venture capital investments, Hoban concluded that the proportion of equity owned by the principals, the stage of development of the product, and the extensiveness of the market evaluation were positively correlated with the rate of return achieved.[6]

Finally, a third study by Van de Ven and colleagues analyzed 13 software firms and found that start-up success was correlated with certain characteristics of the founders: education, experience, and internal locus of control.[7] This study also concluded that success was positively correlated with having a single person in command, the active involvement of top management and board members in the decision making, and implementing the start-up on a small scale with incremental expansion over time.

In another group of studies, researchers have explored those criteria used by venture capitalists in evaluating venture proposals.[8] However, these studies do not directly relate those factors that are considered critical by the investors to the subsequent success or failure of the company.

Clearly, a high-tech start-up depends on many factors to increase its likelihood of being successful. Like a newborn child, the infant high-technology company requires nurturing and constant attention.[9] But even more important to the infant's successful development are the factors related to the genes of the parents and to the nature of the gestation cycle. In a similar way, the likelihood of success of a high-technology company is often influenced by certain inherent characteristics of the firm's founders and other prefunding factors.

Except for the handful of studies mentioned above, the prefunding factors that influence the success or failure of technological start-ups have received little attention by management scientists.[10]

This article is part of a larger study concerning the key prefunding factors that influence the success of high-technology companies funded through venture capital. In this paper, we will report on the initial results of our exploratory research, which also includes an analysis of the first eight ventures examined within the larger research sample.

METHODOLOGY

Our exploratory research was conducted in three stages. First, we reviewed the relevant literature on technological start-ups (including related work on entrepreneurship), product development, and venture capital. Secondly, we arranged semistructured interviews with eight venture capitalists who collectively have over 100 years of investment experience with high-tech firms. In these interviews, we obtained their general views on those factors which they believe influence the success and failure of high-tech start-ups. The following cri-

---

[6] J. P. Hoban, "Characteristics of Venture Capital Investments," unpublished doctoral dissertation, University of Utah, 1976.

[7] A. H. Van de Ven, R. Hudson, and D. M. Schroeder, "Designing New Venture Start-Ups: Entrepreneurial, Organizational, and Ecological Considerations," *Journal of Management* 10, no. 1 (1984), p. 87.

[8] W. A. Wells, "Venture Capital Decision Making," unpublished doctoral dissertation, Carnegie-Mellon University, 1974; J. B. Poindexter, "The Efficiency of Financial Markets: The Venture Capital Case," unpublished doctoral dissertation, New York University, 1976; T. T. Tyebjee and A. V. Bruno, "A Model of Venture Capitalist Investment Activity," *Management Science* 30, no. 9 (September 1984); I. C. MacMillan, et al., "Criteria Used by Venture Capitalist to Evaluate New Venture Proposals," Babson College Entrepreneurship Research Conference, 1985.

[9] A. C. Cooper and A. V. Bruno, "Success among High-Technology Firms," *Business Horizons,* April 1977, pp. 16–23.

[10] M. A. Maidique, "Key Success Factors in High-Technology Ventures," State of the Art in Entrepreneurship Research, IC Institute, University of Texas, Austin, February 1985.

teria were used in selecting the venture capitalists interviewed for our preliminary study: (1) affiliation with a recognized venture capital firm, and (2) experience in the venture capital business for at least five years. Finally, our third step was to analyze eight venture capital financed start-ups, identifying the factors which differentiate successful ventures from unsuccessful ones.

For the purpose of this study, we define a successful venture as a firm that: (1) has been incorporated for more than three years, (2) has reached a sales level over $20 million, and (3) has attained after-tax profits greater than 5 percent of sales. On the other hand, a venture that has been: (1) discontinued within five years, or (2) never reached $3 million in sales during the same period, was considered unsuccessful. This set of criteria is similar to the success-failure criteria employed by many institutional venture investors.

We chose our eight ventures by asking two major West Coast venture capital firms through their managing partners to choose from their portfolios matched pairs of companies that met our success/failure criteria. By studying companies that were part of professionally managed portfolios, we were insured of the availability of business plans, closing documents, and other reference data. All eight high-tech companies were from the electronics industry and were funded after 1974 (see Table 1). All eight companies were located on the West Coast (five in northern California) and in geographical areas with a supportive local infrastructure.

TABLE 1 Types of Businesses Selected by Venture Capitalists

Company	Business
A	Semicustom semiconductors manufacturer
B	Disk drives
C	Computer-aided engineering work stations
D	Fault tolerant computers
E	Specialized computer terminals and software
F	Computer-aided design work stations
G	Color graphics hardware
H	Marine instruments

The four successful firms, all currently public companies, ranged in annual sales from $65 to $500 million, with after-tax profits of around 10 percent. Of the four failures, two firms were dissolved within five years of their founding, and the other two firms did not reach sales of $3 million during the five years after founding.

ANALYSIS OF RESULTS

We focused on five categories of factors that, based on our exploratory interviews and the previous field research discussed earlier, we hypothesized could be important determinants of success in high-tech ventures:

1. Founders' track records.
2. Characteristics of founding team.
3. Target market.
4. Technological strategy.
5. Deal structure.

The following sections report on the preliminary findings in each one of these categories.

1. Founders' Track Records

We define as founders those members of the original team who were expected: (1) to play a key role in the development of the firm, (2) to become full-time employees within the first year after the initial funding date, and (3) to share in the ownership of the company in a significant manner. A founder's track record comprises previous positions held, tenure, and the host organization where he/she had work experience.

Virtually all the founders of both the successful and unsuccessful companies had five or more years of experience in the particular industry that their firm competed in. This result is consistent with the findings of Cooper and Bruno's longitudinal study.[11] What did seem to differentiate between success and failure, however, is the nature of the founders' relevant experience and the characteristics of the organizations in which they obtained their experience. Founders of the successful companies typically had prior experience of two years or more in the same functional position as in the new company. They also had a successful and fast-rising career in previous organizations, and had worked in units of "large" companies (over 500 people) which were characterized by high growth (more than 25 percent increase in sales or employees per year).

In contrast, the founders of the firms that failed had significantly less experience in similar positions to those in the new venture. In addition, founders of unsuccessful companies had a moderate career pace in a mixture of both large and small companies. Moreover, these companies often were characterized by a rate of growth much lower than the companies of their successful counterparts. These results are supported by the Van de Ven et al. study which found that start-up success and company stage of development were positively related to the skills and expertise that the entrepreneurial team brought to the new venture.[12]

2. Characteristics of Founding Team

Founders of high-technology companies often form groups to start new companies. In three studies carried out in different U.S. regions, the authors found that the percentage of new firms started by two or more founders was 48 percent in Austin, Texas; 61 percent in Palo Alto, California; and 59 percent in a study of 955 geographically diversified firms.[13] Cooper and Bruno in their study of Silicon Valley firms found that groups of two or more founders were involved in 83.3 percent of the successful high-growth companies but only 53.8 percent of discontinued firms, suggesting that groups of founders are more successful than individuals.

Notwithstanding these important findings, a review of the literature did not reveal any research on the characteristics of the original founder team. Our analysis of eight cases of success and failure persuaded us that there are two principal characteristics of the team that seem to influence the success of technological start-ups. The first one is the *degree of team completeness,* or the percentage of relevant functions in the new company that are already filled by the founders' team. We define a "complete" team as a group that included as original founders the chief executive officer and the executives responsible for marketing, engineering, finance, and operations. Occasionally, for a specific company, one of these functional areas was not viewed as critical by the venture capitalists for the start-up phase. In this case, we did not include that function in our completeness analysis. In the same way, if in a certain company it was considered that both hardware and software experts were required, we included both of these functions as required for the team completeness.

The successful companies had larger and more complete teams at the time of funding. In two of the four successful companies, the team was 100 percent complete and in the other two

[11] Cooper and Bruno, "Success."

[12] Van de Ven et al., "Designing."

[13] J. C. Susbauer, "The Technical Company Formation Process: A Particular Aspect of Entrepreneurship," unpublished doctoral dissertation, Texas University, Austin, 1969; A. C. Cooper, "The Founding of Technology-Based Firms," The Center for Venture Management, Milwaukee, Wis., 1971; A. Shapero, "An Action Program for Entrepreneurship" (Austin, Texas: Multi-Disciplinary Research, Inc., 1971).

the team was 80 percent complete. On the other hand, while one of the unsuccessful companies had an 80 percent complete team, another had a 75 percent complete team, and the other two included only 50 percent of the relevant functions (see Table 2). In addition, one of these latter companies did not have the future president on board when it was initially funded.

The second important characteristic of the founders' team that we found is highly correlated with the success of a new firm is *the degree of prior joint experience;* that is, the extent to which founders had previously worked together in the same organization for at least six months (the minimum amount of time considered to be relevant as stated by the founders in their resumes). We measured this variable by determining the actual number of prior relationships among pairs of founders and dividing it by the total number of possible relationships among them. For example, if the entrepreneur starts a new company with three additional founders, but he has worked before as a team with only two of them, and the third one is new to the team, the number of existing pairwise relationships in this case would be three; however, the total possible would be six. Thus, for this hypothetical firm, the joint experience factor would be 50 percent.

Of the four successful ventures, two had 100 percent level of joint experience, and the other two had 50 percent levels. In contrast, only one of the failures had 50 percent, while two of them had 33 percent, and in the other no prior joint experience (see Table 2). This result is not surprising. Entrepreneurs during the start-up period have a limited period of time and a limited amount of capital to achieve clear results. Therefore, any time spent developing relationships among the group significantly detracts from other vital activities.

TABLE 2 Founding Team Characteristics

Company	Number of Members	Completeness	Joint Experience
S1	5	100%	50%
S2	4	80%	50%
S3	4	100%	100%
S4	4	80%	100%
Average	4.25	90%	75%
U1	3	75%	33%
U2	3	50%	33%
U3	3	50%	0%
U4	4	80%	50%
Average	3.25	64%	29%

Note: S = Successful Company; U = Unsuccessful Company.

3. Characteristics of the Target Market

The first characteristic of the venture's target market that we evaluated was whether the market selected by the new business was existing, expanded, or new. This categorization was chosen as an extension of the widely used "product-market matrix" originally proposed by Ansoff.[14] In three of the cases, the entrepreneurs targeted already existing markets, while five targeted expanded markets and none targeted new markets. However, it was not possible to link these differing market approaches with the subsequent success and failure of the ventures. As expected, no cases were found in which venture capitalists had invested in companies targeting truly new markets. Indeed, the entrepreneur's product-market choices clustered closely on a market-technology matrix with a slight tendency for successes to be further down the risk diagonal.

All start-ups in our sample targeted high-growth markets. The rate of growth of the market targeted by the eight ventures ranged between 35 and 50 percent annually at the time of funding (see Table 3). We did not, however,

[14] H. I. Ansoff, *Corporate Strategy* (New York: McGraw-Hill, 1965).

find this characteristic to be a factor in the success or failure of a start-up.

A third market factor that we evaluated was the target market share of the new firm. This data was found either in the company's business plan, where it was sometimes explicitly given, or in notes taken by the venture capitalists when they conducted their due diligence investigation. If this data was not readily available, we divided the company's fifth- (or latest available) year sales projection by the overall relevant market size for the venture in that year. We found that the successful companies targeted significantly higher market shares than the unsuccessful ones (see Table 3).

Based on these exploratory results, it seems reasonable to speculate that successful companies try to carve out focused product market segments within which they can play an important role. This finding is partially supported by a study done on 177 start-ups of PIMS (Profit Impact of Market Strategy) corporate ventures, in which the level of market share achieved and the level of return on investment were correlated.[15]

If our follow-up statistical study confirms these results regarding market share, they could serve to refute conventional wisdom evaluating new ventures; that is, the larger the market the better. What does this imply regarding optimum target market size? Assuming the venture capitalist has a goal of participating in ventures that can reach $100 million in sales in five years and address markets with growth rates of 40 to 50 percent per year, the size of the relevant market at the time of funding should be in the range of $100 to $200 million, assuming target market shares of 20 and 10 percent, respectively.

TABLE 3 Characteristics of the Target Market

Companies	Growth (percent)	Market Share (percent)	Buyer Concentration
S1	50	10	High
S2	40	15	High
S3	50	9	High
S4	45	13	High
Average	46	12	High
U1	50	2	Low
U2	35	8	Low
U3	40	6	High
U4	35	3,4,20*	Low
Average	40	5	Low–Medium

* The company had three different products addressing three different markets.

Another market characteristic that we hypothesized might be important was the level of buyer concentration. We considered a venture to have a "low" level of buyer concentration if it served a relatively large number of potential customers with limited buying capacity. In contrast, a venture had a "high" level of buyer concentration if it served a relatively small number of potential customers which had a large buying capacity.

All the successful companies in our sample were characterized by high levels of buyer concentration. On the other hand, only one of four failures had a high level of buyer concentration (see Table 3). This result emphasizes the necessity in building a distribution system during the start-up period, and the complexity of maintaining effective communications with a varied and dispersed set of customers.

Finally, we analyzed the nature of competition in the venture's targeted market segment. Three of the successful companies targeted markets where, because of their early entry, no competitor (at the time of the funding) enjoyed a dominant position and where large companies were supplying only a generic product. The fourth one targeted a market where there

[15] M. A. Maidique and B. J. Zirger, "A Study of Success and Failure in Product Innovation: The Case of the U.S. Electronic Industry," *IEEE Transactions on Engineering Management* EM-31, no. 4 (November 1984); S. Myers and D. G. Marquis, "Successful Industrial Innovation," *National Science Foundation*, NSF 69-17 (1969), p. 11.

were two competitors and one of them was clearly the leader. The company did not, however, challenge the existing leader. Rather it adopted a follower strategy by becoming a second source for the leader's products.

In contrast, two of the unsuccessful ventures targeted markets where a clear leader existed, plus some lesser competitors. In both situations, the new companies tried to compete head-to-head with the leaders. The two other failures targeted markets where several competitors, including some major firms, were already vying for leadership, and where still other major firms were planning to enter in the immediate future. Therefore, the successful companies, contrary to the unsuccessful ones, targeted market segments relatively uninhabited by strong competitors, thereby trying to avoid head-on competition with firms already established in that market.

4. Technological Strategy

As an extension of the already mentioned "product-market matrix," we broke down the core technology employed by the venture into the categories existing, improved, or new. In all eight cases, the core technology used in the company's products was primarily an improvement on the current state of the art and therefore not truly "new." (see Figure 1). Thus, all of our subject companies used already existing technologies as their base and added minor improvements which they hoped, when embodied into their new product concepts, would yield strategic technological advantages.

We measured the level of the strategic technological advantage of the products by comparing projected performance improvements and cost reductions to those of competing existing products. Successful companies targeted major or significant improvements in performance and/or cost. That is, they aimed at substantially increasing the value of their product to the customer (see Table 4). These results are consistent with the findings of studies done on success and failure of new products.[16]

TABLE 4 Level of Strategic Technological Advantage

Companies	Performance Improvement*	Cost Reduction*
S1	Major	None
S2	Major	Minor
S3	Significant	Significant
S4	Major	Major
U1	Minor	Minor
U2	Significant	None
U3	None	Significant
U4	None	Significant

* Scale: Major, Significant, Minor, None.

Another variable that we studied was the gross margin aimed for by the new companies. Gross margin is especially important because it can be viewed as a proxy for the proprietary technological edge of the venture. We found that the gross margin percentages targeted by the eight companies (during their fifth year of operations) ranged between 40 and 60 percent. This factor, however, did not discriminate between success and failure.

Lastly, we analyzed the effort required to complete product development in each of the eight ventures. A study done in 1976 found that the probability of higher returns is improved by investing in ventures where product development is quite advanced.[17] Our data, however, did not support these findings. The projected remaining product development time for our start-ups ranged between 12 and 18 months, but this measure did not seem to be a factor in the success or failure of a new business.

---

[16] R. Rothwell, C. Freeman, A. Horsley, V. T. P. Jervis, A. B. Robertson, and J. Townsend, "SAPPHO Updated—Project SAPPHO, Phase II," *Research Policy* 3 (1974), pp. 258–91.

[17] Maidique, "Key Success Factors."

On the other hand, we did find a qualitative difference between successes and failures. Successful venture teams, in contrast with unsuccessful ones, had planned the product development in greater detail, identifying the main milestones. This finding coincides with one of the main conclusions of the Stanford Innovation Project (SINPRO) on new product success and failure.[18]

5. Deal Structure

Hoban found that the proportion of equity owned by venture capitalists was negatively related to the success of the new venture.[19] The results of our interviews with venture capitalists suggest that both entrepreneurs and external investors must be highly motivated in order for a venture to succeed. Therefore, the equity should be structured in a way that the founders and the venture capitalists should both benefit significantly in the event that the venture is successful. However, from our analysis of the eight cases, we were not able to confirm this hypothesis. The range of the equity percentage owned by venture capitalists after the first round of financing varied between 15 and 73 percent, but did not appear to discriminate between success and failure.

Finally, we analyzed the composition of the board of directors after the first round of financing. One of the more efficient ways that a venture capitalist can influence the direction of a new company is by becoming a member of its board of directors. Thus, we hypothesized that the presence of venture capitalists and entrepreneurs with previous experience on the board of start-ups could be correlated with the success of the venture. The firms in our sample, however, *all* had two or three partners from venture capital firms on their boards, and three of the companies had a successful entrepreneur as board member. However, no relation was detected between the composition of the board and the success of the company. In order to obtain useful data on this variable, we would have to compare our sample with companies that do not have experienced venture capitalists on their boards, that is, with companies funded by sources other than venture capitalists.

SUMMARY

Of necessity, a study such as ours, due to its limited sample size and exploratory nature, must restrict itself to the articulation of hypotheses that can be subsequently tested by either our own follow-up research or that of others. The most salient of the preliminary hypotheses that have come out of our analysis deal with the characteristics of the entrepreneurs and the founding team, certain key market factors, and the strategic cost and/or performance advantages provided by the product to be introduced by the venture.

Our preliminary study confirms the findings first by Cooper and Bruno[20] and lately by Van de Ven et al.[21] regarding the importance of the founder's prior experience. In addition, our study sheds some light on the nature of the founding team experience. Successful entrepreneurs had prior experience in the same roles they fulfilled in the new venture, and had fast-rising careers in high-growth units of medium to large companies. Successful companies had a higher percentage of critical functions filled at the time of first financing and had a higher degree of joint experience in the founding team. The successful start-ups in our sample targeted market segments uninhabited by strong competitors and with a limited number of potential customers. Also, similar to the study done on start-ups of PIMS corporate

[18] Maidique and Zirger, "A Study of Success."

[19] Hoban, "Characteristics of Venture Capital Investments."

[20] Cooper and Bruno, "Success."

[21] Van de Ven et al., "Designing."

ventures,[22] we found that successful companies targeted focused product market segments where they could expect a higher market share.

The successful companies developed a technological strategy that resulted in a significant performance to cost improvement compared with other companies serving the same markets. Also, these successful ventures had prepared more detailed technology development plans. These results support the findings of prior studies of success and failure patterns of new products.[23]

Finally, we examined the deal structure and its effect on the success of the venture. We hypothesized that founders and venture capitalists should both be strongly motivated financially if the venture is to be successful. We, however, did not find confirmation for this hypothesis in our small sample. Hoban, on the other hand, concluded that the proportion of equity owned by the venture capitalists is negatively related to the success of the new company.[24]

Further testing of these hypotheses, as well as others that we have derived from our initial study, represent a timely research endeavor, in view of the recent surge in high-technology ventures and their importance to U.S. international competitiveness.[25] Our next step will be to test the hypotheses we have developed here as part of our ongoing large sample study of high-technology start-ups.

CONCLUSION

Our findings all point in the same direction. They point towards a resource conservation model of high-technology start-ups. The start-up, much more so than the established firm, is short on capital, human resources, and the time to make its mark. For this reason, resources must be very carefully selected, deployed, and aimed. There is little time for costly mistakes. It follows then that the successful teams would be long in experience of all types—joint, industry, growth company, senior management. It also follows that the translation of this experience base into high-value products must be carefully planned every step of the way. And finally, focused, embryonic, limited customer base market targets must be chosen such that neither intense competition and a diffuse customer base will not dilute the company's fragile resource base.

The task of the next stage of our research is to test, enhance, and refine the validity of this resource-limited view of the new venture formation process. For the venture capitalist, this model would allow a more effective selection of investments and thus higher returns. For the entrepreneur, the knowledge would provide a basis for the design of a successful venture structure and plan, and a blueprint for early corrective action to enhance the success of the venture or simply the avoidance of the pain and waste of resources that accompany failure.

---

[22] Maidique and Zirger, "A Study of Success"; and Myers and Marquis, "Successful Industrial Innovation."

[23] Rothwell et al., "SAPPHO."

[24] Hoban, "Characteristics."

[25] Joint Economic Committee, Congress of the United States, *Location of High-Technology Firms and Regional Economic Development* (Washington, D.C.: U.S. Government Printing Office, 1982).

IV Designing and Managing Innovative Systems

In Part IV of this book, we present a series of cases and readings which allow an examination and discussion of issues and problems related to the design and management of innovating systems at the *corporate level.* Corporate management is faced with difficult decisions concerning the development and maintenance of a central R&D capability whose output can serve as a substrate for the identification and inception of new, technology-based business opportunities. Another series of decisions concerns the design of structures and processes that will enhance the development of new ideas into new, viable businesses, and will facilitate the emergence and support of internal entrepreneurs. Some of the key issues and problems addressed in this part of the book are:

- How to link corporate R&D to corporate strategy.
- How to insure that the output of corporate R&D is transferred to other parts of the firm.
- Under what conditions a New Venture Division should be established and how it should be managed.
- Under what conditions innovative activities can be adequately institutionalized and what the trade-offs are.
- What the role of internal entrepreneurs can be and how to manage internal entrepreneurs as "strategists."
- The need for new types of arrangements between extremely talented individuals and corporate entities.

Increasing the capacity for innovation and entrepreneurship is one of the most important challenges facing many firms in the 80s. Established firms must compete for talent with newly emerging firms and can only do so effectively if young professionals can be convinced that the prospective employer provides opportunities for doing psychologically exciting *and* financially rewarding work.

Another, perhaps even more important, reason is that many large, established firms are struggling to achieve real growth as their markets reach saturation and their growth decelerates. Achieving real growth has traditionally been an important corporate objective. Not only does it create additional wealth for the stockholders, it also provides the room for promoting capable employees, an important factor in keeping as well as attracting talented people. This is especially important since during the last decade even some of the best managed firms have experienced difficulties in achieving real growth.

Growth can be achieved through *acquisitions* and/or through *internal development.* In this part of the book, we will be concerned primarily (but not exclusively) with growth through internal development.

In high-technology companies, internal development is fueled through ideas for new products and businesses which often originate in the R&D capability of the firm. *Designing and managing the corporate R&D function* is therefore a central topic in this part of the book. The pressure to innovate has made firms increasingly concerned about the effectiveness of the vast amounts of resources that they spend on R&D. Organizations want to improve the linkages between their strategies and the corporate R&D efforts. They want to get the most out of the talent they assemble by carefully designing the organization structure and systems within which researchers and technicians operate. They want to facilitate the transfer of knowledge and technology from the pure research-oriented corporate R&D group to other, more development-oriented R&D groups in the corporation.

Ideas, as has been pointed out in Part II of this book, do not constitute "innovation." For innovation to take place, *entrepreneurs* must take the ideas and develop them into marketable products. In Part III, we considered examples of how to achieve this goal for individual new products and product families. In Part IV, we are more generally concerned with how companies attract and encourage their entrepreneurial talent.

Different companies have been successful with different approaches. Some firms have attempted to institutionalize entrepreneurship and innovation by making it an ongoing concern for all general managers through finely tuned and very explicit *structural* arrangements. Others have evolved (over a long period of time) a more subtle and soft "entrepreneurial culture" within which managers almost instinctively adopt an entrepreneurial orientation.

Still other companies have established *separate* units for incubating and nurturing fledgling new products and businesses. Such units usually comprise a mixture of projects that were developed "in-house" and smaller firms (usually with complementary technologies) that were acquired from outside the firm.

In general, there does not seem to exist any kind of arrangement or approach that is "always the best." Some scholars argue that a contingency type of approach may be most appropriate. This means that one should consider the large array of possible arrangements (internal and external ones) that firms have experimented with and attempt to determine which of these is most appropriate given a certain set of objective (and subjective) conditions. In this part of the book, we have the opportunity to explore such a contingency approach in greater depth.

The closing case in the book—on the Control Data Corporation—provides an example of a company that has, more than perhaps any other firm, experimented with a wide array of arrangements to accommodate its internal entrepreneurs. A provocative way to conclude the study of the materials presented in this book is to ask whether Control Data is "ahead of the times" and is a model for the future Innovative Corporation, or whether they are perhaps eroding their advantage by dissipating too much of their energy in attempting to stimulate entrepreneurship.

Managing Corporate R&D

Case IV–1
Aerospace Systems (D)

A. Ruedi and P. Lawrence

Dr. Roger Simon had just passed up an opportunity to teach at Yale, preferring instead to stay on at Aerospace and set up a new corporate research lab. Looking forward to his new role as director, he knew it would be a big job; one which would be not only profitable but challenging as well. In his short time at the Atomic Energy Division, Simon had convinced both his division president and the president of the corporation, Al Douglas, of the need for a central research facility staffed by top-notch scientists working at the frontiers of their disciplines. He now had to draw upon all his previous experience working at the Manhattan Project and in the Zeta Labs to build a research organization which would enable Aerospace to remain competitive in the future.

Aerospace had adapted successfully to previous changes in business strategy, but over the past 20 years most of the technical accomplishments were in the area of classical physics applied to large aircraft moving at high velocities. With the recent advances in quantum mechanics and relativity, the nature of the industry was changing dramatically, and, to keep abreast of these changes, Simon felt that Aerospace had little choice but to embark on this expensive undertaking. Only within a central research facility could a company as large and diverse as Aerospace attract and maintain a critical mass of fundamental scientists working at the forefront of their fields.

Simon had been recruited initially as a research scientist, but in less than a year he found himself planning for a $3.5 million technical center with a projected annual budget at $5 million. This represented a major commitment by the board to the role of basic research in its future business strategies, but money alone would not guarantee success. Simon had to confront the generic difficulty of establishing a research group working in areas of high uncertainty within a business environment where investments must be justified by performance.

AEROSPACE SYSTEMS: A COMPANY IN TRANSITION

In 1959, Aerospace was one of the largest companies in the industry, with contract sales to the U.S. government of well over a billion dollars per year. From its founding in 1933, Aerospace grew rapidly to become one of the largest producers of aircraft in the world until the close of World War II, when—almost overnight—90 percent of its contracts with the Defense Department were canceled. In adjusting to this, Aerospace adopted a policy of diversification with less emphasis upon production and more upon contracts for complex systems requiring greater technical competence. By the 1950s, Aerospace was again one of the largest competitors in the field. With the change in business strategy, the older production-oriented structure, in which the heads of the major functions all reported to the chief executive, soon became unwieldy. To provide better support for its diversified activities, Aerospace decentralized its operations into eight separate divisions.[1] Each was given greater autonomy, being responsible to corporate headquarters

Harvard Business School case 9-474-164

[1] These were the Radar, Aircraft, Space Systems, Rocket Engines, Electronic Systems, Atomic Energy, Information Systems, and Submarine Systems Divisions.

solely on a profit basis, and during the decade of the 1950s many became the largest, or second largest, producer of their product or system in the world.

For a time, decentralization proved to be an effective organizational solution to the problems of diversification and growth, but by the late 1950s there was a growing tendency within the divisions toward parochialism and interdivisional competition leading to redundance and a concentration on short-run problems. Each division had its own research staff, but they tended to work on rather short-range projects. The one exception was in the Atomic Energy Division which employed the largest number of Ph.D. scientists, but the work here was done primarily for the Atomic Energy Commission, and this tended to limit both the scope of research activities and opportunities to draw upon this classified work in the other divisions. Several previous attempts had been made to establish a corporate research lab using AE scientists, but the resistance by other divisions stymied these efforts until the arrival of Dr. Roger Simon in July 1960.

Simon was well-known in scientific circles as a theoretical chemist with numerous publications to his credit. An eminent scientist, he had nonetheless become increasingly interested in management. It was this which caused Simon to leave Zeta Labs on two occasions for research positions elsewhere. Speaking of his desire to become a manager, Dr. Simon said:

> I had been doing science for a good many years, and although I still had a certain amount of ambivalence, I felt that I ought to try my hand at a little bit of organization, to organize the world about me. The opportunity at Alfa Steel seemed interesting, something called an assistant director of research. But it turned out that except for a small amount of fundamental work, they did not even do applied research; they put food in cans to see how long it would take for the cans to erode. I decided that this was simply not the place for me. I had the option of returning to Zeta with continuity of service, and I did.
>
> At Zeta, I spent my time totally involved with scientific matters, and I suppose nobody ever thought of me as a manager, although I did begin to express some of these views before I left. I thought—well, maybe I could knit together a chemical physics group for them. I was in fact the leader of that group without the title, simply because I was able to generate sufficient excitement of ideas—and most theoretical chemists are motivated by such ideas. To this day, the theoretical chemists at Zeta do not interact much with the rest of the lab, which is primarily physics oriented. They are a resource which is not fully utilized, and they can be tolerated because the laboratory is so successful anyway. My proposal was to make use of this resource for the company without destroying its scientific competence. They were willing to think about it, but they were not willing to do anything about it. I decided to leave again, since there was this interesting opportunity at AS.
>
> While I was at the Gordon conference,[2] a fellow from AE invited me to stop by and present a paper there. While I was there I was offered a job. I spoke to Howard Elliott, the director of research, and to the vice president of AE. What I noticed at AE was what I thought was a remarkable group of scientists. There were a lot of poor ones too—but nevertheless a remarkably high density of good scientists. But they were unconnected, they weren't interacting even among themselves, and they were unrecognized. I thought they were, as a group, as high quality as you could find at Zeta. Now there are a very few industries that have an opportunity like that. Some of these fellows were actually world authorities in their fields, like Bruce Nelson and Carl Nadel.
>
> I told Howard Elliott, the director of research, that I would only be interested in coming if I had the chance to do some organization, particularly with these resources—it seemed like an exciting idea to bring them together somehow. I really had no firm idea how to do it. I had never thought of that sort of thing. I guess it was sort of rash to think I could do it. I finally joined AE as a

[2] A series of yearly scientific conferences devoted to solid-state physics and solid-state chemistry.

research advisor, with the promise that I would become associate director of research after about six months. Howard Elliott was not interested in me as a manager, he was interested in hiring me as a scientist—but he could only get me as a scientist by permitting me to do some organization.

At that time, the research department was divided into four subdivisions: physics, chemistry, electron physics, and metallurgy. Each one was headed by a department head—all were competing for Howard Elliott's job. There was no coherence; they were all busy building up their empires by acquiring more contracts.

Supervisors and group leaders also spent much of their time doing paperwork and had little time to do science. That is the main reason why some of the better scientists didn't want to be bothered with becoming group supervisors.

The average scientist behaved as if he were working for one of the national laboratories; he did research essentially for the AEC and was responsible for getting out the data—that was all. Some of the research was good, some of it wasn't so good. Nobody thought they were working for AE, to say nothing of AS. Each man had a special relationship with some sponsor at the AEC. There was a great deal of brochuremanship, of time being wasted in getting contracts. Each scientist was an entrepreneur in his own right; he had control over consultants, travel, recruiting, and equipment. He was not called upon to think about the relation of his scientific problem to anything else. It was a more pure researchy attitude than you would find at a university.

An executive no longer with AE described the situation as follows:

Howard Elliott was a great guy, but he had to evaluate everything from three angles before he could make a decision. He was very soft spoken and technically very good. At meetings he would criticize things technically and everybody loved him and respected him. Roger Simon with his extremely dynamic approach to life was getting in there, muddying up the waters like crazy, and stirring things around to find out where the mud was going to travel.

When Howard Elliott had a skiing accident and was sent to the hospital, Roger Simon became acting director of research. In the meantime, Howard Elliott had received several offers from various universities and he confided to Roger Simon that he was thinking of leaving. Roger Simon recalls:

That was a bit disconcerting to me because I thought it was unfair of him to have recruited me a few months before and then to run out before my feet were wet. I told him that I would have preferred him to stay on, while I myself was thinking of going because this was such a mess. At that time, I had a nice offer from Yale University at $24,000 a year.

At this time, Sinclair Reed (the division president) was attempting to get a million dollars of internal funds to replace a like amount withdrawn by the AEC. Howard Elliott was still director of research, and I told Sinclair that my decision as to whether or not I would go to Yale hinged upon whether or not they got that million dollars. Without it, I was not interested in staying because there was too much instability.

To me, Sinclair was the ultimate; if I put enough pressure on him, then he could settle the problem. What I did not realize then was that he was not the ultimate, that putting pressure on him just squeezed him. Howard Elliott felt crushed more than anyone else, and after some discussion with Sinclair Reed he finally quit and I became acting director of research without any obligation to stay on.

With the issue of the million dollars still pending, Roger Simon proceeded to tackle what he considered his two main problems. The first was to "weed out the deadwood."

The lab was in considerable turmoil as a result of Roger Simon's promotion. The section heads were disgruntled because their chance for promotion seemed to have vanished, and the bench scientists were apprehensive about the future of their jobs. These anxieties were not relieved when Simon promoted Bruce Nelson and Carl Nadel to associate directorships. Nelson had never held an administrative position, although he had a worldwide reputation as a scientist; Nadel, who had been in and out of management positions at the group leader level, was also well known as a scientist.

The unrest and anxiety produced by these rapid changes was utilized by Simon to get rid of many people he thought incompetent or useless. Those people who, in concert with Nelson and Nadel, Simon thought were valuable, were privately reassured of their positions and future with the company. In several instances, Simon promoted people to positions of group leader based primarily on the criterion of scientific accomplishment. In like fashion, Bod Nordson, an ex-physicist turned administrator who had been with the research department for only a few months as administrative liaison, was promoted to research administrator. Nelson, Nadel, and Nordson became, with Roger Simon, the management team for AE's research department.

The second problem facing Roger Simon was to get the better scientists to relate their activities to the corporation and interact with others:

> I had already begun to motivate some scientists to relate to the overall mainstream of effort at AE. For example, Bill Tresbon and his group were working quietly on the electronic properties of metals and conductivity. Tresbon told me that he discovered a very interesting high-field super conductor. I woke up one morning a few days after this and said to myself, "Why the devil don't they make a magnet?"
>
> Of course, this was unheard of as far as they were concerned. They were not interested in this. They were going on to something else. So I went down to see one of the men in Tresbon's group, as Tresbon was away, and said, "Look, why don't you try to make a magnet? But first we'll have to draw a wire," and he said, "Tresbon won't like that . . . it's such an applied thing," and I said, "Well, you know you might learn something from this."
>
> "Besides," he said, "you haven't got any money."
>
> "How much will it take?"
>
> "About $50,000."
>
> "You start making that magnet and I'll get you the $50,000."
>
> Very reluctantly he began, and we had the first superconductive wire of this type in the country. I said to Sinclair, "I'm going to make you very happy," and I told him about the discovery. "You do make me happy," he said; and I said, "Well, if you're so happy, get me $50,000 so that these fellows can go ahead," and he did.
>
> When Tresbon came back, he was ready to go through the ceiling, but the day after the first sample of wire came in, it was interesting to see the reaction. It carried 10 to 15 times as much current as the original slab, and it went to much higher fields. The point is—by having taken the first technological step, they added to their store of scientific knowledge. Then they became very interested and they got into the magnet race. And these two guys, all by themselves, successfully competed with much larger efforts at Ipsylon, Zeta, and Gamma Labs.
>
> There were several instances like this and apparently Sinclair Reed developed a confidence in me as a manager. Reed came to the conclusion that this was the time to ask Douglas not only for the million dollars, but to establish some sort of central research effort, and he asked me to write a report telling Douglas what he would have to do to establish a research facility of this kind.

Roger Simon had two weeks to prepare his proposal. He achieved this by working together with Nelson, Nadel, and Nordson. The four men would discuss the issues, then Simon dictated his ideas which were subsequently edited by Nadel and Nordson. The brochure (Appendix) suggested a research laboratory geographically separate from AE, although organizationally part of AE, reporting to Reed.

Though the line of argument and the tone that the brochure would take had been discussed by Simon with Reed before publication, Reed did not read the final document until the day before it was to be presented to Mr. Douglas. Reed was in agreement with the general content and took the lead in presenting the idea of a geographically separate, corporate-oriented research center.

During the course of the meeting, Mr. Douglas's primary concern appeared to be how one might obtain the necessary interdivisional acceptance and utilization of the research effort if this corporate-oriented effort was to be a part of a single division. Reed responded with the key suggestion that the central research function be removed from his jurisdiction and placed under that of the general office. Although Reed's extemporaneous offer was well received by Douglas, no conclusions were reached at the meeting. In recalling the next few days, Mr. Simon said:

> Things went on for a couple of weeks; there was no comment from Douglas and no report as to what his decision was. I asked Sinclair how things were going and he said, "I haven't heard anything so it looks pretty grim," so I said, "Well, Sinclair, summer is coming along and I think I'm going to accept the job at Yale University." Sinclair said, "Let's go down and see Douglas before you make up your mind." So we went down to see Douglas and talked with him. This is the longest talk I ever had with Douglas. And then I detected after some minutes that there was a rapport between Douglas and me and right then and there I decided, "Well, this is a big company and they are good scientists. It will not only be challenging, but profitable, to establish a lab in an organization like this."

Alan Douglas recalled the meeting with Dr. Simon and events that led up to it:

> There were two thoughts in my mind. The first and foremost was the need to upgrade the technical excellence of the corporation and the fact that I, myself, did not possess the requisite technical background to accomplish this without a staff. Secondly, the intrinsic value of a science effort in the corporation during this period.
>
> Then I started hearing about Roger Simon. Sinclair Reed talked to me about Roger Simon quite often, mentioning his intellectual capabilities and his abilities to form a lab, and I began to get interested. I began to talk to Roger Simon because his background was so different from mine that we had to go through an adjustment period before we could understand each other. In fact, he gave me a series of lectures which gave me some feeling about what the science that he had in mind was about. It developed at this point that he was also a good administrator since he was running that lab at AE.
>
> The mental cycle that I went through was something like this: Here was this group working under contract for the AEC, working on their own problems because you can't get a first-rate scientist to work on anything unless it is a problem he is interested in. But the restraining interest of government was considerable and I started to give some thought about funding this research out of internal funds. The question was how to set up the lab. The idea of taking any of our research and development funds and putting them into that kind of an organization was not very popular with the divisional people who are product and development oriented.

Dr. Simon recalled his relationship with Mr. Douglas in this period as follows:

> In view of the urgency connected with the upgrading of science in the corporation and because of the variety of unbalanced pressures to which I thought Douglas would be subjected, what I had to do was to learn to motivate him. I thought, "How can I motivate Douglas? I can't really motivate him because of my management experience, because I have none. I may have some brash ideas, but I can't really represent myself as a distinguished manager." The only thing I had to go on was my hopefully distinguished reputation as a scientist—a reputation which very few people at AS had the equivalent of; the only other people might be a few in the research department at AE—Nelson, Nadel, and Tresbon. So I made no effort; in fact, I told him point blank I was no manager but that I was a good scientist. I told him I was fed up with the way science was being managed at AS and for that reason I was going to Yale University. At this point he said, "Well, I want you to think about it a little longer."
>
> I thought about it a little longer and went back a couple of days later, and he said, "We have been trying to establish a central research activity

> for a long time, but we have not had the man whom we could count on to do it. We now think we have." I sat there very quietly. Then I warned him again that I had no experience as a manager, that he was gambling, that he would have to give me the authority to do this thing the way I wanted to do it, not to make me a shunt to him or anybody else. And he agreed! It was a startling thing, he agreed!

By July, the Aerospace System's board of directors had approved the establishment of a corporate research center as outlined in Dr. Simon's proposal and recommended by Al Douglas. The research center was to become a separate division in October 1962 and in the interim period remain dependent on the Atomic Energy Division for facilities and administrative support. Prior to the board meeting, Simon was spending most of his time in the day-to-day operations of the AE lab, but with this go-ahead decision things began to move fast.

In preparation for the early discussions with architects, Simon and his associate directors had to spend many late nights discussing what the laboratory would look like in terms of size and disciplines so they could pin down an appropriate physical layout. Building upon his initial proposal, Simon wanted to construct a lab based upon key scientific disciplines, for only in this way did he feel he could combine both specialization and flexibility in a single research unit. Each researcher would be eminent in his field, but the fields, being basic, could combine in numerous ways, depending upon the interests generated. To support this, each senior staff member would have his own lab, but in a facility which was compact to facilitate interaction. As the lab evolved in these meetings, other factors such as the size and location of offices, the library, and even the cafeteria were all designed to maintain a stimulating collegial atmosphere.

Beyond these architectural considerations there were a great many other issues raised but not yet fully resolved. The whole problem of financial support was only one of these. Based on their experience in AE, they knew of the dangers associated with government contracts, but they were still not certain what the actual mix of outside to inside support should be. Related to this was the issue of whether divisions should be encouraged to pay for work in the laboratory on a "contract" basis, or whether all research should be corporately funded. This in turn raised the question of how the interface between the lab and the divisions could be arranged to promote autonomy of the lab while still having it in touch with the divisions.

Another problem they talked about dealt with the administrative components of the lab. To what extent should the day-to-day operation and liaison responsibilities be centralized and kept from the scientists? Since they were planning to spend in the order of $100,000 a man in each of 15 key scientific areas and they hoped to recruit top persons in each field, they wanted to avoid bureaucratic red tape and too many levels of management, which would constrain this talent. One possibility was for Simon, Nelson, and Nordson to have the key scientists report directly to them as a group. However, they felt that key personnel might end up spending too much of their time in meetings rather than in their labs. Simon and his group talked often about this problem in an effort to get a balance between autonomy for the individual scientists and some unity of purpose in the lab.

In addition to these questions on financing and direction, Simon and his team also were beginning to grapple with several related issues. Where would they recruit these scientists? Would they try to attract researchers from the other divisions as well as go outside? What type of compensation scheme would they offer to attract and motivate these people? And how would this be tied into performance evaluations?

APPENDIX A MASTER RESEARCH PLAN FOR AEROSPACE SYSTEMS

DEFINITION

It is best to begin with the definition of terms. By "research" I shall mean something distinct from "development," though the separation can never be complete. With this in mind, I classify under the heading of "research" those activities which are devoted to the creation of new understanding of both phenomena and materials.

NEED FOR A RESEARCH PROGRAM

Aerospace Systems, in consort with the rest of the industry, has as yet not demonstrated outstanding capacity in research in the sense in which I have defined it. It is important to note that research is a different activity in the sense that it is involved with the creation of new knowledge, and very often this is offered in terms of the molecular and submolecular mechanisms which underlie materials and phenomena. The developments at AS cited above have, with minor exception, all been based on the sound application of scientific principles which were thoroughly established at the time the developments were initiated. One might say that the science of yesterday is the engineering development of today. The converse is obvious. The engineering which we will do tomorrow will be the direct descendent of the research we do today.

Why is it that we can no longer rely solely on engineering applications? Even more important, the need for research stems from the fact that today's accelerated pace has drastically reduced the delay time between the moment when new science is created and the instant when it is utilized as a tool of development. It has become customary for the same organization to create the science and to exploit the development.

FUNCTIONS OF A PROFITABLE RESEARCH LABORATORY

If it is granted that the corporation must, literally, pursue research, there must next follow an understanding as to how a group of individuals, synthesized into a "laboratory," can come up with the necessary knowledge that the corporation seeks to exploit.

Maintaining lines of communication with the external scientific community. It is vital to become aware of new and exciting developments in the world of science as soon as they happen. In today's highly competitive atmosphere, time—even a little time—is a valuable commodity, and only early cognizance of important developments can provide the necessary time. There is no better way to establish broad lines of communication with the world of science than through the natural personal relationships which occur among scientists who are peers. If the laboratory has gained the respect of the scientific community and its scientists are similarly respected, then they will be welcomed within the most elite professional circles. The normal give-and-take of information which occurs within these groups will provide for the early transmission of news.

Many modern developments are so specialized that they can only be interpreted by a man who has a working familiarity in a new field. Thus it may sometimes be of value to support a specialist and his small research effort in order to keep an eye on the new field.

Stimulating other laboratories. An important function of a research laboratory is that of stimulating the community of science to work on problems which are of interest to the parent firm. This can, perhaps, best be indicated by the example of Bell Telephone Laboratories and the transistor. When the transistor effect was first discovered by Brattain and Bardeen, it was realized immediately that the effect could be converted into a useful device which Bell could use to replace costly, short-lived, and less reliable vacuum tubes which were found by the millions in the Bell System. On the other hand, there were grave materials problems of a fundamental nature, and even Bell's large staff could not be expected to solve them within a reasonable period of time. By inducing some of their most outstanding people to work and publish in this area, they succeeded in stimulating the entire world to concentrate on the field of solid-state physics. So popular did the subject become that the solid-state sessions at the Physical Society meetings permitted standing room only. The scientific community published the results of its work and the literature was available to Bell. In a sense, Bell achieved immense amplification of its own effort, and in fact they were able to actually use transistors in the Bell systems within a very few years. That other firms manufactured and sold transistors was of secondary importance. Bell only wanted to use them.

It is clear that this sort of strategy cannot be pursued with a second-rate staff. To nucleate interest, it is not enough to publish work in the field. The work must be imaginative, exciting, and clearly indicative of the possibilities for further inquiry.

Advising and consulting with development groups. This is a fairly obvious function of a research laboratory. Sometimes a development man's problem can be solved as soon as he contacts the appropriate research man. The latter may have sufficient knowledge concerning the recent literature in the field so that the solution involves little more than the location of a specific reference. Alternatively, the knowledge may be available but not published and yet the research man may know of it. Sometimes the development man may be able to enlist the collaborative effort of the research man in the development problem itself.

THE RESEARCH SCIENTIST

When the need for research is accepted, there must be a simultaneous acceptance of the fact that the research scientist is a different species from the engineer. Because of this, a whole new host of problems arises for which the solutions developed to fit the engineer are inappropriate. These problems are discussed here, but first a few words are in order concerning the motivation and attitudes of the truly outstanding and creative research scientist. Analyses of the scientist have appeared in many reports and are well documented. These remarks combine our own experiences with those of other well-known and well-established research organizations. *AS has been attempting to deal with the research scientist with methods applied previously to engineers.* An immediate result is that AS is not regarded by scientists as an especially desirable place to work.

It must be accepted from the very beginning that most scientists are profession oriented and not company oriented. It is not very far from the truth that scientists regard the company partly as a means to an end, that is, a means to advance scientific research. But the company wishes to use science as a means to its end, that is, to earn profits. Therefore, it is necessary that a compromise be made. The company must make peace with the world of science in order to extract information from it and to secure maximum allegiance from its workers.

To attract these top-notch research scientists, the company must display an enthusiasm for good science as an end in itself. It is generally true that the degree of company-oriented enthusiasm shown by the scientist is directly proportional to the degree of science-oriented enthusiasm displayed by the company. The surest way to alienate a talented research worker is to tolerate good science passively while exhibiting enthusiasm actively for the technical activities which are clearly directed toward a commercial goal. The enthusiasm must be felt; it cannot be produced by edict. But if senior management cannot honestly display enthusiasm for science, then it is probably fair to say that only among a few members of senior management does enthusiasm for science per se exist.

The research scientist considers himself a professional person. There is a large complex of small indignities in existence at AS which militates against providing the scientist with the dignity and atmosphere he seeks. Each one of these items considered by itself seems so trivial as to be of no consequence, but the entire complex of annoyances makes for a complaint of some significance. It is also true that some of the individual annoyances are not annoyances at all, but are merely regarded so by the research workers.

In this instance, we have an example of the importance of recognizing the personal characteristics of research workers as distinct from those of engineers.

In his personal nonscientific behavior, the scientific worker is no more a logical individual than is any other person. For example, capable scientists are often completely unimpressed by an appeal for economy from management. This, in spite of the irrefutable argument that all funding, no matter how generous, is limited. One might say that this is an aspect of the real world which is not real to scientists.

If a company means to establish a research laboratory, it must be prepared to put up with all the seemingly illogical intangibles that go along with it.

MANAGERIAL AUTHORITY REQUIRED BY A RESEARCH LABORATORY

An atmosphere in which creative ideas will flourish can only be attained with certain managerial authorities granted to the laboratory by the corporation's senior management. Two such areas are the control and direction of the research program. One cannot employ a creative individual with a national reputation and then tell him what to do; his creative talent is often intimately bound to an independent nature. The principal elements of control available to the research director are: (1) the careful selection of personnel at the time of employment, and (2) those influence methods associated with the personal rather than business relationship with the scientist. For example, it is only reasonable to select as employees workers whose natural interests are in those fields upon which it is felt that the company should concentrate. One also attempts to employ responsible individuals who will understand the nature of their obligation to the company. But once employed, the scientist must be allowed to work on that task to which his enthusiasm drives him. That extra spark that is needed to inspire a new idea will only come from the drive for understanding that the scientist finds within himself.

Another attribute of the better scientist helps channel the laboratory's energies into the desired fields. This is the fact that very few scientists enjoy working in isolation. Ordinarily they like to discuss their problems with colleagues whose interests are similar. Thus, if some central theme of scientific interest pervades the laboratory, the new employee will tend to gravitate towards it.

The important thing is to make everyone desirous of being a member of the team. The personality of the director may be used as a control element by appropriate use of enthusiasm or by development of interest through the active collaboration of the director himself in a new field.

If under this system workers occasionally pursue lines of inquiry which are not of immediate use to the company, then this is the price which must be paid in order to establish an atmosphere in which creative thinking will flourish, and to which outstanding personnel will be attracted.

The company must also make it clear that they will not be perpetually anxious about how their funds are being applied to research. They must be confident in the judgments of the research management. Constant detailed accounting of the manner in which funds are utilized should not be necessary.

Those limitations which the corporate management wishes placed on its research effort should be clearly and publicly delineated. Outside of these, research management should be given a free hand to create the most fertile environment that its ingenuity can provide.

It is clear that the principal emphasis must be placed on the acquisition of outstanding personnel. The success of the entire venture will stand or fall on the company's ability to recruit talented people. The more creative scientist (measured by some suitable standard) may be 100 times as creative as the average. On this basis, it is well worthwhile for the company to spend two or three times the average in support of such a person; the return per dollar is 30 to 50 times greater.

MANAGEMENT'S ATTITUDE TO THE RESEARCH PROGRAM

I have noted previously that one of the essential requirements for a productive research laboratory is a display of enthusiasm for good science on the part of the senior corporate and divisional managements. Displaying enthusiasm for science is an expensive thing, in connection with which more often than not direct financial return cannot be visualized. For example, it may be necessary at times to sponsor research projects having no direct bearing on current company interests. It will be necessary to have scholarly academic visitors whose consultant fees are high and who do little more than raise the morale of the organization. It may be necessary to provide scientists with paid sabbaticals during which they can visit other research institutes or universities. At times it will be necessary to sponsor topical scientific conferences, even to the extent of publishing the proceedings. It may even be necessary to employ scientists on a full-time basis who spend only a part of their time in the company laboratory; for example, some of their time may be devoted to teaching.

THE RETURNS OF RESEARCH

Nobody can guarantee in advance just what return, if any, will be realized on the very appreciable sum of money which must be invested to establish an adequate research program. It therefore requires faith on the part of the investor that, given first-class personnel surrounded by an atmosphere within which creative activity can flourish, the chances for profit are excellent. Faith is in a large measure familiarity. That the value of such a very unconventional mode of business function as has been described should be real can only become clear with sufficient familiarity.

Many businessmen, although highly intelligent and extremely sensitive to reasonable arguments, find it difficult to bring themselves to risk funds in a research effort because they have not acquired the necessary intimate familiarity with the details of the research process.

Another bitter pill must be swallowed. This is the fact that once having mounted the best possible kind of effort, no major advantage over competitors who have mounted similar efforts can be assured. The best that can be hoped for is the ability to remain abreast of the leaders, not to outdistance them. The aspects here are negative rather than positive in the sense that the firm without a first-class research effort will almost surely fall far behind.

Probably the best that can be done in the way of assuring a profitable research effort is to employ an outstanding research director and to place one's faith in him. The chances are, if he is competent and it is possible to provide an atmosphere in which he can recruit outstanding scientists, the effort will be profitable given enough time. This is at once the very most and very least that one can do.

I have deliberately emphasized the scientific rather than the applications aspects of some activities at AE. This is not to belittle the importance of an effective applied research effort. The scientific advances which AE scientists are apparently capable of making would have little significance for the company if there did not exist an effective product development department to which the research effort could be eventually coupled. I emphasize these specific activities in the realm of pure science because this is an exceedingly valuable effort currently owned by the company which has attracted too little attention.

The research discussed is of a caliber and significance equal to that found at the best universities or at an outstanding research institute. The fact that such work can be done indicates that AS has a nucleus of talent in these fields. In spite of the outstanding work which I have just described, this nucleus has not fulfilled its potential and in fact is in danger of dissolution. If we lose this group of scientists, it is questionable whether the company will ever mount a similar research effort in the future. In today's competitive market, it will be almost impossible to recruit such talent again.

THE NEED FOR CENTRALIZATION

What is the trouble? Why has not the group fulfilled its potential? The answer is many-faceted. In the first place, the whole of a laboratory should be greater than the sum of its parts. There should be an integration of effort. Scientists should interact spontaneously with one another. There should be a feeling of teamness. The laboratory should reach a kindling temperature. Interaction is instrumental in achieving this kindling temperature, and it is also necessary for the research effort to attain a critical mass.

At AE, there has not been much spontaneous interaction and I suspect the same is true of the other divisions of AS. Certain things catalyze interaction. Instead of the things necessary for such catalysis, it seems as if just those which act to quench interaction have been present. For example, most of AE's research is supported by external sponsors. The resulting project system has compartmentalized the scientist's time and equipment. If he is stimulated by a colleague to investigate a problem outside the scope of his immediate project, how will he charge his time? More important, where will he find the equipment to pursue his interest? In the end, his mind also becomes compartmentalized and he is no longer susceptible to stimulation. There are several individuals who exemplify this effect at AE.

To achieve an integrated effort, there must be a centralized research effort within AS. This will have to come to pass before the kind of research I have been discussing can be realized. As it stands, there are not only costly redundancies within the company, but there is obvious interference between divisions. It seems unreasonable in the light of the scarcity of topflight research personnel that AS should further scatter its talents among its own several parts. Unfortunately there is an unsymmetrical distribution of both profitability and research talent among the various subdivisions of AS. For example, Rocket Engines and Electronic Systems are capable of large earnings, but there is little question that the largest concentration of research talent, as opposed to development talent, is lodged at AE. Therefore, AE would have to play a major role in the process of centralization. Of course, opposition to this concept may arise, but this has little bearing on the fact that it is required on the grounds of logic.

Consider the case of the high-field superconducting program. It was a real stroke of luck that we, with our very small effort (supported entirely by AEC funds), were able to effect a breakthrough which our large and capable competition had not achieved. Because AS had not supported this research, our proprietary position was somewhat compromised, but worse yet we had neither equipment, nor funds, nor personnel to capitalize on this breakthrough and to fully exploit the time advantage which had been given us.

Any research laboratory, in order to be successful, ought to have one or at the most only a few themes at a time. The development of high-field superconductors seemed to provide excellent opportunity to bring the integration of purpose which can be initiated by so powerful a central theme, but very little of this plan could be effected as there are no uncommitted funds for the support of a diversified program of this sort. In this instance, time is a critical element since there are very large research efforts functioning in the hands of our competitors. Worse yet, the capital equipment is not available.

I have quoted at length the high-field superconductor to emphasize how ineffective our system is from the point of view of sponsoring a mobile and highly competitive research program. There is so much inertia associated with the ponderous mechanism for making up our collective mind that whatever advantage in time our research effort may have given us erodes until no advantage remains. I would say that all of this stems from the lack of a master plan for research, a commitment in advance to do certain things, and a placing of confidence in the person of some trusted and highly competent research director. The establishment of a corporate research center would undoubtedly provide some of the basis for the elimination of such difficulties. It would at least give us a master plan to which we could refer all decisions which have to be made in a hurry.

A master plan would have some further advantages. Where AEC or DOD sponsorship ceases for a project of scientific excellence before such work can be completed and published, the company should see to it that such work is completed. This is a part of the principle that the company should make peace with the world of science to establish an atmosphere and reputation which will attract the more outstanding personnel.

THE OPPORTUNITY EXISTS

In spite of the above criticisms, AS is situated at a point in time and space (circumstance and geography) which is especially opportune for the establishment of the best industrial research laboratory on the West Coast.

West Coast industrial laboratories are all of the same type, heavily involved with government contracts and committed to cost-plus research. There is little project security—even job security is uncertain—and in fact, a large class of what might be called migrant scientific workers exists who follow the contracts from company to company. Each laboratory has a few competent people, but no one is able to acquire the critical mass necessary to mount a really outstanding research effort.

Most of the personnel engaged by these laboratories continue on their jobs because, among other reasons, they prefer to live on the West Coast. If any one company would give the clear sign that it meant to conduct its affairs in a manner similar to the outstanding East Coast laboratories (Bell, GE, IBM), it would very rapidly accumulate the best personnel now scattered among the many similar mediocre firms. For AS to give this sign, it will have to give public demonstration to the acceptance of those precepts mentioned above. The establishment of this laboratory will require considerable initial expense. AS will have to let it be known publicly that such expense has not been spared and that the research is being heavily financed by company funds. Furthermore, it would have to be made clear that it was research, not development, which would be sponsored, and that the company was willing to be quite patient in connection with the time schedule for results, perhaps looking 5 to 10 years into the future.

A MASTER RESEARCH PLAN

Development of a research theme. A master plan for research at AS requires, first, the statement of a theme or themes. An industrial research laboratory ought to be concerned with but a few all-consuming central research themes at any time.

To be appropriate to AS, a research theme should fulfill two criteria. First, it should be the frontier of science. Second, it should evolve naturally from the corporation's fields of interest.

By tradition, AS has been a physics-based organization. It is, therefore, doubly appropriate that AS should concentrate primarily on physical research.

Fields of research. It is appropriate that the phase of the master plan which indicates the areas of research of interest to the corporation be indicated in terms of those scientific disciplines for which we desire coverage. It will be the job of research management to integrate these diverse interests into a unified effort, now supporting one theme, and later, perhaps, another. In examining the list of disciplines, it must not be assumed that the people assigned to one discipline will be limited to that particular field. People of the proper caliber will be able to work in a number of the areas contained within the list. One attribute of a good scientist is interest in allied disciplines. Through such cross-fertilization, progress is made.

Many of the subjects can overlap; many of the people who are working at one time in one area might very easily work at another time in another area. It is only through juxtaposing people with such overlapping interests that it is possible to achieve the interaction and integration of purpose which is required to advance a currently important research theme. What is important is to have all of these people on board all of the time. It may seem expensive if at times we are not using all of them in connection with fields which are of primary interest to the company, but we cannot tolerate the lag which is involved in acquiring them only when we need them.

Personnel for the central research laboratory. Since the research center is to be representative of the entire corporation, it ought to draw upon personnel from all divisions as required. The staff, however, should not come exclusively from the present divisions and a large fraction of the required staff will have to be recruited from universities and other laboratories. The contribution of manpower from each of the divisions of AS to the initial staffing of the research center should depend on the qualifications of the scientists themselves, and no requirements should be imposed regarding proportionality to a division's size, solvency, or seniority. Any conflict based on this issue is liable to immobilize the plan at the outset. Any compromise to questions of personality is liable to do the same. If a large sum of money is to be invested, we ought to be sure that it is put to the best possible use. The corporate director of research should report directly to the general offices and be charged with the responsibility of showing no special bias towards any of the operating divisions.

We will need to recruit for the several disciplines not now adequately covered. Such recruiting must place very strong emphasis on the acquisition of creative people. Recruiting will be slow because of the lack of availability of adequate personnel. It should be reiterated that such personnel will be attracted to the corporate research center only if the proper image can be created. Everything must be done to produce and maintain this image. Once we acquire outstanding personnel and are provided with proper funding and adequate working

Physics	
Theoretical physics	8
Metal physics	5
Nonmetallic solids	7
Plasma physics	5
Geophysics	2
Nuclear physics	5
Field physics	5
Space physics	4
Surface physics	1
Biophysics	1
High-temperature materials	6
Device physics	4
	53

Physical Metallurgy	
Mechanical properties	6
Structure	4
	10

Chemistry	
Electrochemistry	4
Chemical kinetics	3
Radiation	2
Chemical thermodynamics	1
Theoretical chemistry	2
Combustion chemistry	2
Polymer chemistry	2
Surface chemistry	2
Biochemistry	1
Analytical chemistry	2
	21

Mathematics	
Statistics	1
Information theory	2
Operational analysis	3
Logic	1
Numerical analysis	2
	9

conditions, the rest will almost take care of itself.

Size of the central research laboratory. The corporate research center should have a minimum size in order to achieve the critical mass necessary for effective functioning. It has to be large enough to cover the wide variety of disciplines listed. It must have breadth in order to be cognizant of the latest developments in these fields. It must have depth in order to make it possible for AS to deploy its forces in the most efficient manner to meet the demands of the moment. It must be large enough to provide the community of science with a purposeful and respected image. It must provide a sufficient variety of fields to attract and interest people of stature. It must promise a top-notch potential staff member an association with colleagues sufficiently learned and versatile to supply him with all desired support information.

How large is large enough? My best estimate gleaned from actual working experience is about 100 scientists of Ph.D. caliber. Each of these must have support to the extent of about one technician per scientist so that the entire research organization should involve about 200 direct people.

The number of people enumerated in the above list represents senior people, all of whom have the Ph.D. degree or equivalent. The total of the above is 93. To allow for imperfect estimates, we propose to round the figure off at 100. In addition to these, and the 100 supporting technicians, there will be need for a staff of service and clerical personnel concerned with administrative service duties, a library, and a shop.

Facility needs. In the previous section of this discussion, examples of impendences have been given, and it has been argued that a certain degree of managerial autonomy and stability of funding is necessary. For these reasons, and others previously noted, a research center separate from the present AS divisions appears desirable and necessary. It would require approximately 120,000 square feet of laboratory and office space to house the number of people previously noted. This figure is, of course, only approximate and is presented here merely as a guide for further study.

Reading IV–1
The Lab that Ran Away from Xerox

Bro Uttal

On a golden hillside in sight of Stanford University nestles Xerox's Palo Alto Research Center, a mecca for talented researchers—and an embarrassment. For the $150 million it has lavished on PARC in 14 years, Xerox has reaped far less than it expected. Yet upstart companies have turned the ideas born there into a crop of promising products. Confides George Pake, Xerox's scholarly research vice president: "My friends tease me by calling PARC a national resource."

Not that the center has been utterly barren of benefits for Xerox. The company's prowess in designing custom chips, to be used in future copiers, comes largely from PARC. So do its promising capabilities in computer-aided design and artificial intelligence. PARC did most of the research for Xerox's laser printers, now a $250-million-a-year business growing at 45 percent annually and expected to turn a profit in 1984.

But Xerox hasn't cashed in on PARC's exciting research on computerized office systems, which was the center's original reason for being. According to Stanford J. Garrett, a security analyst who follows Xerox for Paine Webber, the company's office systems business lost a horrific $120 million last year and will probably drop $80 million in 1983. "Xerox has got a lot out of PARC," says Garrett, "but not nearly as much as it could have or should have."

Why has Xerox had trouble translating first-rate research into money-making products? Partly because the process takes time at any large company—often close to a decade. Sheer size slows decision making, and the need to concentrate on existing businesses impairs management's ability to move deftly into small, fast-changing markets. This is a special problem for Xerox, still overwhelmingly a one-product company whose copiers accounted for three quarters of last year's $8.5 billion in revenues and almost all the $1.2 billion in operating profits.

Serious organizational flaws, acknowledged by high Xerox executives, have also proved a handicap. PARC had weak ties to the rest of Xerox, and the rest of Xerox had no channel for marketing products based on the researchers' efforts. The company has revamped office equipment marketing five times in the last six years. "Xerox has creaked, twisted, and groaned trying to find out how to use PARC's work," says an insider. While Xerox has groaned, disgruntled researchers have left in frustration. These Xeroids, as they call themselves, have showered PARC's concepts—for designing personal computers, office equipment, and other products—on competing companies.

PARC's influence outside the walls of Xerox is an ironic tribute to the ambitious vision of the man who founded the center in 1969. C. Peter McColough, then Xerox's president, charged PARC with providing the technology Xerox needed to become "an architect of information" in the office. The new center, in a mutedly elegant three-story building whose rock-garden atria foster meditation, quickly lured many of the nation's leading computer scientists, offering what an alumnus calls "a blank check and 10 years without corporate interference."

Roughly half of PARC's money went for research in computer science and half for research in the physical sciences. Most of the glamour radiated from the computer crew. Members were notorious for long hair and beards and for working at all hours—some-

Reprinted with permission from the September 5, 1983, issue of *Fortune* magazine.

times shoeless and shirtless. They held raucous weekly meetings in the "bean-bag room," where people tossed around blue-sky concepts while reclining on huge pellet-filled hassocks. PARC's hotshots were not just playing at being geniuses. Before long, computer scientists recognized PARC as the leading source of research on how people interact with computers.

The hands-off policy at Xerox's headquarters in Stamford, Connecticut, proved a double-edged sword. PARC researchers used their freedom to explore concepts for personal computing that have since swept the industry. All sorts of computers, including some from Apple and IBM, now offer "bit mapped" displays, which PARC championed 10 years ago. Such displays link each of the thousands of dots on a video screen to a bit of information stored in computer memory, thus allowing the computer to change each dot and create very fine-grained images. Apple's new, easy-to-use Lisa flaunts a display that can be divided into "windows" for viewing several pieces of work at once, as well as a pointing device, or "mouse," for giving commands. PARC did the lion's share of work on both ideas.

But Xerox's loose management also encouraged PARC to overstep its charter, which was to do research, not nuts-and-bolts product development. By the mid-1970s, the center was hard at work on the Alto, an expensive machine with some of the attributes of a personal computer, which was supposed to serve as a research prototype. Alto and its software became so popular inside Xerox, where PARC installed a couple of thousand of the systems, that some renegade researchers began to see them as commercial products. Out of top management's sight, they slaved like distillers of moonshine whiskey to develop the Alto for the market.

Product development, however, was the turf of another Xerox group, which was championing a rival machine called the Star, later to reach the market as Xerox's 8010 workstation. Unlike a personal computer, which generally relies on its own processing power and memory, the Star worked well only when linked with other Xerox equipment. (See "Xerox Xooms toward the Office of the Future," *Fortune*, May 18, 1981.)

PARC rebels not only took on the development group, but also dominated a Xerox unit set up to test-market research prototypes. This group got over 100 Altos installed in the White House, both houses of Congress, and a few companies and universities. Unwilling to support rival machines, Xerox guillotined the Alto and in 1980 liquidated the whole test-marketing group.

Veterans of that group have been the chief evangelists of PARC technology. John Ellenby, one of the unit's managers, later founded Grid Systems. His Compass computer approximates some prescient PARC concepts first used in the Alto. It's portable, uses a bit-mapped display, and easily hooks up with remote computers. At $8,000 to $12,000, the Compass sounds too costly to be popular, but Grid expects revenues of more than $28 million in 1983, its first full year of operations; in August, Grid said it was on the verge of profitability.

Another manager of the test-marketing unit, Ben Wegbreit, had previously been one of PARC's brightest technical talents. Convergent Technologies of Santa Clara, California, founded in 1979 to make workstations, picked off Wegbreit and two colleagues to design software. Convergent's word-processing program shows some of its origins in the form of a "piece table," a type of software developed at PARC. It allows computers with fairly small memories to process long documents. It does this by storing only the changes made when editing, along with the original version, instead of the original plus a full-length edited version, as other programs do. Conveniences like that have helped Convergent land contracts that could produce some $450 million in sales to big computer companies that haven't developed their own desktop systems.

Charles Simonyi, who defected from Hungary at 17, styles himself "the messenger RNA of the PARC virus." He worked at the center for seven years, mostly on Bravo, a text-editing software program for the Alto that never reached the market. "We weren't supposed to do programs like that," he confesses, "so Bravo started out as a subterfuge. But when people at Xerox saw it, they wanted to use it inside the company. Bravo was why people used Altos, just as VisiCalc was the reason people bought the Apple II." Simonyi expected some brilliant executive to see his product's market potential. "That wasn't dumb," he says, "but it was naive to assume such a person would come from Xerox." Simonyi found a warmer welcome at Microsoft Corporation, based in Bellevue, Washington, which rang up $50 million in sales of personal-computer software in the year ended last June. A big chunk of this year's sales, which should approach $100 million, will come from Microsoft Word, a streamlined version of Bravo.

Lisa is the unkindest cut of all. In December 1979, Steve Jobs, then Apple's vice chairman, visited PARC with some colleagues to poke around. They saw Smalltalk, a set of programming tools. "Their eyes bugged out," recalls Lawrence Tesler, who helped develop Smalltalk. "They understood its significance better than anyone else who had visited." Seven months later, Jobs hired Tesler, having decided to use many Smalltalk features in the Lisa.

The Lisa had to be priced at $10,000, two to four times Job's earlier estimates. But it seems to be taking off. Apple claims to have shipped as many Lisas in July, the first month they were available, as Xerox has shipped Stars, or 8010s, in 19 months of availability. The Star, which embodies many concepts used in the Lisa, has been ill-starred. The influential *Seybold Report on Professional Computing* calls it "a jack-of-all-trades which does none really well." Sales suffered initially because some of the Star's software was late in coming to market.

Office equipment analysts have started referring to PARC-style systems as "Lisa-like," not "Star-like." Apple's next computer, Macintosh, scheduled to ripen into a commercial product by the end of this year, could further identify Apple with PARC's ideas. The engineering manager for Macintosh came from PARC, where his last big project was a personal computer.

From this, Xerox might appear to have muffed the chance to make it big in personal computers with PARC's creations. Some Xeroids are sure the company could have been an early winner if only it had launched a less expensive Alto in the late 1970s. Unlike the Star, the Alto was an "open" computer, easy for outsiders to program. Independently written software has helped touch off the personal computer explosion, so the dissidents have a point. Because the Star is "closed," outsiders can't write programs for it.

To mourn the Alto, though, is to blame unfairly those who killed it. Xerox was out to produce office equipment, and no office equipment supplier, including IBM, foresaw that personal computers would compete with their wares. It was inconceivable that the cost of computer memory would decline 31 percent a year, as it has for the last five years, or that today's microcomputers would be as powerful as yesterday's mainframe computers. Xerox and its ilk concentrated not on freestanding personal computers but on clusters of workstations that share the use of computer hardware. That way, customers could spread high hardware costs across many workers. And suppliers could defray the costs of their prized sales forces with big-ticket orders.

Besides, Xerox had, and still has, ulterior motives in the office. Competition in the copier market keeps growing, and the company's chief aim has been to protect copier installations by strengthening its control of large, lucrative accounts. Companies that can sell complete office systems—workstations with reliable software, printers, and data-storage devices, all linked into a network—have a stronger lock on their customers than do suppliers of stand-alone equipment. Thus the Star, which works well only when hooked up with other Xerox gear, seemed to fit the company's strategy better than free-standing little computers would.

The complete-system approach, moreover, was more compatible with Xerox's expansive ways of thinking than the alternative of making piecemeal improvements on an individual machine like the Alto. Big companies often can't make the modest efforts needed to probe emerging markets. "It's a problem when you're getting your feet wet in a new business," says Jack Goldman, formerly Xerox's research chief. "In a large company, every product must be a home run to justify the costs of marketing and development."

That has been especially true at Xerox, which owes its existence to xerography, one of the longest homers on record. Top management "followed the big-bang strategy," says one veteran. "They wanted to build absolutely the best office system instead of taking things bit by bit." At PARC, the company's urge to build the best at the expense of the merely better, like an Alto, had its own name: biggerism.

Biggerism could pay off in some ways, to be sure. Xerox has big hopes for Ethernet, a PARC-invented network that uses a cable and translating devices to connect different types of office equipment. By souping up the performance of PARC's original version of Ethernet, Xerox drastically raised the cost of hooking up, to as much as $5,000 per connection. That move discouraged sales and deterred other equipment makers from adapting their machines to talk through Ethernet. But now, improved chip technology has sliced the cost of connecting by about two thirds. Over 70 office equipment makers are using Ethernet or plan to, including Apple. The temporary setback helped keep Ethernet from becoming *the* industry standard, but it is *a* standard. (The only other company likely to set a standard is IBM.)

Xerox still thinks PARC's work can produce some big hits. No one is more convinced than John Shoch, a remarkably hard-boiled former PARC researcher who became the company's office systems chief last October. His first priority is to expand the number of Xerox products that will communicate over Ethernet (20 do now, including laser printers and facsimile machines). Making a winner out of the Star will take more effort. Because the technology is old and the system tries to do so many things, the workstation seems expensive and inept in many functions, especially compared to Lisa.

Shoch wants to bring out a less costly version of the $15,000 Star, which he sees as one claw of a pincer's movement to narrow the Lisa's potential market share. The other claw, in his view, will be IBM's personal computer armed with a Lisa-like set of programs written by VisiCorp. Priced at some $7,000, that system won't compete directly with the Star but will be far cheaper than the Lisa. It will also tap into Ethernet—thanks to a helping hand from Xerox. Says Shoch: "There's going to be a squeeze between the lower priced Star and commodity-type computers that run better software. It'll be a tough place to compete."

The company's support of PARC has never wavered. This year's budget of $35 million or so will set a record. But changes have taken place. Last March, Xerox appointed a new director of PARC, William Spencer. A veteran of two decades at Bell Labs, Spencer admires AT&T's ability to transfer technology out of the lab by attaching satellite labs to major manufacturing plants. "PARC's main shortcoming," he feels, "has been a lack of management attention. We started things that didn't match what was going on in other parts of Xerox."

Spencer is trying to produce a better fit by meeting a couple of times a year at PARC with Xerox's division managers, some of whom haven't visited for years. Every three weeks or so he breakfasts with Shoch, and they've started a joint hiring program: some new researchers will spend their first year or so at PARC, then join the office systems group.

Time is on Spencer's side. Having taken its lumps in the office systems business, Xerox has a better fix on what kinds of products make sense. While Shoch's division still struggles to discover a successful way of selling office systems, PARC, having created much of the technology McColough sought, is stepping up its work on a new frontier: very large-scale integrated circuits used for everything from diagnosing copier breakdowns to connecting personal computers with mainframes. "The foundation for our future will be the next generation of chips," says Spencer, who originally came to PARC to set up a line for making them. "Office systems is a smaller part of our work now."

When a company wants to make it big in a new business, a solid base of technology is necessary. But it's hardly sufficient. Without a clear understanding of corporate strategy and pressure from a hungry marketing group, even the best technologists can get out of hand. The tricky part is to strike a balance between encouraging creativity and getting your money's worth.

Managing the Research Professional

Case IV–2
Duval Research Center

M. A. Maidique

"Duval has a dual ladder system," explained Jim Breitmaier, vice president of R&D, Duval Plastics Corporation, "however, it's not working well."

> Two years ago, we created the Fellow position because there was a feeling, particularly among the younger scientists, that our dual ladder didn't go far enough. But the technical people continue to complain that the managers have more status in the hierarchy than they do.
>
> Recently, I've begun to question the effectiveness of the dual ladder, at least our version of it. Can the dual ladder upgrade the status of the technical people? And, more importantly, is it a practical way to get scientists more involved in decision making? I have a hunch that our system can be significantly improved but I don't yet know what changes to make.

Harvard Business School case 9-680-046

DUVAL PLASTICS CORPORATION

Duval traced its origins to a chemical business established in 1900 by Alphonse Duval in Wilmington, Delaware. From the original chemical business, the company had branched out to plastics after World War II. Although by now Duval manufactured a wide variety of other products, the Duval Plastics Corporation is principally a specialty plastics manufacturer. Duval leadership in the specialty plastics industry is seen by management as strongly influenced by the firm's early commitment to search for better materials and more efficient ways of forming them into useful products. Jacques Duval, the company chairman, described the Duval team as committed to "making our science useful."

Research and development was a deeply rooted tradition at the Duval Plastics Corporation. The origin of Duval's R&D center was one of the first plastics research labs in the nation, established in 1918 by Thomas Murphy.

The commitment to research evidenced by Dr. Murphy's pioneering labor paid off. Over the decades following the First World War, the company had introduced a wide variety of new plastic materials. The company also pioneered

EXHIBIT 1 Sales and Net Income 1968–1977 (in millions)

	Sales	Net Income	R&D
1977	560	46	27
1976	512	42	24
1975	470	15	21
1974	525	24	19
1973	472	35	17
1972	357	27	NA
1971	302	17	
1970	304	22	
1969	270	27	
1968	240	24	

SOURCE: Duval Plastics Corporation, Annual Report, 1977.

in plastics manufacturing. The first automated plastics processes had been developed at Duval.

Research concepts had been successfully turned into sales and profits at the Duval Plastics Corporation. Net profits in 1977 reached $46 million on sales of $560 million while R&D expenditures increased to $27 million, a new record (see Exhibit 1). Less than 5 percent of these R&D expenditures were financed by the government.

The company's 30,000 products fall into three broad categories: *Consumer products* that include a wide variety of plastic tablewares and housewares for preparing, serving, and storing food and plastic components for other kitchen utensils; *electrical and electronic products* used widely in consumer, industrial, aerospace, military, and telecommunications equipment; *technical and other* products that include such wide applications as laboratory plastic equipment, clinical chemistry instruments, and specialized diagnostic testing systems. (See Exhibit 2 for a product breakdown by industry segment.)

THE PLASTICS RESEARCH CENTER

From its modest beginning 60 years earlier, Duval's Research Center had grown to a complex of several buildings located in Wilmington, Delaware, and surrounded by over 500 acres of meadowlands and hills. In late 1975, the center extended its activities to include a laboratory on the outskirts of Munich, Germany.

The Research Center's organization included 150 scientists and engineers with bachelor's and advanced degrees in 30 disciplines, including 62 Ph.D.s. Two hundred fifty technicians, secretaries, and other supporting personnel made up the remainder of the center's organization.

Several of the Plastics Research Center's Ph.D.s had gone on to senior management po-

EXHIBIT 2 Product Breakdown by Industry Segment* (in millions)

	Consumer Products	Consumer Durable Components	Capital Goods Components	Health and Science	General Corporate
Net sales	149	168	132	110	—
Income from operations	28	37	22	13	43*
Total assets	74	86	64	75	240†

* Corporate research and development projects which are designed to benefit a wide range of products and processes.

† Includes cash, short-term investments, and investments in associated companies.

sitions at Duval. The president of Duval, Dr. Jonathan Glasgow, had started out as a chemist in one of the Plastics Research Center's laboratories in 1960. Dr. Selden Loring, vice chairman of the board and director of the Research Division, had begun his career as a researcher in 1945. Dr. James Breitmaier, vice president R&D, had likewise begun his career as a research scientist in 1965.

Research, development, and engineering activities at the Plastics Research Center covers a very wide range. Five broad areas, however, account for most of Duval's R&D activities:

1. Materials sciences.
2. Physical sciences.
3. Life sciences.
4. Product development.
5. Process research and development.

To accomplish its broad range of R&D goals, Duval had over the years assembled a cadre of scientists with impressive credentials and diverse backgrounds. The Plastics Research Center includes scientists with degrees in analytical, physical, and organic chemistry; ceramics; metallurgy; and crystallography. There are also electrical and mechanical engineers, computer scientists and product designers, optical physicists and mathematicians. During the last decade, Duval has also recruited a team of biochemists, biomedical engineers, immunologists, and microbiologists.

Duval's management could point to several examples to demonstrate the wide freedom that Duval scientists enjoyed in their research pursuits. The curiosity and enterprising abilities of Duval's scientists had led the company far afield, from plastics materials research to fields such as biochemistry, immunology, and blood analysis.

ORGANIZATION

The Plastics Research Center is one of the three main organizational blocks of the Duval Plastics Corporation that report directly to the chairman, Jacques Duval. President Jonathan Glasgow is primarily responsible for domestic operations and has responsibility for most of the company's 45 manufacturing plants and its 15,000 employees. Vice Chairman Charles Duval, the chairman of the board's cousin, has responsibility for international operations. About one third of Duval's sales were made to customers outside the United States. Vice Chairman Selden Loring is responsible for the Research Division. (See organization chart in Exhibit 3.)

Neither of the Duvals were technical people. They had both, however, graduated in business from Stanford University. Thus, President Glasgow, a Ph.D. physical chemist, in conjunction with Selden Loring, exerted a great deal of influence over technical decisions.

Jon Glasgow explained his role:

> It would be difficult for me to play a key role in the company's technological decisions if I had not had the technical training that I have. At Duval, I headed up the physics and electronics lab before assuming general management responsibilities. My technical background is not absolutely essential but it does give me credibility with the technical people. But my real role is that of a "creative intermediary" who brings together the technology and the market opportunity.
>
> As a minimum, the manager of a technology firm must have enough highly qualified technical people around him so that he can have access to sound judgments regarding technology. The key to this access is a management style that allows the right communications to come up through the organization to the president's office.

Some of these technical inputs had been provided by many years at Duval by Dr. Selden Loring, who had distinguished himself in the company by his technical work as well as by building an outstanding research team. While Selden Loring directed Duval's technical efforts, he also took care to groom a successor. Loring had decided many years before that Jim Breitmaier would succeed him.

EXHIBIT 3 Duval Plastics Corporation Organization

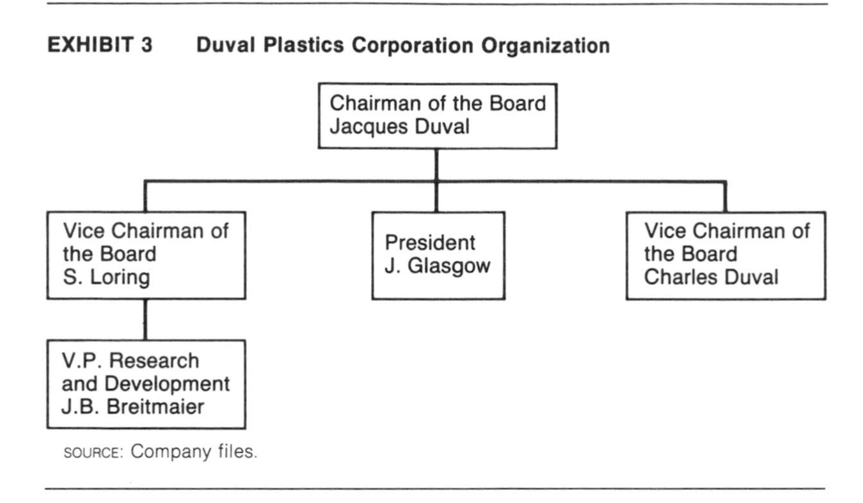

SOURCE: Company files.

Dr. James Breitmaier was now responsible for all of the line operations of the Plastics Research Corporation, including the European lab. Reporting to Dr. Breitmaier were four line, director-level departments:

1. Research.
2. Development (process and product).
3. Bubble memories.
4. European R&D.

and two staff departments:

1. Planning.
2. Administrative and technical services.

The Dual Career Ladder

The first rung of a parallel career ladder for scientists and engineers, the Senior Associate, had existed for 20 years at Duval (Exhibit 4).[1] However, over the last few years a feeling had developed, particularly among the younger scientists, that the ladder didn't go far enough. Partially in response to this sentiment Dr. Breitmaier had established the Fellow position in 1977. "I guess you could call me the 'champion' of the Fellow concept," said Breitmaier.

"The first two Fellows, Dave Marein and Bernie Dante, were chosen without formal guidelines," Breitmaier explained.

> They are outstanding scientists, however, and certainly no one would question their appoint-

EXHIBIT 4 Parallel Ladder Structure

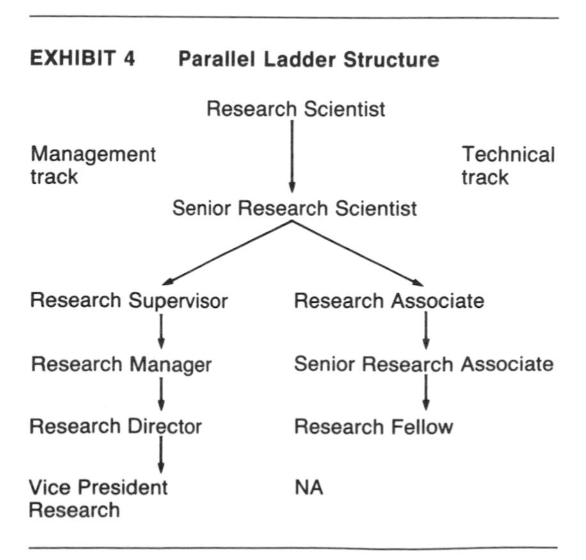

[1] The parallel ladder is a dual hierarchy system, widely employed by U.S. technology-based systems since the early 1960s, which aimed at improving the status and the compensation levels of distinguished scientists and engineers.

ments. But in the future we must be more careful, more systematic about the appointments. That's why we have prepared a document to establish ground rules for future appointments to the Research Associate and Fellow positions (Exhibit 5).

I think that the creation of the Research Fellow position has been a positive step. However, there remain a lot of unanswered questions about the parallel ladder concept and how it should be implemented. I've still to resolve a number of difficult issues.

Two years ago, for example, research, development, and engineering were one organization.[2] Thus the issue of whether there should be an Engineering Fellow, or a Development Fellow, wouldn't have come up. We are now discussing whether we shouldn't create such positions (see Exhibit 6 for proposal). There are clearly pros and cons. We want to stimulate first-class technical work throughout the company. But will this dilute the significance of appointment to a Fellow position?

Then there is the question of how many Fellows and Associates to appoint. Right now we have 2 Fellows, 8 Senior Associates, and 15 Associates out of a pool of 300 technicians and scientists. But technical people point out that there are over twice as many managers, 20, as there are Senior Associates, the equivalent ladder position. With more opportunities available, this signals to some that we are still emphasizing managers.

[2] Now most of the company's engineering and development is done at the operating divisions.

EXHIBIT 5 Research Associate/Research Fellow Position Definition

Objective

To provide for the professional career advancement of those members of the Plastics Research Center who have made significant research contributions and whose talents and skills lie particularly in the scientific and technological fields.

Background

Duval has established a series of research positions and titles for senior members of the technical staff, ranging from the Associate level to the post of Fellow. This memorandum outlines the procedures, qualifications, and guidelines to be used in considering candidates for promotion to the Associate position or higher in this series.

Procedures

Proposals for promotion of a candidate will be prepared by the immediate supervisor and presented to the Research Center Staff, who will make a recommendation to J. B. Breitmaier for final decision. The supervisor should present a written proposal two weeks in advance of meeting with the staff and should expect to present and discuss the proposal personally.

Qualifications

Advancement to Associate and to higher posts will depend on identification and recognition of scientific and technical stature of the candidate and contributions made. These qualities should be considered from the viewpoint of the company as well as from the viewpoint of the scientific community worldwide.

Strict quantitative measures are not adequate and can only serve as guidelines together with qualitative factors in considering the merits of a proposal. Some suggested guidelines and factors are included below. A candidate will seldom meet all of the qualifications fully, but should meet most.

Business Contribution

- Should be identified as key contributor to inventions and/or developments which are important to the company business.
  - At the *Associate* level, these contributions may not yet be reflected in significant new sales volume.
  - At the *Senior Associate* level, the candidate will usually be associated with significant sales volume contribution and with several invention/development areas.

EXHIBIT 5 (*concluded*)

- At the *Fellow* level, the candidate will usually be identified as introducing an important discontinuity in the flow of corporate technology, resulting in the opening of new business(es).

■ In considering business contribution qualifications, it will be useful to judge:

- How far away from traditional Duval technology has the candidate taken us successfully?
- What degree of professional risk was assumed?
- How *consistently* has the candidate demonstrated ability to originate ideas and solutions over time?
- What degree of autonomy has been accorded the candidate in the choice of program?

Technical Stature

The *Associate* candidate will usually have several patent disclosures, may have one or more issued patents and a few published technical papers of recognized quality.

A Senior Associate is usually recognized by the organization as an expert in a limited field and is consulted by peers and by marketing specialists on occasion.

The *Fellow* candidate will normally be widely known in the scientific community, with more than 20 issued patents, of which several are in commercial use.

A Fellow is recognized as the corporate expert in his field, and is consulted by peers, by marketing, and by management. He is looked to as setting a standard of technical excellence for the corporation.

In considering the technical stature qualifications, it may be useful to list:

■ Publications and their significance.
■ Patents and their impact.
■ Professional society prizes/awards.
■ Editorial activities—professional journals/books.
■ Invited-speaker role at professional meetings.
■ NSF reviewer role.

3-10-79.
SOURCE: Company files.

EXHIBIT 6 Engineering Fellow/Associate Proposal

TO: Research Center department heads
FROM: J. B. Breitmaier and A. B. Johnson*
DATE: December 15, 1978
SUBJ: Engineering Fellow/Associate Advisory Committee

Attached is the proposal for the procedure and qualifications for the positions of Engineering Fellow, Senior Engineering Associate, and Engineering Associate. This proposal has been approved by us and the appropriate people in the Personnel organization and now needs to be made ongoing.

We would like each of you to be a member of the first Advisory Committee for Engineering Fellow/Associate. We have asked Mike Cassandino† to be the chairman and he will be contacting you regarding your first meeting.

We feel this is a definite step forward in recognizing and rewarding exceptional engineering talents as well as more closely tying together the Research Center to the operating divisions.

* Vice President, Engineering Division.
† Director, Personnel Department.
SOURCE: Company files.

EXHIBIT 6 (*concluded*)

Proposal

To formalize between the Plastics Research Center and the Operating Divisions the process for selection of Engineering Associate and Senior Engineering Associate, and to recognize a new level of corporate engineering contributions through the new position of Engineering Fellow.

System's Objective

To recognize and reward those individuals employed by Duval who have made significant-measurable-outstanding engineering contributions to the company.

Qualifications

These positions are considered a reward for past accomplishments rather than anticipated future behavior. The guidelines are as follows:

	Associate		*Fellow*			
Qualification	A1	A2	F1	F2	F3	F4
Contribution to D.P.C. profitability (in millions of dollars)	1–2	2–10	10–30	30–60	60–100	>100
Contributions (multiple) recognized outstanding in his field by:						
Division	Yes	Yes	Yes	Yes	Yes	Yes
Corporation (WW)	No	Yes	Yes	Yes	Yes	Yes
USA	No	No	Yes	Yes	Yes	Yes
World	No	No	No	No	Yes	Yes
Recognized as a consultant by:						
Peers	Yes	Yes	Yes	Yes	Yes	Yes
Division	Yes	Yes	Yes	Yes	Yes	Yes
Corporation (WW)	No	Yes	Yes	Yes	Yes	Yes
Recipient of individual outstanding contributor award or equivalent	0	1	2	3	4	5
Number of patents/awards	0–5	5–10	10–15	15–20	20–25	>25
Performance rating last three years	EX/OS	EX/OS	EX/OS	EX/OS	EX/OS	EX/OS

Qualifier

1. Should not be in the "mainstream" of management, but instead have a specific technical orientation.
2. Individual can have people reporting to him.
3. Contributions can be accumulative and it would be possible to move progressively higher in group number as additional contributions are made.
4. The first three guideline factors will represent 80 percent of the qualifying needs.
5. Since all of these individuals contribute more towards future growth than short-term profitability, we did not feel additional compensation on a regular basis would be desirable. However, they should be considered for such rewards as stock options.

Approval Process

1. All recommendations for these positions should be screened first by the Advisory Committee consisting of three Engineering Division Directors, two Plastics Research Center Directors, and one Fellow.
2. At least 50 percent of the Advisory Committee should be rotated every three years.
3. The Advisory Committee should meet formally three times a year with candidates for all levels proposed by the appropriate director (sponsor).
4. All recommendations from the Advisory Committee must be unanimous and will need the final approval of Dr. Breitmaier and Mr. Johnson.

But the crux of the problem is improving the quality of our technological decisions, finding better ways to integrate decision making and technology.

Although I enjoy technical work, I generally don't make technical decisions myself. I don't get involved in, accept, or reject decisions on projects that have less than a $5,000 sales potential, and 20 percent of the programs we're involved in are below this level.

The burden of decision making falls on the six department directors and their supporting managers. But I try to stay abreast of what is going on technically. I have often said to Fellows and Associates, "Drop in and talk whenever you want." And quite a few do come in.

It's very important for us to maintain the highest quality technical communications. In the early stages of a project development, using traditional business tools such as IRR and ROA are not very useful. It's really technical judgment that counts. You can't analyze the project in numerical terms until it gets into the very latest stages of development. (Duval had defined five new project development stages.)

Stage	Objective
1. Basic research	Basic knowledge
2. Applied research	Technical feasibility
3. Exploratory development	Commercial feasibility
4. Scheduled development	Production and initial sales
5. Commercial development	Profit

At present, technical requests are evaluated by our technical directors in conjunction with managers from the line divisions. Perhaps some of the dual ladder people should be automatically included in these committees.

INTERVIEWS WITH TECHNICAL STAFF'S DIVISION SCIENTISTS

Several scientists on the technical ladder and younger aspirants were questioned by the casewriter regarding their attitudes towards the Duval ladder system.

Bernie Dante, Research Fellow, Ph.D. physical chemistry, MIT, 25 years at Duval, 18 patents, 27 publications.

I've been a Research Fellow since the program began in 1977, along with Dave Marein. Earlier I had been the manager of a small chemistry department. Now I have less administrative responsibility; otherwise, I do the same kind of work.

The Fellow position was created in part as a response to the criticisms of younger scientists. They argued that the dual ladder didn't go far enough. The ladder now, for the Fellow position, has no salary limit.

But people still complain that it is tough to get ahead in the technical ladder. There are 8 Senior Research Associates, and this may be too few compared to the over 20 managers.

However, when you get there the dual ladder positions are fully equivalent. And I haven't heard anyone else say otherwise. Of course this doesn't mean that the technical people control the management decisions.

In principle, the dual ladder system is modeled on the achievements of Dr. W. George. Dr. W. George, who started out as a research scientist, became the "Dean of the Fellows" at Duval. He has one of the most impressive technical records in the history of the company. Altogether he has about 40 patents. He was a research director until he retired a couple of years ago.

Patents, in part for historical reasons, are an important part of the Duval corporate strategy. For many decades, Duval has thrived on a number of unusual material and process innovations.

Recently David Marein, the only other person thus far appointed to the Fellow position, coinvented bubble memories. (Marein and a Bell Labs scientist are considered the coinventors of bubble memories.) This was a major development, though quite a bit outside of our main plastics business, where I have done most of my own work.

Every year, Duval is granted about 50 patents. We generally follow a policy of patenting first, to assure protection, and then publishing. Sometimes, however, to get the benefit of outside reaction, we will publish before a patent issues.

It's tough, however, to evaluate the value of patents. Material patents, for instance, are easier to defend. In general, it's very difficult to evaluate the potential of an R&D idea. It's far easier to reward product development and sales.

I'm delighted to have been selected a Research Fellow, but in a sense I am not absolutely clear about the need for the position. To me the achievements that lead to the position are a great reward in themselves.

On the other hand, maybe I'm too influenced by the old times when any of us could drop in and talk to Glasgow and Loring. Then the labs used to be next to the administrative building. Now we're at the Center, 15 miles away. I don't know how the younger guys feel about this, that is, the ones that don't know the top management people personally.

Charlie Lucas, Senior Research Associate, Ph.D., Cornell, physics, 17 years at Duval, 40 papers, 9 patents.

The second ladder is a bunch of nonsense. I don't care what they call me even if it's "Lord King of Research." Let me point out at the outset that at least half of my peers would disagree with my views on this.

The key issue to me is who tells you what to do. I have no more say in what goes on now than before. I'm likely to report to someone 10 or 12 years my junior. I control no funds, no people. I can only sign for $100. But, on the other hand, I see no way that this could be changed. That is, no way short of anarchy.

There is a lot of pressure around here to be a manager. It's a very prestigious and powerful position. It's the goal of many of our research people. Some of them don't really belong in research. They are interested in management, not science. They'd just as soon manage an underwear factory.

Yet often they make technical decisions. But when a project fails, doesn't make a buck, the failure trickles back to us and we bench scientists shoulder the criticism.

But don't get me wrong. Being a manager is difficult. For me it would be almost impossible. I don't like to tell others what to do. I have an inferiority complex about ordering others around.

Becoming a Research Fellow is to a certain extent a matter of luck. You can't judge a guy's technical ability, so what do you do? You make the appointment depend on the commercial success his developments have had.

In the physics group, for instance, there is a small chance of a big commercialization success so there is a small chance of becoming a Fellow. It helps, however, if you're working on one of the big projects (projected ultimate market of $5 million or more). Then you have the best access to funds, space, people, and any other resources you might need. But a lot of exploratory research still winds up being aborted.

It's the old, paradoxical tension between basic research and commercialization. If you're Bell Labs—and everyone aspires to be Bell Labs—you presumably can ignore it. Duval can't. We are not Bell Labs.

Yet no one would disagree with the appointments to the Fellow level that have thus been made. Most people believe that more should be made.

There should be a Senior Fellow position but there is no one at this point to appoint at that level. Only Dr. George, who is now retired, would have qualified. He was responsible for more important inventions than anyone else in the company. He has over 40 patents.

But why worry so much about the dual ladder? All that people should be told is, "Look, there are two ways to go, management or individual contributor. And you should think very hard about this."

Bill Osell, Research Scientist, Ph.D. immunogenetics, Columbia, 30 years old, three years at Duval.

I've had a unique opportunity here at Duval to set up a new program in immunology. The funding has been fantastic. Basically they allowed me to do my own thing. The support for my program, in general, has been excellent. This is very important to all scientists.

In the long term, I have a choice between climbing the technical or management ladder. The question is how long will I be able to retain my present productivity, that is, my output in terms of publications and patents. To advance in the Duval parallel ladder, you need to develop an international reputation. Several of our scientists, for instance, are members of the National Academy of Sciences. But getting to the top of the ladder has its rewards. Bernie Dante, for instance, might make more than the president of Duval. His job is also very, very secure.

On the other hand, promotions in the scientific end are basically conservative, that is, progress up the ladder is slow. We are top heavy in managers.

But nonetheless, management is a real alternative for me. I like working with people and organizing. If I run out of technical ideas, I'll give it a try.

Robert Richards, Senior Research Associate, chemist, M.S. protein chemistry, RPI, Ph.D. work (complete except for thesis), Stanford University, 28 patents, 32 published papers, 14 years at Duval, Fundamental Life Sciences Department. Present projects: waste conversion, tissue culture fermentation.

The parallel ladder is a great concept, that is, it has the potential to be a great system. But to work, it has to be an equal partnership between scientists and management.

We are still far from this ideal at Duval. But things are better than they were a few years ago. At one time, the parallel ladder was simply a dumping ground for scientists that had failed as managers. This is no longer the case. But management and economics still dominate the technical people.

Ideally, the managers should make the budget, planning, and personnel decisions while the scientists make the decisions on the technical programs. In short, there should be an equal partnership. This is the only way the positions on the ladder could be fully equivalent.

Right now they aren't. Take compensation, for instance. Equivalent positions on the ladder have equivalent salary ranges. But it takes much longer in elapsed time for a technical person to get to the equivalent managerial level. You're penalized for being on the technical side.

The differences manifest themselves in other ways, too. The managers have larger, better furnished and located offices, and easier access to clerical and secretarial assistance.[3] What all this adds up to is that the "equivalent" management positions have more status and recognition in the Duval community.

Managers can also rise higher in the hierarchy than technical people. The highest position I can aspire to on the technical track is a Research Fellow, which is rated on the ladder as a director-level equivalent. However, there is no equivalent Senior Research Fellow that could be, at least theoretically, equivalent to a corporate vice president. But I recognize that it has to stop there. There can only be one president. But why shouldn't he have one or more VP-level technical advisors?

Duval has actually gone backwards in this regard. The Fellows used to report to our vice president, Jim Breitmaier, but now they report to the director of research. It's been like that, one step forward, a half-step backward.

But maybe the most significant problem is that the judgment of technical people is not yet weighed sufficiently. If you're in a meeting, it's always the management people that have the final say, even on technical issues. It's simply not an equal partnership.

But the problems, I believe, are solvable. Our present system can be improved. One thing that we have going for us is that Jim (Breitmaier) listens and people are willing to speak their mind to him. I am basically optimistic that future changes will be for the better.

CONCLUSION

Jim Breitmaier had recently done a good deal of thinking about the Duval dual career structure. He was intimately familiar with the prevailing attitudes and reservations regarding the Duval dual ladder system at the Plastics Research Center. He thought some changes could be constructive but he hadn't yet decided what these might be and when they should be implemented. Jim had recently obtained access to

[3] Parking for the R&D Center was on a first-come, first-serve basis from technicians to James Breitmaier.

an excerpt from a consultant's report that discussed the recent literature on dual career ladders (Appendix A). He wondered how relevant these academic findings might be to his own operation.

APPENDIX A
DUVAL RESEARCH CENTER

PARALLEL LADDERS: THE MODERN VIEW

To the modern researcher the parallel ladder, in MIT's Ed Roberts's words, "is an oversimplified solution to a complex problem." Thus the focus of management research has in the last decade shifted to the study of the overall problem of the organizational implications of technical careers.

The groundwork for understanding the dual ladder dilemma is laid out in Schoner and Harrell's study.[1] Schoner and Harrell surveyed the attitudes of 100 engineers and managers in an electronics company that had implemented the dual ladder concept. They designed nine questions that measured morale, attitudes towards the dual ladder, and attitudes towards recognition. Only in two questions are the differences in response between the technical and management groups statistically significant:

1. The technical personnel—much more so than the management group—felt that they were underpaid.
2. A significant minority of technical personnel were dissatisfied with being on the technical ladder.

However, surprisingly, morale was high *throughout the firm*. There were no significant morale differences between the managerial and the technical groups. This data led the authors to conclude that:

> High morale among technical personnel does not *necessarily* depend on their having equal prestige in the company, or on being paid on the same scale (author's emphasis),

for technical people,

> tend to look to their professional colleagues (which may be outside the firm) for recognition.

Thus to Schoner and Harrell the paradoxical finding of universally high morale in a situation in which the dual ladder had failed to confer equal prestige on the managerial and technological groups,

> suggests that the dual ladder policy is based on a misconception of what engineers and scientists really want from their jobs.

However, despite these caveats, the dual ladder has proliferated. One survey of 22 technology-based companies found that 75 percent had "some kind of parallel ladder system,"[2] while another study of 10 similar firms found that 70 percent had "certain dual hierarchies."[3] The widespread character of the dual ladder combined with the contradictions pointed out by the Schoner and Harrell study highlight the need for a more sophisticated examination of the dual ladder dilemma.

Lotte Baylyn, in a continuing study of MIT graduates, attempts such an analysis by segmenting the mid-career technically trained professional into three groups:[4]

---

[1] B. Schoner and T. W. Harrell, "The Questionable Dual Ladder," *Personnel,* January/February 1973.

[2] Stanford University, *Motivation of Scientists and Engineers,* Stanford Graduate School of Business, Stanford University, 1959, pp. 14–16.

[3] John W. Riegel, *Administration of Salaries and Intangible Rewards for Engineers and Scientists* (Ann Arbor: Bureau of Industrial Relations, University of Michigan, 1958), p. 23.

[4] Lotte Baylyn, "An Analysis of Mid-Career Issues and Their Organizational Implications," to appear in J. E. Paap (ed.), *New Dimensions in Management of Human Resources* (London: Prentice-Hall International).

1. Technically oriented.
2. People oriented.
3. Nonwork oriented.

These groups are then further subdivided into "high" and "low" (or ordinary) organization potential. This results in the six cells given in Chart I. According to Professor Baylyn, each of these cells has its own characteristics and must be considered individually. Baylyn notes that "most organizations are geared to cell 3, to people-oriented employees who will rise to top positions in the company." The difficulties that dual ladder systems face are clarified by the Baylyn model:

> The trouble with the technical ladder is that it has so often been used for cell 4 employees that it has lost its value for those in cell 1 for whom it is really intended. In other words, the technical ladder has been implemented in most companies in such a way that it has not differentiated between high and low potential employees. Companies must find a place for the "plateaued" manager—the people-oriented employees who will not make it at the top; they must also find ways of allowing their potential technical people to expand in influence and express their particular competences. But the same mechanism cannot easily serve both these needs, and the technical ladder has too often been asked to do just that.[5]

The dual ladder decision is an important one for a technically based company. But it is subtle enough to demand a careful analysis of precisely what group it is that is to be rewarded. And it must not be allowed to become "a 'booby prize' for those that fail to make the more prestigious administrative ladder," since for most engineers "success seems to consist in winning a place in management."[6]

But perhaps most important for high-level technical contributors—dual ladder or not—is that they in Baylyn's words, "must not be isolated from the decision-making top management group." A corollary to this idea is that top

[5] Ibid.

[6] Leonard Sayles and G. Strauss, *Managing Human Resources,* p. 386.

CHART I Organizational Roles for Technically Trained Personnel at Mid-Career

Orientation at Mid-Career		Organizational Evaluation of Potential: High	Organizational Evaluation of Potential: Low ("ordinary")
	Technical	Independent contributor Policy specialist "Idea innovator" "Internal entrepreneur" (1)	Technical support Expert on "formatted" tasks (2)
	People	Top management Sponsor Development as policy (3)	Mentor Individual development functions (4)
	Nonwork	Specialist Internal consultant (5)	(6)

managements that pursue this policy—quite apart from the motivational benefit—will doubtlessly be better able to make sound, well-informed technological decisions.

Reading IV–2 Variations of Individual Productivity in Research Laboratories (excerpt)

William Shockley

Summary

It is well known that some workers in scientific research laboratories are enormously more creative than others. If the number of scientific publications is used as a measure of productivity, it is found that some individuals create new science at a rate at least 50 times greater than others. Thus differences in rates of scientific production are much bigger than differences in the rates of performing simpler acts, such as the rate of running the mile, or the number of words a man can speak per minute.

On the basis of statistical studies of rates of publication, it is found that it is more appropriate to consider not simply the rate of publication but its logarithm. The logarithm appears to have a normal distribution over the population of typical research laboratories. The existence of a "log-normal distribution" suggests that the logarithm of the rate of production is a manifestation of some fairly fundamental mental attribute. The great variation in rate of production from one individual to another can be explained on the basis of simplified models of the mental processes concerned. The common feature in the models is that a large number of factors are involved so that small changes in each, all in the same direction, may result in a very large change in output. For example, the number of ideas a scientist can bring into awareness at one time may control his ability to make an invention and his rate of invention may increase very rapidly with this number.

A study of the relationship of salary to productivity shows that rewards do not keep pace with increasing production. To win a 10 percent raise, a research worker must increase his output between 30 and 50 percent. This fact may account for the difficulty of obtaining efficient operation in many government laboratories in which top pay is low compared to industry, with the result that very few highly creative individuals are retained.

I. INTRODUCTION

Everyone who has been associated with scientific research knows that between one research worker and another there are very large differences in the rate of production of new scientific material. Scientific productivity is difficult to study quantitatively, however, and relatively little has been established about its statistics. In this article, the measure of scientific production I have used is the number of publications that an individual has made.

The use of the number of publications as a measure of production requires some justification. Most scientists know individuals who publish large numbers of trivial findings as rapidly as possible. Conversely, a few outstanding contributors publish very little. The existence of such wide variations tends to raise a doubt about the appropriateness of quantity of publication as a measure of true scientific productivity. Actually, studies quoted below demonstrate a surprisingly close correlation between quantity of scientific production and the achievement of eminence as a contributor to the scientific field.

The relationship between quantity of production and scientific recognition has been studied recently by Dennis, who considered a number of scientists who have been recognized as outstanding.[1] As a criterion of eminence for American scientists, he has used election to the National Academy of Sciences; his study is based on 71 members of the National Academy of Sciences who lived to an age of 70 or greater and whose biographies are contained in the Biographical Memoirs of the Academy. He finds that all of these people have been substantial contributors to literature with the range of publications extending from 768 to 27, the median value being 145. (Based on a productive life of approximately 30 years, this corresponds to an average rate of publication of about 5 per year, a number to which I shall refer in later parts of this discussion.) Dennis concludes that relatively high numbers of publications are characteristic of members of the National Academy of Sciences. He conjectures that of those who have achieved the lesser eminence of being listed in American Men of Science, only about 10 percent will have a publication record exceeding the 27 which represents the minimum publisher of the 71 listed in Biographical Memoirs of the National Academy of Sciences. He has also studied eminent European scientists and comes to essentially the same conclusion. In fact, his study goes further and shows that almost without exception heavy scientific publishers have also achieved eminence by being listed in the *Encyclopedia Britannica* or in histories of important developments of the sciences to which they contributed.

It should be remarked that in Dennis's work, he includes more routine types of contributions (such as popular articles) than are generally associated with scientific eminence. However, it may still be appropriate to quote a few of the statistics obtained by Dennis for people who certainly classify in the genius class of the scientific publishers. Among these Dennis refers to: Pasteur with 172 publications, Faraday with 161, Poisson with 158, Agassiz with 133, Gay-Lussac with 134, Gauss with 123, Kelvin with 114, Maxwell with 90, Joule with 89, Davy with 86, Helmholtz with 86, Lyell with 76, Hamilton with 71, Darwin with 61, and Riemann with 19. Riemann, who was the least productive, died at the age of 40. At his rate of publication, he would probably have contributed at least another 10 or 20 publications had he lived to the age of 70. Even with 19, he was in the top 25 percent of the 19th-century scientists referred to in Dennis's study.

The chief conclusion reached in this article is that in any large and reasonably homogeneous laboratory, such as, for example, the Los Alamos Scientific Laboratory and the research staff of the Brookhaven National Laboratory, which are included in this study, there are great variations in the output of publication between one individual and another. The most straightforward way to study these variations is to list the number of individuals with zero, one, two, etc., numbers of publications in the period studied. This compilation may then be plotted as a distribution graph (see Figure 2 for an example). In some cases, however, the data are too meager for a smooth trend to be seen easily and another form of presenting the data is more convenient.

The form used for most of the data presented in this paper is the *cumulative distribution graph*.

Such a graph can be illustrated in terms of the distribution of the height of a regiment of men. If the men are lined up in order of increasing height at a uniform spacing, then, as shown in Figure 1(*a*), there will be a steady increase in height from the shortest man to the tallest man. There will usually be a few men who are exceptionally short, a few men who are exceptionally tall. For the majority of the men, the height will vary relatively uniformly along the line of the men. In general, one

---

[1] Wayne Dennis, "Bibliography of Eminent Scientists," *Science Monthly* 79 (September 1954), pp. 180–183.

should thus expect an S-shaped curve with an inflection point near the middle of the distribution.

Such a curve is closely related to the distribution in height shown in Figure 1(*b*), which represents the number of men whose height lies in any particular interval of height. This can be obtained from Figure 1(*a*), as is represented there, by drawing two lines bracketing a certain interval in height and counting the number of men lying in this range. Figure 1(*b*) represents a smooth curve drawn through such a distribution. It can, in fact, be obtained from Figure 1(*a*) by drawing a smooth curve through the distribution in height and differentiating the number of men as a function of the height.

For many natural phenomena and in particular for those in which the measured quantity varies due to the additive effects of a large number of independently varying factors of comparable importance, a Gaussian or normal distribution, like that of Figure 1(*b*), is obtained. Conversely, if distribution is normal, then the cumulative distribution graph will have the symmetrical S-shaped characteristic in Figure 1(*a*), the middle flat portion corresponding to large numbers of cases in the central range, and the rapid convergence of the extremes to their asymptotes corresponding to

FIGURE 1

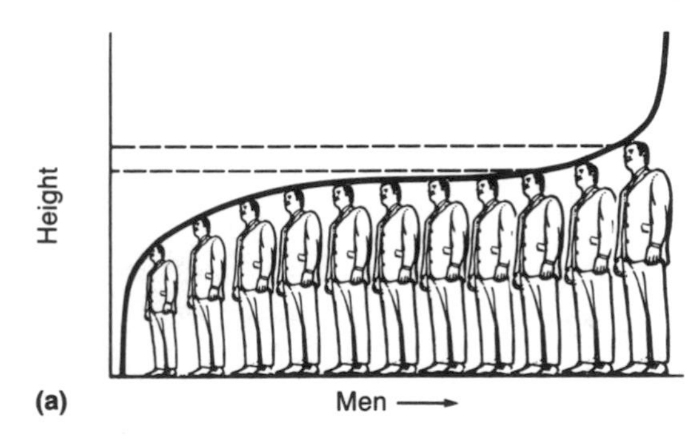

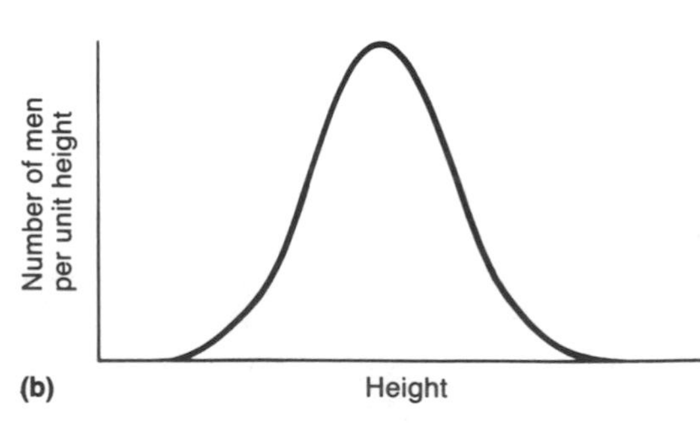

The cumulative-distribution graph and the normal-distribution curve. (*a*) The cumulative-distribution graph represented by men arranged in order of height at uniform spacing. (*b*) A "smoothed" distribution curve, of normal form, such as might be obtained from (*a*) by finding the number of men in each small increment of height.

FIGURE 2

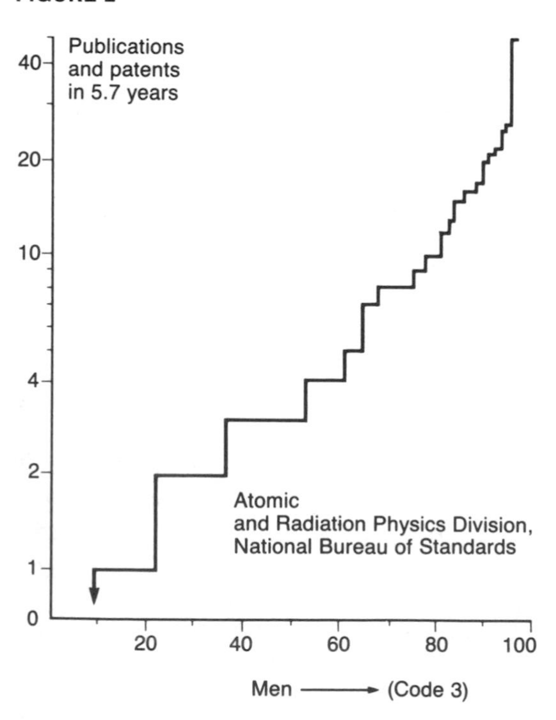

Cumulative distribution on logarithmic scale for publications and patents for Atomic and Radiation Physics Division, National Bureau of Standards, for a period of 5.7 years.

the scarcity of cases which deviate much from the mean value.

One of the new results of this study is that the data on rates of publication can be well represented by a normal distribution (Figure 2).

Transfer from Research to Development

Case IV–3 *The Transfer of Technology from Research to Development*

H. Cohen, S. Keller, and D. Streeter

In this paper, we will discuss some observations we have made in our own laboratories on the transfer of technology from research to development. We have tried to assemble "data" on transfers or attempted transfers that have occurred over the past 15 years. We have inspected these findings to see whether some common features could be recognized.

IBM's Research Division is a separate division of the company, independent of product groups, and reporting directly to the chief operating executives in the corporate office. The division is not, however, a staff advisory group but is charged with two major functions: to contribute to the technologies required for the product line by supporting current technologies and by finding new alternatives; and to contribute to those fields of science which underlie present product technologies and which may provide future ones. Product development is carried out in the laboratories of the groups' development divisions. There are 27 of these throughout the world. Thus our Research Division, with its three laboratories in Yorktown Heights, San Jose, and Zurich, faces a development community about 10 times our size, well-spread geographically, and covering a very large range of technical areas. In only one case, at San Jose, are research and development laboratories located at the same site.

Since completing the data taking and analysis reported here, we have used some of the notions to help guide research managers and project leaders with new transfers as they have come about. In doing this, we have begun to perceive other aspects that we hope will bear further generalizations, perhaps in later reports. In addition, we have added to our divisional staff a full-time marketing representative as program manager for technology transfer, and a full-time cost estimator. The former serves to make the corporate and divisional marketing representatives knowledgeable about our work and to bring their requirements to us. The latter helps us prepare our case with our development colleagues.

METHODOLOGY

First, let us set out some terminology. Transfers will be called successful if the technology has moved from research to a development laboratory and then has become a product or a part of a product or an important enhancement of a production process. A nonsuccessful transfer will be one in which the technology has left research but has not appeared as a product. A nontransfer refers to research projects that were intended for transfer but were never accepted in development.

We began our study with two parallel steps: first, we wrote down the "well-known" lore in the company having to do with the transfer problem; and second, we examined a long list

Research Management vol 22, No. 3, May 1979, pp. 11–17. Reprinted with permission.

of all of the projects in the laboratories over a 15-year period that we felt were intended to be transferred (remember that part of our mission has always been to work in science as well as in applied projects).

The prejudices about what was required for a good transfer were collected from research and development managers and staff members who had been involved in transfers. Here are some of them:

1. There must be an advanced technology group in the receiving organization to enable transfer to take place. (Advanced technology is the term used in the company for an advanced effort in the development laboratory not directly supporting a currently planned product but aimed at follow-on or replacement products.)
2. Advanced technology competes with research, often blocking transfer.
3. Transfer occurs when "outsiders" recognize the value of the technology. These outsiders may be external to the company, or they may be internal users of the technology but not prospective developers.
4. The external marketplace can play an effective role in pressuring a transfer to take place.
5. Once a project has transferred, it is useful to maintain some level of work in research, overlapping and complementing the newly initiated work in the division.
6. There should be joint participation, by research and development, when the project is still in research. The transfer can be most easily facilitated with the transfer of people from the joint program to the receiving organization.
7. Physical proximity of research to the receiving organization is important.

We examined the long list of projects and divided it into new lists of successful, nonsuccessful, and nontransfers. From these, we chose projects that represented the functional areas of our research programs: logic and memory, storage, input/output and communications, systems and programming, and computer applications. We actually reviewed only 18 projects so that we can certainly make no claims to completeness even within our own laboratories. A description of each of the case histories is given in the Appendix. Furthermore, we conducted our examination in an "anthropological" mode of observation and discussion through interviews without trying to quantify results. We have, therefore, arrived at some views and suggestions and not at hard-and-fast conclusions or rules of conduct.

From the original prejudices or "lore" noted above, we produced an interviewing guide and then interviewed more than 50 people who were involved in the transfers. Most, but not all, of these are or were in the research division so that we have more a research view than a general one. Case histories were prepared for each of the projects. The case histories and key factor suggestions were reviewed, testing the original prejudices, sometimes replacing or confirming them, sometimes adding new factors of interest. We finally replaced the original list with a new set of factors or ingredients of a transfer to which we gave an ordering of importance. In the following two sections, we will discuss these technology transfer factors in order of relative importance. In the discussion, we will point out examples from specific projects to help explain the factors.

Primary Factors

Technical understanding

It is essential that research understand the main technical issues of the technology before passing it on. This may seem obvious. However, in some cases we studied this did not seem to be true, and this is why we believe the technical base for each project must be considered carefully. In the germanium project, while the materials and processing problems were understood, the limitations in the advantages of germanium over silicon only became apparent after several years of research activity. These had less to do with devices that were created and more to do with device implementation in packaging and circuitry.

In addition, the target of achieving a very high-speed device, a rather restricted goal, could also be achieved with silicon whose development was continuing to make progress. Another example can be drawn from the beam addressable file project. There were problems in obtaining the laser arrays for addressing and problems in obtaining a material with the desired properties. At the same time, research tried very early in the game to obtain development assistance in the program. It turned out that this was premature. In addition, we had not successfully evaluated the benefits of this technology over what was available in the conventional magnetic recording. Since we didn't have the technology in hand (lack of addressing arrays or appropriate materials) and since we had not fully assessed the advantages over existing technology and the latter's limitations or lack of them, the project was destined to be unsuccessful insofar as the transfer process was concerned.

When a research project is aimed at transferring to development toward product status, it is important to understand where it will fit in the product line and what requirements must be met to reach that fit. While research cannot do its own marketing, it cannot waste time solving problems that don't exist or producing technology that cannot be sold by IBM. A basic ingredient of a technology is its cost. This can be considered at least in a preliminary estimate fashion for hardware but is, obviously, very difficult for programming. Both of the large systems projects we referred to failed in this respect.

Fortunately, for devices, circuits, and other hardware, the research work itself requires that at least one possible means of manufacturing be exhibited. Alternatives and improvements can be left for development and manufacturing engineering. For software, especially systems programming, "manufacture" or implementation in development is not well understood. Research results in software have not seemed to have directly affected eventual implementation methods.

Feasibility

Several projects never demonstrated the feasibility of research concepts because the time pressures forced transfer before demonstration could be accomplished. One thing that we learned was that we have to sometimes bridle our enthusiasm to keep from pushing an idea before we understand it well enough and can demonstrate its feasibility. Research and the receiving division must reach an agreement on what constitutes feasibility. Clearly, this will depend on the topic. For an algorithm such as the Fast Fourier Transform, the requirement is a running program that does better than existing methods. For hardware it will be a working device or even a system of components, performing a function, together with a demonstrated fabrication methodology. The program in magnetic films for memories is one in which feasibility was shown. However, magnetic films never became a product due to advances made in core memories and the quick growth in semiconductor memory technology.

In some cases, there are entire application systems or languages where the feasibility implies acceptability to the end user. This might be an end user in, say, medical diagnosis or the airline business, and must involve some kind of joint study with real users before feasibility can be demonstrated.

Advanced development overlap

For those projects that are transferred out, research must determine whether to maintain activity, either to support development or to defend its concepts, or to explore advanced or related technologies. The successful research development of FETs was followed by an abrupt discontinuation of almost all research in semiconductors for a short period. This was a mistake. The difficulty experienced in getting the development division to pick up the one device memory cell in 1967 may have been partly due to a relative disinterest by research management in semiconductors just at that time as it turned its attention to other areas.

In planning applied projects, and especially as they near transfer, careful preparations have to be made for the proper kind of overlap program. For certain research projects, in particular, systems work, the creation of a special, advanced development effort is often the answer to problems of scaling-up, or to answer questions of marketability or economic feasibility before making a full product commitment. This may require bringing to the research laboratories new kinds of people. For example, with APL, a complete working system with customers was running before transfer.

Growth potential

There have been several research programs that suffered at being too narrowly aimed at a specific need and not having clear paths to technical growth and to growth in product applicability. Examples of this are the germanium and the beam addressable file programs.

Unless there is a prospect of technical advancement, the transfer may not be successful due to the fact that existing technologies "stretch" themselves. The challenge of a new technology forces an existing one to extend itself, to advance its goals, to expand its potential in the face of competition. Frequently, this stretching removes the advantages of this challenge, resulting in its demise. This was exactly the case with the beam addressable file. The germanium project and the magnetic thin film program afford other examples. The cryogenic computer work of the early 60s is another example. The thin superconducting film memories developed at that time lost out to the advances being made by magnetic films and the cryotron logic lost out to the constant advances made by silicon circuitry. In all of these cases, not enough attention was paid to the growth potential in the new technology.

While research could perhaps take credit for stimulating or forcing advances in the existing technologies by offering competing alternatives, it certainly can't be the organization's major ambition or goal. To avoid being caught itself, Research has to constantly and carefully look over its shoulder at what is coming along.

Existence of an advocate

No matter how elegant the research results are or how much benefit they appear to have for the company, someone in research must take the responsibility to see that the results reach the right place. Our study indicates that a strong proponent actively selling the research results is necessary for transfer. It is obviously not sufficient; the several projects that failed to transfer or failed to become products also had research champions. Properly timed seminars for publicizing and explaining transferable research concepts have been helpful when used. The effectiveness of the research champion has been enhanced in several cases via a push-pull provided by an external champion.

Advanced technology activities in a development laboratory

The major conclusion is that advanced technology programs in the development laboratories are helpful and often necessary for transfer from research. In a very clear-cut fashion, the presence of "ad tech" groups aided in our moving electron beam technology and magnetic bubbles work from research to development. Both of these transfers have and are taking place continuously. Materials and knowledge of materials processing, device and circuit invention, and application techniques have all moved out. Interest in electron beam fabrication methods was high enough amongst individuals in ad tech and their connection to research was useful in smoothing the way. With respect to bubbles, the ad tech lab in development had the talents and experience to pick up its technology. Thus, without formal contracting or negotiating, the presence of skilled and research-minded people in the development labs, and their relative freedom from close-in product demands, made it easier to effect the transfer.

In other hardware projects, divisional ad tech has served a critical function and has often looked to be competitive or even obstructional to research. Our case studies show that most of the time, the higher hurdles created by ad tech skepticism or resistance were, in the end, beneficial. For LSI–FET, research had to do the work and carry materials processing, device and circuit design, and design automation very far along, further than the research image of itself was comfortable with. However, the results were convincing, the corporation took on FETs in a confirmed fashion, and research benefited by having seasoned people in silicon technology, ready for subsequent efforts.

There is a similar record for magnetic films. The initial hurdles put up by the relevant development organization required very solid results and a thorough involvement of research people, not only in technology but also in systems usage and its economy. In the cases of the already mentioned beam addressable file and germanium programs, ad tech groups were correctly negative. By and large, having an ad tech activity in a divisional development laboratory is a positive asset to technology transfers.

All of the above is relevant to hardware technology. In the case of software activities, the picture is not as clear. In one case, that of APL, it was protected and developed slowly inside research. While it created its technology, it also created a user audience and usage patterns. Keeping it in research longer than might have been thought desirable had the benefit of producing a new language and tools for its operation, not seriously reduced in effectiveness by having to comply with then current marketing philosophy. Thus, research, willing or not, provided the advanced technology phase for APL.

There is a class of programming results that really do not require very much further development. The FFT, VM Monitor, and ASTAP contributions were passed on fairly directly, but in each case the successful transfer came about by negotiation as to what level of programming would be acceptable.

In general, in the software area it is a little more difficult to define the role of ad tech groups in the development sector.

External pressures

For many of the hardware projects that were transferred and for some of the software ones, the presence of some form of the same technology in a competitor's laboratory or a product announcement has helped transfer. For our work in LSI–FETs, most of the industry was beginning preparation for FET componentry while our components development groups were still concentrating on bipolars. Research was able to draw attention to the competition when it was needed. For magnetic thin films, a competitor had announced a memory product before our product was close to announcement, but it was clear from the published work of research activities in several laboratories that a number of companies were working on magnetic films. At the time we were urging a development lab to become interested in electron beam technology, some manufacturers were publicly talking about methods and preliminary results. Reports of work on bubbles at other laboratories have kept product development people interested enough to make the continuous transfer that is occurring easier.

In other projects, however, there was no competitive pressure. In the applications area, cryptography and Fast Fourier Transforms were unique to research and IBM. The beam addressable file was not specifically pushed by competitors although their people had similar projects underway. In the S.S.A., there was an immediate development following the research work in order to fulfill a government contract.

In general, for hardware, parallel activity elsewhere has helped research to transfer to development. It has also created an external standard against which to judge the research progress and achievement. When there is no outside activity we can expect greater difficulty in making judgments ourselves and in transferring.

For the applications results, external pressure has not played an important role. For the systems and programming transfers, competition has played a part in the past and may again in the future.

Joint programs

Joint programs can have several forms. They can involve support by money or by people. They may involve research people in development laboratories or development people in our laboratories. The most interesting observation from our case studies is that there was no joint activity in any of the systems and programming projects. There may have been a number of reasons, but for the most part this seemed to have been because of an inability by research to convince development managers that our ideas were any better or might be more productive than their own.

There were (or are) activities involving jointly planned programs or lending of people in LSI–FET, magnetic bubbles, and in magnetic thin films. In the germanium and electron beam projects, research took on development people in training or as a mode of entry hiring into the company. In general, we conclude that joint programs are good to have but do not ensure success.

Secondary Factors

Timeliness

Timeliness may enter in several ways. For one, research may try to provide a new or unique technology early because there are other candidates for a new product. More often research will have to be concerned about product cycles and when entry of improvements or even a new technology in a conventional product area is feasible. Good timing is important but it is not sufficient for successful transfer. If what we have is good enough, timeliness may not even be necessary.

Internal users

In addition to the useful pressure that external competition may create in helping to move a technology to the development laboratories, in some cases internal IBM users can play a similar role. A demand from hardware systems people, in one of the development labs, for low-cost, high-density FET circuitry helped in getting our divisional components people to pick up the research results. Internal use of APL helped create pressure on the sales side of the business. Hopefully, internal users of magnetic bubbles will grow and augment the market for device and circuit manufacture. If such internal users demands do not naturally arise, perhaps research labs should stimulate them.

Government contracts

In one of the cases studied, magnetic bubbles, the presence of a government contract was useful in furthering the research work itself and in providing a good stimulus for transfer. In the early days of magnetic thin film research, contracts were helpful in getting started in the technology. Another contract supported some early work in sparse matrices. The difficulty of government's requirements for the Social Security Page Reader forced a collaboration between research and development that probably would not have taken place otherwise. The collaboration produced technical advances that were useful in subsequent products. In effect, the stiff external requirements forced the development groups to look to research for advanced work.

High-level involvement

Occasionally research has turned to corporate management for help in transfer. This was true in the case of LSI–FETs. At other times, staff committees were involved. In general, however, this has not been an important or even an effective mechanism for research to use.

Individual corporate responsibility

In one case, cryptography, an individual with a corporate watchdog role was useful. In general, this is rare and this may be important only when there is a totally new area of technical endeavor, such as cryptography, for us to deal with.

Proximity

In practically no case was the proximity of a development laboratory to a research laboratory an important factor. At times, being close was convenient and saved money, but no transfer failed because of distance.

APPENDIX—THE 18 CASE STUDY PROJECTS

LARGE-SCALE INTEGRATION—N–CHANNEL FIELD EFFECT TRANSISTORS (SUCCESSFUL)

In 1963, the Yorktown Heights laboratory began work on integrated silicon circuitry. This included silicon processing techniques, involving considerable physics and chemistry, device and circuit design, light table development for mask making and other optical lithographic requirements, and a design automation program. The devices were primarily FETs, field effect transistors, but the methodology evolved proved to be useful for bipolar devices and circuits as well. The general idea of large-scale integration was transferred to the component development laboratories in 1966, but it was not until 1968 that the specific technology for FET was finally adopted. This transfer provided the basis for IBM's main memories of the early 1970s and for the logic in most of the company's terminals and small machines in the same period.

ELECTRON BEAM FABRICATION METHODS (SUCCESSFUL)

In the early 1960s, the Yorktown Heights laboratory used electron beams to produce an optically read storage disk. Its original use was as the dictionary in a Russian translation system. Some of this early "photostore" technology was transferred as early as 1963 into special-purpose storage products. From that time on, there were a number of parallel research activities: the beam column itself including an improved filament, the software to automatically run circuit patterns and, importantly, efficient sensitive resists for the lithographic processing. These were transferred continuously into the component development laboratories beginning about 1966. Electron beam fabrication methods are now in use in the lithographic processes of circuit chips.

THE GERMANIUM PROGRAM (NONTRANSFER)

Germanium has a higher mobility than silicon and, in the early days of transistors, was widely used for point contact and junction transistors. With the coming of integration in the early 60s, there was a brief period of competition between germanium and silicon for use in integrated circuits. The Yorktown laboratory started a program in 1964 which was supported by funds from the components development division. With this support, the program grew to a rather large size. As the silicon technology advanced, the germanium studies experimented with low-temperature environments (liquid nitrogen) to gain further speed and other advantages. Both were aimed at a high-speed circuit requirement for a large computing system which was in design at the time. By 1968, however, it had become apparent that, although germanium might meet the requirements of speed for this particular computer project, the power required to attain these speeds was very high. Although this was also true of silicon, silicon had much more attractive characteristics at medium and low speeds and appeared to have greater growth and extendibility prospects. The project was terminated in 1968, and the use of germanium in computer circuitry has disappeared.

ONE DEVICE MEMORY CELL (UNSUCCESSFUL)

Until 1966, integrated circuits for memory in the research laboratory and in development in IBM had used a number of transistors for each memory cell. A cell is the physical location of memory bit storage. At this time, a research staff member invented a memory circuit, which required only one device and a patent was issued in 1968. Attempts were made to interest the development laboratories in this circuit which gave a very large decrease in cell area and, therefore, represented a primary means of increasing memory density on a chip, and increasing speed. Unfortunately, other designs had already been adopted for current development in 1967 and little headway was made. Eventually, cell designs of this kind did appear in IBM memory technology but an early lead was lost.

MAGNETIC THIN FILM MEMORY (UNSUCCESSFUL)

Early work on using thin magnetic films to form memories was carried on in IBM, Lincoln Laboratories, Univac, and other laboratories. In 1960, joint preliminary studies by IBM research and a development group resulted in a research project in the Zurich laboratory. This was successful to the extent that the technology was brought from Zurich to Yorktown Heights and with further work was transferred to a components development laboratory. By 1964, a product design for a very fast memory was completed. Plans were made to use the memories in a large computer. In research, further activity produced new technical ideas for other versions of thin film memories. While all of this was happening in magnetic thin films, the major memory product was ferrite cores which were being continuously improved as to size and speed. Also, the first transistor memories were being considered. In the end, only one computer system with a fast magnetic thin film memory was shipped. It had made its goals but the technology, by 1968, had been overtaken by transistor memories. In 1969, efforts were terminated by both research and development.

BEAM ADDRESSABLE STORAGE (NONTRANSFER)

In the mid-60s, before the serious advent of magnetic bubbles as a storage candidate, much thought was given to replacing magnetic induction recording with a beam addressable storage system. To gain high-bit density, magnetic domain sizes on disks and tapes have been continuously reduced, and hence requiring the magnetic head to move closer to the disk surface. As the head-to-surface gap becomes smaller, design and operational control become more difficult and more costly. Beam addressed disk storage did not have this limitation and, therefore, looked interesting. A research project was underway in the San Jose laboratory by 1968. It used a magneto-optic effect: originally a europium oxide coated disk was written on by a light beam produced by low-temperature injection lasers. At first, the disk surface materials also had to be operated at liquid nitrogen temperatures. New disk coatings were found and plans were made to do the work necessary to bring continuously operating room temperature semiconductor lasers into the system. A deeper understanding of the physical mechanisms involved in the transduction of light energy through a thermal phase to a change in magnetic phase was studied. As all of this was being done, the magnetic induction recording technology in the neighboring development laboratory was spurred to significant improvements. Higher densities of magnetic bits and dramatically smaller head-to-disk gaps were found feasible. The projected densities and costs of magnetic induction recording became equal to or better than those set out as goals for the beam addressable project, and it was terminated.

MAGNETIC BUBBLES (SUCCESSFUL)

Although IBM researchers had worked with the interesting garnet crystal materials and were aware of magnetic effects themselves and those observed at the Bell Laboratories and Philips Eindhoven laboratory, it was not until the announcement by the Bell Labs of its bubble technology and its patents in the area that interest was really spurred. Research groups were formed in 1969, and a small NASA contract was accepted in 1971 calling for a simple operating chip with bubbles of rather large diameter. The contract was completed in 1972. New materials, including an amorphous substitute for the garnet crystals, began to come out of the research activities. Inventions of new bubble devices and of a new system concept, the bubble lattice file, appeared. However, efforts to interest the component development laboratories and computer system development groups in the company were not successful. Research then undertook a campaign to interest not only the technology developers in the company but also future systems users. Finally, the storage development laboratory became interested and early research work was transferred. Research continues to work on advanced concepts in bubble storage.

COPIER TECHNOLOGY (SUCCESSFUL)

In the early 1960s, relatively basic work was started in the San Jose Research laboratory on organic photoconductors. Although there was not a specific product goal in mind, it was thought that microfilms or perhaps copiers might require such photosensitive materials. The early studies led to the discovery of a very high-sensitivity photoconductor just at the time when technology for an office copier was required. A robot model was built to show that the new material would work. Since this was a new product area, there was no development group to accept the work. Eventually some of the research people carried the technology into development while others created an advanced technology group for the development division.

SOCIAL SECURITY PAGE READERS (SUCCESSFUL)

Character and pattern recognition had been a research field in the computer sciences all throughout the 1950s. In the early 1960s, the Yorktown Heights group developed a system for character recognition with multiple scanners and software and hardware for processing recognition logic flexible enough to operate on a wide variety of fonts. In product development, however, character recognition concentrated on special single fonts such as might be employed in a bank check reader. When the Social Security Administration requested a multifont page reader in 1963, the research facility and its processing experience was used to show feasibility. A joint effort was carried on by research and the development laboratory for two years involving transfers of people both ways. When the page reader product was delivered to the Social Security Administration, a large number of the research concepts were included.

SYSTEM Y (UNSUCCESSFUL)
SYSTEM A (UNSUCCESSFUL)

These were two large projects—one in the mid-60s, the other in the early 70s. One dealt with an advanced hardware design for a computer and the other a software architecture. We cannot discuss these projects in any detail because some of the results are still sensitive. However, they were similar in the following respect: in each there were some extremely interesting and potentially powerful concepts developed while they were in research (Yorktown Heights). In both cases, this was only a short period of time, one year, and before these concepts could be worked to any degree of feasibility, the projects were moved almost intact into a development program. In hindsight, it appears now that not enough understanding was provided during the research period.

APL (SUCCESSFUL)

The concepts of the APL language were brought to IBM by the research staff member who conceived of them at Harvard. The language was unique in that it developed a new notation and syntax and among other attractive features allowed for the powerful operators on vectors and matrices that are desired by people in many kinds of mathematical applications in science. After a trial as a batch system, a time-sharing implementation was created, nominally for use in the Yorktown Heights laboratory. Classes were taught and very quickly a large number of researchers began to use the system. Other users came on to the system from other parts of IBM. All of this was carried on relatively informally, and as the user set grew and the language became well known it served as a proof to the development and marketing groups in the company that APL deserved to become a product. This finally happened in 1970.

M-44 (SUCCESSFUL)

This was the local name at the Yorktown Heights laboratory for a project in the early 60s that tested concepts for virtual memory and virtual machines. An older computer was physically modified and a new operating system created to try out the ideas. For example, the notion of paging, bringing blocks of data from disk or drum to main memory in an ordered fashion so as to give the user the impression of an enhanced or virtual memory, was tested by literally coding algorithms and trying them out. The virtual machine concept was first used in this experimental system. The research results were positive and were quickly transferred to development groups for use in time-sharing systems in the late 60s and virtual memory and machine systems in the 70s.

VM
MONITOR/STATISTICS-GENERATING PACKAGE (SUCCESSFUL)

These are two related software programs that enable users of VM, one of IBM's main operating systems, to measure the performance of their workload on the systems. The programs were developed in Yorktown Heights for use on the local computing systems to help understand computing efficiency and improvements. They were transferred relatively smoothly to a development division and have become a part of the VM system provided for customers.

CRYPTOGRAPHY (SUCCESSFUL)

Data security became an issue in IBM in the late 60s. Corporate responsibility was assigned to an individual who stimulated interest and activities amongst the mathematicians at Yorktown Heights. Simultaneously, others in the laboratory were coding and designing hardware for some new encryption methods. Attempts were made in 1970 to interest advanced technology groups in the terminal development laboratories, but there were no takers. However, in 1971, a special product was produced by the same development laboratory for a banking customer. The cryptographic code developed by that laboratory was sent to Yorktown Heights for testing and it was easily broken. The new technology, ideas, hardware, and software that had been underway at research was quickly put into use instead, and the transfer was effected. An enhanced version of these codes has now become the federal cryptographic standard.

FAST FOURIER TRANSFORM (SUCCESSFUL)

This now well-known algorithm came into being in its present easily computed form through the joint efforts of two IBMers in research and a staff member of the Bell Labs in 1963. The algorithm was suggested to solve a particular problem in low-temperature physics and programmed at Yorktown Heights. Its amazing usefulness was publicized and propagated to IBM customers and scientists by reports, papers, newsletters, and a large number of personal contacts. Within four years' time, programs were available, special hardware was under development, and the algorithm was on its way to becoming one of the most widely used in all of scientific computing. Important extensions are still being made.

ASTAP (SUCCESSFUL)

This is an acronym for an internal IBM circuit analysis program. Between 1963 and 1975, mathematicians in the Yorktown Heights laboratory made a number of contributions. Two of these, methods for handling "stiff" differential equations and for dealing efficiently with sparse matrices, have made huge improvements in circuit analysis running times. They have also led to a large number of independent mathematical investigations by workers in the field in a number of other institutions.

GRAPHIC DOCUMENT SYSTEM (UNSUCCESSFUL)

This project began as a possible solution to the problem of mapping electric utility holdings. It was stimulated by a known customer need and it allowed field maps, roughly sketched on the job, to be easily and swiftly transformed into properly dimensioned, annotated, and rectified maps. The system used special hardware and required new software. It was used in a test with one of the major regional utility companies and proved effective in this trial. Using the mapping system as a base, a drafting system was also evolved and tested in one of IBM's development laboratories. Both projects have since wandered through a number of development projects in both domestic and European development laboratories, but no products have resulted.

Integrating Entrepreneurship and Strategy

Case IV–4 Texas Instruments Shows U.S. Business How to Survive in the 1980s

The 1980s loom as a bloody battlefield for U.S. industry. America, which so often led the world in its ability to mass-produce and market innovative products, is fast losing its edge, many experts feel. These same experts worry about a disappearance of innovation and the increasingly minuscule gains in U.S. productivity. But no U.S. company is working harder than Texas Instruments, Inc., to foster innovation and to focus an entire corporation on boosting productivity—a crucial factor in an era of seemingly endemic inflation.

Today the Dallas electronics giant leads the world in such fast-moving, high-technology markets as semiconductors, calculators, and digital watches. And it is now building its production machine to move even faster in the 1980s. TI is doing this with a very complex system designed to stimulate and manage innovation.

Under Its 10-Gallon Hat, a Japanese-Style Culture

The overpowering culture of Texas Instruments, Inc., so vital to the success of the Dallas company's management systems, has its roots buried deep in a soil of Texas' pioneer work ethic, dedication, toughness, and tenacity. Says one former TI manager: "Everybody in that organization is either from Texas or just out of school. And they honestly believe—I used to believe it myself—that the company can do anything."

"The TI culture is a religion," pronounces one TI vice president. As such, "the climate polarizes people—either you are incorporated into the culture or rejected," declares Arnoldo C. Hax, who has studied the company at the Sloan School of Management of the Massachusetts Institute of Technology. Some of his students who have gone to work at TI fit right in, but others quickly bail out.

Glenn E. Penisten, president of American Microsystems, Inc., and a former TI executive, agrees. "I've seen people brought in at reasonably high levels and not survive . . . the culture tends to reject 'strange' individuals."

Involvement Teams

TI figures that it takes five years to train a full-fledged TI manager. For those who survive the course, it is as a cog in a demanding, no-nonsense world. The management takes itself and the company very seriously. "There's not a helluva lot of frivolity over there," says another former TI executive. In fact, the company is developing its culture in a way that causes both competitors and admirers to compare TI to Japanese companies.

When J. Fred Bucy, TI president, described his Japanese competition recently, he could have just as easily been describing what he wants in his own company. "Japan has a culture and society well suited to achieving increased productivity and the growth that results from it," said Bucy. "They are hard-working, dedicated people . . . and are highly motivated, in part, because of a culture that assigns personal responsibility for the quality of work." And, he added, "there is a strong tendency in the Japanese culture to align personal goals with goals set by their companies."

Looking at TI, the similarities can be startling. More than 83 percent of all TI employees, for example, are now organized into "people involvement teams" seeking ways to improve their own productivity. At TI, "the employee is subservient to the success of the corporate entity," says a former TI manager. Adds another, "The company looks at its people as being completely interchangeable—kind of like auto parts."

But just as in Japan, TI employees do not normally get fired—particularly those who have worked at the company for five years or more. "There's lots of yelling and screaming, but not much ripping off of badges," says one former TI employee. TI is not compassionate toward the manager who is not meeting the goals that he set for himself. "He'll be moved to the side, or down, or put on special assignment," comments a former TI manager. "But TI doesn't cut a guy's throat and put him out on the street."

The Work Ethic

Corporate loyalty is big at TI. In "Silicon Valley" south of San Francisco, where many of TI's semiconductor competitors are concentrated, "you've got a group of very bright people whose loyalty is focused on the industry, not on the company they work for," says Jack R. Yelverton, a veteran San Francisco executive recruiter for the industry. "At TI, it's the other way around."

Most of the experienced professionals that TI hires from other companies do not have a great deal of success. Says one who left after two years: "TI doesn't want experienced people; they want to hire them young and train them." Over the past five years, in fact, TI has hired 5,604 graduates right out of college.

The work ethic is a cornerstone of the TI culture. "If you didn't work overtime you were ostracized—at least in the early days," comments Bruce D. Henderson, president of the Boston Consulting Group. And it has not changed much since then. "When you're a professional and work for TI, long hours come with the territory," says one competitor. "They demand, and get, a lot of mileage out of their people."

TI still works a 42½-hour week, and 5:30 P.M. meetings "are rather common," says one former employee. "The office is certainly a good place to meet people on Saturday mornings. People show up because it's expected."

Keeping Employees Happy

Seniority is another common denominator with the Japanese. "Seniority is all important at TI," says Jerry Wasserman, an industry consultant at Arthur D. Little, Inc. "Other companies have badges where color denotes levels of authority, but at TI, your badge color shows your years at the company."

And like the Japanese, TI works to keep its employees happy. "TI tries hard to keep you in the fold by covering all aspects of life," says one former employee who is still impressed with the company's efforts. "They have this fantastic 'rec' center, with a gymnasium and baseball diamonds; they have a rod and gun club; and they have 75 acres on Lake Texoma, where a lot to build a cabin, or put a trailer on, costs you $43—and a nickel a year."

TI's MAGIC IN MANAGING INNOVATION

Running any $2 billion company with 68,000 workers in 45 plants spread throughout 18 countries is a tough management task. But when such a company is expected to grow 15 percent annually in a worldwide business keyed to rapid technology change and declining unit costs, the job of managing might seem all but impossible. To meet that challenge, Texas Instruments, Inc., has successfully evolved one of the most formal planning systems in existence.

"If we hit $10 billion [by the late 1980s], it will be because we planned every foot of the way," emphasizes TI President J. Fred Bucy. Innovation is the lifeblood of the company, and the key to TI's success has been its novel, highly complex system to stimulate and manage innovation.

"It requires an extraordinary amount of coordination to work," says Arnoldo C. Hax, who follows TI at the Sloan School of Management at the Massachusetts Institute of Technology. But, he adds, "I think it has worked at TI."

Indeed it has. TI has grown at an average rate of 15 percent annually for the past 15 years and yet has kept the innovation juices flowing. "I would think that TI would be getting old and a little bit creaky at this point," comments Richard L. Petritz, an early research and development manager at TI who now heads Inmos, Ltd. "But it is not happening."

George H. Heilmeier, who saw a large amount of cutting-edge technology during the three years that he ran the Pentagon's Defense Advanced Research Projects Agency, got his first, close-up look at TI recently when he joined the company as a vice president, the first corporate officer TI has ever elected from the outside. "This is a large company that still has the esprit of a small company, and I'm not sure I understand that," he says. "It may be because they're all Texans," Heilmeier adds, "and it may be because it's a company run by engineers rather than by lawyers and accountants."

That kind of environment was precisely what Patrick E. Haggerty, then president and now honorary chairman, was aiming for when he decided in 1962 that TI was getting too big for all of its managers to sit down together and hammer out strategies. He formalized his own strategic planning style, but it took more than a decade of hard work before this "objectives, strategies, and tactics" (OST) system became an integral part of the TI culture. "Haggerty understood what it took to motivate people, and he worked hard to keep the company from becoming insensitive as it got bigger," says Glenn E. Penisten, a 16-year TI veteran who is now president of American Microsystems, Inc. (AMI).

It was Mark Shepherd, Jr., Haggerty's successor as president, who really made the OST system work. In the late 1960s, he split TI's annual expenditures into separate strategic (OST) and operating budgets, placing the strategic funds under the control of a corporate committee that was to set the guidelines for spending them. Under OST, a project-oriented management structure focuses entirely on tomorrow's growth, while a more conventional operating hierarchy concentrates entirely on today's profitability. Shepherd set up this dual reporting structure to ensure that managers did not underplay or postpone long-range strategic programs in favor of short-range profits.

OST is a highly decentralized "bottom-up" planning system, where more than 250 funded projects called tactical action programs—TAPs in TI jargon—drive more than 60 strategies that support the company's dozen business objectives. The objectives are now set up to build TI into a $10 billion company in 10 years. One TAP manager, for example, is responsible for developing a new liquid-crystal display (LCD) watch for the watch strategy manager, who, in turn, reports to the consumer objectives manager.

Management by Objective

This OST pyramid is overlaid across TI's operating hierarchy of 32 divisions (ranging in annual sales from $50 million to $150 million each) and more than 80 product-customer centers ($10 million to $100 million each). The centers (PCCs) are roughly equivalent to departments, except that they are more self-sufficient. Many of them have their own engineering, manufacturing, and marketing units. TI pulls together a tactics program from whatever PCCs are required. "The real power of OST," says Inmos' Petritz, "is that it enables you to get into a new business without reorganizing your company to do it."

One 10-year TI veteran, now a major competitor, liked TI's system so much he patterned his own management structure after it. But he adds: "Frankly, TI's OST program is a label for management by objective, Haggerty's naval strategy system." Unlike military program management systems, however, at TI most TAP managers also manage PCCs, and strategy managers typically are division heads. In this way, TI gets away from the "handoff" problem between a separate development or planning group and the operating organization.

OST is a highly visible program where all managers are constantly measured against documented goals. There are monthly reviews, and TI's computerized scheduling, or PERT, system, is updated by status reports on every TAP. That gives top management a window deep into the company. "I don't believe in the hierarchy system where you have to funnel information through several management levels," says Bucy.

TI's Heilmeier observes: "It's impossible to bury a mistake in this company; the grass roots of the corporation are visible from the top. There's no place to hide in TI—the people work in teams, and that results in a lot of peer pressure and peer recognition." Adds Bucy: "An outside manager joining TI would be surprised at how much upper management knows. We communicate so thoroughly through OST that [managers] don't need day-to-day guidance."

On the other hand, Wilfred J. Corrigan, chairman of Fairchild Camera & Instrument Corporation, a TI competitor, declares that "the middle manager [at TI] often finds that the system is really a tool to manage him, not the other way around." But Bucy says: "As long as the guys are on course, they don't get interfered with."

Even so, some managers do find the tight control stifling. "[Bucy] sets policies and procedures, and the consequences of not following them are quite severe," notes one longtime TI manager. While this executive believes that TI's management system is overdone, and at times serves to work against the company, he also sees it as a real strength of TI. "Companies need to be well managed and disciplined," he says, "and though TI has gone overboard, I'd rather err that way than be like the many companies that go the opposite route."

AMI's Penisten is currently trying to transplant much of the TI planning system to his Santa Clara (California) semiconductor company. But he worries that "you can quickly spend too much time on detail and spend all your energies on planning and reporting." Some of the questions that Penisten is still wrestling with: How does he get control without stifling creativity and interfering with decision making, and how does he avoid creating a bureaucracy?

A Constant Search for Ideas

TI officials believe that OST can be transplanted to other companies, but they warn that the system has grown in the TI culture, and its success depends partly on that structure. Professor Hax of MIT does not think it can be transplanted. "TI's formal structure and management systems can't be understood apart from the culture," he says.

Former TI managers generally give the company's structure high marks. "There are a lot of middle managers running around TI talking about 'growth share matrices' and 'learning-curve pricing,'" says one. "You don't get that depth of planning in other companies." Says a 17-year TI veteran who is now a competitor: "TI has numerous systems for getting people to think and to make their ideas known—management is constantly seeking ideas."

Just about now, TI managers are submitting more than 400 tactical proposals for OST money next year that range in thrust from product development and feasibility testing to cost reduction and new marketing techniques. Because TI uses zero-based budgeting—it was the first company to employ the concept—managers must rank their proposals in order of priority. Next March, the process will culminate in a full week's strategic planning conference in Dallas attended not only by 500 managers but also by TI's board of directors. Together they will decide on the corporate plan and allocate funds.

TI is constantly refining the OST. In 1975, "we realized we had a problem; we were slighting the more speculative development efforts," says Charles H. Phipps, an assistant vice president heading OST. So TI now asks its objectives managers to decide how much OST money TI should spend on such programs and to rank those proposals separately. Haggerty called them "wild hare" ideas, and the name has stuck.

Wild hare seems to be solving TI's problem. It funded the company's highly successful portable computer terminal, the first product on the market using magnetic-bubble memory devices.

Such funding has even started new businesses. TI's entry into the marine electronics market last fall was kicked off two years earlier by a wild hare grant. The result was a navigation receiver that was introduced at $2,095, a full $1,000 below its closest competitor.

Like any strategic planning effort, the original OST thrust was aimed externally at developing and building new technologies, products, and businesses. But TI has successfully turned the system inward as well. "We're using OST to look hard at internal funding programs that will impact productivity," Phipps says. Called people and asset effectiveness (P&AE), the system is aimed not only at reducing manufacturing costs but at paring indirect costs as well.

Like the OST program, P&AE took an entire decade to get fully accepted into the TI culture. But now the company has 83 percent of its employees organized into teams to participate in the planning and control of their own work to improve productivity. TI believes that P&AE has played a major role in good employee relations. One measure of its effectiveness, as well as that of TI's broad benefits package, was that TI is the "third largest, nonunionized corporation in the United States—after IBM and Kodak," Senior Vice President James L. Fischer noted recently.

The OST spinoff is spawning a host of successful productivity improvements. Last year, TI gave one of its $300 programmable calculators, along with six hours of training, to each of 8,000 technical and administrative personnel. In six months, the $3 million P&AE program had paid for itself by boosting the productivity of those people by 3.5 percent to 4 percent.

But most of the P&AE programs, which have to compete with other OST programs for funding, are closely tied to advanced manufacturing techniques. Factory automation is high on TI's list, and programs are underway to automate the assembly of calculators, large-scale integrated (LSI) circuits, and a host of other products.

Passing Out $25,000 Grants

Any large company, especially one with formal management systems, sooner or later finds that it is freezing out some innovation, and TI is no exception. But in 1973, the company started a program dubbed IDEA to further encourage innovation. TI splits up $1 million annually among 40 IDEA representatives—usually senior technical staffers, not managers—who pass out grants of up to $25,000 to employees with ideas for a product or process improvement.

If an employee is turned down by one IDEA representative, he can take his idea to another. "We've found that about a third to a half eventually get funded," says Vice President Bernard H. List. "And once the guy gets his money, no one—not even Bucy—can take it away from him." Half of the ideas funded end up paying off. "But aside from the payoff," List says, "the motivational effect has been very positive."

With the IDEA program, TI employees avoid the massive presentation and documentation that OST requires. "We're just learning how to use it at the working and management levels," says Hector A. Cardenas, consumer technology head. "Like many institutionalized things, it was slow to get started."

But already, IDEA is starting an amazing number of innovative products, particularly in the consumer area. The $19.95 digital watch that tore apart the market in 1976 got its start as an IDEA program in the Semiconductor Group. "The people running the watch division figured it wouldn't work," Phipps recalls. "They were convinced that the watch was a jewelry business."

And in June 1977, Cardenas himself went to an IDEA representative for money after his managers turned him down for OST funding because they did not believe that it was possible to manufacture a new type of watch. For less than $25,000, Cardenas and his team proved that TI could. And a year later, they were in New York announcing it—the first all-electronic, analog watch that uses a new type of liquid-crystal display to show the traditional hands.

Gene A. Frantz's idea was a low-cost speech synthesizer built on one tiny chip with voice quality equal to that of the telephone system. His team won a $25,000 award and "then spent a month convincing corporate research that we could implement its speech technology with our semiconductor technology," he says. The Frantz team did it, and the first product using their breakthrough technology was announced in June—a talking, learning aid to teach spelling. The $50 Speak and Spell is the first of what is expected to be a flood of products over the next several years using the revolutionary chip.

With these kinds of innovative products spilling out of TI now, it seems clear that the company is solving much of the problem of stimulating and managing innovation in a large company. There will always be room for many management styles, and other companies may not have to emulate the TI strategy to survive in the 1980s. "But TI seems to have discovered the style that allows a multibillion-dollar corporation to grow at 15 percent a year," points out TI's Heilmeier. "To match that, companies are going to have to share information with more people. They're going to have to maintain entrepreneurial spirit, and management is going to have to have good visibility into all of the company."

The Entrepreneurial Organization

Case IV–5
The Technical Strategy of 3M: Start More Little Businesses and More Little Businessmen

It sounds simple, but it's not. To Minnesota Mining & Manufacturing Co., it means finding novel technologies, protecting them vigorously with patents and plenty of company security, and stimulating entrepreneurs to champion them through to the marketplace. Senior editor Ford Park explores the ways 3M uses this strategy to keep getting richer, more diversified, more innovative.

It's probably because the winters are so very cold—20 below and lower, they say. But there it sits, a loudspeaker-microphone, on a pole of its own at curbside, as you are driven up to the administration building.

The car stops. Your driver-escort rolls down

his window. "Yes?" queries the disembodied voice of the guard sitting in his shelter six feet away. "This is Warren Feist, public relations, with two editors from *Innovation* magazine." A pause, presumably to check names and the day's schedule, and then the satisfied response: "Okay, proceed." The car rolls forward, and you're "inside."

But not quite. Behind a massive counter in the modern lobby, next to a graceful spiral stairway flanked by the healthiest of potted plants, sits a pleasant but firm receptionist. "Names please." A schedule is consulted. Cameras slung on shoulders eyed a trifle unbelievingly. Attache case . . . hmmmm. But the name cards (escort required) are there and ready, and the check-in is swift. Up you move for the beginning of two days of interviews with top brass.

A missile base? An electronics company doing work for the Defense Department? An aerospace company? A government R&D lab? None of these. It's the headquarters of the Minnesota Mining & Manufacturing Co. in St. Paul, Minnesota.

Now wait a minute. That's 3M. They're . . . let's see . . . all those products with Scotch on the front: transparent tape, reflective signs for highways, water and stain repellent chemicals for clothes, carpets that athletes play on, slide projectors, magnetic tape, sandpaper, tape recorders, cook-in bags for food, floor-polishing pads, copying machines, adhesives, electrical insulating tape . . . and, oh yes, that new one on TV—hair-setting tape. Well, maybe they all don't have "Scotch" in their brand names, but they have a lot of Scotch plaid on them, don't they?

You're a sales manager's dream: Out of the 35,000-odd products that 3M makes, you've identified only a handful. But you did it almost without thinking and, what's better yet, the products you reeled off come from one or another of most of 3M's six major product groups. That's customer acceptance for you!

But what about all this red tape (not a 3M proprietary product!) and security stuff getting in? You've just encountered one of several startling anomalies of this prodigiously productive and profitable company: As freewheeling as 3M is at producing items found all over American homes and businesses, it's at the same time one of the most security conscious companies in the country. More about this oddity later, but what about those words "prodigiously productive and profitable?" That's mighty strong language.

Let's take the last of these first: 3M is considered one of the best-managed industrial companies in the country in the sense that it has compiled a most remarkable growth record over the last 30 years. Consider the following:

- 1968 was the 30th consecutive year that 3M has increased its revenue; during this period its sales have increased over 100 times—from $14 million in 1938 to $1.4+ billion in 1968.
- 1968 was the 17th consecutive year in which earnings per share increased over the previous year; only three times during the 30-year growth period we are considering have earnings per share declined.
- Sales and earnings have grown at rates of 16.6 percent and 13.6 percent, respectively, since 1938.
- 3M stock has grown in market value at a rate of roughly 18 percent per year since 1938.

Investors have been enthusiastic over this growth record, but their enthusiasm—as reflected in the price-earnings ratio of 3M stock—has been tempered in recent years. Nineteen sixty: the future of the Thermofax copying process looked assured, so the stock sold at a high of $88 per share, and was traded at a multiple of 64 times earnings. With the introduction by Xerox of its 914 copier, even though 3M's growth pattern continued, the bloom went slightly off the investment rose; it wasn't until 1967 that the company's stock exceeded the 1960 price.

The idea that 3M could be hurt in the marketplace, that its seemingly endless stream of products didn't ensure invulnerability, took hold. As a consequence, the price-earnings ratio has held in the low 30s for the last few years. Investors, in their paradoxical way, seem to be saying: 3M knows how to grow, but it's not necessarily a growth company.

But our concern here is not with the fickleness of the stock-buying public, but with the fact that the company does, indeed, know how to grow—the first half of the paradox. And this brings us to one (but only one) of the key ingredients of 3M's growth: its productivity in the realm of new technical ideas that, in turn, lead to new products that can be imaginatively packaged and marketed to the buying public.

In 3M's recent annual reports to its stockholders, you find a recurring phrase: "Approximately 25 percent of last year's sales were in products marketed in the last five years." This kind of statement is commonly made by many a company with a diversified product line. But when you examine that line, or part of it, in 3M's Hall of Products, you begin to believe the statement is an accurate one.

Like most displays of company products these days, the 3M Hall of Products at once intrigues and disappoints. The array is at first bewildering: the latest in tape recorders side-by-side with industrial abrasives; roofing granules cheek-by-jowl with tarnish-inhibiting silver polish; a novel slide projector snuggled up to reflective signs. And, above all, tape, tape, and more tape—for every use you can think of.

But then, eye and mind synchronize, and you begin to associate things you see with your own immediate concerns. You wonder: Is that small tape recorder any lighter or better than the Japanese one in your briefcase? What about that crepe-paperlike tape that clings to itself and holds the surgical dressing in place on the model? Wouldn't that have held the bandage on your son's knee, after he ripped it open falling off a bike (nine stitches), better than all those various-sized Band-Aids? And that slide projector over there, the one with the soundtrack running around each transparency? Say, I'll bet we could use that somehow back at the office. Oh, here's that carpet the pro-football players use; but you're a tennis player . . . what about the bounce?

Impressions . . . of needs, desires, unfulfilled wishes. But in the end—somewhat disappointing, frustrating. Why? Ultimately because the display is remote, under glass, untouchable. What fun it would be—and how stimulating to the consumer in you—to be able to feel, examine, and play with this somehow sterile array of products. To see how sticky the tape is (or nonsticky, for there are tapes like that); to record your voice on the recorder; take your own picture; combine the two on the "talking" slide projector; to see and hear the result onscreen; to unroll that bandage, polish some silver, bounce that tennis ball . . . Mr. President of 3M, take note: Read Marshall McLuhan and visit the Smithsonian Institution.

Still, you shouldn't cavil about detail. The products are there, and there are more where these came from. Mr. President of 3M, Harry Heltzer, is the man chiefly responsible for seeing to it that the creative well runs clear and full. It doesn't happen by accident, and the reasons are interesting to explore.

Harry Heltzer himself is not the least of these reasons. Heltzer is a metallurgical engineer who began at 3M in the depths of the depression, in the nitty-grittiest of jobs—shoveling abrasives in one of the company's sandpaper factories. The company had been founded on paper coated with abrasives in the early 1900s, and in the early years one was likely as not to begin by pushing the basic product—abrasive grit—around. Heltzer's husky shoulders atop the short, chunky frame hint at the arduousness of this labor, even after all these years.

The president of 3M proves an excellent listener, the steady gaze of brown eyes "reading" you as you ask your questions, the impeccably tailored chest leaning forward against the edge of his desk. But the round, rather pixieish face really comes alive when he talks—especially about his first love, Scotchlite, the reflective material you find brightening highway signs all over the land. And here you begin to learn something of a key ingredient to 3M's success: the product champion.

"Scotchlite," says Heltzer, "is a romantic story going way back."

> Someone asked the question, "Why didn't Minnesota Mining make glass beads, because glass beads were going to find increasing use on the highways?" Just before this, I had done a little work in the mineral department lab on trying to color glass beads we'd imported from Czechoslovakia, and had learned a bit about their reflecting properties. And, as a little extracurricular activity, I'd been trying to make luminous house numbers—and maybe luminous signs as well—by developing luminous pigments. With mediocre success, I must admit.
>
> Well, this question and my free-time lab project combined to stimulate my lab director, Ed Clark, and myself to search out where glass beads were being used on highways. We found a place where glass beads had been sprinkled on the highway and we saw that they did provide a more visible line at night than the ordinary painted line did. From there, it was only natural for us to conclude that since we were a coating company, and probably knew more than anyone about putting particles onto a web than anybody, we ought to be able to coat glass beads very accurately on a piece of paper.
>
> So that's what we did. The first reflective tape we made was simply a double-coated tape—glass beads sprinkled on one side and an adhesive on the other. We took some out here in St. Paul and, with the cooperation of the highway department, put some down. After the first frost came, and then a thaw, we found we didn't know as much about adhesives under all weather conditions as we thought. The line on the road came loose in places and would kind of weave down the center. I must admit that the phones rang at a pretty good clip!

What happened next seems to be quite typical of product development at 3M. Recalls Heltzer,

> We looked around inside the company for skills in related areas. We tapped knowledge that existed in our sandpaper business on how to make waterproof sandpaper. We drew on the experience of our roofing people who knew something about exposure. We reached into our adhesive and tape divisions to see how we could make the tape stick to the highway better.
>
> The product was still "iffy" enough that not one of the divisions wanted to take it on, so we put it into what was then called the New Products Division and brought it along there. After only modest success, the question arose: Are we more skilled in the production of those glass beads (which we were trying to make by then, rather than import, so as to get a better index of refraction), or should we concentrate on the coating aspects, because of the problem of adhesion to the pavement? Dick Carlton, who was a vice president of 3M then, suggested we concentrate on the sign market and put pavement marking on the shelf for a while. This was sound judgment, I believe, since reflective signs made with glass buttons were quite expensive and there clearly was a market for a better way. So we focused on making highway signs from our paper sprinkled with glass particles—the product that became known as Scotchlite—and came back to pavement marking later on, a market we're heavily into at present.

Now this little vignette, interesting as it is in its own right, sparkles with a number of themes that illuminate 3M's technical style. One of these—the concept of the product champion—was mentioned at the outset. Harry Heltzer was Scotchlite's product champion, the man who was fascinated by a technical idea, convinced of its market potential, and determined to make it work.

This key coalescence of man and idea—Heltzer and reflective sheeting—is vividly remembered by a man who watched it take place. No longer with 3M, he nonetheless recalls it as a classic case. "Harry," he says, "would get some time on a sandpaper machine and make a roll of this stuff, put it in the back of his car, and go out and try to peddle it someplace locally. Bert

Cross, then boss of the New Products Division (and now board chairman), became intrigued, and divided the country into two regions—east of the Mississippi and west of it. Heltzer took the west and another chap the east. They'd make this reflective paper and then get in their cars like a pair of drummers. They'd sell it and come back and make more. Or they'd work to improve on it when the customers had complaints—and they did. The two were competing to see who would get to be general manager of the division that would be spun off if the product went. Heltzer made it, but there wouldn't have been any competition if Cross hadn't set it up and continued to have faith that one of the two would bring the product off."

3M is a company, then, where you very quickly get the sense that the product champion is king—or can at least aspire to a throne. That it is less so today than it was a decade or so ago is a possibility that deserves looking at a bit later. But first it's necessary to look more closely at this philosophy so central to 3M's technical style.

Most people you talk to about 3M—present- or ex-employees—pinpoint the tradition of the product champion as stemming from the succinct maxim of the company's first president, W. L. McKnight. A canny Scots (whence the Tartan Brand for 3M's products?) accountant, with a boundless interest in new ways of doing things and an intuitive sense that applied research could uncover them, McKnight pounded a constant refrain into his people: "Look for uninhabited markets." He was forever looking himself, and one of his questions—"Why can't we invent a track surface that horses can race on when it rains?"—led to the development of 3M's Tartan surfacing material. (As the owner of several racehorses, including the famed Dr. Fager, this invention satisfied McKnight's personal interest in "new ways of doing things.")

But this is essentially the story of 3M in the 60s and beyond, and so a present-day articulation of the philosophy of W. L. McKnight (now retired) is perhaps more in order. For this we can profitably turn to Bob Adams, newly appointed vice president for R&D at 3M.

Adams, the very picture of a Ph.D. chemical engineer-cum-marketing specialist-cum-R&D director, answers the question succinctly: "The technical style of 3M? Start little and build."

The idea is provocative, but needs expansion.

> We don't look to the president, or the vice president for R&D, or the director of our central research lab to say, all right, now Monday morning 3M is going to get into such-and-such a business. Rather, we prefer to see someone in one of our laboratories, or marketing, or manufacturing, or new products bring forward a new idea that he's been thinking about. Then, when he can convince the people around him, including his supervisor, that he's got something interesting, we'll make him what we call a "project manager"—with a small budget of money and talent—and let him run with it.
>
> In short, we'd rather have the idea for a new business come from the bottom up than from the top down. Throughout all our 60 years of history here, that's been the mark of success: Did you develop a new business? The incentive? Money, of course. But that's not the key. The key, the carrot out in front of each of these budding entrepreneurs, is becoming the general manager of a new business . . . having such a hot project that management just has to become involved whether it wants to or not.

How involved is "involved"? You find rather quickly at 3M that, except for the big, round, overall figures, folks are pretty coy. So you ask around outside, among ex-3M people who have seen the project manager scheme work. One such, now a high officer in a diversified chemical company, puts it this way:

> Usually, someone won't be picked to be a project manager until a new product has begun to generate perhaps one half to two million dollars in sales. Then, someone in the nucleus of people that has nurtured it may be selected as the person responsible for all phases. He's really a tiny general manager accountable for sales, manufacturing, and product development activities, with his

> own P&L responsibility. He may be the one who originated the idea, or he may not be. He may even be brought in from one of the divisions—as I was. When the sales get closer to the $15 million mark, the company begins thinking of making it a department, or even a small division. That's the hierarchy—first a project, then a department, then a division. But for a company to be able to say, at a very low level of sales, "Okay, Charlie, if you can make it go, it's yours . . . but we're going to be looking to you for growth and profits," that's a pretty strong motivator.

It is, indeed, but it does convey a tone of loneliness, of Horatio at the Bridge battling against overwhelming odds. But 3M isn't interested in chucking people in the water just to see whether they can swim. It wants each enterprise, big or small, to meet three financial criteria: a 20 to 25 percent pre-tax return on sales; a 20 to 25 percent return on stockholder's investment; and an earnings growth of 10 to 15 percent per year. To meet these average requirements means considerable attention to the budding ventures, to insure that as many of them as possible flourish in an above-average fashion.

This sense of parental concern for adolescent growing pains is nicely conveyed by Bob Purvis, an ex-3M chemist who was technical manager of the company's industrial finishing department in the early 1960s. Recalls Purvis,

> When Harry Heltzer was a group vice president, our department was a new and very small part of his group of divisions, contributing maybe one third of 1 percent to his total sales. And yet once a month, on the same Monday morning, at 8 A.M., there would be Heltzer, on the scene, for a meeting with our key management group—the department manager, the sales manager, the manufacturing manager, and myself—asking: "How can I help you fellows?"

Powerful as this project manager scheme is, it sounds a little pat: someone gets an idea; a nucleus of talent is formed to develop it and try it out in the market; sales grow; one person is picked to manage all aspects; sales grow further; and finally a decision is made whether to make it into a separate division (of which there are now 21 at 3M) or to keep it as a product line in an existing division. You find yourself hungering for an example, to make the whole thing real. Find me a typical project manager, you say. People smile and nod consent. They say okay because they want you to understand—up to a point—how things work at 3M. They smile because they know, and tell you directly, that there's no universal mechanism whereby 3M directs ideas into project status. Like Topsy, you quickly learn, they just kind of grow.

Still, you press for a name, a face, a real person grappling with a real project. Hmmmm. We've talked about things like Scotchlite, and tapes—pretty simple products, really, once the technology was mastered. But the company's growing. Products are getting more complex. Maybe we'd better look at one of these. Well, there's Dr. Marshall Hatfield over in the Computer Graphics Project . . . Great, you say, let's go!

Going means leaving the tall administration building for one of the low, yellow-brick lab buildings across the expanse of greensward and parking lots that knits together the 3M complex. The company station wagon that constantly circulates has departed, so you decide to walk. After soggy, rainy New York, the sun is a joy. Fleecy clouds edge across a crystalline Minnesota sky. You see them reflected in the blue-green windows of the new administration-annex buildings, which sparkle rather like facets on the Emerald City of Oz.

But what's going on? You see the cloud reflections, but that's all. No lights. No people. Just clouds on a hundred windows. Your omnipresent guide explains: "Those windows? They're glass, with a layer of 3M Scotchtint, a new product that filters out the harsher wavelengths of the sun." And filters out all the wavelengths of light from inside, you think to yourself. It gives the building a closed, shuttered look. Keep out. Don't look. No peeking. Like those millions of Americans who seem to want to hide their real selves behind dark glasses.

You're taken on a shortcut, so you come in the back way, where the 3M-ers go in, and the products go out on special 3M trailer-trucks. But back door or not, the ubiquitous guards are there. Your name tags are in order. Your public relations guide says with a smile, "I guess you don't want to see my identification, do you?" The guard, with a glance at your cameras, reacts like an airport cop to a passenger who's just asked, "Does this plane go to Havana?" "Yes sir, I do," he responds, and waits, steady-eyed, until proper identification has been produced. Once again, you're "inside."

Inside is primrose yellow, doubtless cheery in the long winter months, but in the summer . . . sort of neo-airport. People abound in the corridors as you are hurried past glass-windowed laboratory doors. The men nearly all suit-coated, in colors lively to the eye; little of commuter-run gray here. The madras sportcoat and crew cut above suntan crops up now and again. A few longhairs around the ears, an occasional side-vent jacket, but no kooks. The miniskirts (alas!) are mere tentative experimentation.

But you feel at home here. As a former divisional R&D director you talked to said, "You'll like the 3M people." And you do. In a few minutes, you're shaking hands with Marshall Hatfield (no relation to the McCoy antagonists of yore, he answers, for what must be the umpteen-millionth time), manager of 3M's Computer Graphics Project.

What's that? you ask. Well, it turns out that 3M is getting embroiled in the computer-peripheral market—what's known as the man-machine interface. Computers have become so fast that the person who buys one can't keep up with the output anymore. He's getting swamped with paper, and the more jobs he finds for the computer, to justify its high cost, the deeper he gets in the paper quagmire. To 3M, here is an "uninhabited market"—or a *relatively* uninhabited market, as one ex-3M technical employee paraphrases W. L. McKnight's favorite slogan.

Marshall Hatfield, a tall man, graying at the temples, who looks as though he should be governor of one of the Rocky Mountain states, explains what 3M is doing in these profitable but potentially dangerous waters.

> Over the recent years, the company has experienced a very attractive and healthy growth based on graphic arts technologies. Our copying programs, high-speed printing plates, microfilms, and the like are products created by expanding, building onto unique 3M proprietary technology.
>
> We've begun to think of ourselves as an "image-forming" company, just as we're a coating and adhesives company. So it was only natural that R&D in the central research laboratory and the Duplicating Products Division lab began to focus on all kinds of ways to form images.
>
> One of these was the use of an electron beam to write on some sort of sensitized material.
>
> You see, an electron beam can be controlled, told what to do, at very high speed by electrical impulses. Our research management had the feeling that if this beam movement could be recorded on some sort of special film, the combination would provide a unique technology that would be important in the computer field. We began to develop such materials and came up with a dry-silver coating on ordinary film that would react wherever the electron beam struck it. To develop the image, all you had to do was apply heat to carry to completion the reaction that the electron impact had begun. That, plus learning how to move the film rapidly in and out of the vacuum that the beam requires, gave us the rudiments of a recording system.

Aha, now the tie-in with computers becomes clearer. Electrical impulses from the magnetic-tape output of the computer drive the beam, the beam writes on the film, which is then heated continuously to develop the image. . . .

"Exactly," nods Hatfield, "and the whole system is what we call the 3M electron-beam recorder. The user can then send the roll of computer film, which is what we call it, wherever he wants to instead of a stack of paper. The receiver can then use our regular microfilm reader-printer to scan the film itself, or he can call for an enlarged copy by pressing a button.

This is where the savings occur: You use film instead of paper to keep the record of your computer output."

This is the product, but what about the market? For 3M, if anything, is a market-oriented company. The market analysis must go hand-in-hand with the evolution of the product idea. Hatfield puts it this way:

> I felt quite strongly that before any of the electron-beam technology we were developing could move into the product development stage, we would have to know more about the computer market than we did at the time. Until then, 3M's activity there was almost exclusively in making and selling magnetic tape to computer manufacturers. We really didn't know the computer customer. So I deliberately hired a successful computer salesman to be our sales manager. We put him right into our laboratory and told him to become familiar with what our people thought could be done with this new technology. Then he took some of these men out to the computer customers, his old friends, so we could begin to evolve an understanding of what they needed and wanted. The system we discussed a moment ago was the result.

From the tone of Hatfield's remarks here you get the feeling of a manager talking—a man talking technics, marketing, product development, all the key aspects. But how did he get to be manager of this project—that's what you came across the sunny plaza, past the guards, up to this windowless but still rather elegant office to find out. The heart of the technical strategy of 3M.

The path is somewhat complicated, as are all human odysseys, but all the more revealing for it. Hatfield began his 3M career as a research chemist in the central research laboratory back in 1950. A few years later he was asked to organize a physical chemistry section, and what better man to ask than a Ph.D. physical chemist from the University of Illinois?

Pretty soon the Thermofax copying project started to take off, and it would need its own research program. This led almost simultaneously to Hatfield's developing a research program for the Duplicating Products Division in which Thermofax was being nurtured. Yet another project reared its head within the division, something called the microfilm reader-printer project—"Nothing more," recalls Hatfield, "than two or three people trying desperately to get a product on the market."

So by this time, Hatfield had two basic responsibilities, one to develop a research program, and the other to organize a laboratory that would oversee quality control, technical service, and the like for the budding microfilm project. Gradually, the microfilm products side of things began to grow to the point where it might split off from the Duplicating Products Division into a division of its own—which it eventually did.

Hatfield found himself faced with a decision: to stick with the research programs in Duplicating, or to stay with the quality control laboratory for microfilm, which included research to help nurture this tender, young plant.

Perhaps indicative of a latent entrepreneurial itch, Hatfield chose the latter and became technical director of the soon-to-emerge Microfilm Products Division. As it turned out, the new division began to pay its own way before too long, and Hatfield began to exercise one of the key functions of any 3M divisional technical director.

"Someone in that position," says Hatfield, "not only has responsibility for R&D, quality control, technical service, production service, and giving help to the selling organization, but he also has to look around for things that might make a new business. The 3M company really doesn't care *where* its new businesses come from, and our own division—the Microfilm Division—wanted one in the worst way. Some general managers don't care, feeling that central research or the New Business Ventures Division will hand them something. It just happened that my general manager had a great desire to create a new business out of his division."

It also just happened that 3M was experiencing some difficulty in coordinating the various divisional R&D efforts—too many people were working on similar things. One of these turned out to be electron beam technology and, since Hatfield had become interested in the coordination problem, this one caught his technical director's eye. He became interested in the computer possibilities of electron beam imaging, and succeeded in bringing in the computer sales specialist mentioned earlier.

Right about here the birth of a 3M project manager can be said to have occurred. Or at least its conception. Listen to Marshall Hatfield:

> All of a sudden I began to realize that this little thing we had could be as big a business as the company has ever seen. So I thought, hell, I'm not interested in microfilm anymore, I'm more interested in this new electron-beam activity we've got going. So I asked to take on the project myself.

Ask and ye shall receive. Not necessarily, at least not at 3M. But if you're a technical director, and you have your division general manager's blessing, you can try. The still-adolescent Microfilm Division didn't have enough money to begin a project of this magnitude, involving, as it did, placing a firm foot on the edge of the computer market. So Hatfield turned to central research for support.

Says Hatfield,

> I sold our vice president for R&D, then Dr. Charles Walton, on the idea that we had to have this project for the company's future health. So he provided some free activity—money from his budget. We also went to some of the other divisions where electron beam work had been going on. This can get touchy, because if the thing goes, who's going to get the profits? But we convinced the Duplicating Products Division to spend some money and provide some talent—people who were eager, by the way, to see their technology blossom into a new business—on the grounds that they would benefit later on. All this, of course, before we'd got the project going to the point where you say to management, okay, now you're going to start losing money in a hurry!

This process by which the 3M entrepreneur seeks out financial underpinnings for his project has been described in somewhat less elegant language by a former 3M technical manager who has watched it take place many times. "Actually," he says, "the typical product champion at 3M turns out to be one-quarter technical person and three-quarters entrepreneur. He pushes a product idea by what amounts to conning the management—somewhat exaggerating its chances for success, hoping to get enough money to make the product fill the bill."

Webster's New Collegiate Dictionary defines *con* thus: (1) swindle, (2) coax, cajole. Taking definition number 2, we can realistically assume that 3M top management has heard (and done) its share of conning. And doesn't mind listening to more.

President Harry Heltzer puts it succinctly and engagingly:

> We're in the business of gambling on individuals. If someone has been a pretty good judge in the past of what he said he was going to do, and has done it, then if he comes in and says I want to start importing moon dust, I guess I'm likely to let him try. In like manner, if a fellow who's been around here a great many years comes forward and says this year we've got a tremendous breakthrough that's going to cause all sorts of things to happen, and he's had four or five years in a row of not delivering on what he said, well, I'm likely to give him an argument.

Heltzer's comment here implies more of a one-to-one link between the person with an idea and the chief executive officer than is really possible in a company the size of 3M. Still, it echoes a theme that ran through Heltzer's own experience years ago when he was struggling to make Scotchlite a viable product. This is the bedrock principle of free communication within the company: If you need help, go find it. Anywhere. This of course requires two correlary principles: Know where to look for the information you need, and have it be made available when you ask.

Many ex-3M people instantly put their finger on these rules of intellectual behavior as one of the key strengths of the company over the years. Recalls one former technical director,

> There was a tremendous exchange of information, and no real holding back. I could get any report on anything I wanted to know about. People were encouraged to move about from lab to lab. If, for example, one of my people heard about something going on in another division, I'd tell him to get over there and find out about it because it might just be the key to our problem. And he would be welcomed over there.

Even though you, as a visitor from "outside," are not able really to gauge this sort of interchange, because you cannot circulate freely among 3M's technical personnel, you do get a feeling for it—logically—in the dining room where you are taken for lunch. You're seated apart, feeling a bit "in Coventry," but you catch snatches of conversation—arrangements being made for later meetings, a product line being discussed, the results of a trip being reported.

Feeling frustrated at the invisible fence around you, you let your attention wander from the remarks of your host and to the surroundings. It's an unexpectedly large room for an executive dining room, richly paneled with 3M's walnut Tartan-Clad vinyl veneer. Your eye drifts to the high ceiling, painted black, and there lurking in the shadows is a row of spotlights of various sizes and shapes. Just like a TV studio.

Well, you've heard 3M wryly described as Minnesota Mining & *Meeting* Co., but then something else you've heard about clicks . . . the 3M Technical Forum, an unusual example of the company's emphasis on internal communication. Like many other technologically based firms, 3M many years ago sought a way to encourage its professional people to mix and exchange ideas. One of its early vice presidents for R&D, Dr. Lloyd Hatch, sponsored the idea of an internal technical society, run by the professionals, which would hold seminars and colloquia on all sorts of technical topics drawn from the work going on in the various company labs. This was broadened, before long, to include management subjects, and the behavioral sciences. Anything the professionals wanted to chew over that might help them in their work.

Eventually, someone got the bright idea of annually displaying to the Technical Forum members the newest goodies from the divisions—new products just about to be introduced to the marketplace. So a few days were set aside—called Tech Forum Days—during which the many divisions would erect displays of their latest wares. Quite naturally, the competition has proved fierce among the divisions as to who will put on the best show, who will come up with the most ingenious products. The result: not only intense communication of who's doing what, but a stimulating race to see who can do it best. The idea is a charming one, and the gregariousness of Nottingham Fair springs instantly to mind. But they didn't have those spotlights back in 12th-century England.

But 3M does have them, and they are more than a little symbolic of the climate there. You get the sense that it's *de rigeur* to jostle a bit for the limelight—especially if you want to become a project manager. Sometimes, at the division level, the spotlight wavers—there are so many things to push along that managerial scrutiny must, perforce, skip lightly over some of the newer ideas. But even then, the person with an idea—especially if he's done his homework and can answer some hard-nosed questions—can bring his dream to an arena where the candlepower is a bit more intense.

This is 3M's New Business Ventures Division, and its new general manager, Lester Krogh, is accustomed to giving hard looks at where the company is going. Formerly director of technical planning, Dr. Krogh has been a division technical director (Abrasives) and was at one time the head of the chemical research lab in central research. In an essentially chemical company, which began in the abrasives business, that's bespoken background, indeed.

You ask whether someone who wants to be a product champion can gain access to advice and maybe even funds from the New Ventures Business Division. Krogh, a tall, rather professional sort of fellow, chooses his words with care:

> It depends a great deal on the climate in the person's division—the attitude, the resources, how the present business is going. If the management of that division is hungry for a new product, he'll do better staying right there, because then he doesn't interrupt a chain of events that could help him build a business. On the other hand, if the division management is pretty well absorbed in other things, then he definitely ought to consider coming to us for corporate sponsorship.
>
> He can come to us directly—after consulting his own management, of course. Say they don't give him much encouragement, but okay his talking with us to find out our enthusiasm for his idea. It's a feeling-out process, really. His division is a bit dubious, so they play a conservative game. He comes to us and gets a fresh set of inputs. This gets fed back to his division management, and they may become more enthusiastic as a result. Or they may not.
>
> Let's suppose not, and we in New Business Ventures are interested. Then there's some more feeling out. The person wants to know what our commitment might be. Do we have in mind a six-months look, a two-year look, or a five-year look at his new-product idea? This is important to him, because if the project fails, he will still be able to go back to his old job, but he will have lost some time. He's got to resolve the conflict between security and opportunity. This is not at all an unhealthy conflict: we've had plenty of people who had the chance to become entrepreneurs but who chose instead to stay where they were. They're better employees because they had the chance and were able to decide what was best for them.

Close observers of the 3M company, people who were active in it in its heyday, would tend to agree. But, they point out, the company has gotten so big that the risks of failure in becoming an entrepreneur have increased. To grow at such past meteoric rates means the company needs bigger new-business ideas—often bigger than one man cares to tackle.

Then, too, management has gotten more diffuse; despite strenuous efforts to keep in touch, top management finds it harder and harder to give the personal encouragement that gave its men such incentive up to the mid-1950s. As a consequence, say these observers, the traditional 3M product champion is perceptibly becoming a rare breed. Rather than lose so much time, with the risks of failure so much higher, the tendency seems to be for talented people simply to wait for the company to grow; they move up by attrition.

One former 3M employee, who has thought long and hard about his old company as it moves into what might be described as a third generation of management, put it this way:

> It seems to me that there are basically four routes by which a talented person can get ahead in 3M these days:
>
> First, and least risky, is to follow the company's natural growth—the odds being that there will be a good slot for you.
>
> Second, a little more risky, tie in with a new acquisition—of which 3M is making more and more nowadays.
>
> The third way is to tie in with an emerging product champion, someone who's willing to put his neck on the block.
>
> Last and most risky of all, try the entrepreneurial route yourself.
>
> The Harry Heltzers and Bert Crosses, men over 50, succeeded by being product champions. But there's an increasing preponderance of people under 50 who are advancing by the other three routes—though most of them prefer to think of themselves as having been product champions.

Les Krogh agrees that 3M faces something of a stand-patter's syndrome. As head of the company effort to get new businesses started outside of divisional channels, he's one of those who has to cope with it. You ask, how? The answer is cautious, and general:

> First, let me say that a person who can do a good job in our existing business is an important cog in the whole establishment. But you have to stimulate people to move out of that by clearly stating that growth is the corporate objective. He has to have before him continually examples of people who have been rewarded for taking the risks inherent in getting us into new business. But more than this, risk-takers in a corporation have to have some sort of security. If his project fails, for no fault of his own, you've got to have a place for him to go. Hopefully, if he's a good employee, one who can handle a risky enterprise, you'll encourage him to try again later on. But if he says, in effect, once was enough, you must have a mechanism for getting him back into the stream of things.

These are clearly sound strategies, ably articulated. Just what tactics will be adopted to see that they succeed are Lester Krogh's abiding concern. And 3M's. One can penetrate no further: only time will tell, really, whether the New Business Ventures Division grows beyond its present modest base into something more than a discreet limbo for new-product ideas which divisional general managers do not want to pursue, or in which they have already sunk too much of their precious capital.

But these are internal problems and, vital as they are, they pale beside Krogh's broader charter—to bring 3M beyond just the making of products and offering of them for sale. Says Krogh: "Our philosophy has changed to the point where we see the establishment of a new business now as a total package, including a sound marketing plan." This sounds a little strange, considering that 3M has always been noted for being quick off the mark in getting its products to the consumer. So you explore further.

"More than ever," explains Krogh, "we try to probe the market as soon as we can with a new development."

> For one thing, you can spend enormous sums in developing a product to perfection, and yet find that no one wants it. We're looking to see, first of all, whether it's the right time in the marketplace. Secondly, we're *imagining* what the customer wants. We may or may not be right. As soon as we get him to say, "yes, that's a good idea, BUT," we have a valuable input to our technical people. If we keep getting the same answer—we like that, BUT—the idea goes on the shelf temporarily. Then we'll dust it off periodically and try it again, until finally the marketplace really is ready for it, or until we have found a new way of doing a job for which the market is ready.
>
> Scotchlite was a good example of this. The highway-department customer thought for a long while that painted signs were good enough. It took a long time to improve the product to the point where we could convince him that the safety aspects of high-reflectivity were worth the extra cost. The same thing applies to reflectorizing *people;* we know all sorts of ways to do it, but it's a question of finding ways of getting the customer to want it. We're still trying.

Besides the care and feeding of internally generated new business ideas, and helping develop a more integrated marketing strategy, Krogh's division also examines the possibilities of growth by acquisition. This is a technique 3M has opted for more and more in recent years. Over the last decade, for example, 3M has entered the photographic field firmly through the acquisition of four companies—Ferrania in Italy; Revere Camera Company and Dynacolor Corporation in this country; and Allied Colour Film Company in Canada. The idea here was to compete with Kodak in this "relatively uninhabited market." (The attempt, thus far, has not been a stunning success.) At present, 3M is negotiating for the acquisition of Riker Laboratories, a modest-sized pharmaceutical house.

Some see these moves as evidence that 3M has concluded it cannot continue to grow primarily through the internal generation of new products. President Heltzer says firmly that he doesn't believe in acquisition for acquisition's sake, which doesn't quite settle the point, and then goes on to say, "It's hard to think of any acquisition in which there wasn't a relationship either in existing technology or potential technology." Which doesn't settle it either.

The photographic company acquisitions certainly relate strongly to 3M's proprietary position in "image forming," and the pharmaceutical company plans relate to the company's strong position in fluorochemicals (an early acquisition in itself) in ways 3M is hesitant to spell out.

Lester Krogh adds a few shadings to the picture: "I think that we're actually building more technologies than we are acquiring, or want to acquire. Our acquisitions are going to be, if you like, chink filling. If there's a hole in the wall, a brick missing, we'll provide that brick so that the wall is sturdy."

The image is still blurred, as through a glass darkly.

Whatever the mix of growth modes 3M chooses in the years ahead, by now you are sure that certain features will continue to stand out in the picture of the company's technical strategy—all of them reinforcing a traditional dictum: "Market products that are novel, patentable, and consumable—each returning as high a profit on investment as is possible."

Research thinking will assuredly play an outstanding role. More inventive than scientific, to be sure, for inventor-type thinking is embedded deep in the company's soul. The uses of *directed* basic research will be appreciated, but its usefulness in impressing Wall Street will not be overlooked.

As 3M's professionals go about their work, they will continue to be encouraged to spend part of their time (up to 15 percent is the commonly accepted figure) working on dreams of their own, thereby uncovering enough nuggets to convince their bosses to stake out a claim.

The ideas they come up with will still be protected to the fullest extent allowed by law, for 3M reputedly has the largest (and most aggressive) staff of patent lawyers working for it of any company its size in the country. For just this reason, competitors will continue to think twice before taking on 3M.

The company's professionals—the scientists and engineers—will probably continue to pursue their careers at this one company longer than most. Low turnover of technical employees (reportedly 2 to 3 percent) has long been a 3M characteristic—in part because the work there is exciting and profitable, in part because 3M has an exceptionally tough employee agreement that each professional must sign. In essence, it prevents an employee from working for a competitor for two years, and with 3M as diversified a company as it is, that can be a long dry spell.

An aggressive sales and marketing force will be looking for more and more subtle ways to make you want 3M's products when (or as close to when as possible) the company's researchers come up with them. Lester Krogh means business.

Product champions will continue to sprout among the stand patters, perhaps fewer of them than before. But each will be encouraged with a sort of messianic fire by the earlier generation of entrepreneurs who know the thrill of having the reins of a new business in their hands.

And finally, the ubiquitous guards will doubtless be there, making sure that scrutiny from "outside" is controlled just the way Minnesota Mining & Manufacturing Co. has always liked it to be.

The guards. That reminds you: Don't forget to turn in your badges on the way out. (They'll be relieved to see those cameras go!) The station wagon waits to carry you to the airport. The driver, Mr. DuBois, is courteous and informative about the features of St. Paul rushing past. And, above all, delighted that you remembered enough of your French to pronounce his name correctly.

As the Boeing 747 lofts its way into a sparkling Minnesota sky, leaping the Mississippi to begin its trajectory back to humid, cranky New

York, you think back on the people you've met—inside 3M and out—over the past few weeks. An insistent question thrusts its way through your fatigue: Would *you* like to work at 3M? You reflect. I would . . . I wouldn't . . . I would . . . I wouldn't . . . I would . . . no, I wouldn't. Hmmmmmm. Maybe you'd better go and get a new daisy, before winter sets in.

Reading IV–3
New Ventures for Corporate Growth

E. B. Roberts

To meet ambitious plans for growth and diversification, corporations are turning in increasing numbers to new venture strategies. However, most new ventures fail. And even when they do succeed, they often take 10 years or more to generate substantial returns on the initial investment of capital and management attention. The question is obvious: Given its uncertain promise, why is corporate venturing proving so attractive?

The odds against its success are enormous. The push toward a venture strategy usually comes when a company, wishing to address customer needs it has not previously served, seeks either to enter new markets or to sell dramatically different products in its existing markets. Second, most ventures involve a new technology—whether that technology is new to the world or only to the company. Third, almost every corporation undertaking a venture has found it both necessary and desirable to establish for it a structure quite different from that in use throughout the rest of the organization.

Reprinted by permission of the *Harvard Business Review*. "New Ventures for Corporate Growth" by Edward B. Roberts (July/August 1980).

Entering unfamiliar markets, employing unfamiliar technology, and implementing an unfamiliar organizational structure—even taken separately, each of these presents a troublesome challenge. Put all three together in a single new venture organization, and it is no wonder that their joint probability of success is rather small.

Nonetheless, venture strategies are increasingly attractive to many companies. My purpose in this article is, first of all, to consider just why this should be so. I then examine and evaluate the various options available to companies embarking on a venture strategy. Finally, I discuss what companies can do to improve the likelihood of venture success.

Though I address all three points, my focus will primarily be on two of the options: the large and small company joint venture, which has the principal virtue of speed of market impact, and the internal venture organization, which is best illustrated by Minnesota Mining and Manufacturing Company (3M) with its long-term record of success in venturing.

WHY ESTABLISH A VENTURE ORGANIZATION?

If the odds against a new venture strategy are high, what makes it so very appealing? The answer is really quite simple: the alternatives are no better. No other strategy for enhancing growth in size or profitability currently offers a higher probability of success. Consider:

- When it was still easy to identify unmet needs in the marketplace, companies could launch products to meet them with every expectation that the markets thus defined would continue to grow and to support continued company growth. Today, however, many such traditional markets have become saturated, incapable of additional sustained growth.
- When there were still few large companies in the developed world and little technological

competition among them, a company could apply its R&D capacities to develop new products, which it could then sustain relatively easily by ongoing efforts at incremental innovation. Today, with technological sophistication diffused throughout the world, for a company to remain competitive—especially with mature high-volume products—requires far more than the uneven performance of traditional R&D.

- When untapped foreign markets were still plentiful, it was a simple matter to enter them. Today, the overwhelming likelihood is that those markets have long since been populated both by domestic competitors and by native companies.
- When interest rates were lower, price/earnings ratios higher, and antitrust regulation less strictly enforced, companies could readily grow and diversify by acquiring other companies. Today, an acquisitions strategy can rarely be pursued on such advantageous terms.

In short, the most common growth strategies of an earlier era are no longer so easy to follow or so likely to succeed. Consequently, venture strategies, even with their low probability of success, have begun to look much better.

THE SPECTRUM OF VENTURE STRATEGIES

Exhibit 1 displays the range of alternative strategies for launching new ventures. At one end are those approaches that feature essentially low company involvement; at the other, approaches that demand high levels of commitment both in dollars and in management time.

Venture Capital

At the far left of the spectrum is venture capital, the investment of money in the stock of one company by another. During the mid- to late 1960s, many major corporations decided to secure entry into new technologies by taking investment positions in young high-technology enterprises. Major companies in a variety of industries—companies such as Du Pont, Exxon, Ford, General Electric, and Singer—sought out a "window" on promising technologies through the venture capital route, but few of them have been able to make the venture capital approach by itself an important stimulus of corporate growth or profitability.

EXHIBIT 1 Spectrum of Venture Strategies

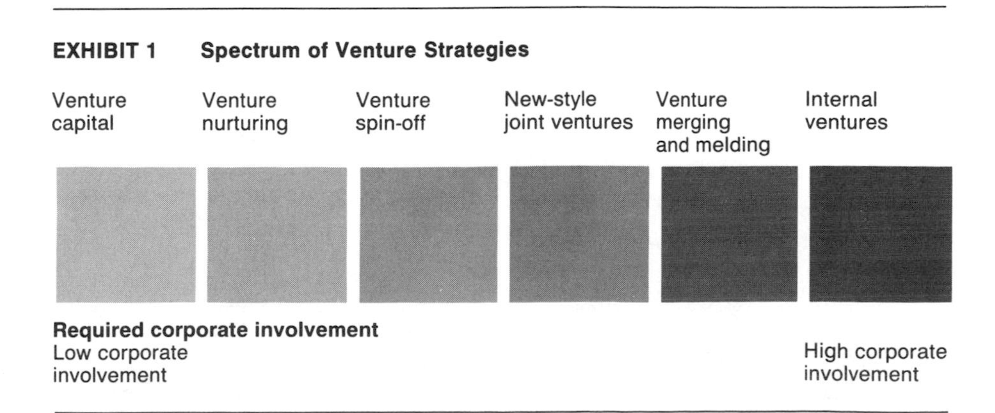

Venture Nurturing

Second along the spectrum, venture nurturing involves more than just capital investment. Here the investing company also gives managerial assistance to the nurtured enterprise in such areas as marketing, manufacturing, and research. Though perhaps a more sensible approach to diversification than just the arm's-length provision of funds, venture nurturing is still unlikely to have a significant impact on the investor's sales or profits. Cabot Corporation, for example, tried this approach but gave up after two years of frustrating experience with several start-up companies.

Venture Spin-Off

As a by-product of its R&D efforts, a corporation may develop an idea or technology that does not fit its mainstream interest, that may entail substantial risks to the parent, or that may be better developed on an independent basis outside the company. The originating company will then spin off the new business as a separate corporation, either seeking to gain market and operational experience in a new field, as Exxon Enterprises did temporarily with its Solar Power Corporation, or to attract outside growth capital, as General Electric did with its formation of Nuclepore and other companies.[1] Venture spin-off may be a good way to hold on to an internal entrepreneur or to exploit a by-product technology, but the limited involvement it allows still promises only limited returns to the parent company.

New-Style Joint Ventures

Because I think this approach of particular importance, I will discuss it in more detail later on. Here large and small companies enter jointly into new ventures. The small companies provide entrepreneurial enthusiasm, vigor, flexibility, and advanced technology; the large ones, capital and, perhaps more important, worldwide channels of marketing, distribution, and service. This combination allows for the rapid diffusion of technology-based product innovations into large national and international markets.[2]

Venture Merging and Melding

Toward the right side of Exhibit 1 is an approach that I call, for lack of a better name, venture merging and melding. This is what Exxon Enterprises is attempting by deliberately piecing together all the various forms of technologically oriented venturing shown in Exhibit 1 into a critical mass of marketing and technological strengths. In turn, these strengths have allowed Exxon to transform itself from a huge—though unglamorous—one-product, narrow-technology oil company to an exciting company that is expanding into computers and communications, advanced composite materials, and alternative energy devices.[3]

Internal Ventures

Finally, on the far right are internal ventures, those situations in which a company sets up a separate entity within itself—an entirely separate division or group—for the purpose of entering different markets or developing radically different products. This approach has great potential but a mixed record to date. Du Pont, for instance, has had a spotty record in internal corporate-level venturing for nearly two decades.[4] Ralston Purina, however, has done rea-

---

[1] Sharon Sabin, "At Nuclepore, They Don't Work for G.E. Anymore," *Fortune*, December 1973, p. 145.

[2] James D. Hlavacek, Brian H. Dovey, and John J. Biondo, "Tie Small Business Technology to Marketing Power," *Harvard Business Review*, January-February 1977, p. 119.

[3] "Exxon's Next Prey: IBM and Xerox," *Business Week*, April 28, 1980, p. 92.

[4] Russell W. Peterson, "New Venture Management in a Large Company," *Harvard Business Review*, May-June 1967, p. 68.

sonably well. For the record, the most consistently effective performance with internal ventures I know of is that of 3M, whose philosophy and methods I will describe in some detail later in this article.

TEAMING THE LARGE WITH THE SMALL

Let us now examine in depth an approach to venturing that has shown itself to be relatively "quick and dirty" and adaptable: the new-style joint venture. You will recall that new-style ventures are those in which large and small companies join forces to create a new entry in the marketplace. The idea here is quite simple. The large company usually provides access to capital and to channels of distribution, sales, and service otherwise unavailable to the small company; in return, the small company provides advanced technology and a degree of entrepreneurial commitment the larger one often lacks. Together the strengths of both add up to a distinct competitive advantage.

Numerous studies on the process of innovation have shown time and again that small companies and individual inventors account for a disproportionate share of commercially successful, technologically based innovations.[5] Whether the explanation lies in their superior commitment, drive, freedom from constraint, flexibility, or closeness to the market, the facts themselves are quite clear. Small entrepreneurially minded companies have been unusually able to come up with technological advances that are competitive in the marketplace.

Balancing Needs with Strengths

But the small company has an obvious problem: its size. It has neither extensive market coverage nor an extensive sales force. It is usually not even a national company. Young entrepreneurial companies are often regional at best, and the obstacles they face—organizational and financial—in becoming national or international are tremendous. The great success stories of corporations such as Polaroid, Xerox, or Digital Equipment are clear exceptions to the rule.[6] In the vast majority of cases, the small technology-based company simply cannot grow from within to a large-scale size with the time and resources available to it.

By contrast, a large company has relatively easy access to capital markets as well as significant capital availability within itself. Moreover, it not only has large sales but a large establishment overall. It has ample manufacturing capacity located near its various national and international markets. It has a distribution and marketing organization that covers all its relevant market territory. It can service its products on a national and international basis.

Competitive Advantage

Now, if the entrepreneurial commitment, innovative behavior, and advanced technological products of the small company were combined with the capital availability, marketing strength, and distribution channels of the large, it stands to reason that the synthesis might well create significant competitive advantage. Indeed, many pairs of differently sized companies have entered into just this kind of venture arrangement. Exhibit 2 lists but a few of the attempts in the Boston area alone.

A typical example of a successful arrangement is the joint venture between Roche Electronics, a division of Hoffmann–La Roche, and the Avco Everett Research Laboratory. Their venture was to produce an inflatable balloon heart assist pump, and it has been both technically and commercially successful. The de-

[5] Five of these studies are referenced in my article, "Technology Strategy for the European Firm," *Industrial Marketing Management,* August 1975, p. 193.

[6] See my article, "Entrepreneurship and Technology," *Research Management,* vol. XI, no. 4, 1968, p. 249.

EXHIBIT 2 Examples of Large/Small Company Joint Ventures

Large Company	Small Company	Area of Joint Venture
American Broadcasting Company	Technical Operations	Black-and-white film transmitted for color TV viewing
American District Telegraph	Solid State Technology	Industrial security systems
Bell & Howell	Microx	Microfilm readers
Dravo	Anti-Pollution Systems	Molten-salt pollution control systems
Elliott (division of Carrier)	Mechanical Technology	High-speed centrifugal compressors
Exxon Nuclear (division of Exxon)	Avco Everett Research Laboratory	High-energy laser uranium isotope separation and enrichment
Ford Motor	Thermo Electron	Steam engines for automobiles
General Electric	Bolt Beranek & Newman	Hospital computer systems
Johnson & Johnson	Damon	Automated clinical laboratory systems
3M	Energy Devices	Updatable microfilm systems
Mobil	Tyco Laboratories	Long-crystal silicon solar conversion technology
Pitney Bowes	Alpex Computer	Electronic "point of sale" checkout systems
Roche Electronics (division of Hoffmann–La Roche)	Avco Everett Research Laboratory	Inflatable balloon heart assist systems
Wyeth Laboratories (division of American Home Products)	Survival Technology	Self-administered heart attack drug and injection systems

velopment of the product came from a combination of the electronics technology and materials capability of Avco Everett with the marketing, distribution, and field service capability of Roche. More recently, Avco Everett has taken over the entire venture as part of its own diversification movement into the medical field.

Characteristic Difficulties

However, no matter how appealing the prospect, the problems with this new-style approach are significant and troubling. Consider, for example, Johnson & Johnson's joint effort with Damon Corporation to develop automated clinical laboratory equipment.

At the time of the joint venture, Johnson & Johnson had annual sales of roughly $3 billion. By contrast, Damon, a spin-off from the MIT Research Laboratory for Electronics, had only $3 million in sales when it started negotiations with Johnson & Johnson and $30 million by the time it successfully concluded negotiations to initiate the joint venture three years later. The intended product was to sell at prices of $100,000 or more to large hospitals for doing clinical analyses of patient fluids.

Though a partial technical success, the product was a commercial failure. Why? Because two kinds of problems often confront new-style ventures.

Misreading

Often both partners misread the appropriateness of the large company's channels of marketing and distribution. It is all too easy for a large company to think that it can sell almost anything through its vast field sales and service organization.

In the Johnson & Johnson–Damon case, Johnson & Johnson could correctly say that it had salespeople regularly calling on every major hospital in the free world. Therefore, it might well have felt it had the representation necessary to sell a Damon-developed clinical laboratory system. But Johnson & Johnson sold largely disposable medical products such as Band-Aids; the people to whom it sold were reorder clerks, inventory supervisors, or head nurses; and the basis on which it sold was a combination of product quality and volume discounts.

To whom, however, does one sell a $100,000 piece of clinical laboratory equipment? Certainly not reorder clerks or inventory supervisors. The director of the hospital's clinical laboratories will be involved, as will the hospital's chief administrator. And in all but the largest hospitals, so will the board of trustees. It is too much to expect that sales personnel used to selling Band-Aids to reorder clerks can switch overnight to such a different level of responsibility and remain effective.

In addition, a major piece of clinical apparatus requires a special level of field service. With its different experience, Johnson & Johnson simply did not have that kind of field operation in place. To be fair, Johnson & Johnson had also been selling small medical instruments, but even this had not prepared it for the service requirements of a clinical laboratory analyzer.

Though it is easy to misread at first glance the appropriateness of a large company's marketing channels, a little careful thought and common sense are often all that are needed. The central question is clear: Do the company's sales and service organizations meet the particular requirements of the new product? If not, can they be made appropriate with only slight modifications—say, expanding an existing service capability or adding a specialty salesperson to an existing field office? Incremental change of this sort, if a realistic alternative, is almost always less expensive than starting from scratch and trying to build a whole new organization.

Impedance mismatch

This is more of a generic problem of new-style ventures than the misreading just discussed. Differently sized companies tend to breathe, play, and act on very different frequencies. They have very different ways of managing themselves and their decision processes. David Kosowsky, the president of Damon, is quoted as saying that he would come to a negotiating session with Johnson & Johnson prepared to bet his company, ready to make decisions as needed. Yet he saw the Johnson & Johnson people as coming to the same meetings prepared to listen, absorb, report, and carry information back to their superiors for further consideration.

The small company entrepreneur is often ready to make a decision based on gut feelings and to commit on the spot whatever is necessary to implement the decision. The large corporation's time scale for making decisions extends for months and sometimes years.

In general, the behavior of a large company is very different from that of a small one. The large company does basic research. The small company does technical problem solving. The large company does market research. The small company executive talks to a few friends in other organizations to get a feeling for how they view a potential product. Such differences in organizational temperament can easily produce strains and misunderstandings.

Despite these various difficulties, I believe that the promise of new-style joint ventures is quite high. More than any other form of venturing, they offer the possibility of reasonably quick market impact and profitability, for they seek to build on competitive strengths already in place.

INTERNAL VENTURING AT 3M

For over 30 years, 3M has primarily based its steady growth in size and profitability on new businesses developed through internal ventures. More than most other major corporations, it has thoroughly organized itself to encourage and support them. Its long-term record of success—ROI increasing at approximately 16 percent compounded annually—speaks for itself, but we may legitimately ask just how 3M goes about venturing so successfully.

What Is a New Business?

To look at 3M's organization on paper (see Exhibit 3) is initially to see a rather ordinary structure. Near the top of the organization are two divisions that report to the vice president of research and development: the Corporate Research Laboratory and the New Business Development Division. The latter, however, has quite a different charter than that of comparable units in most other companies. Here is where the real distinctions begin.

EXHIBIT 3 3M Structure for New Ventures

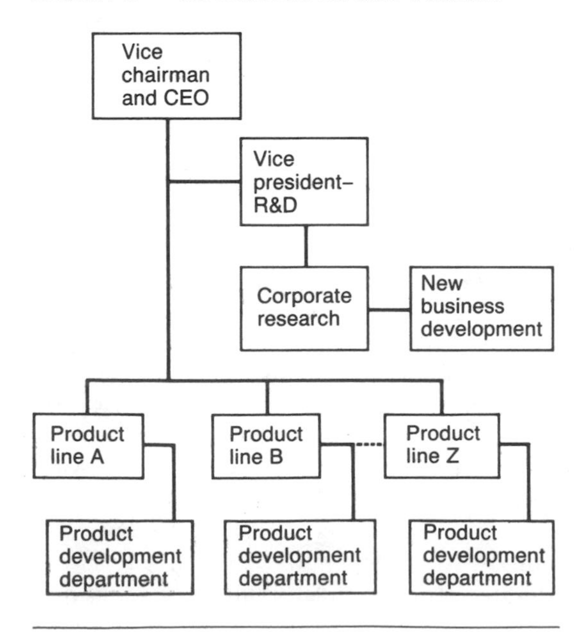

In 3M, a new business is defined behaviorally as one that has not yet reached critical mass in the marketplace, although it has perhaps as much as $20 million in annual sales. This means that the corporate New Business Development Division is charged with the responsibility for evolving, nurturing, and maintaining diverse business activities at various stages of development. It is an internal venture nurturing organization, an operation that not only gives birth but also support and sustenance. When new products are big enough to be self-sustaining, it spins them down the organizational chart as part of an existing division or as a new product line division.

Product Development

The second point worth noting about 3M's structure is that it is for the most part built on product line organizations, which have doubled in number since 1970, each with its own product development department. By itself this structure is not unusual. What is unusual is the charter of the product development departments. Each of them is charged with assisting the division that it serves by coming up with new products for that division's present line of business, with incremental improvements in old products, and with useful process changes. All of this is conventional. What is unconventional is that each of the product development departments is also charged with the responsibility for new venture development—new ventures without product line or business area constraints.

It is perfectly acceptable for any of these departments to develop products in any line of business, even if the new product competes with the output of another product division. Put simply, the 3M philosophy is, "We would rather have one of our own new products competing with an existing product line of 3M than have a competitor's new product competing." To the argument, "Surely that creates dissent and competition; isn't it bad?" 3M responds, "No, not from our way of thinking. We think that it can be good."

Does this philosophy create duplication of resources? Of course it does. Does it create efficiently used resources? I suspect that it does because the model of efficient competition is applied not just to the external but also to the internal marketplace. Competition for money, ideas, people, and market dominance keeps all participants in fighting trim.

The Eleventh Commandment

From top to bottom, 3M's management provides active, spirited encouragement for new venture generation. Many at the company even speak of a special 11th commandment: "Thou shalt not kill a new product idea." And they follow it seriously in practice. Contrary to the situation in many other companies, those in 3M who want to stop the development of a new product are saddled with the burden of proof. Benefit of the doubt goes of right to those who propose projects, not those who oppose them.

Of course, pushing a new product idea does not immediately throw open an endless bank account, but it does guarantee a chance to succeed. With the burden of proof on those who wish to kill ideas, the work environment within the company is distinctly favorable to entrepreneurial activity. In part, this environment is the result of promoting top management from within, frequently from successes in venture management. But it is everywhere reinforced. As one longtime observer has remarked, "You can't even talk for 10 minutes to a janitor at 3M without the conversation turning to new products."

Sources of Funding

Another important kind of support for new ventures is the multiple sources of venture capital within the company. Corporate groups can provide funding for new ventures without regard to source, and each product line department can provide funding for its employees' ideas no matter what market they are aimed at.

Say someone engaged in product development or marketing approaches his boss with an idea for a new product. If, despite all of the pressures on the boss to be supportive, he still answers, "We really don't have the money; we can't afford to handle it; we can't support your activity," the proposer is not then shut off permanently inside 3M. He is free to go elsewhere in the company to seek support for his idea, and a real market exists for the potential support of these ideas.

Moreover, if he can convince someone else to support his idea, then his idea does not go alone; he goes with it. The individual must be able to move with his idea and to join with his sponsor in undertaking the product development work. Then he and his sponsor quite properly share the blame if it fails and the benefits if it succeeds.

Product Teams

3M also gives special attention to the formation of product teams, entrepreneurial mini-business groups that 3M calls business development units. At an early stage of developing a new product idea, 3M tries to recruit individuals from marketing, the technical area, finance, and manufacturing to come together as a team, each member of which is committed to the further development and movement of this particular product into the market.

To make the team more effective, 3M does not *assign* people to such activities; the team members are recruited. This makes a very big difference in results. In most companies, a marketing person assigned to evaluate a technical person's idea can get off the hook most easily by saying that the idea is poor and by pointing out all of its deficiencies, its inadequate justification, and its lack of a market. Given the usual incentive systems, why should the marketing person share the risk? But instead of assigning him or her to evaluate the idea, 3M approaches Marketing and says, "Is anyone here interested in working on this?"

Here is a good instant test of a new product idea. If no one in the organization wants to join the new team, the idea behind it may not be very good. More important, whoever says, "I want in," becomes a partner, not a subordinate. He or she shares both in the risk and in the commitment and enthusiasm that go along with it. Team members are not likely to say, "This cannot be produced. It can never break even. It will never sell." They are involved as a team because they want to be, and they have a lot invested in making the idea work.

3M then supports its teams by saying to them in effect, "We are committed to you as a group. You will move forward with your product into the marketplace and benefit from its growth. But we cannot promise to keep you together forever as a new venture team. We will do our best to keep the team going so long as you meet our standard financial measures of performance throughout the life cycle of the product. If you fail, we will give you a backup commitment of job security at the level of job you left to join the venture. We cannot promise any specific job. But if you try hard and work diligently and simply fail, then we will at least guarantee you a backup job."

And some 3M ventures do get canceled. Although the company will not reveal its success/failure data, it has said that its ratio of success is comparable to that of other organizations. The key difference, of course, is that 3M starts many more new ventures.

Measures of Performance

The financial measures that 3M applies to new ventures are simple and to the point: ROI, profit margin, and sales growth rate. Only if a product team meets these criteria will the company make every effort to keep it going. These measures are straightforward and objective.

What sets 3M's standards apart, however, is what they do *not* include. First, they do not require a minimum "promised" size in sales volume for any given product idea. Instead, 3M says something like this to its product teams:

"Our experience tells us that prior to its entry into the market, we do not really know how to anticipate the sales growth of a new product. Consequently, we will make market forecasts that stick after you have entered the market, not before. We will listen to your ideas, argue with them, and do all kinds of analyses and estimates, but we will not say at the outset, 'The idea must be capable of generating $50 million or $100 million per year in sales.' Of course, we prefer larger businesses, but we will accept smaller businesses as entries into new fields."

Further, 3M's standards do not place area-of-business constraints on the generation of new product ideas. Unlike most other companies, it does not say to its teams, "Whatever ideas you come up with are fine, so long as they fall within business areas where we are strong." Nor does it say, "We want your new ideas—provided, of course, that the resulting products can be manufactured in our existing plants out of our existing stocks of raw materials and can be sold through our existing sales and distribution channels."

Reward Systems

The final element of 3M's approach to new ventures is its handling of rewards. All individuals involved in a new venture will have more or less automatic changes in their employment and compensation categories as a function of the sales growth of their product. Moreover, because the stimulation and sponsorship of new products is a responsibility of management at all levels, 3M has established special compensation incentives for those managers who are able to "breed" new ventures or departments.

KEY LESSONS FROM VENTURE STUDIES

I can make with confidence only three summary generalizations about successful new venture strategies.

They require long-term persistence. How long is long term? At the bare minimum, if a corporation is not willing to commit itself to a five- to seven-year involvement, then it should not even think of undertaking new ventures. What is needed is "patient money"—money in the hands of an executive group that is centrally concerned with the future growth and development of the company, money that need not generate payoffs in the next few years. In fact, 10 to 12 years is a more reasonable time span.

They depend on entrepreneurial behavior. The basis of every venture strategy is the attempt by a large company either to link up with or to emulate a small entrepreneurial company. In a sense this is surprising because it violates many of the textbook arguments for economies of scale. Yet in increasing numbers multimillion- and multibillion-dollar corporations are trying to scale down their manner of operating when they want to enter new business areas. They have rediscovered the special virtues of building an entrepreneurial organization and of harnessing entrepreneurial energy.

No single strategy works for all. What works for 3M will not necessarily work for every company. There are no magic formulas, and it is dangerously misleading to mimic the particular success of others. The current state of knowledge about venturing supports a far more modest conclusion: a variety of possible venture strategies is available, and it is up to each company's management to assess its own special needs, abilities, and personnel. This is simple common sense, but—like much sound managerial wisdom—it is all too often forgotten.

Corporate Venture Divisions

Case IV–6
Du Pont Company Development Department: Evolution from 1960 to 1976

N. D. Fast

In 1976, Du Pont Company was the largest U.S. chemical producer with revenues of over $7 billion annually. In 1976, the company's broadly diversified product line included both industrial and consumer products and included synthetic fibers, paint, agricultural chemicals, and medical instruments.

During the period 1950–74, Du Pont's sales increased more than fourfold and net income doubled. Most of this growth occurred in markets in which the company had attained a leading position through the introduction of innovative new products—either internally developed or purchased through the acquisition of small companies.

COMPANY HISTORY

To understand the evolution of Du Pont's new venture activities, it is necessary to take a historical perspective of the company's development.

Du Pont had been a leading company in the United States chemical industry since the early 1900s. During World War I, the company more than tripled in size and was very profitable. In the 1920s, the expansion of the previous decade was absorbed and consolidated and Du Pont began to diversify into new areas within the chemical industry.

From *The Rise and Fall of Corporate New Venture Division,* UMI Research Press, Ann Arbor, MI. Research in Business Decisions Series No. 3, 1978.

The company was family owned and managed at this time and top management was young. This was significant as it allowed them to develop a long-range plan for growth and diversification with the expectation that they would be in control to see this plan carried out over the following two decades.

In the diversification program of the 1920s, several small companies with promising new technologies were acquired. The technologies acquired during this period were seen as the levers for opening up broad potential markets which could be developed and expanded in the future. Most of these companies remained small for more than a decade before blossoming into the growth businesses of the late 40s and 50s. On the average at Du Pont, it had taken 23 years for a new product to reach sales of $50 million.

During the 1940s, the technologies acquired in the 20s were integrated into the company. "Pioneering" research was centralized and this brought about much cross-fertilization of process and chemical knowledge.

The 50s was a very profitable "growth decade" for the company as it realized the payoff from the innovative products acquired and developed in the 20s. Increasing demand for these products, the development of closely related product line extensions, and the discovery of new applications for products and technologies already existing in the company all contributed to the growth of the 50s.

CORPORATE CULTURE

Within Du Pont there was a strong corporate culture and sense of identity which strongly influenced the evolution of both the company and its new venture activities. Du Pont's corporate culture was characterized by the following:

Research and Development Orientation

Historically, the company had achieved market penetration and dominance by having technologically innovative products. Du Pont's approach to diversification in the past had been to acquire technology and/or to internally develop innovative products. During the late 60s, approximately $250 million annually was spent on research and development. The company had a very high percentage of individuals with advanced degrees in the sciences, its salaries for researchers were above average, and it was known to have one of the best research capabilities in American industry. Its compensation system supported this R&D orientation by having a system of bonuses which rewarded innovation—particularly that achieved by a "product champion" who successfully overcame the skepticism and resistance of others.

"People" Orientation

The company had a strong concern for its employees. There was an unusually large number of employees who had spent their whole career with the company—many serving 30 or more years. For the most part, its plants were not organized by national unions, its on-the-job accident record was the safest in the industry, and it was the first chemical company to have a major concern about safety. It is important to note that this concern stemmed from an interest in employees' welfare rather than just profit considerations.

Corporate Responsibility

A third core value at Du Pont was the company's strong sense of social responsibility and high standards of conduct. These were reflected in the statements of its top management as well as a system of high product standards it set for itself. One member of the executive committee described this characteristic of the company as a "sense of ethical obligation."

Historical Perspective

Among top management, as well as employees at lower levels of the corporation, there was an unusually strong awareness of the company's history. There was an intangible notion of "what would be proper" for Du Pont (i.e., technologically innovative products), and this historical perspective exerted an influence on decision makers. It supported and complemented the other core values. For example, Du Pont's R&D orientation was reinforced by the belief that Nylon, a product which played a key role in Du Pont's growth, had developed out of a "basic" research discovery in the corporate laboratory.

This corporate culture and system of values were shaped to a large extent by the fact that the company was family owned and managed for much of its history. It was not until the 1970s that the first chief executive outside of the founding family was selected. The values of the Du Pont family, which were ahead of their time, were ingrained in the company.

NEW VENTURE DIVISIONS

From 1960 through 1976 (the time of this writing), Du Pont had an organizational unit which could be considered a "new venture division"—a department whose primary task was the initiation of new ventures and the management of their early commercialization.

1960–1970: Development Department

From 1960 to 1970, this function was carried out by the corporate development department. Traditionally, the role of this department had been to serve as the "staff arm of the Executive Committee"—as a department to provide it with information and special studies as input for policy-making and often to take the necessary actions to carry out its policies as well. In the course of Du Pont's history, the development department had thus assumed diverse and varied roles, often combining several functions. During the 1960s, the development department's primary task was to launch new ventures.

1970–1976: New Business Opportunities Division (NBO)

In 1970, a subunit was created within the development department and this new department performed the functions of a "new venture division" from 1970 to 1976. This unit, called the New Business Opportunity Division (NBO), was given primary responsibility for the development department's new venture activities. As will be described below, the creation of the NBO was more than simply a reorganization. It constituted a fundamentally different approach to launching new ventures, brought about in response to a changing corporate situation and the experience gained from Du Pont's venture activities of the 60s.

The evolution of the "new venture divisions"—both the development department and the NBO—will be described in two sections below. In the first section, the role of the development department from 1960 to 1970 will be discussed. In the second section, the new venture activities of the 1970s will be discussed, focusing on the New Business Opportunities Division which was created within the development department to handle this function.

1960–1970: DEVELOPMENT DEPARTMENT

In 1960, Du Pont was in a very favorable position. It was the dominant company in several segments of the chemical industry. Both sales and earnings were in an upward trend. There was an atmosphere of confidence with positive expectations for the future.

Du Pont's financial position was strong in 1960. It had been conservatively managed financially and the high profits of the 1950s had generated funds for which "a significant number of major new investment opportunities were not immediately forthcoming." However, it was becoming apparent that many of the company's businesses were maturing. The chairman and Executive Committee recognized that Du Pont was beginning to "saturate its markets; its growth was declining and it had to diversify."

Increase in R&D Expenditures

One response to this situation was to increase expenditures for research at both the corporate and divisional levels. The activities of the central research department were expanded with the objective of making technological breakthroughs from which new businesses could be developed. The logic of this approach was to use research breakthroughs (as opposed to its management skills, financial strength, or existing technological capabilities) as the primary resource the company was going to build on in entering new businesses. Top management clearly saw Du Pont as a technologically based company with research the driving force for growth. Comments such as the following were typically found in the annual reports of the 60s:

- "Research has been the foundation of Du Pont's record of innovation in the chemical industry."
- "As a technically based company, Du Pont relies primarily on its own staff of scientists and engineers for the discovery and development of new lines of business."

Diversification

A more basic response to the surplus of funds and maturing of Du Pont's businesses was a conscious shift in corporate strategy. This took two forms. The company launched a major international expansion of existing operations and parallel to that it initiated a program of diversification. With respect to the latter, a member of the Executive Committee explained:

> Much like during the 1930s, we put a greater emphasis on diversification. The Executive Committee decided that in addition to the stream of investment opportunities from the divisions, there might be merit in a corporate diversification effort to lead us into new fields and provide investment opportunities.

There was no strong desire to get out of the chemical industry but rather a commitment to look in new directions and to compare what was found in relation to opportunities in existing businesses.

Diversification looked especially attractive because of the company's success with its earlier diversification program. A second factor was the atmosphere of confidence and optimism in the company at the time.

The domestic diversification program was to be carried out through internal development rather than through acquisition. There were two main reasons for this. First, it was felt that because of Du Pont's size there would be antitrust hazards with acquisitions. In the late 50s, the courts had ruled against the company in a major antitrust case. No violation of law was found but the courts ruled that there was a "potential" violation. This case had a major influence on top management thinking and made them "a little gun-shy of large acquisitions." One member of the Executive Committee explained: "When you go through a long case like that, it chews up quite a bit of corporate assets and time. After that, you are leary of getting tangled up again."

A second reason for avoiding major acquisitions was that the company's past growth had been primarily through internal development and the acquisition of small companies with technological innovations. Du Pont's past success with this approach weighed in favor of it.

Initiation of New Venture Program

The internal development program was carried out by both the operating divisions and the corporate development department which took on the function of a new venture division.

The operating divisions were encouraged to launch "new ventures." Efforts to develop new products were labeled "new ventures" if they met certain established criteria, the most important of which were size and uniqueness—that it be a sizable effort to develop and introduce a product which did not already exist in the marketplace. General managers of operating divisions could elect to have their major development efforts labeled "new ventures." These would thus get special accounting treatment with their costs carried below the bottom line of the division's profit and loss statement. This advantage to the division was offset by the increased exposure to the Executive Committee that the venture received (rather than being buried in the operating division's total activities). Increased exposure limited the flexibility of the divisional general manager in handling the venture.

The purpose of this arrangement was twofold. It encouraged the divisional general managers to become venturesome, since the development and commercialization costs of their new ventures would not penalize the performance of their existing operations. Secondly, it helped the Executive Committee keep track of how much venture activity was being undertaken and how the ventures were performing.

The assignment of the function of a new venture division to the development department was a step consistent with that department's traditional role as the staff arm of the Executive Committee. In the four decades prior to 1960, the development department had assumed different functions in accordance with the needs of the Executive Committee. One member of the committee observed:

> The development department's roles have ebbed and flowed with what the committee saw

> as the corporate's needs which were not readily attainable through the operating divisions. In fact, if you look at the changing roles of the development department, you would get a good feel for how Du Pont's corporate strategy has changed in the course of our history.

In the early 20s, the development department had been given the task of proposing new directions which the company should move in. The department came up with a plan for diversification and carried this out during the next decade. From the end of World War II through 1960, the development department took on various "trouble shooting" types of functions. These included corporate planning, licensing of technology, acting as government liaison, and special studies for the Executive Committee. The latter included studies of the corporate organization (for example, considering the desirability of combining departments) and more specific projects including analyzing the company's patent position.

In 1960, an assistant director of the development department was appointed and made responsible for the department's new venture activities. The development department continued its other activities, including corporate planning and internal consulting, with these responsibilities assigned to another assistant director.

Organization and Charter

The organization of the departments in 1960 is shown in Figure 1.

FIGURE 1

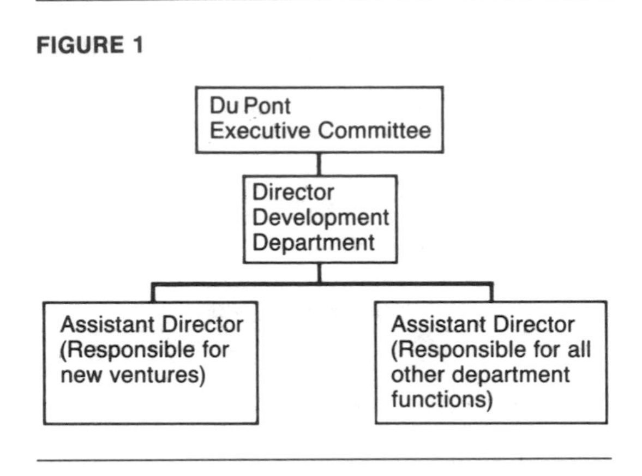

The assistant director responsible for new ventures was given little initial direction or guidance for how to carry out this function, nor did he have an organization. His first priorities were thus to "put together a staff and determine the bounds of this assignment." He presented the Executive Committee a written paper offering seven ways that the venture activity might be carried out—alternatives such as different types of start-up ventures, minority equity investments, and acquisitions of small companies. It was expected that only one or two of these alternatives would be selected. However, the Executive Committee instructed him to pursue all seven approaches. This very positive response was attributed to the Executive Committee's desire to experiment with various approaches and the fact that "things in the business world are faddish" and the approaches offered were somewhat novel.

The corporatewide venture effort launched in the early 60s was characterized by a lack of specific direction. An individual in an operating department at the firm observed:

> Top management saw a need for ventures and said, "Go ahead and do it." Nobody really managed it or directed it. So the whole company began to get into ventures but there was no clear direction or purpose.

Development Department Approach

The diversification program which was implemented consisted of two main thrusts:

- "Embryonic investment program," consisting of minority equity investments in small, high-technology companies.
- Large-scale start-up ventures.

High-technology companies to which Du Pont could make a contribution in the area of technology were sought. The logic behind this program was that if these companies were successful, they would provide one or more of the following:

- Investment opportunities for Du Pont.
- Businesses to be acquired by Du Pont.
- A synergistic effect through the exchange of technology.

In the start-up venture effort, the development department sought out "investment opportunities which would use our existing skills and give above-average return." There was the hope and expectation that these ventures would have a major impact on corporate earnings. Therefore, the size of the venture and its potential growth were key criteria for selecting ventures. With the limited staff of the development department, it was not feasible to initiate a sufficient number of small ventures to make an impact.

The development department attempted to generate venture ideas by "looking to the future to determine where there would be abnormal rates of growth and asking how could we get into these areas with our skills and technology." In addition, they kept an "open door" for ideas people wanted to submit from both within the company and outside of it. As part of this, they looked to the operating departments for venture ideas that they failed to back because they were too far out, radical, or unrelated to their existing businesses.

Although the objective was to build on existing company skills—primarily in-house technology and markets they were already serving—the development department could not pursue ventures which fell within the province of an operating department. Since the company was already well diversified and the operating departments were in quite different fields, the development department was forced into areas which were "even more speculative," taking on the riskiest, least related projects.

The typical venture launched by the development department in the 1960s was described by one individual as a "large-scale 'frontal assault' on a new market." A venture often consisted of several products aimed at the same market. Most often, these required heavy front-end investment and when they ran into obstacles, the typical reaction was to "call another division in" and "pour more money in." The result was that when ventures failed, they were large-scale failures, well into the tens of millions of dollars.

Relations with Other Departments

The ventures launched were "more or less fully self-contained" with large staffs of technical personnel and the venture's own facilities. The development department's relations with operating departments were thus minimal.

Nevertheless, in the operating departments there was some dissatisfaction with the development department's ventures in the 60s. Some individuals felt that they lacked the people resources and close disciplined management and attention which were found in the operating departments.

As the ventures were self-contained, they did not rely on other departments for assistance. There was little competition for resources as funds were not constrained. An individual in one of the operating departments explained:

> Everybody had just about all the money they wanted for ventures. All the departments were encouraged to do new things. We were not competing with each other. Management didn't put an upper limit on what would be spent. If you had a good project, it was funded.

The development department's interactions with the central research department were also minimal during the 1960s. This was attributed to the completely different orientations and objectives of the departments. The central research department conducted "academic, basic research." This orientation developed during the 1930s and 1940s when the operating departments grew to a size where they could support their own research organizations. As research pertinent to existing businesses was farmed out to the operating departments, central research became more "basic," pursuing projects of a much longer range nature.

During the 60s, the assistant director of the development department felt that the skills in the central research department were not relevant to the task of his department. Thus, a new organization was built "from the ground up" within the development department. It was observed that much of the money spent on the department's ventures really went into building this organization, serving to inflate the cost of these ventures. Although much smaller than central research, which was the largest research organization in the company, this department grew to be substantial in size.

Management Philosophy

The philosophy and atmosphere in the company that prevailed during this period shaped the venture activities. The dominating idea was an extreme confidence in the company's capabilities. This was reflected in the ventures in the following ways. There was great optimism in discussing ventures, "everything always looked good, problems were solvable, and the venture was always heading in the right direction." Second, there was a reluctance to recognize or accept that a venture could fail. Thus, when one ran into difficulty, the corporate response was to try to overcome the problem by committing additional resources. Third, it was believed that the time required to develop a venture into a large profitable business could be collapsed. This idea of reducing the time required to launch a venture was central to the "frontal assault" approach of the 60s.[1]

A fourth characteristic of the 60s was that the products developed were of the highest quality. One individual observed: "We offered nothing but the best—no crass commercialization." It was felt that during the 60s, several ventures priced themselves out of the market because of a tendency to resist compromise in product quality or performance even when economically warranted.

The assistant director of the development department had almost total freedom in investigating new opportunities, determining the strategy for ventures, and guiding their operations. The degree of Executive Committee involvement was determined by the size of the venture and amount of funds it was consuming. The assistant director of the development department observed: "The Executive Committee only closely watched a venture when it became a big consumer of capital. Capital was the trigger for their attention."

The Executive Committee established a Subcommittee on Planning composed of three vice presidents. They followed the ventures more closely than the total committee meeting with the director or assistant director of the development department every few weeks. The development department reported on ventures formally to the full Executive Committee only quarterly. The Executive Committee rarely directed that a venture be discontinued. Instead, it would provide that continuation of a venture was contingent on meeting certain established goals and targets.

Growth and Evolution

From 1960 to 1966, new venture activities were of increasing importance to the development department and gradually took an increasing share of its time. The number of staff people working on ventures grew from 12 to over 200 people.

In 1966, the director of the development department retired and the assistant director was promoted to that position. The organization structure of the department was changed to re-

[1] It ultimately became apparent that although product development could be accelerated, market acceptance of truly innovative products often could not be predicted or speeded up.

flect the increasing importance of its venture activities. These were carried out directly by the department director with the other department functions reporting to his assistant. The organization is shown in Figure 2.

In 1968, the director of the development department was promoted to general manager of an operating department. He was replaced by another individual who headed up the department for only one year before becoming an assistant general manager of an operating department.

By 1969, several large start-up ventures were being discontinued or spun off. These ventures had been initiated in the early 60s and had come through the pipeline in parallel. One successful venture had been transferred to an operating department. Successful and promising products in other ventures were beginning to be spun off to operating departments.

At the peak of Du Pont's venturing activity, 25 "new ventures" were underway. The bulk of these were in the operating departments, with only five in the development department. By 1969, several of these had run up losses in the tens of millions of dollars and were being shut down. Minority equity positions in six small companies had been acquired under the "embryonic investment program" and in 1969, the company was making plans to divest these.

A member of the Executive Committee commented on the change in attitude toward ventures that occurred in about 1969:

> At no time was there a precipitous withdrawal of support for ventures. No one woke up one day and said, "Let's not do it anymore." It was a gradual thing and a variety of factors contributed to it.

FIGURE 2

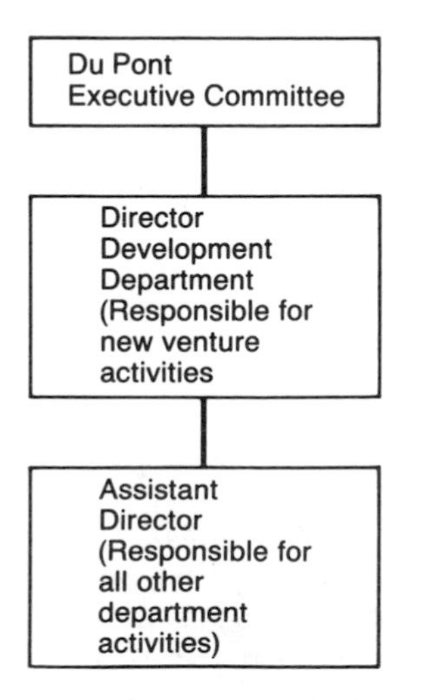

NEW BUSINESS OPPORTUNITIES DIVISION: 1970–1976

Corporate Situation

In the late 60s, Du Pont's corporate situation began to change. The general economic situation was becoming unfavorable due to rising inflation and, in 1970, the first recession since the early 1960s.

Du Pont's earnings had been in a downtrend since 1965. Many of the company's patents had expired, resulting in overcapacity and pressure on prices. The Kennedy Round tariff cuts had hit the chemical industry particularly hard, increasing foreign competition especially in textiles from the Far East. Sales in 1970 were down slightly from a year earlier and profits had declined by about 10 percent. In the decade 1960 to 1970, sales had increased by close to 60 percent; however, earnings had increased the first half of the decade and declined the second half, so that 1970 earnings were at a level equal to 1963's, and only about 20 percent higher than 1960. Although Du Pont's balance sheet remained very strong, the company was entering a capital short period.

The outlook for the company's existing businesses was becoming more favorable. Its product portfolio included a larger percentage of new and growing businesses than a decade earlier. In the period 1960 to 1972, more than 50 new products (not including product modifications or improvements) had been introduced, and in 1972, more than one quarter of sales were from products less than 15 years old.

Corporate Strategy

The changing corporate situation brought about a shift in corporate strategy.

Corporate diversification slowed and the diversification of departments into related areas was given priority over corporate diversification into businesses farther afield. There was a basic change in the philosophy and atmosphere within the company. A member of the Executive Committee explained: "No one suddenly turned a valve on this. We are talking about degrees of emphasis."

This transition from a phase of expansion to one of consolidation is clearly reflected in the annual reports of 1969, 1970, and 1971, as shown below:

1969 Annual Report

> Newly developed products are expected to account for half of the company's earnings growth over the next five years. . . . The company will continue to broaden its business base by entering new markets.

1970 Annual Report

> The long-term outlook remains excellent. The markets for chemical industry products are expanding with growth in physical volume expected to average 6 percent to 7 percent a year throughout the 70s. Du Pont . . . is particularly strong in those products having high growth potential worldwide. . . . A number of new ventures which have been a drain on earnings are moving to the commercial phase. . . . The company is making intensive efforts to improve profits by improving productivity and reducing costs. In 1970, economies were achieved through systematic companywide reevaluation of activities and programs.

1971 Annual Report

> During the past two years, the company has pursued a rigorous program to reduce costs and increase profitability. Manufacturing operations have been consolidated in several product areas. A number of unprofitable operations have been dropped. . . . Our work force has been cut 10 percent during the last two years chiefly through normal attrition and reduced hiring. Major organizational changes have been made, including an extensive realignment of product lines among industry divisions to facilitate the development and marketing of related products.

Research Policy

In about 1970, there was also a change in the direction of research and development. There was greater emphasis on research and development in support of the existing businesses through product improvement and product line extensions and a decline in "basic" or "pioneering" research. Thus, departmental research organizations were directed to support their existing products and channel their pioneering research to the central research department. Two factors influenced this. One was a shortage of funds and a desire to increase the return on dollars invested in research. The other factor was that the environment for research was getting tougher with increasing government regulation of new product introductions and longer lead times for their development. As its name implies, the research "establishment" at Du Pont was not increased and decreased in size "with every pop in the business cycle." Rather, changes in corporate policy were reflected by shifting the focus of the departmental research organizations which consumed approximately 90 percent of total research expenditures.

Support for Ventures

Support for the company's venture activities diminished in the period 1969 to 1971. One major reason for this was the changing corporate situation and strategy. One member of the Executive Committee summed it up stating:

> In a capital short period, you clearly finance first those businesses you are already in. The urge to diversify is less. In fact, you simply don't diversify when you are short of capital unless your existing businesses are in trouble and ours were not.

The second factor bringing about a decline in support for venturing was the experience of the 60s. The five start-up ventures launched by the development department had run up total precommercial costs of approximately $80 million, almost $50 million in one venture alone. And this was only a fraction of the company-wide total.

The time frame of management's expectations contributed to the growing dissatisfaction with the performance of ventures. A member of the development department explained:

> Top management believed that you could develop a $50 million business in 5 to 10 years if sufficient resources were committed. After a couple of years, it became apparent that we couldn't speed up new business development. Management became disillusioned with the progress of our ventures. The irony is that some of those ventures which looked so bad in 1969 turned out okay. We just expected results too soon.

In fact, one venture was a major success growing into a business with over $100 million in sales by 1976. A former director of the development department felt that this single venture justified the entire development department program.

Attitudes towards new ventures (which ultimately affected the development department) were influenced more by the ventures launched in the operating divisions than by those of the development department. The operating departments were "where the action was." Many in the operating departments felt that the company's overall ventures program was a failure. The comments below were typical assessments:

> It is awfully easy to *launch* diversification efforts in several attractive directions and build these up in scale, but you incur a heck of a lot of expense before you get the first buck back. In the late 60s, it became apparent that this could become pretty expensive and far reaching without the necessary management attention. The Executive Committee got disenchanted with the concept.
>
> There is nothing wrong with rolling the dice on a big venture if it's obviously a winner, but if you roll them many times on those that are "iffy" and chew up money in chunks of $50 million at a crack, that gets to be a bloody game. We did some of that in the 60s.
>
> The experience of the 60s put a large damper on the concept that all you have to do is take off in 20 directions. Management became particularly disenchanted with $50 million and $100 million holes.

The third factor which brought a change in new venture policy in about 1970 was that many of the ventures launched were moving in parallel toward commercialization. This facilitated the winding down of the venture program since a large percentage of them could be disposed of either by being spun off to a division or being discontinued.

Creation of NBO

In 1970, the New Business Opportunities Division (NBO) was created within the development department. The decline of the development department's venture activities had become evident in 1968 and 1969. The appointment of a new director of the department in 1969 was a key event because the individual selected favored a "MICRO" approach to launching ventures rather than that of the development department. He implemented this new approach by creating the NBO. The manager who headed up the NBO observed:

> There are several ways to change a department like the development department. If you have enough power, you can modify its formal charter,

> or an alternative is to destroy it and build a new department. The new development department director didn't have the kind of power to achieve the former. So he started up the NBO and let the development department's old venture program die out.

The decline of Du Pont's venture program could have resulted in the development department simply being disbanded. However, instead it was reduced in scope and carried out on a smaller scale with a different approach. The decision to continue the new venture division function on a smaller scale rather than discontinuing it was attributed to two factors. First, Du Pont's culture highly valued research-based new ventures. Historically these accounted for most of the company's leading products. Thus, there was a reluctance to completely discontinue this activity. A member of the Executive Committee explained:

> We are a research-based company. Our most profitable products were developed from research. Thus, we always want a stable of ventures. We try to keep it appropriately sized for our business situation.

The second factor was that Du Pont was conservatively managed financially. The company had minimal long-term debt and did not "spend up to the peg." Thus, it could "suffer for a couple of years" while adjusting the level of its venture activities to the appropriate size.

Change in Philosophy

The approach and philosophy of the NBO represented almost a complete reversal from that of the development department in the 60s. The director of the development department who established the NBO commented on this change in orientation:

> The NBO evolved as a change in organization within the development department, but the significant change was in its way of carrying out new business development. This change in approach resulted from our disappointing experience in the 1960s.
>
> The approach we adopted puts more emphasis on finding a way to run a meaningful market test at an earlier stage while trying to concentrate on not making any significant early capital investment. We believe it is possible to be innovative in using equipment and facilities which already exist in the company. Our objective is to try to sell a product (which is the only true measure of value) before asking for any sizable commitment of funds.

The controller of the development department also discussed this change in philosophy:

> In the 15 years I have been here, our philosophy has changed markedly. In the early 60s, we had a big block of ventures in which we came out with several major products aimed at the same market. These did not really materialize as quickly or easily as we had expected. We now attempt to make a venture largely self-supporting early on. We don't invest a large amount of money in a venture until there are indications of market acceptance and evidence of a financial return. This more recent approach appears to be more successful. If I had to say why, I would say because when you simultaneously launch several products you don't really know whether the market needs them. In our present approach, we live with the market for a while and then decide what additional needs we can fill.

The manager of the NBO observed that there was a clear business justification for the NBO approach versus that of the 60s:

> When we learned that new ventures take a much longer time to develop, it made us rethink our approach. If it takes 20 years instead of 10 for a venture to develop, you have to be more careful how you spend your money and more selective about opportunities to pursue. The time value of money has become a more important consideration and we have thus sought to get a positive cash flow as early as possible.

There were five main aspects of the approach taken by the NBO which differed from that of the development department in the 60s.

Resyntheses of resources and capabilities

The NBO sought ventures which were more closely related to the company's existing resources and capabilities. They sought ventures which were less risky and "far out." The department director explained this as follows:

> One of the things we do in a very disciplined manner is to look 5 or 10 years down the road and say to ourselves, "Suppose this venture turns out to be successful—what will it look like then, and what are *all* the resources we would need to get from here to there?" Then we ask, "Of all of these required resources, which do Du Pont already have?" I believe that we are better off leading from some strength. If we have few or none of the necessary resources and other companies have them, it would be difficult to convince me to undertake the venture. This says something about the degree of diversification I am willing to support. There are innumerable opportunities we can pursue. I give those that we have the resources for highest priority.

Thus, the NBO sought to develop ventures which represented recombinations or resyntheses of existing company resources and capabilities.

Relations with other departments

The second major difference was that while the ventures of the 60s were essentially self-contained, the NBO sought to have lean low-budget ventures which drew heavily, either formally or informally, on the people and other resources of the various operating departments. The NBO manager explained:

> Our job is basically to identify the resources that are needed for a venture, determine whether Du Pont has these resources in house, or at least enough of them to be successful, and then to pull them together and try to manage them. This differs from our role during the 1960s. Then, we would have tried to create those same resources ourselves.
>
> You do better when the business you're aiming at is close to what you already know. My job then becomes to organize and to tap the necessary resources to test our assumptions and to carry out that plan.

For example, in one venture which involved developing a device to aid textile designers, researchers in the photo products department worked on applying a technology for reproducing colors which existed in that department while an individual in the fibers department surveyed the potential market for the device.

Beachhead ventures

A third difference in approach was that the NBO attempted to launch small ventures targeted at a particular market niche as compared to the "frontal assault" ventures of the 60s in which a broader market was attacked with several parallel product introductions. The five ventures launched during the 60s had precommercial costs averaging almost 10 times the average for the 6 ventures launched by the NBO. The latter averaged under $2 million. In the approach adopted by the NBO, the notion of "testing" is the key. The manager of the NBO stated:

> Our entry strategy is to try to develop a "beachhead"—an approach that will give us a niche in the marketplace. Our aim is to make money in establishing that niche. The first step is to test out all of the assumptions with regard to our resource base and the niche we're aiming at. Many times those assumptions are incorrect but we have the flexibility to make changes. Nevertheless, ultimately it may turn out that the niche is not a business.

This approach is clearly intended to avoid the large-scale failures that were experienced in the 60s.

Long-term impact

The fourth major difference was that the mission of the NBO was not to redirect Du Pont or make a substantial short-term impact on its diversity, but to establish financially sound businesses in future growth markets. An important element of this approach was to defer major financial commitments until the new business was proven.

Although ventures were initially small, the NBO still had the objectives of impacting on the company. The critical difference from the approach of the 60s was the time frame considered. The NBO did not expect its ventures to have a major impact for at least 15 or 20 years.

The director of the development department also felt that it would be unrealistic to attempt to significantly increase the product market diversity of Du Pont. There were few markets that Du Pont did not sell to in one way or another, and even though it was a chemical company, its diversity in terms of products and markets was comparable to that of a conglomerate.

An individual in one of the operating departments had the following comment about the role of the NBO:

> The NBO is one of the charming things we have around. It is a nice luxury, but the NBO is not Du Pont. It's a nice little organization, but in Du Pont the real action is in the operating departments, even in new product development. The NBO only takes what does not fit into a department or what they can generate themselves.

Learning philosophy

The fifth difference between the development department of the 60s and the NBO of the 70s was one of philosophy. In the 60s, ventures were seen as "the implementation of a plan" for entry into new markets. Venture strategies were adapted and modified when they ran into difficulty, but the inclination was to attempt to overcome the problems encountered by committing additional resources. In contrast, the NBO's philosophy was to "learn and adapt." The idea was to test business ideas and modify them in response to the feedback from the experiments carried out. The terms *success* and *failure* which were appropriate for the approach of the 60s were considered inappropriate when applied to the NBO's ventures. The NBO manager explained:

> *Success* and *failure* are terms which bother me. I do not think they are appropriate for what we do. For example, two of our ventures, Cronel and Viaflow, had similar objectives: to develop a technology into a product, develop a plan to test market that product, and get the test market results. The test market results in Viaflow were that people were buying. In Cronel people weren't buying. From my perspective, we ran two experiments and got different answers. It's like trying to decide between investing in the stocks of two companies. If you investigate both and you decide not to buy one, you haven't "failed."

ORGANIZATION

The organization structure of the department in 1970 is shown in Figure 3.

The activities of the development department included corporate planning, internal consulting, licensing technology which was not used internally, the accounting and control functions for the department, and the NBO.

The licensing manager interfaced closely with the NBO. When new opportunities arose, the decision had to be made whether to pursue it internally or license it. When a venture failed, the licensing manager would often become involved and attempt to license what had been developed.

The organization of the NBO itself was described as very fluid. It changed to meet the needs of the individual projects and ventures at a particular point in time. The group responsible for searching out new ventures and screening ideas varied in size from year to year and its activity ranged from "almost dormant" to "very active." The activity of this group depended to a large extent on the amount of money the Executive Committee appropriated.

The individuals in that group were very mo-

FIGURE 3

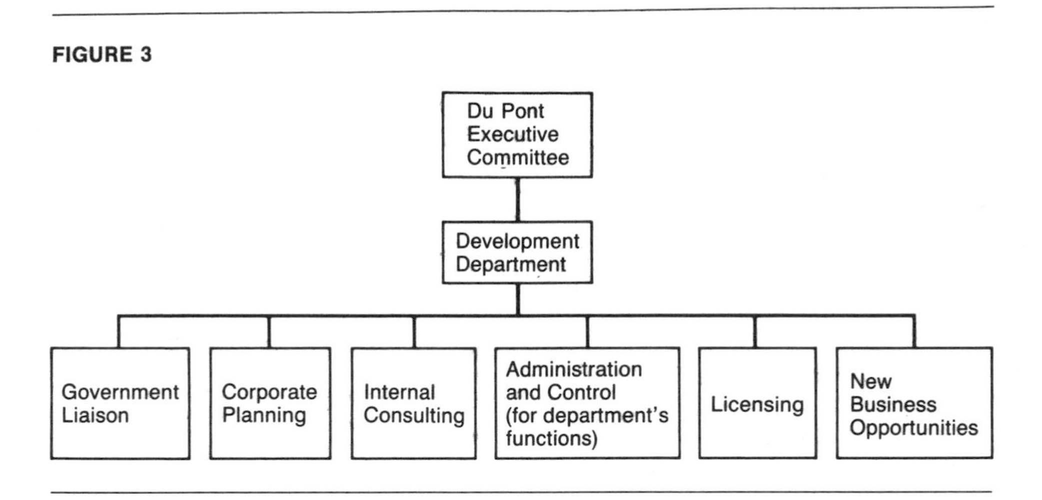

bile, and when a venture idea was approved, the individual who studied it initially often moved out of this group to run the venture. He might not be replaced for a long period of time, but then several individuals would be added to this group and the cycle begun again.

Staffing

The individual chosen to head up the NBO had begun his career at Du Pont in the plastics department 12 years earlier. He had spent three years in research in another department and then nine in sales and marketing before coming to the NBO.

Most of the individuals in the NBO had technical backgrounds and many had worked on the development department ventures during the prior decade. One member of the department described a typical individual within the NBO as "someone with commercial savvy and a background from different departments in the company."

Generally, individuals who came into the NBO from an operating department followed one of three paths. After a few years, they either returned to a new position in an operating department, returned to an operating department as part of a venture that was spun off from the NBO, or remained permanently in the NBO. In 1974, a decision was made to hire M.B.A.s into the NBO as one of the entry-level positions in Du Pont. The rationale for this was that it would give them an opportunity to get a relatively quick, broad overview of the company before they were offered employment in an operating department. In their work with the NBO, their lack of contacts in Du Pont was considered a disadvantage, but this was partly overcome by their use of the more experienced, long-term Du Pont people in the NBO to facilitate access to other departments.

NBO Approach

The NBO's activities were limited to start-up ventures. The department evaluated potential new business ideas that were generated both within Du Pont and outside of it. When an outside party came to Du Pont with a venture idea, the host department would contact the NBO if the idea did not fit into an operating department. The key factor in determining who would develop a venture was where the required resources were located.

The NBO did not carry out any internal idea

generating or brainstorming, but rather served as a clearinghouse for the ideas of others.

When a potential venture was brought to the attention of the NBO, it was evaluated by a committee of several department members, usually including the manager. If a decision was made to pursue the opportunity, two individuals were usually assigned to manage the initial phase of the venture. This gave the NBO manager flexibility in choosing a venture manager. Usually within a year, one of these individuals was selected as the venture manager.

One of the first responsibilities of the venture manager was to prepare a business plan. The development department director stressed the importance of planning:

> Our aim is to shorten the time and dollars necessary to get a true market appraisal of a potential new business. To achieve this, we put a great deal of emphasis on well thought-out planning and accurate estimation of financial costs.

The business plan was updated as the venture went along and more information was gained. There was no pressure on a venture manager to adhere to an earlier plan if he had gained information which warranted changing it.

Resource Board

Early in a venture, a "Resource Board" was established. Members of other departments who could contribute to the business were asked to serve on this board, usually aside one or two members of the NBO and its manager. The Resource Board was viewed as a prestigious appointment—"it gave the NBO an opportunity to cater to their egos somewhat" and build support. The purposes of the Resource Board were to assess the progress of the venture, make suggestions and recommendations, to facilitate communication and cooperation with other departments in the company, and to plan for the ultimate disposition of the venture. The development department director discussed the rationale for having a board for each of the ventures:

> Resource Boards are working boards of people who can contribute and have access to the required resources in the operating departments. One of the things I feel strongly about is that it would be a mistake to try to build completely self-contained venture groups in this department. You lose a lot by transferring a person out of his department to this venture. When you contract out development work to other departments, you get not only the expertise of the individual who works on it but also the help of his supervisor and others around him. The Resource Board compensates in some way for not having a completely self-sufficient venture group, and it gives us access to greater resources.

A former board member discussed what he saw as its function:

> The role of the Resource Board is to provide the varied inputs that the venture manager needs to round out his experience. Its members operate both as individual specialists and as a group to support, criticize, or modify the venture manager's proposals. The Resource Board provides a sounding board for the venture manager. To a large degree, this is the way Du Pont manages its businesses. We like to get all points of view. Very few decisions in Du Pont are made unilaterally.

When the venture was spun off to an operating department, the Resource Board was usually disbanded.

Management Control

A venture manager reported to the Resource Board quarterly. He determined the nature of the reports he would provide to both the Resource Board and the manager of the NBO. The only required report was a monthly budget for the year for all expense and revenue categories.

The relationship between the NBO manager and the venture manager was likened to that of a venture capitalist and the entrepreneurs he had invested in. It was generally observed that the NBO manager served more of an advisory role than a decision-making role with respect to the ventures.

In contrast to the approach of the 1960s, the initial objectives for a venture were limited. They were often stated in terms of questions that the venture manager would attempt to answer, knowledge that he would attempt to gain, and the activities that would be carried out towards these ends.

The controller for the development department contrasted tracking the performance of ventures to that of ongoing businesses:

> In your existing businesses, you evaluate return on investment. We do not attempt to evaluate return on investment for ventures because normally we haven't invested much. We will use someone else's plant, equipment, and what have you at least to the point where the venture is a viable business. We are more concerned that the venture is going toward a profitable operating position and that we are achieving it within our budget constraints.
>
> With a venture, you have to go beyond the bottom line because very often it hasn't developed to a point where there is anything even close to a profit. In fact, losses may be increasing and the venture may still be healthy as long as you have a picture that it will turn profitable and you'll get a good return.

The controller's staff did the accounting for individual ventures. Individuals from his staff were assigned to specific ventures and reported on a "very heavy" dotted line to the venture manager. They developed an accounting and information system for their venture and reported to the development department controller at least monthly on the progress on the venture. These were typical accounting reports with sales, expenses, and profits versus budget and explanations for variances. These reports did not go to the Resource Board, because a venture's performance varied widely from month to month.

The monthly accounting reports were distributed to the manager of the NBO and the director of the development department. However, the director of the development department did not get involved in the details of individual ventures. He explained:

> I intentionally try not to get into the substantive details of a venture. I like to understand them, but I feel it is critical to have the decisions made at the level of the venture manager. My job is to set the proper environment, decide on the type of approach we will take, and assure that the right people will run the ventures. I am here to have ideas bounced off of and give the venture managers authority and control over the ventures. My responsibility to top management is to control the money and to know how we are progressing.

The development department reported to a "liaison" vice president on the Executive Committee. He had a fairly detailed knowledge of the ventures. The rest of the Executive Committee received an annual plan for the whole department including the NBO and individual ventures, as well as quarterly reports. However, they were primarily interested in the "direction" individual ventures were going in and what the total cost of the NBO was.

In general, there was a lower level of Executive Committee attention devoted to individual ventures of the NBO than those of the development department in the 60s. This was attributed to the smaller size of the ventures. The NBO manager felt that this gave the venture group flexibility:

> There is little resistance to redefining a venture's strategy if the argument makes sense. A low profile is important because you don't have to continually answer to people who don't know what you're doing. The other reason is that it takes a lot of time and energy to have a high profile, and that energy should be spent carrying out the objectives of the venture.
>
> If you've got a high profile, it is harder to change your approach—the organization becomes committed to that approach and there is resistance to changing the plan. Exposure to the organization produces commitment to an ap-

> proach and that is difficult to reverse. We have tried to structure the organization so that there are as few bosses to go through as possible. My boss (the head of the development department) reports to the Executive Committee. There isn't much space between the guys who head the ventures and the top.

The decision to discontinue a venture was made by either the NBO manager or the venture manager. Individuals higher up in the organization generally did not get involved. This was in contrast to the 60s when the Executive Committee was sometimes instrumental in the decision to discontinue a venture. The NBO manager stated that if he felt a venture should be continued, the development department director will usually "let it run." The same attitude prevailed in the relationship between the NBO manager and the venture manager. The former would sometimes let a venture run for several months after he felt that it should be discontinued so that the decision would come from the venture manager.

When a venture was "discontinued," it was not a large-scale failure as in the 60s but only the beginning of another phase in its development. It would come under the jurisdiction of the licensing manager who would attempt to find opportunities for other companies which were either in a different direction or on a different scale.

> The word *salvage* might describe my role, but it has the wrong connotation. It implies something of little value when very often a situation I am attempting to license represents an excellent business opportunity for someone else.

For most ventures, there was usually a successful business that could be salvaged.

Venture Spin-Off

A venture which was "successful" was transferred out of the NBO when its potential had been determined. The actual timing depended on subjective factors such as the individuals involved, the progress of the business, and the organizational fit in its new home. The transfer of a business to an operating department was often a difficult step, and one NBO member pointed out:

> If you were truly looking for new businesses and new markets, then by definition there is no easy fit with an existing Du Pont department.

There appeared to be little desire within the NBO to retain profitable ventures. One department member explained:

> Success for us is in placing the ventures. That is our ultimate goal—to have somebody want the venture—to demonstrate enough potential, profitability, or whatever so that someone wants it. There is no doubt about the genuine satisfaction that people get in seeing a venture placed well. There would be reluctance to give a venture up if it was going to the wrong place, but that's not usually a problem. From the very beginning, we're planning for the ultimate placement of the business.

One of the functions of the Resource Board was to plan for a smooth transfer of the venture of the NBO to an operating department. The individuals from the operating department who had served on the Resource Board had been with the venture through its growing pains. Thus, they were able to give their management an inside view of what it was. This facilitated a smooth transfer of the venture and reduced the possibility of conflict afterwards.

The director of the development department seemed to have little desire to retain ventures after they were commercialized. He seemed to prefer the initiation of ventures to "empire building." He commented:

> We hope that ventures will move to an industrial department when their feasibility has been tested. We don't want to become an industrial department by retaining ventures because it is not our charter. We would lose our ability to start new ones if that happened. The best way to ensure that the transfer is made smoothly is to have an industrial department involved in a working

way from the beginning on the Resource Board or otherwise.

The head of the NBO also shared these views. He stated:

> The only thing that is certain is that a successful venture will leave this department—because the business of this department is to start new ones. If we continue to run old ones, then we will cease to start new ones. The only time it bothers me to turn over a venture to an operating department is when it is done before our objectives have been realized. Then if the thing doesn't succeed, it reflects on me and the NBO. The pressure to turn it over is greatest when there is a close fit with an operating department. There have even been some ventures identified here, where the conclusion has been that all the resources reside somewhere else and they have been immediately turned over to the proper department.

1975 Reorganization

In 1975, there was a major change in the organization of the development department. It was combined with central research into a single department. Several of its functions including internal management consulting, serving as the Executive Committee's staff arm, and corporate planning were placed elsewhere in the company where top management felt they would be better served. The historical mission—serving as the staff arm of the Executive Committee—was assigned to a new department called Corporate Planning. This was attributed to the reluctance of the Executive

FIGURE 4

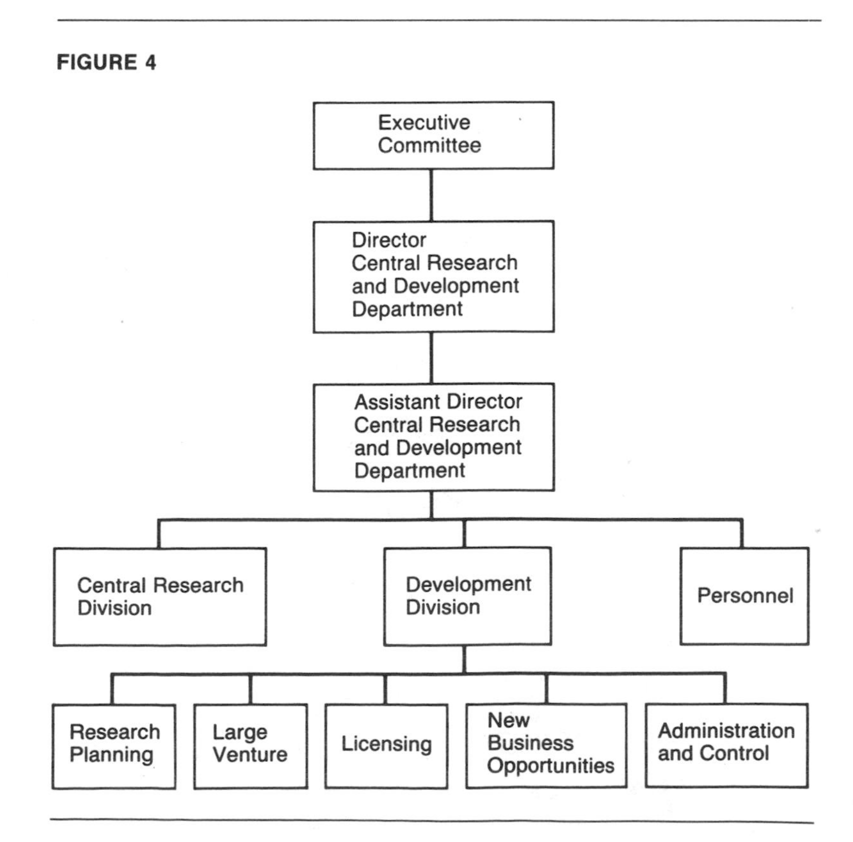

Committee to really use the development department as its staff arm in recent years. The change in organization is shown in Figure 4.

In 1976, the NBO manager was promoted to director of the Development Division. Prior to that, the assistant director of the central research and development department had held both titles. The effect of this promotion was to have the Development Division assume the functions of the NBO.

Relations with Central Research

The combination of the central research and development departments brought about some degree of increased coordination and cooperation between the two departments. This had been an objective since the creation of the NBO. An integral part of the philosophy was to attempt to use the existing corporate resources rather than duplicating them in the department.

One of the NBO manager's first responsibilities was to begin working more closely with central research. Since the two departments had different objectives and orientations, there was limited overlap.

The 1975 reorganization and changes which accompanied it facilitated cooperation and coordination between the research and NBO departments in several ways, including the following:

- The directors of the NBO and research were peers, reporting to the same individual. They dealt with each other directly rather than through intermediaries.
- Both individuals were on the research and development department staff. This body, composed of eight key individuals in the R&D department, met formally at least once per month.
- A Research Planning Group, which previously was in the Research Division, was transferred to the Development Division. This group of 12 people helped the Research Division evaluate research projects in terms of their business potential before they were initiated. Its role, however, was strictly advisory.

Continued Evolution

In considering the future of new venture activities in Du Pont, a member of the Executive Committee observed:

> The changes in our new venture activities are not irreversible. Our approach will change as the intermediate term business outlook does. We could start up large-scale ventures again like we did a decade ago. Even though they may be ill advised, all we need is extra money to get into the bad habits of the 60s.

EXHIBIT 1 Du Pont Company Financial and Operating Record, 1957–1966

Year	Sales	Average Operating Investment (*a*)	Construction Expenditures	Operating Income Net		Total Net Income	Per Share of Common Stock			
							Net Income			
				Amount	As Percentage of Average Operating Investment (*a*)		Du Pont Sources	General Motors Dividends (*b*)	Total	Cash Dividends Paid
		(millions)				(million)				
1966	3,158	4,643	531	367	7%	389	8.23	—	8.23	5.75
1965	2,999	4,267	326	384	9	407	8.63	—	8.63	6.00
1964	2,761	3,910	290	337	8	471	7.58	2.06	10.03	7.25
1963	2,554	3,604	370	310	8	472	6.84	3.21	10.05	7.75
1962	2,406	3,341	245	305	9	451	6.73	2.87	9.60	7.50
1961	2,191	3,120	204	258	8	418	5.72	3.16	8.88	7.50
1960	2,142	2,933	213	248	8	381	5.57	2.53	8.10	6.75
1959	2,114	2,745	174	288	10	418	6.38	2.54	8.92	7.00
1958	1,829	2,581	231	212	8	341	4.71	2.54	7.25	6.00
1957	1,964	2,421	220	265	11	396	5.93	2.55	8.48	6.50

(*a*) Operating Investment is the sum of "Total Current Assets" and "Plants and Properties" before deduction of accumulated depreciation and obsolescence, as shown in the company's consolidated balance sheets; the average is based on investment at the beginning and end of the year.

(*b*) In compliance with a U.S. District Court judgment, the company disposed of its entire investment of 63 million shares of General Motors Corporation common stock. Distributions made to holders of Du Pont common stock in July 1962, January 1964, and January 1965 aggregated 1.36 shares of GM common stock for each share of Du Pont common stock. In addition, 447,847 GM shares were sold in 1964.

EXHIBIT 2 Du Pont Company Financial and Operating Record 1967–1975

	1975(*a*)	1974(*a*)	1973	1972	1971	1970	1969	1968	1967
Sales	$ 7,222	$ 6,910	$5,964	$4,948	$4,371	$4,118	$4,133	$3,931	$3,519
Cost of goods and other operating charges	5,410	5,052	3,879	3,262	2,867	2,682	2,587	2,410	2,220
Interest expenses	126	62	35	24	18	18	15	11	10
Earnings before income taxes and minority interests	453	682	1,077	774	652	623	772	824	622
Income taxes	177	267	480	353	289	283	393	433	293
Earnings before minority interests	276	415	597	421	363	340	379	391	329
Percent of sales	3.8%	6.0%	10.0%	8.5%	8.3%	8.3%	9.2%	9.9%	9.3%
Percent return on average total investment (*b*)	2.5%	4.2%	6.8%	5.3%	4.9%	4.9%	5.8%	6.4%	5.7%
Net income	272	404	586	414	357	334	369	380	321
Net income earned on common stock	262	393	576	404	347	324	359	370	311
Per common share (*c*)	5.43	8.20	12.04	8.50	7.33	6.86	7.62	7.99	6.73
Percent return on average common stockholders' equity	7.4%	11.5%	18.0%	13.7%	12.4%	12.1%	14.1%	15.5%	13.7%
Dividends paid per common share	4.25	5.50	5.75	5.45	5.00	5.00	5.25	5.50	5.00
Dividends as percent of amount earned on common stock	78%	67%	48%	64%	68%	73%	69%	69%	74%
Working capital	1,276	1,607	1,435	1,399	1,330	1,189	1,218	1,099	970
Plants and properties (*d*)	8,585	7,669	6,786	6,155	5,731	5,366	4,963	4,668	4,426
Total investments (*b*)	11,418	10,521	9,215	8,308	7,696	7,167	6,784	6,373	5,906
Long-term debt	889	793	250	241	236	162	147	114	99
Common stockholders' equity	3,596	3,514	3,355	3,029	2,856	2,725	2,615	2,458	2,319
For the year:									
Capital expenditures	1,066	1,038	781	561	474	499	417	355	466
Depreciation	580	506	450	418	399	372	351	324	302

(*a*) Du Pont adopted changes in 1974 and 1975. See relevant annual reports for details.
(*b*) Total investment is the sum of all assets as shown in the company's consolidated balance sheets, before deduction of accumulated depreciation and obsolescence.
(*c*) Based on the average number of common shares outstanding.
(*d*) Before deduction of accumulated depreciation and obsolescence.

Institutionalizing Entrepreneurship

Case IV–7
PC&D, Inc.

E. T. Christiansen, R. G. Hammermesh, and J. L. Bower

When we promoted you to the presidency five years ago, we expected that there would be changes, but we never expected you to diminish the importance of the old line businesses to the extent that you have. I think you have erred in doing so. . . .

The new entrepreneurial subs are certainly dynamic and have brought positive press to the company. But, by investing all new resources in them, you are jeopardizing the health of the company as a whole. . . .

My division's reputation has been built over the past 50 years on the superior quality of its products and sales force. But, as the leadership of our products begins to erode, my salesmen are beginning to leave. Without resources, I cannot stop this trend, and, as much as it saddens me to say so, I am losing my own motivation to stay with the company.

These were some excerpts from a letter that the senior vice president and head of the Machinery Division, George McElroy, 58, sent to John Martell, president of PC&D, Inc., in February 1976. McElroy was highly respected in both the company and the industry, a member of the board of directors, and a senior officer of the company for 20 years. Therefore, Martell knew that it was important to respond and resolve the issues with McElroy successfully. At the same time, Martell had no intention of giving up his own prerogatives to direct the company.

Harvard Business School case 9-380-072

HISTORY OF PC&D, INC.

The Payson & Clark Company

Payson & Clark, the forerunner of PC&D, Inc., was founded during the merger movement around the turn of the century. Four regional machinery companies merged to form a national industrial machinery manufacturing corporation named after the two largest enterprises in the merger, Payson and Clark. With the growth of industry across the country at the time, the demand for heavy machinery took off. The new company benefited from economies of scale, both in production and distribution, and grew and prospered.

By 1965, Payson & Clark Company was an old, stable company, still producing machinery. With revenues of $300 million and net after-tax profits of $6 million, it was still the largest firm in the industry. (See Exhibits 1 and 2 for additional financial information.) The company offered the most complete line of heavy industrial equipment in the industry, the different available configurations of standard and custom models filling a large, encyclopedic sales manual. The consistently high quality and unusual breadth of the product line had made attracting high-caliber salesmen relatively easy. These people were highly knowledgeable in the applications of the product line and saw themselves as consultants to their industrial customers.

While Payson & Clark was the leader in quality and breadth of its product line, it was not the leader in innovations. It left expensive R&D to others, copying products after they were widely accepted. It could afford to follow others primarily because the industry itself was slow moving. In 1965, the business was essentially the same as when the company was founded. Its growth depended on the general growth of industry in the United States, effi-

EXHIBIT 1 PC&D, Inc.

PAYSON & CLARK COMPANY
Income Statement
1956–1965
(in millions)

	1956	1957	1958	1959	1960	1961	1962	1963	1964	1965
Sales	177.6	190.7	205.0	220.5	237.2	247.9	259.1	273.3	288.1	302.7
Cost of goods sold	136.1	145.8	157.6	171.0	184.4	192.4	202.1	218.7	230.8	243.6
Gross profit	41.5	44.9	47.4	49.5	52.8	55.5	57.0	54.6	57.3	59.1
Depreciation	5.0	5.0	5.0	4.0	4.0	4.0	4.0	4.0	3.5	3.5
Marketing and general and administrative	18.2	19.7	20.5	22.2	25.6	27.5	28.4	28.0	30.0	33.3
Engineering and product development	8.1	8.6	9.9	10.1	10.6	11.0	11.4	8.8	9.2	7.1
Total expenses	31.3	33.3	35.4	36.3	40.2	42.5	43.8	40.8	42.7	43.9
NBIT	10.2	11.6	12.0	13.2	12.6	13.0	13.2	13.8	14.6	15.2
Interest	3.0	4.0	4.0	4.0	3.0	3.0	3.0	3.0	3.0	3.0
Profit before tax	7.2	7.6	8.0	9.2	9.6	10.0	10.2	10.8	11.6	12.2
Tax	3.6	3.8	4.0	4.6	4.8	5.0	5.1	5.4	5.8	6.1
Profit after tax	3.6	3.8	4.0	4.6	4.8	5.0	5.1	5.4	5.8	6.1
Earnings per share	$1.29	$1.36	$1.44	$1.65	$1.72	$1.80	$1.83	$1.94	$2.08	$2.19
Average stock price	$18	$22	$19	$30	$29	$29	$27	$31	$35	$33

ciencies in purchasing raw materials, and the scale and automation of production. Indeed, the company's major innovation came in the early 50s with the introduction of plastics in some of the models.

The company was structured in 1965 as it had been in the 20s, with a standard functional organization and highly centralized chain of command. Its top executives were old-time managers, the average age being 55. Many had spent their entire careers with the firm and could remember the days when old Mr. Payson had kept tight reins on the company in the 30s and 40s. Harold C. Payson IV, aged 53 in 1965, was president of the company from the late 40s and president and chairman since 1955. Although the company was publicly held, the Payson family still owned a considerable amount of the stock.

In the early 1960s, Harold Payson began to consider succession. He wanted to leave the company in good condition not only for his own personal pride but for the betterment of his heirs. From discussions with his investment bankers and friends in the business world, Mr. Payson had recognized that an association with a high-technology, high-growth industry would strengthen Payson & Clark's image. One way in which Mr. Payson sought to implement this suggestion was to use some of the excess capital thrown off by the machinery business to enter into joint ventures with young, new companies developing high-technology, innovative products. Several such investments were made in the early 1960s, including one with the Datronics Company in 1962.

Datronics Company

In 1965, the Datronics Company was 10 years old with revenues of $50 million. (See Exhibits 3 and 4 for additional financial information.)

EXHIBIT 2 **PC&D, Inc.**

PAYSON & CLARK COMPANY
Balance Sheet
1956–1965
(in millions)

	1956	*1957*	*1958*	*1959*	*1960*	*1961*	*1962*	*1963*	*1964*	*1965*
Assets										
Cash and securities	6	7	3	1	2	2	2	1	1	1
Accounts receivable	33	36	38	39	41	43	45	47	51	55
Inventories	56	61	64	66	69	74	78	82	88	91
Total current assets	95	103	105	106	112	119	125	130	140	147
Plant and equipment	65	60	60	61	63	67	65	65	64	65
Investments in joint ventures	—	—	—	—	—	—	5	10	11	14
Total assets	160	163	165	167	175	186	195	205	215	226
Liabilities and Net Worth										
Accounts payable	31	33	36	38	46	54	62	65	70	75
Accrued liabilities	7	9	10	11	13	17	22	25	31	36
Long-term debt due	6	6	6	6	6	6	6	6	6	6
Total current liabilities	44	48	52	55	65	77	86	96	107	117
Long-term debt	52	47	41	35	29	23	18	12	6	—
Total liabilities	96	95	93	90	94	100	104	118	113	117
Common stock	27	27	27	27	27	27	27	27	27	27
Retained earnings	37	41	45	50	54	59	64	70	75	82
Total liabilities and net worth	160	163	165	167	175	186	195	205	215	226

EXHIBIT 3 **PC&D, Inc.**

DATRONICS COMPANY
Income Statement
1956–1965
(in millions)

	1956	*1957*	*1958*	*1959*	*1960*	*1961*	*1962*	*1963*	*1964*	*1965*
Contracts	1.2	6.4	8.2	7.5	8.0	7.9	6.0	4.3	3.4	2.4
Sales	—	—	.2	2.1	4.4	8.1	14.3	22.5	34.2	48.1
Revenues	1.2	6.4	8.4	9.6	12.4	16.0	20.3	26.8	37.6	50.5
Cost of goods sold	1.0	4.5	6.0	6.9	8.9	11.5	14.7	19.6	27.8	37.9
Gross profits	.2	1.9	2.4	2.7	3.5	4.5	5.6	7.2	9.8	12.6
Expenses	.5	.6	.7	.7	.7	.7	.9	.9	1.0	1.1
R&D	—	.7	.8	1.0	1.2	1.5	2.2	3.0	4.0	5.1
Profit before tax	(.3)	.6	.9	1.0	1.6	2.3	2.5	3.3	4.8	6.4
Tax	(.15)	.2	.4	.5	.8	1.1	1.2	1.6	2.4	3.2
Net profit	(.15)	.4	.5	.5	.8	1.2	1.3	1.7	2.4	3.2
Earnings per share	($1.50)	$4	$5	$5	$8	$12	$10.40	$13.60	$19.20	$25.60

EXHIBIT 4 PC&D, Inc.

DATRONICS COMPANY
Balance Sheet
1956–1965
(in millions)

	1956	1957	1958	1959	1960	1961	1962	1963	1964	1965
Assets										
Cash	.05	.10	.10	.40	.20	.60	.60	.65	1.56	.70
Inventories	.20	2.60	2.70	3.70	5.20	6.20	6.80	10.15	15.22	20.10
Accounts receivable	—	.30	.50	1.00	2.00	2.20	3.00	4.00	5.12	6.00
Total current assets	.25	3.00	3.30	5.10	7.30	9.00	10.40	14.85	21.90	26.80
Plant and equipment	.50	1.00	1.20	1.40	2.00	3.10	5.10	7.50	8.5	9.0
Total assets	.75	4.00	5.50	6.50	9.30	12.10	15.50	22.35	30.40	35.8
Liabilities and Net Worth										
Accounts payable	.10	2.15	2.20	2.60	3.65	4.75	5.50	8.78	12.10	14.25
Accrued liabilities	.10	1.00	1.05	1.25	1.65	2.25	1.70	2.77	3.80	3.85
	.20	3.15	3.25	3.85	5.30	7.00	7.20	11.55	15.90	18.10
Notes payable	.60	.50	1.40	1.30	1.85	1.75	2.50	2.2	3.5	3.5
Total liabilities	.80	3.65	4.65	5.15	7.15	8.75	9.70	13.75	19.45	21.60
Additional paid-in capital	—	—	—	—	—	—	1.125	2.225	2.225	2.225
Common stock ($1 par)	.10	.10	.10	.10	.10	.10	.125	.125	.125	.125
Retained earnings	(.15)	.25	.75	1.25	2.05	3.25	4.55	6.25	8.65	11.85
Total liabilities and net worth	.75	4.00	5.50	6.50	9.30	12.10	15.50	22.35	30.40	35.80

The company had started as an engineering firm subsisting on government research grants and contracts. As a by-product of the government projects, the company also developed several types of sophisticated electronic equipment with wide applications to industry. The company concentrated its efforts on R&D, however, and subcontracted the production and bought marketing services for its commercial products. The lack of control over marketing and production and the lost profits passed to the marketers and subcontractors displeased the company's young president, John Martell. In his opinion, the growth of the company was limited until the right product emerged to justify going to a full manufacturing and marketing company.

Following Payson & Clark's investment in 1962, Datronic's engineers developed an exciting new product toward the end of 1964 which promised to sell extremely well due to its increased capacity and lower cost. John Martell saw the promise of the new product as the waited-for opportunity to expand the company. It was clear, however, that a major influx of capital was needed to bring the product to the market, build a sales force, and begin volume production. Therefore, Martell began a search for external capital that included a presentation to the joint venture partner, Payson & Clark, which already owned 20 percent of Datronics' stock.

Meanwhile, Harold Payson had been following the activities at Datronics closely and was quite aware of the growth potential of the company before John Martell's visit. Further, he recognized that Datronics, once its manufacturing operations started, would have a continual need for new capital. If Payson & Clark invested once, it would not be long until another request for resources came from Datronics. With these factors in mind, Payson decided that the most beneficial arrangement for both parties would be for his company to acquire Datronics. Martell agreed to this offer and negotiations for a friendly takeover were consummated. Payson & Clark acquired Datronics for $42 million in November 1965. John Martell himself received $8.4 million in cash, notes, and securities.

The acquisition provided an opportunity for the Payson & Clark Company to update its image. Patterning itself after other successful growth companies of the time, it changed its name to PC&D, Inc., to denote the beginning of a new era in the company.

PC&D, INC., 1965–1970

After the acquisition, Harold Payson restructured the company with the help of consultants, setting up a divisional organization. The old Payson & Clark Company now became the Machinery Division, headed by George McElroy, formerly vice president, manufacturing. The Datronics Company became the Electronics Division, headed by John Martell.

The Electronics Division

At the time of the acquisition, the Datronics Company consisted of several scientific labs, some test equipment, 10 professional engineers, administrative staff, and John Martell.

Martell, an electrical engineer by training, was a man in his mid-30s. He was energetic and a risk taker by nature, and even as a child in Iowa could not imagine working for someone else all his life. After college at M.I.T., he worked for eight years at a large, scientific equipment company in the Boston area. Initially, he was hired for the research group, but he was more attracted to the management positions in the company. He transferred first to the Corporate Planning Office and then became plant manager for one of the divisions. With his technical competence and management experience, it was not surprising that he was approached by several of the more innovative of the company's research engineers to invest in and head up a new, independent R&D company. Martell bought in for 25 percent of the founding stock and, thus, began the Datronics Company.

During his term as president of Datronics, Martell was highly regarded by the small group of employees. While he had a respectable command of the technology, he left the research to the engineers, devoting his time to developing sources of challenging and lucrative contracts.

After the acquisition by the Payson & Clark Company, Martell retained full control of the operations of his old organization that was now the Electronics Division. He hired an experienced industrial marketer from a large technical firm to set up the marketing operations and a friend of his from his old employer to head up the production operations. As expected, the demand for the division's new product was very high. Five years later, by 1970, the division was a successful growing enterprise, having expanded into other electronics fields. It had 700 employees, marketing offices established or opening throughout the United States, Europe, and Japan, plants at three different sites, and revenues of over $160 million. The business press reported these activities very favorably, giving much credit to the leadership of Martell.

The Machinery Division

Meanwhile, the Machinery Division continued to be the stalwart of the industry it always had been, retaining its structure and activities of the earlier time. George McElroy, division manager and senior vice president, was considered the mainstay of the division. He had joined the company in the early 1950s and was primarily responsible for the plastics innovations of that time. Advisor and confidant of Payson, McElroy was thought by his subordinates to be the next in line for the presidency.

As for Harold Payson himself, he limited his involvement in the company's internal affairs to reviewing budgets and year-end results, and spent most of his time with community activities and lobbying in Washington. He felt justified in this hands-off policy because of the quality of both his division vice presidents, McElroy and Martell. PC&D's performance further supported Payson's approach. Revenues climbed to $530 million, and profits after tax to $14 million by 1970. The solid 26 multiple of its stock price reflected the confidence in PC&D's prospects. (See Exhibits 5 and 6.)

The compensation schemes reflected the extent to which Harold Payson allowed the division managers to be autonomous. McElroy's compensation was 90 percent salary, with a 10 percent bonus based on ROI. Martell received two thirds of his pay as a bonus based on growth in revenues. Compensation policies within each division were entirely at the discretion of either Martell or McElroy. In general, Martell made much greater use of incentive compensation than McElroy.

EXHIBIT 5 PC&D, INC.

Income Statement
1966–1970
(in millions)

	1966	1967	1968	1969	1970
Sales:					
Machinery	315.1	327.5	340.2	354.1	368.2
Electronics	66.1	84.7	106.7	132.3	161.4
Total	381.2	412.2	446.9	486.4	529.6
Cost of goods sold:					
Machinery	251.7	264.3	271.8	284.7	297.9
Electronics	49.6	63.0	79.6	96.8	118.5
Total	301.3	327.3	351.4	381.5	416.4
Gross margin	79.9	84.9	95.5	104.9	113.2
Expenses:					
Marketing general and administrative expense	46.1	48.3	50.3	51.6	53.1
Product development—Machinery	6.9	4.6	4.7	4.1	4.5
R&D—Electronics	4.2	5.3	10.3	17.8	27.3
Total	52.2	58.2	65.3	73.5	84.9
NBIT	24.7	26.7	30.2	31.4	28.3
Interest	3.0	3.0	0.2	0.2	0.2
Profit before tax	21.7	23.7	30.0	31.2	28.1
Taxes	10.8	11.8	15.0	15.6	14.0
Net profit	10.9	11.9	15.0	15.6	14.1
Earnings per share	$3.63	$3.97	$5.00	$5.20	$4.70
Average stock price	$94	$111	$145	$146	$103

EXHIBIT 6 PC&D, INC.

Balance Sheet
1966–1970
(in millions)

	1966	1967	1968	1969	1970
Assets					
Cash and securities	2	5	9	7	11
Accounts receivable	67	71	77	87	101
Inventories	118	128	145	166	180
Total current assets	187	214	231	260	292
Plant and equipment	83	95	97	108	120
Investments in joint ventures	10	11	12	12	10
Goodwill	6	6	5	5	5
Total assets	286	320	345	385	427
Liabilities and Net Worth					
Accounts payable	90	96	103	111	127
Accrued liabilities	31	33	31	32	35
Long-term debt due	1	1	1	2	3
Total current liabilities	122	130	135	145	165
Long-term debt	16	30	35	49	57
Total liabilities	138	160	170	194	222
Common stock and paid-in capital	55	55	55	55	55
Retained earnings	93	105	120	136	150
Total liabilities and net worth	286	320	345	385	427

1970 Change at PC&D

Toward the end of 1970, Harold Payson decided that it was time to limit his involvement to that of chairman of the board and to name a new president of PC&D. He, himself, supported the appointment of George McElroy as the next president. McElroy was the next senior officer in the company and, after years of working with Harold Payson, held many of the same views as to the traditional values of PC&D. However, Payson agreed with the school of thought that chief executives should not choose their own successors. He therefore established a search committee, consisting of three outside members of the board of directors. (See Exhibit 7 for a list of board members.) A thorough job was done. The committee interviewed several candidates within PC&D, including John Martell and George McElroy. Outside candidates were also considered. The committee utilized executive search firms and consultants to identify candidates and carefully compared external and internal prospects. The result was the nomination of John Martell. While his relative youth was a surprise to some, the search committee's report explained the thinking behind the choice that "during the past five years, PC&D has experienced an exciting and profitable period of growth and diversification. But it is essential that the company not become complacent. One of our major criteria in choosing a new president was to find a person with the energy and

EXHIBIT 7 PC&D, Inc.

Members, Board of Directors, 1970

Harold Payson IV, President—PC&D
George McElroy, Senior Vice President—Machinery Division, PC&D
John Martell, Vice President—Electronics Division, PC&D
Carl Northrup, Treasurer—PC&D
David S. Curtis, Partner—Barth & Gimbel, Wall Street Brokerage Firm
Elizabeth B. Payne, Partner—Payne, Bartley, & Springer, Washington Law Firm
Charles F. Sprague, President—Forrest Products, Inc. (a large manufacturing firm)
Gardner L. Stacy III, Dean—Business School, State University
James Hoffman, Vice President—Baltimore Analysts Association (an international firm)

vision to continue PC&D's growth and expansion." The board unanimously approved the selection of John Martell as president and CEO.

Martell began his new position with the board's mandate in mind. He planned to continue the diversification of PC&D into high-growth industries. He expected to follow both an acquisition mode and a start-up mode, using the excess funds from the Machinery Division and PC&D's rising stock to finance the growth. For start-ups, Martell planned to use joint ventures supporting newer companies, much as the old Payson & Clark Company had supported his venture in its early days.

Martell brought to his position a very definite management style. He was a strong believer in the benefits accruing from an opportunistic, entrepreneurial spirit and he wanted to inject PC&D with this kind of energy. However, he was concerned that the kind of people with this kind of spirit would not be attracted to work with PC&D because of the stigma, real or imagined, of being attached to a large company.

As Martell commented:

> It was my experience that there are two worlds of people, some of whom are very secure and comfortable and satisfied in their career pursuits in large institutionalized companies, and others of whom are, I think, wild ducks, and who are interested in perhaps greater challenges that small companies present in terms of the necessity to succeed or die.
>
> In many work environments, the constraints placed upon the individual by the nature of the institution are such as to sometimes make people uncomfortable.
>
> The decision-making process is long and involved, sometimes not known, in the sense that the people who act upon decisions are not in close proximity to those who benefit or suffer from the effects of those decisions.
>
> The formalization of the decision-making process is frequently an irritant, and for people who are unusually energetic and demanding, in the sense of desiring, themselves, to take action and to have their actions complemented by the actions of other people upon whom they are dependent, I would characterize these people as perhaps being wild ducks rather than tame ducks. In that sense, I wanted more "wild ducks" in our company.

Martell himself credited the success of the Electronics Division to Payson's willingness to turn the reins completely over to him. The secret, Martell thought, was in spotting the right person with both ability and integrity. Corporate headquarters' role should be to provide resources in terms of both money and expertise as needed, to set timetables, to provide measurement points and incentive, and then to keep hands off.

While the board's directives were clear to

Martell, the specifics for implementation were not. Not only were the larger questions of which way to diversify or how to encourage innovation unanswered, but how to plan and who to involve were also unclear. Martell was not given the luxury of time to resolve these issues. Within the first week in his new position, three professionals from the Electronics Division called on Martell. Bert Rogers and Elaine Patterson were key engineers from the Research Department and Thomas Grennan was head of marketing, western region. They had been working on some ideas for a new product (not competing with any PC&D current lines) and were ready to leave the company to start their own business to develop and market it. Indeed, they had already had a prospectus prepared for their new venture. They were hoping either Martell personally or PC&D, Inc., might be able to provide some venture capital. The president particularly liked these three and admired their willingness to take such personal risks with a product as yet unresearched as to market or design. Indeed, with his energy and "can do" aggressive style, Tom Grennan reminded Martell of himself just a few years ago when he left to start the Datronics Company.

Martell liked the product and saw the idea as a possible route for continuing the diversification and growth of PC&D. But there was a problem. It was clear from the presentation of the three that much of their motivation came from the desire to start their own company and, through their equity interest, to reap the high rewards of their efforts if successful. Martell did not fault this motivation, for it had been his as well. He could not expect PC&D's managers to take large personal risks if there was no potential for a large payoff. Further, a fair offer to the group, if in salary, required more than PC&D could afford or could justify to the older divisions. Martell told Rogers, Patterson, and Grennan that he was very interested and asked if he could review the prospectus overnight and get back to them the next day. That night, he devised a plan of which he was particularly proud. The major feature of the plan Martell called the "Entrepreneurial Subsidiary." Martell presented this proposal to Grennan, Rogers, and Patterson the next day. They readily accepted and a pattern for most of PC&D's diversification over the next five years was begun.

The Entrepreneurial Subsidiary

Martell's plan was as follows:

When a proposal for a new product area was made to the PC&D corporate office, a new (entrepreneurial) subsidiary would be incorporated. The initiators of the idea would leave their old division or company and become officers and employees of the new subsidiary. In the current example, the new subsidiary was the Pro Instrument Corporation with Grennan as president and Rogers and Patterson as vice presidents.

The new subsidiary would issue stock in its name, $1 par value, 80 percent of which would be bought by PC&D, Inc., and 20 percent by the entrepreneurs involved—engineers and other key officers. This initial capitalization, plus sizable direct loans from PC&D, Inc., provided the funds for the research and development of the new product up to its commercialization. In the case of Pro Instruments, Patterson and Rogers hired 10 other researchers, while Grennan hired a market researcher and a finance/accounting person. These 15 people invested $50,000 together and PC&D invested another $200,000.

Two kinds of agreements were signed between the two parties. The first was a research contract between the parent company and the subsidiary, setting time schedules for the research, defining requirements for a commercializable product, outlining budgets, and otherwise stipulating obligations on both sides. In general, the sub was responsible for the R&D and production and testing of a set number of prototypes of a new product, while the parent company would market and produce the product on an international scale. Pro Instruments' agreement stipulated two phases, one lasting 18 months to produce a prototype, and another lasting 6 months to test the product in the field and produce a marketing plan. Detailed budget and personnel needs were outlined, providing for a $900,000 working capital loan from PC&D during the first phase and $425,000 during the second.

While PC&D, Inc., had proprietary rights on the product and all revenues received from marketing it, the agreement often included an incentive kicker for the key engineers in the form of additional stock to be issued if the finished product produced certain specified amounts of revenue by given dates. Indeed, this was the case for Pro Instruments: 5,000 shares in year 1, to be issued if net profits were over $250,000; 20,000 shares in year 2 if profits were over $1 million, and 10,000 in year 3 if profits were over $3 million.

The second agreement specified the financial obligations and terms for merger. Once the terms of the research contract were met, PC&D, with board approval, had the option for a stated period of time (usually four years) to merge the subsidiary through a one-for-one exchange of PC&D stock for the stock of the subsidiary. The sub was then dissolved. To protect the interests of entrepreneurs, PC&D was required to vote on merger of the sub within 60 days if the sub met certain criteria. For Pro Instruments, the criteria were (1) the product earned cumulative profits of $500,000 and (2) if the earnings of PC&D and the sub were consolidated, dilution of PC&D's EPS would not have occurred over three consecutive quarters. If PC&D did not choose to merge during the 60 days, then the sub had a right to buy out PC&D's interest.

Since PC&D's stock was selling for $103 in 1970 and subsidiary stock was bought for $1 per share, the exchange of stock represented a tremendous potential return. Depending on the value of PC&D's stock at the time of merger, the net worth of the "entrepreneurs" who originally invested in the sub multiplied overnight. Indeed, as subs were merged in ensuing years, typical gains ranged from 100 to 200 times the original investments in the entrepreneurial sub. For example, PC&D exercised its option to merge Pro Instruments when its product was brought to market in 1972. Thomas Grennan, who had bought 6,000 shares of Pro Instruments stock, found his 6,000 shares of PC&D valued at $936,000 (PC&D common selling for $156 on the New York Stock Exchange at the time). By the end of 1974, Pro Instruments' new product had earned $50 million in revenue and $4.8 million in profits, thus qualifying the original entrepreneurs for stock bonuses. Grennan received another 4,200 shares valued at $684,600. Thus, in four years, he had earned about $1.6 million on a $6,000 investment.

By setting up entrepreneurial subs like Pro Instruments, Mr. Martell had several expectations. In the process of setting up a subsidiary with the dynamics of a small, independent group, Martell hoped to create the loyalty, cohesion, and informal structure conducive to successful research and development efforts. The sub would have a separate location and its own officers who decided structure and operating policies. Further, it provided the opportunity to buy into and reap the benefits of ownership in the equity of a company. In Mr. Martell's words:

> I think the concept of the entrepreneurial subsidiaries was the outgrowth of the insight that in many industrial corporations the system of rewards is perhaps inverted from what many people think it should be; that the hierarchy of the institution commends itself to those people who are capable of managing other people's efforts, and those people at lower echelons who are unusually creative and who, as a result of their creativity and innovation and daring in the technical sense or perhaps in a marketing sense, are unusually responsible for the accomplishments of the business, are very frequently forgotten about in the larger rewards of the enterprise.

> I, on the other hand, recognized that such persons are frequently, perhaps by training, inclination, or otherwise, not capable of marshaling the financial resources or organizing the manufacturing and marketing efforts required to exploit their creativity. Without the kind of assistance that PC&D was capable of lending to them—an assured marketing capability was often a key concern—they are wary of undertaking new ventures.

Further, it was Mr. Martell's opinion that the organizational and incentive structure of the entrepreneurial sub would attract the best engineers from older, more secure firms to PC&D—the so-called "wild ducks."

More important, Martell hoped to encourage the timely development of new products with minimal initial investment by PC&D. If Pro Instruments, for example, did not meet its timetable with the original money invested, its officers would have to approach PC&D for new money just as if they were an outside company. PC&D would then have multiple opportunities to review and consider the investment. If the entrepreneurial sub failed or could not get more money from the parent, PC&D was under no obligation to keep the company alive or to rehire its employees. If loans were involved, PC&D could act as any other creditor. As Martell observed,

> The benefit to PC&D shareholders was in the rapid expansion of PC&D's products, the size of the company, the ability of the company to compete in the marketplace in a way which PC&D, dependent upon only internal development projects, could never have achieved, or could have achieved only at much greater costs and over a longer period of years.

However, Martell felt the stock incentives would properly reward the genius of creative engineers for the service performed without having to pay high salaries over a long potentially unproductive period after the initial product was developed. Employees did not have to be rehired, nor were they obligated to continue employment, even if the sub was merged. Those that were rehired would be paid at the normal salary levels of comparable people at PC&D. The reasoning here was that:

> There were two criteria for establishing an entrepreneurial subsidiary. The first criterion was that the R&D objectives of the subsidiary could not be reached except under the aegis of the subsidiary, because it involved people who were not involved in PC&D's main lines of business.
>
> The other criterion was that considerable career risk must exist for the people who would leave their established positions within the management structure of PC&D to undertake the entrepreneurial venture of the new subsidiary. Also, the people, in some part, had to be new talent who came from outside PC&D. When I refer to career risk, I mean for example that if a director of engineering at PC&D left his or her post to join an entrepreneurial subsidiary, a new director of engineering would be appointed, and given the lack of success of the entrepreneurial subsidiary, there would in effect be no position of director of engineering to which the person could return. Moreover, it is probable that we would not want the individual to return.

The stock incentive also motivated the engineers to produce without having to commit any resources of the parent company for the future, since the corporation was not required to merge the sub or to produce and market the new product. The incentive kicker, moreover, would insure quality. A product that was rushed through development would be more likely to have problems and not reach revenue goals.

Another advantage of the entrepreneurial sub was its effect on decision making. Without the need to go through the entire corporate hierarchy, decisions would be made closer to the operating level. This would enhance the quality of decisions because managers performed best, according to Martell, when given objectives and resources from top managers but with operating decisions left unfettered.

Finally, Martell expected that the entrepreneurial sub would be the training and proving ground of PC&D's future top managers. By providing the means for these executives to gain great personal wealth, Martell expected to gain their loyalty and continued efforts for both himself and PC&D.

PC&D: 1970–1975

During the first five years of Martell's presidency, PC&D's growth was quite impressive. With revenues topping the billion dollar mark in 1975, growth had averaged about 15 percent in revenues and 35 percent in profits after tax during the five years. (See Exhibits 8 and 9 for financials.) Such growth had been achieved, to a large extent, from new products developed in entrepreneurial subsidiaries. In 1975, sales of $179.2 million and profit before taxes and interest of $22.1 million came from these new products.[1] All together, 11 entrepreneurial subsidiaries had been organized during the 1970–75 time frame. Of these, four had successfully developed products and had been merged into PC&D—one in 1972, one in 1973, and two in 1974. The other seven were younger and work was still in process. None had failed so far.

Most subsidiaries grew out of needs of the Electronics Division or Pro Instruments. Competitors in the electronics equipment industry were beginning to integrate backward, lowering costs by producing their own semiconductors. The need to remain cost competitive caused PC&D to establish entrepreneurial subs to develop specialized components including semiconductors, assuming that these could be used both by PC&D and sold in outside markets. In the process of selling semiconductors to outside customers, ideas for new products using PC&D components were stimulated, and new subs were formed to develop these equipment products. The cost of merging the two types of subs, components or equipment, differed, however. Equipment subs were cheaper in so far as they could share the already existent sales force of the Electronics Division; many parts could be standard ones already utilized in other products; and the processes were similar to other Electronics products. But with semiconductors, new plant, new sales channels, new manufacturing processes, and new skills at all levels had to be built. While to Martell the move into semiconductors promised a large cash flow in the future in a booming industry, some in the company were concerned that the current cash drain was not the best use of scarce cash resources.

When Martell first became president, he made few changes in PC&D's organization structure. McElroy continued as vice president, Machinery Division, and retained control over that division's structure and policies. Martell himself retained his responsibilities as manager of the Electronics Division. This he did reluctantly and with all intentions of finding a new executive for the job; however, the unexpected nature of his promotion left Martell without a ready candidate.

As the subs began to be merged, beginning with Pro Instruments in 1972, questions of organization began to arise. In typical fashion, Martell wanted to pass involvement in these decisions down to the appropriate managers. There was also no question that Pro Instrument's president, Tom Grennan, had proven himself with the new subsidiary. So in 1972, Martell appointed Grennan to division vice president, electronics, based on Grennan's superlative performance. Further, because the products were complementary, all of the subs that were merged in this period were placed in the Electronics Division. Moreover, in recognition of the increased number of products, Grennan did reorganize the Electronics Division. He appointed his Pro Instruments col-

[1] Of PC&D total assets in 1975, approximately 40 percent were devoted to the Machinery Division, 35 percent to the traditional Electronics Division, and 25 percent to the entrepreneurial subsidiaries.

EXHIBIT 8 PC&D, INC.

Income Statement
1971–1975
(in millions)

	1971	1972	1973	1974	1975
Sales:					
Machinery	382.9	397.8	412.5	426.9	440.6
Electronics*	193.6	235.6	300.1	397.4	561.4
Total	576.5	633.4	712.6	824.3	1,002.0
Cost of goods sold:					
Machinery	311.3	322.6	338.2	350.9	359.1
Electronics	145.2	174.3	216.1	282.2	421.1
Total	456.5	496.9	554.3	633.1	780.2
Gross margin	120.0	136.5	158.3	191.2	221.8
Expenses:					
Marketing general and administrative expense	54.7	56.3	59.1	63.3	67.7
Development—Machinery	5.0	5.1	5.2	5.2	5.3
R&D—Electronics	28.4	29.5	30.7	31.9	33.5
Total	88.1	90.9	95.0	100.4	106.5
Profit before interest and taxes	31.9	45.6	63.3	90.8	115.3
Interest	0.2	3.0	3.0	7.0	11.0
Profit before tax	31.7	42.6	60.3	83.8	104.3
Taxes	15.8	21.3	30.1	41.9	52.1
Net profit	15.9	21.3	30.2	41.9	52.2
Earnings per share	$5.30	$6.45†	$8.39	$10.47	$13.05
Average stock price	$106	$156	$158	$163	$238

* Sales figures for Electronics include both sales by the original division plus sales of new subsidiaries after they are merged. Thus in 1975, the $561.4 million in sales for Electronics includes $179.2 from products developed in subsidiaries. Profit before interest and taxes from new products was $22.1 million.

† Number of shares increased in 1972 by .3 million from the merger of Pro Instruments. They increased in 1973 by .3 million from merger of Sub #2, and again by .4 million in 1974 from the merger of Subs #3 and #4. Thus in 1974, there was a total of 4 million shares outstanding. In late 1973, there was a secondary offering of 1 million shares.

league, Bert Rogers, to director of research which was organized by product area. Manufacturing, also organized by product, reflected the development by subsidiary as well. Marketing, on the other hand, was organized by region as it had been previously. Until they were merged, however, subsidiary presidents went directly to Mr. Martell for resolution of problems that arose. (See Exhibit 10 for an organization chart in 1975.)

By 1975, the Electronics Division's enlarged marketing and production departments employed 4,000 people with production plants in three different locations. Electronics now had sales of $561.4 million as compared to Machinery's $440.6 million.

EXHIBIT 9 PC&D, INC.

Balance Sheet
1971–1975
(in millions)

	1971	1972	1973	1974	1975
Assets					
Cash and securities	10	5	2	2	3
Accounts receivable	117	131	155	171	213
Inventories	200	223	270	327	401
Total current assets	327	359	427	500	617
Plant and equipment	122	124	125	178	232
Investments in joint ventures	10	8	10	9	6
Investments in subsidiaries	5	10	21	16	25
Goodwill	4	4	3	3	2
Total assets	468	505	586	706	882
Liabilities and Net Worth					
Accounts payable	151	160	179	193	243
Accrued liabilities	37	41	46	51	65
Long-term debt due	4	4	4	6	7
Total current liabilities	192	205	229	250	315
Long-term debt	55	58	84	138	193
Total liabilities	247	263	313	388	508
Common stock and paid-in capital	55	55	56	57	57
Retained earnings	166	187	217	261	317
Total liabilities and net worth	468	505	586	706	882

While successful development projects from subsidiaries had been largely responsible for the sales growth at PC&D, this result had not come without costs. First, the subsidiaries required funds—$60 million by the end of 1975. Some of these funds came from retained earnings, but much was new money raised in the form of long-term debt. Further, stock issued to capitalize subs and pay bonuses to "entrepreneurs" had a diluting effect on PC&D's shares. If all subsidiaries were merged and successful, the number of new shares could be significant. While raising such a sizable amount of new funds was not particularly difficult for a company as large as PC&D, the needs arising from the subsidiaries left little new money for the core businesses of PC&D. The Machinery Division, for example, had not had their development budget increased at all during the five years ending 1975.

Current Concerns

Despite PC&D's recent successes, Mr. Martell was not without worries. Several problem areas had appeared in both the Electronics and Machinery Divisions.

In Electronics, personnel and products originating in subsidiaries now equaled or surpassed those from the original division. It had been part of the strategy of the entrepreneurial subsidiaries to use them as devices to attract

EXHIBIT 10 PC&D Inc.

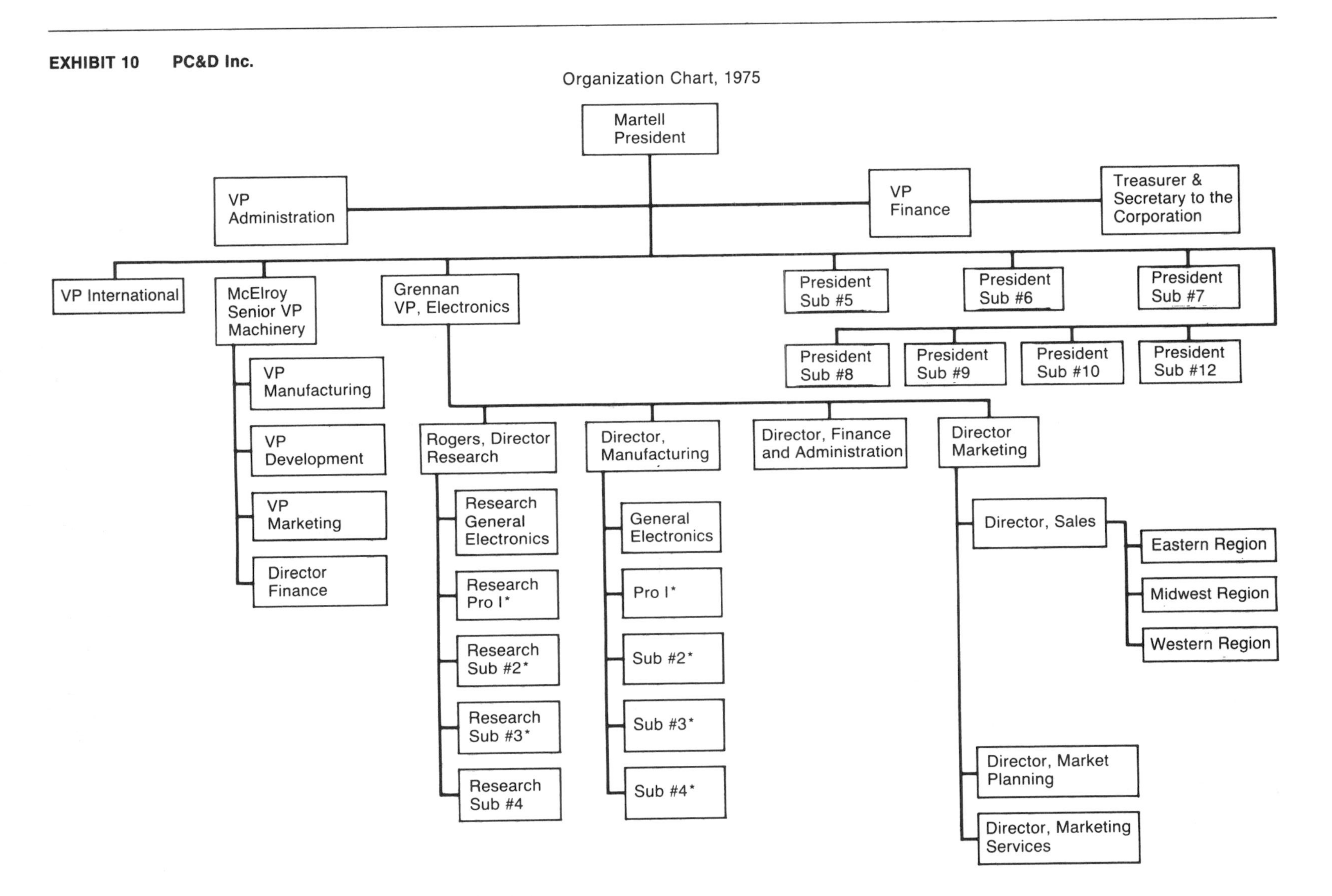

* Reference to subsidiaries indicates origin of personnel and product.

talent from other firms. A key researcher hired from outside was encouraged to hire, in turn, the best of his or her former colleagues. Thus, the loyalty and friendships between key "entrepreneurs" and their staffs were often strong and of long standing. As the entrepreneurial subsidiaries were merged, their personnel tended to retain this loyalty to the president or key officers of the old sub rather than transferring it to PC&D. Thus, several warring spheres of influence were developing in the division, particularly in the research department and between research and other departments. Martell was concerned that such influences and warring would lead to poor decisions and much wasted energy in the division.

Turnover in Electronics was also increasing. This was of particular concern to Martell for it was just those talented engineers that the entrepreneurial subsidiaries were meant to attract that were beginning to leave. For example, Elaine Patterson, formerly of Pro Instruments, left during 1975 to start her own company, taking 20 research engineers with her. The source of the turnover was unclear but possible factors included distaste for the kind of warring atmosphere mentioned above and the inability to be a part of a large corporate R&D department with its demand for budgets and reports.

For many employees, the sudden absence of monetary incentives changed the climate drastically, however. This lack of incentive, coupled with the discovery that the most challenging projects were taken on by newly formed subsidiaries which favored hiring outside expertise, caused dissatisfaction. For Martell, such turnover was of greatest concern in the long run, for the inability to create a strong central R&D department in Electronics created a continuing need for more entrepreneurial subs. These subsidiaries were still too new an idea for Martell to want to risk his entire future R&D program on their successes. Further, most of the new products were in highly competitive areas. Without continuing upgrades, these products would soon become obsolete. A strong central R&D department was needed for follow-up development of products started by subsidiaries.

Finally, Martell was concerned by recent indications of rather serious operating problems in the Electronics Division. This was particularly disturbing in that Martell had placed complete faith in Grennan's managerial ability. The most recent cost report, for example, indicated that marketing, G&A, and engineering expenses were way out of line in the division. Further, the marketing and production departments reported problems in several products originating in the subs. One product, with expected obsolescence of four years, now showed a six-year breakeven just to cover the engineering and production costs. Another product, completing its first year on the market, had been forecasted by the subsidiary to achieve $20 million in sales in its first two years. However, during the first six months, losses had been incurred because of customer returns. A report on the causes of the returns showed a predominance of product failures. The chances for breakeven on this product looked bleak. While none of these problems had affected operating results yet, Martell was especially concerned that these operating problems would have a negative impact on first-quarter 1976 earnings.

Martell had not confronted Grennan with these operating problems as yet. He had wanted to see how the division itself was attacking these issues through its long-range plan. Martell had requested Grennan to prepare a Long-Range Plan (five years) as well as the usual one-year operating plan. The product of this effort had only arrived recently (February 1976) and Martell had not had a chance to study it. (Table of Contents is reproduced in Exhibit 11.) Its 100-page bulk loomed on Martell's desk. Quick perusal had indicated maybe four pages of prose scattered through the plan, and dozens of charts, graphs, and tables of numbers, every one of which manifested an upward trend.

EXHIBIT 11 PC&D, Inc.
Electronics Division
1976 Operating Plan
1977–1980 Long-Range Outlook

Table of Contents

EXHIBIT 11 *(concluded)*

In an attempt to get employee feedback on all of these problems, Martell had contracted an outside consulting firm to carry out confidential interviews with personnel in the Electronics Division. The interviews found middle managers quite concerned over the "confusion in the division" which was causing a loss of morale there. The consultant's report cited concrete problems, including lost equipment, missed billings, and confusion in the plant. Typical comments from lower level personnel included:

> Either upper management is not being informed of problems or they don't know how to solve them.
>
> Morale is very poor, job security is nil.
>
> There is little emphasis on production efficiencies.
>
> Scrap is unaccounted for.
>
> Market forecasts are grossly inaccurate.
>
> Production schedules have a definite saw-tooth pattern. There is very little good planning.
>
> There are no systematic controls.

These were not the sort of comments Martell expected from the division responsible for the major portion of PC&D's future growth. His concern, at this time, was not so much the problems themselves, but what was being done about them. His preferred policy was obviously to stay out of day-to-day operating problems. He wondered how long it was prudent to allow such problems to continue without some intervention on his part.

Meanwhile, the Machinery Division had its own problems. The last major construction of new plant had been in the early 1950s. Since that time, McElroy had upgraded production methods, which succeeded in checking rising costs. However, since 1965, resources for such improvements had not been increased, and, with inflation in the 1970s, less and less could be done on a marginal basis. McElroy was currently of the opinion that capacity was sufficient for the short term, but that it was impossible to remain state-of-the-art.[2] Indeed, the Machinery Division's products were beginning to fall behind the new developments of competitors.

---

[2] McElroy suspected that the Machinery Division would require an investment of $100 to $125 million over two to three years to revitalize the product line and plant and equipment. McElroy felt that in the long term the return on this investment would match the division's historic ROI.

Further, the costs of Machinery's products were beginning to inch up. As the production line aged, quality control reported an increasing percentage of defective goods. In contrast to the situation in Machinery, the rather extensive investment in new plant for the production of semiconductors did not sit too well with McElroy who was concerned with the lack of flexibility that could result from backward integrating and thought component needs should be farmed out to the cheapest bidder from the numerous small component firms. Martell was concerned how long he could keep McElroy satisfied without a major investment in Machinery and how long he could count on the cash flow from Machinery for other users.

Also, turnover, a problem never before experienced in the Machinery Division, had appeared. Here, however, it was the salespeople who were leaving. Martell worried over this trend, for the sales force was the strength of the division. According to the head of marketing, the salespeople considered themselves the best in the industry, and they did not wish to sell products which were not the best. They saw Machinery's products no longer as the best in quality or state-of-the-art. Further, they did not wish to work for a company where they felt unimportant. Whether true or not, the sales force certainly appeared less aggressive than in previous times.

Thus, Martell was not overly surprised to receive McElroy's letter nor was he certain that some of McElroy's anger concerning the Electronics Division was not justified. Martell knew he had to do something about McElroy, as well as Grennan and the Electronics Division. He also had to decide whether entrepreneurial subsidiaries should continue to be part of PC&D's research and development strategy. Finally, all of Martell's decisions concerning the divisions and subsidiaries needed to be consistent with a strategy that would continue PC&D's growth.

Reading IV–4
Designs for Corporate Entrepreneurship in Established Firms

R. A. Burgelman

As firms grow large, their capacity to maintain a certain growth rate, based on pursuing opportunities in their mainstream areas of business, eventually diminishes. Sooner or later, firms—Apples and IBMs alike—have to find and exploit opportunities in marginally related, even unrelated, areas through internal corporate venturing and/or acquisition.

Systematic research shows that such diversification is both difficult and risky. Not surprisingly, various authors have argued that firms should maintain the "common thread"[1] and "stick to the knitting."[2] This may be good advice for firms which have not sufficiently exploited their incremental opportunities. It does, however, assume away the fundamental problem and offers little in terms of how firms could improve their capacity to engage in *corporate entrepreneurship:* extending the firm's domain of competence and corresponding opportunity set through internally generated new resource combinations.[3]

In the light of received theory of strategy and organization, the term *corporate entrepreneur-*

---

[1] H. Igor Ansoff, *Corporate Strategy* (New York: McGraw-Hill, 1965).

[2] T. J. Peters and R. H. Waterman, *In Search of Excellence* (New York: Harper and Row, 1983).

[3] For a discussion of the theoretical foundations of corporate entrepreneurship, see R. A. Burgelman, "Corporate Entrepreneurship and Strategic Management: Insights from a Process Study," *Management Science* 29 (1983), pp. 1649–64.

ship seems oxymoronic. This article presents a new model of strategic behavior in large, established firms which identifies entrepreneurial activity as a natural and integral part of the strategic process. The model sheds more light on why the strategic management of entrepreneurial activities constitutes a real challenge for corporate management. This article also proposes a conceptual framework which corporate management may find useful for improving its capacity to deal effectively with entrepreneurial initiatives. This, in turn, provides the basis for discussing conditions under which various organization designs for corporate entrepreneurship may be appropriate and raises some issues and problems associated with implementing such designs.

FIGURE 1 A Model of the Interaction of Strategic Behavior, Corporate Context, and the Concept of Strategy

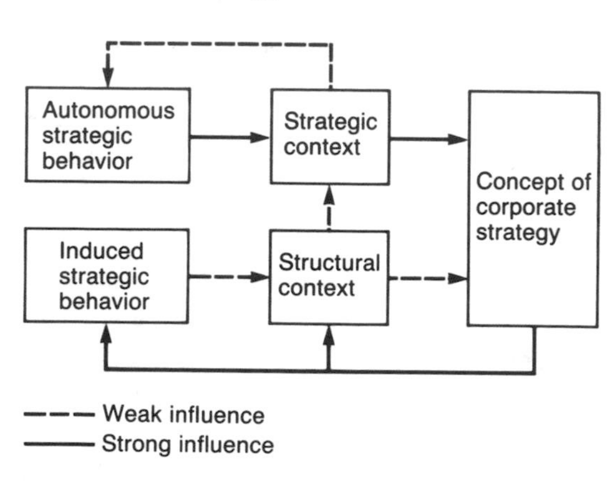

A NEW MODEL OF THE STRATEGIC PROCESS

Based on a field study of the internal corporate venturing process in one large, established firm and on the reanalysis of landmark studies in the field of strategic management, I have proposed a new model of the strategic process (see Figure 1).[4] Figure 1 shows two fundamentally different strategic processes going on simultaneously in large, complex firms.

Induced Strategic Behavior Loop

The bottom loop of the model corresponds to the traditional view of top-driven strategic management. The current *concept of strategy* represents the more or less explicit articulation of the firm's theory about the basis for its past and current successes and failures. It provides a more or less shared frame of reference for the strategic actors in the organization, and provides the basis for corporate objective-setting in terms of its business portfolio and resource allocation.

The concept of strategy induces strategic activity in the firm. *Induced* strategic behavior fits in the existing categories used in the firm's strategic planning and takes place in relation to its familiar external environments. Examples of such strategic behavior emerge around, among others, new product development projects for existing businesses, market development projects for existing products, and strategic capital investment projects for existing businesses.

The current concept of strategy corresponds to a *structural context* aimed at keeping strategic behavior at operational levels in line with the current concept of strategy. Structural context refers to the various administrative mechanisms which top management can manipulate to influence the perceived interests of the strategic actors at the operational and middle levels in the organization. It intervenes in the relationship between induced strategic behavior and the concept of strategy, and operates as a selection mechanism on the induced strategic behavior.

[4] R. A. Burgelman, "A Model of the Interaction of Strategic Behavior, Corporate Context, and the Concept of Strategy," *Academy of Management Review* 8 (1983), pp. 61–70.

The so-called "excellent companies"[5] all seem to have found *their* way of making the induced strategic behavior loop work exceedingly well. Operational and middle-level managers in these firms understand what strategic actions are required in view of the corporate development needs, even though there may be very little explicit attention given to formal "strategy." Managers in such companies identify with the corporate ways and means, yet maintain an element of creative independence. Marks and Spencer, 3M, and Hewlett-Packard are examples of excellent implicit management of the induced strategic behavior loop. General Electric with its SBU system and, until recently, Texas Instruments with its OST system are examples of excellent explicit management of the induced strategic behavior loop.

Autonomous Strategic Behavior Loop

During any given period of time, the bulk of strategic activity in large, complex firms is likely to be of the induced variety. The present model, however, proposes that large, resource-rich firms are likely to possess a reservoir of entrepreneurial potential at operational levels that will express itself in autonomous strategic initiatives. *Autonomous* strategic behavior introduces new categories for the definition of opportunities. Entrepreneurial participants, at the product market level, conceive new business opportunities, engage in project championing efforts to mobilize corporate resources for these new opportunities, and perform strategic forcing efforts to create momentum for their further development. Middle-level managers attempt to formulate broader strategies for areas of new business activity and try to convince top management to support them. This is the type of strategic behavior encountered in the study of internal corporate venturing.[6] Such autonomous strategic initiatives attempt to escape the selective effects of the current structural context, and they make the current concept of corporate strategy problematical. They lead to a redefinition of the corporation's relevant environment and broaden the scope of its business portfolio.

Autonomous strategic behavior takes shape outside of the current concept of strategy. Yet, to be successful, it needs eventually to be accepted by the organization and to be integrated into its concept of strategy. The process through which this can be achieved has been identified as the process of *strategic context* determination. Strategic context determination intervenes in the relationship between autonomous strategic behavior and the concept of strategy. It encompasses the activities through which middle-level managers question the current concept of strategy and provide top management with the opportunity to rationalize, retroactively, successful autonomous strategic behavior. This, in turn, changes the basis for the further inducement of strategic behavior.

The model thus suggests that corporate entrepreneurship is typically constituted by the interlocking strategic activities of managers at multiple levels in the firm's organization. It subsumes two important findings from the literature on innovation in organizations:

- Different processes are involved in generating and implementing innovations.[7]
- There may be a conflict between being excellent at incremental innovation and the capacity for more radical innovation.[8]

---

[5] Peters and Waterman, *In Search of Excellence.*

[6] R. A. Burgelman, "A Process Model of Internal Corporate Venturing in the Diversified Major Firm," *Administrative Science Quarterly* 28 (1983), pp. 223–44.

[7] J. Q. Wilson, "Innovation in Organization: Notes Toward a Theory," in J. D. Thompson, ed., *Approaches to Organizational Design* (Pittsburgh: University of Pittsburgh Press, 1966), pp. 195–218.

[8] W. Abernathy, *The Productivity Dilemma* (Baltimore: The Johns Hopkins Press, 1978).

It also provides a theoretical explanation for the seemingly contradictory findings of Peters and Waterman that excellent companies seem to have *both* an extraordinarily strong CEO whose influence pervades the entire organization *and* independent mavericks who engage in activities outside the regular channels of hierarchical decision making.[9] Finally, it provides a preliminary conceptualization of the experimentation-and-selection process through which diversity becomes transformed into a new order in large, complex organizations.

The Managerial Challenge Posed by Autonomous Strategic Behavior

Consider the following vignettes:

> In 1966, calculators were largely mechanized. A young man working for one of the calculator companies brought a model for an electronic calculator to Hewlett-Packard. His own firm was not interested in it because they didn't have the electronic capability. In spite of unfavorable market research forecasts, William Hewlett personally championed the project.[10]
>
> Back in 1980, Sam H. Eletr, a manager in Hewlett-Packard's labs, tried to persuade the company's new product people to get into biotechnology. "I was laughed out of the room," he said. But venture capitalists didn't laugh. They persuaded Mr. Eletr to quit Hewlett-Packard and staked him to $5.2 million to start a new company. Its product: gene machines, which make DNA, the basic material of the genetic code—and the essential raw material in the burgeoning business of genetic engineering. Now, three years later, Hewlett-Packard has formed a joint venture with Genentech, Inc., to develop tools for biotechnology. One product it is considering: gene machines.[11]

How should corporate management deal with such autonomous strategic behavior? Clearly, not every new idea or proposal can or should be adopted and developed. Yet, it is not a gratuitous exercise in Monday morning quarterbacking to ask whether the managements of the firms involved in the vignettes had made a *strategic* analysis of the proposals and whether they did indeed make a *strategic* decision not to pursue the proposals of their internal entrepreneurs.

From a strategic management perspective, it does not seem adequate to reject the electronic calculator because "we don't have an electronics capability" or to reject the gene-making machine because "we are not in biotechnology." It seems likely that there must have been some important relevant capabilities in place which allowed the internal entrepreneurs to come up with the proposal and perhaps even develop a prototype in the first place! Even if there is no apparent significant relationship with current capabilities and skills, it is still important to consider the strategic opportunities and/or threats potentially implied by the entrepreneurial proposal. It is precisely these efforts—to extend the firm's domain of competence, to elaborate and recombine the current capabilities, and to define new, unanticipated opportunities—which make internal entrepreneurial activity a vital part of the strategic process in large, established firms.

From a strategic management perspective, the problem is *how* could corporate management improve its capacity to deal with autonomous strategic behavior, given that, by definition, it does not fit with the current corporate strategy. What is now needed is an analytical framework which can be used by corporate management to assess entrepreneurial proposals and which also leads to tentative conclusions about the use of various organization design alternatives to structure the relationships between entrepreneurial initiatives and the corporation.

---

[9] Peters and Waterman, *In Search of Excellence.*

[10] R. M. Atherton and D. M. Crites, "Hewlett-Packard: A 1975–1978 Review" (Boston: Harvard Case Services, 1980).

[11] This quote was taken from, "After Slow Start, Gene Machines Approach a Period of Fast Growth and Steady Profits." *The Wall Street Journal,* December 13, 1983.

A FRAMEWORK FOR ASSESSING INTERNAL ENTREPRENEURIAL PROPOSALS

The conceptual framework focuses on two key dimensions of strategic decision making concerning internal entrepreneurial proposals. One is the expected *strategic importance* for corporate development. The other is the degree to which proposals are related to the core capabilities of the corporation, i.e., their *operational relatedness.*

Assessing Strategic Importance

How can management assess, as accurately as possible, the strategic importance of an entrepreneurial initiative? Even though this is one of the most important top management responsibilities, it is also one for which top management is often not well equipped. Corporate-level managers in large, diversified firms tend to rise through the ranks, having earned their reputation as head of one or more of the operating divisions. By the time they reach the top management level, they have developed a highly reliable frame of reference to evaluate business strategies and resource allocation proposals pertaining to the main lines of business of the corporation. By the same token, their substantive knowledge of new technologies and markets is limited.[12]

There is a tendency for top management to rely on corporate staffs, consultants, and informal interactions with "peers" from other companies to make assessments of new business fields. Such information sources have merit, but they are no substitute for efforts to understand the deeper substantive issues involved in a specific proposal. The latter efforts should be based on requiring middle-level managers to "educate" corporate management and to encourage middle-level managers to "champion" new proposals based on their own substantive assessments. Such *substantive* interaction between different levels of management is likely to improve top management's capacity to make strategically sound assessments, making them rely less on purely quantitative projection.[13]

It would be useful for top management to have a checklist of critical issues and questions in these substantive interactions. Examples of these are:

- How does this initiative maintain our capacity to move in areas where major current or potential competitors might move?
- How does this help us to find out where *not* to go?
- How does it help us create new defensible niches?
- How does it help mobilize the organization?
- To what extent could it put the firm at risk?
- When should we get out of it if it does not seem to work?
- What is missing in our analysis?

Strategic assessment will sometimes result in a classification of a proposal as "very" or "not at all" important. In other cases, the situation will be more ambiguous and lead to assessments like "important for the time being" or "maybe important in the future." Key to the usefulness of the analysis is that such assessments are based on specific, substantive factors.

Assessing Operational Relatedness

The second key dimension of strategic decision making concerns the *degree* to which the entrepreneurial proposal requires capabilities and skills that are different from the core capa-

[12] Henry Kissinger has made the interesting observation that top policy-makers are, basically, strategies-in-action whose fundamental strategic premises are a *given* by the time they reach their positions. See H. A. Kissinger, *White House Years* (Boston: Little Brown, 1979).

[13] See also R. A. Burgelman, "Managing the Internal Corporate Venturing Process," *Sloan Management Review* (Winter 1984), pp. 33–48.

bilities and skills of the corporation. This is relevant for a number of reasons. First, new business proposals may either be driven by newly developed capabilities and skills or they may drive the development of new capabilities and skills, both of which have very different implications.[14] Second, entrepreneurial proposals typically are based on new combinations of corporate capabilities, and this may reveal potential opportunities for positive synergies (or threat of negative synergies). Often, internal entrepreneurs will weave together pieces of technology and knowledge which exist in separate parts of the organization and which would otherwise remain unused.

In order to be able to make the required assessments of the dimension of operational relatedness, corporate management should rely on substantive interactions with middle-level managers who champion entrepreneurial projects. To guide these interactions, corporate management again needs a checklist of critical issues and questions in these substantive interactions. Some examples are:

- What are the key capabilities required to make this project successful?
- Where, how, and when are we going to get them if we don't have them yet, and at what cost?
- Who else might be able to do this, perhaps better?
- How will these new capabilities affect the capacities currently employed in our mainstream business?
- What other areas may possibly require successful innovative efforts if we move forward with this project?
- What is missing in our analysis?

A useful tool to help corporate management with this assessment is to develop an accurate inventory of current capabilities and skills in various functional areas and to spell out in detail how each area of business activity uses these capabilities and skills. Such a capabilities/businesses *matrix* should be regularly updated and should provide a reference frame for each new entrepreneurial proposal. In light of this, new proposals will sometimes be classified as "very" or "not at all" related. In other cases, the situation will again be somewhat unclear and lead to a "partly related" assessment. In the context of the matrix, these assessments should be made in specific, substantive terms for each proposal.

The assessment framework can now be used to discuss the use of various organization design alternatives for the different types of entrepreneurial proposals.

DESIGN ALTERNATIVES FOR CORPORATE ENTREPRENEURSHIP

Having assessed an entrepreneurial proposal in terms of its strategic importance and operational relatedness, corporate management must choose an organization design for structuring the relationship between the new business and the corporation which is commensurate with its position in the assessment framework. This involves various combinations of *administrative* and *operational* linkages.

Determining Administrative Linkages

The assessment of strategic importance has implications for the degree of *control* corporate management needs to maintain over the new business development. This, in turn, has implications for the administrative linkages to be established.

If strategic importance is high, strong administrative linkages will be in order. This means, basically, that the new business must be folded into the existing structural context of the firm. Corporate management will want a say in the strategic management of the new business

[14] A. R. Fusfeld, "How to Put Technology into Corporate Planning," in Technology Review, *Innovation* (1976), pp. 53–57.

through direct reporting relationships, substantive involvement in planning/budgeting processes, and involvement in trade-offs between the strategic concerns of the new and existing businesses. Measurement and reward systems must reflect clearly articulated strategic objectives for the new business development. Low strategic importance, on the other hand, should lead corporate management to examine how the new business can best be spun off.

In more ambiguous situations, where strategic importance is judged to be somewhat unclear as yet, corporate management should relax the structural context, and allow the new business some leeway in its strategic management. In such situations, the strategic context of the new business remains to be determined. This requires mechanisms facilitating substantive interaction between middle and corporate levels of management, and measurement and reward systems capable of dealing with as yet unclear performance dimensions and strategic objectives.

Determining Operational Linkages

The degree of operational relatedness has implications for the *efficiency* with which both the new and the existing businesses can be managed. This, in turn, has implications for the required operational linkages.

If operational relatedness is judged to be high, strong coupling of the operations of the new and existing businesses is probably in order. Corporate management should ensure that both new and existing capabilities and skills are used well through integration of work flows, adequate mutual adjustment between resource users through lateral relations at the operational level, and free flows of information and know-how through regular contacts between professionals in the new and existing businesses. Low operational relatedness, on the other hand, may require complete decoupling of the operations of new and existing businesses to avoid interferences and concomitant wasteful (because unnecessary) communications and negotiations.

In situations where operational relatedness is partial and not completely clear, loose coupling seems most adequate. In such situations, the work flows of new and existing businesses should remain basically separate, and mutual adjustment should be achieved through individual integrator roles or through task force type mechanisms, rather than directly through the operational-level managers. Information and know-how flows, however, should remain as uninhibited as is practical.

Choosing Design Alternatives

Various combinations of administrative and operational linkages produce different design alternatives. These correspond to choices which corporate management has to make regarding the different situations identified in the assessment framework. Figure 2 shows nine such design alternatives.

The design alternatives discussed here are not exhaustive, and the scales for the different dimensions used in the assessment framework remain rudimentary. Much room is left for refinement through further research. By the same token, the framework represented in Figure 2 allows for a preliminary conceptual underpinning for a number of practices encountered in today's business environment.

1. Direct integration

High strategic importance and operational relatedness require strong administrative and operational linkages. This means that there is a need to integrate the new business directly into the mainstream of the corporation. Such integration must anticipate internal resistance for reasons well documented in the organizational change literature. The role of "champions"—those who know the workings of the current system very well—are likely to be important in such situations. The need for direct integration

FIGURE 2 Organization Designs for Corporate Entrepreneurship

Operational relatedness	Very important	Uncertain	Not important
Unrelated	3 Special business units	6 Independent business units	9 Complete spin-off
Partly related	2 New product business department	5 New venture division	8 Contracting
Strongly related	1 Direct integration	4 Micro new ventures department	7 Nurturing and contracting

Strategic importance

is perhaps most likely to occur in highly integrated firms, where radical changes in product concept and/or in process technologies could threaten the overall strategic position of the firm.[15]

2. *New product business department*

High strategic importance and partial operational relatedness require a combination of strong administrative and medium-strong operational linkages. This may be achieved by creating a separate department around an entrepreneurial project in that part (division or group) of the operating system where potential for sharing capabilities and skills is significant. Corporate management should monitor the strategic development of the project in substantive terms and not allow it to be folded (and "buried") into the overall strategic planning of that division or group.[16]

3. *Special business units*

High strategic importance and low operational relatedness may require the creation of specially dedicated new business units. Strong administrative linkages are necessary to ensure the attainment of explicit strategic objectives within specified time horizons throughout the development process. It will often be necessary to later on combine and integrate some of these business units into a new operating division in the corporate structure.[17]

[15] An example of the need for direct integration is documented by Twiss's account of the development of "float glass" at Pilkington Glass, Ltd. See B.Twiss, *Managing Technological Innovation* (London: Longman, 1980); see also A. C. Cooper and D. Schendel, "Strategic Responses to Technological Threats," *Business Horizons* (1976), pp. 61–69, for a discussion of the difficulties firms face when confronted with radical changes in their mainstream operations.

[16] An example where the proposed approach might have been useful is provided by one of the major, diversified automotive supplier's handling of electronic fuel injection development. In spite of having the required technology, there was strong resistance from the firm's carburetor division, and the automotive group management level did not support the development either. Only after a new group-level manager took charge of the strategic management of the project and brought in additional operational capabilities and skills did the project take off.

[17] IBM's use of the Special Business Unit design to enter the personal computer business is an example. See "Meet the New Lean, Mean IBM," *Fortune*, June 13, 1983, p. 78.

4. *Micro new ventures department*

Uncertain strategic importance and high operational relatedness seem typical for the "peripheral" projects which are likely to emerge in the operating divisions on a rather continuous basis. For such projects, administrative linkages should be loose. The venture manager should be allowed to develop a strategy within budget and time constraints but should otherwise not be limited by current divisional or even corporate-level strategies. Operational linkages should be strong, to take advantage of the existing capabilities and skills and to facilitate transferring back newly developed ones. Norman Fast has discussed a "micro" new ventures division design,[18] which would seem to fit the conditions specified here.

5. *New venture division (NVD)*

This design is proposed for situations of maximum ambiguity in the assessment framework.[19] The NVD may serve best as a "nucleation" function. It provides a fluid internal environment for projects with the potential to create major new business thrusts for the corporation, but of which the strategic importance remains to be determined as the development process unfolds. Administrative linkages should be fairly loose. Middle-level managers supervising a few ventures are expected to develop "middle-range" strategies for new fields of business: bringing together projects which may exist in various parts of the corporation, and/or can be acquired externally, and integrating these with some of the venture projects they supervise, to build sizable new businesses. Operational linkages should also be fairly loose, yet be sufficiently developed to facilitate transferring back and forth relevant know-how and information concerning capabilities and skills. Long time horizons—8 to 12 years—are necessary, but ventures should not be allowed to languish. High-quality middle-level managers are crucial to make this design work.

6. *Independent business units*

Uncertain strategic importance and negligible operational relatedness may make external venture arrangements attractive. Controlling ownership with correspondingly strong board representation may provide corporate management with an acceptable level of strategic control without administrative linkages.[20]

7. *Nurturing plus contracting*

In some cases, an entrepreneurial proposal may be considered unimportant for the firm's corporate development strategy, yet be strongly related to its operational capabilities and skills. Such ventures will typically address "interstices" in the market which may be too small for the company to serve profitably but which offer opportunities for a small business.[21] Top management may want to help such entrepreneurs spin off from the corporation and may, in fact, help the entrepreneur set up his or her business. This provides a known and, in all likelihood, friendly competitor in those interstices, keeping out other ones. Instead of administrative or ownership linkages, there

---

[18] N. D. Fast. *The Rise and Fall of Corporate New Venture Divisions* (Ann Arbor, Mich.: V.M.I. Research Press, 1979).

[19] For a discussion of major problems associated with the NVD, see R. A. Burgelman, "Managing the New Venture Division: Research Findings and Implications for Strategic Management," *Strategic Management Journal* (forthcoming). See also Fast, *The Rise and Fall.*

[20] IBM's use of Independent Business Units is one example where the corporation keeps complete ownership. See *Fortune*, June 13, 1983. An example of joint ownership is provided by how Bank of America has organized its Venture-Capital Business. See "Despite Greater Risks, More Banks Turn to Venture-Capital Business," *The Wall Street Journal,* November 28, 1983.

[21] For a discussion of the concept of "interstice," see E. T. Penrose, *The Theory of the Growth of the Firm* (Oxford: Blackwell, 1968).

may be a basis for long-term contracting relationships in which the corporation can profitably supply the entrepreneur with some of its excess capabilities and skills. Strong operational linkages related to these contracts may facilitate transfer of new or improved skills developed by the entrepreneur.

8. *Contracting*

The possibilities for nurturing would seem to diminish as the required capabilities and skills of the new business are less related. Yet there may still be opportunities for profitable contracting arrangements and for learning about new or improved capabilities and skills through some form of operational linkages.

9. *Complete spin-off*

If strategic importance and operational relatedness are both low, complete spin-off will be most appropriate. A decision based on a careful assessment of both dimensions is likely to lead to a well-founded decision from the perception of both the corporation and the internal entrepreneur.

Implementing Design Alternatives

In order to implement designs for corporate entrepreneurship effectively, three major issues and potential problems need to be considered. First, corporate management and the internal entrepreneur should view the assessment framework as a tool to clarify—at a particular moment in time—their community of interests and interdependencies and to structure a non-zero sum game. Second, corporate management must establish measurement and reward systems which are capable of accommodating the different incentive requirements of different designs. Third, as the development process unfolds, new information may modify the perceived strategic importance and operational relatedness, which may require a renegotiation of the organization design.

To deal effectively with the implementation issues and potential problems, corporate management must recognize internal entrepreneurs as "strategists" and perhaps even *encourage* them to think and act as such. This is necessary because the stability of the relationship will be dependent on both parties feeling that they have achieved their individual interests to the greatest extent, given the structure of the situation. On the part of corporate management, this implies attempts to appropriate benefits from the entrepreneurial endeavor, but only to the extent that they can provide the entrepreneur with the opportunity to be more successful than if he or she were to go it alone. This, in turn, requires simultaneously generous policies to help internal entrepreneurs based on a sound assessment of their proposals and unequivocal determination to protect proprietary corporate capabilities and skills vigorously.[22]

CONCLUSION

Until recently, the Schumpeterian distinction between entrepreneurial and administrated ("bureaucratic") economic activity could be considered adequate. However, in the light of the turbulence of recent industrial developments, this distinction loses much of its relevance. Large, established corporations and new, maturing firms alike are confronted with the problem of maintaining their growth, if not their existence, by exploiting to the fullest the unique resource combinations they have assembled.

Increasingly, there is an awareness that internal entrepreneurs are necessary for firms to achieve growth. The internal entrepreneur,

---

[22] For an account of some examples, see "Spin-Offs Mount in Silicon Valley," *The New York Times,* January 3, 1984.

like the external entrepreneur, enacts new opportunities and drives the development of new resource combinations or recombinations. As a result, new forms of economic organization—a broader array of arrangements—are necessary.

In turn, new theories of the firm and a more nuanced view of the role of hierarchies, contracts, and markets are required. The conceptual foundations of these theories are currently being laid in such fields as the economics of internal organization, agency theory, the theory of legal contracts, and theories of organization design and change. Usually, practitioners are already ahead, experimenting with new organizational forms and arrangements. In the process, they generate new data and raise the basis for new research questions.

A better understanding of the process of corporate entrepreneurship will facilitate the collaboration between firms and their internal entrepreneurs.

Maintaining an Entrepreneurial Environment

Case IV–8 Hewlett-Packard: A 1975–1978 Review

R. M. Atherton, D. M. Crites, and G. Greenberg

In May 1978, John Young was appointed chief executive of Hewlett-Packard. He was simultaneously handed the difficult task of charting a path through a rapidly growing and increasingly complex competitive jungle. Forbes reported that Young would be paid $280,000 to lead the classy electronics company into the rapidly changing, new computer market in which the biggest competitor (IBM) had vastly greater resources.[1] Although handpicked by the company's two founders, Bill Hewlett and Dave Packard, who between them owned 39 percent of the stock, he would be watched carefully. Whether he could continue the growth, the success, and the same egalitarian leadership style of his predecessors was a real question. Whether he should even try to adopt the same general strategy and tactics of recent years was also a real question.

An earlier case described the industry, the company's objectives, strategies, policies, structure, and performance from 1972 to 1975.[2] This case depicts some of the major facets of the 1975–78 transitionary period for Hewlett-Packard. In summary, the period appears to have been marked by a continuation of impressive growth; by repeated affirmation of, and only slight changes in, the company's basic objectives and policies; by a smooth transfer of top executive responsibilities; and by a changing product mix and marketing strategy which had brought Hewlett-Packard into increasingly more competitive markets and direct confrontation with IBM and other major computer companies.

HEWLETT-PACKARD: A BRIEF SKETCH

Innovative products have been the cornerstone of Hewlett-Packard's growth since 1939, when Hewlett engineered a new type of audio

[1] *Forbes*, "Welcome to the Hot Seat, John Young," July 24, 1978, pp. 62–63.

[2] Hewlett-Packard (A), ICCH 9-376-754.

oscillator and, with Packard, created the company in Packard's garage. The product was cheaper and easier to use than competitive products, and it was quickly followed by a family of test instruments based on the same design principles. Hewlett-Packard has since become one of the giants of the high-technology electronics industry. Their products include electronic test and measuring systems; medical electronic products; electronic instrumentation for chemical analysis; and solid-state components. According to company sources, Hewlett-Packard has remained a people-oriented company with management policies that encourage individual creativity, initiative, and contribution throughout the organization. It has also tried to retain the openness, informality, and unstructured operating procedures that marked the company in its early years. Each individual has been given the freedom and the flexibility to implement work methods and ideas to achieve both personal and company objectives and goals.

Both Hewlett and Packard have indicated that their Corporate Objectives, first put into writing in 1957 and modified occasionally since then, have served the company well in shaping the company, guiding its growth, and providing the foundation for its contribution to technological progress and the betterment of society. Last updated in 1977, the Corporate Objectives were, according to company sources, remarkably similar to the original versions developed from management concepts formulated by Hewlett and Packard in the company's early years.[3]

The following is a brief listing of the Hewlett-Packard objectives in 1978.

1. Profit Objective: To achieve sufficient profit to finance our company growth and to provide the resources we need to achieve our other corporate objectives.
2. Customer Objective: To provide products and services of the greatest possible value to our customers, thereby gaining and holding their respect and loyalty.
3. Fields of Interest Objective: To enter new fields only when the ideas we have, together with our technical, manufacturing, and marketing skills, assure that we can make a needed and profitable contribution to the field.
4. Growth Objective: To let our growth be limited only by our profits and our ability to develop and produce technical products that satisfy real customer needs.
5. People Objective: To help HP people share in the company's success, which they make possible; to provide job security based on their performance; to recognize their individual achievements; and to help them gain a sense of satisfaction and accomplishment from their work.
6. Management Objective: To foster initiative and creativity by allowing the individual great freedom of action in attaining well-defined objectives.
7. Citizenship Objective: To honor our obligations to society by being an economic, intellectual, and social asset to each nation and each community in which we operate.

Except for slight changes in wording, the objectives were the same as in 1975.

THE 1973–74 REDIRECTION

Adversely affected by computer and aerospace downturns in 1970, Hewlett-Packard had at first welcomed the 30 percent increase in sales in 1972 and the 40 percent increase in 1973. Problems arose, however, as inventories and accounts receivable increased substantially. A 32 percent increase in employees to handle the increased sales, administrative, and

[3] *Measure,* "Revised Corporate Objectives," May 1977, pp. 7–10.

manufacturing activities required extensive training efforts and organizational readjustments. Some products were put into production before they were fully developed. Prices were sometimes set too low for an adequate return on investment. Short-term borrowing increased substantially to $118 million and management seriously considered converting some of its short-term debt to long-term debt, a practice the company had traditionally avoided, preferring to operate on a pay-as-you-go basis.

In 1973–74, top management decided to avoid adding long-term debt and to reduce short-term debt by controlling costs, managing assets, and improving profit margins. As Packard made clear to the management at all levels, they had somehow been diverted into seeking market share as an objective. So both he and Hewlett began a year-long campaign to reemphasize the principles they developed when they began their unique partnership. Packard toured the divisions to impose this new asset-management discipline. In addition, while other companies dropped prices to boost sales and cut research spending to improve earnings, Hewlett-Packard used quite different tactics. It raised prices by an average of 10 percent over the previous year, and it increased spending on research and development by 20 percent, to an $80 million annual rate. These two strategies were intended to improve company profitability, to slow the rate of growth that had more than doubled sales in the previous three years, and to enable it to compete primarily on the basis of quality and technological superiority.

The improvements in 1974 performance compared with 1973 were quite dramatic. During fiscal 1974, inventories and receivables increased about 3 percent while sales grew 34 percent to $884 million. The effect of this better asset control, combined with improved earnings, resulted in a drop in short-term debt of approximately $77 million. Earnings were up 66 percent to $84 million and were equal to $3.08 per share compared to $1.89 per share. Only 1,000 employees were added compared to 7,000 in the previous year.

Both Hewlett and Packard were reportedly dismayed that they had been forced to initiate and personally lead the efforts to get the company back on the track. It was particularly disconcerting to them because they believed the issues were fundamental to the basic strategy of the company. They had also had to intervene directly in day-to-day operational management, which was counter to their basic philosophy of a decentralized, product-oriented, and divisionalized organization structure.

GROWTH, 1975–78

The dramatic growth that followed the 1973–74 "redirection" was in essence maintained through the 1978 fiscal year. The sales increase from $981 million in 1975 to $1.73 billion in 1978 averaged almost 21 percent per year. Net earnings, growing from $84 million in 1975 to $153 million in 1978, averaged 22 percent per year with an 8 percent increase in 1975–76, a 33 percent jump in 1976–77, and a 1977–78 growth of 26 percent. A four-year consolidated earnings summary, strategic ratios, financial ratios, contributions to sales and earnings by business segments, and a percent of sales analysis are presented in Exhibits 1, 2, 3, and 4. Total employees, about 29,000 at the beginning of the 1975 fiscal year, grew about 11 percent a year to a level of about 42,400 at the end of the 1978 fiscal year. The total number of products increased from roughly 3,400 in late 1975 to over 5,000 in mid-1978. The number of new product introductions increased significantly from about 90 major new products in 1975 to 130 in 1978. In keeping with its traditional attention to research and development, these expenditures grew from $90 million in fiscal 1975 to $154 million in fiscal 1978. Data on growth are given in Exhibit 5.

EXHIBIT 1 Hewlett-Packard Company

Four-Year Consolidated Earnings Summary

	1975	1976	1977	1978
Net sales	$981.2	$1,111.6	$1,360.0	$1,728.0
Other income, net	8.3	12.0	13.9	23.0
Total revenues	989.5	1,123.6	1,373.9	1,751.0
Costs and expenses:				
Cost of goods sold	462.7	535.6	622.2	805.0
Research and development	89.6	107.6	125.4	154.0
Marketing	162.0	176.6	207.5	264.0
Administrative and general	124.5	139.1	185.4	226.0
Interest	2.2	4.1	4.2	6.0
Total costs and expenses	841.0	963.0	1,144.7	1,455.0
Earnings before taxes on income	148.6	160.6	229.2	296.0
Taxes on income	65.0	69.8	107.7	143.0
Net earnings	$ 83.6	$ 90.8	$ 121.5	$ 153.0
Per share:				
Net earnings	3.02	3.24	4.27	5.27
Cash dividends	.25	.30	.40	.50
Common shares outstanding at year-end	27.6	28.0	28.5	29.0

In millions of dollars; for fiscal years ending October 31.
Note: Figures may not add exactly due to rounding.
SOURCE: Hewlett-Packard *Annual Reports* (1975–78).

STRUCTURE

In a 1978 statement of philosophy, HP emphasized as a basis for high-level achievement their provision of a realistic and simple set of long-term objectives on which all could agree and on which people could work with a minimum of supervision and a maximum of responsibility.[4] They stated that to attain such a participative working environment requires special attention to the basic organizational structure of the company. At Hewlett-Packard, a product division was an integrated self-sustaining organization with a great deal of independence that performed in much the same way as the company had 22 years ago. The fundamental responsibilities of a division, extending worldwide, were to develop, manufacture, and market appropriate products. Acting much as an independent business, each division was responsible for its own accounting, personnel activities, quality assurance, and support of its products in the field. Coordination of the divisions was achieved primarily through the product groups. Group management had overall responsibility for the operations and financial performance of its divisions. Each group had a common sales force serving all of its product divisions. To keep an atmosphere that encour-

[4] *Measure*, "Working Together: The Hewlett-Packard Organization," June 1978, pp. 10–11.

EXHIBIT 2 Hewlett-Packard Company

Strategic Ratios

	Net earnings / Total revenues	×	Total revenues / Average assets	=	Net earnings / Average assets	×	Average assets / Average net worth	=	Net earnings / Average net worth
1975	8.5%	×	1.39	=	11.8%	×	1.39	=	16.4
1976	8.1%	×	1.32	=	10.6%	×	1.38	=	14.7
1977	8.8%	×	1.31	=	11.6%	×	1.40	=	16.2
1978	8.7%	×	1.34	=	11.7%	×	1.44	=	16.8

Financial Ratios

	1975	1976	1977	1978
Current ratio	2.51	2.62	2.58 (1)	2.29
Acid test	1.36	1.59	1.62	1.44
Collection period (days)	75	76	72 (2)	77
Accounts payable T/O (days)	25	21	27	32
Inventory T/O	2.31	2.42	2.41	2.54
Debt to net worth	.33	.35	.37 (3)	.42
Interest coverage	66.9	38.2	53.6	48.3
Gross profit margin	.53	.52	.54	.53
Net profit to net sales	.086	.082	.089 (4)	.089

Note: Dun's Review (December 1977) figures are the averages of the Electronic Component and Scientific Instrument business lines. Comparable figures for 1978 were not published.

(1) *Dun's Review* indicated the industry median was 2.64.
(2) *Dun's Review* indicated the industry median was 65.
(3) *Dun's Review* indicated the industry median was .85.
(4) *Dun's Review* indicated the industry median was .047.

SOURCE: Developed by casewriters from Hewlett-Packard *Annual Reports* (1975–78).

aged the making of problem-solving decisions as close as possible to the level where the problem occurred, HP has striven over the years to keep its basic business units—the product divisions—relatively small and well defined.

SELECTED STRATEGIES AND RELATED POLICIES

Hewlett-Packard's product-market strategy has concentrated on developing quality products, which make unique technological contributions and are so far advanced that customers are willing to pay premium prices. Products originally limited to electronic measuring instrument markets have expanded over the years to include computers and other technologically related fields. Customer service, both before and after the sale, has been given primary emphasis. Their financial strategy has been to use profits, employee stock purchases, and other internally generated funds to finance growth. They have avoided long-term debt and have resorted to short-term debt only when sales growth exceeded the return on net worth. Their growth strategy has been to attain a position of technological strength and leadership by continually developing innovative products

EXHIBIT 3 Hewlett-Packard Company

Contributions to Sales and Earnings

	1975	1976	1977	1978
Sales:				
Test, measuring, and related items	$ 453	$ 501	$ 593	$ 740
Electronic data products	395	453	580	761
Medical electronic equipment	99	119	135	163
Analytical instrumentation	53	58	76	98
Total	1,000	1,131	1,384	1,762
Less: Sales between business segments	(19)	(19)	(24)	(34)
Net sales to customers	$ 981	$1,112	$1,360	$1,728
Earnings:				
Test, measuring, and related items	$ 94	$ 103	$ 134	$ 180
Electronic data products	68	69	106	124
Medical electronic equipment	13	21	22	26
Analytical instrumentation	8	7	12	16
Operating profit	183	200	274	346
Less: Eliminations and corporate items	(34)	(39)	(45)	(50)
Earnings before taxes on income	$ 149	$ 161	$ 229	$ 296

In millions of dollars; for fiscal years ending October 31.
SOURCE: Hewlett-Packard *Annual Reports* (1975–78).

and by attracting high-caliber and creative people. Their motivational strategy has consisted of providing employees with the opportunity to share in the success of the company through high wages, profit sharing, and stock-purchase plans. They have also provided job security by keeping fluctuations in production schedules to a minimum by avoiding consumer-type products and by not making any products exclusively for the government. Their managerial strategy has been to practice "management by objective"[5] rather than management by directive; they have used the corporate objectives to provide unity of purpose and have given employees the freedom to work toward these goals in ways they determine best for their own area of responsibility. The company has exercised its social responsibility by building plants and offices that are attractive and in harmony with the community, by helping to solve community problems, and by contributing both money and time to community projects.

Division Review

A principal vehicle for effecting communication between corporate management and the basic operating units has been the division review conducted annually at almost every division and sales region. Described as the natural outgrowth of the personal interest and hands-

[5] "Management by objective" is Hewlett-Packard's phrase for using corporate objectives primarily as a framework for coordination, decision making, and planning rather than for performance appraisal as in typical MBO programs.

EXHIBIT 4 Hewlett-Packard Company

Percent of Sales Analysis

	1975	1976	1977	1978
Earnings:				
Cost of goods sold	46.8	47.7	45.3	46.0
Research and development	9.1	9.6	9.1	8.8
Marketing	16.4	15.7	15.1	15.1
Administrative and general	12.6	12.4	13.5	12.9
Interest	.2	.4	.3	.3
Total costs and expenses	85.0	85.7	83.3	83.1
E.B.I.T.	15.0	14.3	16.7	16.9
Taxes	6.6	6.2	7.8	8.2
Net earnings	8.5	8.1	8.8	8.7
Sales by business segment (percent of total):				
Test, measuring, and related	45.3	44.3	42.8	42.0
Electronic data products	39.5	40.1	41.9	43.2
Medical electronic equipment	9.9	10.5	9.8	9.3
Analytical instrumentation	5.3	5.1	5.5	5.6
Earnings by business segment (percent of total):				
Test, measuring, and related	51.4	51.5	48.9	52.0
Electronic data products	37.2	34.5	38.7	35.8
Medical electronic equipment	7.1	10.5	8.0	7.5
Analytical instrumentation	4.4	3.5	4.4	4.6

Developed by casewriters from Exhibits 1 and 3 in the case.
Note: Figures may not add exactly due to rounding.

on style so characteristic of HP, reviews by 1978 were covering a full range of business matters: financial performance for the past year; outlook for orders, shipments, and facilities for the next three years; detailed presentations on product development strategy and key programs; and a look at people management including training, recruiting, and affirmative action goals and results. A very broad cross section of division personnel as well as a visiting group of reviewers were involved in organizing, presenting, and participating in the reviews. The visiting reviewers generally included several members of the corporate executive committee, corporate staff heads such as personnel and controller, appropriate group and related division managers, and on occasion even outside directors.

MBWA

Another concept has received considerable attention at HP as "an extra step that HP managers needed to take in order to make the HP open-door policy truly effective." Developed by John Doyle, vice president, personnel, earlier in his career at HP, it has been termed "management by wandering around" or MBWA.[6] It has been described as friendly, unfo-

[6] *Measure,* "What Is This Management by Wandering Around?" April 1978, pp. 8–11.

EXHIBIT 5 Hewlett-Packard Company

Selected Growth Indicators, 1975–1978*

	1975	1976	1977	1978
Employees:				
Domestic	22,000	22,800	25,400	31,000
International	8,200	9,400	9,700	11,400
Total	30,200	32,200	35,100	42,400
Total customers	35,000	N.A.†	over 50,000	N.A.
Domestic orders (millions)	$500.4	$592.4	$768.8	$977.0
International orders (millions)	$501.3	$557.6	$664.1	$898.0
Backlog of orders (millions)	$145	$175	$252	N.A.
R&D expenditures (millions)	$ 89.6	$107.6	$125.4	$154.0
Patents held/and pending	770/151	837/158	850/165	N.A.
Number of products	~3,400	~3,600	~4,000	~5,000
Major new products introduced	~90	~100	~115	~130
Capital expenditures (millions)	$ 66	$103.4	$115.5	$159.0
Increases in plant capacity (sq. ft.)	760,000	768,000	696,000	741,000
Increases in sales and service (sq. ft.)	N.A.	175,000	183,000	253,000

* For fiscal years ending October 31.
† Not available.
SOURCE: Hewlett-Packard *Annual Reports* (1975–78), Form 10-Ks, and correspondence with HP.

cused, unscheduled, and—to any employee at their work with whom a wandering manager stops to chat—an invitation to repay the visit and walk through that open door whenever they choose. To encourage MBWA, it has been the subject of management briefings and seminars. A two-part video program on MBWA has been taped and made available to all HP organizations. The three corporate personnel administrators have also begun to encourage it wherever they go on their liaison missions. One division general manager said of MBWA, "It's really a body chemistry kind of thing. You've got to really want to wander around and communicate at all levels." A manufacturing manager, talking about MBWA, indicated that "management by involvement" was more descriptive of the HP way than would be "management by overview." A sales region personnel manager, however, citing their communication problem as "a certain sense of isolation," noted that a "manager can't do much spontaneous wandering around" a sales territory.

CORPORATE ORGANIZATION AND LEADERSHIP TRANSITION, 1975–78

The April 1975 restructuring which led to speculation on who would later be taking the corporate reins had three main parts: (1) it expanded the product groups from four (Test and Measurement, Data Products, Medical Equipment, and Analytical Instrumentation) to six (Instruments, Computer Systems, Components, Medical, Calculators, and Analytical); (2) it added a new management level of top vice presidents; and (3) established an Executive Committee to oversee day-to-day operations of the company. The June 1978 corporate structure—except for some changes in the person-

EXHIBIT 6 Hewlett-Packard Corporate Organization, April 1975

BOARD OF DIRECTORS
Dave Packard, Chairman

CHIEF EXECUTIVE OFFICER
Bill Hewlett, President

Special Assistant
Ed Porter, Vice President

Corporate Development
John Doyle, Director

Research & Development
Barney Oliver, Vice President

HP Laboratories
- **Director** Barney Oliver
- **Administration** Dan Lansdon
- **Electronics Research** Paul Stolt
- **LSI** Bob Grimm
- **Physical Electronics** Don Hammond
- **Physical Research** Len Cutler
- **Solid State** Paul Greene
- **Corporate Libraries** Mark Baer

ADMINISTRATION
Bob Boniface, Vice President, Corporate Administration

Corporate Staff
- **Corporate Engineering** Eb Rechtin, Chief Engineer
- **Corporate Services** Bruce Wholey, Vice President
- **Finance** Ed van Bronkhorst, Vice President
- **Government Relations** Jack Beckett, Director
- **Legal** Jean Chognard, General Counsel; Jack Brigham, General Attorney
- **Personnel** Ray Wilbur, Vice President
- **Public Relations** Dave Kirby, Director
- **Secretary** Frank Couter, Vice President
- **Marketing** Al Oliverio, Vice President
- **International** Bill Doolittle, Vice President

OPERATIONS
John Young, Executive Vice President
Ralph Lee, Executive Vice President

Product Groups

INSTRUMENT
Bill Terry, Vice President and General Manager
Ray Demere, Vice President

Divisions
- **Boeblingen, Germany** David Rose
- **Civil Engineering** (Loveland, Colorado) Bill McCullough
- **Colorado Springs** Hal Edmondson
- **Delcon** (Mountain View, California) Brian Moore
- **Loveland Facility** Ed Shideler
- **Loveland Instruments** Don Schutz
- **Manufacturing (Loveland)** Don Cullen
- **Manufacturing (Palo Alto)** Jim Farrell
- **New Jersey** John Blokker
- **San Diego (California)** Dick Moore
- **Santa Clara (California)** Al Dagley
- **Santa Rosa (California)** Doug Chance
- **So. Queensberry, U.K.** Peter Carmichael
- **Stanford Park (California)** Rod Carlson

Instruments/Civil Engineer Sales/Service

COMPUTER SYSTEMS
Paul Ely, General Manager

Divisions
- **Automatic Measurement (Sunnyvale, California)** Al Seely
- **Boise (Idaho)** Ray Smolek
- **Data Systems (Cupertino, California)** Dick Anderson
- **Grenoble (France)** Karl Schwarz

Computer Systems Sales/Service

COMPONENTS
Dave Weindorf, General Manager

Divisions
- **HPA (Palo Alto, California)** Dave Weindorf
- **Singapore/Malaysia** Tom Lauhon

Components Sales/Service

MEDICAL
Dean Morton, Vice President and General Manager

Divisions
- **Andover (Massachusetts)** Burt Dole
- **Boebitngen, Germany** Karl Grund
- **Brazil** Guenter Warmbold
- **McMinnville (Oregon)** Wall Dyke
- **Waltham (Massachusetts)** Law Platt

Medical Sales/Service

CALCULATORS
George Newman, General Manager

Divisions
- **Advanced Products (Cupertino, California)** Ray King
- **Brazil** Guenler Warmbold
- **Singapore** Tom Lauhon
- **Loveland Calculators** Tom Kelley

Calculator Sales/Service

ANALYTICAL
Emery Rogers, General Manager

Divisions
- **Avondale (Pennsylvania)** Mason Byles
- **Grotzingen, Germany** Peter Hupe
- **Scientific Instruments (Palo Alto, California)** Ed Truitt

Analytical Sales/Service

U.S. and Canada Sales Administration
Eastern: Rick Weaver • **Midwest:** Wall Wallin • **Southern:** Gene Stiles • **Western:** Phil Scalzo • **Canada:** Chuck Williams

International Sales and Subsidiary Administration
Europe: Dick Alberding, Managing Director — **Northern Area:** Fred Schroeder • **Southern Area:** Doug Hardt • **Germany:** Eberhard Knoblauch • **United Kingdom:** Dennis Taylor
Intercontinental: Alan Bickell, Director — **Asia/Africa:** Lee Ting • **Australasia:** John Warmington • **Brazil** (Manufacturing): Guenter Warmbold • **Japan:** Kenzo Sasaoka • **Latin America:** Marc Gumucio • **Southeast Asia:** Tom Lauhon

SOURCE: *Measure,* "Working Together: The HP Organization," April–May 1975, pp. 16–17.

EXHIBIT 7 Hewlett-Packard Corporate Organization, June 1978

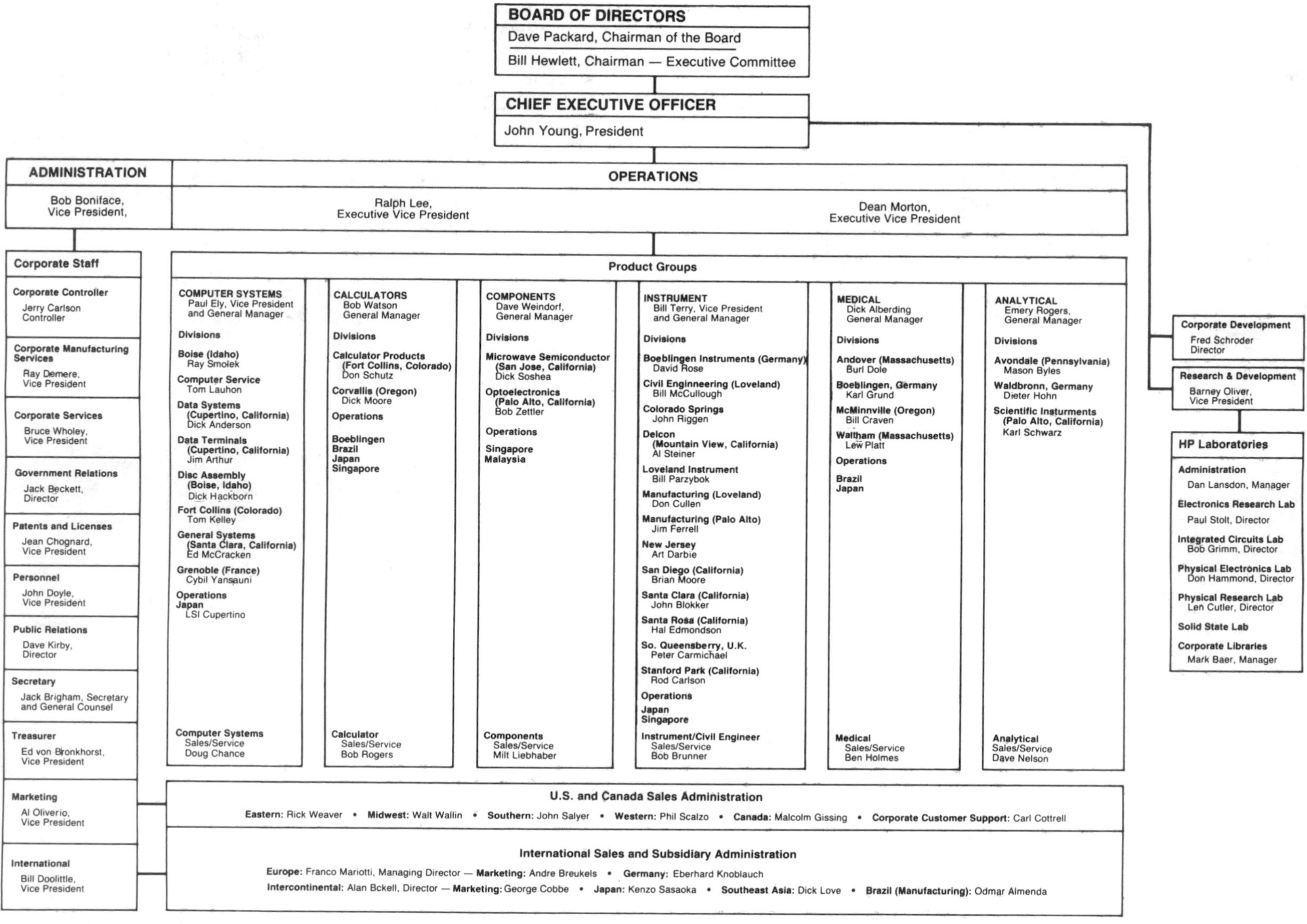

SOURCE: *Measure*, June 1978, pp. 8–9.

nel holding various positions, a growing number of divisions within the product groups, and an increasing emphasis on computers and calculators—was basically the same structure as in 1975. (See Exhibits 6 and 7.) The company magazine *Measure*, introducing the 1978 organization, wrote that, "Except for an official transfer of titles and responsibilities plus a birthday celebration, you would hardly have known that HP made a rather significant change in its organizational character last month."[7] One day before his 65th birthday, Bill Hewlett's resignation as chief executive officer was made official in a brief announcement; thereupon, John Young, who in 1974 had been designated as "the leading contender," became CEO as well as president. Elevated to one of the then-new executive vice presidencies and to the board of directors in 1974, Young had fulfilled the numerous predictions made during the 1974–77 period by succeeding Bill Hewlett as president and chief operating officer in November 1977. Thus, by June 1978, John Young had completed a four-year preparation for the top spot wherein HP for the first time in its 39-year history would be managed by a team of managers developed within the organization rather than by its original founders.

Although Bill Hewlett, as chairman of the Executive Committee, and David Packard, as chairman of the board of directors, were still spending about half their time at HP, it was John Young who had been handed the tough task of taking Hewlett-Packard deeper and deeper into the unfriendly territory of computational technology.

COMPUTATIONAL TECHNOLOGY

Hewlett-Packard has always been heavily engaged in electronic technology. Even as recently as 1977, a special section of their annual report indicated that nowhere else did technological innovation show more momentum than in electronics and its offspring, electronic computation. The environment, as pointed out by *Forbes,* is friendly indeed for HP in the field of measuring instruments, where the company has made a big name for itself and the competition was comparable in size or more often specialized and smaller (e.g., Beckman, Tektronix, and Varian).[8] But the instrument business had slowed in rate of growth; *Forbes* claimed the company, in order to keep its growth record intact, has had to move into a more competitive environment where the opposition is bigger and tougher (e.g., Digital Equipment, Texas Instruments, and IBM). See Exhibit 8 for asset size, debt position, and financial strength for typical instrument, electronic, and computer companies. See Exhibit 9 for key performance data for selected companies.

HP first became involved in the use of computational technology in the early 1960s when its engineers began to design instruments that could work together automatically in computer-controlled systems. The company carried the concept one step further in the mid-1960s with the introduction of a computer designed specifically to work with its instruments. The principal contribution offered by HP in that first computer was ruggedness—the ability to function outside a controlled environment, exposed to wide variations in temperature, humidity, and pressure. In subsequent years, HP products have been prominent in engineering and scientific applications, where there was a high premium on advanced instrumentation to solve complex problems of instrumentation and measurement, in widely varying environmental conditions.

More recently, the need for precise measurement and computation had become widespread in many different industries, businesses,

[7] *Measure*, "The HP Organization: Reaching a Landmark Quietly," June 1978, p. 7.

[8] *Forbes*, "Welcome to the Hot Seat."

EXHIBIT 8 Selected Financial Position Data on Selected Firms

Company	Total Assets 1978 (millions of dollars)	Total Debt 1978 (millions of dollars)	Short-Term Debt as a Percent of Total Investment Capital 1978	Long-Term Debt as a Percent of Total Investment Capital 1978	Common Equity as Percent of Total Investment Capital	Stock Price as Percent of Book Value per Share 9/22/78
Beckman Instruments	277.0	79.3	17.7	17.8	64.4	244.4
Data General	322.1	59.6	0.0	25.4	74.6	349.2
Digital Equipment Company	1,436.5	119.2	3.4	10.6	86.1	237.0
Fairchild Camera	387.9	91.0	8.4	23.4	68.2	99.1
General Instrument	363.5	72.7	0.1	26.1	69.1	141.9
Hewlett-Packard	1,295.8	105.2	9.4	1.0	89.6	278.9
International Business Machines	19,114.1	428.2	1.3	2.0	96.7	313.6
National Semiconductor	278.9	24.9	15.0	1.0	84.1	276.8
Raytheon	1,966.1	97.1	2.3	11.2	86.5	219.5
Texas Instruments	1,350.7	78.8	6.0	3.6	90.4	246.2
Tektronix	491.1	47.4	2.8	9.9	87.3	255.6
Varian Associates	312.6	62.7	13.7	14.5	71.8	86.2

SOURCE: *Business Week*, "A Significant Swing to Short-Term Debt," October 16, 1978, pp. 122–36.

EXHIBIT 9 Selected Performance Data on Selected Firms, 1974–1978

Sales Growth (Percent of Change)

	1974–75	1975–76	1976–77	1977–78*
Beckman Instruments	17	6	18	18
Data General	30	49	58	47
Digital Equipment	27	38	44	36
Fairchild Camera	−24	52	4	15
General Instrument	−11	24	8	8
Hewlett-Packard	11	13	22	24
IBM	14	13	11	13
National Semiconductor	10	38	19	28
Raytheon	16	10	14	16
Tektronix	24	9	24	32
Texas Instruments	−13	21	23	21
Varian Associates	6	10	3	14
Average	8.9	23.6	20.7	22.7

Net Profit Margin (Percent)

	1974	1975	1976	1977	1978*
Beckman Instruments	4	4	5	6	7
Data General	12	12	12	11	11
Digital Equipment	11	9	10	10	10
Fairchild Camera	7	4	3	2	5
General Instrument	3	3	4	5	6
Hewlett-Packard	10	9	8	9	9
IBM	15	14	15	15	15
National Semiconductor	8	7	6	3	5
Raytheon	3	3	4	4	4
Tektronix	8	8	8	10	10
Texas Instruments	6	5	6	6	6
Varian Associates	3	3	3	4	3
Average	7.5	6.8	7.0	7.1	7.6

Earnings on Net Worth (Percent)

	1974	1975	1976	1977	1978*
Beckman Instruments	8	9	10	13	15
Data General	21	14	17	20	21
Digital Equipment	13	12	12	15	16
Fairchild Camera	17	6	7	6	13
General Instrument	7	7	9	12	14
Hewlett-Packard	18	15	13	15	15
IBM	18	17	19	21	21
National Semiconductor	35	25	20	10	17
Raytheon	14	15	16	18	20
Tektronix	12	13	13	16	17
Texas Instruments	17	11	15	16	16
Varian Associates	6	6	6	8	8
Average	15.5	12.5	13.1	14.2	16.1

* Estimated by *Value Line Investment Survey* (Arnold Bernhard & Co., July 7, 1978, p. 187).

and professions. Among the company's newest customers were those involved in business data processing. The first HP product aimed exclusively at this market was a hand-held calculator for financial analysis. At the other end of the size scale was the development in the early 1960s of HP's first minicomputer-based time-share system which found wide use in science and engineering and was particularly well received in the educational market. The next generation of computers, introduced in the early 1970s, also found a ready market in the educational field because it could accommodate many different programs and computer languages. HP has steadily upgraded this computer as a result of applying the computer to HP's own business problems. This development has proved particularly useful to HP customers with similar worldwide manufacturing operations.

The relative success, however, of HP's excursions into hand-held calculators and minicomputers have been quite different. Erratic market conditions and heavy competition characterized both industry segments. There were marked differences, however, in the ability and willingness of the company to adapt and respond to these product/market changes.

Hand-Held Calculators

More widely known to college students and the general public than its broad line of basic products was the company's line of hand-held calculators. David Packard described HP's entry into this field in *The AMBA Executive* newsletter in September 1977: "Actually we got into the electronic calculator business by accident. We hadn't planned it at all."[9] In 1966, calculators were largely mechanical; a young man working for one of the calculator companies brought to HP a model for an electronic calculator. His own company was not interested in it because they didn't have the electronic capability. An HP team was put together and the first electronic calculator, with a great deal of power, was designed for the engineers at HP. It was, however, a large device about one foot square. Coincidentally, HP was also doing research on large-scale integrated circuits and on light-emitting diodes. Bill Hewlett realized that these technologies could be combined into a calculator, that these light-emitting diodes would make it possible to have a small readout, and that the result would be something that could be put into a pocket. A year later, the HP-35, the first hand-held calculator, was introduced.

Forbes has reported that for a brief period, HP made itself the leader in the business and scientific hand-held calculator field, which in 1974 was estimated to have yielded roughly 30 percent of company profits.[10] Shortly after, HP's high-priced, high-quality calculators fell before the competition led by Texas Instruments. Rather than compete across the board, HP decided to remain in the specialized upper end of the market. In 1978, the division was reputed to be barely profitable, but with relatively stable sales.

Minicomputers

HP had become, by 1978, a well-integrated minicomputer manufacturer, competing with International Business Machines, Digital Equipment Corporation, and Data General. This business had long been characterized by high technological risk and erratic earnings. During the late 1960s, HP successfully directed sales efforts toward the educational, scientific, and engineering markets, where it was an established supplier of instruments. Subsequently, in entering the minicomputer market, the company chose to service the time-sharing sector,

[9] *The AMBA Executive,* "Hewlett-Packard Chairman Built Company by Design, Calculator by Chance," September 1977, pp. 1 ff.

[10] *Forbes,* "Welcome to the Hot Seat."

which fell apart in the 1970s, causing profit reversals.

Recently, however, the picture has improved and HP's electronic data processing product category has contributed over 40 percent of sales and almost the same proportion of profits, despite the drag from hand-held calculators. (See Exhibit 4.) HP has expanded its computer line into the area where others hold strong positions. HP's minicomputer line consisted basically of two products, one for business and one for scientific/technical use. Big customers often bought several systems at a time complete with peripherals terminals, disk drives, printers, and even instruments that could be attached. A single sale could easily exceed $1 million. The company was well aware of the dangers of its thrust into computers. Many big and smart companies had tried to take on IBM and lost. Hewlett-Packard has mounted its effort carefully. The division's domestic sales force has been almost doubled in the previous year to 500 people. The sales force has been split between business and engineering systems. Mr. Young has reportedly spent 10 percent of his time making sales presentations to customers' top management, since commitments in the $1 million range typically require board-of-directors' approval. The company has also limited its marketing efforts by foregoing well-covered markets like banks and insurance companies in favor of large manufacturing companies which could use systems that HP had developed initially for its own operations. Such firms could take a whole computer line from the technically slanted machines on the factory floor and near engineers' desks to business systems for payrolls and customer billing.

To effectively compete in minicomputers, the company has had to continue to be extremely innovative, and creative, as well as efficient. The minicomputer environment was difficult, rapidly changing, and extremely competitive. In this market, Hewlett-Packard has started to kick at the shins of IBM, which was 15 times larger (see Exhibit 8). A June 1976 article in *Business Week* quoted a former HP marketing executive, then president of Tandem Computers, Inc., as saying, "The first rule of this business is not to compete with IBM."[11] And in October 1978, *Business Week* described "an incredibly fast adjustment in (HP's) marketing strategy."[12] It also noted some rough spots in the road that HP had already traveled in the field of computational technology: (1) its early reliance on techniques that worked well with sales to engineers, but not with the applications-oriented commercial EDP customers; (2) the difficulty of selling the idea of distributed processing, a concept involving pushing data processing out of the central computer room, HP's primary strategic difference from IBM; (3) a period in 1973 when the HP 3000 had to be taken off the market and redesigned because its software was too powerful for the hardware; (4) the different requirements, buyer attributes, and decision processes that characterized the larger and more fragmented market of commercial systems; (5) the tough task of meeting systems repair and maintenance response standards set by the mainframe companies it was now up against; and (6) the hard push by its customers for more applications software that would allow customers to perform specific tasks. Included in the same *Business Week* article were two items that must have intrigued long-time observers of HP and the computer industry. The product manager for HP's new HP 3000, Series 33, noting how, since 1974, they had concentrated on expanded capability for the 3000 at about the original price, was quoted, "Now let's use the technology to drive down the price." *Business Week* also indicated that HP is likely "to see more competition in distributed processing, especially from IBM, which is expected to an-

[11] *Business Week*, "Hewlett-Packard Takes on the Computer Giants," June 7, 1976, pp. 91–92.

[12] *Business Week*, "Hewlett-Packard Learns to Sell to Business Managers," October 26, 1978, pp. 62B and 62G.

nounce a powerful new series of low-cost mainframe computers this fall."

Case IV–9
Can Hewlett-Packard Put the Pieces Back Together?

J. W. Wilson and C. Harris, with G. Bock

Almost from the day he took over as chief executive officer at Hewlett-Packard Company in 1978, John A. Young realized he had a serious problem. The company was the world's largest maker of electronic instruments, but computers were about to take over as its main product. The trouble was that the company's efforts to develop new computers were badly fragmented, and HP was becoming a technological laggard in a highly competitive field.

HP needed a well-orchestrated line of machines for the 1980s. Instead, it had three separate computer divisions, all pumping out products that were incompatible. Watching the company fall dangerously behind archrivals Digital Equipment, Data General, and IBM, Young looked for a drastic remedy. "We had to do something, or we'd sink," he remembers.

BLAZING SPEEDS

He did something big. In a move that ran counter to HP's tradition of autonomous, entrepreneurial divisions, the unflappable Young set out to centralize HP's computer research efforts. The project—the biggest in the company's 47-year history—was called Spectrum, and it was risky. Young was putting all the company's eggs in one basket: He killed HP's other computer design projects. If he didn't succeed in building a unified computer organization and creating a central technology that could be used in all the company's machines, HP would be consigned to the computer industry's backwaters.

On February 25, after spending an estimated $200 million on Spectrum, Young introduced his first two machines, built around a new and largely unproven approach known as RISC, for Reduced-Instruction-Set Computer. Its more efficient design lets a computer gain blazing speed by drastically reducing the complexity of its central processor.

Many analysts and customers think Young is playing a winning hand. His first Spectrum machine, code-named Indigo during development and now called the HP 3000 Series 930, sounds impressive. It's supposed to perform better than competing $450,000 minicomputers from Digital Equipment Corporation and International Business Machines Corporation—for about half the price. That could help HP steal accounts from the two giants. With its new machines, notes Grant S. Bushee, executive vice president of the market research firm InfoCorp, "HP can go after any market."

The company needs a boost. Its overall share of the worldwide minicomputer market slipped to 5 percent last year from 5.5 percent in 1983, according to InfoCorp (chart). In the $37 billion market for commercial minis, it dropped from a 4.2 percent share in 1983 to 3.9 percent last year, while DEC nudged up from 5 percent to 7 percent. In the $15 billion market for technical minis, HP sank to 7.8 percent, from more than 12 percent five years ago. DEC's inroads mostly caused that, too, but so did competition from IBM and AT&T. Says George F. Colony, president of Forrester Research, Inc., in Cambridge, Massachusetts: "It's damage-control time for HP in the minicomputer market."

Not everyone is convinced that Young's RISC strategy will be a winning one. One critic is Stephen K. Smith, a PaineWebber Incorporated computer analyst who warns that many claims

---

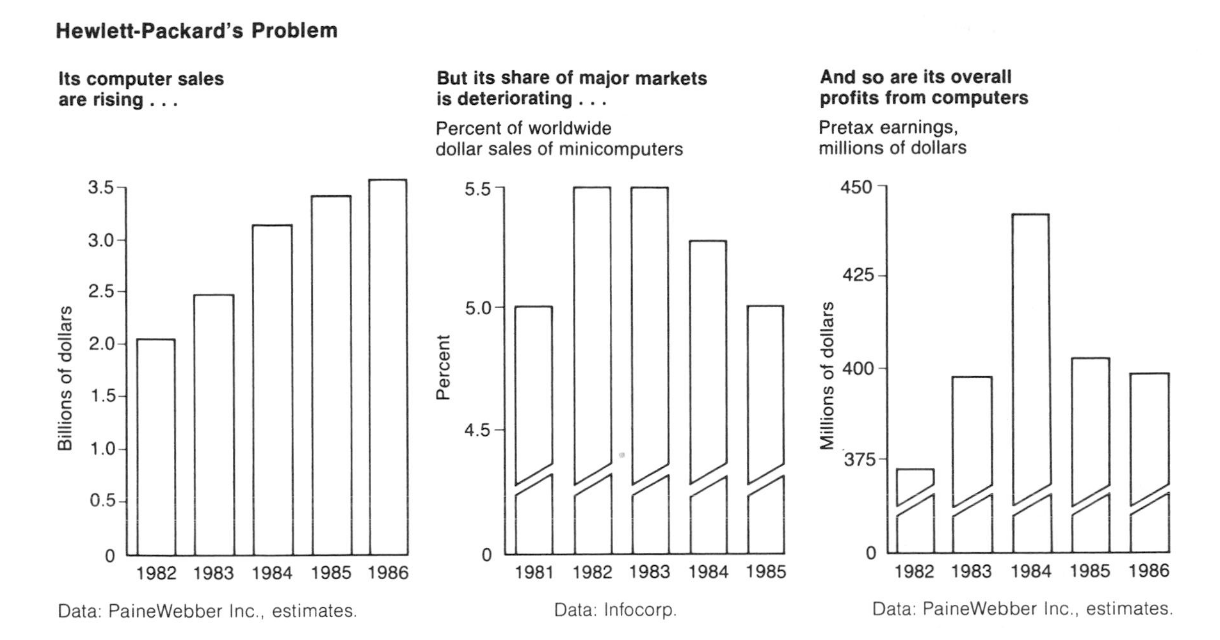

about RISC's advantages "have yet to be substantiated outside an academic environment."

Even if Spectrum does succeed, it will be a while before Young's moves bolster HP's flagging computer profits or improve its overall bottom line. Spectrum "isn't going to show great payoffs in 1986 or 1987," says David C. Moschella, director of systems research for International Data Corporation. That's bad news for investors, who bid up the company's shares from 29 in the fall to 44 recently in what analysts say was anticipation of the Spectrum line.

"BAD SMELL"

Spectrum's payoff will be slow because the Series 930 machine isn't scheduled for shipment until the end of this year, and its big brother, the powerful Series 950, code-named Cheetah, won't ship until 1987. HP is moving slowly to make sure the machines' software works. "HP was built on a reputation of quality and dependability, and they don't want to monkey with that," says William C. Rosser, a vice president of Gartner Group, Inc. "If it comes out, and somebody gets a bad smell, that would hurt them more than any delay."

Such thoroughness has been largely responsible for HP's success in the past. But the delay could push away customers who have grown restless for more computing power. "We've been holding on by our toenails," says Wayne E. Holp, a Seattle-based consultant to HP customers. "The fingernails gave way a long time ago." Some of his clients are disappointed because HP has been using 100 of its Spectrum-line machines internally for a year—and they expected to have at least a few by now.

This problem coincides with two others. Instrument sales, accounting for nearly half the Palo Alto (California) company's revenues, dropped 3 percent last year, estimates Kidder, Peabody, & Co. And there's the industrywide slump, which caused HP's overall revenues to grow only a modest 8 percent, to $6.5 billion, in the fiscal year ended in October. That compares with a double-digit growth rate for most other computer makers. HP's earnings in the same period fell for the first time since 1975, by 10 percent, to $489 million. In August, Young had to impose 10 percent salary cuts, since trimmed to 5 percent for everyone but top executives.

Although Spectrum's delay has created problems for Young, the project has given him a chance to put his stamp on a company molded by two of Silicon Valley's most revered figures. Before Spectrum, Young had the titles of president and CEO. But he still toiled in the shadows of William R. Hewlett and David Packard, who chaired the executive committee and board, respectively.

The founders, now semiretired, were set in their ways, and HP needed dramatic change. The small, entrepreneurial divisions they had used so successfully to invent instruments were a big problem when it came to computers. Customers wanted products that could work together, and that made a unified, companywide approach essential. Young was reluctant to tamper with the founders' formula, and it was several years before he centralized product development, as IBM and DEC had done. By finally conceiving a more coordinated approach and building a consensus among his computer division managers to go with Spectrum, Young put his imprint on HP.

In doing so, he radically reshaped a management system long held up as a paragon of excellence. Not only did he have to squelch at least four competing approaches to building high-powered computers, but he also had to sell the untried RISC technology to the losing divisions and then harness them to the task of making it work. Division managers, who once had a virtually free hand at HP, now spend much of their time coordinating their activities with counterparts in other units. Says a former manager: "HP is now organized as a John Young company."

While Hewlett and Packard both were decisive, creative entrepreneurs, Young, 53, is more a detail man who relies on committees and consensus-building before acting. Since getting his M.B.A. from Stanford University, Young has spent his entire 28-year career with HP, mostly on the instruments side. "His style is to lead people where he wants them to go rather than tell them what to do," says an insider. To some, that's a weakness. Complains a former HP manager: "I never saw him make a decision."

He did make a key one in 1980. One of HP's weaknesses was that Young and fellow officers had little experience with computers. "People at the decision-making level didn't have a good gut feel for what made sense technically," says a former employee. "So they tended not to make decisions." To beef up research and development, which was then still focused on test-and-measurement technology, Young launched a recruiting campaign. His prize catch: Joel S. Birnbaum, who was for five years the head of computer sciences at IBM's Thomas J. Watson Research Center. Taking over as HP's director of computer research, Birnbaum quickly proposed a program to bring order out of the divisional chaos. "I felt it was hopeless unless we could do this," he recalls.

MISTRUST

Loquacious, quick-witted, and sociable, the new research chief was a veteran of IBM's early work on RISC designs. He felt that HP should be making simpler computers, not the more complex ones many divisions were designing. Realizing that HP was run by engineers, not the marketing specialists who dominate IBM, he set out to discover what customers wanted HP's computers to do. His research created the impetus for a prototype RISC machine he called LESS, for Low-End Spectrum System. "Our motto was: LESS is more," Birnbaum says. "And even that little machine was much faster than what was then our top of the line." RISC design strips away from a computer's central processor all but its essential functions, so it can process information faster.

Once Young developed the consensus he wanted to give Spectrum a final go-ahead in February 1983, he needed someone who could make sure that all the divisions agreed on the details of how the project should proceed. That job fell to Douglas C. Chance, general manager of HP's Information Systems Group. Young saw Chance as a strong strategist who could build a team and develop confidence. This was crucial because many technical decisions had to be made quickly before the machine could be handed over to a development team.

Chance set up a committee of top officers and division managers that met monthly over dinner. As often as not, Chance learned, the issues that worried them most weren't technical. "The real concerns had to do with whether they could trust the others to do what we had all agreed to do," he recalls. Such fears "typically surfaced late at night after several bottles of wine."

Eventually, as Spectrum grew, HP's divisions learned to cooperate under a new umbrella called the Information Technology Group. At its helm was veteran engineer George E. Bodway, a hearty, rumpled man with a knack for administration. "Nothing escapes him," an admirer says. Formed in May 1984, ITG quickly built a staff of 1,000 and for the first time brought into one group the people who develop such key elements of computers as operating systems, programming languages, and integrated-circuit designs. Some 20 divisions chipped in specialized software, peripheral devices, and other elements of Spectrum, with disagreements arbitrated by Bodway and a series of specialized councils.

LOSING VERVE?

Young is keeping the centralization in place and shifting the role of the once-autonomous divisions dramatically. The divisions continue as marketing arms for specific groups of customers. But when it comes to product development, they can no longer "do any damn thing they want," he says. Young hopes new reporting and performance-measuring systems will help develop links among divisions "without getting swamped in a lot of bureaucracy."

Some former employees fear that HP is in danger of losing its entrepreneurial verve. Now, "HP looks a lot more like DEC or IBM," says one. "That creates a different work environment." But it is clear to Young that the company's old environment didn't work well in the high-stakes computer game. The combination of new technology and a new corporate structure looks like his only bet.

Case IV–10
Joel Birnbaum, the Odd Man in at Hewlett-Packard

J. W. Wilson with M. A. Harris

Joel S. Birnbaum may be the closest thing to a Renaissance man at Hewlett-Packard Company. An amateur photographer, nuclear physicist, computer visionary, and former Ivy League baseball pitcher, he is an anomaly in a California company run largely by an ingrown crowd of test-and-measurement engineers.

It might seem that an Easterner with Birnbaum's varied credentials would have trouble fitting in at a place like HP. But Birnbaum, 48, won a heated intramural battle to switch the company's minicomputers to a drastically simplified architecture known as RISC. In doing so, he has emerged as a corporate vice president and director of HP Laboratories, which he is rapidly turning into one of the computer industry's hottest research facilities.

KINDRED SPIRITS

After attending Cornell on a baseball scholarship and picking up a physics Ph.D. at Yale, Birnbaum spent 15 years at International Business Machines Corporation. But as director of computer research there, he became frustrated by the difficulty of getting ideas to market. "I had the feeling I didn't make a difference," he says. When a recruiter for HP called in 1980, Birnbaum found: "I could talk to the HP guys in a different way. The top guys at IBM were not technical people."

It was at IBM, however, that Birnbaum ran across RISC technology—during a mid-1970s research project for a digital phone switch that could route a million calls an hour. John Cocke, another IBM researcher, came up with an idea for special software, plus a simplified processor that eliminated the circuitry needed to decode seldom-used instructions. It could thus run much faster than existing processors. The phone switch was dropped, but Birnbaum built a model processor. Although crude, it worked—and became the basis for the RT PC technical workstation, a RISC machine that IBM introduced in January.

The prototype for Spectrum, the machine designed for HP by Birnbaum and William S. Worley, Jr., another ex-IBM researcher, fell short of pure RISC technology. HP wanted to minimize risks, so it used well-proven chips instead of the hottest ones available. That limits the speed of its new machines. But they're still faster than most conventional ones, and Birnbaum plans to soup them up. That might make it possible to "domesticate" computers by building in artificial intelligence that makes them much easier to use.

His work at HP Labs leaves Birnbaum with little time for athletics or his darkroom, where he develops his own portrait and landscape shots. But his real love, he says, is ideas. And tossing those around isn't a bad occupation for an ex-pitcher.

Reading IV–5
The Art of High-Technology Management

M. A. Maidique and
R. H. Hayes

Over the past 15 years, the world's perception of the competence of U.S. companies in managing technology has come full circle. In 1967, a Frenchman, J.-J. Servan-Schreiber, expressed with alarm in his book, *The American Challenge,* that U.S. technology was far ahead of the rest of the industrialized world.[1] This "technology gap," he argued, was continually widening because of the *superior ability of Americans to organize and manage technological development.*

Today, the situation is perceived to have changed drastically. The concern now is that the gap is reversing: the onslaught of Japanese and/or European challenges is threatening America's technological leadership. Even such informed Americans as Dr. Simon Ramo express great concern: In his book, *America's Technology Slip,* Dr. Ramo notes the apparent inability of U.S. companies to compete technologically with their foreign counterparts.[2] Moreover, in the best seller *The Art of Japanese Management,* the authors use as a basis of comparison two technology-based firms: Matsushita (Japanese) and ITT (American).[3] Here, the Japanese firm is depicted as a model for managers, while the management practices of the U.S. firm are sharply criticized.

Reprinted from *Sloan Management Review,* Vol. 25, Winter 1984, pp. 18–31, by permission of the publisher.

[1] See J.-J. Servan-Schreiber, *The American Challenge* (New York: Atheneum Publishers, 1968).

[2] See S. Ramo, *America's Technology Slip* (New York: John Wiley & Sons, 1980).

[3] See R. Pascale and A. Athos, *The Art of Japanese Management* (New York: Simon & Schuster, 1981).

Nevertheless, a number of U.S. companies appear to be fending off these foreign challenges successfully. These firms are repeatedly included on lists of "America's best-managed companies." Many of them are competitors in the R&D intensive industries, a sector of our economy that has come under particular criticism. Ironically, some of them have even served as models for highly successful Japanese and European high-tech firms.

For example, of the 43 companies that Peters and Waterman, Jr., judged to be "excellent" in *In Search of Excellence,* almost half were classified as "high technology," or as containing a substantial high-technology component.[4] Similarly, of the five U.S. organizations that William Ouchi described as best prepared to meet the Japanese challenge, three (IBM, Hewlett-Packard, and Kodak) were high-technology companies.[5] Indeed, high-technology corporations are among the most admired firms in America. In a *Fortune* study that ranked the corporate reputation of the 200 largest U.S. corporations, IBM and Hewlett-Packard (HP) ranked first and second, respectively.[6] And of the top 10 firms, 9 compete in such high-technology fields as pharmaceuticals, precision instruments, communications, office equipment, computers, jet engines, and electronics.

The above studies reinforce our own findings, which have led us to conclude that U.S. high-technology firms that seek to improve their management practices to succeed against foreign competitors need not look overseas. The firms mentioned above are not unique. On the contrary, they are representative of scores of well-managed small and large U.S. technology-based firms. Moreover, the management practices they have adopted are widely applicable. Thus, perhaps the key to stimulating innovation in our country is not to adopt the managerial practices of the Europeans or the Japanese, but to adapt some of the policies of our *own* successful high-technology firms.

THE STUDY

Over the past two decades, we have been privileged to work with a host of small and large high-technology firms as participants, advisors, and researchers. We and our assistants interviewed formally and informally over 250 executives, including over 30 CEOs, from a wide cross section of high-tech industries—biotechnology, semiconductors, computers, pharmaceuticals, and aerospace. About 100 of these executives were interviewed in 1983 as part of a large-scale study of product innovation in the electronics industry (which was conducted by one of this article's authors and his colleagues).[7] Our research has been guided by

---

[4] See T. J. Peters and R. H. Waterman, Jr., *In Search of Excellence* (New York: Harper and Row, 1982). For purposes of this article, the high-technology industries are defined as those which spend more than 3 percent of sales on R&D. These industries, though otherwise quite different, are all characterized by a rapid rate of change in their products and technologies. Only five U.S. industries meet this criterion: (1) chemicals and pharmaceuticals; (2) machinery (especially computers and office machines); (3) electrical equipment and communications; (4) professional and scientific instruments; and (5) aircraft and missiles. See National Science Foundation, *Science Resources Studies Highlights,* NSF81-331, December 31, 1981, p. 2.

[5] See W. Ouchi, *Theory Z: How American Management Can Meet the Japanese Challenge* (New York: John Wiley & Sons, 1980).

[6] See C. E. Makin, "Ranking Corporate Reputations," *Fortune,* January 10, 1983, pp. 34–44. Corporate reputation was subdivided into eight attributes: quality of management, quality of products and services, innovativeness, long-term investment value, financial soundness, ability to develop and help talented people, community and environmental responsibility, and use of corporate assets.

[7] See: M. A. Maidique and B. J. Zirger, "Stanford Innovation Project: A Study of Successful and Unsuccessful Product Innovation in High-Technology Firms," *IEEE Transactions on Engineering Management,* in press; M. A. Maidique, "The Stanford Innovation Project: A Comparative Study of Success and Failure in High-Technology Product Innovation," *Management of Technological Innovation Conference Proceedings* (Worcester Polytechnic Institute, 1983).

a fundamental question: what are the strategies, policies, practices, and decisions that result in successful management of high-technology enterprises? One of our principal findings was that no company has a monopoly on managerial excellence. Even the best run companies make big mistakes, and many smaller, lesser regarded companies are surprisingly sophisticated about the factors that mediate between success and failure.

It also became apparent from our interviews that the driving force behind the successes of many of these companies was strong leadership. All companies need leaders and visionaries, of course, but leadership is particularly essential when the future is blurry and when the world is changing rapidly. Although few high-tech firms can succeed for long without strong leaders, leadership itself is not the subject of this article. Rather, we accept it as given and seek to understand what strategies and management practices can *reinforce* strong leadership.

The companies we studied were of different sizes ($10 million to $30 billion in sales); their technologies were at different stages of maturity; their industry growth rates and product mixes were different; and their managers ranged widely in age. But they all had the same unifying thread: a rapid rate of change in the technological base of their products. This common thread, rapid technological change, implies novel products and functions and thus usually rapid growth. But even when growth is slow or moderate, the destruction of the old capital base by new technology results in the need for rapid redeployment of resources to cope with new product designs and new manufacturing processes. Thus, the two dominant characteristics of the high-technology organizations that we focused on were growth and change.

In part because of this split focus (growth and change), the companies we studied often appeared to display contradictory behavior over time. Despite these differences, in important respects, they were remarkably similar because they all confronted the same two-headed dilemma: how to unleash the creativity that promotes growth and change without being fragmented by it, and how to control innovation without stifling it. In dealing with this concern, they tended to adopt strikingly similar managerial approaches.

THE PARADOX: CONTINUITY AND CHAOS

When we grouped our findings into general themes of success, a significant paradox gradually emerged—which is a product of the unique challenge that high-technology firms face. Some of the behavioral patterns that these companies displayed seemed to favor promoting disorder and informality, while others would have us conclude that it was consistency, continuity, integration, and order that were the keys to success. As we grappled with this apparent paradox, we came to realize that continued success in a high-technology environment requires periodic shifts between chaos and continuity.[8] Our originally static framework, therefore, was gradually replaced by a dynamic framework within whose ebbs and flows lay the secrets of success.

SIX THEMES OF SUCCESS

The six themes that we grouped our findings into were: (1) business focus; (2) adaptability; (3) organizational cohesion; (4) entrepreneur-

[8] A similar conclusion was reached by Romanelli and Tushman in their study of leadership in the minicomputer industry, which found that successful companies alternated long periods of continuity and inertia with rapid reorientation. See E. Romanelli and M. Tushman, "Executive Leadership and Organizational Outcomes: An Evolutionary Perspective," *Management of Technological Innovation Conference Proceedings* (Worcester Polytechnic Institute, 1983).

ial culture; (5) sense of integrity; and (6) "hands-on" top management. No one firm exhibits excellence in every one of these categories at any one time, nor are the less successful firms totally lacking in all. Nonetheless, outstanding high-technology firms tend to score high in most of the six categories, while less successful ones usually score low in several.[9]

1. Business Focus

Even a superficial analysis of the most successful high-technology firms leads one to conclude that they are highly focused. With few exceptions, the leaders in high-technology fields, such as computers, aerospace, semiconductors, biotechnology, chemicals, pharmaceuticals, electronic instruments, and duplicating machines, realize the great bulk of their sales either from a single product line or from a closely related set of product lines.[10] For example, IBM, Boeing, Intel, and Genentech confine themselves almost entirely to computer products, commercial aircraft, integrated circuits, and genetic engineering, respectively. Similarly, four fifths of Kodak's and Xerox's sales come from photographic products and duplicating machines, respectively. In general, the smaller the company, the more highly focused it is. Tandon concentrates on disk drives; Tandem on high-reliability computers; Analog Devices on linear integrated circuits; and Cullinet on software products.

Closely related products

This extraordinary concentration does not stop with the dominant product line. When the company grows and establishes a secondary product line, it is usually closely related to the first. Hewlett-Packard, for instance, has two product families, each of which accounts for about half of its sales. Both families—electronic instruments and data processors—are focused on the same technical, scientific, and process control markets. IBM also makes two closely related product lines—data processors (approximately 80 percent of sales) and office equipment—both of which emphasize the business market.

Companies that took the opposite path have not fared well. Two of yesterday's technological leaders, ITT and RCA, have paid dearly for diversifying away from their strengths. Today, both firms are trying to divest many of what were once highly touted acquisitions. As David Packard, chairman of the board of Hewlett-Packard, once observed, "No company ever died from starvation, but many have died from indigestion."[11]

A communications firm that became the world's largest conglomerate, ITT began to slip in the early 1970s after an acquisition wave orchestrated by Harold Geneen. When Geneen retired in 1977, his successors attempted to redress ITT's lackluster performance through a far-reaching divestment program.[12] So far, 40 companies and other assets worth over $1 billion have been sold off—and ITT watchers believe the program is just getting started. Some analysts believe that ITT will ultimately be restructured into three groups, with the com-

---

[9] One of the authors in this article has employed this framework as a diagnostic tool in audits of high-technology firms. The firm is evaluated along these six dimensions on a 0–10 scale by members of corporate and divisional management, working individually. The results are then used as inputs for conducting a strategic review of the firm.

[10] General Electric evidently has also recognized the value of such concentration. In 1979, Reginald Jones, then GE's CEO, broke up the firm into six independent sectors led by "sector executives." See R. Vancil and P. C. Browne, "General Electric Consumer Products and Services Sector" (Boston: Harvard Business School Case Services 2-179-070).

[11] Personal communication with David Packard, Stanford University, March 4, 1982.

[12] After only 18 months as Geneen's successor as president, Lyman Hamilton was summarily dismissed by Geneen for reversing Geneen's way of doing business. See G. Colvin, "The Re-Geneening of ITT," *Fortune,* January 11, 1982, pp. 34–39.

munications/electronics group and engineered products (home of ITT semiconductors) forming the core of a "new" ITT.

RCA experienced a similar fate to ITT. When RCA's architect and longtime chairman, General David Sarnoff, retired in 1966, RCA was internationally respected for its pioneering work in television, electronic components, communications, and radar. But by 1980, the three CEOs who followed Sarnoff had turned a technological leader into a conglomerate with flat sales, declining earnings, and a $2.9 billion debt. This disappointing performance led RCA's new CEO, Thorton F. Bradshaw, to decide to return RCA to its high-technology origins.[13] Bradshaw's strategy is to now concentrate on RCA's traditional strengths—communications and entertainment—by divesting its other businesses.

Focused R&D

Another policy that strengthens the focus of leading high-technology firms is concentrating R&D on one or two areas. Such a strategy enables these businesses to dominate the research, particularly the more risky, leading edge explorations. By spending a higher proportion of their sales dollars on R&D than their competitors do, or through their sheer size (as in the case of IBM, Kodak, and Xerox), such companies maintain their technological leadership. It is not unusual for a leading firm's R&D investment to be one and a half to two times the industry's average as a percent of sales (8 to 15 percent) and several times more than any individual competitor on an absolute basis.[14]

Moreover, their commitment to R&D is both enduring and consistent. It is maintained through slack periods and recessions because it is believed to be in the best, long-term interest of the stockholders. As the CEO of Analog Devices, a leading linear integrated circuit manufacturer, explained in a quarterly report which noted that profits had declined 30 percent, "We are sharply constraining the growth of fixed expenses, but we do not feel it is in the best interest of shareholders to cut back further on product development . . . in order to relieve short-term pressure on earnings."[15] Similarly, when sales, as a result of a recession, flattened and profit margins plummeted at Intel, its management invested a record-breaking $130 million in R&D, and another $150 million in plant and equipment.[16]

Consistent priorities

Still another way that a company demonstrates a strong business focus is through a set of priorities and a pattern of behavior that is continually reinforced by top management: for example, planned manufacturing improvement at Texas Instruments (TI); customer service at IBM; the concept of the entrepreneurial product champion at 3M; and the new products at HP. Belief in the competitive effectiveness of their chosen theme runs deep in each of these companies.

A business focus that is maintained over extended periods of time has fundamental consequences. By concentrating on what it does well, a company develops an intimate knowledge of its markets, competitors, technologies, employees, and of the future needs and opportunities of its customers.[17] The Stanford Innovation

[13] See "RCA: Still Another Master," *Business Week,* August 17, 1981, pp. 80–86.

[14] See "R&D Scoreboard," *Business Week,* July 6, 1981, pp. 60–75.

[15] See R. Stata, Analog Devices *Quarterly Report,* 1st Quarter, 1981.

[16] See: "Why They Are Jumping Ship at Intel," *Business Week,* February 14, 1983, p. 107; M. Chase, "Problem-Plagued Intel Bets on New Products, IBM's Financial Help," *The Wall Street Journal,* February 4, 1983.

[17] These SAPPHO findings are generally consistent with the results of the Stanford Innovation Project, a major comparative study of U.S. high-technology innovation. See M. A. Maidique, "The Stanford Innovation Project: A Comparative Study of Success and Failure in High-Technology Product Innovation," *Management of Technology Conference Proceedings* (Worcester Polytechnic Institute, 1983).

Project recently completed a three-year study of 224 U.S. high-technology products (half of which were successes, half of which were failures) and concluded that a continuous, in-depth, informal interaction with leading customers throughout the product development process was the principal factor behind successful new products. In short, this coupling is the cornerstone of effective high-technology progress. Such an interaction is greatly facilitated by the longstanding and close customer relationships that are fostered by concentrating on closely related product-market choices.[18] "Customer needs," explains Tom Jones, chairman of Northrop Corporation, "must be understood *way ahead of time*" (authors' emphasis).[19]

2. Adaptability

Successful firms balance a well-defined business focus with the willingness, and the will, to undertake major and rapid change when necessary. Concentration, in short, does not mean stagnation. Immobility is the most dangerous behavioral pattern a high-technology firm can develop: technology can change rapidly, and with it the markets and customers served. Therefore, a high-technology firm must be able to track and exploit the rapid shifts and twists in market boundaries as they are redefined by new technological, market, and competitive developments.

The cost of strategic stagnation can be great, as General Radio (GR) found out. Once the proud leader of the electronic instruments business, GR almost singlehandedly created many sectors of the market. Its engineering excellence and its progressive human relations policies were models for the industry. But when its founder, Melville Eastham, retired in 1950, GR's strategy ossified. In the next two decades, the company failed to take advantage of two major opportunities for growth that were closely related to the company's strengths: microwave instruments and minicomputers. Meanwhile, its traditional product line withered away. Now all that remains of GR's once dominant instruments line, which is less than 10 percent of sales, is a small assembly area where a handful of technicians assemble batches of the old instruments.

It wasn't until William Thurston, in the wake of mounting losses, assumed the presidency at the end of 1972 that GR began to refocus its engineering creativity and couple it to its new marketing strategies. Using the failure of the old policies as his mandate, Thurston deemphasized the aging product lines, focused GR's attention on automated test equipment, balanced its traditional engineering excellence with an increased sensitivity to market needs, and gave the firm a new name—GenRad. Since then, GenRad has resumed rapid growth and has won a leadership position in the automatic test equipment market.[20]

The GenRad story is a classic example of a firm making a strategic change because it perceived that its existing strategy was not working. But even successful high-technology firms sometimes feel the need to be rejuvenated periodically to avoid technological stagnation. In the mid-1960s, for example, IBM appeared to have little reason for major change. The company had a near monopoly in the computer mainframe industry. Its two principal products—the 1401 at the low end of the market and the 7090 at the high end—accounted for over two thirds of its industry's sales. Yet, in one move, the company obsoleted both product lines (as well as others) and redefined the rules of competition for decades to come by simultaneously introducing six compatible

[18] See: Maidique and Zirger, "Stanford Innovation Project." Several other authors have reached similar conclusions. See, for example, Peters and Waterman, *In Search of Excellence*.

[19] Personal communication with Tom Jones, chairman of the board, Northrop Corporation, May 1982.

[20] See W. R. Thurston, "The Revitalization of GenRad," *Sloan Management Review,* Summer 1981, pp. 53–57.

models of the "System 360," based on proprietary hybrid integrated circuits.[21]

During the same period, GM, whose dominance of the U.S. auto industry approached IBM's dominance of the computer mainframe industry, stoutly resisted such a rejuvenation. Instead, it became more and more centralized and inflexible. Yet, GM was also once a high-technology company. In its early days when Alfred P. Sloan ran the company, engines were viewed as high-technology products. One day, Charles F. Kettering told Sloan he believed the high efficiency of the diesel engine could be engineered into a compact power plant. Sloan's response was: "Very well—we are now in the diesel engine business. You tell us how the engine should run, and I will . . . capitalize the program."[22] Two years later, Kettering achieved a major breakthrough in diesel technology. This paved the way for a revolution in the railroad industry and led to GM's preeminence in the diesel locomotive markets.

Organizational flexibility

To undertake such wrenching shifts in direction requires both agility and daring. Organizational agility seems to be associated with organizational flexibility—frequent realignments of people and responsibilities as the firm attempts to maintain its balance on shifting competitive sands. The daring and the willingness to take "you bet your company" kind of risks is a product of both the inner confidence of its members and a powerful top management—one that either has effective shareholder control or the full support of its board.

[21] See T. Wise, "IBM's 5 Billion Dollar Gamble," *Fortune,* September 1966; "A Rocky Road to the Marketplace," *Fortune,* October 1966.

[22] See A. P. Sloan, *My Years with General Motors* (New York: Anchor Books, 1972), p. 401.

3. Organizational Cohesion

The key to success for a high-tech firm is not simply periodic renewal. There must also be cooperation in the translation of new ideas into new products and processes. As Ken Fisher, the architect of Prime Computer's extraordinary growth, puts it, "If you have the driving function, the most important success factor is the ability to integrate. It's also the most difficult part of the task."[23]

To succeed, the energy and creativity of the whole organization must be tapped. Anything that restricts the flow of ideas, or undermines the trust, respect, and sense of a commonality of purpose among individuals is a potential danger. This is why high-tech firms fight so vigorously against the usual organizational accoutrements of seniority, rank, and functional specialization. Little attention is given to organizational charts: often they don't exist.

Younger people in a rapidly evolving technological field are often as good—and sometimes even better—a source of new ideas as are older ones. In some high-tech firms, in fact, the notion of a "half-life of knowledge" is used; that is, the amount of time that has to elapse before half of what one knows is obsolete. In semiconductor engineering, for example, it is estimated that the half-life of a newly minted Ph.D. is about seven years. Therefore, any practice that relegates younger engineers to secondary, nonpartnership roles is considered counterproductive.

Similarly, product design, marketing, and manufacturing personnel must collaborate in a common cause rather than compete with one another, as happens in many organizations. Any policies that appear to elevate one of these functions above the others—either in prestige or in rewards—can poison the atmosphere for collaboration and cooperation.

[23] Personal communication with Ken Fisher, 1980. Mr. Fisher was president and CEO of Prime Computer from 1975 to 1981.

A source of division, and one which distracts the attention of people from the needs of the firm to their own aggrandizement, are the executive "perks" that are found in many mature organizations: pretentious job titles, separate dining rooms and restrooms for executives, larger and more luxurious offices (often separated in some way from the rest of the organization), and even separate or reserved places in the company parking lot all tend to establish "distance" between managers and doers and substitute artificial goals for the crucial real ones of creating successful new products and customers. The appearance of an executive dining room, in fact, is one of the clearest danger signals.

Good communication

One way to combat the development of such distance is by making top executives more visible and accessible. IBM, for instance, has an open-door policy that encourages managers at different levels of the organization to talk to department heads and vice presidents. According to senior IBM executives, it was not unusual for a project manager to drop in and talk to Frank Cary (IBM's chairman) or John Opel (IBM's president) until Cary's recent retirement. Likewise, an office with transparent walls and no door, such as that of John Young, CEO at HP, encourages communication. In fact, open-style offices are common in many high-tech firms.

A regular feature of 3M's management process is the monthly Technical Forum where technical staff members from the firm exchange views on their respective projects. This emphasis on communication is not restricted to internal operations. Such a firm supports and often sponsors industrywide technical conferences, sabbaticals for staff members, and cooperative projects with technical universities.

Technical Forums serve to compensate partially for the loss of visibility that technologists usually experience when an organization becomes more complex and when production, marketing, and finance staffs swell. So does the concept of the dual-career ladder that is used in most of these firms; that is, a job hierarchy through which technical personnel can attain the status, compensation, and recognition that is accorded to a division general manager or a corporate vice president. By using this strategy, companies try to retain the spirit of the early days of the industry when scientists played a dominant role, often even serving as members of the board of directors.[24]

Again, a strategic business focus contributes to organizational cohesion. Managers of firms that have a strong theme/culture and that concentrate on closely related markets and technologies generally display a sophisticated understanding of their businesses. Someone who understands where the firm is going and why is more likely to be willing to subordinate the interests of his or her own unit or function in the interest of promoting the common goal.

Job rotation

A policy of conscious job rotation also facilitates this sense of communality. In the small firm, everyone is involved in everyone else's job: specialization tends to creep in as size increases and boundary lines between functions appear. If left unchecked, these boundaries can become rigid and impermeable. Rotating managers in temporary assignments across these boundaries helps keep the lines fluid and informal, however. When a new process is developed at TI, for example, the process developers are sent to the production unit where the process will be implemented. They are allowed to return to their usual posts only after that unit's operations manager is convinced that the process is working properly.

---

[24] At Genentech, Cetus, Biogen, and Collaborative Research, four of the leading biotechnology firms, a top scientist is also a member of the board of directors.

Integration of roles

Other ways that high-tech companies try to prevent organizational, and particularly hierarchical, barriers from rising is through multidisciplinary project teams, "special venture groups," and matrixlike organizational structures. Such structures, which require functional specialists and product/market managers to interact in a variety of relatively short-term problem-solving assignments, both inject a certain ambiguity into organizational relationships and require each individual to play a variety of organizational roles.

For example, AT&T uses a combination of organizational and physical mechanisms to promote integration. The Advanced Development sections of Bell Labs are physically located on the sites of the Western Electric plants. This location creates an organizational bond between Development and Bell's basic research and an equally important spatial bond between Development and the manufacturing engineering groups at the plants. In this way, communication is encouraged among Development and the other two groups.[25]

Long-term employment

Long-term employment and intensive training are also important integrative mechanisms. Managers and technologists are more likely to develop satisfactory working relationships if they know they will be harnessed to each other for a good part of their working lives. Moreover, their loyalty and commitment to the firm is increased if they know the firm is continuously investing in upgrading their capabilities.

At Tandem, technologists regularly train administrators on the performance and function of the firm's products and, in turn, administrators train the technologists on personnel policies and financial operations.[26] Such a firm also tends to select college graduates who have excellent academic records, which suggest self-discipline and stability, and then encourages them to stay with the firm for most, if not all, of their careers.

4. Entrepreneurial Culture

While continuously striving to pull the organization together, successful high-tech firms also display fierce activism in promoting internal agents of change. Indeed, it has long been recognized that one of the most important characteristics of a successful high-technology firm is an entrepreneurial culture.[27]

Indeed, the ease with which small entrepreneurial firms innovate has always inspired a mixture of puzzlement and jealousy in larger firms. When new ventures and small firms fail, they usually do so because of capital shortages and managerial errors.[28] Nonetheless, time and again they develop remarkably innovative products, processes, and services with a speed and efficiency that baffle the managers of large companies. The success of the Apple II, which created a new industry, and Genentech's genetically engineered insulin are of this genre. The explanation for a small entrepreneurial firm's innovativeness is straightforward, yet it is difficult for a large firm to replicate its spirit.

Entrepreneurial characteristics

First, the small firm is typically blessed with excellent communication. Its technical people are in continuous contact (and oftentimes in cramped quarters). They have lunch together,

---

[25] See, for example, J. A. Morton, *Organizing for Innovation* (New York: McGraw-Hill, 1971).

[26] Jimmy Treybig, president of Tandem Computer, Stanford Executive Institute Presentation, August 1982.

[27] See: D. A. Schon, *Technology and Change* (New York: Dell Publishing, 1967); Peters and Waterman, *In Search of Excellence.*

[28] See S. Myers and E. F. Sweezy, "Why Innovations Fail," *Technology Review,* March–April 1978, pp. 40–46.

and they call each other outside of working hours. Thus, they come to understand and appreciate the difficulties and challenges facing one another. Sometimes they will change jobs or double up to break a critical bottleneck; often the same person plays multiple roles. This overlapping of responsibilities results in a second blessing: a dissolving of the classic organizational barriers that are major impediments to the innovating process. Third, key decisions can be made immediately by the people who first recognize a problem, not later by top management or by someone who barely understands the issue. Fourth, the concentration of power in the leader/entrepreneurs makes it possible to deploy the firm's resources very rapidly. Lastly, the small firm has access to multiple funding channels, from the family dentist to a formal public offering. In contrast, the manager of an R&D project in a large firm has effectively only one source, the "corporate bank."

Small divisions

In order to recreate the entrepreneurial climate of the small firm, successful large high-technology firms often employ a variety of organizational devices and personnel policies. First, they divide and subdivide. Hewlett-Packard, for example, is subdivided into 50 divisions: the company has a policy of splitting divisions soon after they exceed 1,000 employees. Texas Instruments is subdivided into over 30 divisions and 250 "tactical action programs." Until recently, 3M's business was split into 40 divisions. Although these divisions sometimes reach $100 million or more in sales, by *Fortune* 500 standards they are still relatively small companies.

Variety of funding channels

Second, such high-tech firms employ a variety of funding channels to encourage risk taking. At Texas Instruments, managers have three distinct options in funding a new R&D project. If their proposal is rejected by the centralized Strategic Planning (OST) system because it is not expected to yield acceptable economic gains, they can seek a "Wild Hare Grant." The Wild Hare program was instituted by Patrick Haggerty, while he was TI's chairman, to insure that good ideas with long-term potential were not systematically turned down. Alternatively, if the project is outside the mainstream of the OST system, managers or engineers can contact one of dozens of individuals who hold "IDEA" grant purse strings and who can authorize up to $25,000 for prototype development. It was an IDEA grant that resulted in TI's highly successful Speak and Spell learning aid.

3M managers also have three choices: they can request funds from (1) their own division, (2) corporate R&D, or (3) the new ventures division.[29] This willingness to allow a variety of funding channels has an important consequence: it encourages the pursuit of alternative technological approaches, particularly during the early stages of a technology's development, when no one can be sure of the best course to follow.

IBM, for instance, has found that rebellion can be good business. Thomas Watson, Jr., the founder's son and a longtime senior manager, once described the way the disk memory, a core element of modern computers, was developed:

> [It was] not the logical outcome of a decision made by IBM management; [because of budget difficulties] it was developed in one of our laboratories as a bootleg project. A handful of men

---

[29] See: *Texas Instruments* (A), 9-476-122, Harvard Business School case; *Texas Instruments Shows U.S. Business How to Survive in the 1980s,* 3-579-092, Harvard Business School case; *Texas Instruments "Speak and Spell Product,"* 9-679-089, revised 7/79, Harvard Business School case.

> . . . broke the rules. They risked their jobs to work on a project they believed in.[30]

At Northrop, the head of aircraft design usually has at any one time several projects in progress without the awareness of top management. A lot can happen before the decision reaches even a couple of levels below the chairman. "We like it that way," explains Northrop Chairman Tom Jones.[31]

Tolerance of failure

Moreover, the successful high-technology firms tend to be very tolerant of technological failure. "At HP," Bob Hungate, general manager of the Medical Supplies Division, explains, "it's understood that when you try something new you will sometimes fail."[32] Similarly, at 3M, those who fail to turn their pet project into a commercial success almost always get another chance. Richard Frankel, the president of the Kevex Corporation, a $20 million instrument manufacturer, puts it this way, "You need to encourage people to make mistakes. You have to let them fly in spite of aerodynamic limitations."[33]

Opportunity to pursue outside projects

Finally, these firms provide ample time to pursue speculative projects. Typically, as much as 20 percent of a productive scientist's or engineer's time is "unprogrammed," during which he or she is free to pursue interests that may not lie in the mainstream of the firm. IBM Technical Fellows are given up to five years to work on projects of their own choosing, from high-speed memories to astronomy.

5. Sense of Integrity

While committed to individualism and entrepreneurship, at the same time successful high-tech firms tend to exhibit a commitment to long-term relationships. The firms view themselves as part of an enduring community that includes employees, stockholders, customers, suppliers, and local communities: their objective is to maintain stable associations with all of these interest groups.

Although these firms have clear-cut business objectives, such as growth, profits, and market share, they consider them subordinate to higher order ethical values. Honesty, fairness, and openness—that is, integrity—are not to be sacrificed for short-term gain. Such companies don't knowingly promise what they can't deliver to customers, stockholders, or employees. They don't misrepresent company plans and performance. They tend to be tough but forthright competitors. As Herb Dwight—president of Spectra-Physics, one of the world's leading laser manufacturers—says, "The managers that succeed here go *out of their way* to be ethical."[34] And Alexander d'Arbeloff, cofounder and president of Teradyne, states bluntly, "Integrity comes first. If you don't have that, nothing else matters."[35]

These policies may seem utopian, even puritanical, but in a high-tech firm they also make good business sense. Technological change can be dazzlingly rapid; therefore, uncertainty is high, risks are difficult to assess, and market opportunities and profits are hard to predict. It

---

[30] Thomas Watson, Jr. Address to the Eighth International Congress of Accountants, New York City, September 24, 1962, as quoted by D. A. Shon, "Champions for Radical New Inventions," *Harvard Business Review,* March–April 1963, p. 85.

[31] Personal communication with Tom Jones, chairman of the board, Northrop Corporation, May 1982.

[32] Personal communication with Bob Hungate, general manager, Medical Supplies Division, Hewlett-Packard, 1980.

[33] Personal communication with Richard Frankel, president, Kevex Corporation, April 1983.

[34] Personal communication with Herb Dwight, president and CEO, Spectra-Physics, 1982.

[35] Personal communication with Alexander d'Arbeloff, cofounder and president of Teradyne, 1983.

is almost impossible to get a complex product into production, for example, without solid trust between functions, between workers and managers, and between managers and stockholders (who must be willing to see the company through the possible dips in sales growth and earnings that often accompany major technological shifts). Without integrity, the risks multiply and the probability of failure (in an already difficult enterprise) rises unacceptably. In such a context, Ray Stata, cofounder of the Massachusetts High-Technology Council, states categorically, "You need an environment of mutual trust."[36]

This commitment to ethical values must start at the top, otherwise it is ineffective. Most of the CEOs we interviewed consider it to be a cardinal dimension of their role. As Bernie Gordon, president of Analogic, explains, "The things that make leaders are their philosophy, ethics, and psychology."[37] Nowhere is this dimension more important than in dealing with the company's employees. Paul Rizzo, IBM's vice chairman, puts it this way, "At IBM we have a fundamental respect for the individual . . . people must be free to disagree and to be heard. Then, even if they lose, you can still marshall them behind you."[38]

Self-understanding

This sense of integrity manifests itself in a second, not unrelated, way—self-understanding. The pride, almost arrogance, of these firms in their ability to compete in their chosen fields is tempered by a surprising acknowledgment of their limitations. One has only to read Hewlett-Packard's corporate objectives or interview one of its top managers to sense this extraordinary blend of strength and humility. Successful high-tech companies are able to reconcile their "dream" with what they can realistically achieve. This is one of the reasons why they are extremely reticent to diversify into unknown territories.

6. "Hands-On" Top Management

Notwithstanding their deep sense of respect and trust for individuals, CEOs of successful high-technology firms are usually actively involved in the innovation process to such an extent that they are sometimes accused of meddling. Tom McAvoy, Corning's president, sifts through hundreds of project proposals each year trying to identify those that can have a "significant strategic impact on the company"—the potential to restructure the company's business. Not surprisingly, most of these projects deal with new technologies. For one or two of the most salient ones, he adopts the role of "field general": he frequently visits the line operations, receives direct updates from those working on the project, and assures himself that the required resources are being provided.[39]

Such direct involvement of the top executive at Corning sounds more characteristic of vibrant entrepreneurial firms, such as Tandon, Activision, and Seagate, but Corning is far from unique. Similar patterns can be identified in many larger high-technology firms. Milt Greenberg, president of GCA, a $180 million semiconductor process equipment manufacturer, stated: "Sometimes you just have to short-circuit the organization to achieve major change."[40] Tom Watson, Jr., (IBM's chairman) and Vince Learson (IBM's president) were doing just that when they met with programmers and designers and other executives in Watson's ski cabin in Vermont to finalize software design

[36] Personal communication with Ray Stata, president and CEO, Analog Devices, 1980.

[37] Personal communication with Bernie Gordon, president and CEO, Analogic, 1982.

[38] Personal communication with Paul Rizzo, 1980.

[39] Personal communication with Tom McAvoy, president of Corning Glass, 1979.

[40] Personal communication with Milt Greenberg, president of GCA, 1980.

concepts for the System 360—at a point in time when IBM was already a $4 billion firm.[41]

Good high-tech managers not only understand how organizations, and in particular engineers, work, they understand the fundamentals of their technology and can interact directly with their people about it. This does not imply that it is necessary for the senior managers of such firms to be technologists (although they usually are in the early stages of growth): neither Watson nor Learson were technical people. What appears to be more important is the ability to ask lots of questions, even "dumb" questions, and dogged patience in order to understand in-depth such core questions as: (1) how the technology works; (2) its limits, as well as its potential (together with the limits and potential of competitors' technologies); (3) what these various technologies require in terms of technical and economic resources; (4) the direction and speed of change; and (5) the available technological options, their cost, probability of failure, and potential benefits if they prove successful.

This depth of understanding is difficult enough to achieve for one set of related technologies and markets; it is virtually impossible for one person to master many different sets. This is another reason why business focus appears to be so important in high-tech firms. It matters little if one or more perceptive scientists or technologists foresees the impact of new technologies on the firm's markets, if its top management doesn't internalize these risks and make the major changes in organization and resource allocation that are usually necessitated by a technological transition.

THE PARADOX OF HIGH-TECHNOLOGY MANAGEMENT

The six themes around which we arranged our findings can be organized into two, apparently paradoxical groupings: business focus, organizational cohesion, and a sense of integrity fall into one group; adaptability, entrepreneurial culture, and hands-on management fall into the other group. On the one hand, business focus, organizational cohesion, and integrity imply stability and conservatism. On the other hand, adaptability, entrepreneurial culture, and hands-on top management are synonymous with rapid, sometimes precipitous change. The fundamental tension is between order and disorder. Half of the success factors pull in one direction; the other half tug the other way.

This paradox has frustrated many academicians who seek to identify rational processes and stable cause-effect relationships in high-tech firms and managers. Such relationships are not easily observable unless a certain constancy exists. But in most high-tech firms, the only constant is continual change. As one insightful student of the innovation process phrased it, "Advanced technology requires the collaboration of diverse professions and organizations, often with ambiguous or highly interdependent jurisdictions. In such situations, many of our highly touted rational management techniques break down."[42] One recent researcher, however, proposed a new model of the firm that attempts to rationalize the conflict between stability and change by splitting the strategic process into two loops, one that extends the past, the other that periodically attempts to break with it.[43]

Established organizations are, by their very nature, innovation resisting. By defining jobs and responsibilities and arranging them in serial reporting relationships, organizations en-

[41] See Wise, "IBM's 5 Billion."

[42] See L. R. Sayles and M. K. Chandler, *Managing Large Systems: Organizations for the Future* (New York: Harper and Row, 1971).

[43] See R. A. Burgelman, "A Model of the Interaction of Strategic Behavior, Corporate Context, and the Concept of Corporate Strategy," *Academy of Management Review* (1983), pp. 61–70.

courage the performance of a restricted set of tasks in a programmed, predictable way. Not only do formal organizations resist innovation, they often act in ways that stamp it out. Overcoming such behavior—which is analogous to the way the human body mobilizes antibodies to attack foreign cells—is, therefore, a core job of high-tech management.

The Paradoxical Challenge

High-tech firms deal with this challenge in different ways. Texas Instruments, long renowned for the complex, interdependent matrix structure it used in managing dozens of product-customer centers (PCCs), recently consolidated groups of PCCs and made them into more autonomous units. "The manager of a PCC controls the resources and operations for his entire family . . . in the simplest terms, the PCC manager is to be an entrepreneur," explained Fred Bucy, TI's president.[44]

Meanwhile, a different trend is evident at 3M, where entrepreneurs have been given a free rein for decades. A recent major reorganization was designed to arrest snowballing diversity by concentrating its sprawling structure of autonomous divisions into four market groups. "We were becoming too fragmented," explained Vincent Ruane, vice president of 3M's Electronics Division.[45]

Similarly, HP recently reorganized into five groups, each with its own strategic responsibilities. Although this simply changes some of its reporting relationships, it does give HP, for the first time, a means for integrating product and market development across generally autonomous units.[46]

These reorganizations do not mean that organizational integration is dead at Texas Instruments, or that 3M's and HP's entrepreneurial cultures are being dismantled. They signify first, that these firms recognize that both (organizational integration and entrepreneurial cultures) are important, and second, that periodic change is required for environmental adaptability. These three firms are demonstrating remarkable adaptability by reorganizing from a position of relative strength—not, as is far more common, in response to financial difficulties. As Lewis Lehr, 3M's president, explained, "We can change now because we're not in trouble."[47]

Such reversals are essentially antibureaucratic, in the same spirit as Mao's admonition to "let a hundred flowers blossom and a hundred schools of thought contend."[48] At IBM, in 1963, Tom Watson, Jr., temporarily abolished the corporate management committee in an attempt to push decisions downward and thus facilitate the changes necessary for IBM's great leap forward to the System 360.[49] Disorder, slack, and ambiguity are necessary for innovation, since they provide the porosity that facilitates entrepreneurial behavior—just as do geographically separated, relatively autonomous organizational subunits.

But the corporate management committee is alive and well at IBM today. As it should be. The process of innovation, once begun, is both self-perpetuating and potentially self-destructive: although the top managers of high-tech firms must sometimes espouse organizational disorder, for the most part they must preserve order.

---

[44] See S. Zipper, "TI Unscrambling Matrix Management to Cope with Gridlock in Major Profit Centers," *Electronic News,* April 26, 1982, p. 1.

[45] See M. Barnfather, "Can 3M Find Happiness in the 1980s?" *Forbes,* March 11, 1982, pp. 113–16.

[46] See R. Hill, "Does a 'Hands-Off' Company Now Need a 'Hands-On' Style?" *International Management,* July 1983, p. 35.

[47] See Barnfather, "Can 3M Find Happiness?"

[48] S. R. Schram, ed., *Quotations from Chairman Mao Tse Tung* (Bantam Books, 1967), p. 174.

[49] See: D. G. Marquis, "Ways of Organizing Projects," *Innovation,* August 1969, pp. 26–33; T. Levitt, *Marketing for Business Growth* (New York: McGraw-Hill, 1974), in particular, ch. 7.

Winnowing Old Products

Not all new product ideas can be pursued. As Charles Ames, former president of Reliance Electric, states, "An enthusiastic inventor is a menace to practical businessmen."[50] Older products, upon which the current success of the firm was built, at some point have to be abandoned: just as the long-term success of the firm requires the planting and nurturing of new products, it also requires the conscious, even ruthless, pruning of other products so that the resources they consume can be used elsewhere.

This attitude demands hard-nosed managers who are continually managing the functional and divisional interfaces of their firms. They cannot be swayed by nostalgia, or by the fear of disappointing the many committed people who are involved in the development and production of discontinued products. They must also overcome the natural resistance of their subordinates, and even their peers, who often have a vested interest in the products that brought them early personal success in the organization.

Yet, firms also need a certain amount of continuity because major change often emerges from the accretion of a number of smaller, less visible improvements. Studies of petroleum refining, rayon, and rail transportation, for example, show that half or more of the productivity gains ultimately achieved within these technologies were the result of the accumulation of minor improvements.[51] Indeed, most engineers, managers, technologists, and manufacturing and marketing specialists work on what Thomas Kuhn might have called "normal innovations,"[52] the little steps that improve or extend existing product lines and processes.

Managing Ambivalently

The successful high-technology firm, then, must be managed ambivalently. A steady commitment to order and organization will produce one-color Model T Fords. Continuous revolution will bar incremental productivity gains. Many companies have found that alternating periods of relaxation and control appear to meet this dual need. Surprisingly, such ambiguity does not necessarily lead to frustration and discontent.[53] In fact, interspersing periods of tension, action, and excitement with periods of reflection, evaluation, and revitalization is the same sort of irregular rhythm that characterizes many favorite pastimes—including sailing, which has been described as "long periods of total boredom punctuated with moments of stark terror."

Knowing when and where to change from one stance to the other, and having the power to make the shift, is the core of the art of high-technology management. James E. Webb, administrator of the National Aeronautics and Space Administration during the successful Apollo ("man on the moon") program, recalled that "we were required to fly our administrative machine in a turbulent environment, and . . . a certain level of *organizational instability was essential if NASA was not to lose control*" (authors' emphasis).[54]

---

[50] Charles Ames, former CEO of Reliance Electric, as quoted in "Exxon's $600 Million Mistake," *Fortune,* October 19, 1981.

[51] See, for example, W. J. Abernathy and J. M. Utterback, "Patterns of Industrial Innovation," *Technology Review,* June–July 1978, pp. 40–47.

[52] See T. Kuhn, *The Structure of Scientific Revolutions,* 2d ed. (Chicago: University of Chicago Press, 1967).

[53] After reviewing an early draft of this article, Ray Stata wrote, "The articulation of dynamic balance, of ying and yang, . . . served as a reminder to me that there isn't one way forever, but a constant adaption to the needs and circumstances of the moment." Ray Stata, president, Analog Devices, letter of November 29, 1982.

[54] Quoted in "Some Contributions of James E. Webb to the Theory and Practice of Management," a presentation by Elmer B. Staats before the annual meeting of the Academy of Management on August 11, 1978.

In summary, the central dilemma of the high-technology firm is that it must succeed in managing two conflicting trends: continuity and rapid change. There are two ways to resolve this dilemma. One is an old idea: managing different parts of the firm differently—some business units for innovation, others for efficiency.

A second way—a way which we believe is more powerful and pervasive—is to manage differently at different times in the evolutionary cycle of the firm. The successful high-technology firm *alternates* periods of consolidation and continuity with sharp reorientations that can lead to dramatic changes in the firm's strategies, structure, controls, and distribution of power, followed by a period of consolidation.[55] Thomas Jefferson knew this secret when he wrote 200 years ago, "A little revolution now and then is a good thing."[56]

[55] See Romanelli and Tushman, "Executive Leadership."

[56] See J. Bartlett, *Bartlett's Familiar Quotations,* 14th ed. (Boston: Little, Brown), p. 471B.

The Future of Entrepreneurship in Established Firms

Case IV–11 Control Data Corporation

M. van den Poel and R. A. Burgelman

Within Control Data Corporation there is a deep commitment to promote entrepreneurship at every level inside and outside. It is part of the culture that has been generated within the company, and it is also based on the knowledge that in a high-technology business, when you lose your creativity, you have lost the company. . . . As a result, the entrepreneurial spirit has always been highly prized and promoted within the company, to the point where some (outside) analysts complain that we are constantly moving away from our bread-and-butter businesses into risky and unusual projects.[1]

Started as a small computer company in 1957, Control Data had by 1983 become a $4.6 billion company of mainframe computers, computer services, computer peripherals, and financial services. One of the few companies to diversify away from mainframes during the 60s, CDC had also invested early on in some unusual projects such as computer-based education programs (Plato-project), launched a number of ventures to create agricultural and inner-city jobs, set up some 150 business and technology centers to stimulate new businesses, and established a CDC technology institute, all guided by CDC basic business strategy "to apply the problem-solving capabilities of its computer and its financial and human resources to those markets that have evolved from society's major unmet needs." Guided by this philosophy, CDC had also over the years facilitated a number of spin-offs in high-technology fields and had recently set up an office to advise employees who wanted to start their own businesses. At the same time, the company was looking for ways to promote innovation within the company—or "intrapreneurship," as CDC called it.

[1] John Pfouts, vice president CDC, Stanford Business School, March 12, 1984.

HISTORICAL OVERVIEW AND RECENT DEVELOPMENTS

After working as a salesman for Westinghouse and later as an engineer for the U.S. Navy, William Norris, a Nebraskan, started in 1946 a small company called Engineering Research Associates (ERA) to exploit an embryonic computer technology. ERA got off the ground with some Navy contracts and a few computer experts (among them Seymour Cray). Remington Rand acquired ERA in 1952 and then merged with Sperry Rand, a company that had centered its computer operations division, Univac, in Minneapolis. In 1957, William Norris, then vice president and general manager of Univac, and some associates spun off of Sperry Univac and set up a new company, Control Data Corporation, to build very large and fast computers. As the *Scientific American* notes:

> William Norris has had a profound impact on Minnesota in two ways. His Engineering Research Associates and its successor companies have spawned dozens of new technology companies. Since 1945, most of them have been set up by former associates or employees of ERA, Sperry Rand, Univac, Control Data, and generations beyond. A recent count showed some 40 companies that can trace their lineage directly to ERA. Seventeen alone were born out of Control Data, itself one of the 40.[2] (Exhibit 1)

In the early 60s, the Control Data group produced the world's largest and fastest computer, the CDC 6600. Ordered by the U.S. Atomic Energy Commission, this computer staggered the industry and forced IBM to revamp its own design and manufacturing to try to assure its supremacy. But Control Data had wider interests than large mainframe computers, and in the late 60s Chairman Norris decided to diversify the company away from its dependence on mainframe computers, making it the first mainframe maker to recognize the futility of competing head-on with IBM.[3]

Control Data diversified into financial services, computer peripherals, and computer services. The strategy seemed successful during the late 60s and the 70s: consolidated revenues tripled between 1973 and 1983, and pretax earnings more than doubled during 1973 and 1981 but were declining in recent years. (Exhibit 2)

By 1983, financial services accounted for 23 percent of total consolidated revenues, while the computer business accounted for 77 percent. Within the computer business, mainframe computers—CDC's original product—only represented around 22 percent of the computer business revenues by 1983 (compared to 35 percent in 1975), while computer peripherals went from 26 percent in 1975 to close to 40 percent in 1982 and computer services remained relatively stable, representing 38 to 40 percent of total computer business between 1975 and 1982. (Exhibit 3) Each business had its own set of problems, however.

Control Data acquired Commercial Credit Corporation in 1968, a *financial services* concern that provided business credit insurance, life insurance, casualty insurances, and financing and leasing services. A reliable and large source of income during the 70s, Commercial Credit was plagued by a number of problems in recent years: its property-casual insurance business was losing money, and the 125 business and technology centers—a favorite Norris project—run by Commercial Credit and set up to help small businesses by selling services and computer hardware to them, were plagued by large losses. In November of 1984, Control Data announced it would sell the Commercial Credit Corporation.[4]

The diversification into *computer peripherals*

---

[2] Peter J. Brennan, "Minnesota . . . Technology Wellspring," *Scientific American,* October 1980.

[3] *Business Week,* October 17, 1983.

[4] *San Jose Mercury News,* October 28, 1984.

EXHIBIT 1 Control Data

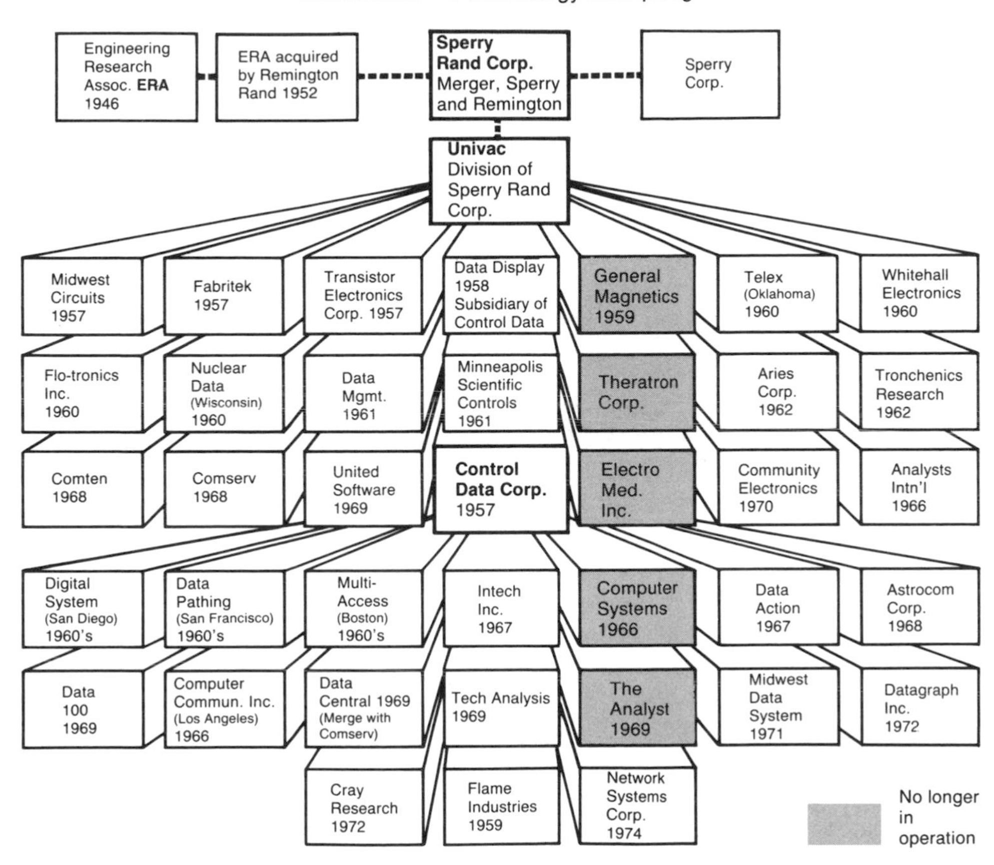

Examples of technical company formations and spin-offs originating with Engineering Research Associates (ERA) in 1946. (All are Twin Cities companies except as noted.)

SOURCE: *Scientific American*, October 1980.

in the late 60s had been a successful move. CDC had become the dominant independent producer of large disk drives for minicomputers and mainframes sold to the OEM (including NCR, Honeywell, and Burroughs). This became the company's traditional cash cow, representing $1.4 billion in 1983. In computer peripherals, the need to expand into new markets, as well as high-technology development costs, had brought CDC to embark on a large number of joint ventures and partnerships. Among these ventures were Computer Peripherals (with NCR and ICL), Magnetic Peripherals (with Honeywell and Sperry Rand), and Optical Peripherals Laboratory (with Philips).

Cooperative ventures became the company's policy: "Over the next three years, CDC's technical expenditures alone will be in excess of $1

EXHIBIT 2 Control Data

	1984	1983	1982	1981	1980	1979	1978	1977	1976	1975	1974	1973
Consolidated revenues (billions):	5.0	4.6	4.3	4.1	3.8	3.2	2.7	2.3	2.0	1.9	1.8	1.5
Percent computer business	75	77	76	75	73	70	68	66	68	65	61	63
Percent financial service	25	23	24	25	27	30	32	34	32	35	39	37
Pretax earnings (millions):	31 (net)	214	229	295	264	210	167	117	92	78	3	120
Percent computer business	N.A.	72	62	66	62	55	53	40	29	36	48	35
Percent financial service	N.A.	28	38	34	38	45	47	60	71	64	52	65
Employees (thousands)	54,000	55,858	56,005	60,627	57,068	57,182	51,430	46,197	42,696			
Earnings per share of common stock	0.81	4.20	4.11	4.48	4.23	3.60	2.59	1.83				
Stockprice (market):												
High	40.75	63.3	42.8	42.1	38.8	28.5	22.3	14.6	13.6	11.8	19.6	31.0
Low	24.30	35.5	21.1	29.8	21.5	14.4	11.4	9.7	8.8	5.3	4.8	15.2

SOURCE: Control Data annual reports; *Value Line; Business Week,* October 22, 1984.

EXHIBIT 3 Control Data

Change in Computer Business Revenue

	1983	1981	1979	1977	1975
Computer systems	22%	20%	21%	26%	35%
Computer peripherals	39%	42%	40%	33%	26%
Computer services	39%	38%	39%	41%	39%
Computer business revenues (billions)	3.5	3.1	2.2	1.5	1.2

SOURCE: Control Data annual reports.

billion," said Robert Price, president and COO, "so any part of that which you can share simply frees up resources to do things unique to CDC." Price estimated that, in general, cooperation could increase the effect of technology spending by 30 percent. "It's going to be necessary to cooperate. That is one thing the Japanese have done for us."[5]

Among a large number of joint ventures (Exhibit 4), CDC had in 1982 also embarked on a partnership with Burroughs Corporation's Memorex subsidiary to build IBM 3380-type disk drive products. Recently, though, product delays and technical problems with its large IBM-compatible disk drives forced CDC to give up that business, which represented estimated sales of $100 million. CDC's peripherals then concentrated on building hard, high-capacity disk drives (8 inch and 14 inch) for which CDC is still the dominant manufacturer, although, even in that segment, Japan's Hitachi, NEC, Fujitsu, and the United States' Quantum were moving in fast.[6]

CDC also plunged heavily into *computer services* after acquiring IBM's Service Bureau Corporation in 1973, again a move which initially paid off. A number of its operations were highly profitable, such as Arbitron Ratings, Inc., a television and radio audience-measuring service, and Ticketron, Inc., which sold tickets for lotteries, sporting, and entertainment events. But CDC's more conventional computer time-sharing services, delivered from remote mainframe computers via the telephone, were floundering, largely because of increased competition from low-priced microcomputers.[7]

The Plato Project

Within the Computer Services Division, the "Plato project" constituted an interesting venture. Started in the early 1960s as a computer-based teaching and education system, it was Bill Norris's pet project. Plato covered education and industrial training programs, farm management courses, and medical services. A very long-term investment—an estimated $900 million was poured into it over the years—Plato became profitable only in 1983, on sales of $125 million, mostly due to its large-scale industrial training programs with big clients, e.g., General Motors, United Airlines, Du Pont. Plato was originally developed when mainframes with central computer processing dominated the market and over 12,000 hours of training courses were being run on mainframes. With the advent of self-contained microcomputers, Plato faced problems and compromised by offering central services to minicomputers and converting some 2,000 hours of training courses to microcomputers, especially those used for secondary school education. The secondary school market for computers, however, was in 1983 already covered by Apple and Tandy computers, who together held more than 50 percent of that market. But Norris believed in the Plato project: "Fifteen years from now," he said, "computer-based education will be the largest source of revenues and profits. . . . I don't think in terms of 10 years. I think in terms of 20 years . . . addressing basic needs in society (such as the Plato project) takes longer to get a return . . . otherwise, hell, everybody would be out there if it were easy. . . ."[8] Very recently, a new CDC subsidiary was formed to sell computer-based education to schools, the most prized and least tapped market for CDC's Plato system.

Mainframe Computers

Finally, Control Data's original core business, *mainframe computers,* severely diminished in importance over the years. By 1983, it

---

[5] *Electronics,* June 16, 1983.

[6] *Business Week,* October 22, 1984.

[7] Ibid.

[8] *Business Week,* October 22, 1984, and *The Wall Street Journal,* November 12, 1982.

EXHIBIT 4 Control Data

Control Data: Joint Ventures or Partnerships

	Present Participants and Percentage Owned	Product Category	Formation Date	Headquarters Location	Activities
Computer Peripherals, Inc.	Control Data Corporation, 60% NCR Corporation, 20% ICL plc, 20%	Tape drives	May 1972	Minneapolis	Manufacturing and development
Magnetic Peripherals, Inc.	CDC, 67% Honeywell, Inc., 17% CII-Honeywell Bull, 3% Sperry Corporation, 13%	Rotating-disk memories	April 1975	Minneapolis	Manufacturing and development
Optical Media Laboratory	NV Philips Gloeilampenfabrieken, 52% CDC, 48%	Optical media	February 1982	Eindhoven, the Netherlands	Development
Optical Peripherals Laboratory	Magnetic Peripherals, 52% Philips Optical Storage Corporation, 48%	Optical-memory drives	February 1982	Colorado Springs	Development
Centronics Data Computer Corporation	CDC, 35% NCR, 4.5% ICL, 4.5% public, 56%	Printers	June 1982	Hudson, N.H.	Manufacturing, development, marketing
Disk Media, Inc.	Memorex/Burroughs, 60% CDC, 40%	3380-type disk media	September 1982	Santa Clara, California	Development
Peripheral Components, Inc.	Magnetic Peripherals, 60% Memorex/Burroughs, 40%	3380-type thin-film heads	September 1982	Minneapolis	Development

SOURCE: *Electronics*, June 16, 1983.

represented only 16 percent of total revenues and 22 percent of the computer business. Originally addressing the market for large, medium, *and* small computers, CDC seemed to concentrate during the 70s on its standard large-scale systems with the successful Cyber 170 series, mostly for the atomic and nuclear industry and the electric utilities and petroleum industries. Development of custom supercomputers suffered some setbacks in the mid-70s (STAR-system), but renewed commitment in the late 70s (with the testing of the Cyber 203 supercomputers) brought to the market in 1981 the new Cyber 205 supercomputer, which represented a reasonable profit potential for the division in the early 80s.

A number of spin-offs in the Computer Systems Division had occurred over the years. Seymour Cray, Bill Norris's close associate and cofounder of Control Data, had decided to resign in 1972 to start his own company, Cray Research, and to develop supercomputers. Cray had been the leading designer of superscale computers in the industry. With CDC, he had developed the first superscale CDC 6600 in the early 60s, and then, subsequently, the 7600 series when he was senior vice president of CDC's Chippewa Laboratories. Then, in 1974, Jim Thornton, the chief engineer of Chippewa Laboratories under S. Cray, had also left CDC to start Network Systems.

In an effort to accelerate the development of new supercomputers, Control Data had also in August 1983 decided to partially "spin out" its next generation Cyber 205 compatible supercomputer into a new venture called ETA Systems, with a CDC equity of 40 percent.

An Important Spin-Off: Cray Research

On March 14, 1972, William C. Norris, chairman of the board and president of Control Data Corporation, and Seymour R. Cray, senior vice president of the company's Chippewa Laboratories, jointly announced that Mr. Cray would be phasing out of a full-time position with Control Data over the next year.

Mr. Cray, who had been with Control Data from its early days and had been the leading designer of superscale computers in the industry, would continue to assist the company on its 8000 computer series. Beyond that, he would serve on a consulting basis.

In making the announcement, Cray issued the following statement:

> I would not be making this change if I did not believe that Control Data's technical strengths in supercomputers are firmly established as a base on which to build for the future. Therefore, I cannot foresee any better time than the present for me to phase out my full-time responsibilities at Control Data. Although Bill Norris and the company have given me unusual support and freedom, I have long believed that the things I want to do and feel most comfortable in doing can best be done outside the environment of a large company. I am contemplating the establishment of a very small basic computer research laboratory funded by myself and perhaps a few personal friends.[9]

"We are naturally very happy that we can continue to benefit from Seymour's services on a consulting basis in the future," Norris said. "Although we are sorry to alter an arrangement that has been so highly productive, we understand Seymour's desires to pursue his personal interests and we also must recognize the need for change."[10]

Apparently it was CDC's refusal to fund Seymour's proposal to develop a supercomputer faster than the 7600, coupled with the seemingly inevitable increase of his corporate duties as senior vice president, that made Seymour Cray move on. Cray left CDC with the company's blessings and with four of his colleagues to start his own company. Cray Research, Inc., was funded in 1972 with $500,000 of Seymour's

---

[9] *Control Data News,* March 14, 1972.

[10] Ibid.

own money, $500,000 from CDC's Commercial Credit Corporation, and $1.5 million from 14 other investors.[11] Seymour Cray noted: "The purpose of my new company is to design and build a larger, more powerful computer than anyone now has."[12]

Ten years later—1982—Cray Research, Inc., had become a $140 million company with over 1,300 employees. (Exhibit 5) The company eyed a very limited market of some 400 customers (governments, universities' research labs, and occasional large corporations) for their custom supercomputers, which ranged in price from $4 million to $14 million. As *Scientific American* noted: "There are not many companies having only 17 sales in 10 years that one could consider successful. One that is is Cray Research. In a sense, Cray is the keeper of the faith since the new company was set up exclusively to do what Control Data set out to do but somewhere along the way got sidetracked—build super-big, super-fast computers."[13]

Rather than trying to be all things to all people, Cray decided to concentrate on a market so small (estimated at a total of 80 users worldwide in 1976) and so demanding that competitors could not easily break into it. The strategy paid off. From a $1.6 million loss in 1976 on a mere $1 million in sales, Cray, Inc., showed a $19 million profit in 1982 on $140 million in sales. By 1984, Cray had installed 88 supercomputers around the world, controlling nearly two thirds of the market. The market was estimated to grow from $260 million a year in 1984 to over $1 billion a year by 1990.[14]

Cray had introduced its first supercomputer, the Cray-1, which was five times faster than the CDC 7600, during 1976. With the introduction of its first machine, Cray quickly dominated the narrow supercomputer market niche. The Cray-1 essentially knocked CDC out of its dominant position. According to a former CDC salesman, few, if any, CDC 7600s were sold after the Cray-1 introduction. CDC introduced its more powerful Cyber 205 in 1981, but by that time 35 Cray-1 systems had been installed.[15]

Continuously segmenting the supercomputer market, Cray then introduced in 1982 the Cray 1-M (for metal oxide silicon memory), a machine with the same capacity as the original Cray-1 but, thanks to the MOS memory, half the price and half the cost, putting the 1-M for the first time in reach of the large corporate customer.[16] At the same time, however, Cray introduced a more powerful supercomputer, the "X-MP," providing two to five times the computing power of its predecessor, the Cray-1. (Exhibit 6) An industry observer notes that "The X-MP widens the distance between Cray's products and those of his original employer, Control Data, which has begun to move into Cray's market niche (with the CDC 205)."[17]

To concentrate fully on the supercomputer development, Seymour Cray had in 1977 passed on his duties as CEO and president to John Rollwagen. He had protected his independence further in 1981 when he became an independent consultant, under contract with Cray Research until 1987. As *Fortune* pointed out:

> Seymour Cray buys his own equipment, hires his own employees, and shares the rights to his results with Cray Research. Rollwagen finds nothing strange about the relationship and says he feels more secure today than he did when Cray was an employee. "Now, at least, we have him under contract until 1987," he says.[18]

[11] Francis J. Aguilar and Caroline Brainard, "Cray Research, Inc.," Boston: Harvard Business School, case 4-385-011 (rev. 12/84), p. 4. Copyright © 1984 by the President and Fellows of Harvard College.

[12] From *Chippewa Falls Herald Telegram,* May 1972.

[13] *Scientific American,* October 1980.

[14] *Fortune,* March 18, 1985.

[15] See note 11.

[16] *Forbes,* September 12, 1983.

[17] Ibid.

[18] *Fortune,* March 18, 1985.

EXHIBIT 5 Cray Research, Inc.

Financial Summary (dollars in thousands)

	1972*	1973	1974	1975	1976	1977	1978	1979	1980	1981	1982	1983
Revenue:												
Sales	—	—	—	—	—	$ 8,816	$ 8,357	$26,496	$37,645	$ 65,207	$ 91,535	$125,008
Leased systems	—	—	—	—	450	2,261	7,349	12,545	17,558	27,604	35,623	26,151
Service fees	—	—	—	—	59	317	1,471	3,674	5,545	8,831	13,991	18,531
Total revenue	$	$	$	$	$ 509	$11,394	$17,177	$42,715	$60,748	$101,642	$141,149	$169,690
Net income (loss)	$ (72)	$ (527)	$ (944)	$ (887)	$ (1,551)	$ 2,027	$ 3,501	$ 7,819	$10,900	$ 18,170	$ 19,000	$ 26,071
U.S. regions:												
Revenue	—	—	—	—	—	—	$14,646	$14,293	$43,784	$ 90,852	$110,486	$ 87,443
Operating profit	—	—	—	—	—	—	7,448	3,259	20,225	49,926	51,191	37,772
International subsidiaries:												
Revenue	—	—	—	—	—	—	2,531	28,422	16,964	10,790	30,663	82,247
Operating profit	—	—	—	—	—	—	1,042	19,817	10,280	2,482	15,531	39,383
Working capital	$2,375	$1,783	$3,314	$5,756	$10,786	$ 9,083	$ 7,498	$13,949	$53,543	$ 61,927	$ 98,600	$129,436
Long-term debt	—	—	—	2,720	2,720	3,321	4,670	3,029	7,876	12,360	33,741	37,612
Stockholders' equity	2,478	1,961	3,664	3,434	12,054	14,636	20,638	28,623	75,803	94,951	144,561	172,385
Operating and financial ratios:												
Return on stockholder's average equity	—	—	—	—	—	15.2%	19.9%	31.7%	25.6%	22.0%	18.6%	17.0%
Net income as percent of revenue	—	—	—	—	—	17.8%	20.4%	18.3%	17.9%	17.9%	13.5%	15.4%
Current ratio	—	21.3 : 1	19.8 : 1	65.7 : 1	18.3 : 1	16.1 : 1	3.6 : 1	2.6 : 1	4.3 : 1	3.8 : 1	4.6 : 1	3.7 : 1
General data:												
Cumulative systems installed	—	—	—	—	1	3	8	13	22	35	50	65
Number of employees at year-end	12	21	30	45	124	199	321	524	761	1,079	1,352	1,551
Earnings (loss) per share	$(.04)	$(.18)	$(.20)	$(.18)	$(.17)	$.19	$.29	$.63	$.85	$1.32	$1.38	$1.77
Stock price:												
High	—	—	—	—	2⅞	4⅛	11	16⅞	48½	48⅜	45¾	57⅛
Low	—	—	—	—	1⅞	2	3½	7¾	12¾	28	20	36⅜
Price/earnings ratio	—	—	—	—	NM†	43/21	51/16	27/12	57/15	37/21	33/14	32/21

* April 6 (date of organization) to December 31.
† Not meaningful.
SOURCE: Reprinted from HBS-Case "Cray Research, Inc.," 4-385-011. Revised 12/84.

EXHIBIT 6 Cray Research, Inc.

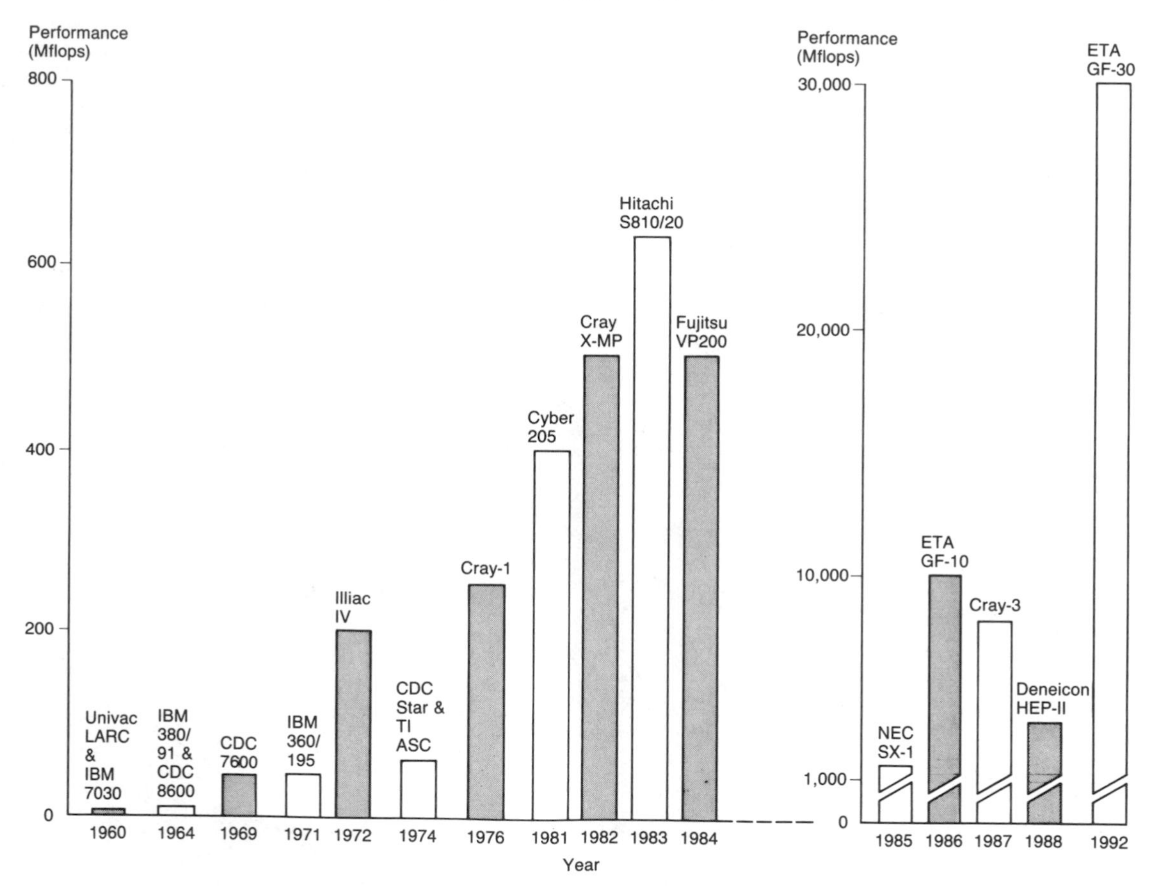

The peak performance of supercomputers has progressed steadily during the past two decades. Performance capabilities will jump dramatically over the next several years, however, if the various projects planned meet their goals. This chart approximates each machine's theoretical peak megaflop (million floating-point operations per second) rating, which is rarely attained in actual operations. For example, the 64-processor Illiac IV never came close to its theoretical peak performance.

SOURCE: *High Technology*, May 1984.
In HBS-Case "Cray Research, Inc.," 4-385-011. Revised 12/84.

To build the Cray-3, Seymour Cray is now using a new component, gallium arsenide, which would make the transistors 5 to 10 times faster than the ones made with silicon. Rollwagen, however, has some reservations about the use of gallium arsenide: "If you ask Seymour, he would say all the problems are solved. . . . If you ask the 100 people that support his effort, they would tell you they are not."[19]

[19] Ibid.

Rollwagen's insurance policy is Steve Chen, a supercomputer designer he hired in 1979. Now vice president of development at Cray Research, Chen aims to extend the design principles of the X-MP (using silicon transistors) to create faster machines than Seymour Cray's. Rollwagen remarked: "We are very pleased with Chen's performance. Pleased enough that we have told Chen he can compete head to head with Seymour if he wants to. Naturally he wants to."[20]

CDC'S EMPLOYEE ENTREPRENEURIAL ADVISORY OFFICE

Reflecting on Seymour Cray's spin-off, CDC Vice President John Pfouts remarked:

> CDC has always had a policy of encouraging employees who felt they had to break off on their own. Thirteen years ago we helped Seymour Cray, and more recently Mel Stuckey, who set up a software business. But somehow we missed Jim Thornton and Network Systems, so Bill Norris figured that as long as CDC was going to lose good people, we might as well do so in a positive way, and in 1979 we formalized the policy into an Employee Entrepreneurial Advisory Office (EEAO).[21]

EEAO

The record of the EEAO is impressive. Since the program has started, 812 CDC employees have been advised and a total of 92 new businesses have been started, offering jobs to over 500 people.

The EEAO function was set up as a corporate office, reporting via a senior staff executive to the chairman of the board. E. Strickland, vice president for corporate growth, was responsible for EEAO. The purpose of the EEAO was "to provide assistance to CD employees who, acting under their own initiatives, are motivated to do their own business." The services provided by the EEAO—confidential if the employee desired—covered a variety of areas: business plans, technology, technology ownership, marketing studies, education, financing, location, legal advice, etc.

The overall intent of the policy was "to provide a process designed to facilitate a favorable outcome in the best interests of the employee and Control Data." Most importantly, the EEAO assisted in the resolution of problems that might develop between employer and departing entrepreneur, for example, in the domain of the use of CD-owned technology. As such, the outcome of a negotiation between EEAO and the employee could be a soundly based new and independent employee venture, a licensing arrangement, a new CD venture, or, finally, an informed conclusion on the part of the employee that his contemplated venture was not viable.

In the early 1980s, a number of spin-offs occurred where CDC looked for ways both to help the entrepreneur and retain some CDC interest in the spin-offs. Funded by a CDC minority participation and a venture capital group, Mel Stuckey set up Microtechnology Source Limited to manufacture software for computers. Zycad was another spin-off. Producing semiconductor simulation equipment, it also was funded with a minority CDC participation and a venture capital group. Another CDC executive set up Edge, to develop a super minicomputer for Cad/Cam applications, where CDC contributed part of the technology and also took a minority participation.

Reflecting on these spin-offs, a venture capitalist—who financed Mel Stuckey's Microsource Technology—remarked:

> Maybe CDC did not make the right decision with Cray Research and Network Systems. But their policy has changed since: Loyal to CDC's entrepreneurial spirit and to Bill Norris's philosophy, CDC now fosters certain types of ventures, then decides not to pursue the venture in-house

---

[20] Ibid.

[21] John Pfouts, vice president CDC, Stanford Business School, March 12, 1984.

(for example, Microsource Technology was software, which is not in CDC's current main product line), but CDC finds ways to keep an interest in the venture either through direct minority investments, through their SBIC–Control Data Capital Corporation set up in 1979 that takes equity, or through technology arrangements.

A former CDC executive, who at CDC worked in the area of new ventures, explained how spin-off decisions were reached:

> First, we looked to see if the project was interesting to CDC. If we thought it was, we then looked at the possibility of the project to attract outside capital. This depended then on the type of individual proposing the project. With an entrepreneurial, aggressive individual, CDC would spin off the project and take a small equity in it. Examples include Cybernetics, a testing-equipment project where CDC took a 25 percent equity, and Magnetic Disk, Inc., a disk repair company that was started in 1981 with a 25 percent CDC participation. Magnetic Disk now has revenues of $20 million and CDC subcontracts their disk repair to them.
>
> If they were "corporate" type individuals who wanted to start a venture, they would be less likely to attract lots of venture capital, so then CDC would put in a larger amount of money and technology, hoping that later on more outside capital could be attracted. An example of this would be the recent creation of ETA Systems for the supercomputer development, where CDC originally planned to take a 40 percent equity but ended up taking 90 percent.

The executive also pointed out that the most successful internal CDC ventures are the joint venture operations in the Peripherals Division, such as Magnetic Peripherals, Optical Storage, Inc., and Centronics.

INTERNAL ENTREPRENEURSHIP AT CDC

While trying to manage the spin-offs in a beneficial way for both the company and the entrepreneur, CDC was also looking for ways to manage innovation *within* the company. As John Pfouts noted:

> The basic idea entrepreneurship addresses is: Do employees really have to leave their corporate positions in order to achieve their goals? Must an organization squeeze out individuals with new ideas instead of creating an environment for them to develop those ideas into thriving new businesses? This is the idea we are working on now. Our approach is a form of what has come to be known as "intrapreneurship."

Unlike the entrepreneur who starts up a venture on his or her own, the intrapreneur develops an idea with the full support of the company. The corporation, in fact, does much the same as the venture capitalist does for the entrepreneur, but the intrapreneur develops a new business *within* the corporation, to the benefit of both.

From the employees' point of view, the advantages of intrapreneurship are obvious. They take a limited risk regarding their individual economic situations. They can utilize the know-how of the corporation. They can use machinery and tools during slack times for design and testing. They have access to the corporate marketing organization, given some special rules. And they are entitled to and expected to look upon the environment with a different perspective. They are endowed with special means to develop new business that the corporation might otherwise have overlooked or started to develop much later. On the other hand, their special status must be reciprocated through some sort of guarantee to the corporation that its interests will be enhanced by the intrapreneur's activity.

To this end, CDC has set up some general rules:

1. The business should be consistent with (but not identical to) the company's own goals. Clearly it should in some way further the company's strategy.
2. The company should take an equity posi-

tion in the new business, thus participating in both the risk and reward.

3. The individual should take a well-defined risk—for example, a 10 percent salary reduction that would apply toward his or her own equity position.
4. The intrapreneur should also have a well-defined reward, based on the development of the business.
5. Since the business is to be independent, there should be rules facilitating the build-up of capital.

It is clear that a variety of problems and questions still need to be addressed if CDC wants to stimulate intrapreneurship, and John Pfouts raised some critical issues:

> How many employee "intrapreneurs" can the company work with? How does it choose? What about people with high expectations who can't be accommodated? How do you structure the benefits to the intrapreneur and how much should the company invest in a given project? What support will the corporation give in terms of flexible scheduling, in-kind services, and relief from other responsibilities? What about the line managers who have worked long and hard and have had a loyalty to the company for years and who now see a bunch of new people getting corporate support to go out and get rich on their own?

One of the recent and more visible examples of CDC's policy to develop intrapreneurship was the formation in 1983 of ETA systems.

ETA Systems: An Internal Corporate Venture?

In August 1983, CDC announced to spin out the development of its next generation of Cyber 205 Compatible Supercomputer—the Cyber 2XX—to a start-up headed by two of its top supercomputer executives and employing some 100 CDC employees.[22]

Control Data transferred capital, equipment, technology, and personnel to the new firm called ETA Systems, based in St. Paul, Minnesota. CDC's equity share was to be 40 percent; additional capital would be searched in the venture capital market and through R&D partnerships. Total capitalization of the firm by 1986 should be $100 million ($40 million by CDC through 1986).

The decision for CDC was an important one. Two of its former executives would head the new firm: Lloyd Thorndyke, Control Data senior vice president for technology development, and Neil Lincoln, CD's executive for supercomputer development architecture, now both new board members of ETA. The new personnel for ETA consisted of 100 CDC employees, who represented 25 percent of CDC's Computer Systems Division personnel.

CDC strongly denied speculation that the move was a prelude to its exit from the supercomputer market, terming the ETA venture one step in a diversified supercomputer development strategy that will produce by the late 1980s at least three different CDC supercomputers for varying applications.[23]

One reason for the creation of ETA seemed to be the use of outside capital to cover the large development cost of supercomputers—an estimated $60 million over the next three years.[24]

In terms of product segmentation, ETA Systems plans to produce a supercomputer, Cyber 2XX, with eight parallel processors capable of maximum computing speeds of 10 billion floating point operations per second (gigaflops) for introduction to the market in 1986.

CDC has committed itself to purchasing several of the initial production units of ETA's systems, both for in-house installation and for remarketing to commercial and government customers.

---

[22] *Electronic News,* August 22, 1983.

[23] Ibid.

[24] *Business Week,* October 17, 1983.

Control Data will continue to enhance their Cyber 205, the most powerful computer so far with maximum computing speed of 400 megaflops, and market it to customers that do not need the Cyber 2XX's 10 gigaflops. Furthermore, Control Data is developing in-house another supercomputer for introduction in the late 80s.

Commenting on the creation of ETA, Chairman William Norris said, "CDC chose to spin out Cyber 2XX development to accelerate its process," and noted, "the plan is entrepreneurial and dedicated. ETA personnel have fortunes to gain from success and bankruptcy to face from failure. There is no greater motivation for hard and creative work."[25]

Reading IV–6 *Epilogue: A New Organizational Revolution in the Making?*

R. A. Burgelman and L. R. Sayles

As we look back on our efforts and attempt to put them into the context of the major changes that have occurred in the way American managers think about their organizations and their own roles within them, we would like to leave the reader with some final themes for further reflection.

TOWARD A THEORY OF CORPORATE ENTREPRENEURSHIP

At the end of our efforts to describe some of the more complex management processes in large, established firms, we feel even more strongly than at the outset that a theory of corporate entrepreneurship is needed. As once-excellent companies lose their luster and new ones are emerging as bright new stars, it seems clear that simply looking for exemplars of success, whose practices can be readily emulated, is not a workable alternative for serious theory-building efforts.

The outlines of the theory of corporate entrepreneurship we propose are still dim. However, we believe that it will be grounded in increased understanding of the evolutionary processes of organizational learning.

In these evolutionary organizational learning processes, we believe, entrepreneurial *individuals* at the operational and middle levels will play an increasingly important role. Our book has provided some evidence of the fact that such individuals elaborate the organization's capabilities and enact the new opportunities that are associated with the elaboration efforts. In a very important sense, they help their organizations to enlarge and embroider their "knitting" rather than just sticking to the existing domains.[1]

The challenge for American business will be to integrate new theoretical insights into its strategic management practices as they become available. This challenge is real, because the Japanese (and perhaps others as well) are already aware of the need to do so.[2]

---

[25] *Electronic News,* August 22, 1983.

From *Inside Corporate Innovation: Strategy, Structure, and Managerial Skills.*

[1] See R. A. Burgelman, "Strategy-Making and Evolutionary Theory: Towards a Capabilities-Based Perspective," Research paper series no. 755, Graduate School of Business, Stanford University, Stanford, CA, 1984.

[2] This is clearly suggested by the following quote:

> If there is a clearly defined goal, we can follow the ordinary decision process. . . . However, when the corporate goals and strategic objectives are often ambiguous, as in the present situation, the ordinary approach does not work properly. We can only take the evolutionary approach; we act first, examine the feedback, and then gradually define the goal.

See "Business Management in the New Industrial Revolution," White Paper, Japan Committee for Economic Development, 1983, pp. 14–15.

INDIVIDUALISM AND BIG BUSINESS: A NEW BEGINNING

This leads to our second theme. We feel that our inquiry into the nature of internal corporate venturing and corporate entrepreneurship provides the basis for a rather optimistic view, but, at the same time, suggests a tremendous challenge for top management of established firms.

Almost continuously, established firms hire from the best and brightest graduates the nation's educational system has to offer. Systematic surveys, as well as our own limited but direct contact with hundreds of students at two major universities, suggest that a significant number of these young professionals have entrepreneurial aspirations and are looking for something more than the job security traditionally offered by large, established firms. On the one hand, this creates a necessity for dramatic changes in the management practices of large, established firms if they want to be able to compete with glittering new firms for entrepreneurial talent. On the other hand, it suggests the existence of an enormous potential for corporate entrepreneurship ready to exert itself and to be channeled in directions beneficial to both the individual and the firm.

It is perhaps worth noting that the existence of this potential was proposed more than 20 years ago by one of the authors of this book, who critically examined the foundations for the widely asserted contradiction between "individualism" and "big business" and found them lacking.[3] Now, it seems that the time is ripe for a new integration of individualism and big business through strategies for the implementation of corporate entrepreneurship.

In view of the challenge posed by the need to integrate individualism and big business, one should perhaps be cautious with respect to the currently fashionable recommendation that a homogeneous and overly integrated "corporate culture" be created. The challenge for established firms, we believe, is not either to be well organized and to act in unison or to be creative and entrepreneurial. The real challenge, it would seem, is to be able to live with the tensions generated by both modes of action. This will require top management's exploitation of existing opportunities to the fullest (because only relatively few will be available), the generation of entirely new opportunities (because today's success is no guarantee for tomorrow), and the balancing of exploitation and generation over time (because resources are limited). Strategic management approaches will have to accomplish all three concerns simultaneously and virtually continuously.

THE CORPORATION OF THE FUTURE

Our last theme, then, concerns the evolving nature of the established firm. It seems to us that the developments currently crystallizing in American business herald an epoch-marking change. We believe the change may well be of the same magnitude as the one that occurred during the first quarter of the 20th century, which led to the organizational innovation represented by the "divisionalized firm" as brilliantly documented in Alfred D. Chandler's landmark study on strategy and structure.[4]

The new wave of organizational innovations involves new types of arrangements between individuals and corporations. It is likely to continue to produce *new organizational forms,* spanning the entire range of combinations of

---

[3] L. R. Sayles, *Individualism and Big Business* (New York: McGraw-Hill, 1964).

[4] A. D. Chandler, *Strategy and Structure* (Cambridge, Mass.: MIT Press, 1962).

markets and hierarchies[5] and involving complex, sometimes protracted *negotiation processes* between individuals and corporate entities. Such negotiation processes, we believe, will be an increasingly pervasive aspect of corporate life and an important mechanism for facilitating the earlier-mentioned new integration of individualism and big business through corporate entrepreneurship.

[5] O. E. Williamson, *Markets and Hierarchies* (New York: Free Press, 1975).

Index